Proceedings of the
Ist World Congress of the
BERNOULLI SOCIETY

Proceedings of the Ist World Congress of the BERNOULLI SOCIETY

Tashkent, USSR
8-14 September 1986

Volume 2
Mathematical Statistics
Theory and Applications

Editors
Yu. A. Prohorov and V. V. Sazonov

VNU SCIENCE PRESS
Utrecht, The Netherlands
1987

VNU Science Press BV
P.O. Box 2093
3500 GB Utrecht
The Netherlands

First published in 1987

ISBN 90-6764-103-0 (set)
ISBN 90-6764-104-9 (Vol. 1)
ISBN 90-6764-105-7 (Vol. 2)

Printed in Great Britain by J. W. Arrowsmith Ltd, Bristol

CONTENTS

ABSTRACT INFERENCE (semi-parametric models. . .)
(Session 1)

Chairman: P.J. Bickel

EFFICIENT TESTING IN A CLASS OF TRANSFORMATION MODELS: AN OUTLINE

by
P.J. Bickel
University of California, Berkeley

Transformation models of the following type have been discussed by Cox (1972), Clayton and Cuzick (1985), Doksum (1985), among others. We observe (Z_i, Y_i) with $Y_i \in J_i$ an open subinterval of R, which are a sample from a population characterized as follows. There exists an unknown transformation τ from J_0 an open subinterval of R onto J_i with $\tau' > 0$ such that $Y = \tau(T)$ where (Z,T) follow a parametric model. The intervals J_i here may be proper or halfrays or R itself. Colloquially, if Y is expressed in the proper unknown scale, i.e. as T, then the joint behaviour of (Z,T) has some nice parametric form. The case considered by previous authors is

$$\log T = \theta^T Z + \varepsilon$$

where ε is independent of Z. The distributions of ε considered so far include:

Cox (1972): e^{ε} has an exponential distribution.

Clayton and Cuzick (1985): e^{ε} has a Pareto distribution with density

$$f(t) = (1+tc)^{-(\frac{1}{c}+1)}, \quad t>0, \quad c \geq 0 \tag{1}$$

where c = 0 is the Cox model. An important special case of (1) considered by Bennett (1983) is the log logistic model, c = 1 which has the attractive proportional odds property.

Doksum (1985): In generalization of the Box-Cox model, ε has a Gaussian distribution.

It seems reasonable in these models to base inference about the parameters of the underlying parametric model such as θ,c above on the maximal invariant of the group of transformations generating this semiparametric model, $\{(z,t) \to (z,\tau(t))\}$. This maximal invariant is just $M = (\underset{\sim}{Z},\underset{\sim}{R})$ where $\underset{\sim}{Z} = (Z_1, \cdots, Z_N)$ and $\underset{\sim}{R} = (R_1, \cdots, R_N)$ is the vector of ranks of the Y_i. The likelihood of M or the conditional likelihood $L(\theta)$ of $\underset{\sim}{R}$ given $\underset{\sim}{Z} = \underset{\sim}{z}$ can in general only be expressed as an N dimensional integral. It can be evaluated explicitly for the Cox model. Clayton and Cuzick propose some ingenious approximations and Doksum proposes that both the value of L and its distribution be calculated approximately by Monte Carlo. So far, however, the asymptotic behaviour of these procedures is not well understood.

In this paper we specialize to $Z = 0,1$ as in Bickel (1985). Moreover we suppose, as did Clayton and Cuzick that the parameter θ governing the conditional density of $T = \tau(Y)$ given $Z = j$, denoted $f_j(\cdot,\theta)$ is real, and in particular that the distribution of ε is assumed known.

In this context, for a subclass of transformation models, we indicate how to construct asymptotically efficient tests of $H: \theta = \theta_0$ vs $K: \theta > \theta_0$. The proofs of our results and a detailed treatment are given in Bickel (1986).

The subclass includes the Pareto model for $c \geq 1$. The testing problem as such is not very interesting save in the case where θ_0 corresponds to independence of Y and Z which is already well understood. However, the solution of the testing problem is a first step in the solution of the estimation problem whose importance is clear. The tests we propose are based on "quadratic rank statistics".

$$T_N = N^{-1}\sum_{i=1}^{N} a\left(\frac{R_i}{N}, Z_i\right) + N^{-2}\sum_{i,j} b\left(\frac{R_i}{N}, \frac{R_j}{N}, Z_i, Z_j\right). \tag{2}$$

We interpret efficiency in this context conditionally on $\underset{\sim}{Z}$, or equivalently the two sample sizes $\sum_{i=1}^{N} Z_i$ and $N - \sum_{i=1}^{N} Z_i$. We show,

i) If $\theta_N = \theta_0 + tN^{-1/2}$, $t \geq 0$,

$$\mathbf{L}_{\theta_N}\left(\frac{T_N}{\sigma_N} \mid \underset{\sim}{Z}\right) \to \mathbf{N}(at,1) \text{ in probability for some } a > 0 \tag{3}$$

where σ_N is a sequence of normalizing constants.

ii) If S_N is any other sequence of statistics not necessarily depending on the ranks only such that

$$\text{p}\overline{\lim}_N \sup_\tau P_{(0,\tau)}[S_N \geq s \mid \underset{\sim}{Z}] = \alpha$$

then, for each τ, θ_N as above,

$$\text{p}\underline{\lim}_N P_{(\theta_N,\tau)}[S_N \geq s \mid \underset{\sim}{Z}] \leq 1 - \Phi(z_{1-\alpha} - at).$$

An important consequence of (i) and (ii) is the following. Let $\Lambda_N(t) = L(\theta_N)/L(\theta_0)$, the conditional rank likelihood ratio statistic for $H: \theta = \theta_0$ vs $K: \theta = \theta_N$ given $\underset{\sim}{Z}$. Then $L(\theta_N)/L(\theta_0)$ is also efficient. That is an asymptotically size α test based on $\Lambda_N(t)$ has power given by (3) and in fact tests based on $\Lambda_N(t)$ for different t are asymptotically equivalent to each other and our quadratic rank test.

We first heuristically derive what turn out to be appropriate a and b. We then state a theorem establishing the existence of our heuristically derived a and b and the asymptotic efficiency of our test.

The locally most powerful rank test statistic for $H: \theta = \theta_0$ vs. $K: \theta > \theta_0$ is given by $N^{1/2}S_N$ where

$$S_N = N^{-1}\sum_{i=1}^{n} Z_{i0}E_{\theta_0}\{c_0(T_i) \mid \underset{\sim}{Z},\underset{\sim}{R}\} + Z_{i1}E_{\theta_0}\{c_1(T_i) \mid \underset{\sim}{Z},\underset{\sim}{R}\}$$

$c_j(t) = \dfrac{\partial \log}{\partial \theta} f_j(t,\theta_0)$ and $Z_{ij} = I(Z_i = j)$. Equivalently, if $\underset{\sim}{D} = (D_1, \cdots, D_N)$ are the antiranks defined by $T_{(j)} = T_{D_j}$ where $T_{(1)} < \cdots < T_{(N)}$ are the order statistics of the sample, then

$$S_N = N^{-1}\sum_{j=0}^{1}\sum_{i=1}^{N}\{Z_{D_i j}E_{\theta_0}(c_j(T_{(i)}) \mid \underset{\sim}{Z},\underset{\sim}{D}). \tag{4}$$

To get an approximation to the scores in (4) we write, $f_j(\cdot,\theta_0)$ as $f_j(\cdot)$ and define,

$$n = \sum_{i=1}^{N} Z_i, \quad m = N - n, \quad \hat{\pi}_0 = \frac{m}{N} = 1 - \hat{\pi}_1.$$

We treat $\hat{\pi}_j$ as deterministic constants in the sequel. Let,

$$h = \hat{\pi}_0 f_0(\cdot) + \hat{\pi}_1 f_1(\cdot)$$

with H the corresponding distribution function. Note that h and H depend on N and are random only through the $\hat{\pi}_j$.

Finally let, for $0 < t < 1$,

$$\lambda_j(t) = c_j(H^{-1}(t)), \tag{5}$$

$$g_j(t) = f_j(H^{-1}(t))/h(H^{-1}(t))$$

the density of $H(T_1)$ given $Z_1 = j$, and

$$\gamma_j(t) = -\frac{g_j'}{g_j}(t). \tag{6}$$

We can rewrite (4) as

$$S_N = S_{N1} + S_{N0}$$

where

$$S_{Nj} = N^{-1}\sum_{i=1}^{N} Z_{D_i j} E(\lambda_j(U_{(i)}) \mid \underset{\sim}{Z}, \underset{\sim}{D})$$

where (Z_i, U_i) are i.i.d with U_1 given $Z_1 = j$ having density g_j and the marginal density of U_1 is uniform,

(7) $$\hat{\pi}_0 g_0 + \hat{\pi}_1 g_1 = 1.$$

The next step is to note that $U_{(i)} \cong \frac{i}{N}$ so that

(8) $$S_{Nj} \cong N^{-1}\sum_{i=1}^{N} \{Z_{D_i j}(\lambda_j(\tfrac{i}{N}) + \lambda_j'(\tfrac{i}{N})E[(U_{(i)} - \tfrac{i}{N}) \mid \underset{\sim}{Z}, \underset{\sim}{D}]\}$$

plus terms we expect to be of order $O(N^{-1})$.

The first term of the approximation is a linear rank statistic. For the second we use a heuristic argument of Clayton and Cuzick who argue that if

$$\bar{y}_i = E(U_{(i)} \mid \underset{\sim}{Z}, \underset{\sim}{D})$$

then $\bar{y}_i$ satisfies approximately the recurrence relation,

(9) $$(\bar{y}_{i+1} - \bar{y}_i)^{-1} - (\bar{y}_i - \bar{y}_{i-1})^{-1} = (1 - Z_{D_i})\gamma_0(\bar{y}_i) + Z_{D_i}\gamma_1(\bar{y}_i).$$

Let

$$\hat{G}_j(t) = (N\hat{\pi}_j)^{-1}\sum_{i=1}^{N} I(U_i \le t) Z_{ij}$$

be the empirical d.f.s of the two subsamples of U_i from g_0, g_1, and let

(10) $$\hat{R}(t) = \pi_N \hat{G}_0 + (1 - \pi_N)\hat{G}_1$$

be the empirical d.f. of the sample $U_1, \cdots, U_N$. Define,

(11) $$\hat{Q}_j(t) = \hat{G}_j \hat{R}^{-1}(t+0), \quad 0 < t \le 1$$

$$\hat{Q}_j(0-) = 0$$

where for any d.f. F, $F^{-1}(t) = \inf\{s : F(s) \ge t\}$. $\hat{Q}_0$ is a distribution function with jumps of size m^{-1} at $\frac{j-1}{N}$ such that $Z_{D_j} = 0$ while $\hat{Q}_1$ jumps $(N-m)^{-1}$ at $\frac{j-1}{N}$ with $Z_{D_j} = 1$. Evidently $\bar{y}_i$ is a function of $\frac{i}{N}$, $\underset{\sim}{Z}$, $\hat{Q}_0$, $\hat{Q}_1$ only. Interpolate smoothly in some way between $\frac{i-1}{N}$ and $\frac{i}{N}$, $1 \le i \le N$ to obtain a function v on (0,1) such that,

$$\bar{y}_i = v(\tfrac{i}{N}).$$

Any solution of (9) must satisfy, for some c,d

$$\overline{y}_i = d + \sum_{j=1}^{i}(c + \sum_{k\geq j} Z_{D_k 0}\gamma_0(\overline{y}_k) + Z_{D_k 1}\gamma_1(\overline{y}_k))^{-1}$$

or for $u = \frac{i}{N} \cong \frac{i-1}{N}$

(12) $$v(u) \cong d + \int_0^u (\frac{c}{N} + \int_t^1 \gamma_0(v(s))\hat{\pi}_0 d\hat{Q}_0(s) + \gamma_1(v(s))\hat{\pi}_1 d\hat{Q}_1(s))^{-1} dt.$$

This is essentially the integral equation of Bickel (1985), save that we make the transformation $H(\cdot)$ and apply (8). Unfortunately, the hopes for analytic approximation of solutions to (12) expressed in Bickel (1985) have so far not been realized. However, suppose we (still formally) extend the definition of (12) to functions $v(\cdot,Q,Q')$ by replacing $\hat{Q}_0$ $\hat{Q}_1$ by arbitrary Q,Q' such that,

$$\hat{\pi}_0 Q(t) + \hat{\pi}_1 Q'(t) = t, \text{ for } t = 0, \frac{1}{N}, \cdots, 1$$

with c,d depending on Q,Q'. Then, if $Q = G_0$, $Q' = G_1$, $\frac{c}{N} = 1$ and $d = 0$, $v(u) = u$ formally satisfies the extension of (12) since by (7)

$$\gamma_0 \hat{\pi}_0 g_0 + \gamma_1 \hat{\pi}_1 g_1 = 0.$$

Therefore, if $\tilde{\Delta}(u) = v(u,\hat{Q}_0,\hat{Q}_1) - u$, $v = v(\cdot\, \hat{Q}_0 \hat{Q}_1)$

$$\tilde{\Delta}(u) = d + \int_0^u \{c + \int_t^1 [\gamma_0(v(s))\hat{\pi}_0 d\hat{Q}_0(s) + \gamma_1(v(s))\hat{\pi}_1 d\hat{Q}_1(s)]\}^{-1} dt$$
$$- \int_0^1 \{1 + \int_t^1 [\gamma_0(s)\hat{\pi}_0 dG_0(s) + \gamma_1(s)\hat{\pi}_1 dG_1(s)]\}^{-1} dt$$

We determine the constants $c(\hat{Q}_0,\hat{Q}_1)$, $d(\hat{Q}_0,\hat{Q}_1)$ formally by smooth fit at the boundaries,

(13) $$\tilde{\Delta}(0) = \tilde{\Delta}(1) = 0.$$

Let,

(14) $$\alpha(s) = \sum_{j=0}^{1} \gamma_j'(s)\hat{\pi}_j g_j(s).$$

Then,

(15) $$\tilde{\Delta}(u) \cong -\int_0^u \{\int_t^1 [\hat{\pi}_0(\gamma_0(v(s))d\hat{Q}_0(s) - \gamma_0(s)dG_0(s)) + \hat{\pi}_1(\gamma_1(v(s))d\hat{Q}_1(s)$$

$$-\gamma_1(s)dG_1(s)]\}dt+(c(\hat{Q}_0,\hat{Q}_1)-1)u+d(\hat{Q}_0,\hat{Q}_1)$$

$$\cong -\int_0^u\{\int_t^1 \alpha(s)\tilde{\Delta}(s)ds+\int_t^1 \gamma_0(s)\hat{\pi}_0 d(\hat{Q}_0(s)-G_0(s))+\gamma_1(s)\hat{\pi}_1 d(\hat{Q}_1(s)-G_1(s))]dt$$

$$+(c(\hat{Q}_0,\hat{Q}_1)-1)u+d(\hat{Q}_0,\hat{Q}_1)$$

$$\cong -\int_0^u\int_t^1 \alpha(s)\tilde{\Delta}(s)ds+\int_t^1\sum_{j=0}^1 \gamma_j(s)\hat{\pi}_j d(\hat{Q}_j-G_j)(s)$$

$$+u\int_0^1(\int_v^1 \alpha(s)\tilde{\Delta}(s)ds+\int_v^1\sum_{j=0}^1 \gamma_j(s)\hat{\pi}_j d(\hat{Q}_j-G_j)(s)).$$

After some algebra, this reduces to,

$$\tilde{\Delta}(u)\cong -\int_0^1 K(s,u)\alpha(s)\tilde{\Delta}(s)ds-\int_0^1 K(s,u)\sum_{j=0}^1 \gamma_j(s)\hat{\pi}_j d\hat{Q}_j(s) \tag{16}$$

where

$$K(s,u) = s\Lambda u - su.$$

Continuing to ignore existence and unicity questions we define $\Delta(u)$ as the solution of the <u>linear</u> integral equation obtained from the approximate equation (16). We introduce a Greens functions solving,

$$\Delta(u,v)+\int_0^1 K(s,u)\alpha(s)\Delta(s,v)ds = K(u,v). \tag{17}$$

Then $\Delta(t)$ is given by

$$\Delta(t) = -\int_0^1 \Delta(t,u)\sum_{j=0}^1 \hat{\pi}_j\gamma_j(u)d\hat{Q}_j(u). \tag{18}$$

We now define,

$$T_N = \int_0^1\sum_{j=0}^1(\lambda_j(t)+\lambda_j'(t)\Delta(t))\hat{\pi}_j d\hat{Q}_j(t) \tag{19}$$

$$= \sum_{j=0}^1\int_0^1 \lambda_j(t)\hat{\pi}_j d\hat{Q}_j(t)-\sum_{j=0}^1\sum_{k=0}^1\int_0^1\int_0^1 \Delta(t,u)$$

$$\lambda_j'(t)\gamma_k(u)\hat{\pi}_j\hat{\pi}_k d\hat{Q}_j(t)d\hat{Q}_k(u)$$

which is of the form (2) with,

$$a(\frac{i}{N},j) = \lambda_j(\frac{i-1}{N})$$

$$b(\frac{i}{N},\frac{i'}{N},j,k) = \lambda_j'(\frac{i-1}{N})\gamma_k(\frac{i-1}{N})\Delta(\frac{i-1}{N},\frac{i'-1}{N})$$

Theorem: Suppose

(20) $$\int_0^1 |\gamma_j'(t)|dt < \infty, \quad j = 0,1.$$

Then, Δ defined by (17) exists and is unique.
If further,

(21) $$\int_0^1 |\lambda_j''(t)|dt < \infty \quad j = 0,1$$

then,

(22) $$T_N = N^{-1}\sum_{i=1}^{N}\sum_{j=0}^{1} A_j(U_i)Z_{ij} + O_P(N^{-\frac{3}{4}+\delta})$$

where,

(23) $$A_j(u) = \lambda_j(u) - \int_0^1 \lambda_j(u)dG_j - \sum_{l=0}^{1}(v_l'(u) - \gamma_j(u)v_l(u))$$

and

(24) $$v_j(u) = \hat{\pi}_j\int_0^1 \Delta(t,u)\lambda_j'(t)dG_j(t).$$

Suppose (20) and (21) hold. If v_g are given by (24) let

(25) $$q(t,\theta) = t - (\theta-\theta_0)\sum_{j=0}^{1} v_j(t), \quad 0 \le t \le 1,$$

if $|\theta-\theta_0| < (\sum_{j=0}^{1}\|v_j'\|_\infty)^{-1}$. Then,

$$q'(t,\theta) > 0$$

and since

$$v_j(0) = v_j(1)$$

q maps [0,1] monotonely onto itself. Let,

(26) $$\tilde{g}_j(t,\theta) = g_j(q(t,\theta),\theta)q'(t,\theta).$$

The test based on T_N is asymptotically most powerful for testing

$H: \theta = \theta_0$ vs $K_N: \theta = \theta_N$ when $T|Z = j$ has distribution $\tilde{g}_j(\cdot, \theta)$.

Extension of these results to the case Z finite say $= \{0, \ldots, p-1\}$ is straightforward. However, θ in such cases is typically multivariate so that one sided hypotheses without nuisance parameters are not very interesting. The extension to such hypotheses and estimation can be carried through by studying, under a fixed θ_0, the family of statistics $T_N(\theta)$, with λ_j, γ_j etc chosen appropriate to θ being true, at least for $|\theta - \theta_0| = O(N^{-1/2})$. Our methods permit this kind of analysis. We intend to report on this subsequently.

It is relatively straightforward to show that for the Pareto family, including the Cox model for $c = 0$, γ_j is continuously twice differentiable on [0,1) but,

$$\gamma_j^{(r)}(t) = \Omega((1-t)^{c-1-r}) \quad \text{as } t \to 1.$$

So,

$$\|\gamma_j'\|_1 = \infty \quad \text{if } c < 1.$$

For the normal model, α blows up and is not integrable at either 0 or 1. It appears that these difficulties can be resolved by considering statistics T_N based on a censored version of the data, Z and $\{R_i : \varepsilon_1 N \le i \le (1-\varepsilon_2)N\}$ with $\varepsilon_2, \varepsilon_1 \downarrow 0$ at a slow enough rate. This analysis which is in progress should be extendable to the case of general right censoring and possibly also to time dependent covariates. Our results so far establish the efficiency of rank likelihood ratio tests. We expect that our extensions will show that estimation by maximizing the rank likelihood is generally efficient in transformation models, not just in the Cox model as was shown by Efron (1977) and Begun et al (1983). This expected conclusion is supported by the results of Doksum (1985).

References

BEGUN, J.M., HALL, W.J., HUANG, W.M. and WELLNER, J.A. (1983). Information and asymptotic efficiency in parametric and non-parametric models, Ann. Statist. 11, 432-452.

BENNETT, S. (1983a). Log-logistic regression models for survival data. Appl. Statist. 32, 165-171.

BICKEL, P.J. (1985). Discussion of papers on semiparametric models. Proc. I.S.I. Amsterdam.

BICKEL, P.J. (1986). Efficient testing in a class of transformation models. (Centrum for Wiskunde en Informatica) Amsterdam.

CLAYTON, D. and CUZICK, J. (1985). The semiparametric Pareto model for regression analysis of survival times. Proc. I.S.I. Amsterdam.

COX, D.R. (1972). Regression models and life-tables (with Discussion). J.R. Statist. Soc. B. 34, 187-220.

DOKSUM, K. (1985). (Preprint) Partial likelihood methods in transformation models.

Bernoulli, Vol. 2, pp. Ø13-Ø25

ABSTRACT INFERENCE IN IMAGE PROCESSING

GRENANDER, U.
Institut Mittag-Leffler, Djursholm, Sweden, and Division of Applied Mathematics, Brown University, Providence, Rhode Island, U.S.A.

AIM OF STUDY. The literature on pattern research, which is at present expanding at an accelerating rate, contains numerous algorithms intended for the recognition of patterns of different types, among them especially pictorial patterns. Common to most of these algorithms is that their construction is based on plausibility arguments and only seldom rests on a firm mathematical foundation. Nevertheless, some of them seem to work well when tried on real or artifically generated patterns.

Actually, it is difficult to evaluate their performance in some exact terms just because of their lack of precisely stated assumptions and mathematical justification. We have argued for a long time that the algorithms intended for pattern inference (restoration, recognition, extrapolation, etc.) should be derived from precisely formulated models of pattern ensembles: the algorithms should be derived as solutions, exact or approximate, to optimality problems formulated in terms of an explicitly stated criterion.

A multitude of pattern theoretic models have been suggested, see Grenander (1976, 1978, 1981) and some of them have been used in special situations, especially for pictorial patterns. We shall illustrate the general discussion in Section 2-4 by two special cases in Sections 5,6.

Viewed from a general statistical perspective it turns out that the problems encountered with our approach are of a type dealt with in abstract inference, since both parameter and sample space will turn out to be of high or infinite dimensionality.

In this paper we shall use exclusively a Bayesian approach, but others are sometimes indicated, for example the method of sieves and penalized likelihood.

We shall deal with patterns that form complex systems in the sense that they exhibit great variability but at the same time also possess a good deal of structure, as is the case for most pictoral patterns in biology and medicine for example. We believe that the approach answers pressing needs for more powerful picture processing methods in the rapidly expanding image industry.

Since our methods admittedly require formidable computing power we have to design the mathematical treatment in such a way that it can be implemented by existing or projected computer technology. We refer not only to the continuing increase in speed and memory size but also, more importantly, to advances in computer architecture (supercomputers, array machines, highly parallel computers). This has influenced our choice of mathematical strategy, and is one of the motivations behind the selection of Markovian measures as priors as will be clear later.

PRIOR MEASURES ON IMAGE ENSEMBLES. The models used in pattern theory are built from configurations, denoted $c = \sigma(g_1, g_2, \ldots, g_n)$, from generators g_i in some generator space G and from connector graphs σ from some set Σ (the connector type). The meaning of c is that for i=1,2,...n a generator g_i is located at site i of the graph σ, and that specified couplings between a generator g_{i_1} and its neighbors g_{i_2}, $(i_1, i_2) \in \sigma$, regulate the pattern structure.

The nature of the generators varies widely from application to application. In case of pictorial patterns the g's express geometric tendencies; the two examples below will make clear what this term is intended to mean.

A generator g sends out signals $\beta_j(g)$; $j=1,2,\ldots\omega$; to its $\omega = \omega(g)$ neighbors in σ. These signals, or bond values, will decide whether the configuration is regular or not. More precisely, c will be regular if, given a truth valued function ρ (the bond relation), the bond relation $\rho(\beta', \beta'')$ holds true for pairs of bonds that meet each other. More formally, the configuration $c = \sigma(g_1, g_2, \ldots g_n)$ is regular iff

$$(1)\qquad \text{TRUE} = \bigwedge_{s=1}^{m} \rho[\beta_{j_1}(g_{i_1}),\beta_{j_2}(g_{i_2})]$$

Here s enumerates all the segments s = j_1th bond from site i_1 to the j_2th bond from site i_2 in σ.

The regular configuration space, i.e. the set of all regular configurations, is denoted $\mathscr{C}(\mathcal{R})$, where $\mathcal{R} = \langle G,\Sigma,\rho\rangle$. It can play the role of parameter space in pattern inference. Often, however, some of the detailed information about c is lost to the (ideal, no-noise) observer, and we can only observe a function of c, I = R(c) taking values in some space $\mathscr{T}$, the image algebra. The elements I of $\mathscr{T}$ are the (pure) images and $\mathscr{T}$ will then be our parameter space.

Still more information is lost if only observe a noisy version I^D of I, applying a stochastic deformation mechanism D to I; $D: \mathscr{T} \to \mathscr{T}^D$.

The above is intended for situations with rigid regularity. This is sometimes too demanding, and we compensate for the lack of rigid structure by replacing the Boolean valued bond relation ρ by an acceptor function A taking non-negative real numbers as values. This will induce a probability measure over $\mathscr{C}$, the set of all configurations over Σ and generator space G; clearly $\mathscr{C}(\mathcal{R}) \subseteq \mathscr{C}$. The measure will be defined by a density p

$$(2)\qquad p(c) = \frac{1}{Z} \prod_{s=1}^{m} A\,[\beta_{j_1}(g_{i_1}),\beta_{j_2}(g_{i_2})]$$

with respect to some fixed measure, usually of product type over G^n. The acceptor function A in (2) couples neighboring generators, and Z is just a normalizing constant, known in statistical mechanics as the partition function.

The measure over $\mathscr{C}$ induces another, denoted P, over the image algebra $\mathscr{T}$ via an identification function $\mathbb{R} : \mathscr{C} \to \mathscr{T}$. This measure will be our prior on the parameter space $\mathscr{T}$ and we shall deduce our algorithms for pattern inference from P.

The space $\mathscr{T}^D$, containing as elements the deformed images, will then be our sample space. Both $\mathcal{T}$ and $\mathcal{T}^{\mathcal{D}}$ will typically be very high or infinite dimensional which motivates the use of abstract inference methods.

The above is an attempt to give the reader a quick overview of ideas used in pattern theory; to obtain a better understanding, with many special cases, the reader is referred to Grenander (1976,1978,1981) and to the more recent but somewhat rhapsodic paper Grenander (1983).

SOME PATTERN INFERENCE. In a practical situation the first step is to choose a regularity $\mathcal{R} = \langle G,\Sigma,\rho\rangle$ for rigid regularity or $\mathcal{R} = \langle G,\Sigma,A\rangle$ for relaxed regularity; the latter will be done below. The choice of $\mathcal{R}$ is usually not obvious and requires inventiveness as well as insight and understanding of the structure of the patterns at hand.

Once $\mathcal{R}$ has been chosen we have a prior measure over $\mathcal{C}$ and/or $\mathcal{T}$ as well as a joint measure over $\mathcal{T} \times \mathcal{T}^{\mathcal{D}}$ when the statistics of $\mathcal{D}$ has been specified. In principle, we can therefore calculate the posterior measure, say a density $p(I|I^{\mathcal{D}})$.

The next question is: what are the inference tasks to be confronted? We may be asked for a restoration function $I^* = I^*(I^{\mathcal{D}}) \in \mathcal{T}$ that minimizes some criteria

$$(3) \qquad E[d(I^*,I)] = \min.$$

In statistical language this amounts to point estimation.

Or, we may have observed a part of the image, with or without noise, and wish to reconstruct the whole image. This is an extrapolation problem.

Or, we may start from a partition

$$(4) \qquad \mathcal{T} = \bigcup_{\alpha} \mathcal{T}^{(\alpha)}$$

where the sets $\mathcal{T}^{\alpha} \subseteq \mathcal{T}$ are the patterns, and we want to determine from which $\mathcal{T}^{\alpha}$ our observed $I^{\mathcal{D}}$ is likely to have originated. This pattern recognition task is therefore a hypothesis testing (multiple decision) problem.

We shall only discuss the first of these, image restoration. As a preparation for this kind of pattern synthesis let us make some comments on pattern synthesis, that is Monte Carlo simulation of the prior measure P. This is needed during the model construction phase but we shall see that it is also closely related to pattern analysis.

Except for extremely simple cases direct simulation of P does not seem computationally feasible. To deal with this difficulty stochastic relaxation was introduced starting from the Metropolis scheme, see Metropolis et al. (1983), originally used for the Ising model.

The basic idea is simple and attractive. Consider the generator g located at site i in σ and with the neighboring sites $i_1, i_2, \ldots i_\omega$. It is then easy to calculate the conditional probability of g given $g_{i_1}, g_{i_2}, \ldots, g_{i_\omega}$ as

$$(5) \qquad p(g|c) = \frac{\prod_{j=1}^{\omega} A[\beta_j(g), \beta_{j'}(g_i')]}{\sum_{g \in G} \prod_{j=1}^{\omega} A[\beta_j(g), \beta_{j'}(g_i')]}$$

Here the ω segments of the connector σ go from site, bond j, to site i', bond j'.

To simulate the prior P over $\mathscr{C}$ let us initialize by choosing an arbitrary configuration $c(0) \in \mathscr{C}$. The main loop in the algorithm is as follows: simulate g distributed as in(5) given the values of the neighbors g_{i_ν}, $\nu=1,2,\ldots\omega$.

Iterate the main loop for different sites i in such a way that each site appears infinitely often. After t iterations we denote the current configuration by c(t). An elementary argument shows that, under mild conditions, c(t) converges in law to P.

<u>Remark 1.</u> The prior in (2) is known to represent a Markov process over σ which shows up in that (5) depends only upon $g_{i_1}, g_{i_2}, \ldots g_{i_\omega}$. Thus we need only fetch that ω values (not n-1 values) from memory, which reduces CPU time - after all, a fetch is also a computation and takes time.

<u>Remark 2.</u> If $|G| < \infty$, as assumed above, and if this cardinality is small or moderate the simulation of (5) can be done by brute

force. If $|G|$ is large or infinite, and if $p(g|c)$ is of well known form then the simulation can be done by well known algorithms.

__Remark 3.__ Unfortunately we know little analytically about the speed of convergence to P. We are therefore unable to give firm advice on when to stop iterating. It is hoped that this baffling problem will receive attention from other mathematicians.

Now let us return to pattern analysis, in particular to image restoration. The joint density of I and $I^{\mathcal{D}}$ (where for simplicity we assume R = identity so that c=I) can be written

$$(6) \qquad p(I, I^{\mathcal{D}}) = p(I)q(I^{\mathcal{D}}|I)$$

so that

$$(7) \qquad p(I|I^{\mathcal{D}}) = \text{constant} \times p(I)q(I^{\mathcal{D}}|I)$$

where the constant may depend upon $I^{\mathcal{D}}$ but not upon I.

An important observation, due to S. Geman, points to the fact that for most $\mathcal{D}$, described by the conditional density q, q is like (2) in that it is of product form over some graph with low arities ω. When this is so we can apply stochastic relaxation to simulate (7) too.

Now it depends upon what optimality criterion we have adopted. Apply stochastic relaxation N times, leading to an i.i.d. sample $c_1, c_2, \ldots c_N$ approximately distributed according to P. We then solve the empirical analog of (3) and have the answer.

A related but different approach is MAP = maximum a posteriori density. We then search for the mode of $p(I|I^{\mathcal{D}})$. It has been shown in Geman and Geman (1984) and Gidas (1985) that this can be done by a version of stochastic relaxation, using not $p(I|I^{\mathcal{D}})$ but the density

$$(8) \qquad \text{constant} \times [p(I|I^{\mathcal{D}})]^{1/T}, \quad T > 0$$

where the parameter T, the "temperature", is made to decrease slowly to T=0. More precisely, the schedule

$$(9) \qquad T = \frac{c}{\log t}$$

is sufficient if c is large enough. Then c(t) converges in probability to the MAP solution.

Other variations of stochastic relaxation have been suggested, see e.g. Grenander (1985), pp. 135-141, but will not be discussed here.

PATTERN THEORETIC LIMIT THEOREMS. An attractive feature of stochastic relaxation is its generality: it can be applied directly to all pattern theoretic models to simulate the prior over the configuration spaces.

As can be expected, unfortunately, we pay a price for the generality, namely the massive computational effort that is sometimes needed for its implementation. This is especially serious if:

(i) the cardinality $|G| = \infty$, for example when G is a continuum

(ii) the couplings are strong so that $A(\cdot,\cdot)$ viewed as the kernel of an integral operator is an approximate identity

(iii) the graph is large, $|\sigma| >> 1$, for example in lattice based models as in Section 5

(iv) the conditional distributions in (5) are not of familiar type.

This objection will be less serious when massively parallel computers become more commonly available. Although the rapid pace of technological progress is likely to continue, we would like to supplement the technological advance with analytical ones. Indeed, it now seems possible to prove limit theorems for the priors in pattern theory that can be used for obtaining computationally feasible approximations. During the last decade such limit theorems have been proven.

Space does not permit any full presentation of them; an example will have to suffice. Say that σ = linear chain of $|\sigma|$ = n sites, so that a configuration consists of n values g_i, where g_i is coupled to g_{i+1}; $1 \leq i < n$. Say that $G = \mathbb{R}$ so that (i) holds, that n is large so that (iii) holds,

$$A(g',g") = Q(g')A_0\left[\frac{g'-g"}{\epsilon}\right]Q(g") \tag{10}$$

where Q is a non-negative weight function, A has the maximum at zero, and ϵ a positive small parameter. When $\epsilon \downarrow 0$ the couplings become stronger, so that (ii) holds. If Q is general and only restricted by mild regularity conditions and with the maximum at 0, case (iv) can also occur. Stochastic relaxation can be expected to require too much CPU-time to be acceptable - analytical help is called for.

Three types of pattern theoretic limit problems can be illustrated through this example. The first limit problem is: what happens to the prior (2) if $\epsilon \downarrow 0$, i.e. couplings are strengthened? This problem is well understood in most interesting situations.

The second limit problem is: what happens to P when $n \to \infty$, i.e. the connector graph is becoming big? The answer is known in some but not all situations.

The third limit problem is: what happens to P when $\epsilon \downarrow 0$, $n \to \infty$ simultaneously? This is the practically most important of the three but also, unfortunately, the hardest one to answer.

In the special case mentioned the following has been shown. Under conditions that can be found in Chow-Grenander (1985), and with the standardization $x_i = g_i/\sqrt{\epsilon}$ the marginal distributions of x_i, $i = [\alpha n]$, $0 < \alpha < 1$, $\epsilon = c/n$, tend to a non-degenerate Gaussian limit.

Actually, one can state more. If we introduce the stochastic process with the argument $t \in (0,1)$ and

$$x_{n,\epsilon}(t) = \begin{cases} x_i & \text{for } t = i/n \\ \text{linear interpolation between successive points } i/n \end{cases} \tag{11}$$

it has been shown that the measure of $x_{n,\epsilon}(t)$ converges weakly to that of a Gaussian, stationary, and Markovian process. This is one of the most thoroughly studied stochastic process and is easy to simulate, which facilitates pattern synthesis.

It ought to be mentioned that this rather special result was based on a conjecture obtained from a computer experiment. The conjecture at first seemed somewhat doubtful, or at least limited to the special case. Recent work by Chow, Grenander, Sethuraman has shown, however, that it extends to much more general graphs; this will be reported elsewhere. We will therefore have access to useful approximations for simulation in pattern inference.

A LATTICE BASED IMAGE MODEL. Both examples will model the same biological shape, two dimensional views of stomachs from different human subjects. The shapes are from a textbook in pathology, they are highly variable although of course resembling each other.

The first model, lattice based, intends to describe only the local statistical properties of the boundary of the shape.

The connector σ will be chosen as an $L \times L$ square lattice (with periodic boundaries), each site of which is connected to its 8 closest neighbors.

The generators will represent geometric tendencies of a site to become an internal point, an external one, or one on the boundary of the shape. In the last case the boundary element can have a tendency to turn left 90°, 45°, be straight, turn right 45°, or 90°. This is formalized in Figure 1.

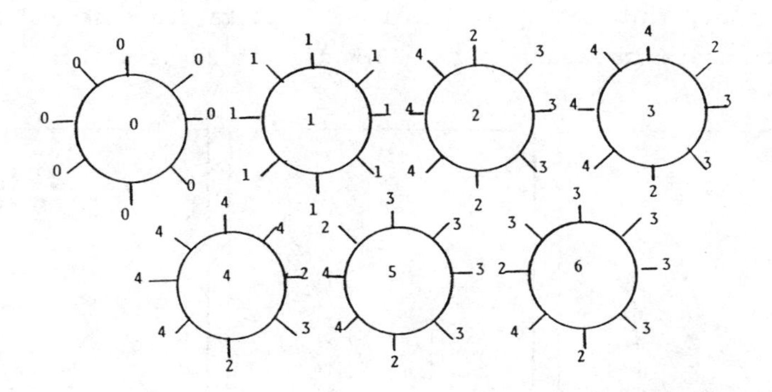

Figure 1

In Table 1 we show the bond values $\beta_j(g)$

Table 1

g	β_0	β_1	β_2	β_3	β_4	β_5	β_6	β_7
0	0	0	0	0	0	0	0	0
1	1	1	1	1	1	1	1	1
2	2	3	3	3	2	4	4	4
3	2	3	3	2	4	4	4	4
4	2	3	2	4	4	4	4	4
5	2	3	3	3	3	2	4	4
6	2	3	3	3	3	3	2	4

We extend this set of 7 generators by rotating each all multiples of 45°. This gives us the generator space G.

The truth valued bond relation function $\rho(\beta',\beta'')$ is given in Table 2.

Table 2

β' \ β''	0	1	2	3	4
0	1	0	0	0	1
1	0	1	0	1	0
2	0	0	1	0	0
3	0	1	0	0	0
4	1	0	0	0	0

We now apply this set up to the image restoration task and show a stomach shape in Figure 2a as a B/W digital 32×32 picture.

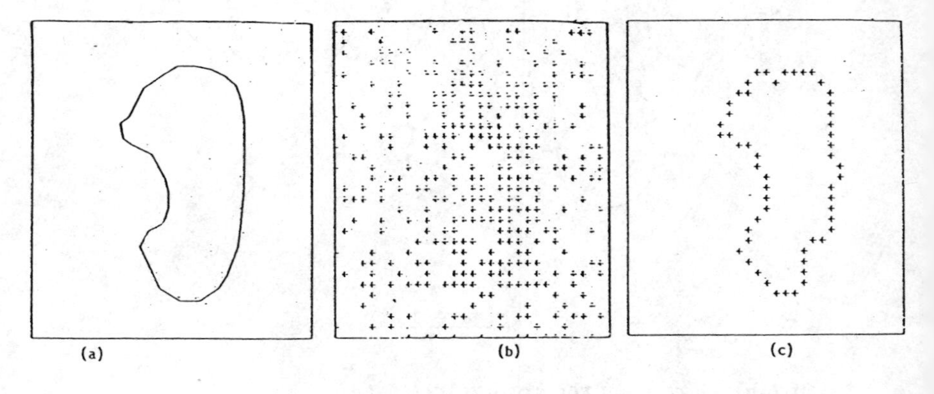

(a) (b) (c)

Figure 2

The picture in (a) becomes the deformed one $I^{\mathcal{D}}$ in 2(b), when $\mathcal{D}$ means symmetric binary noise with the error rate 20%. Stochastic relaxation applied to $I^{\mathcal{D}}$ gives us the restored picture I* represented by its boundary shown in 2(c).

A CONTINUUM BASED MODEL. In our second example we shall not let the pure image I mean a digital picture on an L×L lattice, but let it reside in the continuum $\mathbb{R}^2$.

Indeed, we shall let the boundaries be n-gons in $\mathbb{R}^2$. In the computer experiment with actual stomach type we used n=32, which shows up as an (almost) smooth outline. For clarity we shall use octagons here, n=8, but the approach is exactly the same.

Let the generators be directed line segments, each generator having one in-bond equal to the start point of the segment. The bond relation ρ will be chosen to mean that $\beta_{out}(g_i) = \beta_{in}(g_{i+1})$, i=0,1,...n-1, with i=n identified with i=0.

Note that the connection type Σ, which in our first example consisted of lattice graphs, now has cyclic graphs as its elements.

The prior is then impressed on the regular configuration space to fit the ensemble of observed shapes. How to do this is a tricky question which has been studied a great deal recently. This is not the right place to discuss this challenging estimation problem which is also of the type encountered in abstract inference: See Grenander-Osborn (1986).

In Figure 3(a) we show one shape I simulated from the prior measure. It means, statistically speaking, the true value of our parameter (shape). In (b) we show a deformed version $I^{\mathcal{D}}$ as a 32×32 B/W picture also obtained from symmetric binary noise with a 20% error rate.

Figure 3(c) exhibits the restored picture I* obtained by using stochastic relaxation.

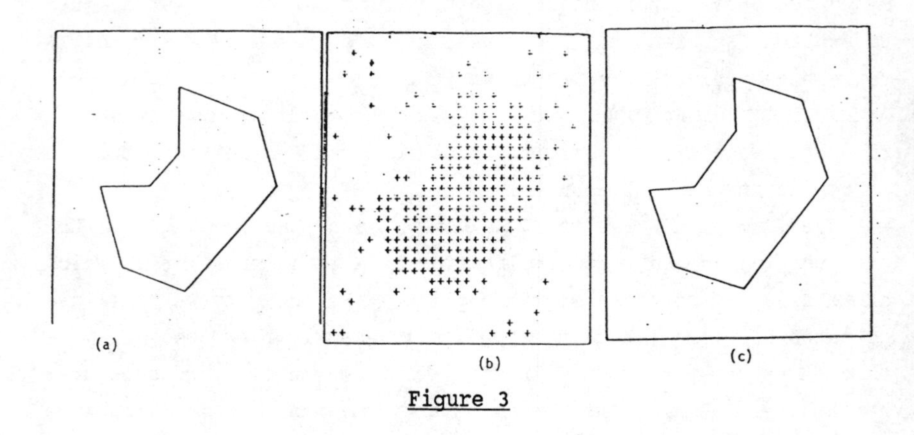

Figure 3

In both of these examples, as well as in a multiple of others treated by the same method, the restoration appears excellent, given the high noise level. But this is not really our point! Many ad hoc techniques also give results that look good. What distinguishes the pattern theoretic approach is that it is model based so that the algorithms are derived as optimal from an explicitly stated model describing the probabilistic variations in shape.

We believe that for shape ensembles with detailed structure continuum based models will turn out to be more powerful than the lattice based ones.

COMPUTATIONAL SET UP. Most of our experiments were at first carried out on a small scale, often with the code written in APL, to be followed by larger ones sometimes with the code in FORTRAN. Execution typically on conventional hardware such as an IBM main frame or a VAX.

Some big and more lifelike computer experiments required more power; we have used occasionally a supercomputer, CRAY. More interestingly, perhaps, we have got excellent performance on our STAR, an array machine which seems nearly ideal for the sort of mathematics we deal with, and will be used by us systematically the next few years.

SUMMARY. A remarkable confluence in time between analytical developments - general pattern theory - and advances in computer technology - parallel and array architectures - makes it possible to construct and execute image processing techniques with a firm theoretical justification.

REFERENCES

1. Besag, J., Spatial interaction and the statistical analysis of lattice systems, J. Royal Stat. Soc. 1974.

2. Chow, Y.S., Grenander, U., A singular perturbation problem, J. Integral Eqns. 1985.

3. Geman, D., Geman, S., Stochastic relaxation, Gibbs distributions, and the Bayesian restoration of images, IEEE Pattern Analysis and Machine Intelligence, 1984.

4. Geman, D., Geman, S., Parameter estimation for some Markov random fields, Brown University Working Papers No. 11, 1983.

5. Gidas, B., Nonstationary Markov chains and convergence of the annealing algorithm, J. Stat. Physics, 39, 1985.

6. Grenander, U., Lectures in Pattern Theory, Vol. I, II, III, 1976, 1978, 1981.

7. Grenander, U., A Tutorial in Pattern Theory, Div. Appl. Math., Brown University, 1983.

8. Grenander, U., Osborn, B., Estimation problems in pattern theory, Preliminary Report, Brown University, 1986.

9. Metropolis, N., Rosenbluth, A.W., Rosenbluth, M.N., Teller, A.N., and Teller, E., Equations of state calculations by fast computing machines, J. Chem. Phys., Vol. 21, 1953.

The unpublished reports mentioned can be obtained by writing to the author.

BAYESIAN INFERENCE IN SEMIPARAMETRIC PROBLEMS

BUNKE O. , Humboldt University Berlin , GDR

We consider a model

(1) $$h_i(X, \vartheta) = e_i \qquad (i=1,\ldots,n)$$

relating the observation X with i.i.d. random vectors e_i having the distribution function F . The k-dimensional parameter and F are assumed to be unknown. The model (1) includes many standard cases, like e.g. location, location-scale and linear models, nonlinear regression models as well as autoregressive time series.

We assume an observation vector $X=(X_1,\ldots,X_n)$, each X_i being uniquely determined by and e_i in some special cases, when

(2) $$h_i(X_i, \vartheta) = e_i \; , \; X_i = k_i(\vartheta, e_i) \quad .$$

The functions h_i defined on $R^n \times \Theta$ are assumed to be measurable as well as also the functions k_i, in case (2) is valid.

Consider a prior distribution, under which ϑ and F are independent, ϑ having a density f and F a Dirichlet distribution D_α .

The index measure is assumed to be absolutely continuous, its normalization $\alpha_1 = \alpha / \alpha(+\infty)$ having the density g .

We need the following assumption A , which is fulfilled e.g. in location-scale and multivariate linear models as well as in autoregressive time series.

A. For all $i,j \in N = \{1,\ldots,n\}$ the set

$$\Theta_{x,ij} = \{ \vartheta \in \Theta \mid h_i(x, \vartheta) = h_j(x, \vartheta) \}$$

has zero Lebesgue measure, if the components x_i of $x = (x_1,\ldots,x_n) \in R^n$ are distinct.

Then we prove, that , if all observations X_i are distinct, the posterior density of ϑ is

(3) $$p(\vartheta \mid X) \propto f(\vartheta) \prod_i g(h_i(X, \vartheta)) \quad ,$$

the same as in the parametric model (1) with known distribution $F = \alpha_1$. Moreover the posterior distribution of F is the mixture

(4) $$\pi(dF) = \int D_{\alpha(h(X, \vartheta))}(dF) \; p(\vartheta \mid X) \, d\vartheta \quad ,$$

using the denotations $h=(h_1,\ldots,h_n)$, $e=(e_1,\ldots,e_n)$ and

(5) $$\alpha(e) = \alpha + \sum_i \delta_{e_i} \quad ,$$

where δ_y is the Dirac measure degenerated at y .
This is known from Diaconis and Freedman (1986) in the location case and has independently been derived in a more general model in Bunke (1985). As a consequence the posterior mean of F , which is a Bayes estimate w.r.t. quadratic loss functions , is

(6) $$\hat{F} = \pi\alpha_1 + \bar{\pi}\int Q_\vartheta \, p(\vartheta \mid X)\, d\vartheta \quad ,$$

where Q_ϑ denotes the empirical distribution function of the "errors" (1) , X and ϑ being fixed , and

(7) $$\pi = a/(a+n) \quad , \bar{\pi} = 1 - \pi \quad , \quad a = \alpha(+\infty) \quad .$$

We notice the surprising fact, that even in the "noninformative" limit case $a \to 0$ the estimate $\hat{F}$ is not the usual empirical distribution of the "residuals"

(8) $$\hat{e}_i = h_i(X, \hat{\vartheta}) \qquad (i=1,\ldots,n) \quad ,$$

$\hat{\vartheta}$ being e.g. a MLE . In the special location case

(9) $$X_i = \vartheta + e_i$$

with "normal noninformative" prior given by

(10) $$\alpha_1 = N(0, \sigma^2) \quad , a \to 0 \quad , \quad f(\vartheta) = \text{const} \quad ,$$

the estimate $\hat{F}$ has a density

(11) $$\hat{F}'(y) = \sum_i \varphi((y - \hat{e}_i)/t_n)/t_n n \quad ,$$

where

(12) $$\hat{e}_i = X_i - \bar{X} \quad , \bar{X} = \sum_i X_i / n \quad , t_n = n^{-1/2} \quad .$$

This is a kernel density estimate with bandwidth t_n, that is, a smoothing of the empirical distribution of the residuals $\hat{e}_i$ is the estimate of F , although F is a discrete distribution (a.s. under the Dirichlet prior) !
Under (2) the distribution F_j of the observation X_j has the estimate

(13) $$F_j = G_j + Q_j$$

with the predictive distribution

(14) $$G_j(y) = \int \alpha_1(h_j(y, \vartheta))\, p(\vartheta \mid X)\, d\vartheta$$

in the model (1) with known $F = \alpha_1$ and the mixture

(15) $$Q_j(y) = \int Q_\vartheta(h_j(y, \vartheta))\, p(\vartheta \mid X)\, d\vartheta \quad .$$

The Bayes estimates are not only interesting in the rare cases, in which a specific prior is available,say, given by normal densities f and g with certain parameters, but also as (admissible) estimates, which moreover have a small average quadratic risk w.r.t. a natural weighting of the unknown and F . Thus one has certain sensible estimates depending on parameters of the prior, which usually are not available , but may be adapted by minimizing some data dependent criterion, say, a bootstrap estimate of the quadratic risk, see Bunke (1986). The estimate (13) may be an essential basis for "smoothed" bootstrap procedures in the model (1) (Efron and Tibshirani (1986)). Although the form of (13) seems to be complicated, it may be easely applied in the simulation of bootstrap samples, as discussed in Bunke (1986).
A restriction to symmetric distributions F may be treated, assuming

$$F(y) = [G(y) - G(-y) + 1]/2 , \tag{16}$$

where G has a Dirichlet prior, but leads to very complicated expressions. In the location case (9) they are relatively simple, as shown in Diaconis and Freedman (1986). An alternative restriction of interest is to distributions with zero mean. Introducing certain sensible priors we may transform the problem to the original unrestricted one.
A restriction to absolutely continuous distributions F may be treated under the same assumptions as assumed in the purely nonparametric model of Lo (1984), namely mixture densities

$$F'(y) = \int K(y, u)\, G(du) , \tag{17}$$

where the kernel K is known and the unknown mixing distribution G has a Dirichlet prior. The resulting estimates have relatively complicated structure and only for small n there is a possibility of numerical calculation. A Monte Carlo approximation may be tried in other cases.

References

1. Bunke O. Bayesian estimators in semiparametric problems.- Preprint Nr. 102 , Humboldt-Univ.,Sektion Mathematik, Berlin 1985.
2. Bunke O. Assessment of the performance of estimators and confidence intervals for parameters in nonlinear regression under nonstandard conditions.- Preprint , Humboldt-Univ., Sektion Mathematik, Berlin 1986.

3. Diaconis P. , Freedman D. On inconsistent Bayes estimates of location.- Ann. Statist.,1986 , Vol. 14 ,p. 68-87.
4. Efron B. , Tibshirani R. Bootstrap methods for standard errors, confidence intervals, and other measures of statistical accuracy.- Statist. Sc. ,1986 , 1 , p. 54-77.
5. Hartigan J.A. Bayes Theory .- New York: Springer-Verlag, 1983.

Bernoulli, Vol. 2, pp. 031-034

SEMI-PARAMETRIC BAYES ESTIMATORS

Nils Lid Hjort
Norwegian Computing Centre, Oslo, Norway

1. INTRODUCTION

This paper discusses a semi-parametric Bayesian approach to fitting a parametric model to data. If the true density for observations X_j is f and the parametric structure is $\{f_\theta;\ \theta \in \Theta\}$, then the maximum likelihood estimator $\hat{\theta}_{ML}$ aims at the parameter value θ_1 that is least false according to the Kullback-Leibler information distance

(1) $$I(f:f_\theta) = \int f \log (f/f_\theta)\, dx.$$

The Bayes solution we will present is consistent for the value θ_2 that minimises another criterion, see (10) below. In general these are different; indeed θ_2 usually lies somewhere between θ_1 and a third variant θ_3 that is least false according to a criterion that is tied to prediction, or probability assessments for certain sets. This result can be rephrased in terms of "inconsistency of Bayes estimators", cf. Diaconis and Freedman (1986) and the discussion following it, and Doss (1985b). It can however be interpreted in a more positive light, in that fitting a parametric model to data from a f that does not belong to the class, as statisticians do every week, necessarily entails some notion of "least false" parameter values, and the Bayes solution and the ML solution just happen to correspond to two different notions of least false.

The main reference for the paper is Hjort (1986), where further motivation and some generalisations are given.

2. BUILDING A SEMI-PARAMETRIC MODEL AROUND A PARAMETRIC ONE.

As a motivating example, suppose the normal distribution is fitted to data X_j, so that $X_j = \mu + \sigma Y_j$, $j = 1,\ldots,n$, where the distribution G for Y_j equals $G_o = N(0, 1)$ in the idealised case. If normality cannot be trusted one might try to model uncertainty in G via a probability distribution on the space of distributions, centred at G_o in some sense. The most convenient way of doing this is by means of the Dirichlet process. Write $G \sim Dir(kG_o)$ to signify that G is such a process, with "prior sample size" parameter k and "prior guess" $EG = G_o$. If G is Dirichlet and $\theta = (\mu,\sigma)$ in addition is given some prior $\nu(\mu,\sigma)d\mu d\sigma$, then it turns out that θ's posterior distribution is exactly the same as the usual one that results from no randomness in G. (We assume that the distribution F for X_j has a density f, in particular there are no ties in data.) While one does get very interesting results about the posterior distribution of G, the distribution of $(X_j-\mu)/\sigma$, our main concern here is estimating the parameters $\theta = (\mu,\sigma)$, and then the following approach is more fruitful: Pin down the Dirichlet process G by conditioning on $G(B_i) = z_i$, $i = 1,\dots,m$, where $B_1,\dots,B_m$ are <u>control sets</u> partitioning the outcome-space for Y. For example, take

$$G(-\infty, -1.645] = .05,\ G(-1.645, 1.645] = .90,\ G(1.645, \infty) = .05.$$

(The z_i's must agree with the $G_o(B_i)$'s.) Then G splits into m separate and independent Dirichlet processes; $G = z_i G_i$ on the set B_i, where $G_i \sim \mathrm{Dir}(kz_i(G_o/z_i))$.

One can now work out the posterior density for $\theta = (\mu,\sigma)$. It becomes

$$\nu_n(\theta|x) = c(x)\, M_n(\theta)\, L_n(\theta)\, \nu(\theta), \tag{2}$$

where $L_n(\theta)$ is the usual likelihood $f_\theta(X_1)\dots f_\theta(X_n)$, and where

$$M_n(\theta) = \prod_{i=1}^{m} z_i^{C_i(\theta)} / \Gamma(kz_i + C_i(\theta)). \tag{3}$$

$C_i(\theta)$ is the number of X_j's that has $Y_j = (X_j-\mu)/\sigma$ in B_i, i.e., $C_i(\theta) = nF_n(\mu + \sigma B_i)$, writing F_n for the empirical distribution function.

Thus the posterior density have to (equally important, as we shall see) contributors; one that peaks in $(\hat\mu_{ML},\hat\sigma_{ML})$ which maximises $L_n(\theta)$, and one that peaks in say $(\tilde\mu,\tilde\sigma)$ which maximises $M_n(\theta)$. The latter one aims at getting $F_n(\mu+\sigma B_i)$ close to z_i for $i = 1,\dots,m$. In the example above this means trying to get about 5% of the $(X_j-\tilde\mu)/\tilde\sigma$ below -1.645, about 90% of them in $(-1.645, 1.645]$, and about 5% of them above 1.645.

The theory outlined above generalises to general parametric families, and in particular, to multivariate observations and to regression models. Derivations and examples will be published elsewhere.

3. WHAT ARE THE BAYES ESTIMATES REALLY ESTIMATING?

One is used to seeing that contributions to the posterior density that stem from the Bayesian's chosen prior are "washed out by the data" as the sample size grows. Work by Diaconis and Freedman and others have shown, however, that this pleasant phenomenon of "Bayesian stableness" must not at all be taken for granted in situations with infinite-dimensional parameters. This is nicely illustrated here: the factor $M_n(\theta)$ that comes from the Dirichlet modelling of the uncertainty of G around G_o proves to be of the same order of magnitude as the likelihood $L_n(\theta)$. A Stirling approximation shows that

$$(4) \qquad M_n(\theta) \doteq \text{const.} \exp\{-n\, I((F_n h_\theta B_i):(z_i))\},$$

which is of the same asymptotic importance as

$$(5) \qquad L_n(\theta) = \exp\{-n \frac{1}{n} \sum_{j=1}^{n} -\log f_\theta(X_j)\}.$$

Here $X_j = h_\theta(Y_j)$ is the appropriate generalisation of $X_j = \mu + \sigma Y_j$, and

$$(6) \qquad I(p:z) = \sum_{i=1}^{m} p_i \log (p_i/z_i)$$

is the Kullback-Leibler distance from probability vector p to probability vector z, cf. (1).

This shows that the posterior density asymptotically behaves like

$$(7) \qquad \nu_n(\theta) = c(x) \exp\{-n\, Q_n(\theta)\}\, \nu(\theta),$$

where $Q_n(\theta)$ is made up as a sum of

$$(8) \qquad Q_{n,1}(\theta) = -\frac{1}{n} \sum_{j=1}^{n} \log f_\theta(X_j)$$

and

$$(9) \qquad Q_{n,2}(\theta) = \sum_{i=1}^{m} F_n(h_\theta B_i) \log \{F_n(h_\theta B_i)/z_i\}.$$

Whereas $Q_{n,1}$ is smooth and well approximated with a quadratic function, $Q_{n,2}$ is very ragged, even if the points of discontinuity it has, since F_n is a step function, are smoothed away. It can still be approximated well enough with a quadratic function, however, to show that the Bayes solution $\hat{\theta}$ (under quadratic loss or other reasonable loss functions) tends to the parameter value θ_2 that minimises

$$(10) \qquad Q(F, \theta) = I(f:f_\theta) + I((Fh_\theta B_i):(G_o B_i)).$$

4. INFLUENCE FUNCTIONS AND ASYMPTOTIC NORMALITY

It is of interest to look into the large sample behaviour of the posterior density (2) and the corresponding Bayes estimates, in terms of limiting distribution results.

This proves to be a very delicate problem. It seems very difficult to calculate the influence function of $\hat{\theta}$ directly, for instance. One may show, however, that $\hat{\theta}$ is $\sqrt{n}$-equivalent to the estimator

(11) θ^* = the minimiser of $Q(F_n, \theta) = \phi(F_n)$,

where Q is the functional in (10), and the influence function $I(F, x) = \lim_{\varepsilon\to 0} \{\phi((1-\varepsilon)F + \varepsilon\delta_x) - \phi(F)\}/\varepsilon$ can be handled. After that the usual heuristic argument is that

$$\sqrt{n}\ (\theta^* - \theta_2) \doteq \frac{1}{\sqrt{n}} \sum_{j=1}^{n} I(F, X_j),$$

suggesting in its turn a limiting distribution result. Further technical arguments can be furnished to prove that the result and the procedure are valid.

In the idealised case when the parametric model happens to be correct i.e. $f = f_{\theta_0}$ for some θ_0, then both of the terms of the distance function (10) are minimised for $\theta = \theta_0$, and the Bayes solution $\hat{\theta}$ as well as its non-Bayesian sister θ^* are consistent for this value, making it meaningful to compare say the limiting covariance matrix with that of the efficient $\hat{\theta}_{ML}$. Limited experience indicates that one never loses much by preferring $\hat{\theta}$ (or θ^*) to $\hat{\theta}_{ML}$, even in the (never-occuring) $f = f_{\theta_0}$ case. In the simple example of Section 2 one gets limiting standard deviations of 1.058 and $1.042/\sqrt{2}$, compared to the optimal possible 1 and $1/\sqrt{2}$. It is a remarkable fact that one cannot lose more than a maximum of 6.066%, in terms of limiting standard deviation, in the normal family case, regardless of choice of control sets $B_1, \ldots, B_m$.

The Bayes estimates also tend to be robust, in that their influence functions are more cautious than those for the ML estimator.

5. SUMMARY

This paper has proposed a way of modelling uncertainty about a parametric model by building a larger semi-parametric model around it, using pinned-down Dirichlet processes and control sets that can be chosen by the statistician to reflect relative importance of predictions to be made after fitting the model. The resulting estimators aim at reasonable least false parameter values, are robust, and nearly efficient.

REFERENCES

Diaconis, P. and Freedman, D. (1986). On the consistency of Bayes estimates (with discussion). Ann. Statist. 14, 1-67.

Doss, H. (1985a,b). Bayesian nonparametric estimates of the median. I: Computation of the estimates. II: Asymptotic properties of the estimates. Ann. Statist. 13, 1432-1444 and 1445-1464.

Hjort, N.L. (1986). Discussion of Diaconis and Freedman (op. cit.). Ann. Statist. 14, 49-55.

ON ESTIMATING IN MODELS WITH INFINITELY MANY PARAMETERS

A.W. van der Vaart
Department of Mathematics
University of Leiden
P.O. Box 9512
2300 RA Leiden
The Netherlands

Important examples of semi-parametric models belong to the class of *mixture models*. Let Θ be an open subset of $\mathbb{R}^k$ and H be a collection of probability measures on a measurable space $(\mathcal{Z},\mathcal{A})$. For each $(\theta,z)\in\Theta\times\mathcal{Z}$ let $\underline{p}(\cdot,\theta,z)$ be a density with respect to a σ-finite measure μ on a measurable space $(\mathcal{X},\mathcal{B})$. Suppose that $\underline{p}(x,\theta,z)$ is measurable as a function of (x,z) and set

(1) $$p(x,\theta,\eta)= \int \underline{p}(x,\theta,z)\, d\eta(z).$$

A *mixture model* is defined by

$X_1,X_2,\ldots,X_n$ are i.i.d. random elements
$(\theta,\eta)\in\Theta\times H$ is unknown
X_j has density $p(\cdot,\theta,\eta)$ of type (1)

Mixture models are sometimes called *structural models* as opposed to *functional models*. The latter type of model is described by

$X_1,X_2,\ldots$ are independent random elements
$(\theta,z_1,z_2,\ldots)\in\Theta\times\mathcal{Z}^\infty$ is unknown
X_j has density $\underline{p}(\cdot,\theta,z_j)$ w.r.t. μ on $(\mathcal{X},\mathcal{B})$

In fact it is possible to embed the two models in a single and more general model

for each $n=1,2,\ldots$

(2)
$X_{n1},X_{n2},\ldots,X_{nn}$ are independent random elements
$(\theta,\eta_{n1},\ldots,\eta_{nn})\in\Theta\times H^n$ is unknown
X_{nj} has density $p(\cdot,\theta,\eta_{nj})$ w.r.t. μ on $(\mathcal{X},\mathcal{B})$
$p(\cdot,\theta,\eta)$ takes the form (1).

If H contains the degenerated distributions δ_z then the functional model is a submodel of (2), since $p(\cdot,\theta,\delta_z)=\underline{p}(\cdot,\theta,z)$.

Next suppose that there exist measurable functions $h(\cdot,\theta):(\mathcal{X},\mathcal{B}) \to \mathbb{R}$ $\psi(\cdot,\theta):(\mathcal{X},\mathcal{B}) \to \mathbb{R}^m$ and $g(\cdot,\theta,\eta):\mathbb{R}^m \to \mathbb{R}$ such that

(3) $\quad \underline{p}(\cdot,\theta,z)= h(\cdot,\theta)\ g(\psi(\cdot,\theta),\theta,z)$.

Then the model determined by (1)-(3) is a special case of the following model

for each n=1,2,...

(4)
$X_{n1},X_{n2},\ldots,X_{nn}$ are independent random elements
$(\theta,\eta_{n1},\ldots,\eta_{nn})\in \Theta \mathrm{x} H^n$
X_{nj} has density $p(\cdot,\theta,\eta_{nj})$ w.r.t. μ on $(\mathcal{X},\mathcal{B})$
$p(\cdot,\theta,\eta)= h(\cdot,\theta)\ g(\psi(\cdot,\theta),\theta,\eta)$

where $g(\cdot,\theta,\eta):\mathbb{R}^m \to \mathbb{R}$ is a density of $\psi(X,\theta)$ with respect to a measure ν_θ on $\mathbb{R}^m$ (if X has density $p(\cdot,\theta,\eta)$) and H may be an arbitrary set. The last condition in (4) means that, for fixed θ, there exists $\psi(X,\theta)$ which is sufficient for η.

In the i.i.d. version of model (4), where $\eta_{nj}= \eta$ for every n and j, it is often possible to construct a sequence of estimators which is efficient for θ in the sense of the convolution and local asymptotic minimax theorem (cf. van der Vaart(1986)). The idea behind the construction of such an estimator is as follows. We suppose that the density $p(\cdot,\theta,\eta)$ is differentiable with respect to θ in the sense that there exists a *score* $\ell(\cdot,\theta,\eta)$ *for* θ satisfying for every $h\in\mathbb{R}^k$

$$\int[t^{-1}(p^{\frac{1}{2}}(x,\theta+th,\eta)-p^{\frac{1}{2}}(x,\theta,\eta))-\tfrac{1}{2}h'\ell(x,\theta,\eta)p^{\frac{1}{2}}(x,\theta,\eta)]^2 d\mu(x) \to 0,$$

as $t \to 0$. In many examples which have the structure (4), *the efficient score function for* θ (cf. Begun et al.(1983)) is given by

$$\tilde{\ell}(\cdot,\theta,\eta)= \ell(\cdot,\theta,\eta)- E_\theta(\ell(X_{11},\theta,\eta)\,|\psi(X_{11},\theta)=\psi(\cdot,\theta)).$$

Hence, setting $\tilde{I}(\theta,\eta)= \int \tilde{\ell}(x,\theta,\eta)\tilde{\ell}(x,\theta,\eta)'\ p(x,\theta,\eta)\ d\mu(x)$, T_n is efficient for θ if

(5) $\quad \sqrt{n}(T_n-\theta)= n^{-\frac{1}{2}}\ \Sigma_{j=1}^n\ \tilde{I}^{-1}(\theta,\eta)\tilde{\ell}(X_{nj},\theta,\eta)+ o_{P_{\theta\eta}}(1)$.

A general method to obtain T_n is to replace $\tilde{\ell}(\cdot,\theta,\eta)$ by an estimated version $\hat{\ell}(\cdot,\theta)$ and to define T_n by the one step method, based on an initial $\sqrt{n}$-consistent estimator $\hat{\theta}_n$,

$$T_n = \hat{\theta}_n + n^{-1} \Sigma_{j=1}^n \hat{I}_n^{-1}(\hat{\theta}_n) \hat{\ell}(X_j, \hat{\theta}_n),$$

where $\hat{I}_n^{-1}(\hat{\theta}_n)$ should estimate $\tilde{I}(\theta,\eta)$. Here, the main problem is to find a suitable estimator for $\tilde{\ell}(\cdot,\theta,\eta)$, for given θ. In the i.i.d. version of the model (4) this is often possible. Indeed, assuming ν_θ to be Lebesgue measure and $g(\cdot,\theta,\eta)$ sufficiently smooth, we have

$$\ell(x,\theta,\eta)= \dot{h}/h(x,\theta)+ \dot{\psi}(x,\theta)\circ\nabla g/g(\psi(x,\theta),\theta,\eta)+ \dot{g}/g(\psi(x,\theta),\theta,\eta),$$

Here $\nabla g(\cdot,\theta,\eta)=(g^{(1)}(\cdot,\theta,\eta),\dots,g^{(m)}(\cdot,\theta,\eta))'$ is the vector of partial derivatives with respect to s of $g(s,\theta,\eta):\mathbb{R}^m \to \mathbb{R}$ and $\dot{h}$, $\dot{g}$ and $\dot{\psi}$ are the vectors, respectively a (k×m) matrix, of which the i-th rows contain partial derivatives with respect to θ. Hence

$$\text{(6)} \qquad \tilde{\ell}(x,\theta,\eta)=\tilde{H}(x,\theta) + \tilde{\psi}(x,\theta)\circ\nabla g/g(\psi(x,\theta),\theta,\eta),$$

where

$$\tilde{H}(x,\theta)= \dot{h}/h(x,\theta)- E_\theta(\dot{h}/h(X_{11},\theta)\,|\psi(X_{11},\theta)=\psi(x,\theta))$$

$$\tilde{\psi}(x,\theta)= \dot{\psi}(x,\theta)- E_\theta(\dot{\psi}(X_{11},\theta)\,|\psi(X_{11},\theta)=\psi(x,\theta)).$$

The key to the construction of an estimate $\hat{\ell}(\cdot,\theta)$ for $\tilde{\ell}(\cdot,\theta,\eta)$ is that (6) depends on η only through $\nabla g/g$. If in (4) $\eta_{nj}= \eta$ for every n and j, then for given θ, $\psi(X_{11},\theta),\dots,\psi(X_{nn},\theta)$ is an i.i.d. sample from the density $g(\cdot,\theta,\eta)$. The *kernel method* gives an estimate

$$\text{(7)} \qquad \hat{g}(s,\theta)= n^{-1}\Sigma_{j=1}^n \sigma_n^{-m}\omega(\sigma_n^{-1}(s-\psi(X_{nj},\theta)))$$

for $g(s,\theta,\eta)$. Then $\nabla\hat{g}/\hat{g}(s,\theta)$ should estimate $\nabla g/g(s,\theta,\eta)$ and plugging this into (6) yields a candidate for $\hat{\ell}(\cdot,\theta)$.

Now consider the behaviour of the estimator in the model with infinitely many nuisance parameters (4). The most important difference is that the sufficient statistics $\psi(X_{n1},\theta),\dots,\psi(X_{nn},\theta)$ are no longer i.i.d. The kernel estimate (7) does not estimate a common density $g(\cdot,\theta,\eta)$, but rather the *average* density

$$\bar{g}_n(s,\theta)= \bar{g}_n(s,\theta,\eta_{n1},\dots,\eta_{nn})= n^{-1}\Sigma_{j=1}^n g(s,\theta,\eta_{nj}).$$

Straightforward calculations show that exactly the same construction should yield an estimator satisfying

$$(8)\qquad \sqrt{n}(T_n-\theta)= n^{-\frac{1}{2}}\Sigma_{j=1}^n \tilde{I}_n^{-1}(\theta)\tilde{\ell}_n(X_{nj},\theta,\eta_{n1},\ldots,\eta_{nn})+ o_{P_{\theta\eta_{n1}\ldots\eta_{nn}}}(1),$$

where

$$\tilde{\ell}_n(x,\theta)= \tilde{H}(x,\theta) + \tilde{\psi}(x,\theta)\circ\bar{Q}_n(\psi(x,\theta),\theta,\eta_{n1},\ldots,\eta_{nn})$$

$$\bar{Q}_n(s,\theta,\eta_{n1},\ldots,\eta_{nn})= \Sigma_{j=1}^n \nabla g(s,\theta,\eta_{nj})/\Sigma_{j=1}^n g(s,\theta,\eta_{nj})$$

$$\tilde{I}_n(\theta)= \tilde{I}_n(\theta,\eta_{n1},\ldots,\eta_{nn})= \int \tilde{\ell}_n(x,\theta)\tilde{\ell}_n(x,\theta)'\ \bar{p}_n(x,\theta)\ d\mu(x).$$

Actually $\tilde{\ell}_n(\cdot,\theta)$ has the interpretation of being the *efficient score for* θ *of the average density* $\bar{p}_n(x,\theta)= n^{-1}\Sigma_{j=1}^n\ p(x,\theta,\eta_{nj})$.

While the estimator T_n is often optimal in the i.i.d. model, it is difficult to make a same statement for the performance in the model (4), due to the fact that it is unclear how to define an optimality concept in models with infinitely many nuisance parameters. However the estimator improves other constructions in the literature.

As an example consider the model (1)-(2) with $Z=(0,\infty)$ and

$$\underline{p}(x,y,\theta,z)= ze^{-zx}\ \theta ze^{-\theta zy}\ 1\{x>0,y>0\}.$$

In the functional form of this model we have a sequence of pairs (X_j,Y_j) of independent, exponentially distributed random variables with hazard rates z_j and θz_j respectively and the problem is to estimate the ratio θ of the hazard rates. Set

$$\bar{\eta}_n= n^{-1}\Sigma_{j=1}^n \eta_{nj}\ .$$

Under the condition that the sequence of measures $\{\bar{\eta}_n\}$ is tight in such a way that all limit point η_∞ have $\eta_\infty(0,\infty)=1$ (no mass should escape to either zero or infinity), the construction sketched above can be made rigorous and gives an asymptotically normal estimator with variance determined by (8). This follows from general results which will appear in van der Vaart(1987).

REFERENCES

Begun, J.M., Hall, W.J., Huang, W.M. and Wellner J. (1983). Information and asymptotic efficiency in parametric-nonparametric models. *Ann. Statist. 2*, 435-452.

Vaart, A.W. van der (1986). *Estimating a Real Parameter in a Class of Semi-Parametric Models*, Report 86-9, Dep. Math., Univ. Leiden.

Vaart, A.W van der (1987). *Statistical Estimation in Models with Large Parameter Spaces.* Thesis.

INFERENCE FOR STOCHASTIC PROCESSES

(Session 2)

Chairman: A.N. Shiryaev

SEMIMARTINGALE CONVERGENCE THEORY AND CONDITIONAL INFERENCE FOR STOCHASTIC PROCESSES

Feigin, P.D.
Technion - Israel Institute of Technology, Haifa, Israel.

1. CONVERGENCE OF EXPERIMENTS

We commence by giving some brief and non-rigorous background to the decision theoretic approach to asymptotics for statistical experiments.

Suppose we have a parameter space Θ and two experiments with observables X and Y and families of measures $\mathcal{P} = \{P_\theta\}$ and $\mathcal{Q} = \{Q_\theta\}$, respectively. Consider procedures (statistics) $\rho_{\mathcal{P}} = \delta_{\mathcal{P}}(X)$ and $\rho_{\mathcal{Q}} = \delta_{\mathcal{Q}} = \delta_{\mathcal{Q}}(Y)$ for each experiment. For the loss function $l(\rho,\theta)$ we may define the risk function as

$$R_{\mathcal{P}}(\rho_{\mathcal{P}},\theta) = \int l(\rho_{\mathcal{P}},\theta)\frac{dP_\theta}{dP_0}dP_0\ ; \tag{1.1}$$

where we assume $P_\theta << P_0$ for all $\theta \in \Theta$, for some $P_0 \in \mathcal{P}$; and similarly for $\mathcal{Q}$. Define $\rho^*_{\mathcal{P}}$ as the "best" procedure for $\mathcal{P}$ with respect to some criterion (e.g. "minimax").

Let $\mathcal{L}[\cdot|P]$ denote the "law of $\cdot$ under P". If, for any procedure $\rho_{\mathcal{Q}}$ for the $\mathcal{Q}$ experiment, there exists a procedure $\rho_{\mathcal{P}}$ such that

$$\mathcal{L}\left[\left(\rho_{\mathcal{P}},\left\{\frac{dP_\theta}{dP_0};\theta\in\Theta\right\}\right)|P_0\right] = \mathcal{L}\left[\left(\rho_{\mathcal{Q}},\left\{\frac{dQ_\theta}{dQ_0};\theta\in\Theta\right\}\right)|Q_0\right] \tag{1.2}$$

then it clearly follows (see (1.1)) that

$$R_{\mathcal{P}}(\rho_{\mathcal{P}},\theta) \equiv R_{\mathcal{Q}}(\rho_{\mathcal{Q}},\theta)\ ,\theta\in\Theta. \tag{1.3}$$

In this case, the "best" $\mathcal{Q}$ procedure can do no better than $\rho^*_{\mathcal{P}}$ with respect to the decided upon criterion for ordering risk functions. If we find a $\rho_{\mathcal{Q}}$ for which $R_{\mathcal{Q}}(\rho_{\mathcal{Q}},\theta)$ achieves the same level of the criterion as does $R_{\mathcal{P}}(\rho^*_{\mathcal{P}},\theta)$, then $\rho_{\mathcal{Q}}$ is optimal for $\mathcal{Q}$ as well.

The fact is that (1.3) is a triviality given (1.2). However, on studying (1.2) we see that the essential fact that makes the experiments comparable is the equivalence of the likelihood ratio processes. If we now introduce a sequence of experiments $\{\mathcal{Q}^t = \{Q_\theta^t\}\}$ and a sequence of corresponding procedures $\{\rho_{\mathcal{Q}}^t\}$, we need to show that

$$\mathcal{L}\left[\left(\rho_{\mathcal{Q}}^t, \left\{\frac{dQ_\theta^t}{dQ_0^t}; \theta \in \Theta\right\}\right) | Q_0^t\right] \to \mathcal{L}\left[(\rho, \{H_\theta; \theta \in \Theta\}) | P_0\right] \quad (1.4)$$

where $\{H_\theta\}$ is a likelihood process (i.e. $E_{P_0}(H_\theta) = 1$, for all $\theta \in \Theta$). This would then allow us to give asymptotic bounds to "best" (e.g. "minimax") procedures for $\{\mathcal{Q}^t\}$. In most cases, by a tightness argument, the joint weak convergence of $(\rho_{\mathcal{Q}}^t, \left\{\frac{dQ_\theta^t}{dQ_0^t}\right\}$ will follow from the weak convergence of the second component, at least along subsequences. Thus the essential aspect of verifying the convergence of experiments is in showing the weak convergence of the likelihood ratio processes.

Remark 1. If we think of t (or call it n) as a sample size index then typically Q_θ^t and Q_0^t will separate as $t \to \infty$. So, for the usual asymptotics to fit into this framework we are required to localize the original experiment about a given value θ_0 and define:

$$\mathcal{P}^t = \left\{P_h^t = Q_{\theta_0 + h\, a_t(\theta_0)}^t\, ; h \in H\right\} \quad (1.5)$$

with $a_t \to 0$ at rate fast enough to prevent separation of P_h^t and P_0^t, and slow enough to make them asymptotically distinguishable.

Remark 2. Given the convergence of $\{\mathcal{P}^t\}$ to $\mathcal{P}$ in the sense of convergence of likelihood ratios, together with weak convergence of procedures $\{\rho^t\}$ to ρ, we define the asymptotic risk of $\{\rho^t\}$ as that of ρ in $\mathcal{P}$. (The loss function l is given and fixed). These will be *locally asymptotic risks* if ρ^t and $\mathcal{P}^t$ are localized versions of some $\rho_{\mathcal{Q}}^t$ and $\mathcal{Q}^t$; see Feigin (1986) for more details.

2. LOCALLY ASYMPTOTIC MIXED NORMAL EXPERIMENTS

When the (localized) likelihood ratio process converges weakly to that of a normal shift experiment with a random variance then we say that the corresponding experiments are *locally asymptotic mixed normal* (LAMN). This behaviour occurs for many families of non-ergodic processes and we will illustrate the counting process case below. The LAMN conditions were discussed by Jeganathan (1982), Basawa and Scott

(1983), and others.

Given $\{Q^t\}$ as in Section 1, we proceed to formally define the LAMN property at $\theta_0 \in \Theta$. We let $\theta_t(h) = \theta_0 + a_t h$ where $\{a_t = a_t(\theta_0)\}$ converges to 0 and $h \in H$. If Θ is some (finite) vector space then $\{a_t\}$ may be a (matrix) linear operator sequence. We define $P^t = \left\{P_h^t \equiv Q_{\theta_t(h)}^t\right\}$ as the local experiment and $\left\{\Lambda_t(h) = \log\left[\frac{dP_h^t}{dP_0^t}\right]\right.$; $h \in H$ } as the logarithm of the likelihood ratio process.

<u>Definition</u>. $\{Q^t\}$ is LAMN at θ_0 if there exists a sequence $\{a_t\}$ as above and sequences of random variables $\{U_t\}$, $\{\sigma_t^2\}$ satisfying:

$$\text{(i) } \Lambda_t(h) + \frac{\sigma_t^2}{2}\left[(U_t - h)^2 - U_t^2\right] \xrightarrow{P_0^t} 0 \tag{2.1}$$

$$\text{(ii) } \mathcal{L}\left[(\sigma_t U_t, \sigma_t^2) \,|\, P_0^t\right] \to \mathcal{L}\left[(Z, \sigma^2) \,|\, P_0\right] \tag{2.2}$$

where Z is N(0,1) and independent of σ^2 under P_0.

Conditions (i) and (ii) ensure the convergence of the likelihood ratio process to that of the experiment

$$\mathcal{P} = \left\{P_h = \mathrm{N}(h, \sigma^{-2}) \times \mathcal{L}\left[\sigma^2 \,|P_0\right]\right\} \tag{2.3}$$

i.e. observe σ^2 with distribution fixed for all P_h (*ancillarity*), and then observe U with normal distribution with mean h and variance σ^{-2}.

The ancillarity of σ^2 for the *local* limit experiment makes a conditional analysis of this experiment most attractive. We show below that a decision theoretic argument can be given for computing conditional risk functions in such cases.

We shall turn to the role of semimartingale stable convergence theory in showing that the LAMN conditions hold. We note that other approaches are also possible - see, for example, Sweeting (1980).

Although the index t may not refer to time for the LAMN definition to make sense we are concerned with the applications to stochastic processes in time. We consider situations in which there is one single history (*filtration*) $\{F_u; 0 \leq u < \infty\}$ of increasing σ-fields and P^t is a restriction to F_t. Martingales will be defined in terms of this history or derived histories $G^t = \{F_{st}; 0 \leq s \leq 1\}$. We denote by P_0^∞ the extension of $\{P_0^t\}$ to $\vee_t F_t$.

A Taylor expansion in the scalar Θ case will demonstrate the semimartingale approach to verifying (i) and (ii). We denote by [M] the "square bracket" process of a martingale M, and by $\dot{f}$ the derivative of f with respect to h, and evaluated at $h = 0$. Then

$$\Lambda_t(h) \approx h\,\dot{\Lambda}_t(0) + (1/2)h^2\ddot{\Lambda}_t(0)\,. \tag{2.4}$$

We may readily show, under regularity conditions, that

$$\dot{\lambda}^t = \left\{\dot{\Lambda}_{st}\,;0 \le s \le 1\right\} \tag{2.5}$$

and

$$R^t = \ddot{\lambda}^t + [\dot{\lambda}^t] = \left\{\ddot{\Lambda}_{st} + [\dot{\Lambda}]_{st}\,;0 \le s \le 1\right\} \tag{2.6}$$

are $G^t = \{F_{st};0\le s\le 1\}$ martingales for each t. Expanding (2.4) we have

$$\Lambda^t(h) \approx h\,\dot{\lambda}^t - (1/2)\,h^2[\dot{\lambda}^t] + (1/2)\,h^2\,R^t \tag{2.7}$$

where $\Lambda^t = \{\Lambda_s^t(h) = \Lambda_{st}(h);0\le s\le 1\}$. Comparing (2.7) and (2.1) we make the identification:

$$\dot{\lambda}_1^t = \sigma_t^2\,U_t\,, \quad [\dot{\lambda}^t]_1 = \sigma_t^2 \tag{2.8}$$

As we show later, based on the convergence of the sequence of $\{[\dot{\lambda}^t]\}$ processes we can conclude that the martingale sequence $\{\dot{\lambda}^t\}$ also converges stably. Moreover, the martingales $\{R^t\}$ are typically asymptotically null because they involve an extra factor of a_t (and so are smaller) due to taking an extra derivative with respect to h or due to squaring a first derivative. One, of course, will often need to determine the behaviour of the approximation error in (2.4). Nevertheless, verifying the LAMN property centers on establishing the convergence in probability of the process $\{[\dot{\lambda}^t]\}$.

We present the relevant theorem in the next section. It provides the key to establishing stable convergence of the sequence of martingales $\{\dot{\lambda}^t\}$ in order to ensure that (ii) of the LAMN condition holds.

3. STABLE CONVERGENCE OF SEMIMARGINGALES

A more general result is given in Feigin (1985) - here we give the part required for the LAMN theory in the current context.

Suppose, for each t, that $Y^t = \{Y_s^t; 0 \le s \le 1\}$ is a martingale adapted to $G^t = \{G_s^t; 0 \le s \le 1\}$; and for which $Y_s^t = a_t M_{st}, G_s^t = F_{st}$ for $0 \le s \le 1$; and such that $E(Y_1^t)^2$ is bounded in t. Then we have the following:

Proposition 1. If

$$\text{(i) } a_t \to 0 \text{ as } t \to \infty \tag{3.1}$$

$$\text{(ii) } \mathrm{E}\left[\sup_{s \le 1} |\Delta Y_s^t|\right] \to 0 \text{ as } t \to \infty \tag{3.2}$$

$$\text{(iii) } [Y^t]_1 \xrightarrow{P_0^\infty} \eta^2 \tag{3.3}$$

then $\mathcal{L}[Y_1^t] \to \mathcal{L}[Z \cdot \eta^*]$ (stably). (Here $Z \sim N(0,1)$ and $\mathcal{L}[\eta^*] = \mathcal{L}[\eta]$, with η^* independent of Z).

The stability (or Renyi mixing) property allows one to deduce that, for example,

$$\mathcal{L}\left[\left([Y^t]_1^{-\frac{1}{2}} Y_1^t, [Y^t]_1\right)\right] \longrightarrow \mathcal{L}[(Z, \eta^*)] \tag{3.4}$$

as required for (ii) of the LAMN condition.

Example. Let $X = \{X_t; t \ge 0\}$ be a counting (point) process with a Q_θ stochastic intensity $\{u_t(\theta); t \ge 0\}$ with respect to Lebesque measure. It follows that

$$\Lambda_t(h) = \int_0^t g_s \, dM_s - \int_0^t [\exp(g_s) - g_s - 1]\, u_s \, ds \tag{3.5}$$

where

$$g_s(h) = \log\left(u_s(\theta_t(h)) \,/\, u_s(\theta_0)\right)$$

and

$$M_t = M_t(\theta_0) = X_t - \int_0^t u_s(\theta_0)\, ds\,. \tag{3.6}$$

Here again $\theta_t(h) = \theta_0 + a_t h$, and a_t is to be determined. We now write X_s^t for X_{st}, u_s^t for u_{st}, and so on. Remember that $\dot{g}$ means the derivative of g with respect to h at $h = 0$.

In this example we therefore have:

$$\dot{\lambda}^t = \int \dot{g}_s^t \, dM_s^t\,, \quad [\dot{\lambda}^t] = \int \left(\dot{g}_s^t\right)^2 dX_s^t \tag{3.7}$$

and

$$\ddot{\lambda}^t = \int \ddot{g}_s^t \, dM_s^t - \int \left(\dot{g}_s^t\right)^2 u_s^t \, ds \tag{3.8}$$

with, after a little manipulation,

$$R^t = \int \left(\ddot{g}_s^t + \left(\dot{g}_s^t\right)^2\right) dM_s^t \tag{3.9}$$

Writing $\dot{g}_s^t = a_t\dot{v}_{st}$ and $\ddot{g}_s^t = a_t^2\ddot{v}_{st}$ and if $\{(\dot{v}_s^2 + \ddot{v}_s)/\dot{v}_s\}$ remains bounded, then whenever $\int_0^{1\cdot t}\dot{g}_s^t\, dM_s^t$ converges so does R_1^t to 0 (since $a_t \to 0$). So we need to choose a_t so that

$$a_t^2 \int_0^1 (\dot{v}_{st})^2\, dX_s^t = a_t^2 \int_0^t (\dot{v}_s)^2\, dX_s \xrightarrow{P_0^\infty} \sigma^2 \; ; \qquad (3.10)$$

then not only will Proposition 1 apply, but in most cases the LAMN condition will follow immediately for these counting processes.

Remark 3. Since there is only one underlying history, the stability (or mixing) of the weak convergence comes virtually automatically using the arguments in Feigin (1985). It does not follow directly from earlier work on weak convergence to processes with conditionally independent increments - see for example Liptser and Shiryayev (1980).

4. A DECISION THEORETIC CONDITIONALITY APPROACH

The limit experiment P of (2.3) has a shift invariance structure and if the loss function l makes the whole problem invariant then Hajek (1965) argued that the *conditional risk* given σ^2 is the appropriate risk function for comparing procedures. He even used terms such as "obligatory" when discussing the use of this conditional risk function.

Since the analysis is for the limit experiment we may now assert that the (locally) asymptotic conditional risk ((L)ACR) is the correct way to evaluate and compare procedures asymptotically for the original sequence of experiments.

We take the example of a fixed level (think of this as a fixed "risk") confidence interval. In this context $l(\rho,h) = 1 - I_\rho(h)$ where I is the indicator function of the subscript set and we choose to only consider symmetric intervals $\rho = (U - g(\sigma), U + g(\sigma))$ for h. The conditional level (risk) is

$$R(\rho,h) = 2\,[1 - \Phi(\sigma\, g(\sigma))]\;, \text{for all } h \in H\,. \qquad (4.1)$$

Since we wanted a "fixed" or given level we must have $\sigma g(\sigma)$ = constant, say $g(\sigma) = \sigma^{-1} z_{1-\alpha/2}$.

To complete the story we need to determine which sequence of confidence interval procedures for the original experiments has local asymptotic level given by (4.1). From the LAMN property, the sequence of intervals (procedures)

$$\{\rho_t = (U_t - g(\sigma_t), U_t + g(\sigma_t))\} \tag{4.2}$$

converges weakly to the ρ described above.

One usually requires one further step - removing the dependence on θ_o from U_t and σ_t by replacing them with asymptotically equivalent statistics depending only on the data. Le Cam (1986) describes the general theory and we will not pursue it further here. Some more details appear in Feigin (1986).

5. OTHER APPROACHES AND PROBLEMS

Bayesian Analysis. Heyde and Johnstone (1979) showed that a Bayesian analysis would lead, asymptotically to the same inference procedures as the conditional approach described above. Because of the smoothing effect of integrating with respect to a positive prior density, many of the technical problems of the conditional approach are averted.

Convergence of Conditional Distributions. The decision theoretic approach that we discussed above avoids any attempt to evaluate the conditional risks for each t, or to show that they converge in some sense to the conditional risk of the asymptotic experiment. For the confidence interval example, we may wish to show that

$$P_h^t(U_t - g(\sigma_t) \leq h \leq U_t + g(\sigma_t) \mid \sigma_t) \tag{5.1}$$

converges in probability to

$$P_h(U - g(\sigma) \leq h \leq U + g(\sigma) \mid \sigma) . \tag{5.2}$$

A result such as this would further confirm the treatment of σ_t as "asymptotically" (or "locally" or "approximately") ancillary. These kinds of results are harder to prove in general. One compromise is to condition on a *smoothed* version σ_t^* - see Feigin (1986) for some more details.

REFERENCES

Basawa, I.V., and Scott, D.J. (1983). Asymptotic Optimal Inference for Non-ergodic Models. Lecture Notes in Statistics, New York, Springer Verlag.
Feigin, P.D. (1985). Stable Convergence of Semimartingales. Stoch. Proc. Appl., V.19, pp. 125-134.
Feigin, P.D. (1986). Asymptotic Theory of Conditional Inference for Stochastic Processes. Stoch. Proc. Appl., V. 22, pp. 89-102.
Hajek, J. (1965). On Basic Concepts of Statistics. Proc. Fifth Berkeley Symp. on Math. Stat. and Probab., V. I, pp. 139-162.
Heyde, C.C., and Johnstone, I.M. (1979). On Asymptotic Posterior Normality for Stochastic Processes. J. Roy. Statist. Soc. B, V.41, pp. 184-189.
Jeganathan, P. (1982). On the Asymptotic Theory of Estimation when the Log-Likelihood Ratios is Mixed Normal. Sankhya, V. A44, pp. 173-212.
Le Cam, L.M. (1986). Asymptotic Methods in Statistical Decision Theory. New York, Springer Verlag.
Liptser, R.Sh., and Shiryayev, A.N. (1980). A Functional Central Limit Theorem for Semimartingales. Th. Probab. Appl., V. XXV, pp. 667-688.
Sweeting, T.J. (1980). Uniform Asymptotic Normality of the Maximum Likelihood Estimator. Ann. Statist., V. 8, pp. 1375-1381.

Bernoulli, Vol. 2, pp. Ø49-Ø54

THE FOUNDATIONS OF FINITE SAMPLE ESTIMATION IN STOCHASTIC PROCESSES - II

V.P. Godambe
University of Waterloo, Waterloo, Ontario, Canada N2L 3G1.

ABSTRACT. This paper provides a generalization of the theory of finite sample estimation for stochastic processes proposed by Godambe (1985). The generalization is in two directions. One, single parameter case is now extended to multiparameter case. Second, the martingale structure and its corresponding standard filtration is extended to absorb some features of general spatial processes.

INTRODUCTION. Let $y_1,\ldots,y_n$ be a discrete stochastic process. Since we study finite sample properties we restrict the process to R^n. For the process $y_1,\ldots,y_n$ we assume the class of possible distributions $F = \{F\}$ on R^n. On F is defined a real parameter θ, such that for some given real functions $h_i(y_1,\ldots,y_i\ ;\ \theta)$ of the arguments shown,

$$E_{i-1,F}[h_i\{y_1,\ldots,y_i;\theta(F)\}] = 0,\ F\varepsilon F,\ i = 1,\ldots,n, \tag{1}$$

$E_{i-1,F}$ denoting the expectation under F, holding $y_1,\ldots,y_{i-1}$ fixed. A special form of h_i common in the literature is $h_i = y_i - E_{i-1,F}(y_i)$. To estimate the unknown parameter θ from the observations $y_1,\ldots,y_n$ we construct a real function $g = g(y_1,\ldots,y_n,\theta)$ of the arguments shown such that

$$E_F[g\{y_1,\ldots,y_n;\theta(F)\}] = 0,\ F\varepsilon F. \tag{2}$$

A solution in θ, assuming it exists, of the equation $g = 0$ is said to provide an estimate of θ. Therefore any function g satisfying (2) is called an *unbiased estimating function* and $g = 0$ is called an *estimating equation*. We now define the class of all unbiased estimating functions

$$G = \{g : E_F[g\{y_1,\ldots,y_n;\theta(F)\}] = 0,\ F\varepsilon F\}. \tag{3}$$

To define an optimum estimating function g^* in G, we assume the underlying class of distributions $F = \{F\}$ mentioned above to be a union of parametric families of distributions. That is for every distribution $F \epsilon \mathcal{F}$ is defined a parameter ν, $\nu(F)$ being called 'the form of the distribution F' where $\nu(F)$ takes values in the set C as F varies over $\mathcal{F}$. Further we assume $F \Longleftrightarrow (\theta(F), \nu(F))$, hence we write $F = F_{\theta,\nu}$, and

$$\mathcal{F} = \bigcup_{\nu \epsilon C} \{F_{\theta,\nu} : \theta \epsilon \Omega\} \tag{4}$$

where the range of θ, namely Ω, is independent of ν. Next we denote by $f_{\theta,\nu}$ the density function of $F_{\theta,\nu}$ w.r.t. the underlying Lebesque measure on $\mathcal{R}^n$. The central assumption of our approach is the existence of the θ-score function.

$$(\partial \log f_{\theta,\nu}/\partial\theta). \tag{5}$$

Definition 1. An estimating function g^* is said to be *optimal* in G in (3) if $g^* \epsilon G$ and for every estimating function $g \epsilon G$,

$$E_{\theta,\nu}(g^* - \mu \frac{\partial \log f_{\theta,\nu}}{\partial\theta})^2 \leq E_{\theta,\nu}(g - \mu \frac{\partial \log f_{\theta,\nu}}{\partial\theta})^2 \tag{6}$$

for all $\theta \epsilon \Omega$ and $\nu \epsilon C$ i.e. for all $F \epsilon \mathcal{F}$ of (4), μ being some real function of θ and ν. Restricting the above definition of 'optimality' to a subclass G' of G in (3) of estimating functions which satisfy some 'regularity' conditions, it can be shown (Godambe & Thompson, 1985) that g^* is 'optimal' in G' if and only if $g^* \epsilon G'$ and

$$E_F[g^{*2}\{y_1,\ldots,y_n;\theta(F)\}] / [E_F\{ \frac{\partial g^*(y_1,\ldots,y_n;\theta)}{\partial\theta}\Big|_{\theta=\theta(F)}\}]^2$$
$$\leq E_F g^2\{y_1,\ldots,y_n;\theta(F)\} / [E_F\{ \frac{\partial g(y_1,\ldots,y_n;\theta)}{\partial\theta}\Big|_{\theta=\theta(F)}\}]^2 \tag{7}$$

for $g \epsilon G'$, for all $F \epsilon \mathcal{F}$ of (4). It follows that the optimum estimating function g^* *satisfying* (7) *and* (6) *is defined only upto a constant multiple.*

It is easy to varify that estimating function g^* satisfying (7) is also the one which in G' is most highly correlated with the (true) underlying score functions:

$$\text{Corr.}_{\theta,\nu}(g^*, \frac{\partial \log f_{\theta,\nu}}{\partial\theta}) \geq \text{corr.}_{\theta,\nu}(g, \frac{\partial \log f_{\theta,\nu}}{\partial\theta}), \tag{8}$$

for all $g \epsilon G'$ and $\theta \epsilon \Omega$, $\nu \epsilon C$.

In general no optimal estimating function g^* exists in G or G' (Godambe & Thompson, 1985). Hence we restrict further the definition of 'optimality' to a subclass G_ℓ (of G' and G) of *linear estimating functions*. With reference to functions h_i in (1), we define a linear estimating function g as

$$g = \sum_{i=1}^{n} h_i \; a_{i-1} \tag{9}$$

where $a_{i-1} = a_{i-1}(y_1,\ldots,y_{i-1},\theta)$ are any real functions of the arguments indicated, $i = 1,\ldots,n$. Note g in (9) is a linear function of h_i, $i = 1,\ldots,n$. With appropriate *regularity conditions* on the functions h_i and a_{i-1}, $i=1,\ldots,n$ if G_ℓ is the class of all *regular* linear estimating functions,

$$G_\ell = \{g\} \tag{10}$$

then because of (1), $G_\ell \subset G' \subset G$. Now in terms of the functions h_i, we define a *subclass* F' *of* F in (4) of the underlying distributions:

$$F' = \{F: \frac{E_{i-1,F}(\partial h_i/\partial\theta)}{E_{i-1,F}(h_i^2)} = r(F)s_i(y_1,\ldots,y_{i-1}\theta(F)), \quad i = 1,\ldots,n\} \tag{11}$$

where r and s_i are some real functions of the arguments shown, $i = 1,\ldots,n$.

Theorem 1. Restricting Definition 1 of 'optimality' to the class G_ℓ in (10) of the regular linear estimating functions and to the class F' in (11) of the underlying distribution functions we have the optimum estimating function, given by,

$$g^* = \sum_i^n h_i\{E_{i-1,F}(\partial h_i/\partial\theta)\}/E_{i-1,F}(h_i^2). \tag{12}$$

For the proof of the above theorem and its application we refer to Godambe (1985). It is interesting to compare the 'finite sample' character of the theorem with the asymptotic methods of estimation. Let $y_0,\ldots,y_n$ be a branching process with $y_0 = 1$ and y_1 having some parametric distribution with $E(y_1) = \theta$ and $E(y_1-\theta)^2 = \sigma^2$. Taking $h_i = y_i - \theta y_{i-1}$ we have, for $i = 1,\ldots,n$, $E_{i-1}h_i = 0$ and $E_{i-1}h_i^2 = \sigma^2 y_{i-1}$. Hence the optimum estimating function

$$g^* = \sum_{i=1}^{n} (y_i - \theta y_{i-1}); \tag{13}$$

'$g^* = 0$' also happens to be the maximum likelihood equation for the parametric distribution put forward by Kendall (1949). But our justification of g^* is, of course, without any distributional assumption. There are some intrinsic difficulties in justifying and/or explaining, in terms of asymptotic properties, the estimation based on $g^* = 0$, or any other estimation for that matter. Suppose the data in hand, namely $y_0, \ldots, y_n$ shows that the 'extinction' has already occurred. How can then one investigate the asymptotic properties of the estimation? Even if the extinction has not occurred upto the nth epoch, to do 'asymptotics' one will have to assume being on a non-extinction sequence. How from a finitie segment, can one tell if it is a non-extinction sequence or otherwise? On the other hand, in our theory, non-extinction or extinction (up to the nth epoch) is the part of the data; in the latter case it gives a reasonable estimate $\hat{\theta} < 1$, a condition for extinction.

Suppose we replace in the branching process discussed above the mean θ by θ_1 and the variance σ^2 by θ_2. Let g_1^* be obtained from g^* in (13) by replacing in it θ by θ_1. For a given (known) θ_1, it can be shown (Godambe, 1985) that an optimum estimating function g_2^* for θ_2 is given by

$$g_2^* = \Sigma_1^n \{(y_1 - \theta_1 y_{i-1})^2 - \theta_2 y_{i-1}\}/(c-3+2y_{i-1}), \tag{14}$$

c being the kurtosis of the distribution of y_1. Thus the estimating functions g_1^* and g_2^* are *separately* 'optimal' for the parameters θ_1 and θ_2 respectively. Now we ask the question, when are g_1^* and g_2^* *jointly* optimal for θ_1 and θ_2? (Answer: if $E(y_1-\theta_1)^3 = 0$!) To answer this question, we develop, a general theory of estimating many parameters of a spatial process.

<u>MANY PARAMETERS</u>. The functions $h_i(y_1 \ldots y_{i-1}, \theta), i=1,\ldots,n$ of Theorem 1 are now replaced by functions $h_j(y_1 \ldots y_n; \theta_1, \ldots, \theta_m) \equiv h_j(\underset{\sim}{y}, \underset{\sim}{\theta})$, $j = 1,\ldots,k$, k arbitrary. The condition $E_{i-1,F} h_i = 0$ in (1) is now replaced by

$$E_{jF} h_j = 0, \quad j=1,\ldots,k \tag{15}$$

where $E_{jF} \equiv$ expectation (F), conditional on $\underset{\sim}{y} \varepsilon (j)$ a specified subset of R^n. Note for $j \neq j'$, the subset (j) may be $\equiv (j')$. We standardize the function h_j as

$$h_j^*(\alpha) = h_j \frac{E_{jF} \dfrac{\partial h_j}{\partial \theta_\alpha}}{E_{jF}(h_j)^2}, \ j=1,\ldots,k; \ \alpha=1,\ldots,m \tag{16}$$

Now corresponding to F' in (11), we introduce the underlying class of distributions

$$F' = \{F: E_{jF} h_j = 0, \ \frac{E_{jF}(\partial h_j/\partial\theta_\alpha)_{\theta_\alpha=\theta_\alpha(F)}}{\varepsilon_{jF}(h_j^2)} = \gamma(F) S_{j\alpha}(\underset{\sim}{y}, \underset{\sim}{\theta}(F)),$$

$\gamma, S_{j\alpha}$ being some functions of the indicated arguments, $(j=1,\ldots,k, \alpha=1,\ldots,m)$, $E(h_j^* h_j^* q_j) = 0$ for $j \neq j'$ (17)
$q_j = q_j(\underset{\sim}{y}, \underset{\sim}{\theta})$, being *any* function, of the arguments shown, which is constant on the set (j), $j,j'=1,\ldots,k$,}

Note the *orthogonality* of h_j^* and $h_{j'}^*$, $j \neq j'$ in (17). A multi-parametric generalization of the class, G_ℓ in (10), of regular, linear, unbiased estimating functions is given by

$\underset{\sim}{G}_\ell = \{\underset{\sim}{g}: \underset{\sim}{g} = (g^{(1)},\ldots,g^{(r)},\ldots,g^{(m)})$ where

$$g^{(r)} = \sum_{\alpha=1}^{m} \sum_{j=1}^{k} h_j^*(\alpha) q_{j\alpha}^{(r)}, \tag{18}$$

$q_{j\alpha}^{(r)}$ being any function of $(\underset{\sim}{y}, \underset{\sim}{\theta})$ which is constant on the subset (j), $j=1,\ldots,k$, $r=1,\ldots,m\}$

<u>AN EXTENTION OF THE OPTIMALITY CRITERION (7).</u> Let corresponding to the estimating function $\underset{\sim}{g}$, matrix $J \equiv ||E_F g^{(r)} g^{(r')}||$, $r,r'=1,\ldots,m$ and $H \equiv ||E_F \partial g^{(r)}/\partial\theta_\alpha||$, $r,\alpha=1,\ldots,m$. Further let J^* and H^* be the corresponding matrices for the estimating function $\underset{\sim}{g}^* = (g^{(1)*},\ldots,g^{(r)*},\ldots,g^{(m)*})$. Now in the class of estimating functions $\underset{\sim}{G}_\ell$ in (18), the estimating function $\underset{\sim}{g}^*$ is said to be *optimal* if for any $\underset{\sim}{g} \varepsilon \underset{\sim}{G}$ in (18),

$$J - H(H^*)^{-1} J^* (H^{*t})^{-1} H^t \tag{19}$$

is positive semidefinite for all $F \varepsilon F'$ in (17).

A similar criterion for a multiparametric set up was proposed by Durbin (1960) and Bhapkar (1972).

Theorem 2. According to the above extended criterion of optimality, the optimal estimating function $\underset{\sim}{g}^*$ is given by

$$g^{(r)*} = \sum_{j=1}^{k} h_j^* (\alpha), \quad r = \alpha=1,\ldots,m. \tag{20}$$

Proof. It is easy to check that the matrix

$$||E_F \partial g^{(r)}/\partial\theta_\alpha|| \equiv ||E_F g^{(r)} g^{(\alpha)*}||, \quad r,\alpha=1,\ldots,m \tag{21}$$

for all $F\varepsilon F'$ in (17) because of the 'orthogonality' condition in (17). As shown in Godambe and Thompson (1986), the equation (21) implies the positive semidefiniteness of (19); hence the proof.

Similar theorems could be proved using appropriate multiparametric extentions of the criteria (6) and (8).

Illustration. In the branching processes with mean θ_1 and variance θ_2, discussed before, we have, for $j=1,\ldots,n,(k=2n,m=2)$,

$$h_j = (y_j - \theta_1 y_{j-1}),$$
$$h_{2j} = \{(y_j - \theta_1 y_{j-1})^2 - \theta_2 y_{j-1}\}.$$

Now $h_j^*(1)=-(y_j-\theta y_{j-1})\frac{1}{\theta_2}$, $h_j^*(2)=0$, $h_{2j}^*(1) = 0$ and $h_{2j}^*(2)=(c-3+2y_{j-1})^{-1}$, c as in (14) being the kurtosis of the distribution of y_1. It is easy to check that the condition of 'orthogonality' in (17) is satisfied if the $E(y_1-\theta_1)^3 = 0$. Hence under the condition, by Theorem 2, the estimating functions g_1^* and g_2^* given by (13) and (14) are *jointly* optimum for estimating θ_1 and θ_2.

REFERENCES.

Bhapkar V.P. (1972). On a measure of efficiency of an estimating equation. Sankhyā 34, 467-472.

Durbin J. (1960). Estimation of parameters in time-series regression models. J.R. Statist. Soc. B22, 139-153.

Godambe V.P. (1985). The foundations of finite sample estimation in stochastic processes. Biometrika, 72, 419-428.

Godambe V.P. and Thompson M.E. (1985). The logic of least squares revisited. (Unpublished).

Godambe V.P. and Thompson, M.E. (1986). Parameters of super-population and survey population: Their relationships and estimation. I.S. Review, 54, 123-134.

Kendall D.G. (1949). Stochastic processes and population growth. J.R. Statist. B., 230-264.

SOME USES OF MAXIMUM-ENTROPY METHODS FOR ILL-POSED PROBLEMS IN SIGNAL AND CRISTALLOGRAPHY THEORIES.

D. DACUNHA-CASTELLE (Université Paris Sud ORSAY France)

Let f a positive function on the torus $\mathbb{T}^k$ and

$$H(f) = -\int_{\mathbb{T}^k} f(x) \log f(x)dx \leqq \infty$$

the entropy of f.

1. Maximum entropy (ME) as a summation procedure.

Let (e_n) be an orthonormal system of $L^2(\mathbb{T})$, for instance $e_n(x)=\exp inx$. Let $f_k=\langle f,e_k\rangle$. Let (S_m) a summation procedure, i.e. $S_m : C^m \to L^2(\mathbb{T})$, in order that for every f, $(S_m[(f_j,|j| \leqslant m])_k = f_k$, $|k| \leqslant m$.

By classical, we mean linear usual summation procedures as Dirichlet, Cesaro, Fejer, De la Vallée Poussin.

If $C_m = \{c_j, j\leqslant|m|\}$ is given, $C_m\in\mathbb{C}^{2m}$, let ε_m the set of functions f such that $f_j=c_j$, $j \leqslant |m|$, and let f_m^{ME} an element of ε_m of largest entropy.

THEOREM. Let $f\geqslant 0$, $\int f(x)dx = 1$, $H(f) < \infty$ and $c_j=f_j$, $|j|\leqslant m$ in the previous notations.

1. Then f_m^{ME} is uniquely defined by

$$f_m^{ME}(x) = Z^{-1}(\lambda_j, |j| \leqslant m) \exp \sum_{|j|\leqslant m} \lambda_j e_j(x)$$

where λ_j are the multipliers of the Lagrange associated problem and Z the normalizing function.

2. If $\log f\in L^2$, then $\lim_m f_m^{ME} = f$ (in L^2).

The proof of (1) is easy, see GASSIAT for results on

the speed in [2]. For instance we have :

$\forall\lambda>0\ \exists n(\lambda)$ such that for

$$n>n(\lambda),\ K(f,f_n) \leqslant (1+\lambda^4 \|f\|^2_{L^2})^{1/2} \quad \|e_n\|_{L^2}$$

when e_n is the tail of the Fourier expansion of $\log f$. K being the Kullback distance.

So we have suggested to use M.E procedure as a summation procedure when $f \geqslant 0$ and m any integer.

Our heuristic, checked by experimentation, is that M.E is much better than classical procedures when f looks like a mixture of a small number of regular approximations of f functions. The improvement can be surprising (see GASSIAT). It seems that however the smoothness of the functions have also some importance.

2. Signal theory.

M.E entropy has been suggested as a method of extrapolation in signal theory for spectra unknown out of a band, or as a method of restoration for images [5].

Of course, the advocates of the use of M.E claim that it is as a "panacée".

Our conviction, built on experiments is that is not a correct position. So it is necessary to justify what kind of situations are suitable for M.E. Our feeling is that it is only the case when the image is very bright for some pixels isolated, and very dark for all the others. We are now trying to give a definitive quantitative form to this heuristic.

In all these ill-posed linear problems M.E seems so interesting (there are of course other non-linear methods which are good competitions).

3. Cristallography.

Let f be the electronic density of a crystal ; i.e. $f>0$ on $\mathbb{T}^3$. By mean of X-rays diffraction we measure some $|f_k|$, $k \in A$, with $f_k = \langle f, e_k \rangle$, $e_k = \exp ik.x$, $k \in \mathbb{Z}^3$. # A is large, from 10^2 to 10^5 for proteins.

$|f_k|$ can be measured exactly, or with some incertitude

$$\sum_{k\in A} \big| |f_k|_{obs} - |f_k|_{true} \big|^2 \leqslant \varepsilon^2 .$$

We know also a few $\varphi_k = \arg\langle f, e_k \rangle$ (initial phasis). The crystallographic problem is how to build f, thus first choose φ_k for $\varphi_k \in A$ and then choose f_k, $k \notin A$. The problem is known as the PHASE-problem. It is clearly a (very) ill-posed problem. Here f is near a δ-function mixture and there is always some noise around.

The results are functions f looking to the one described in 2. Of course, stereo chimical constraint helps to the choices, and it is not necessary to have a very accurate precision for φ_k (10 % can be consider as good). G. BRICOGNE [1] has proposed to choose f as f_A^{ME} where f_A^{ME} maximizes the entropy under the constraints $|f_{A,k}^{ME}| = |f_k|$, $k\in A$. This problem has a high degree of complexity for the following reasons :

1. It is a macro-problem, which has to be splitted in a large number of ordinary optimisation problem.
2. It is a non linear problem.
3. The lack of convexity implies that solutions are not unique. Combined with splitting, it furnishes a tree of iterative solutions.
4. With G. BRICOGNE, we suggest the use of a stochastic model in order to help to the solution of the previous point. Let us consider the electronic density f as the density of $X_1, \ldots, X_n$ i.i.d., where the parameter n is interpreted as a degree of liberty and is unknown. Suppose $f_k = 1/n \sum_{k=1}^{n} \exp i\langle k, X_j \rangle$.
   Let $A = \bigcup_k A_k$, A_k increasing, $B_k = A_k \setminus A_{k-1}$.
   If at the k-step, $\{\varphi_j, j\in A_k\}$ has been chosen (the phase-problem has been solved up to the k-steps), we shall use the conditional likelihoods of $(\varphi_j, j\in B_{k+1})$ with respect to $(|f_j|, j\in A_{k+1})$ in order to choose the M.E solution at step (k+1) [See Dacunha-Castelle [3] for details].

Another interesting point of view on the use of M.E in cristallography is NAVAZA [5].

CONCLUSION.

ME as other non linear methods is a good way to solve ill-posed problems. But all situations are not suitable for it, and it is necessary to developp a mathematical theory of stochastic extrapolation.

BILIOGRAPHIE.

[1] G. BRICOGNE. Maximum entropy and the foundations of direct methods (Acta Cryst.) 1984.

[2] S.F. BURCH, S.F. GULL, J. SKILLING. Computer vision Graphics and Image Processing n°23, 1983, p.113-128.

[3] D. DACUNHA-CASTELLE. Reconstruction des phases en cristallographie par maximum d´entropie. Semin. Bourbaki 36e année 1983-1984, n° 628.

[4] E. GASSIAT. Problème sommatoire par maximum d´entropie C.R. Acad. Sc. Paris, t. 303, Série I, 14, 1986.

[5] J. NAVAZA. On the maximum entropy estimate of the electron density function. Acta Cryst. 1985, A 41, p. 232-244.

On Asymptotic Efficiency of the Cox Estimator

K. Dzhaparidze
Centre for Mathematics and Computer Science,
P.O. Box 4079, 1009 AB Amsterdam, The Netherlands

Let r_n, $n=1,2, \cdots$ be a non-decreasing sequence of integers. Let

$$N^n = (N_t^n, F_t^n, P_{\alpha,\beta}^n) \quad \text{and} \quad A^n = (A_t^n, F_t^n, P_{\alpha,\beta}^n),$$

$$A_t^n = \int_0^1 col\{Y_s^{in} e^{\beta Z_s^{in}}, \ i=1, \cdots, r_n\} d\alpha_s$$

be an r_n-variate counting process and its compensator, resp., defined on a stochastic basis which is n-th in the sequence

$$(\Omega^n, F^n, \{F_t^n, \ 0 \leqslant t \leqslant 1\}, P_{\alpha,\beta}^n), \quad n=1,2, \cdots, \tag{1}$$

where F_t^n-predictable censoring factors Y_t^{in} and the covariate processes Z_t^{in} satisfy the conditions I-III (cf. [1], p.1105). The cumulative hazard rate $\alpha = \alpha_t$ is nuisance parameter, and β is the scalar-valued parameter of interest to be estimated from an observed sample path of N^n. It is assumed that α_t, $0 \leqslant t \leqslant 1$, is a continuous nondecreasing function with $\alpha_1 < \infty$, and that β takes its values from a finite open set $\mathfrak{B} \in R^1$.

I (Asymptotic boundedness). The censoring factors Y_t^{in}, $i=1, \ldots, r_n$ take values 0 or 1 (so N_t^{in} only jumps when $Y_t^{in}=1$), and for sufficiently large values of n

$$P_{\alpha,\beta}^n \text{ a.s. } \ sup\{Y_t^{in} |Z_t^{in}|; \ 0 \leqslant t \leqslant 1, \ i=1, \ldots, r_n\} < \infty$$

II (Asymptotic stability). Let k_n, $n=1,2, \cdots$ be a sequence of unboundedly increasing numbers. There exists a family $\{\phi^{(0)}(\beta), \ \beta \in \mathfrak{B}\}$ of deterministic functions $\phi^{(0)} = \phi_t^{(0)}$, $0 \leqslant t \leqslant 1$ such that the difference

$$d_t^n(\beta) = \frac{1}{k_n} \sum_{i=1}^{r_n} Y_t^{in} e^{\beta Z_t^{in}} - \phi_t^{(0)}(\beta)$$

and its first two derivatives with respect to β satisfy the following asymptotic relations: for each $\delta > 0$

$$P_{\alpha,\beta}^n (sup\{ \left| \frac{\partial^j}{\partial \beta^j} d_t(\beta) \right|, \ 0 \leqslant t \leqslant 1\} > \delta) \to 0, \quad j=0,1,2,$$

as $n \to \infty$ uniformly in $\beta \in \mathfrak{B}$.

III (Asymptotic regularity). The function $\phi^{(0)}$ and its first two derivatives $\phi^{(1)} = (\partial / \partial \beta) \phi^{(0)}$ and $\phi^{(2)} = (\partial^2 / \partial^2 \beta) \phi^{(0)}$ are continuous in $\beta \in \mathfrak{B}$ uniformly in $t \in [0,1]$; they are bounded on $\mathfrak{B} \times [0,1]$ and $\phi^{(0)}$ is bounded away from zero. Besides,

$$\sigma^2 \equiv \sigma^2(\alpha, \beta) = \int_0^1 \{\phi^{(2)}(\beta) - \frac{[\phi^{(1)}(\beta)]^2}{\phi^{(0)}(\beta)}\} d\alpha > 0.$$

Define the Cox estimator $\hat{\beta}_n$ for β by the condition

$$\sup_{\beta \in \mathfrak{B}} \int_0^1 \ln^T \Psi_s^n(\beta) d\mathbb{N}_s^n = \int_0^1 \ln^T \Psi_s^n(\hat{\beta}_n) d\mathbb{N}_s^n, \quad \ln \Psi^n = col\{\ln \Psi^{in},\ i=1,\ldots,r_n\} \tag{2}$$

with

$$\Psi_t^{in}(\beta) = Y_t^{in} e^{\beta Z_t^{in}} / \sum_{i=1}^{r_n} Y_t^{in} e^{\beta Z_t^{in}}.$$

Before characterising asymptotic properties of $\hat{\beta}_n$ we give the following definitions:

Definition 1. Let $H^n(\alpha,\beta) = (H_t^n(\alpha,\beta), F_t^n, P_{\alpha,\beta}^n)$ be an r_n-variate predictable process such that

$$\mathfrak{L}\{\kappa_n^{-\frac{1}{2}} \int_0^1 (H^n, S^n)^T dM^n \,\Big|\, P_{\alpha,\beta}^n\} \Rightarrow N(0, [c_{ij},\ i,j=1,2]) \tag{3}$$

where $M^n = M^n(\alpha,\beta) = N^n - A^n(\alpha,\beta)$, while for each $b \in R^1$ and $a \in L^2(\phi^{(0)} d\alpha)$ $S^n = S^n(a,b) = col\{bZ^{in} + a,\ i=1,\ldots,r_n\}$. Therefore concerning the second component solely the above requirement is met under the Conditions I-III with the limiting variance $c_{22} = \int_0^1 \{b^2\phi^{(2)} + 2ba\phi^{(1)} + a^2\phi^{(0)}\} d\alpha$. Also the limiting covariance matrix in (3) is nonsingular (with entries dependent on α, β, a and b, of course).

An estimator $\beta_n^* = \beta_n^*(H^n, c_{11})$ is called asymptotically linear and asymptotically normal if for certain H^n and c_{11} as above and for each $\delta > 0$

$$P_{\alpha,\beta}^n\{\, |(\kappa_n c_{11})^{\frac{1}{2}} (\beta_n^* - \beta) - (\kappa_n c_{11})^{-\frac{1}{2}} \int_0^1 H^n(\alpha,\beta) dM^n(\alpha,\beta)| > \delta\} \to 0 \tag{4}$$

Obviously, under $P_{\alpha,\beta}^n$ $\zeta^n = (\kappa_n c_{11})^{\frac{1}{2}} (\beta_n^* - \beta)$ has the standard normal limiting distribution.

Remark 1. Suppose that the filtration in (1) is minimal: $F_t^n = \sigma(\omega\colon N_s^n,\ s \leq t)$. Define on the n-th space of events the probability measure P_{α^n,β^n}^n giving to N^n the compensator $A^n(\alpha^n,\beta^n)$ where $\beta^n = \beta + \kappa_n^{-\frac{1}{2}} b \in \mathfrak{B}$, $b \in R^1$ and α^n is a function of the same type as α such that $d\alpha^n / d\alpha = 1 + \kappa_n^{-\frac{1}{2}} a$, $a \in L^2(\phi^{(0)} d\alpha)$.

Proposition 6.2 in [2] allows us to apply here the third LeCam's lemma according to which the limiting distribution of ζ^n under P_{α^n,β^n}^n gets the bias equal to $c_{12} c_{11}^{-\frac{1}{2}}$.

Definition 2. Retain the special situation introduced in Remark 1. An estimator β_R^n is called regular in Hajek's sense (at "point" α and β) if for some nondegenerate distribution function G the following weak convergence takes place: for each a and b as above

$$\mathfrak{L}\{\kappa_n^{\frac{1}{2}} (\beta_R^n - \beta^n) \,|\, P_{\alpha^n,\beta^n}^n\} \Rightarrow G.$$

Remark 2. According to Remark 1 the estimator $\beta_n^*(H, c_{11})$ is Hajek's regular iff $c_{12} = bc_{11}$.

Definition 3. Remove now the condition that the filtration is minimal. As in this case Hajek's definition of regularity loses its meaning, we define the regular in wide sense asymptotically linear and asymptotically normal estimators $\beta_n^*(H^n, c_{11})$ by requiring that in (3) $c_{12} = bc_{11}$.

Now we formulate the statement about asymptotic optimality of $\hat{\beta}_n$ (for the proof consult [1,2]).

Theorem 1. Under the Conditions I-III

(1) $\hat{\beta}_n$ is asymptotically linear and asymptotically normal wide sense regular estimator $\hat{\beta}_n = \beta_n^*(\frac{\partial}{\partial \beta} \log \Psi^n, \sigma^2)$; it attains a lower bound for the asymptotic variances of such estimators $\beta_n^*(H^n, c_{11})$, for $c_{11} \leq \sigma^2$.

(2) Suppose in addition that $F_t^n = \sigma\{\omega: N_s^n, s \leqslant t\}$. Then $\hat{\beta}_n$ is Hajek's regular; it attains the lower bound for the risks of such estimators: for any continuous loss function w allowing a polynomial majorant

$$\liminf_{n\to\infty} E^n_{\alpha,\beta} w((\kappa_n\sigma)^{\frac{1}{2}} (\beta^n_R - \beta)) \geqslant \frac{1}{\sqrt{2\pi}} \int_{-\infty}^{\infty} w(x) e^{-\frac{1}{2}x^2} dx$$

$$= \lim_{n\to\infty} E^n_{\alpha,\beta} w((\kappa_n\sigma)^{1/2}(\hat{\beta}_n - \beta))$$

References

1. Andersen P.K., Gill R.D., Cox's regression model for counting processes: a large sample study. Ann. Statist., 1982, Vol. 10, No. 4, p. 1100-1120.

2. Dzhaparidze K., On asymptotic inference about intensity parameters of a counting process. in *Papers on semiparametric models at the ISI centenary session (with discussion)*, R.D. Gill and M. Voors (eds.), Report MS-R86XX, CWI, Amsterdam.

ASYMPTOTIC PROPERTIES OF THE MAXIMUM LIKELIHOOD ESTIMATOR, ITO-VENTZEL'S FORMULA FOR SEMIMARTINGALES AND ITS APPLICATION TO THE RECURSIVE ESTIMATION IN A GENERAL SCHEME OF STATISTICAL MODELS

Lazrieva N.L., Toronjadze T.A.
Tbilisi Razmadze Mathematical Institute of the Academy of Sciences of the Georgian SSR, Tbilisi, USSR

The paper consists of three sections.

In 1° asymptotic properties of the maximum likelihood estimator are studied in a general scheme of statistical models by the LAN technique (Le Cam, Hajek, Ibragimov, Khas'minsky, etc.) with the conditions expressed in terms of predictable characteristics of the transformation of martingales which determine the densities.

In 2° generalization of Ito-Ventzel's formula is given.

In 3° from an "exact" equation for MLE obtained by Ito-Ventzel's formula an "approximated" equation (for the recursive estimator) is derived and some sufficient conditions of the asymptotic proximity of the solutions of these equations are given, from which we conclude that the recursive estimator (RE) has the same asymptotic properties as the MLE.

1°. Asymptotic properties of the MLE. Let

$$\mathcal{E}_n = (\Omega^n, \underline{F}^n, \underline{F}^n, P_\theta^n, P^n), \quad n \geq 1, \quad \theta \in R^1,$$

be a sequence of statistical models where $(\Omega^n, \underline{F}^n, \underline{F}^n, P^n)$ is a space with filtration $\underline{F}^n = \{\underline{F}_t^n\}_{0 \leq t \leq T}$ satisfying the usual conditions, $P_\theta^n \sim P^n$, $P_\theta^n \neq P_{\theta'}^n$ when $\theta \neq \theta'$. $P_\theta^n | \underline{F}_0^n = P^n | \underline{F}_0^n$. Let $\rho_\theta^n(t)$ be a local density of the measure P_θ^n

w.r.t. P^n . It is known that $\rho_\theta^n(t) = \mathcal{E}_t(M_\theta^n)$ where $M_\theta^n \in \mathcal{M}_{loc}(\underline{F}^n, P^n)$ and $\mathcal{E}(\cdot)$ is Dolean's exponential curve.

Assume that the usual regularity conditions are satisfied. Then $d/d\theta\,(\ln \rho_\theta^n) = L(\dot{M}_\theta^n, M_\theta^n)$ where $\dot{M}_\theta^n$ is a strong derivative of the martingale M^n_θ and

$$L(m, M) = m - \langle m^c, M^c\rangle - \sum \frac{\Delta m \, \Delta M}{1 + \Delta M}.$$

By virtue of Girsanov's theorem, $L(\dot{M}_\theta^n, M_\theta^n) \in \mathcal{M}_{loc}(\underline{F}^n, P_\theta^n)$. Also suppose that $L(\dot{M}^n_\theta, M_\theta^n) \in \mathcal{M}^2(\underline{F}^n, P_\theta^n)$.

Asymptotic properties of the MLE are given by the following theorem.

Theorem 1. Let the following conditions be satisfied: uniformly in θ on each compactum we have

(a) $\lim_{n\to\infty} \varphi_n(\theta) = 0$ where $\varphi_n^{-2}(\theta) = E_\theta^n \langle L(\dot{M}_\theta^n, M_\theta^n)\rangle_T$,

(b) $P_\theta^n\text{-}\lim_{n\to\infty} \varphi_n^2(\theta) \langle L(\dot{M}_\theta^n, M_\theta^n)\rangle_T = 1$,

(c) $P_\theta^n\text{-}\lim_{n\to\infty} \varphi_n^2(\theta) I_{\{|x| > \varepsilon \varphi_n^{-1}(\theta)\}} x^2 * \nu_\theta^n = 0$ where ν_θ^n are compensators of jump measures of the process $L(\dot{M}_\theta^n, M_\theta^n)$ w.r.t. the measure P_θ^n ,

(d) $\lim_{n\to\infty} \sup_{\theta\in K} \sup_{|\theta - y| < \varphi_n^\delta} \varphi_n^2(\theta) E_\theta^n \langle L(\dot{M}_y^n - \dot{M}_\theta^n, M_\theta^n)\rangle_T = 0$,

(e) a constant C_0 exists such that for any $n>0, \theta, \theta_1, \theta_2$

$$\varphi_n^2(\theta_1)\, E_\theta^n \langle L(\dot{M}^n_{\theta_2}, M_\theta^n)\rangle_T < C_0,$$

(f) constants $\gamma > 0, \chi > 0, c > 0$ exist such that for any $\theta \in K$ and $n > 0$ $h_\theta^n(u) \geqslant c|u|^\gamma$ where $h_\theta^n(u) = -\ln E_\theta^n\{\exp -\chi(\langle L^c\rangle + \sum^p (1 + \Delta L)^\beta - 1 - \beta \Delta L + \sum(\ln^p K - {}^pK + 1)$, $K = (1 + \Delta L)^\beta$, $0 < \beta < 1$, $L = L(M^n_{\theta + u\varphi^n(\theta)} - M_\theta^n, M_\theta^n)$.

Then:

1) the family P_θ^n satisfies the uniform LAN condition; uniformly w.r.t. θ on any compactum,

2) $P_\theta^n\text{-}\lim_{n\to\infty} \hat{\theta}^n = \theta$, 3) $\lim_{n\to\infty} \mathcal{L}_{P_\theta^n}\{\varphi_n^{-1}(\theta)(\hat{\theta}^n - \theta)\} = \mathcal{N}(0,1)$

<u>2°. Ito-Ventzel's Formula for Semimartingales.</u> Let on some stochastic basis $(\Omega, \underline{\underline{F}}, \underline{F}, P)$ a semimartingale ξ an[d] a family of semimartingales $F(t,x) = M(t,x) + A(t,x)$, $x \in R^1$,

$M(\cdot,x)\in\mathcal{M}^2(\underline{F},P)$, $A(\cdot,x)\in\mathcal{A}(\underline{F},P)$ be given. Assume that:

1) the mapping $F: x\to F(\cdot,x)$ is twice continuously differentiable in the sense of the norm $\|\cdot\|_T$ (if S is a semimartingale and $S=M+A$ then $\|S\|_T=E\int_0^T|dA_s|+E^{1/2}[M]_T$) with the second derivative F_{xx} being continuous in x for all t and ω, the processes $\sup_{|x|\le k}|F_{xx}(t,x)|=\varphi_k(t)$, $F_x(t,0)$, $F(t,0)$ are locally bounded.

Denote by $\Phi(t,x)$ one of the functions $F(t,x)$ and $F_x(t,x)$. By virtue of Assumption 1), $\Phi(t,x)=M_\Phi(t,x)+$ $+A_\Phi(t,x)$ where $M_\Phi(\cdot,x)\in\mathcal{M}^2(\underline{F},P)$, $A_\Phi(\cdot,x)\in\mathcal{A}(\underline{F},P)$, $[M_\Phi(\cdot,x)]_t=\int_0^t f_\Phi(s,x)\,dB_s$, $A_\Phi(t,x)=\int_0^t a_\Phi(s,x)\,dA_s$ with some a_Φ, f_Φ, A and B.

Now assume that for any $K>0$

2) $E\int_0^T\sup_{|x|\le K}|a_\Phi(s,x)||dA_s|<\infty$, $E\int_0^T\sup_{|x|\le K}|f_\Phi(s,x)||dB_s|<\infty$

Then Ito-Ventzel's formula is valid:

$$F(t,\xi_t)=F(0,\xi_0)+\int_0^t F_x(s-,\xi_{s-})d\xi_s+\frac{1}{2}\int_0^t F_{xx}(s-,\xi_{s-})\,d\langle\xi^c\rangle_s+$$
$$+\sum_{s\le t}(F(s-,\xi_s)-F(s-,\xi_{s-})-F_x(s-,\xi_{s-})\Delta\xi_s)+\sum_{s\le t}\Delta F(s,\xi_s)-\Delta(F(s,\xi_s)-$$
$$-\Delta F_x(s,\xi_{s-})\Delta\xi_s)+\left[\int_0^\cdot F_x(ds,\xi_{s-}),\xi\right]_t. \tag{1}$$

The first term of (1), the so called stochastic line integral w.r.t. the family $F(t,x)$, $x\in R^1$, along the curve ξ (Gikhman, Skorokhod (1968), Chitashvili, Mania (1982) substitutes the derivative w.r.t. t in the usual Ito formula, the last two terms are "Ventzel effect". Ito-Ventzel's formula was obtained by various authors (Venzel (1965), Rozovky (1975), Kunita, Bismut (1980), Trofimov, Mikulevichus (1983) under the assumption that the semimartingales F allow an integer representation w.r.t. the components of the semimartingale independent of x.

3°. Recursive Estimation. The MLE $\hat\theta^n_t$, $0\le t\le T$ can be obtained from the equation $F^n(t,\theta)\equiv L_t(\dot M^n_\theta,M^n_\theta)=0$. Applying Ito-Ventzel's formula to $F^n(t,\hat\theta^n_t)$ we obtain an equation for $\hat\theta^n$ and omitting some infinitesimal terms we obtain the following equation for the RE $\tilde\theta^n$

$$d\tilde\theta^n_t=-\frac{F^n(dt,\tilde\theta^n_{t-})}{\tilde F^n_x(t,\tilde\theta^n_{t-})}+\frac{\Delta F^n(t,\tilde\theta^n_{t-})\Delta\tilde F^n_x(t,\tilde\theta^n_{t-})}{\tilde F^n_x(t,\tilde\theta^n_{t-})\,\tilde F^n_x(t-,\tilde\theta^n_{t-})}, \tag{2}$$

where $\tilde{F}_x^n(t,\theta) = -\langle \dot{M}^n(\theta)\rangle_t - 1$, $\Delta F^n(t, \tilde{\theta}^n_{t-}) = \Delta F^n(t,\theta)|_{\theta=\tilde{\theta}^n_{t-}}$

We confine ourselves to consideration of the continuous case. First assume that $F_x(t,\theta) \le -1$. Then the equation for the MLE will have the form

$$d\hat{\theta}^n_t = -\frac{F^n(dt,\hat{\theta}^n_t)}{F_x^n(t,\hat{\theta}^n_t)} - \frac{1}{F_x^n(t,\hat{\theta}^n_t)}\left(\frac{1}{2}\frac{F_{xx}^n(t,\hat{\theta}^n_t)}{(F_x^n(t,\hat{\theta}^n_t))^2}\times\right.$$
$$\left.\times k^n_{1,1}(dt,\hat{\theta}^n_t,\hat{\theta}^n_t) + \frac{1}{F_x^n(t,\hat{\theta}^n_t)} k_{2,1}(dt,\hat{\theta}^n_t,\hat{\theta}^n_t)\right), \quad (3)$$

where $k_{i,j}(t,x,y) = \langle M^{(i)}(\cdot,x), M^{(j)}(\cdot,y)\rangle_t$, hence

$$d\tilde{\theta}^n_t = -(\tilde{F}_x^n(t,\tilde{\theta}^n_t))^{-1} F^n(dt,\tilde{\theta}^n_t), \quad (4)$$

We assume that usual conditions of boundedness and Lipschitz type hold, which guarantee the existence and uniqueness of the strong solution of (3) and (4), as well as the conditions of Theorem 1 allowing to prove the asymptotic "proximity" of the solutions of these equations, i.e.

$$\lim_{n\to\infty} E^n_\theta\, \varphi_n^{-2}(\theta)(\hat{\theta}^n_t - \tilde{\theta}^n_t)^2 = 0, \quad 0 \le t \le T$$

Further we apply the smoothness technique which allows to reduce the general case to the consideration of such $F^{n,*}(t,\theta)$ for which the assumption $F_x^{n,*} \le -1$ holds. If, in addition, the condition of uniform ergodicity

$$\lim_{n\to\infty} P^n_\theta\left(\sup_{\theta-c\le y\le\theta+c} F_x(t,y) > -\varepsilon\right) = 0, \quad \forall c>0,$$

is satisfied for any $c>0$ and some $\varepsilon>0$ it can be easily shown that $\lim_{n\to\infty} P^n_\theta\{\hat{\theta}^n_t \ne \theta^{*,n}_t\} = 0$ where $\theta^{*,n}_t$ is a solution of the smoothed equation $F^{n,*}(t,\theta)=0$ and thus we can conclude that the asymptotic behaviour of $\theta^{*,n}$ is similar to $\hat{\theta}^n$.

In conclusion we shall give a simple example. Let $M_n(t,\theta) = f(\theta) M_n(t)$, where $f(\theta)$ be a twice continuously differentiable function, $\dot{f}(\theta) \ne 0$, $M^n \in \mathcal{M}^{2,c}$. Then $\tilde{\theta}^n_t$ is consistent, asymptotically normal and efficient.

Bernoulli, Vol. 2, pp. 067-070

MAXIMUM ENTROPY SELECTION OF SOLUTIONS TO ILL-POSED MARTINGALE PROBLEMS.

ROLANDO REBOLLEDO.
UNIVERSIDAD CATOLICA DE CHILE
FACULTAD DE MATEMATICAS.
CASILLA 6177.SANTIAGO.CHILE.

The aim of this note is to illustrate a canonical method to select solutions to martingale problems,maximazing a certain entropy functional

We take the usual conventions for the definition of a martingale problem:we denote by D the canonical space of *càdlàg* functions from $\mathbb{R}_+$ to $\mathbb{R}^d$ endowed with the Skorohod's topology;$\mathfrak{D}$ is the corresponding Borelian σ-algebra;$X_t(w):=w(t)$, for all w in D and t in $\mathbb{R}_+$,and $\mathfrak{D}_t:=\bigcap_{u>t}\sigma(X_s,s\leq u)$ for all $t\geq 0$.

Let L be an integro-differential operator defined as follows:

$$Lf(x):=\frac{1}{2}\sum_{i,j=1}^{d}a_{ij}(x)D^iD^jf(x)+\sum_{i=1}^{d}b_i(x)D^if(x)+$$

$$\int_{R^d\setminus\{0\}}(f(x+u)-f(x)-I_{\{|u|\leq 1\}}\langle u,\text{grad } f(x)\rangle)S(x,du)$$

(where $D^i=\partial\cdot\partial x_i$,i=1,...,d).

The coefficients $a_{i,j}$ and b_i are supposed to be continuous and bounded;the matrix $(a_{i,j}(x))$ being semi-elliptic;S(x,du) is a borelian kernel on $\mathbb{R}^d\times(\mathbb{R}^d\setminus\{0\})$,integrating the function $|u|^2I_{\{|u|\leq 1\}}+|u|I_{\{|u|>1\}}$, uniformly in x, and such that $x\mapsto\int h(u)S(x,du)$ be continuous for all bounded and continuous function h.

Thus the martingale problem associated with L is expressed as follows:

Prob$(s,x,L)=\{P$ probability on $(\mathbf{D},\mathfrak{D})$ such that P-a.s. $X_u=x$ for all u

in $[0,s]$ and the process $f(X_t)-f(x)-\int_s^t Lf(X_{u-})du$ is a

$(P,(\mathfrak{D}_t))$-martingale for all f infinitely differentiable with compact support in $\mathbb{R}^d\}$

This problem has no unique solution with the hypothesis assumed for L,but it can be proved that solutions there exist.We describe now a procedure to choose a Markovian solution maximazing a certain entropy functional.Extending TAKAHASHI(1984)'s results one can introduce a suitable entropy functional as follows.Take μ to be a probability measure on $(\mathbf{D},\mathfrak{D})$ dominating all solutions to the martingale problem.Consider then a tight family $(P^n_{s,x})$ of probabilities such that for all n, $P^n_{s,x}$ is a **Markovian solution** to a discretization of the martingale problem (keep in mind the classical way to prove existence of solutions).For all open set G in the weak topology of measures on $(\mathbf{D},\mathfrak{D})$ define

$$C_s(G):=\limsup_n \frac{1}{n} \operatorname{Log} \mu(\{w\in D: \frac{1}{n}\sum_{k=0}^{n-1} P^k_{s,X_s(w)} \in G\}) \quad ,s\geq 0.$$

Then the **entropy for a probability P** on $(\mathbf{D},\mathfrak{D})$-relative to the sequence $(P^n_{s,x})$ and the measure μ-is defined by the expression:

$$H_s(P):=\inf\{C_s(G):\ G \text{ open set containing } P\},\quad s\geq 0.$$

This is a lower semicontinuous functional taking values in$[-\infty,0]$,for all s in a full Lebesgue-measure subset T of $\mathbb{R}_+$.

The main result is then:

THEOREM 1.

There exists a procedure determining for each (s,x) in $\mathbb{R}_+\times\mathbb{R}^d$ a unique probability $P_{(s,x)}$ in Prob(s,x,L) such that:

(1) The map $(s,x) \mapsto P_{(s,x)}$ is measurable;

(2) $H_s(P_{(s,x)})=\sup\{H_s(Q): Q\in \text{Prob}(s,x,L)\}$, $s\geq 0$;

(3) The family $(D, \mathfrak{D},(\mathfrak{D}_t),(X_t),P_{(0,x)})$ is strong Markovian.

SKETCH OF THE PROOF.

The proof is mainly based on BOBADILLA's Selection Theorem allowing to make "Markov Selections" in a more general framework(c.f.BOBADILLA(1986a),(1986b)).

The Selection Theorem works as follows:one chooses a *determining class* of functionals ($g^n; n \in \mathbb{N}$) on the set of probability measures (i.e. $g^n(P)=g^n(Q)$ for all n implies P=Q).Together with this choice one must prove that H_s is a "Regenerating Functional" in the sense of BOBADILLA,which means-loosely speaking-a well behavior with respect to shifts and conditioning.This is the more technical point of the proof.The conclusion follows then straigtforward by Theorem 9 in BOBADILLA (1986b).

REMARK.

As the sketch of the prooof shows,one has an *infinite* number of selection procedures since they depend on the choice of the determining family of functionals.

Let u be an element of the space $C_b(\mathbf{D})$ of all bounded and continuous real-valued functions defined on **D**.We introduce the **pressure** of u as the functional:

$$\Pi_s(u):=\lim\sup_n \frac{1}{n} \operatorname{Log} \int_D \exp(-\sum_{k=0}^{n-1} \int_D u(v) P^k_{s,X_s(w)}(dv))\mu(dw),\ s\geq 0.$$

And define the **free energy** of a probability measure P on $(\mathbf{D},\mathfrak{D})$ as:

$$F_s(P):=\inf\{\Pi_s(u)+\int_D u\,dP;\ u\in C_b(\mathbf{D})\},\ s\geq 0.$$

One has then the following result by adapting the method of TAKAHASHI to martingale problems:

THEOREM 2.

Functionals F_s and H_s are lower semicontinuous for Lebesgue almost all $s\geq 0$;F_s is concave and $H_s \leq F_s$.

Furthermore,for all $u\in C_b(D)$,all (s,x) in $\mathbf{R}_+\times\mathbf{R}^d$,

$$\Pi_s(u)=\sup\{F_s(P)-\int_D u\,dP\ ;\ P\in Prob(s,x,L)\}$$

$$=\sup\{H_s(P)-\int_D u\,dP\ ;\ P\in Prob(s,x,L)\}$$

With the notations of Theorem 1 we have then the following

COROLLARY.

For all (s,x) in $\mathbb{R}_+\times\mathbb{R}^d$,

$$H_s(P_{(s,x)})=\sup\{F_s(Q)\ ;\ Q\in Prob(s,x,L)\}=F_s(P_{(s,x)})=0,$$

Thus the procedure presented in Theorem 1 gives the maxima of both the entropy and the free energy together with the Markovian property.

ACKNOLEDGEMENTS.

The author wish to express his gratitude to the Academy of Sciences of the USSR.

This research was partially supported by UNDP-UNESCO grant and FONDECYT grant #1087/86.

REFERENCES.

BOBADILLA,Gladys (1986a).Problemas de Martingalas sin condiciones de unicidad.Existencia y Aproximación de soluciones Markovianas Fuertes.Doctoral Thesis,Universidad Católica de Chile.

________________(1986b).Une méthode de sélection de probabilités Markoviennes.C.R.Acad.Sci.Paris,t.303,Série I,#4,147-150.

TAKAHASHI,Yoichiro(1984).Entropy Functional for Dynamical Systems and their Random Perturbations.In:Stochastic Analysis,Proceedings of theTaniguchi Int.Symp.Katata and Kyoto,K. Itô (ed.),North-Holland,Amsterdam-N.Y.-Oxford,437-467.

ASYMPTOTIC INFERENCE FOR THE GALTON-WATSON PROCESS WITHOUT RESTRICTION TO THE SUPER-CRITICAL CASE

David J. Scott

Department of Statistics, La Trobe University, Bundoora, Australia.

Consider the Galton-Watson process $\{Z_n; n\geqslant 0\}$ with offspring distribution p_θ given by

$$p_\theta(k) = \Pr(Z_n=k \mid Z_{n-1}=1) = a_k\theta^k/A(\theta)$$

$$\text{for } k=0,1,2,\ldots, \quad \theta \in \Theta .$$

Then

$$\mu(\theta) = E(Z_1 \mid Z_0=1) = \theta A'(\theta)/A(\theta)$$

and

$$\mu_2(\theta) = E[(Z_1-\mu(\theta))^2 \mid Z_0=1] = \theta\mu'(\theta) .$$

Set

$$Y_n = Z_0 + Z_1 + \ldots + Z_n .$$

The maximum likelihood estimator (MLE) of $\mu(\theta)$ is

$$\mu(\theta_n) = (Y_n-Y_0)/Y_{n-1}$$

and $\hat{\theta}_n$ the MLE of θ is the solution of

$$\mu(\hat{\theta}_n) = (Y_n-Y_0)/Y_{n-1} .$$

Let the prior distribution of θ be denoted by $\pi(\theta)$, the posterior distribution by $\pi(\theta \mid Z_0,\ldots Z_n)$ and define

$$\Phi(x) = (2\pi)^{-\frac{1}{2}} \int_{-\infty}^{x} e^{-u^2/2}\, du ,$$

$$\bar{\sigma}_n = (\mu_2(\hat{\theta}_n)/Y_{n-1})^{\frac{1}{2}}$$

and

$$\mu(a,b) = \{\theta : \mu(\hat{\theta}_n) + a\,\bar{\sigma}_n < \mu(\theta) < \mu(\hat{\theta}_n) + b\,\bar{\sigma}_n\} \ .$$

Heyde (1979) proved the following.

<u>Theorem</u>. If $\hat{\theta}_n \to \theta_0$, $Y_{n-1} \to \infty$, $\pi(\theta)$ is continuous at θ_0 and p_{θ_0} is non-degenerate then

$$\int_{\mu(a,b)} \pi(\theta|Z_0, \ldots Z_n)\,d\theta \to \Phi(b) - \Phi(a) \ .$$

From this Theorem, for n and Y_{n-1} large, letting P_π denote the probability assuming the prior Π ,

$$P_\pi(\mu(\theta) > 1) = \int_{\mu(1,\infty)} \pi(\theta|Z_0, \ldots Z_n)\,d\theta$$
$$\simeq 1 - \Phi([1-\mu(\hat{\theta}_n)]/\bar{\sigma}_n)$$

which allows calculation of the posterior probability that the process is supercritical.

Heyde's result is notable in that in contrast to the well-known results using a frequentist approach, no restriction to the supercritical case is required. A corresponding result may be obtained using a frequentist approach, the essential change being that the asymptotics are as Y_{n-1} converges to infinity.

<u>Theorem</u>. For the Galton-Watson process if θ_0 is the true value of θ ,

$$\Pr(|\hat{\theta}_n - \theta_0| > \varepsilon \,|\, Y_{n-1} \geqslant N) \to 0 \quad \text{as} \quad N \to \infty$$

and

$$\sup_{x \in R} |\Pr([\mu(\hat{\theta}_n) - \mu(\theta_0)]/\bar{\sigma}_n \leqslant x \,|\, Y_{n-1} \geqslant N) - \Phi(x)| \to 0$$
$$\text{as} \quad N \to \infty \ .$$

This result justifies an asymptotic test of

$$H_0 : \mu(\theta) \leqslant 1 \quad \text{vs} \quad H_1 : \mu(\theta) > 1$$

which is to reject H_0 if

$$\mu(\hat{\theta}_n) \geqslant 1 + \bar{\sigma}_n\, \Phi^{-1}(1-\alpha) \ .$$

The P-value of the observation $\mu(\hat{\theta}_n)$ for this test is

$$1 - \Phi([\mu(\hat{\theta}_n) - 1]/\bar{\sigma}_n)$$
$$= \Phi([1 - \mu(\hat{\theta}_n)]/\bar{\sigma}_n$$

which is the same as the posterior probability as calculated using Heyde's result.

Heyde's result is a form of asymptotic posterior normality and it is of interest to compare it to the usual Bernstein-von Mises Theorem which also gives asymptotic posterior normality.

Consider observations $X_1, X_2, \dots X_n$ from a stochastic process with density $p_n(x_1, x_2, \dots x_n|\theta)$. Let $\hat{\theta}_n$ be the MLE of θ and let $\ell n(\theta) = \log p_n(X_1, \dots, X_n|\theta)$ be the log-likelihood. Suppose $\pi(\theta)$ is a continuous prior density for θ and $\pi(\theta|X_1, X_2, \dots, X_n)$ is the posterior density.

A fairly recent version of the Bernstein-von Mises Theorem was given by Heyde and Johnstone (1979).

Theorem. Under regularity conditions

$$\int_{\hat{\theta}_n + a\sigma_n}^{\hat{\theta}_n + b\sigma_n} \pi(\theta|X_1 \dots X_n)d\theta \to \Phi(b) - \Phi(a)$$

(where $\sigma_n = [-\ell_n''(\hat{\theta}_n)]^{-\frac{1}{2}}$) in P_{θ_0}-probability.

It is important to note that the regularity conditions involve convergence of various quantities in P_{θ_0}-probability and the result also gives convergence in P_{θ_0}-probability. The Bernstein-von Mises Theorem can actually be viewed as being a composition of two statements. The first is analytic, and is that for certain sequences of observed values asymptotic posterior normality holds. The second is that with P_{θ_0}-probability approaching one, the observed values of the process are such that asymptotic posterior normality holds. Then Heyde's Theorem concerning asymptotic posterior normality of the Galton-Watson process consists of the analytic part of the Bernstein-von Mises Theorem only. This suggests it can be obtained by stripping off the probabilistic aspects of Heyde and Johnstone's proof of the Bernstein-von Mises Theorem. This does indeed work, producing a new result from which Heyde's Theorem may be obtained.

<u>Theorem</u>. If $\hat{\theta}_n \to \theta_0$, $Y_{n-1} \to \infty$ and $\pi(\theta)$ is continuous at θ_0

$$\int_{\hat{\theta}_n + a\sigma_n}^{\hat{\theta}_n + b\sigma_n} \pi(\theta | Z_0, \ldots, Z_n) d\theta \to \Phi(b) - \Phi(a)$$

where $\sigma_n = \hat{\theta}_n / (Y_{n-1}\mu_2(\hat{\theta}_n))^{\frac{1}{2}}$.

The connection between this result and Heyde's Theorem is quite simple. The theorem above states that asymptotically

$$\theta \sim N(\hat{\theta}_n, \sigma_n^2) \ .$$

Thus asymptotically

$$\mu(\theta) \sim N(\mu(\hat{\theta}_n), \bar{\sigma}_n^2)$$

where

$$\bar{\sigma}_n^2 = \sigma_n^2(\mu'(\hat{\theta}_n))^2 \ ,$$

which is Heyde's result.

References

Heyde, C.C. (1979). On assessing the potential severity of an outbreak of a rare infectious disease: a Bayesian approach. Austr. J. Statist. 21, 282-292.

Heyde, C.C. and Johnstone, I.M. (1979). On asymptotic posterior normality for stochastic processes. J. Roy. Statist. Soc. B, 41, 184-189.

CROSS-VALIDATION

(Session 3)

Chairman: D.V. Hinkley

ON RESAMPLING METHODS FOR CONFIDENCE LIMITS

David V. Hinkley
Center for Statistical Sciences
and
Department of Mathematics
The University of Texas at Austin

SUMMARY

Some recent research on bootstrap resampling methods is reviewed. Topics include: Monte Carlo and theoretical approximation as efficient alternatives to naive simulation; construction of approximate pivots; inversion of bootstrap tests; and conditional bootstraps. The majority of the discussion is addressed to statistics based on homogeneous random samples.
Key words and phrases: ancillary statistic, bootstrap, double bootstrap, likelihood, Monte Carlo, pivot, saddlepoint method.

1. INTRODUCTION

This is a review of some bootstrap techniques associated with confidence limit calculations. The objective is to be reasonably comprehensive, and to introduce some topics of current research interest. Our starting point is a summary and illustration of the basic bootstrap method for homogeneous, independent data; see Efron (1982).

Suppose that $\underline{x} = (x_1, \ldots, x_n)$ is a random sample of fixed size n from an infinite population for which $\Pr(X \leq x) = F(x)$ is the cumulative distribution function of a randomly sampled datum. The population quantity θ, which is a differentiable functional $t(F)$, is of interest. It is assumed that θ is estimated by $T = T(x_1, \ldots, x_n) = t(\tilde{F})$, where $\tilde{F}$ is the empirical distribution function; defined by $n\tilde{F}(x) = \text{card}\{i : x_i \leq x\}$, and the functional $t(\cdot)$ is assumed regular.

For the purpose of calculating confidence limits, distributions of quantities D such as $D = T - \theta$ are required. Because F will be unknown, although possibly belonging to a known family indexed by θ and nuisance parameters, it will be usual to estimate this distribution by $\hat{F}$, say, and thence to estimate the distribution of $T - \theta$. If the latter step is not amenable to theoretical calculation, then we can approximate the distribution by a Monte Carlo simulation method.

To be specific, consider $D = T - \theta$ and its distribution function $G(d) = G(d, F) = \Pr(T - \theta \leq d \mid F)$, which is to be estimated by

$$\hat{G}(d) = G(d, \hat{F}) = \Pr(T - \theta \leq d \mid \hat{F}) \ . \tag{1}$$

The simplest Monte Carlo technique for approximating $\hat{G}$ is as follows:

Step 1°. Use a Monte Carlo simulation method to generate a random sample $\underline{x}^* = (x_1^*, \ldots, x_n^*)$ from $\hat{F}$.

Step 2°. Calculate $T^* = T(x_1^*, \ldots, x_n^*)$ and thence the simulated value $T^* - T$, which is to $\hat{F}$ what $T - \theta$ is to F.

Step 3°. Perform Steps 1° and 2° a total of B times and approximate $\hat{G}(d)$ by

$$\hat{G}_B(d) = B^{-1} \text{ freq}\{T^* - T \leq d\} \ . \tag{2}$$

Superscript $*$ will always denote a random variable generated from $\hat{F}$.

When $\tilde{F}$ is used for $\hat{F}$ in Step 1°, $\underline{x}^*$ is obtained by uniform random sampling with replacement from $\underline{x}$ – hence the name "resampling method." While this nonparametric case is of most interest, many theoretical aspects of bootstrap methods are most easily discussed in the parametric case where F belongs to a known family.

The estimated distribution $\hat{G}$ yields confidence limits in the usual way. Thus if $d_p = \hat{G}^{-1}(p)$, then the lower $1-\alpha$ confidence limit for θ is $T - d_{1-\alpha}$ and the upper $1-\alpha$ limit is $T - d_\alpha$. If the Monte Carlo approximation (2) is used, then $\hat{G}_B^{-1}(p)$ is taken to be the $[(B+1)p]^{\text{th}}$ ordered value of $T^* - T$.

Example 1. Suppose that the first row of Table 1 is a random sample from a population whose mean is $\theta = \int x dF(x) = \mu$, and that we estimate μ by $T = \overline{x} = n^{-1}\Sigma x_j$, whose value is 17.87. We wish to estimate the distribution of $\overline{x} - \mu$ and hence obtain an upper 90% confidence limit for μ. (Superior alternatives to use of $\overline{x} - \mu$ will be discussed later.)

If F is assumed to be a normal distribution, then we estimate F by the $N(T, \hat{\sigma}^2)$ distribution with $\hat{\sigma}^2 = n^{-1}\Sigma(x_i - \overline{x})^2$, and thence calculate $\hat{G}$ theoretically to be $\hat{G}(d) = \Phi(\sqrt{n}\, d/\hat{\sigma})$. Because $\hat{\sigma}^2 = 46.53$, $\hat{G}(d) = \Phi(0.46d)$ for these data. Then the 90% upper confidence limit for μ is $T - \hat{G}^{-1}(0.10) = 17.87 - (0.46)^{-1}\Phi^{-1}(0.10) = 20.63$.

If nothing is assumed about F and we take $\hat{F} = \tilde{F}$, then the estimate (1), now written $\tilde{G}(d) = \Pr(\overline{x}^* - \overline{x} \mid \tilde{F})$, will be approximated by (2) using the simulation resampling procedure. A very small application with $B = 9$ is illustrated in Table 1, wherein each sample $\underline{x}^*$ is given in the equivalent form of frequencies f_i^* for data values x_i. We approximate $d_{0.10}$ by the 1st ordered value of $\overline{x}^* - \overline{x}$, namely $15.75 - 17.87 = -2.12$, and thence calculate the 90% confidence limit to be $T - \tilde{G}_B^{-1}(0.10) = 17.87 - (-2.12) = 19.99$.

Table 1. A Random Sample and a Small Bootstrap Analysis of Its Mean

data :	9.6	10.4	13.0	15.0	16.6	17.2	17.3	21.8	24.0	33.8	$\overline{x}$ 17.87
bootstrap sample no.	frequencies of datum values										$\overline{x}^*$
1	1	0	0	1	3	1	1	0	2	1	19.07
2	1	0	1	1	1	1	0	3	2	0	18.48
3	0	0	2	1	2	0	2	0	3	0	18.08
4	1	1	1	2	0	1	1	1	0	2	18.69
5	1	0	1	1	3	1	1	1	1	0	16.77
6	1	1	2	0	0	1	1	2	1	1	18.19
7	0	1	3	1	0	1	3	0	1	0	15.75
8	2	1	0	0	2	1	0	0	2	2	19.56
9	1	1	1	2	0	0	1	1	1	2	19.37
					sample average of bootstrap $\overline{x}^*$						18.22
					sample variance of bootstrap $\overline{x}^*$						1.54

estimated quantiles for $\overline{x} - \mu : d_{.10} = 15.75 - 17.87 = -2.12$, etc.

Several questions now arise, which we state briefly here and then discuss in subsequent sections. (i) When resampling from $\tilde{F}$ is done, how large should be B? Are Monte Carlo methods other than direct simulation available? How do such Monte Carlo approximations for $\hat{G}$ compare with theoretical expansions using estimated coefficients? (ii) In Example 1, $\hat{G}$ does not give exact confidence limits, because the coverage $E\Phi(\hat{\sigma}\,\Phi^{-1}(1-\alpha)/\sigma)$ does not equal $1 - \alpha$. We should use the pivot $(\overline{x} - \mu)/\hat{\sigma}$, for which $\hat{G} = G$ and exact confidence limits are obtained. Are there analogs of pivots, or pivotal confidence limit methods, when resampling from $\tilde{F}$? (iii) In Example 1, if F were in a non-normal location-scale family, there would be conditional confidence limits derived from the conditional distribution of $\overline{x} - \mu$ given the ancillary configuration of residuals. Do concepts of ancillarity and conditional inference apply with resampling methods?

2. MONTE CARLO AND THEORY

The bootstrap estimate (2) includes two errors, the Monte Carlo error $\hat{G}_B(d) - \hat{G}(d)$ and the data error $\hat{G}(d) - G(d)$. The first of these can be reduced by (i) increasing B, (ii) calculating $\hat{G}$ theoretically, possibly by approximate methods, or (iii) using an efficient Monte Carlo method.

The merit of reducing the simulation error will depend upon the magnitude of the data error. In typical applications, B would be of the order of 1000 for accurate approximations of probabilities, unless a normal approximation is reasonable for $\hat{G}$. Exact calculation of $\hat{G}$ will rarely be possible in the nonparametric case $\hat{F} = \tilde{F}$. This leaves theoretical approximation to $\hat{G}$ and improved Monte Carlo simulation for consideration, and both approaches involve the structure of T.

Attention is now restricted to the nonparametric case $\hat{F} = \tilde{F}$. It is assumed that $T = t(\tilde{F})$ possesses the expansion

$$t(\tilde{F}) = t(F) + n^{-1}\Sigma L(x_j, F) + \tfrac{1}{2}n^{-2}\Sigma\Sigma Q(x_j, x_k, F) + \cdots, \tag{3}$$

where L is the first functional derivative, or influence function, and Q is the second functional derivative. Similar expansions exist for derived quantities such as $(T - \theta)/S$, where S has the form $s(\tilde{F})$.

Consider theoretical approximation to $\tilde{G}(d) = \Pr(T^* - T \leq d \mid \tilde{F})$. From (3) we have

$$T^* - T = n^{-1}\Sigma L(x_j^*, \tilde{F}) + \tfrac{1}{2}n^{-2}\Sigma\Sigma Q(x_j^*, x_k^*, \tilde{F}) + O_p(n^{-\frac{3}{2}}). \tag{4}$$

One approach is to use an Edgeworth series for $\tilde{G}(d)$, as discussed by Hall (1983) and Hinkley & Wei (1984), with joint moments of L, Q and other derivatives calculated in the usual way. The normal approximation will have zero mean and variance

$$n^{-1}\sigma^2(\tilde{F}) = n^{-2}\Sigma\{L(x_j, \tilde{F})\}^2; \tag{5}$$

the one-term Edgeworth correction will involve Q, and the two-term Edgeworth correction will involve the third functional derivative.

An alternative approach is to calculate a saddlepoint approximation for $\tilde{G}$, following Daniels (1987). The simplest approximation of this type ignores the Q and higher-order derivatives in (4), so that the cumulant generating function $n\tilde{K}(\lambda)$ of $n(T^* - T)$ is approximated by $n \log_e[n^{-1}\Sigma \exp\{\lambda L(x_j, \tilde{F})\}]$. If $Q \equiv 0$, then the relative error of the saddlepoint approximation is uniform and $O(n^{-1})$; but if Q is nonzero, then the approximation for $\tilde{K}(\lambda)$ carries an error $O(n^{-1/2})$. In Example 1 with $T = \overline{X}$, Q is zero, and the saddlepoint method gives very accurate answers: the exact 1% and 99% points -4.42 and 5.47 for $\overline{X} - \mu$ are approximated by -4.43 and 5.48, whereas the normal approximation gives -5.02 and 5.02. The saddlepoint method also applies with $O(n^{-1})1$ error to solutions of estimating equations. Modification of the method to use the full approximation (4) appears difficult.

There is a variety of Monte Carlo methods that could be considered; see Therneau (1983). Two methods are described briefly here.

The control function method combines theory and simulation, the latter being required only for deviations of T^* from its linear approximation $T + n^{-1}\Sigma L(x_j^*, \tilde{F}) = T_L^*$. Thus to estimate the variance of T^* we write

$$\text{Var}\,(T^* \mid \tilde{F}) = \text{Var}(T_L^* \mid \tilde{F}) + 2\,\text{Cov}(T_L^*, T^* - L_L^* \mid \tilde{F}) + \text{Var}(T^* - T_L^* \mid \tilde{F}). \tag{6}$$

The first term on the right of (6) is the variance (5), and only the last two terms need to be approximated by simulation, so that the simulation error is considerably reduced. This technique can be applied to approximation of higher cumulants, and would be useful with normal and Edgeworth approximations; see Davison *et al.* (1986). Theoretical terms involving $L(x_i, \tilde{F})$ can be approximated by numerical differencing.

The second Monte Carlo method has origins in both classical design theory and numerical analysis. One approach is to replace $(x_1^*, \ldots, x_n^*)$ by the equivalent frequencies $(f_1^*, \ldots, f_n^*)$ of values $(x_1, \ldots, x_n)$, as in Table 1. The idea is to restrict the B vectors $\underline{f}^*$ so that they mimic the multinomial sample space of possible values of $\underline{f}^*$. First order balance is accomplished by randomly permuting a string of B copies of $(x_1, \ldots, x_n)$ and then reading off successive blocks of n. Davison *et al.* (1986) discuss this in some detail. A second approach uses a different representation for x_i^*, in which we write

$x_i^* = x_{\xi_i}$, with $\xi_i \in \{1,\ldots,n\}$. Then each bootstrap sample is determined by a vector $\xi = (\xi_1,\ldots,\xi_n)$ in the n-dimensional lattice cube $\{1,\ldots,n\}^n$. The simple balance described above is achieved for $B = n$ by making the n ξ-vectors form a randomized block design, with ξ_i the treatment label in the ith block. The theory underlying this approach is described and illustrated by Ogbonmwan & Wynn (1985). Higher order balance can be attained by correspondingly more complex designs.

In more complicated situations involving nonhomogeneous data, as in regression, corresponding extensions of the theoretical and Monte Carlo methods will be required.

3. PIVOTAL CONFIDENCE LIMITS

In Example 1, the estimated distribution $\hat{G}$ of $\overline{x}-\mu$ does not equal the true distribution G and so there is error in confidence limits based on $\hat{G}$. When F is known to be normal, exact confidence limits can be calculated by using the pivot $(\overline{x}-\mu)/\hat{\sigma}$ in place of $\overline{x}-\mu$. In general, for a parametric family $\mathcal{F}$ of distributions, a pivot is a quantity $D(\theta,\tilde{F})$ with the same distribution G for all F in $\mathcal{F}$, so that in particular $\hat{G} = G$ when $\hat{F}$ is an estimated member of $\mathcal{F}$. If $D(\theta,\tilde{F})$ is monotone decreasing in θ, as is often the case, then the upper $1-\alpha$ confidence limit for θ is

$$\theta^{1-\alpha} = \max\left\{\theta : D(\theta,\tilde{F}) \geq d_\alpha = \hat{G}^{-1}(\alpha)\right\} . \tag{7}$$

In the non parametric case with sampling from $\tilde{F}$, we say that $D(\theta,\tilde{F})$ is pivotal if

$$\tilde{G}(d) = \Pr\{D(T,\tilde{F}^*) \leq d \mid \tilde{F}\} = G(d) . \tag{8}$$

In fact it is impossible to satisfy (8) exactly. We shall suppose that $D = \Omega(1)$, so that $\tilde{G}(d)-G(d) = O_p(n^{-1/2})$. Then D is called *approximately pivotal* if $\tilde{G}(d)-G(d) = O_p(n^{-r})$ for some $r \geq 1$. Empirical evidence shows that use of non-pivotal $T-\theta$ in resampling methods gives poorly-behaved confidence limits. Chapman & Hinkley (1986) show how to use double-bootstrap analysis to check whether or not an arbitrary $D(\theta,\tilde{F})$ is approximately pivotal. Beran (1985) and Efron (1986) have described different methods for automatic calculation of approximately pivotal confidence limits, which we now review.

In Beran's method, we begin with a quantity $D_0(\theta, \tilde{F})$, such as $T - \theta$, whose distribution G_0 is estimated by $\hat{G}_0$. Then define D to be the estimated probability integral transform of D_0,

$$D(\theta, \tilde{F}) = \hat{G}_0\big(D_0(\theta, \tilde{F})\big) , \qquad (9)$$

whose distribution G is estimated by $\hat{G}$. This D is approximately pivotal in considerable generality, in the sense referred to earlier, because $\hat{G} - G = O_p(n^{-1})$. Therefore the confidence limit formula (7) will give errors of order n^{-1}, whereas use of $\hat{G}_0$ would generally give errors of order $n^{-1/2}$; see Beran (1985) for technical details.

Example 1. (ctd.) Take $D_0(\theta, \tilde{F}) = \overline{x} - \mu$, and consider the parametric case where F is assumed normal. Then $\hat{G}_0$ is the $N(0, \hat{\sigma}^2/n)$ distribution and (9) gives $D(\theta, \tilde{F}) = \Phi(\sqrt{n}(\overline{x} - \mu)/\hat{\sigma})$, which is a monotone decreasing function of the pivotal Student t-statistic. Therefore confidence limit formula (7) gives the familiar Student t confidence limit for μ.

In the nonparametric case when $\hat{F} = \tilde{F}$, calculation and use of (9) can be effected by "double bootstrapping." The simple bootstrap resampling method defined earlier, now applied to D_0, will yield approximation (2) for $\hat{G}_0$; and this whole procedure can be simulated by resampling C times, say, for each $\underline{x}^*$ to obtain an approximation to $\hat{G}$.

If $D_0(\theta, \tilde{F})$ is monotone decreasing in θ, so will be $D(\theta, \tilde{F})$. Then limit (7) corresponds to estimate $\hat{G}_0^{-1}\big(\hat{G}^{-1}(\alpha)\big)$ for the α quantile of $D_0(\theta, \tilde{F})$, in the same sense that the Student-t limit is obtained for $\overline{x} - \mu$ in the preceding example. In fact this double bootstrap method applied to $D_0(\theta, \tilde{F}) = T - \theta$ gives confidence limits close to those obtained from the standard, single bootstrap method applied to $D = \sqrt{n}(T - \theta)/\sigma(\tilde{F})$, with $\sigma^2(\tilde{F})$ as in (5). Both methods give confidence correct to $O_p(n^{-1})$, as Beran shows using arguments based on Edgeworth series expansions.

The double bootstrap procedure is very costly, since the total number of resamples will be very large, *e.g.*, 10^5. Efficient Monte Carlo methods (Section 2) are needed for practical application.

Efron's confidence limit procedure is developed from an earlier attempt to account automatically for transformable non-pivotality. The initial assumption was that for some unknown monotone transformation h, $Z = h(T) - h(\theta)$ is a normally distributed pivot. One should think of h as the variance-stabilizing transformation, which will not necessarily remove the effect of skewness, typically of order $n^{-1/2}$. But Efron observes that this order of skewness can be removed by transforming Z to $g(T, A, \theta) = \{\exp(AZ) - 1\}/A$ where $A = a(\tilde{F})$ is proportional to the standardized skewness of the $L(x_j, \tilde{F})$ values. The approximately normal pivot $g(T, A, \theta)$ may then be used, without specifying $h(\cdot)$, to define a confidence limit procedure which is correct to second order, thus accounting for $n^{-1/2}$ deviations from the normal approximation; a fairly general proof of this fact has been given by Hall (1986). However, in the normal parametric case the procedure yields the $N(0, \hat{\sigma}^2)$ approximation for $\overline{X} - \mu$, which is inadequate. Presumably Efron's procedure should, and could, be adjusted to incorporate studentization.

4. OTHER CONFIDENCE LIMIT METHODS

An alternative to the pivotal method of calculating confidence limits is Neyman's method of inverting hypothesis tests. This method can be extended directly to results from the double bootstrap procedure (Section 3.2) applied to estimate T.

Suppose that θ is the only unknown parameter for a parametric family. Then an upper $1 - \alpha$ confidence limit is the largest value of θ_0 not rejected in a level α test of $H_0 : \theta \geq \theta_0$ versus $\overline{H}_0 : \theta < \theta_0$ based on T. That is, for observed value t_0 of T, the upper limit $\theta^{1-\alpha}$ satisfies $\text{pr}(T \leq t_0 \mid F_{\theta_0}, \theta_0 = \theta^{1-\alpha}) = \alpha$; see Cox & Hinkley (1974, Section 7.2).

In the double bootstrap analog, with no parametric model, variation in θ_0 is replaced by variation in T^*. Probabilities are approximated by empirical distributions of T^{**}, which are values of T in resamples $\underline{x}^{**}$ drawn from $\underline{x}^*$. Then the upper $1 - \alpha$ confidence limit $\theta^{1-\alpha}$ satisfies $\text{pr}(T^{**} \leq t_0 \mid \tilde{F}^* : T^* = \theta^{1-\alpha}) = \alpha$, whose solution must be interpolated from simulation estimates of the probabilities. The procedure can be generalized, for example to studentized values $(T - \theta)/S$; see Chapman & Hinkley (1986).

Several ideas have been proposed for calculating distribution-free likelihood functions for θ, but no satisfactory general theory has been given. In particular, the conventional likelihood methods of calculating confidence limits (Cox & Hinkley, Section 9.4) are not known to apply. Therefore only brief outlines of two ideas will be given here.

One useful idea explored by Ogbonwan & Wynn (1986) applies only to contrast parameters, unless some type of symmetry is imposed on underlying distributions. One assumes that the sampled nonhomogeneous random variables X_i can be (non-uniformly) transformed into i.i.d. variables $Z_i(\theta) = z_i(X_i, \theta)$. Denote the corresponding estimate of θ based on observations $z_i(\theta) = z_i(x_i, \theta)$ by $t(\theta)$. Under the distribution of $Z_i^*(\theta)$ induced by sampling X^*s from $\tilde{F}$, the density of $T^*(\theta)$ is denoted by $\tilde{g}(t \mid \theta)$, and the likelihood is then taken to be $\tilde{g}(t(\theta) \mid \theta)$. This approach has strong similarities to randomization methods.

A second idea would be to fit a smooth density $\tilde{g}(t^{**} \mid T_b^*)$ to double bootstrap values T_{bc}^{**}, $c = 1, \ldots, C$ as defined above, and to define $\tilde{g}(t_0 \mid T_b^*)$ to be the likelihood at $\theta = T_b^*$. Interpolation on $\tilde{g}(t_0 \mid T_b^*)$ would give a continuous likelihood.

5. CONDITIONAL BOOTSTRAPS

In principle, inference should be conditioned on ancillary statistics when these exist, and the paradigm for parametric models is well developed. The challenging case is where resampling from $\tilde{F}$ and its generalizations is used. One definition of ancillarity is that $A = a(\tilde{F})$ is ancillary when its distribution is the same under sampling from F and $\tilde{F}$; this mimics a property of parametric ancillarity, and can be checked using the double bootstrap mentioned earlier.

Usually A would be suggested by analogy with parametric models, but should be robust. One detailed example of conditional bootstrapping is described by Hinkley & Schechtman (1987).

6. CONCLUDING COMMENTS

The examples and general discussion in this paper have been made elementary to clarify the ideas. On the more mathematical side, the referenced papers by Beran, Efron and Hall provide many important theoretical results, as well as additional references. A useful current review with advanced applications is by Efron & Tibshirani (1986). An expanded version of the present paper, with additional examples, is available (Hinkley, 1986).

ACKNOWLEDGEMENTS

Financial support of the National Science Foundation is gratefully acknowledged.

REFERENCES

Beran, R.J. (1985). Prepivoting to reduce level error of confidence sets. Unpublished, University of California at Berkeley.

Chapman, P.L. & Hinkley, D.V. (1986). The double bootstrap, pivots and confidence limits. Unpublished, University of Texas at Austin.

Cox, D.R. & Hinkley, D.V. (1974). Theoretical Statistics, London: Chapman & Hall.

Daniels, H.E. (1987). Tail probability approximations. Inter. Statist. Rev. 55, to appear.

Davison, A.C., Hinkley, D.V. & Schechtman, E. (1986). Efficient bootstrap methods. Biometrika 73, to appear.

Efron, B. (1982). The Jackknife, the Bootstrap and Other Resampling Plans. In: CBMS-NSF Conference Series in Applied Mathematics, 38. SIAM, Philadelphia.

Efron, B. (1986). Better bootstrap confidence intervals. J. Amer. Statist. Assoc. 81.

Efron, B. & Tibshirani, R. (1986). Bootstrap measures for standard errors, confidence intervals, and other measures of statistical accuracy. Statistical Science 1, 54–77.

Hall, P. (1983). Inverting an Edgeworth expansion. Ann. Statist. 11, 569–576.

Hall, P. (1986). Theoretical comparison of bootstrap confidence intervals. Australian National University, unpublished.

Hinkley, D.V. (1986). A review of resampling methods for confidence limits. Unpublished, University of Texas at Austin.

Hinkley, D.V. & Schechtman, E. (1987). Conditional bootstrap analysis of a mean-shift model. Biometrika 74, to appear.

Hinkley, D.V. & Wei, B.C. (1984). Improvements of jackknife confidence limit methods. Biometrika 71, 331–339.

Ogbonmwan, S.M. & Wynn, H.P. (1985). Accelerated resampling codes with low discrepancy. The City University, London, unpublished.

Obgonmwan, S.M. & Wynn, H.P. (1986). Resampling generated likelihoods. The City University, London, unpublished.

Thernau, T.M. (1983). Variance Reduction Techniques for the Bootstrap. Unpublished Ph.D. thesis, Stanford University.

THE INTERPLAY BETWEEN CROSS-VALIDATION AND SMOOTHING METHODS

B. W. Silverman
School of Mathematical Sciences
University of Bath
BATH BA2 7AY
United Kingdom

SPLINE SMOOTHING FOR NONPARAMETRIC REGRESSION

Consider the nonparametric regression problem of estimating a curve g given data pairs $(t_i, Y_i), i=1, \cdots, n$ that are assumed to satisfy $Y_i = g(t_i)+\varepsilon_i$. Assume that the design points t_i are known and are not necessarily evenly spaced, and that the ε_i are random errors. The conventional assumptions that the ε_i are uncorrelated with mean zero and equal variances will also be made.

The most widely used approach to curve fitting is, of course, least squares. If we place no restrictions at all on the curve to be estimated then we can reduce the sum of squares $\sum\{Y_i - g(t_i)\}^2$ to zero by choosing g to be any curve that actually interpolates the data points (provided the t_i are all distinct). Such an interpolant would usually be rejected by the statistician on the grounds that its rapid fluctuations were implausible. The most commonly used device for avoiding such "implausible" estimates is to restrict attention to curves g which fall into some parametric class. Another approach is to quantify the competition between the two conflicting aims in curve estimation, which are to produce a good fit to the data but to avoid too much rapid local variation.

A measure of the rapid local variation of a curve can be given by a *roughness penalty* (see Good and Gaskins, 1971, and Boneva, Kendall and Stefanov, 1971) such as the integrated squared second derivative. Using this measure, define the *penalized sum of squares*

$$S(g) = \sum\{Y_i - g(t_i)\}^2 + \alpha \int g''(x)^2 dx \ ; \tag{1}$$

the *smoothing parameter* α represents the rate of exchange between residual error and local variation. Minimizing $S(g)$ over the class of all (twice-differentiable) functions g will yield an estimate $\hat{g}$ which, for the given value of α, gives the best compromise between smoothness and goodness of fit to the data. It can be shown (see for example Reinsch, 1967) that the curve $\hat{g}$ is a cubic spline with knots at the design points. DeBoor (1978) gives details of algorithms for finding $\hat{g}$ in a number of operations that depends linearly on n. Many other authors have discussed smoothing splines of this kind; for reviews, see, for example, Wegman and Wright (1983) and Silverman (1985).

APPROXIMATING THE EIGENVALUES OF THE HAT MATRIX

The quadratic nature of the penalized sum of squares implies that the spline smoother $\hat{g}$ is linear in the observations Y_i in the sense that there exists a *weight function* $G(s,t)$ such that

$$\hat{g}(s) = n^{-1} \sum_{i=1}^{n} Y_i \, G(s,t_i). \tag{2}$$

The weight function depends on the design points $t_1, \cdots, t_n$ and also on the smoothing parameter α. We can, in fact, obtain the asymptotic form of the weight function; hence an approximate explicit form of the estimate can be derived, thus solving the major conceptual problem that the spline smoother is defined *implicitly* as the solution to a minimization problem rather than as an *explicit* formula involving the data values. It turns out that $\hat{g}(s)$ can be approximated by a variable kernel regression estimator, where the local bandwidth depends in an attractive way on the local density of design points; for details and discussion see Silverman (1984a).

Let $A(\alpha)$ be the matrix

$$A_{ij}(\alpha) = n^{-1} G(t_i, t_j). \tag{3}$$

The matrix $A(\alpha)$ is called the *hat matrix* because it maps the data vector Y_i into the vector $\hat{Y}_i$ of predicted values $\hat{g}(t_i)$. Under suitable conditions the eigenvalue of $A(\alpha)$ can be approximated in a way that will be of great use in developing rapid methods of cross-validation for choosing the smoothing parameter α. The matrix $A(\alpha)$ is symmetric and positive-definite; denote its eigenvalues, in decreasing order, by $\delta_{1n}, \cdots, \delta_{nn}$.

Suppose that the design points all lie in an interval $[a,b]$ on which the estimation of g is of interest. Suppose further that n is large and that the design points can be assumed to have *local density* f, in the sense that, for each t in $[a,b]$, the number of design points in $[t,t+dt]$ is approximately $f(t)dt$. The density f is assumed to be bounded above and below away from zero. For precise statements and proofs of results discussed here, see Silverman (1984b) and the papers by other authors referenced there, especially Utreras (1981). There is no requirement that the t_i form a random sample from f nor need they fall at regularly-spaced quantiles of f, though of course both of these cases are permitted.

Define constants ρ_i by

$$\rho_1=\rho_2=0 \text{ and } \rho_i = \pi^4(i-1{\cdot}5)^4 \text{ for } i = 3,\cdots,n. \tag{4}$$

It can be shown that the eigenvalues of $A(\alpha)$ are given approximately by

$$\delta_{in} \approx (1 + \kappa^4 n^{-1}\alpha\rho_i)^{-1} \tag{5}$$

where the constant $\kappa = \kappa(f)$ depends on f and is defined by

$$\kappa(f) = \int_a^b f(t)^{1/4}\,dt. \tag{6}$$

Let $\mu_1(\alpha) = \text{trace } A(\alpha)$ and $\mu_2(\alpha) = \text{trace } A(\alpha)^2$; from (5) we have

$$\mu_1(\alpha) = \sum_i \delta_{in} \approx \sum_i (1 + \kappa^4 n^{-1}\alpha\rho_i)^{-1} \tag{7}$$

and

$$\mu_2(\alpha) = \sum_i \delta_{in}^2 \approx \sum_i (1 + \kappa^4 n^{-1}\alpha\rho_i)^{-2}. \tag{8}$$

The key to all these approximations lies in the eigenvalue problem that occurs in classical mechanics in the study of the vibrations of a rod of density $f(t)$ with free ends.

ESTIMATING THE FUNCTIONAL κ OF THE DESIGN DENSITY

In practice, the density f of design points is not known, except in some very special cases. All we are given is a set of data. Therefore it is necessary to provide an estimate of the constant $\kappa(f)$ as defined in (6) above. This constant is interesting because it is a *functional* of the unknown density f, and so a natural way to estimate it is to construct a *density estimate* $\hat{f}$ of f and to substitute this estimate in (6). A general discussion of density estimation is provided, for example, by Silverman (1986), whose Section 6.5 provides other contexts in which density estimates are used in order to estimate quantities that depend on unknown densities, rather than densities themselves.

Since the theory summarized above requires f to be bounded below away from zero on its support, the support $[a,b]$ is first estimated by setting

$$a = t_{\min} - \tfrac{1}{2}n^{-1}(t_{\max} - t_{\min}) \tag{9}$$

$$b = t_{\max} + \tfrac{1}{2}n^{-1}(t_{\max} - t_{\min}). \tag{10}$$

Then $\hat{f}$ is estimated by the kernel method using reflection at the boundaries a and b, as defined by Boneva, Kendall and Stefanov (1971) and discussed by Silverman (1986, Section 2.10). For reasons given in Silverman (1986, Section 3.4.2), the kernel estimate is constructed using the Gaussian kernel with bandwidth $1{\cdot}06sn^{-1/5}$, where s is the sample standard deviation of the t_i.

Finally the functional $\kappa(f)$ is estimated by a simple quadrature

$$\hat{\kappa}(f) = \frac{1}{32}\sum_{r=1}^{32} \hat{f}\{a + (r-\tfrac{1}{2})(b-a)/32\}^{1/4}; \tag{11}$$

this estimate may appear a little crude, but it gave excellent results in the cross-validation contexts described below. Substituting $\hat{\kappa}$ for κ in (7) and (8) above yields approximations for $\mu_1(\alpha)$ and $\mu_2(\alpha)$ that can be calculated from a given data set extremely rapidly once $\hat{\kappa}$ is known. It should be stressed that the estimate $\hat{\kappa}$ is computed once and for all for each data set of interest. Because of the fast algorithms available for density estimation using the fast Fourier transform (for example Silverman, 1982) the evaluation of $\hat{\kappa}$ involves trivial computational effort.

APPLICATIONS TO CROSS-VALIDATION

Several methods have been suggested for the automatic choice of the smoothing parameter in nonparametric regression. Probably the most popular automatic methods are those related to cross-validation. Of course, no automatic method should be used blindly, and the choice of which automatic method to use is to some extent arbitrary (as are many other choices in statistics, such as the choice of parametric model in classical statistics). For this reason I use the word *automatic* rather than *objective* to describe methods that do not require direct choice of the smoothing parameter.

The basic principle of cross-validation is to leave the data points out one or more at a time and to choose that value of α under which the omitted points are best predicted by the remainder of the data. Let g_α^{-i} be the smoothing spline calculated from all the data pairs except (t_i, Y_i), using the value α for the smoothing parameter. The cross-validation choice of α is then the value of α which minimizes the cross-validation score

$$\mathrm{XVSC}(\alpha) = n^{-1} \sum \{Y_i - g_\alpha^{-i}(t_i)\}^2. \tag{12}$$

A standard argument in regression theory, given for example by Craven and Wahba (1979) shows that (12) has the easier computational form

$$\mathrm{XVSC}(\alpha) = n^{-1} \sum_{i=1}^{n} \frac{\{Y_i - \hat{g}(t_i)\}^2}{\{1 - A_{ii}(\alpha)\}^2}. \tag{13}$$

Craven and Wahba also suggested the use of a related criterion, called *generalized cross-validation*, obtained from (13) by replacing $A_{ii}(\alpha)$ by its average value, $n^{-1}\mathrm{trace}\, A(\alpha)$. This gives the score

$$\mathrm{GXVSC}(\alpha) = n^{-1}\mathrm{RSS}(\alpha)/\{1 - n^{-1}\mathrm{trace}\, A(\alpha)\}^2 \tag{14}$$

where $\mathrm{RSS}(\alpha)$ is the residual sum of squares $\sum\{Y_i - \hat{g}(t_i)\}^2$. Theoretical and simulation results reported by Craven and Wahba (1979) and in other papers demonstrate that generalized cross-validation usually behaves well in choosing the smoothing parameter. Its computation requires the evaluation of the trace of $A(\alpha)$; until the recent work of Hutchison and deHoog (1985), who gave a linear time algorithm for this calculation, this still represented a considerable computational task, though by no means as substantial as that involved in calculating (13).

The approximations derived above for the trace of $A(\alpha)$ suggest a simple approximation for GXVSC(α). Let AGXVSC(α) be the quantity obtained by substituting into the definition of GXVSC the approximation (7) for trace $A(\alpha)$ with κ estimated by $\hat{\kappa}$ from (11). The choice of smoothing parameter given by minimizing AGXVSC(α) is called *asymptotic generalized cross-validation* and is investigated by Silverman (1984b). In that paper a detailed simulation study is reported using a variety of models for the curve g, for the design points, and for the signal-to-noise ratio. On data sets where the design points are regularly spaced, the performances of AGXV and GXV were found to be almost identical; this is because the approximation (5) is extremely accurate in this case. In the more general case of irregularly spaced data, AGXV was superior to GXV in that the fairly small proportion (approximately 5%) of cases where GXV drastically undersmoothed was reduced to less than 1%. The reason for this lies in the quality of the approximation (5), which underestimated the large eigenvalues of $A(\alpha)$. When α is very small, the effect of this underestimation is to increase AGXVSC(α); hence AGXV will penalize more heavily for undersmoothing than will GXV.

In summary, asymptotic generalized cross-validation has both computational and statistical advantages over generalized cross-validation. Some practical examples of its use are given in Silverman (1985).

ROBUST CROSS-VALIDATION

An alternative approach to cross-validation is introduced by Robinson and Moyeed (1986). This is motivated by the notion that any reasonably robust curve estimate should not be influenced excessively by the omission of any particular data point. The overall amount of such influence for the data point (t_i, Y_i) is measured by $\sum_j \{\hat{g}(t_j) - g_\alpha^{-i}(t_j)\}^2$, and averaging over all points i yields the score

$$\mathrm{SSC}(\alpha) = n^{-1} \sum_i \sum_j \{\hat{g}(t_j) - g_\alpha^{-i}(t_j)\}^2 . \qquad (15)$$

It can be shown from (15) that

$$\mathrm{SSC}(\alpha) = \sum_i \sum_j A_{ij}(\alpha)^2 \left[\frac{Y_i - \hat{g}(t_i)}{1 - A_{ii}(\alpha)} \right]^2 . \tag{16}$$

The direct calculation of (16) involves the evaluation of the entire hat matrix and so is very laborious. In the same spirit as GXV, Robinson and Moyeed (1986) replace $A_{ii}(\alpha)$ and $\sum_j A_{ij}(\alpha)^2$ by their respective average values over i, $n^{-1}\mu_1(\alpha)$ and $n^{-1}\mu_2(\alpha)$, to obtain, after some algebra,

$$\mathrm{GSSC}(\alpha) = n^{-1}\mu_2(\alpha)\mathrm{GXVSC}(\alpha) \tag{17}$$

as a measure of the robustness of the procedure with smoothing parameter α.

In order to robustify cross-validation, it is natural to add the two scores $\mathrm{GSSC}(\alpha)$ and $\mathrm{GXVSC}(\alpha)$ to obtain the *robust generalized cross-validation* score

$$\mathrm{RGXVSC}(\alpha) = \{1 + n^{-1}\mu_2(\alpha)\}\mathrm{GXVSC}(\alpha) \tag{18}$$

making use of the expression (17). The final stage of the Robinson-Moyeed procedure is to replace GXVSC in (18) by AGXVSC and to approximate $\mu_2(\alpha)$ using the expression (8) with κ replaced by $\hat{\kappa}$ from (11). This yields an easily calculated score function $\mathrm{ARGXVSC}(\alpha)$. A simulation study conducted by Robinson and Moyeed (1986) shows that ARGXV improves on AGXV and hence on GXV in the case of regularly-spaced data and is practically equivalent to AGXV for irregularly spaced data. Thus, whatever the design points, ARGXV appears virtually to eliminate the occasionally bad results obtained using GXV. Some applications to real data sets indicate that, as hoped, ARGXV is also less affected by outlying observations.

REFERENCES

Boneva, L.I., Kendall, D.G., and Stefanov, I. (1971). Spline transformations. *J. Roy. Statist. Soc. Ser. B*, **33**, 1-70.

Craven, P., and Wahba, G. (1979). Smoothing noisy data with spline functions. *Numer. Math.*, **31**, 377-403.

DeBoor, C. (1978). *A Practical Guide to Splines*. Springer, New York.

Good, I.J., and Gaskins, R.A. (1971). Nonparametric roughness penalties for probability densities. *Biometrika*, **58**, 255-277.

Hutchison, M.F., and deHoog, F.R. (1985). Smoothing noisy data with spline functions. *Numer. Math.*, **47**, 99-106.

Reinsch, C. (1967). Smoothing by spline functions. *Numer. Math.*, **10**, 177-183.

Robinson, A., and Moyeed, R.A. (1986). Variants of cross-validation in spline smoothing regression. University of Bath preprint, submitted for publication.

Silverman, B.W. (1982). Kernel density estimation using the fast Fourier transform. Statistical Algorithm AS 176. *Appl. Statist.*, **31**, 93-97.

Silverman, B.W. (1984a). Spline smoothing: the equivalent variable kernel method. *Ann. Statist.*, **12**, 898-916.

Silverman, B.W. (1984b). A fast and efficient cross-validation method for smoothing parameter choice in spline regression. *J. Amer. Statist. Assoc.*, **79**, 584-589.

Silverman, B.W. (1985). Some aspects of the spline smoothing approach to non-parametric regression curve fitting. *J. Roy. Statist. Soc. Ser. B*, **47**, 1-52.

Silverman, B.W. (1986). *Density Estimation for Statistics and Data Analysis.* Chapman and Hall, London and New York.

Utreras, F. (1981). Optimal smoothing of noisy data using spline functions. *SIAM J. Sci. Stat. Comput.*, **2**, 349-362.

Wegman, E.J., and Wright, I.W. (1983). Splines in statistics. *J. Amer. Statist. Assoc.*, **78**, 351-365.

Bernoulli, Vol. 2, pp. Ø95-Ø98

BOOTSTRAP OF THE MEAN IN THE INFINITE VARIANCE CASE

K.B. Athreya
Indian Statistical Institute, Bangalore, 560059, INDIA
Dept. of Statistics, Iowa State University, Ames, Iowa,50011,U.S.A.

Abstract: Let $X_1,X_2,\ldots,X_n$ be i.i.d.r.v. belonging to the domain of attraction of a stable law of order α, $0 < \alpha < 2$. Given $X_1^n \equiv (X_1,X_2,\ldots,X_n)$ let $Y_1,Y_2,\ldots,Y_n$ be i.i.d. with distribution $P(Y_1 = X_j \mid X_1^n) = n^{-1}$ for $j = 1,2,\ldots,n$. This paper shows that there exist a_n and b_n such that iff $H_n(x,\omega) \equiv P(n(\bar{Y}_n - b_n) \le a_n x \mid X_1^n)$ then there is a random infinitely divisible distribution function $H(x,\omega)$ such that $H_n(.,\omega)$ converges to $H(.,\omega)$ in the Skorohod space $D(-\infty,\infty)$. This shows that the bootstrap method for the sample mean fails when $\alpha < 2$. The case $\alpha = 2$ and other related results are also discussed.

Keywords & phrases : Bootstrap, random i.d. law, stable law.
AMS (1980) Classification : 62G05, 62E29.

1. Introduction : Let $X_1,X_2,\ldots,X_n$ be independent, indentically distributed random variables belonging to the domain of attraction of a stable law of order α. Given $X_1^n \equiv (X_1,X_2,\ldots,X_n)$ let $Y_1,Y_2,\ldots,Y_m$ be independent identically distributed with $P(Y_1 = X_j \mid X_1^n) = n^{-1}$ for $j = 1,2,\ldots,n$. It was shown by K.Singh [6] and P.J.Bickel and D.Freedman [3] independently that when $EX_1^2 < \infty$ and $m = n$ the random distribution function $H_n(x,\omega) = P(n(\bar{Y}_n - \bar{X}_n) \le s_n x \mid X_1^n)$ where

$$\bar{X}_n = n^{-1} \sum_1^n X_i,\ \bar{Y}_n = n^{-1} \sum_1^n Y_i, \text{ and } s_n^2 = n^{-1} \sum_1^n (X_i - \bar{X}_n)^2$$ converges to

the standard normal distribution function $\Phi(x)$ in the supremum norm with probability one. Here and in what follows ω represents a generic element of the underlying sample space (Ω,B,P) on which all the random variables to be encountered here are supposed to be defined. The goal of this paper is to seek the existence of constants (possibly random) $a_{m,n}$, $b_{m,n}$ and conditions on the rate m,n go to ∞ such that the random distribution function

$$H_{m,n}(x,\omega) = P((\sum_1^m Y_j - b_{m,n}) \leq a_{m,n}\, x | X_1^n)$$

converges to an appropriate limit. This problem arises in the theory of bootstrap introduced by B.Efron [4]. Our results indicate that the bootstrap method works well only when $\alpha = 2$.

2. The results : Let $F(x) = P(X_1 \leq x)$, $\mu(t) = E(X_1^2 : |X_1| \leq t)$. It is known (see Feller [5]) that X_1 belongs to the domain of attraction of a stable law or order α iff $\mu(t) \sim t^{2-\alpha}L(t)$ as $t \to \infty$ where $L(t)$ is slowly varying at ∞. For $\alpha < 2$ this is equivalent to $F(-x) \sim C_1 x^{-\alpha}L(x)$ and $1 - F(x) \sim C_2 x^{-\alpha}L(x)$ as $x \to \infty$ for some nonnegatve numbers C_1, C_2. Without loss of generality we assume that $C_2 = 1$ and set $C_1 = C \geq 0$

Case i) $\alpha < 2$, $m = n$. Let

(1) $H_n(x,\omega) = P(n(\bar{Y}_n - L_n) \leq X_{nn}\, x | X_1^n)$

Where $X_{nn} = \max(X_1, X_2, \ldots, X_n)$ and where for $1 < \alpha < 2$ $L_n = \bar{X}_n$ and for $0 < \alpha \leq 1$, L_n is the unique solution of the equation

$$\sum_{j=1}^{n} \tau\left(\frac{X_j - L_n}{X_{nn}}\right) = 0 \quad \text{with}$$

(2) $\tau(x) = x$ in $(-1, +1)$, $= 1$ for $x > 1$, $= -$ for $x < -1$.

The statistic $n(\bar{Y}_n - L_n)X_{nn}^{-1}$ is the bootstrap analog of $n(\bar{X}_n - \tilde{L}_n)a_n^{-1}$ where $\tilde{L}_n = EX_1$ if $1 < \alpha < 2$ and is the unique solution of $\sum_{j=1}^{n} \tau\left(\frac{X_j - \tilde{L}_n}{a_n}\right) = 0$ and a_n is any sequence of constants going to ∞ such that $na_n^{-2}\mu(a_n) \to 1$. It is known that $n(\bar{X}_n - \tilde{L}_n)a_n^{-1}$ coverges in distribution to $G_\alpha(.)$, a stable law of order α. A question of some importance in the theory of bootstrap is whether $H_n(.,\omega)$ also converges to $G_\alpha(.)$. It turns out that the answer is no and thus the method of bootstrap of the mean fails when $\alpha < 2$. However, $H_n(.,\omega)$ does converge to a limit that is a random distribution function $H(.,\omega)$. To describe this limit we introduce a Poisson random measure $N(A,\omega)$ defined for all Borel sets $B(R)$ of R such that for $A_1, A_2, \ldots, A_k$ disjoint $N(A_i,\omega)$ $i = 1,2,\ldots,k$ are independent and $EN(A,\omega) = \lambda_\alpha(A)$ where λ_α is a measure on $B(R)$ such that

(3) $\lambda_\alpha[x,\infty) = x^{-\alpha}$ and $\lambda_\alpha(-\infty, -x] = c\, x^{-\alpha}$ for $x > 0$.

Let $\phi(t,\omega) = \exp(\int (e^{itx} - 1 - it\, \tau_\alpha(x))\ N'(dx,\omega)$ where $\tau_\alpha(x) \equiv x$ if $1 < \alpha < 2$ and equals to $\tau(x)$ of (2) for $0 < \alpha \leq 1$ and $N'(a,\omega) = N(A\tau^{-1},\omega)$ where $\tau = \inf\{t : N((t,\omega),\omega) = 0\}$. It can be shown that $\phi(t,\omega)$ is the characteristic function of an infinitely divisible distribution $H(x,\omega)$ so that

(4) $\phi(t,\omega) = \int e^{itx}\, dH(x,\omega)$.

Theorem 1 : Let H_n and H be as in (1) and (4) respectively. Then the sequence of satochastic processes $\{H_n(.,\omega)\}$ converge weakly to $\{H(.,\omega)\}$ in the Skorohod space $D(-\infty,\infty)$. In particular, for any $-\infty < a < b\ \infty$, $H_n(a,\omega) \xrightarrow{d} H(a,\omega)$ and $H_n(b,\omega) - H_n(a,\omega) \xrightarrow{d} H(b,\omega) - H(a,\omega)$ where $\xrightarrow{d}$ stands for convergence in distribution.

Case ii) $\alpha = 2$, $m = n$, $EX_1^2 = \infty$.

Theorem 2 : Let H_n be as in (1) and let $\Phi(.)$ be the standard normal c.d.f.

(5) Then, $\sup_x |H_n(x,\omega) - \Phi(x)| \to 0$ in probability.

We remark, in passing, that when $EX_1^2 < \infty$ this can be strengthened to convergence w.p.l. as was shown by K.Singh [6] and P.J.Bickel and D.Freedman [3].

Case iii) $\alpha < 2$, m,n go to ∞ such that $m\,n^{-1} \to 0$.

Let $a_{m,n}$ and $b_{m,n}$ be such that

$$mn^{-1} \sum_{j=1}^{n} \chi_{[a_{m,n},\infty)}(X_j) \to 1. \quad \text{and} \quad \sum_{j=1}^{n} \tau_\alpha\left(\frac{X_j - b_{m,n}}{a_{m,n}}\right) = 0$$

where $\tau_\alpha(x) \equiv x$ if $1 < \alpha < 2$ and $\tau(x)$ of (2) for $0 < \alpha \leq 1$.

(6) Let $H_{m,n}(x,\omega) = P((\sum_{j=1}^{m} Y_j - b_{m,n})\, a_{m,n}^{-1} \leq x | X_1^n)$

Theorem 3 : Let $\alpha < 2$ and $m, n \to \infty$ such that $mn^{-1} \to 0$. Let $H_{m,n}(\cdot,\cdot)$ be as in (6) . Then

$$\sup_x |H_{m,n}(x,\omega) - G_\alpha(x)| \xrightarrow{P} 0$$

where $G_\alpha(\cdot)$ is the distribution function of a stable law of order α whose characteristic function $\phi(t)$ is given by

$$\phi(t) = \exp\left(\int (e^{itx}-1-it\,\tau_\alpha(x))\lambda_\alpha(dx)\right) \text{ where } \lambda_0(\cdot) \text{ is as in (3).}$$

Notice that $(\sum_{j=1}^{m} Y_j - b_{m,n})\, a_{m,n}^{-1}$ is the bootstrap version of the statistics $(\sum_{j=1}^{n} X_j - \tilde{L}_n) a_n^{-1}$ where $\tilde{L}_n$ and a_n are as in case (i).

This indicates that the bootstrap method works when $\alpha < 2$ provided the resample size in is small campared to the original sample size n.

Details of the proof of the results of this paper may be found in [1,2].

REFERENCES:

1. Athreya, K.B. (1986) Bootstrap of the mean in the infinite variance Case-I and II, Technical Reports 86-21, 86-22 of the Department of Statistics, Iowa State University, Ames, Iowa, 50011.

2. Athreya, K.B. (1986) Bootstrap of the mean in the infinite variance case To appear in the Annals of Statistics (1987).

3. Bickel, P.J. and Freedman, D. (1981). Some asymptotic theory for the bootstrap. Annals of Statistics, 9 1196-1217.

4. Efron, B. (1979) Bootstrap methods - another look at the Jack knife. Annals of Statistics, 7, 1-26.

5. Feller, W (1971) An Introduction to Probability Theory and Applications. John Wiley, N.Y.

6. Singh, K (1981) On the asymptotic efficiency of Efron's bootstrap, Annals of Statistics, 9, 1187-1195.

AUTOMATIC CURVE SMOOTHING

Wolfgang Härdle
Institut Wirtschaftstheorie II
Universität Bonn
Adenauerallee 24-26
D-5300 Bonn,
Federal Republic of Germany

1. INTRODUCTION

Regression smoothing is a method for estimating the mean function from observations $(x_1,Y_1),\ldots,(x_n,Y_n)$ of the form

$$Y_i = m(x_i) + \varepsilon_i, \quad i=1,\ldots,n,$$

where the observation errors are independent, identically distributed, mean zero random variables. There are a number of approaches for estimating the regression function m. Here we discuss nonparametric smoothing procedures, which are closely related to local averaging, i.e. to estimate m(x), average the Y_i's which are in some neighborhood of x. The width of this neighborhood, commonly called bandwidth or smoothing parameter, controls the smoothness of the curve estimate. Under weak conditions (bandwidth shrinks to zero not too rapidly as n increases) the curve smoothers consistently estimate the regression function m. In practice, however, one has to select a smoothing parameter in some way. A too small bandwidth, resulting in high variance, is not acceptable and so is *oversmoothing* which creates a large bias. It is therefore highly desirable to have some *automatic curve smoothing* procedure.

Proposed methods for choosing the window size automatically are based on estimates of the prediction error or adjustments of the residual sum of squares. It has been shown by Härdle, Hall and Marron (1986) (HHM) that all these proposals are asymptotically equivalent but can be quite different in a practical situation. In this paper we highlight these difficulties with automatic curve smoothing and construct situations where some of the proposals seem to be preferable.

2. AUTOMATIC CURVE SMOOTHING

To simplify the presentation, assume the design points are equally spaced, i.e. $x_i = i/n$, and assume that the errors have equal variance, $E\varepsilon^2 = \sigma^2$. We study *kernel smoothers*

$$\hat{m}_h(x) = n^{-1}h^{-1}\sum_{i=1}^{n} K\left(\frac{x-x_i}{h}\right)Y_i$$

where h is the bandwidth and K is a symmetric kernel function. It is certainly desirable to tailor the automatic curve smoothing so that the resulting regression estimate is close to the true curve. Most automatic bandwidth procedures are designed to optimize the averaged squared error (ASE)

$$d_A(h) = n^{-1}\sum_{i=1}^{n} [\hat{m}_h(x_i) - m(x_i)]^2 w(x_i),$$

where w is some weight function. These automatic bandwidth selectors are defined by multiplying $p(h) = n^{-1}\sum_{i=1}^{n}(Y_i - \hat{m}_h(x_i))^2 w(x_i)$ by a correction factor $\Xi(n^{-1}h^{-1})$. The examples we threat here are

General Gross-Validation (Craven and Wahba 1979),

$$\Xi_{GCV}(n^{-1}h^{-1}) = (1 - n^{-1}h^{-1}K(0))^{-2}.$$

Akaike's Information Criterion (Akaike 1970),

$$\Xi_{AIC}(n^{-1}h^{-1}) = \exp(2n^{-1}h^{-1}K(0)).$$

Finite Prediction Error (Akaike 1974),

$$\Xi_{FPE}(n^{-1}h^{-1}) = (1 + n^{-1}h^{-1}K(0))/(1 - n^{-1}h^{-1}K(0)).$$

A model selector of Shibata (1981),

$$\Xi_S(n^{-1}h^{-1}) = 1 + 2n^{-1}h^{-1}K(0).$$

The bandwidth selector T of Rice (1984),

$$\Xi_T(n^{-1}h^{-1}) = (1 - 2n^{-1}h^{-1}K(0))^{-1}.$$

Let $\hat{h}$ denote the bandwidth that minimizes $(p\cdot\Xi)(h)$. The *automatic curve smoother* is defined as $\hat{m}_{\hat{h}}(x)$. This automatic curve smoothing procedure is asymptotically optimal for the above Ξ in the sense that

$$\frac{d_A(\hat{h})}{d_A(\hat{h}_o)} \xrightarrow{p} 1,$$

where $\hat{h}_o$ denotes the minimizer of d_A. The relative differences are quantified in the

<u>Theorem</u>. Let $\hat{h}_o \sim n^{-1/5}$ then

$$n^{3/10}(\hat{h} - \hat{h}_o) \to N(0,\sigma^2)$$

$$n[d_A(\hat{h}) - d_A(\hat{h}_o)] \to C\cdot\chi_1^2.$$

in distribution, where σ^2 and C are defined in HHM.

A very remarkable feature of this result is that the constants σ^2 and C are *independent of* Ξ. In a simulated example we generated 100 samples of size n=75 with σ=0.05 and $m(x) = \sin(\lambda 2\pi x)$. The kernel function was taken to be $K(x) = (15/8)(1-4x^2)^2 I(|x|\le 1/2)$. Table 1 shows the number exceedances by ratios of error criteria d_A and $E\{d_A\} = d_M$, for 100 data sets of size n=100.

PARAMETER LAMBDA = 1

	1.05	1.1	1.2	1.4	1.6	1.8	2	4	6	8
RICE										
DA	50	34	15	4	2	1	1	0	0	0
DM	18	8	0	0	0	0	0	0	0	0
GCV										
DA	51	34	18	12	9	7	7	1	0	0
DM	30	17	11	7	3	2	2	0	0	0
FPE										
DA	75	63	52	49	47	40	37	10	2	0
DM	65	56	52	47	40	40	40	0	0	0
AIC										
DA	91	86	85	83	81	69	63	13	3	0
DM	86	84	84	82	79	79	79	0	0	0
SHIBATA										
DA	100	100	100	100	99	90	81	15	3	0
DM	100	100	100	100	100	100	100	0	0	0

Table 1

Rice's proposal T shows a quite good performance. Note that T has a slight bias towards oversmoothing, since this Ξ has a pole at $2n^{-1}K(0)$ whereas all the other selectors have no pole or a pole at $n^{-1}K(0)$, the "no smoothing point". By increasing λ to 2 (Table 2) the T-selector looses its good performance, and is clearly outperformed by Generalized Cross-Validation (GCV).

PARAMETER LAMBDA = 2

	1.05	1.1	1.2	1.4	1.6	1.8	2	4	6	8
GCV										
DA	37	24	15	4	3	3	2	0	0	0
DM	31	10	10	2	0	0	0	0	0	0
RICE										
DA	53	26	6	0	0	0	0	0	0	0
DM	25	5	0	0	0	0	0	0	0	0
FPE										
DA	92	85	75	57	36	19	9	0	0	0
DM	84	80	80	80	0	0	0	0	0	0
AIC										
DA	99	96	90	69	43	23	11	0	0	0
DM	97	95	95	95	0	0	0	0	0	0
SHIBATA										
DA	100	100	95	74	46	23	11	0	0	0
DM	100	100	100	100	0	0	0	0	0	0

Table 2

The reason for that is the tendency of T to oversmooth; this behavior is penalized in a situation where the reduction of bias becomes more important than the reduction of variance. This becomes apparent in Table 3 where λ was 3.

PARAMETER LAMBDA = 3

	1.05	1.1	1.2	1.4	1.6	1.8	2	4	6	8
GCV										
DA	51	28	8	1	0	0	0	0	0	0
DM	69	4	0	0	0	0	0	0	0	0
RICE										
DA	69	36	15	1	0	0	0	0	0	0
DM	99	17	1	0	0	0	0	0	0	0
SHIBATA										
DA	89	70	33	4	2	1	1	0	0	0
DM	100	100	0	0	0	0	0	0	0	0
FPE										
DA	89	70	33	4	2	1	1	0	0	0
DM	100	100	0	0	0	0	0	0	0	0
AIC										
DA	100	100	95	51	19	4	3	0	0	0
DM	100	100	100	100	0	0	0	0	0	0

Table 3

References

Akaike, H. (1970). Statistical predictor information. Annals of the Institute of Statistical Mathematics, 22, 203-217.

Akaike, H. (1974). A new look at the statistical model identification. IEEE Transactions on Automatic Control, AC19, 719-723.

Craven, P. and Wahba, G. (1979). Smoothing noisy data with spline functions. Numerische Mathematik, 31, 377-403

Härdle, W., Hall, P. and Marron, J.S. (1986). How far are automatically chosen regression smoothing parameters from their optimum? (with discussion). J. Amer. Stat. Assoc., to appear.

Rice, J. (1984). Bandwidth choice for nonparametric regression. Annals of Statistics, 12, 1215-1230.

Shibata, R. (1981). An optimal selection of regression variables. Biometrika, 68, 45-54.

NON-PARAMETRIC SMOOTHING OF THE BOOTSTRAP

Young A.

Statistical Laboratory, University of Cambridge, 16 Mill Lane, Cambridge, U.K.

This communication describes a non-parametric smoothing modification to Efron's bootstrap estimation procedure. Much of the discussion summarises technical results contained in the paper Silverman and Young [3], which has been submitted for publication elsewhere.

Efron [1] introduced the bootstrap as a non-parametric approach to the assessment of errors and related quantities in statistical estimation. A typical context in which the bootstrap is used is in assessing the sampling mean square error $\alpha(F)$ of an estimate $\hat{\theta}(X_1,\ldots,X_n)$ of a parameter $\theta(F)$, based on an independent sample of size n, with empirical distribution F_n, drawn from some unknown r-variate distribution F. The bootstrap idea may be applied to any functional $\alpha(F)$: the bootstrap estimates $\alpha(F)$ by its empirical version $\alpha(F_n)$, this in general being itself estimated by repeated resampling, with replacement, from F_n. Efron [1] suggested the smoothed bootstrap as a modification which avoids samples with discreteness properties. Assuming F has a smooth density f, a convenient smoothed bootstrap is obtained from the kernel estimate $\hat{f}_{h,s}$ of f defined by

$$\hat{f}_{h,s}(x) = (1+h^2)^{\frac{1}{2}}\,\hat{f}_h\{(1+h^2)^{\frac{1}{2}}\,x\},$$

$$\hat{f}_h(x) = |V|^{-\frac{1}{2}}\,n^{-1}\,h^{-r}\sum_1^n K\{h^{-1}\,V^{-\frac{1}{2}}\,(x-X_i)\}.$$

Here K is a symmetric probability density function of an r-variate distribution with unit variance matrix, V is the sample variance matrix of the data and h is a parameter defining the degree of smoothing. Notice that realisations generated from $\hat{f}_h$ have expectation equal to $\overline{X}$, the mean of the observed sample data, but that smoothing inflates the variance. The kernel estimate $\hat{f}_h$ is therefore "shrunk" to give an estimate $\hat{f}_{h,s}$ with first and second moment properties the same as those observed in the sample $(X_1,\ldots,X_n)$.

Efron [2] considered application of a smoothed bootstrap to estimation of the standard error of the transformed correlation coefficient constructed from a sample of size 14 drawn from a particular bivariate normal distribution. Direct simulation shows that in this case a suitable smoothed bootstrap gives better estimates of standard error than the unsmoothed bootstrap. Despite Efron's finding, there has, to date, been little systematic investigation of the smoothed bootstrap. Particular questions which arise are: (i) when is it worth smoothing, and (ii) how should the degree of smoothing be chosen in circumstances where smoothing is worthwhile? So far, theoretical work has concentrated on the first of these questions (Silverman and Young [3]), though this communication considers also ideas on the second.

The simplest case is that of estimation of a linear functional $\alpha(F) = \int a(t)dF(t)$. For such functionals we have the theorem (Silverman and Young [3]):

<u>Theorem</u>. Suppose $V = [V_{ij}]$ is a fixed positive definite symmetric matrix. The mean square error of the smoothed bootstrap estimate $\hat{\alpha}_h(F) = \int a(t)\, \hat{f}_{h,s}(t)\, dt$ may be reduced below that of the unsmoothed estimate $\hat{\alpha}_0(F) = \int a(t)\, dF_n(t)$, by choosing suitable $h > 0$, provided $a(X)$ and $a^*(X)$ are negatively correlated. Here $a^*(X) = D_V a(X) - X.\nabla a(X)$, with $D_V a(x) = \sum_i \sum_j V_{ij}\, \delta^2 a(x) / \delta x_i\, \delta x_j$.

The appropriate calculus for statistical functionals and computer algebraic manipulation extend the result to analysis of more general functionals $\alpha(F)$, by local linear approximation. Silverman and Young [3] consider a number of examples which illustrate use of the result and the importance of shrinking the kernel estimate in defining the smoothed bootstrap. Note that the theorem only indicates whether smoothing, with fixed V , is worthwhile or not, and gives no guidance in the choice of h . Also, direct application of the result requires knowledge of the underlying distribution F . From an operational viewpoint, of course, V is not fixed, but is taken as the variance matrix of the observed sample.

The above theorem is proved by considering an asymptotic expansion for the mean square error of the smoothed bootstrap estimate $\hat{\alpha}_h(F)$

in the form

$$\text{MSE}\ \{\hat{\alpha}_h(F)\} = C_0 + C_1h^2 + O(h^4)\ ,$$

and by calculating the coefficient C_1 as $C_1=n^{-1}\int\{a(x)-\mu\}a^*(x)f(x)dx$, where $\mu = \int a(x)f(x)dx$. Guidance on the choice of smoothing parameter h may be obtained by considering the h^4 term in this expansion for MSE. Considering the one-dimensional case for simplicity, we find that, for the Gaussian kernel K, the coefficient of h^4 is given by

$$C_2 = n^{-1}[2\int\{a(x) - \mu\}a^{**}(x)f(x)\,dx + \frac{1}{4}\int a^*(x)^2f(x)\,dx + (n-1)\{\int a^*(x)f(x)\,dx\}^2]\ .$$

Here, as before, $a^*(x) = Va''(x) - xa'(x)$ and

$$a^{**}(x) = \frac{1}{8}V^2a^{(iv)}(x) - \frac{1}{4}Vxa'''(x) + \frac{1}{8}x^2a''(x) - \frac{1}{2}Va''(x) + \frac{3}{8}xa'(x)\ .$$

If $C_1 < 0$ some small degree of smoothing at least is worthwhile. If $C_2 < 0$ some larger degree of smoothing may be appropriate. If both $C_1 > 0$ and $C_2 > 0$ the appropriate bootstrap estimate is the unsmoothed estimate $\hat{\alpha}_0(F)$. Otherwise, the optimal smoothing parameter, in the sense of minimising the approximate MSE $C_0 + C_1h^2 + C_2h^4$, is given by $h = (2|C_1| / 4C_2)^{\frac{1}{2}}$.

The quantities C_1 and C_2 depend on the unknown underlying distribution F. Data-based (essentially unsmoothed bootstrap) estimates are given by

$$\hat{C}_1 = n^{-2}\sum\{a(X_i) - \bar{a}\}\ a^*(X_i)\ ,$$

$$\hat{C}_2 = n^{-2}[2\sum\{a(X_i)-\bar{a}\}\ a^{**}(X_i) + \frac{1}{4}\sum a^*(X_i)^2] + (n-1)n^{-3}\{\sum a^*(X_i)\}^2,$$

where $\bar{a} = n^{-1}\sum a(X_i)$. A possible strategy for choosing the degree of smoothing might be to take $h = 0$ if $\hat{C}_1 \geq 0$, $h = \infty$ if $\hat{C}_1 < 0$ and $\hat{C}_2 < 0$, and $h = (2|\hat{C}_1| / 4\hat{C}_2)^{\frac{1}{2}}$ otherwise. The case $h = \infty$ amounts to bootstrap sampling from the parametric $N(\bar{X},V)$ distribution.

Though it is known (Silverman and Young [3]) that a choice of smoothing suitable for estimation of the underlying density f may be unsuitable for bootstrap estimation of a population functional $\alpha(F)$, an alternative, simplified, strategy might be to take $h = 0$ if $\hat{C}_1 \geq 0$ and $h = 1.06V^{\frac{1}{2}}n^{-1/5}$ otherwise. This latter value for

h is a data estimate of the asymptotically optimal smoothing parameter for estimation of the density f, in the case when the population is Gaussian.

As a simple application of the above, consider estimation of the fifth moment EX^5 of a univariate density, on the basis of an independent sample $X_1,\ldots,X_{10}$ of size 10. In the case when X has a standard normal distribution, the unsmoothed bootstrap estimate $\hat{\alpha}_0(F)$ has mean square error 94.5, while $C_1 = -262.5$ and $C_2 = 562.5$. A (locally) optimal degree of smoothing is therefore provided by $h = 0.5$. A series of simulations was carried out to compare the two strategies for choosing h. A series of 4000 datasets of size 10 were drawn from the $N(0,1)$ distribution: the mean square error of the bootstrap estimates obtained by choosing h by the first procedure was 80.4. A similar series of bootstrap estimates choosing h by the simplified strategy gave mean square error 51.3. This figure represents a considerable improvement over both the unsmoothed bootstrap, and the smoothed bootstrap for $h = 0.5$, for which a mean square error of 73.5 was observed by simulation. In all cases smoothed bootstrap estimates were calculated on the basis of 300 bootstrap resamples.

REFERENCES

1. Efron, B. (1979). Bootstrap methods: another look at the jackknife. Ann. Statist. 7, 1-26.
2. Efron, B. (1982). The Jackknife, the Bootstrap and Other Resampling Plans. SIAM, Philadelphia.
3. Silverman, B.W., and Young, A. (1987). The bootstrap: to smooth or not to smooth? To appear in Biometrika.

DATA ANALYSIS (projection pursuit, curve estimation . . .)

(Session 4)

Chairman: E. Diday

ON CONSTRUCTING A GENERAL THEORY OF AUTOMATIC CLASSIFICATION

S.A.Aivazyan
Central Economics Mathematical Institute of the USSR Academy of Sciences, Moscow

INTRODUCTION

This approach to constructing a general theory of automatic classification (AC) is based on an idea suggested earlier by a number of authors (Aivazyan et al.(1974), Aivazyan (1979), Bukhshtaber and Maslov (1977), (1980)). It was suggested that the basic problems of applied statistics (data analysis) should be stated and solved as special optimization problems. The constructive realization of the current approach is an extention of the well known 'k-means' method which was almost simultaneously suggested by Sebestyen (1962), McQueen (1967), and Shlezinger (1965). An important contribution to this development based on a generalization of the class kernel was made by the authors of the so-called "méthode des nuées dinamiques"(the method of dynamic clusters), Diday et al (1962). Those authors posed various problems of statistical data analysis within a unified optimization AC problem by interpreting the notion of kernel in the broader sense (the centroid of a class, the subset of a class, the axis, the random variable, etc) and seeking to find a partition of the set under consideration such that the given 'kernel' is most representative relative to every class (using a certain representativeness criterion).

As enabling us to generate a broad class of AC algorithms, the nuées dinamiques method does not possess, however, a set of controlled parameters which would provide the possibility of linking and comparing various algorithms. Moreover, there are important classes of AC algorithms which fall beyond the scope of the nuées dinamiques method (for example, FOREL, see Aivazyan et al. (1974).

The suggested approach (also, see Aivazyan and Bukhshtaber (1979)) makes it possible to surmount the abovementioned disadvantage of the nuées dinamiques method. According to this approach the totality of AC algorithms is represented as a hierarchical structure. At the uppermost level there is a general mathematical model whose components are used to formulate AC problems in a uniform manner and to describe the algorithms for solving the problems. The lowermost level consists of the 'movements' of specific algorithms (the definition of the 'movement' is given in the report). The descent to lower levels results from a specialization which fills the model components with information concerning the nature of data, the final cause of classification, the a priori trial hypotheses and so on.

In particular, this construction enables us to conduct

(a) the study into the model of AC algorithms as a mathematical structure so as to obtain the convergence conditions for the algorithms and to describe the class of functionals to be optimized by these algorithms;

(b) the comparison between various available algorithms based on the descent and ascent over the hierarchical structure;

(c) the construction of the parametric families of algorithms so as to implement the algorithms as program systems which are, in particular, intended for the adaptive choice (the 'tuning') of the free parameters of an AC algorithm.

The mathematical construction suggested in the report is described in terms of Aivazyan et al. (1974).

1. THE GENERAL DESCRIPTION OF THE MATHEMATICAL AC MODEL

We begin with describing the components of the general mathematical AC model.

E is a set of the objects x_1, x_2,... to be classified. The form of object specification is determined in the course of problem formulation. Basically, an object x_n is described by a p-dimensional vector whose components are the characters of this object. Specifically, if the characters are numerical in nature, E belongs to a Euclidean space R^p.

S=S(E) is a set of the states whereat the sample is used by the algorithm. Qualitatively, this component represents the admissible classifications resulting from the algorithm operation. In terms of Shlezinger (1965) S(E) is the space of coverings, but we make this component more specific by assuming that every element $s \in S$ is a mapping from E into Z, where Z is the set of the classification values. The composition and structure of Z define the type of the problem to be solved, i.e., provide a means for answering the question what a classification problem is to be solved? For instance, if $Z=[1, 2,\ldots,k,\ldots,K]$ is the set of class numbers, then the set $S=\{s: E \rightarrow Z\}$ coincides with the set of the partitions of E into K disjoint classes $s \Leftrightarrow (E(1),\ldots,E(K))$, where $E(k)= s^{-1}(k)$. If Z is an ordered set, this corresponds to a preference or hierarchical classification. The fuzzy classification problem corresponds to $Z=(z_1,\ldots,z_k)$ where z_k are real numbers such that $z_k \geqq 0$, $\sum_{k=1}^{K} z_k=1$, or, formally, in this case Z is a simplex whose vertices correspond to the numbers of classes and the coordinates of points to the probabilities that an object belongs to a class. Obviously, the availability of a priori information concerning the type of an AC problem and the admissible classifications leads to constraints which pinpoint a set S(E) in the set of all mappings. For example, if there is a training (verified) subsample $E_1=\bigcup_{k=1}^{K} E_1(k)$ in E, then one obtains the conditions $s(x)=k$ for $x \in E_1(k)$. Thus to specify S, one needs to formalize the problem, i.e., to set Z and to state the conditions which single out S in the set of all the mappings $E \rightarrow Z$.

L(E) is the set of the descriptions of E in the framework of a given algorithm. This component relates to the choice of the classification means and corresponds to the representativeness subspace which is understood in a broader sense than in Diday et al.(1979). The set L is regarded as a subset of the set of all mappings $Z \rightarrow Y$, where Y is the set of the values which represent the classification results. For example, if $E \subset R^p$ and $Z=[1,\ldots,K]$ and if every class is supposed to be described by the sample mean, then $Y=R^p$ and $L(E)=\{Z \rightarrow Y\}=R^p \times \ldots \times R^p$. If a class is described by a standard and if the standard of the kth class is known to be in a δ_k-neighborhood of the representative y_k^0, specifically, δ_k may be zero for some k then $L(E)=\{(y_1,\ldots,y_k),\ y_k \in R^p,\ \| y_k-y_k^0 \| \leq \delta_k\}$. If the classes are described by standard sets, then the corresponding part of the set of all subsets of E is taken as Y. The set Y can be of a completely different nature than the space of the characters to be measured. Moreover, Y itself can consist of spaces which differ in nature as required by the standards of different classes. Note that on this way we can incorporate many well-known algorithms which cannot be described by the above-mentioned nuées dinamiques method (for example, FOREL, see Aivazyan et al.(1974)) and construct some new algorithms.

R(E) is the set of certain finite subsets of E, the so-called 'portions', into which E is subdivided for classification. This component is introduced in order that one can treat algorithms both in parallel (R(E) consists of a single element symbolizing the entire set E), and in series (for example, when objects are classified in a one-at-a-time manner, $R(E) \equiv E$). Note that although Diday et al.(1979) describes only parallel procedures, the nuées dinamiques method for serial procedures was developed in Diday(1975).

$\mathcal{K}$ is an operator from $S \times L \times R$ into S called a classifier since it shows how to apply the available means Y to the AC problem of type Z in order to pass from the state s_n of sample E with the description l_n to the state s_{n+1}, given the current portion $r_{n+1} \subset E$, i.e., how to carry out a reclassification. The affectation function $f: L \rightarrow S$ (see Diday et al.(1979), Chapter 1) is a specific case of this operator. Usually, the operator $\mathcal{K}$ is obtained from a functional $\mathcal{F}$ estimating the classification performance in accordance with the a priori expert concept of a 'good' classification.

D is an operator from $S \times L$ into L called a descriptor since it is used to obtain a new more accurate description from the previous description l_n of E and the available classification s_{n+1}. The representativeness function $g: S \rightarrow L$ in Diday et al.(1979) is a specific case of the operator D. The standard-based algorithms of partitioning into K classes construct D from a functional $\mathcal{F} : S \times L \rightarrow R^1$ of the form

$$\mathcal{F}(s, l) = \sum_{k=1}^{K} F_k(E(k), l(k))$$

by the formula

$$D(s, l) = \arg\min_{y \in Y} F_k(E(k), y).$$

Thus the computation of D reduces to minimization of $F_k(E(k),y)$ over Y. Note that here the basic idea of the nuées dinamiques method encounters a severe difficulty. Nontrivial examples of choosing the representatives of classes show that Y can be a rather complicated variety (see, for example, Y for the typological factor analysis). The conventional functionals F_k, which allow efficient calculation of a descriptor in the framework of the k-means algorithms for $Y=R^p$, fail to provide a unique solution to the minimization problem when considered on the varieties of the affine subvarieties of R^p. Diday et al.(1979) had to use artificial techniques for selecting one of several possible descriptor values so that the termination rules of the algorithm became complicated. In this connection it should be noted that Bukhshtaber and Maslov (1977),(1980) developed optimization methods for a broad class of functionals on the varieties of the affine subvarieties of a Euclidean space R.

G is an operator from $R \times S \times L \times N$ into R (where N is th set of natural numbers) which can be interpreted as a po tion generator.

Definition. The model of the AC algorithm (ACA) is the set (E, S, L, R, $\mathcal{K}$, D, G) of the above components.

It is shown in Diday et al.(1979) how to obtain many well-known automatic classification algorithms including those of fuzzy classification by appropriate choice of the ACA-model parameters.

2. MOST IMPORTANT DIRECTIONS OF AUTOMATIC CLASSIFICATION DEVELOPMENT

We shall describe the achievements of automatic classifi cation connected with the necessity of tackling the 'bottlenecks' in this area and based on the drastic grow of computer facilities.

2.1. Interactive classification using the elements of computer-based assistance (for example, the special purpose expert systems).

This division deals with the automation of man-machine interaction in the course of solving a classification problem by efficiently using the data visualization methods (the projection persuit, see Friedman (1974),

Huber (1985), Enyukov (1986), and the tomography methods Bukhshtaber and Maslov (1985)) and the properly formalized knowledge and intuition of the experts experienced in both statistics and the area to which the objects under classification belong.

This man-machine interaction results in trial hypotheses concerning the probabilistic and geometric nature of the multivariate observation to be classified, the number of classes, the general form of the metric in the space under consideration and so on.

The currently available expert systems for automatic classification (Demonchaux et al.(1985), Hahn (1985)) can be thought of as the first modest steps along this direction.

2.2 The construction of a general statistical classification theory covering the following particular cases: classification with training (pattern recognition, discriminant analysis); classification with partial (or quasi-) training; classification without training (automatic classification, cluster analysis).

This division deals with the construction of and statistical investigation into a general probabilistic model of a mixture of distributions when the set of observations making up one, several, or even all the training samples can be empty (see the results obtained in Aivazyan et al.(1974) and Enyukov (1986)).

2.3 Treatment of large-scale data arrays and models

If the data dimension p is comparable with the sample size m and the two values are sufficiently large (i.e., $m \to \infty$, $p \to \infty$, and $p/m \to c > 0$), then some basic statistical assumptions of the classical asymptotic analysis (which are valid for $m \to \infty$ and p=const) are violated.

In situations like this it is required that the statist cal properties of the employed rules and procedures are analyzed under the conditions of the above-mentioned asymptotics (which is often called the Kolmogorov asymp totics). Specifically, the paper of Tsibel' (1987) is devoted to these problems.

In the theory and practice of automatic classificatio it is important that the dimension of a mathematical mo del l is chosen correctly depending on the sample size The formulation and solution of such problems can be found in Enyukov (1986).

2.4 The methods of constructing partitions stable to variation in the controlled free parameters of an AC algorithm

The idea of multiple solution (by different methods) of the same problem and subsequent selection of most frequ variants have largely been used in statistics. This ide underlies the approach (see Aivazyan et al.(1983) and Aivazyan (1980)) to developing the statistic methods which enable us to obtain inference stable to variation in initial conditions on the nature and accuracy of dat Specifically, it is suggested that an AC problem (in it optimization statement) should be multiply solved for different objective functions, for example, for a parametric family of objectives. As a result we obtain the set of partitions into classes: every objective is associated with its best method and conversely every best method corresponds to its partition. In the set obtaine one has to select one or several partitions which are relatively stable to changing the objectives. Obviously the change in values of the controlled free parameters of an AC algorithm, in particular, the parameters which determine the specific form of an objective, is equivalent to varying the initial conditions concerning the nature and accuracy of the data under classification.

2.5 The employment of training elements in the choice of an appropriate metric for AC problems

The definition of the distance between the objects (or the groups of objects) to be classified is a 'bottleneck' of AC theory. Usually, a priori information on the probabilistic and geometrical nature of the multivariate observation is lacking. In such a case the successful choice of a metric depends on the statistician skill of formalizing the professional knowledge and intuition of the expert in the area where the AC problem arises.

An interesting approach to using training elements for the 'adjustment' to an appropriate metric is suggested, for example, in Diday and Moreau (1984).

2.6 Estimation of the number of classes in AC problems

The problems related to estimating an integer parameter are traditionally difficult in mathematical statistics (for example, estimating the number of factors in factor analysis, the number of basic functions in regression analysis, etc). In automatic classification the problem of estimating the unknown number of classes can be stated (in probabilistic terms) as a problem of determining the number of the modes of a multivariate density (in the nonparametric formulation) or the number of components in the mixture of distributions characterizing the multivariate observation to be classified (in the parametric formulation). Some interesting results both asymptotic (the amount of observations grows infinitely) and nonasymptotic have been obtained in the latter case (see Orlov (1983), Tsibel' (1987)).

REFERENCES

Aivazyan, S.A. (1979). Extremal formulation of the basic problems of applied statistics. In: National School on Applied Multivariate Statistical Analysis Algorithms and Software. Computer Centre of the Planning Committee of the Armenian SSR, Erevan, pp. 24-49.

Aivazyan, S.A. (1980). Statistique mathématique appliquée et problème de la stabilitée des inférences statistique. In: Data Analysis and Informatics. North-Holland Publ. Comp.

Aivazyan, S.A., Bezhaeva, Z.I., and Staroverov, O.V.(1974 Classification of Multivariate Observation . Statistika, Moscow.

Aivazyan, S.A., and Bukhshtaber, V.M.(1985). Data analysis, applied statistics, and constructing a general theory of automatic classification. In: Diday et al. (1979)(Russian translation), pp. 5-22.

Aivazyan, S.A., Enyukov, I.S., and Meshalkin, L.D.(1983). Applied statistics: introduction to modelling and primary data processing. Finansy I Statistika, Moscow.

Bukhshtaber, V.M., and Maslov, V.K.(1977). Factor analysis and extremum problems on Grassmann varieties. In: Mathematical Methods of Solving Economic Problems, N 7, pp. 87-102.

Bukhshtaber, V.M., and Maslov, V.K.(1980). The problems of applied statistics as extremum problems on irregular domains. In: Algorithms and Software of Applied Statistical Analysis. Uchenye Zapiski po Statistike. v.36, Nauka, Moscow, pp. 381-395.

Bukhshtaber, V.M., and Maslov, V.K.(1985). Tomography methods of multivariate data analysis. In: Statistics. Probability. Economics. Nauka, Moscow, pp.108-116

Demonchaux, E., Quinqueton, J., and Ralambondrainy, H. (1985). CLAVECIN: Un systeme expert en analyse de donnees. Rapports de Recherche, N 431. Institut National de Recherche en Informatique. Le Chesnay.

Diday, E., et al.(1979).Optimisation en classification automatique. Institut National de Recherche en Informatique et en Automatique, Le Chesnay.

Diday, E., and Moreau, J.V.(1984). Learning Hierarchical Clustering from Examples. Rapports de Recherche N 289. Institut National de Recherche en Informatique et en Automatique, Le Chesnay.

Enyukov, I.S.(1986). Methods, Algorithms, and Programs of Multivariate Statistical Analysis. Finansy I Statistika, Moscow.

Enyukov, I.S.(1986). Projection pursuit in reconnaissance data analysis. Reviews of the First World Congress of the Bernoulli Society for Mathematical Statistics and Probability. Tashkent.

Friedman, J.H., and Tukey, J.W.(1974). A projection pursuit algorithm for exploratory data analysis. IEEE Transaction on Computers, C-23, pp. 881-890.

Girko, V.L.(1985). 'Struggle against dimension' in multivariate statistical analysis. In: Application of Multivariate Statistical Analysis in Economics and Quality of Product. Tartu, pp. 43-52.

Hahn, G.J.(1985). The American Statistician, v.39, N 1, pp. 1-16.

Huber, P.J.(1985). Projection pursuit. The Annals of Statistics, v.13, N 2, pp. 435-475.

McQueen, J.(1967). Some methods for classification and analysis of multivariate observation. Proc. Fifth Berkley Symp. Math. Stat. and Probab.,v.1, pp.281-297.

Orlov, A.I.(1983). Some probabilistic aspects of classification theory. In:Applied Statistics. Nauka, Moscow, pp. 166-179.

Sebestyen, G.S.(1962). Decision Making Process in Pattern Recognition. The McMillan Company.

Shlezinger, M.I.(1965). On spontaneous pattern recognition. In: Reading Automata. Naukova Dumka,Kiev,pp.88-106

Tsibel', N.A.(1987). Statistical investigation into the properties of the estimates of multivariate analysis model dimension. Ekonomika I Matematicheskie Metody, the USSR Acad. of Sci. (in print).
Vapnik, V.N.(1979). Restoration of Dependences from Empirical Data. Nauka, Moscow.

DATA ANALYSIS : GEOMETRIC AND ALGEBRAIC STRUCTURES

Fichet B.
Laboratoire de Biomathématiques - Faculté de Médecine -
Université d'Aix - Marseille II.

Here we present a survey of mathematical structures which arise in data analysis. After recalling the fundamental triple of data analysis, we investigate the usual representations of this triple : Euclidean embeddings, Lp-embeddings, hierarchies, pyramidal representations, additive trees, star graphs... Each representation corresponds to special dissimilarities and the set of these dissimilarities is shown to be a cone in a finite dimensional vector space. We examine the respective inclusions of the cones, as well as their geometric nature, especially convexity and closure. Then, approximation problems may be studied : least squares approximation with respect to a given norm, subdominant (or submaximal) and superdominated (or superminimal) approximation, additive constants... For all these mathematical aspects, many problems remain unsolved and some conjectures have been made.

Finally, we pay particular attention to monotone transformations on data. They play an important role in data analysis and we will present their impact on the afore-mentioned representations.

1 - DATA STRUCTURES

The main basic concept in data analysis is dissimilarity. A dissimilarity d on a finite nonempty set I is a nonnegative real function defined on I^2 such that : $\forall i \in I,\ d(i,i)=0$; $\forall (i,j) \in I^2,\ d(i,j)=d(j,i)$. The finite dimensional vector space of real functions which satisfy the same conditions will be denoted by D, and a dissimilarity is an element of the positive orthant D^+. In practice, I represents individuals, quantitative variables, categories of a qualitative variable, logical or presence-absence characters... Let us assign a mass m_i (i.e. a positive number) to each unit i in I. These masses arise essentially in view of approximation problems. The fundamental triple $(I,d,\{m_i, i\in I\})$ will be called a data structure.

Here we recall some common data structures.

For a set I of individuals and a set J of quantitative variables, let x_i^j be the observation of the variable j on the individual i. Denoting by σ_j, $\rho_{jj'}$ respectively the standard deviation of the variable j and the correlation coefficient of the variables j and j', and supposing $\sigma_j > 0$ for every j, we generally consider the data structures $(I,d_I,\{m_i, i\in I\})$ and $(J,d_J,\{n_j, j\in J\})$ such that :

$$\forall (i,i') \in I^2 ,\quad d_I^2(i,i') = \sum_j \left[\frac{x_i^j - x_{i'}^j}{\sigma_j} \right]^2 \quad ; \quad \forall i \in I,\ m_i = 1/|I|$$

$$\forall (j,j') \in J^2 ,\quad (1/2)d_J^2(j,j') = 1 - \rho_{jj'} \quad ; \quad \forall j \in J,\ n_j = 1$$

Let I and J be two qualitative variables and $\{f_{ij}, (i,j)\in I\times J\}$ be a frequency table (derived from a contingency table). We use the following notations : $\forall i \in I,\ f_{i.} = \sum_j f_{ij}$; $\forall j \in J,\ f_{.j} = \sum_i f_{ij}$

Supposing that the previous quantities are all strictly positive, we generally consider on I the data structure $(I,d_I,\{f_{i.}, i\in I\})$, where :

$$\forall (i,i') \in I^2 ,\quad d_I^2(i,i') = \sum_j \frac{1}{f_{.j}} \left[\frac{f_{ij}}{f_{i.}} - \frac{f_{i'j}}{f_{i'.}} \right]^2 \qquad (\chi^2 \text{ metric})$$

and a symmetrical data structure $(J,d_J,\{f_{.j}, j\in J\})$ on J.

For a set I of individuals and a set J of presence-absence characters, let x_i^j be the observation of the character j on the individual i (x_i^j equals 1 (or 0) for presence (or absence) of j). We use the following notations : $\forall(i,i') \in I^2$, $m_{ii'} = \sum_j x_i^j x_{i'}^j$; $\forall i \in I$, $m_i = \sum_j x_i^j$.
Then, we may consider on I the data structure $(I, d_I, \{m_i, i \in I\})$, where :
$\forall(i,i') \in I^2$, $(1/2)d_I^2(i,i') = 1 - m_{ii'}/\sqrt{m_i m_{i'}}$ (Ochiaï's metric)
and a symmetrical data structure $(J, d_J, \{n_j, j \in J\})$ on J.

A dissimilarity d on I is said to be :
proper iff $d(i,j) = 0$ implies : $i = j$
semi-proper iff $d(i,j) = 0$ implies : $\forall k \in I$, $d(i,k) = d(j,k)$.

If d is semi-proper, an equivalence is introduced as follows : $i \sim j$ iff $d(i,j) = 0$.
Then, a proper dissimilarity $\hat{d}$ may be constructed on the quotient space $\hat{I}$; we have $\hat{d}(\hat{\imath},\hat{\jmath}) = d(i,j)$, where i (resp. j) is in $\hat{\imath}$ (resp. $\hat{\jmath}$).
Moreover, if masses are assigned to units i in I, we put :
$\forall \hat{\imath} \in \hat{I}$, $m_{\hat{\imath}} = \Sigma\,\{m_i \mid i \in \hat{\imath}\}$.
Then, $(\hat{I}, \hat{d}, \{m_{\hat{\imath}}, \hat{\imath} \in \hat{I}\})$ is called the induced quotient data structure.

It is usual practice to aggregate units which are equivalent. In a mathematical sense, a property of d which is preserved on the quotient space, has only to be proved when d is proper.

2 - REPRESENTATIONS IN DATA ANALYSIS

Given a data structure, different graphical representations may be proposed. Each of them corresponds to a particular dissimilarity. Here we recall some usual representations and their associated dissimilarities.

- L_p-spaces

A dissimilarity d on I is said to be L_p iff there exist an integer N and a family of real numbers $\{x_i^j, j = 1,\dots,N ; i \in I\}$ satisfying :
$\forall(i,i') \in I^2$, $d^p(i,i') = \sum_j |x_i^j - x_{i'}^j|^p$.

It admits an embedding in an L_p-space, and the set of such dissimilarities will also be denoted by L_p.

Two particular cases an noteworthy.
Euclidean dissimilarities (i.e. dissimilarities in L_2) which yield a Euclidean representation, and the city-block semi-distances (i.e. dissimilarities in L_1) which yield an L_1-representation .

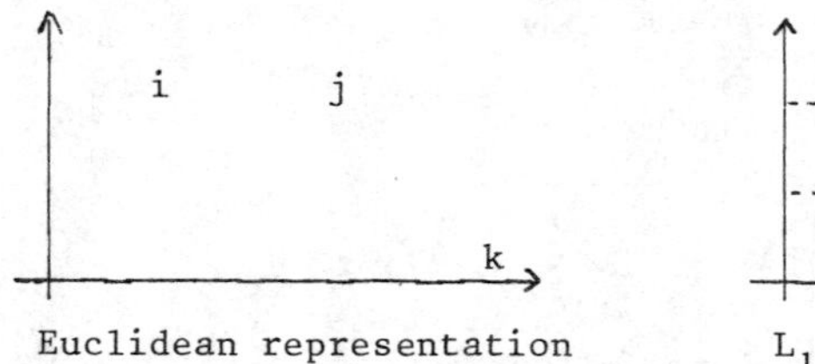

Euclidean representation

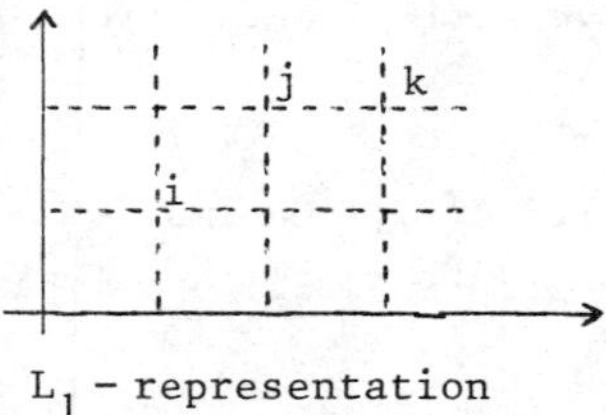

L_1 - representation

Mathematically, it is useful to consider the sets L_p^p , obtained from L_p by the transformation $d \mapsto d^p$. In particular, L_2^2 is the set of squared Euclidean dissimilarities.
Moreover, every finite metric space is shown to be embedded in an L_∞-space. According to this property, the set of semi-distances on I will also be denoted by L_∞ .

- Hierarchical representation

A hierarchy on I is a class $\mathcal{H}$ of non-empty subsets, satisfying :

i) $I \in \mathcal{H}$;

ii) $\forall (H,H') \in \mathcal{H}^2$, $H \cap H' \in \{H,H',\emptyset\}$

iii) $\forall H \in \mathcal{H}$, $\cup\{H' \in \mathcal{H}$, $H' \subset H$, $H' \neq H\} \in \{H,\emptyset\}$

If $f : \mathcal{H} \mapsto \mathbb{R}_+$ is such that :

iv) $(H \subset H', H \neq H') \Rightarrow f(H) < f(H')$

v) H minimal $\Longleftrightarrow f(H) = 0$

$(\mathcal{H},f)$ is said to be an indexed hierarchy.

An indexed hierarchy yields the following representation, called a dendogram :

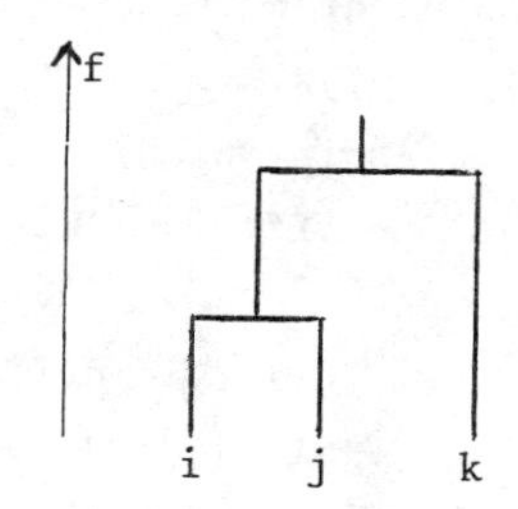

It is well known that there exists a bijection between indexed hierarchies and ultrametrics, i.e. dissimilarities satisfying :
$\forall (i,j,k) \in I^3,\ d(i,j) \leqslant \max [\, d(i,k), d(j,k)\,]$ (ultrametric inequality).
The set of these dissimilarities will be denoted by U.

- Pyramidal representations

These representations extend hierarchical representations.
A pseudo-hierarchy on I is a class of non-empty subsets satisfying the conditions i) and iii) of a hierarchy, as well as :
ii)' $\forall (H,H') \in \mathcal{H}^2,\ H \cap H' \in \mathcal{H} \cup \{\emptyset\}$.
Moreover, we assume the existence of an order on I such that :
$\forall H \in \mathcal{H},\ (i \in H,\ j \in H,\ i \leqslant k \leqslant j)$ implies : $k \in H$.

An indexed pseudo-hierarchy is defined like an indexed hierarchy.
For a weakly indexed pseudo-hierarchy, f satisfies v) and
iv)' $(H \subset H',\ H \neq H') \Rightarrow f(H) \leqslant f(H')$, with the added condition that when $f(H) = f(H')$: there exist H_1, and H_2 in $\mathcal{H}$, different from H, and such that $H = H_1 \cap H_2$.

From a pseudo-hierarchy, we have the following pyramidal representations with overlapping clusters :

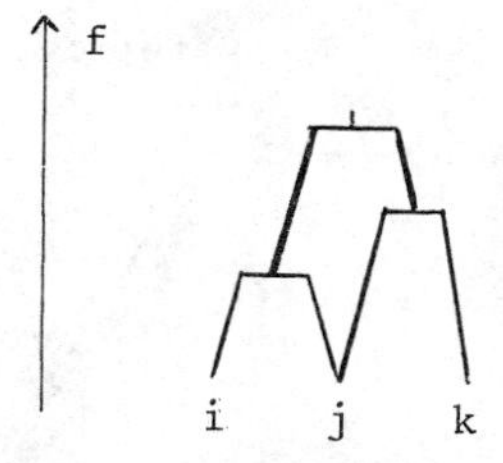

indexed pseudo-hierarchy

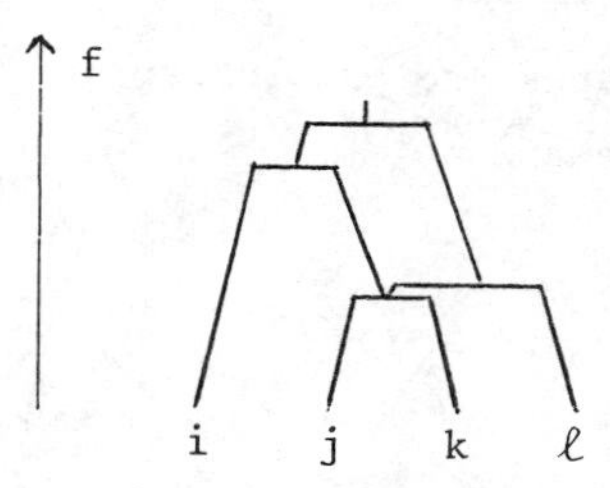

weakly-indexed pseudo-hierarchy

A dissimilaty d is said to be Robinsonian iff there exists an order on I such that : $(i \leqslant j \leqslant k)$ implies $d(i,k) \geqslant \max[d(i,j), d(j,k)]$ (the associated dissimilarity matrix is Robinsonian).
It is said to be strongly Robinsonian iff, in addition to the previous condition, it satisfies for $i \leqslant j \leqslant k$:

$d(i,k) = d(j,k)$ implies : $\forall \ell \geqslant k, \; d(i,\ell) = d(j,\ell)$

$d(i,k) = d(i,j)$ implies : $\forall \ell \leqslant i, \; d(\ell,j) = d(\ell,k)$

Then, a bijection is established between indexed (resp. weakly-indexed) pseudo-hierarchies and strongly-Robinsonian (resp. semi-proper Robinsonian) dissimilarities. The set of such dissimilarities will be denoted by SR (resp. R).

- Addtitive trees

An additive tree is a weighted tree in which the distances between two nodes is the length of the path joining them, i.e. the sum of the corresponding weights. An additive tree yields the following representation :

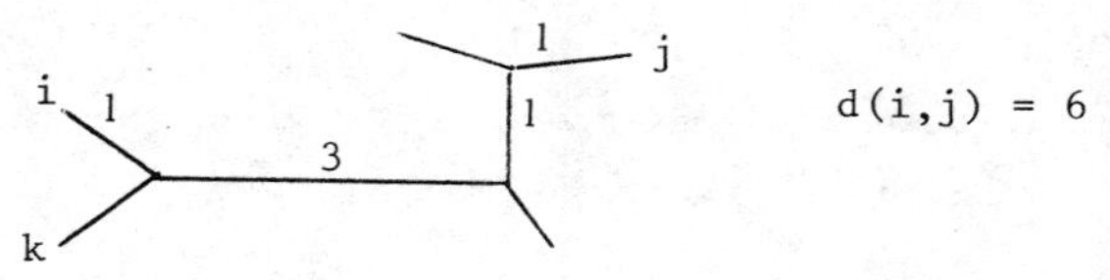

There exists a one-to-one correspondence between additive trees and quadripolar semi-distances, i.e. semi-distances satisfying the following four-point condition :
$\forall (i,j,k,\ell) \in I^4, \; d(i,j) + d(k,\ell) \leqslant \max[d(i,k) + d(j,\ell), d(i,\ell) + d(j,k)]$
The set of these semi-distances will be denoted by Q.

- Star graphs

Star graphs are particular additive trees and yield the following representation :

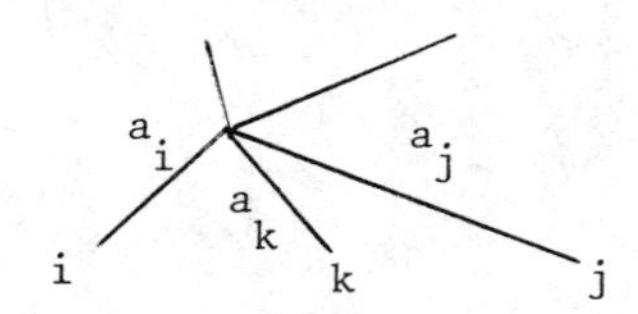

There exists a one-to-one correspondence between these graphs and star semi-distances, i.e. semi-distances such that there exists a family of nonnegative real numbers $\{a_i, i \in I\}$ satisfying :
$\forall (i,j) \in I^2$, $d(i,j) = a_i + a_j$.
The set of such semi-distances will be denoted by S.

3 - COMPARISON OF REPRESENTATIONS

Given a dissimilarity d, it is easy to discover whether d is ultrametric, strongly Robinsonian, semi-proper Robinsonian, quadripolar or is a star semi-distance. However for L_p-spaces we need a characteristic property. To our knowledge, such a property has been found only for L_1 and L_2-spaces. The dissimilarity is (or is not) in L_1 according to the solution of a linear programming problem (with very high dimensionality !) ; d is in L_2 iff the Torgerson's matrix $T(d^2)$ is p.s.d. Here we recall the general term of $W = T(d)$.
$\forall (i,j) \in I^2$, $W_{ij} = (1/2)[-d(i,j) + d(i,.) + d(j,.) - d(.,.)]$
(where : $\forall i \in I$, $d(i,.) = (1/|I|)\sum_j d(i,j)$; $d(.,.) = (1/|I|)\sum_i d(i,.)$).
All the sets mentioned in section 2 are shown to be cones in the space D. Moreover, except for L_1, L_2^2, L_∞ and S, none of these cones is convex, and except for SR and R, all these cones are closed (with respect to any given norm on the finite dimensional space D).
The following strict inclusions have been proved and are helpful in comparing the different representations.

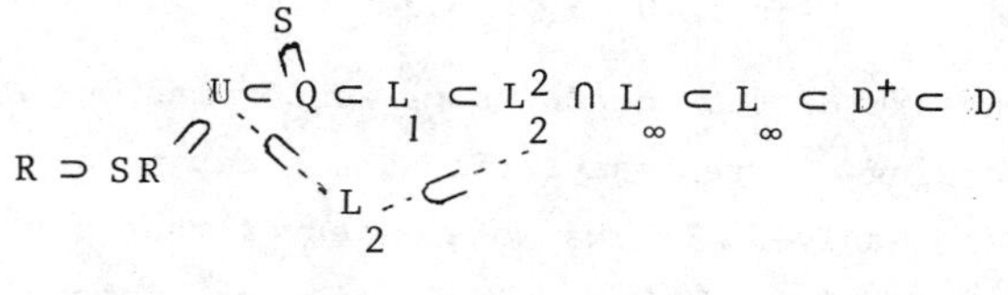

$$\begin{array}{ccccccc}
 & & S & & & & \\
 & & \cap & & & & \\
 & & U \subset Q \subset L_1 & \subset & L_2^2 \cap L_\infty \subset L_\infty \subset D^+ \subset D & & \\
 & \cap & \quad\subset & & \subset & & \\
R \supset SR & & & L_2 & & &
\end{array}$$

Taking the convex hull of cones, some inclusions become equalities.
We have : $\mathrm{conv}(U) = \mathrm{conv}(Q) = L_1$.
However, for L_p-spaces, many unsolved problems arise. Here is a conjecture : it may be shown that L_1 in included in $\mathrm{conv}(L_2)$; do the following strict inclusions hold :
$L_1 \subset \mathrm{conv}(L_2) \subset L_2^2 \cap L_\infty$?

4 - APPROXIMATION PROBLEMS

Given a data structure $(I,d,\{m_i,i\in I\})$ and having chosen a graphical representation corresponding to a special cone C in D, we have to approximate d by an element d' of C. To this end, the geometric and topological properties of C are of capital importance. The existence and uniqueness of the solution obviously depend on the approximation chosen, but they also depend on convexity and closure properties for the cone C. We may imagine the difficulty of approximation problems for pyramidal representation : the cones SR and R are non-convex and non-closed !

Here we recall some classical approximations.

- Projection of d on C according to a given scalar product on D. The (weighted) least squares approximation is an important particular case.
- The additive constant (of order p) : find the minimal constant c such that : $(d^p + cd_1)^{1/p} \in C$ (where d_1 is the dissimilarity equalling 1 for every couple).
- The subdominant or superdominated approximation : find the greatest element (or the maximal elements) in C among the dissimilarities of C which are less than d, or find the smallest element (or the minimal elements) in C among the dissimilarities of C which are greater than d.
- Find the greatest constant α $(0 < \alpha \leqslant 1)$ such that $d^\alpha \in C$.

According to the nature of the cone C, some approximations are well-known. But, considering all representations, a great number of approximations remain unsolved. Let us note moreover that certain approximations have been given a theoretical solution, which because of the combinatory nature of the algorithm, is of no practical use. Examples are least squares problems for hierarchies and for additive tree.

Now we discuss some usual approximations.

The least squares approximation is solved for star graphs and is studied for additive trees in some recent papers, but only for a given tree. For Euclidean dissimilarities, the projection is introduced in terms of squared dissimilarities. Indeed, as opposed to L_2, the cone L_2^2 is convex. Then, an algorithmic method yields the least squares approximation. But the more usual projection derives from principal component analysis by taking into account only positive eigenvalues of the inertia operator. This projection corresponds to a very special norm on D ; we have :

$$\|d\|^2 = (1/4) \sum_i \sum_j [-d(i,j) + d(i,.) + d(j,.) - d(.,.)]^2$$

The additive constant problems of order 1 and 2 are solved for L_2-spaces as well as the additive constant problem of order 1 for L_1-space. The constant c for L_1 is the solution of a linear programming problem. For Euclidean dissimilarities, the constant of order 2 equals $2|\lambda|$, where λ stands for the smallest eigenvalue of Torgerson's matrix $T(d^2)$; the constant of order 1 is the greatest real eigenvalue of a certain matrix.

For hierarchies there exist a subdominant approximation and superminimal approximations. But note that algorithms give only some superminimal approximations.

For pyramidal representation, some results are established only for an order on I compatible with the approximation. Then, according to this order, we have a subdominant and a superdominated approximation for semi-proper Robinsonian dissimilarities, and a unique submaximal approximation for strongly-Robinsonian dissimilarities.

5 - PROXIMITY ANALYSIS

In proximity analysis we use monotone transformations of a dissimilarity d. Indeed, in this approach only the order on data is significant. On a mathematical plane, we consider preordonances, i.e. classes given by the following equivalence : $d \sim d'$ iff :

$d(i,j) \leqslant d(k,\ell) \Longleftrightarrow d'(i,j) \leqslant d'(k,\ell)$.

Let us note that a preordonance may be directly defined by considering a preorder on couples.

Now, we have to study whether the previous structures are preserved when a dissimilarity is replaced by a preordonance. For hierarchical or pyramidal representation, we obtain stratified hierarchies or stratified pseudo-hierarchies. They are defined by a preorder on clusters. However, for additive trees or Euclidean embeddings, the result depends on the dissimilarity in the preordonance, i.e. on the transformation chosen. Many problems then arise, such as the fundamental problem of multidimensional scaling, for which algorithmic and even heuristic methods have been proposed : find the best monotone transformation which gives a Euclidean embedding in the lowest dimension. Several theoretical results show the difficulty of the problem. Here is a surprising result of this kind. For every integer p, we can find a set I with cardinality $n > (p+2)$ and a preordonance on I such that there exist figures satisfying the inequalities of the preordonance only in spaces with dimension p and $(n-1)$.

We end here our survey. Obviously, many other problems could have been raised in this field. However, we hope to have shown that, even within the context evoked here, a great number of questions remain open. A note on bibliography : the field is so vast that we have been obliged to suppress all bibliographical references.

Bernoulli, Vol. 2, pp. 133-136

GENERALIZED CANONICAL ANALYSIS

Michel TENENHAUS, Centre HEC-ISA
1, rue de la Libération, 78350 JOUY-EN-JOSAS, France

Introduction

Canonical analysis has been considered for a long time as a method having a real theoretical interest, but few practical applications. Situation is changing as we can see from the new book of R. Gittins (1985) : "Canonical analysis : a review with applications in Ecology". Generalized canonical analysis (Mc Keon, 1965 ; Carroll, 1968 ; Kettenring, 1971) includes canonical analysis as a particular case (and, consequently, multiple regression, analysis of variance, discriminant analysis, correspondence analysis, ...) and also principal component analysis and multiple correspondence analysis. This method is also useful to study a population described by several numerical or nominal variables and evolving in time. So Generalized Canonical Analysis represents a remarkable synthesis of multivariate linear methods. We present in this paper Generalized Canonical Analysis from a geometrical point of view, showing its relationship with Principal Component Analysis (Saporta, 1975 ; Pages, Cailliez, Escoufier, 1979). This permits many numerical simplifications useful for the writing of a computer program.

I - GENERALIZED CANONICAL ANALYSIS

1. The data

We consider p data tables $X_1, \ldots, X_t, \ldots, X_p$. Each table X_t is formed with n rows corresponding to the same n subjects, the columns represent standardized numerical variables or dummy variables associated with the categories of nominal variables. In other words the data can be numerical or nominal variables. If the variables are nominal they are transformed in binary tables.

2. The problem

We look for standardized and uncorrelated variables $z_1, \ldots, z_m$ maximizing the quantity

$$\sum_{h=1}^{m} \sum_{t=1}^{p} R^2 (z_h, X_t) \qquad (1)$$

where $R^2 (z_h, X_t)$ represents the coefficient of determination between variable z_h and table X_t.

3. The solution

The centering operator P_o is equal to $I - (1/n)\, u\, u'$ where u is a column vector of n ones. We denote by P_t the projection operator into the subspace $L(X_t)$ of R^n generated by the columns of table X_t. The coefficient of determination $R^2(z_h, X_t)$ being equal to $(1/n)\, z'_h P_t z_h$, quantity (1) may be written

$$(2) \qquad (1/n) \sum_{h=1}^{m} z'_h \; P \; z_h ,$$

where $P = \sum_{t=1}^{p} P_t$. The standardized and uncorrelated variables $z_1, \ldots, z_m$ maximizing (2) are obtained as the eigenvectors of the matrix $P_o P$ associated with the m largest eigenvalue $\lambda_1, \ldots, \lambda_m$ ranked in a decreasing order. Then the maximum of (2) is equal to $\sum_{h=1}^{m} \lambda_h$.

In effect it will be enough to diagonalize the matrix P. Vector u is an eigenvector of P associated with the eigenvalue λ_o equal to the number of tables X_t containing a binary table. Consequently the eigenvectors of the matrix $P_o P$ are the eigenvectors of the matrix P different from the eigenvector u associated with the eigenvalue λ_o.

4. The associated principal component analysis

We denote by $X = [X_1, \ldots, X_t, \ldots, X_p]$ the data table obtained by horizontally adjoining the several X_t, M the block diagonal matrix formed with the generalized inverses $n(X'_t X_t)^-$ of the matrices $(1/n) X'_t X_t$, $N = (1/n)\, I$. Generalized canonical analysis of the tables $X_1, \ldots, X_p$ is a principal component analysis of the triplet (X, M, N), (Saporta, 1975). The standardized principal components of the triplet (X, M, N) are the canonical components $z_1, \ldots, z_m$. They are independent from the chosen generalized inverses $n(X'_t X_t)^-$. The duality diagram (Pages, Cailliez, Escoufier, 1979) associated with the triplet (X, M, N) (figure 1) gives immediatly useful relations for numerical calculations : going from calculations in R^n to calculations in R^k, where k is the number of columns of X.

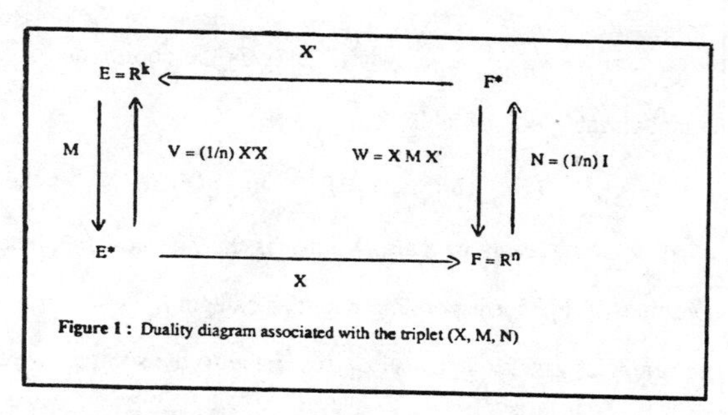

Figure 1 : Duality diagram associated with the triplet (X, M, N)

The projection operator P_t into the subspace $L(X_t)$ may be written $P_t = X_t (X'_t X_t)^- X'_t$ and so we obtain $WN = P$. The following relations are useful for the writing of a generalized canonical analysis computer program.

1) $z_h = \lambda_h^{-1/2} X\varphi_h$ where the factors φ_h are the M^- - orthonormal eigenvectors of the matrix MV.

2) $\varphi_h = M u_h$ where the principal axes u_h are the M - orthonormal eigenvectors of the matrix VM.

3) $u_h = \lambda_h^{-1/2} (1/n) X'z_h$. So the principal axes u_h are independent from the chosen generalized inverses.

4) The chosen generalized inverse may be the Moore - Penrose inverse. The block diagonal matrix A formed with the $(1/n) X'_t X_t$ may be written $A = \sum_{t=1}^{r} \alpha_t \upsilon_t \upsilon'_t$, where r is the rank of A and the υ_t are the eigenvectors of A associated with the non null eigenvalues α_t. So we have :

$$M = A^- = \sum_{t=1}^{r} (1/\alpha_t) \upsilon_t \upsilon'_t \;, M^{1/2} = \sum_{t=1}^{r} (1/\alpha_t)^{1/2} \upsilon_t \upsilon'_t , M^{-1/2} = \sum_{t=1}^{r} (\alpha_t)^{1/2} \upsilon_t \upsilon'_t .$$

We deduce the principal axes u_h and the principal factors φ_h from the orthonormal eigenvectors a_h of the symetric matrix $M^{1/2} VM^{1/2}$ associated with the eigenvalues λ_h :

$u_h = M^{-1/2} a_h$ and $\varphi_h = M^{1/2} a_h$.

5) Some quantities useful for the interpretation of the results can be calculated as functions of the a_h or the u_h :

a) $R^2(z_h , X_t) = \lambda_h a'_{ht} a_{ht} = \lambda_h u'_{ht} n (X'_t X_t)^- u_{ht}$ where a_{ht} (resp. u_{ht}) is the subvector of a_h (resp. u_h) corresponding to table X_t.

b) cor $(x_{tj}, z_h) = \lambda_h^{1/2} u_{htj}$, where u_{htj} is the coordinate of u_h corresponding with the numerical variable x_{tj} of table X_t.

c) cor $(X_{tj}, z_h) = n (\lambda_h /(n_{tj} (n - n_{tj})))^{1/2} u_{htj}$ where X_{tj} is the indicator variable of the category j which appears in table X_t, n_{tj} is the non null frequency of this category, u_{htj} is the coordinate of u_h corresponding with this category.

d) The center of gravity c_{htj} of z_h restricted to the subjects which are in category j of table X_t is equal to $\lambda_h^{1/2} (n/n_{tj}) u_{htj}$.

e) The square of the correlation ratio $\eta^2 (z_h, X_{tk})$ between the canonical component z_h and the nominal variable X_{tk} is equal to $\sum_j (1-(n_{tkj}/n)) \text{cor}^2 (X_{tkj}, z_h)$, where X_{tkj} is the indicator variable of category j of X_{tk}.

References

Carroll, J.D. (1968). A generalization of canonical correlation analysis to three or more sets of variables. Proceeding of the 76th Annual Convention of the Americal Psychological Association, 3, 227-228.

Gittins, R. (1985). Canonical analysis, a review with applications in ecology, Berlin : Springer-Verlag.

Kettenring, J.R. (1971). Canonical analysis of several sets of variables. Biometrika, 58, 433-451.

Mc Keon, J.J. (1965). Canonical analysis : some relations between canonical correlation, factor analysis, discriminant function analysis and scaling theory. Psychometric Monograph 13. University of Chicago Press, Chicago.

Pages, J.P., Cailliez, F., Escoufier, Y. (1979). Analyse factorielle : un peu d'histoire et de géométrie. Revue de Statistique Appliquée, 27,5-28.

Saporta, G. (1975). Liaison entre plusieurs ensembles de variables et codage de données qualitatives. Doctorat de 3è cycle, Université Pierre et Marie Curie, Paris.

DETECTING OUTLIERS AND CLUSTERS IN MULTIVARIATE DATA BASED ON PROJECTION PURSUIT

I.S. Yenyukov
Central Economics Mathematical Institute of the USSR Academy of Sciences, Moscow

1. INTRODUCTION

Let we have a sample $X^{(n)}=(X_1,...,X_n)$ of p-variate observation. We can use the projection pursuit (PP) technique (Friedman, Tukey (1974), Huber (1985)) to detect any singularities of the data, for instance the presence of clusters. The PP is based on extracting "interesting" linear projections $z=U^tX$ of the initial data (X $X^{(n)}$, U is a p-variate vector). More formally, let $Q(U,X^{(n)})$ be a projection index (PI). The step-wise approach usually is used for obtaining several projections vectors $U_1,...,U_q$. At first the vector U_1 is determined by maximizing the PI. Each following vector U_i is found in the same way but on additional condition that the influence all before found vectors is excluded in a suitable way.

Choosing the suitable PI we can pick up the projections useful for detecting the data singularities that are interested for us.

2. PP FOR DETECTING OUTLIERS

As the PI suitable for finding the projections useful for detecting outliers we propose the ratio

$$Q(U,X^{(n)})= s^2(U)/s^2_{rb}(U) \qquad (1)$$

where $s^2(U)$ is the usual estimate of the variance of the projection with the vector U and $s^2_{rb}(U)$ is a robust estimate of it.

It is knouwn that the usual estimate of the variance is rather sensitive to outliers and if they are present its value increase as a rule. And so the projection where the value of the PI (1) is maximal may be reasonable conside-red as one where the influence of outliers is the most expressed. An approximate maximization of the PI (1) can be achived by the solution of the generalized eigenvec-tors problem

$$(\hat{S} - h\hat{S}_{rb})U = 0$$

where $\hat{S}$ is the usual estimate of the covariance matrix of the X and $\hat{S}_{rb}$ is a robust estimate of it. As the ro-bust estimate we used for the programme realisation (Yenyukov (1986)) the estimate of the M-estimate type (Maronna, 1974).

3. PROBABILITY MODEL FOR DESCRIBING A CLUSTER STRUCTURE

We consider that the underlying distribution of the sam-ple $X^{(n)}$ is a k - component mixture with the density

$$P(X) = \sum_{i=1}^{k} a_i c(d,p,W) d((X - M_i)^t W^{-1} (X - M_i)) \quad (2)$$

where d(y) is a positive monotonously decreasing functi of the y (y o) ; c(d,p,W) is normalizing constante; $a_i > o$ is the weight of the i-th component ($\sum a_i = 1$); M_i is the means vector of the i-th component; W is the within-component scatter (covariances) matrix. So the P(X) is the mixture of unimodal ellipsoidally sym-metric densities and may be considered as some model of the cluster structure. The P(X) has k modes (if the mix-ture components are enough scattered) and the points sur rounding some mode may be regarded as the members of the same cluster.

Now let the $z = U^t X$ be an onedimensional projection. The density f(z) of the z is the k-components unimodal symme tric densities mixture

$$f(z) = \frac{1}{w} \sum_{i=1}^{k} a_i e((z - m_i)/w)$$

where

$$e(x) = c(d,p,I) \int d(x^2 + \sum_{j=1}^{k-1} y_j^2) dy_1 \ldots dy_{k-1}$$

$$m_i = M_i^t U \ , \ w^2 = U^t W U \ .$$

Now define the discriminant subspace (Rao, 1965). It is good known that the covariance matrix of the X may be represented as the sum of the matrix W and between-component scatter matrix B, i.e $S=B+W$, where $B=\sum_{i=1}^{k} a_i (M_i-M)(M_i-M)^t$, $M=\sum_{i=1}^{k} a_i M_i$ is means vector of the X. In according to it the variance $s^2(U)$ of the onedimensional projection is $s^2(U)=b^2(U)+w^2(U)$, where $b^2(U)=U^tBU$. Regarde ratio

$$t^2(U)=b^2(U)/w^2(U)$$

The value $t^2(U)$ is a measure of dissimilarity of the mixture components for the projection with vector U. Maximizing the $t^2(U)$ leads to the solution of the generalized eigenvectors problem $(B-tW)V=0$. There are q^+ eigenvectors $V_1,\dots,V_{q^+}$ with corresponding positive eigenvalues $t_1,\dots,t_{q^+}>0$ ($q^+\leqslant \min(p,k-1)$, $t_i=t^2(V_i)$).

The discriminant subspace is defined as subspace

$$R^+=\mathrm{span}(V_1,\dots,V_{q^+})$$

The R^+ contains the complete information about the differences among the mixture (2) components. The PP gives the way to get the R^+ without any information about the matrix B and the W.

4. PI SUITABLE FOR TESTING THE PRESENCE OF CLUSTERS

We consider the one-parametric family of the PI

$$Q_\beta(U,X)=s^\beta(U)E_f f^\beta(z) \quad (\beta>0) \quad (3)$$

where E_f is the averaging operator with the density $f(z)$. Here the "theoretical" quantity of the PI is designated by $Q_\beta(U,X)$ in contrast to the sampling value $Q_\beta(U,X^{(n)})$.

Give without the proof the inequalities connecting the (3) with the ratio $t^2(U)$

$$g(e,\beta)\left(\sum_{i=1}^{k} a_i^{1+\beta}\right)(1+t^2(U))^{\beta/2}\leqslant Q_\beta(U,X)\leqslant g(e,\beta)(1+t^2(U))^{\beta/2}$$

where $g(e,\beta)=E_e e^\beta(z)$ does not depend on the U.

When $t^2(U)=0$ then $Q_\beta(U,X)=g(e,\beta)$, it is minimal value of this PI. Now we formulate the following Lemma (without proof).

<u>Lemma</u>. Let the model (2) be true. Assume that the vec-

tors $U_1,\ldots,U_{q^+}$ are found by step-wise maximizing of the $Q_\beta(U,X)$ and so that the vector U_i is S-orthogonal (i.e. $U_i^t S U_j = o$ if $i=j$) to subspace span$(U_1,\ldots,U_{i-1})$. Then every vector U_i belongs to the discriminant subspace R^+ and what is more R^+=span$(U_1,\ldots,U_{q^+})$,i.e. these vectors define some basis in the disriminant subspace.

Now we note that the value q^+usually is unknown.However, even this case from Lemma it is possible to obtain some useful conclusions. For example, if $q^+=2$ the first two vectors U_1,U_2 extract the complete information about dissimilarities between the clusters. When $q^+>2$ but the eigenvalues $t_i (i>2)$ are small enough in comparison with t_2, then from the reasons of the continuity the same two vectors extract the main part of such information. On the other hand, if all the eigenvalues t_i are approximately equal it is indifferent which vectors take for projecting if only they belong to the R^+, but it is satisfied.

REFERENCES

Friedman J.H.,Tukey J.W. (1974) A projection pursuit algorithm for exploratory data analysis. IEEE Trans. Comput., C-23, 881-889

Huber P.J. (1985) Projection Pursuit. Ann. Statist., 12, 435-474

Rao C.R. (1965) Linear Statistical Inferences and its Applications .New York, Wiley

Yenyukov I.S. (1986) Methods,algorithmes and programmes of the multivariate statistical analysis (in russian). Finances and Statistics, Moscow

THE INVERSE PROBLEM OF THE PRINCIPAL COMPONENT ANALYSIS

Zhanatauov S.U., Computing Center, Novosibirsk, USSR

In data analysis it is usually impossible to connect a single real multivariate sample with one of the theoretical distribution functions and to obtain additional samples from the same population. In this situation one obtains on the computer the artificial samples, which in some way or other are similar to a real multivariate sample. Let $N_s(\bar{x},W)$ be a set of multivariate samples $X^0_{mn} = \{X^0_{i\cdot}\}_{i=\overline{1,m}}$, $X^0_{i\cdot} = (x^0_{i1},\dots,x^0_{in}) \in E^n$, generated from independent observations over dependent multivariate random values with dependent components and having a vector of sampling mean values $\bar{x}_{1n} = (\bar{x}_1,\dots,\bar{x}_n)$, $\bar{x}_j = (1/m)\cdot \sum_{i=1}^{m} x^0_{ij}$ and the sampling covariant matrix $W = (1/m)\cdot$ $\cdot(X^0 - \bar{x})^T(X^0 - \bar{x})$ given; $\mathcal{R}_\Lambda$ is a set of correlation matrices (c.m.) with the spectrum $\Lambda_{nn} = \mathrm{diag}(\lambda_1,\dots,\lambda_n)$ given. The functions $f_1(\Lambda) = \mathrm{tr}(\Lambda) = n$, $f_2(\Lambda) = \mathrm{tr}(\Lambda^2)$, $f_4(\Lambda,1) = (\sum_{j=1}^{1} \lambda_j)/n$ will be called main functional f--parameters of the spectrum Λ c.m., which are stable and reliably calculated statistics.

<u>Problem</u>: to obtain a multivariate sample of the volume $m > n > 2$, satisfying one of the following requirements:

a) a sample should have a sampling c.m., exactly equal to the c.m. given;

b) a sample should have a sampling c.m., whose spectrum is either exactly equal to the spectrum given, or the values of its main f-parameters with given accuracy are equal to the values given.

Here everywhere it is necessary that samples have the vectors of mean values and dispersions given.

For solving this problem it is sufficient to obtain on a computer the samples with sampling c.m., having one and the same given spectrum as well as spectra with values of main f-parameters, with given accuracies of the equal values given.

<u>Theorem</u> (Zhanatauov S.U., 1980). Let $m>n>2$, the elements of the diagonal matrix Λ_{nn} satisfying the relations: $\Lambda_{nn}=\mathrm{diag}(\lambda_1,\dots,\lambda_n)$, $\mathrm{tr}(\Lambda)=\lambda_1+\dots+\lambda_n=n$, $\lambda_1\geqslant\dots\geqslant\lambda_k>\lambda_{k+1}=\dots=\lambda_n=0$, $1\leq k\leq n$. Then there exist infinite sets:

a) of the orthogonal matrices $C_{nn}^{(l)}$, number $l=1,2,\dots$;

b) the correlation matrices $R_{nn}\in\mathcal{R}_\Lambda$ having the spectrum Λ and eigenvectors located in columns of the matrix $C_{nn}^{(l)}$: $R_{nn}^{(l)}=C_{nn}^{(l)}\Lambda C_{nn}^{(l)T}$;

c) of the multivariate samples $U_{mn}^{(t)}\in N_s(0,I)$, $Y_{mn}^{(t)}\in N_s(0,\Lambda)$, number $t=1,2,\dots$, the matrices Λ_{nn}, $C_{nn}^{(l)}$, $R_{nn}^{(l)}$, $Y_{mn}^{(t)}$, $Z_{mn}^{(t,l)}$ satisfying all the relations of the direct model of principal components by H. Hotelling (DM PC), and the Λ-samples of the inverse model of principal components (IM PC) $Y_{m_t n}^{(t)}$, $Z_{m_t n}^{(t,l)}$ having the properties:

1) with number l fixed, number $t=\overline{1,k_t}$, $M=m_1+\dots+m_{k_t}$ if $Y_{m_t n}^{(t)}\in N_s(0,\Lambda)$, $Z_{m_t n}^{(t)}\in N_s(0,R)$ then $Y_{Mn}=\left[Y_{m_1n}^{(1)T}\vdots\dots\vdots Y_{m_{k_t}n}^{(k_t)T}\right]^T\in N_s(0,\Lambda)$, $Z_{Mn}=\left[Z_{m_1n}^{(1)T}\vdots\dots\vdots Z_{m_{k_t}n}^{(k_t)T}\right]^T\in N_s(0,R)$;

2) with number t fixed, number $l=\overline{1,k_l}$, $N=k_l\cdot m$, if $Y_{mn}\in N_s(0,\Lambda)$, $Z_{mn}^{(l)}\in N_s(0,R^{(l)})$ then $Y_{Nn}=\left[Y_{mn}^T\vdots\dots\vdots Y_{mn}^T\right]^T\in N_s(0,\Lambda)$, $Z_{Nn}=\left[Z_{mn}^{(1)T}\vdots\dots\vdots Z_{mn}^{(k_l)T}\right]^T\in N_s(0,\bar{R})$, where $\bar{R}=\{\bar{r}_{ij}\}\in\mathcal{R}_\Lambda$, $\bar{r}_{ij}=(1/k_l)\sum_{l=1}^{k_l} r_{ij}^{(l)}$, $R_{nn}^{(l)}=\{r_{ij}^{(l)}\}\in\mathcal{R}_\Lambda$, $i,j=\overline{1,n}$;

3) with numbers $l=t=\overline{1,k_l}$, $M=m_1+\dots+m_{k_l}$, if $Y_{m_l n}^{(l)}\in N_s(0,\Lambda)$, $Z_{m_l n}^{(l,l)}\in N_s(0,R_{nn}^{(l)})$ then $Y_{Mn}\in N_s(0,\Lambda)$,

$Z_{Mn}^{(\cdot,\cdot)} = \left[Z_{m_1 n}^{(1,1)T} \vdots \cdots \vdots Z_{m_{k_l}}^{(k_l,k_l)T} \right]^T \in N_s(0, \sum_{l=1}^{k_l} \beta_l \cdot R^{(l)})$, where $0 < \beta_l = m_l/M < 1$, $\sum_{l=1}^{k_l} \beta_l = 1$.

For computation of the spectrum with the above properties, algorithms have been developed, making use of the following relations. Let $n \geq k > l > 1$ be integers, $a_i \geq 1$, $i=\overline{2,k}$ be real numbers. Then the elements of the spectrum of c.m. are uniquely defined: $\lambda_{i-1} = \lambda_i a_i$, $\lambda_k = \bar{f}_1/B(k,k)$, $\bar{f}_1 = n$, $\lambda_j = (\prod_{i=j+1}^{k} a_i) \cdot \lambda_k$, $j=\overline{1,k-1}$, $\bar{f}_1 = n$, where $B(t,k) = \sum_{i=1}^{t} (\prod_{j=i+1}^{k} a_j)$. f-Parameters of the spectrum are of the form: $f_1(\Lambda) = B(k,k) \cdot \lambda_k$, $f_2(\Lambda) = D(k,k) \cdot \lambda_k^2$, $f_3(\Lambda) = B(1,k)$, $f_4(\Lambda,l) = B(l.k)/B(k,k)$, $f_5(\Lambda) = E(k) \cdot \lambda_k^k$, $f_6(\Lambda) = \sum_{i=2}^{k} a_i$, where $D(t,k) = \sum_{i=1}^{t} (\prod_{j=i+1}^{k} a_j^2)$, $E(k) = \prod_{i=2}^{k} a_i^{i-1}$.

These algorithms compute monotonous successions of values of f-parameters, obtained after increments: $\tilde{a}_{i+1} = a_{i+1} \gamma_{i+1}$, $\gamma_{i+1} = 1 + \varepsilon_{i+1}$, $\tilde{B}(l,k) = B(l,k) + (\gamma_{i+1} - 1) \cdot B(i,k)$, $\tilde{D}(k,k) = D(k,k) + (\gamma_{i+1}^2 - 1) \cdot D(i,k)$, $\tilde{E}(k) = E(k) \cdot \gamma_{i+1}^i$.

With the use of IM PC there are developed a nonparametric algorithm of interval estimation of statistics, characterizing interrelations between variable samples (including missing values in variables) (Zhanatauov S.U., 1985a,b) and an algorithm of the point estimate of missing values in the multivariate sample (Zhanatauov S.U., 1981). The degree of adequacy of Λ-samples in a real sample and the accuracy of estimates of the algorithm presented further are practically independent of the law of distribution of the standard population, whose samples are transformed into Λ-samples. A multidimensional Gaussian distribution and the uniformly distributed in a unit hypercube are presented as standard distribution. Λ-samples were used both for comparison and elucidation of the domains of the preferable application of methods of incomplete data analysis. As this takes place, algorithms of calculation of the spectrum of the c.m. with given algebraic properties which along with procedures of the IM PC and its applications are a part of the package program "Spectrum", which is the package of modeling of multivariate samples (adequate to those real) of testing of data analysis methods and data processing using the IM PC.

REFERENCES

Zhanatauov S.U.(1980). Technique of computation of the sample with given eigenvalues of its sampling correlation matrix. - In: Matematicheskie voprosy analiza dannykh, Drobyshev Ju.P.(Ed.). VC SOAN SSSR, Novosibirsk, pp. 62-76.

Zhanatauov S.U.(1985a). A nonparametrical algorithm of interval estimations. - In: Vses.simp."Metody i programmnoe obespechenie obrabotki informacii i prikl.stat.analiza dannykh na EVM", BGU, Minsk, pp. 53-54.

Zhanatauov S.U.(1985b). Determination of confidence intervals for estimates of missing values of a real sample - In: Struktury i analiz dannykh. Drobyshev Ju.P.(Ed.). VC SOAN SSSR, Novosibirsk, pp. 111-122.

Zhanatauov S.U.(1981). The method of incomplete data analysis. VC SOAN SSSR, Novosibirsk. (Preprint No 257, 15 p.).

DESIGN OF EXPERIMENTS (nearest neighbour designs . . .)
(Session 5)

Chairman: H.P. Wynn

NUMERICAL METHODS OF OPTIMAL DESIGN CONSTRUCTION

V.V. Fedorov
International Institute for Applied Systems Analysis, Laxenburg, Austria

1. INTRODUCTION

In this paper numerical approaches for the construction of optimal designs will be considered for experiments described by the regression model

$$y_i = \vartheta^T f(x) + \varepsilon_i, \quad i = \overline{1,N} \tag{1}$$

where $f(x)$ is a given set of basic functions, $x \in X$, and X is compact; at least some of the variables x can be controlled by an experimenter, $\vartheta \in R^m$ are estimated parameters, $y_i \in R^1$ is the i-th observation, and $\varepsilon_i \in R^1$ is the random error, $E[\varepsilon_i] = 0$, $E[\varepsilon_i \varepsilon_j] = \delta_{ij}$. In practice, technically more complicated problems could be faced (for instance, y_i could be a vector or errors could be correlated) but usually the methods are straightforward generalizations of the methods developed for problem (1).

The most elegant theoretical results and algorithms were created for a continuous (or approximate) design problem when a design is considered to be a probabilistic measure defined on X, and an information matrix is defined by an integral $M(\xi) = \int f(x)^T(x)\xi(dx)$. In this case, the optimal design of the experiment turns out to be the optimization problem in the space of probability measures:

$$\xi^* = \underset{\xi}{\mathrm{Argmin}} f\, \Phi[M(\xi)], \int_X \xi(dx) = 1, \tag{2}$$

where Φ is the objective function defined by an experimenter.

The first ideas on numerical construction of optimal designs can be found in the pioneer works by Box and Hunter (1965) and Sokolov (1963), where some sequential designs were suggested. These procedures can be considered as very particular cases of some iterative procedures for optimal design in construction, but nevertheless they implicitly contain the idea that one can get optimal design through improving intermediate designs by transferring a finite measure to some given point in X at every step of the sequential design.

This idea was developed and clarified by many authors and the majority of algorithms presented in this survey (which does not pretend to be a historical one) are based on it.

2. FIRST-ORDER ITERATIVE PROCEDURES

It will be assumed that

(a) the functions $f(x)$ are continuous on compact X,

(b) $\Phi(M)$ is a convex function,

(c) there exists q such that

$$\{\xi : \Phi[M(\xi)] = q < \infty\} = \Xi(q) \neq \phi,$$

and

(d) for any $\xi \in \Xi(q)$ and any other $\bar{\xi}$

$$\Phi[(1-\alpha)M(\xi)+\alpha M(\bar{\xi})] = \Phi[M(\xi)]+\alpha\int_X \varphi(x,\xi)\bar{\xi}(dx)+o(\alpha). \quad (3$$

If these assumptions hold, then the following iterative procedure will converge to an optimal design:

$$\xi_{s+1}=(1-\alpha_s)\,\xi_s+\alpha\ \ \xi\ (x_s), \quad (4$$

where $\xi(x_s)$ is a design with the measure totally concentrated at the point x_s,

$$x_s = \mathrm{Argmin}[\varphi\ (x_s^+,\xi_s), -\varphi(x_s^-,\xi_s)], \quad (5$$

$$x_s^+ = \underset{x \in X}{\mathrm{Argmin}}\,\varphi(x,\xi_s), x_s^- = \underset{x \in X_s}{\mathrm{Argmax}}\,\varphi(x,\xi_s),$$

X_s is the supporting set of the design ξ_s, $\alpha_s = \gamma_s$, when $x_s = x_s^+$, an $\alpha_s = -\min[\gamma_s, p_{st}/(1-p_{st})]$, p_{st} is a measure for point x_{st} of design ξ_x.

To provide weak convergence, the sequence $\{\gamma_s\}$ has to satisfy, fo instance, the following condition: $\gamma_s \to 0$ and $\Sigma\gamma_s \to \infty, s \to \infty$. In addition, som other alternatives for the sequence $\{\gamma_s\}$ can be found in Ermakov, 1983 Fedorov, 1972; Fedorov and Uspensky, 1975; Fedorov, 1981; Silvey, 1982; W and Wynn, 1978. The iterative procedures (4), (5) comprise practically all th first-order methods widely discussed in the statistical literature since the lat nineteen sixties. It should be pointed out that the iterative procedure can b realized in practice if the optimization problem (5) is not very difficult from computational point of view, i.e., if the dimension of X is not too high. It especially difficult to work with cases when the controllable variables belon to some functional space (see section IV).

There is a simple idea behind the iterative procedure (4), (5). If one wishes to move along the "best" direction

$$\xi_{s+1} = \mathrm{Arg}\min_{\xi} \Phi[(1-\alpha_s)M(\xi_s)+\alpha_s M(\xi)] \tag{6}$$

then for sufficiently small α_s (see (3)):

$$\min_{\xi} \Phi[(1-\alpha_s)M(\xi_s)+\alpha_s M(\bar{\xi})]=\Phi[M(\xi_s)]+\alpha_s \min_{\xi}\int_X \varphi(x,\xi_s)\bar{\xi}(dx)$$
$$=\Phi[M(\xi_s)]+\alpha_s \min_{x \in X}\varphi(x,\xi_s).$$

Therefore $\bar{\xi}_s^* = \mathrm{Arg}\min_{\xi} \int_X \psi(x,\xi_s)\bar{\xi}(dx)$ can be chosen from the set of point measures $\xi(x), x \in X$ and has to be concentrated at the point $x_x^+ = \mathrm{Arg}\min_{x \in X}\psi(x,\xi_s)$. The same idea is behind "deleting" some points from design ξ_s. These points $\{x_s^-\}$ are "worst" in the sense (6).

For fulfillment of this fact, assumption (d) is crucial. The majority of optimality criteria used in practice satisfy this assumption. But for some quite natural criteria, for instance, $\Phi[M(\xi)] = f^T(x_0)M^-(\xi)f(x_0)$ where x_0 is given, Φ is the variance of $\hat{\vartheta}^T f(x_0)$ and "-" stands for pseudoinversion, formula (3) is not generally valid. One can still use the iterative procedure (4), (5) applying to some regularized version of the initial problem:

$$\Phi_\rho[M(\xi)]=\Phi[(1-\rho)M(\xi)+\rho M(\xi_0)],$$

where $M(\xi_0)$ is regular matrix ($M(\xi_0)\neq 0$). Then

$$\lim_{s\to\infty} \Phi[M(\xi_s)]-\Phi[M(\xi^*)] \le \rho\{\Phi[M(\xi_0)]-\Phi[M(\xi^*)]\}. \tag{7}$$

To adjust the iterative procedure (4), (5) to particular optimality criteria, the following formulae can be useful:

$$\psi(x,\xi)=f^{T}(x)\dot{\Phi}(\xi)f(x) - tr\,\dot{\Phi}(\xi)\,M(\xi),\quad \dot{\Phi} = \frac{\partial \Phi}{\partial M}$$

where the existence of a corresponding derivative is assumed,

$$\frac{\partial M^b}{\partial M_{\alpha\beta}} = \sum_{a=0}^{b-1} M^a E_{\alpha\beta} M^{b-a-1},\quad E_{\alpha\beta\gamma e} = \delta_{\alpha\gamma}\delta_{\beta e},$$

$$\frac{\partial M^{-b}}{\partial M_{\alpha\beta}} = -\sum_{a=0}^{b-1} M^{-a-1}E_{\alpha\beta}M^{-b+a},$$

$$\dot{\Phi} = -M^{-1}\frac{\partial \Phi}{\partial M^{-1}}M^{-1},\quad \frac{\partial \ln|M|}{\partial M} = -M^{-1},\quad \frac{\partial trAM^{-1}}{\partial M} = -M^{-1}AM^{-1},$$

$$M^{-1}(\xi_{s+1})=(1-\alpha_s)^{-1}\left[I-\frac{\alpha_s M^{-1}(\xi_s)f(x)f^T(x)}{1-\alpha_s+\alpha_s d(x,\xi_s)}\right]M^{-1}(\xi_s) ,$$

$$|M(\xi_{s+1})|=(1-\alpha_s)^{m-1}[1-\alpha_s+\alpha_s d(x,\xi_s)]|M(\xi_s)| , d(x,\xi_s)=f^T(x)M^{-1}(\xi_s)f(x) .$$

The convergency rate of the above mentioned algorithms decreases in the vicinity of the optimum. As in general optimization theory, attempts were made to develop second order methods. These methods are based on quadratic approximations of the function $\Phi[M]$ and it is necessary to assume the existence of derivatives $\partial\Phi/\partial p_i, \partial^2\Phi/\partial p_i\partial p_j$, where p_i is the weight for supporting point x_i. Second-order algorithms have at least two features which are handicaps for their use in practice: first, at every step it is necessary to invert the matrix $\{\partial^2\Phi/\partial p_i\partial p_j\}$, and second, all existing modifications can handle only the discrete operability region X (see Ermakov (ed), 1983 Ch 4, Wu, 1978).

In the late seventies, some attention was paid to algorithms which work in the space of information matrices $M(X)$; they are computationally effective if one can easily find the mapping $\Xi(X)\to M(X)$ (see, for example, Gribik and Kortanek, 1977). But usually it is very difficult to realize this mapping numerically.

3. CONSTRUCTION OF OPTIMAL DESIGNS UNDER CONSTRAINTS

In (2) there is only one constraint $\int \xi(dx)=1$. If one considers additional constraints, say $\int \psi(x)\,\xi\,(dx)\le c$, where $\psi(x)$ is a vector of given functions ($c,\psi \in R^k$), then the iterative procedure becomes technically more complicated, although based on the same ideas, see (Fedorov and Gaivorovsky, 1984). In this case, (6) became equivalent to the following optimization problem

$$\bar{\xi}_s = \operatorname*{Argmin}_{\xi} \int_X \varphi(x,\xi_s)\bar{\xi}(dx), \tag{8}$$

subject to $\int_X \psi(x)\bar{\xi}(dx)\le c$.

Due to the classical theorem of Caratheodory, design $\bar{\xi}_s$ can be found within the set of designs containing no more than $k+1$ supporting points. The optimization problem (8) is essentially more complicated than (5). From a computational point of view, the dual problem (see Karlin and Studden, 1966, ch. XII):

$$\max_{u\in U}\min_{x\in X}[\varphi(x,\xi_s)+u^T\psi(x)] \tag{9}$$

where $U=\{u:u_i\ge 0, i=\overline{1,k}\}$, can be useful to define the location of the supporting points of $\bar{\xi}_s$. They have to coincide with the solutions $x_1^*, x_2^*, \ldots$ of (9). The corresponding measures can be found by any linear programming algorithm.

In some applications (see Wynn, 1982), designs have to be restricted in the following sense

$$\int_A \xi(dx) \le \int_A \xi_0(dx) \quad \text{for all } A \subset X \tag{10}$$

where $\int_X \xi_0(dx) = c$, $c \ge 1$, $A \subset X$ and a measure ξ_0 is atomless (see Karlin and Studden, 1966, p. 233). In this case (6) leads to the following problem

$$\bar{\xi}_s = \operatorname*{Argmin}_{\xi} \int_X \varphi(x, \xi_s) \bar{\xi}(dx), \tag{11}$$

where a probability measure $\bar{\xi}_s$ has to satisfy (10).

It is evident that $\bar{\xi}_s$ has to coincide with ξ_0 on any subset of X, where $\varphi(x, \xi_s) \le 0$ and has to be equal 0 otherwise (compare with theorem 1 from Wynn, 1982). Computationally, the search for these sets can be realized through discretization of X. The idea of the iterative procedure (4), (5) will apply once again if one will additionally delete from design ξ_s, some sets where $\varphi(x, \xi_s) > 0$. Thus

$$\xi_{s+1} = (1 - \alpha_s)\xi_s + \alpha_s [\xi_0(E_s, dx) - \xi_0(D_s, dx)], \tag{12}$$

where E_s is the set of new included points and D_s is the set of deleted points.

The procedure similar to (12) (but without the operation of deleting) was considered by Gaivorovsky (1985), and its weak convergency was proven under rather mild conditions. Usually, deleting "bad" points essentially improves the quality of the iterative procedures (compare with the traditional case, Atwood, 1973, Fedorov and Uspensky, 1975).

Let $\Xi(\xi_0)$ be a class of probability measures ξ with supporting sets $A \subset X$, and $\xi(dx) = \xi_0(dx)$, when $x \in A$ and equal to 0 otherwise. For any design problem (2), (10), there exists an optimal design $\xi^* \in \Xi(\xi_0)$, see, for instance, Wynn, 1982, whose results have their origin in the classical moment space theory, particularly in the Liapunov Theorem, see Karlin and Studden, 1966, Ch. VIII). Therefore, it is reasonable to demand that $\xi_s = \xi(A_s, dx) \in \Xi(\xi_0), s = 1, 2, \ldots$ Iterative procedure (12) does not satisfy the latter demand. Instead of this type of iterative procedure (which repeats the idea of (4), (5)), one can apply an iterative procedure of the exchange type:

$$\xi_{s+1} = \xi_0(A_{s+1}, dx) = \xi_0(A_s, dx) + \xi_0(E_s, dx) - \xi_0(D_s, dx), \tag{13}$$

where $$\int_{E_s} \xi_0(dx) = \int_{D_s} \xi_0(dx) = \delta_s , \; A_s \cap E_s = 0 , \quad D_s \subset A_s , \; E_s \cap D_s = 0. \tag{14}$$

If

(e) derivatives $\dot{\Phi}$ exist,

(f) ξ_0 has a continuous density $\mu_0(x)$,

and assumptions (a)–(c) hold, then for sufficiently small δ_s, the approximation

$$\Phi[M(\xi_{s+1})] = \Phi[M(\xi_s)] + [\gamma(x_s^+, \xi_s) - \gamma(x_s^-, \xi_s)]\delta_s + o(\delta_s), \tag{15}$$

where $x_s^+ \in E_s$, $x_s^- \in D_s$ and $\gamma(x,\xi) = f^T(x)\dot{\Phi}(\xi)f(x)$ can be used. If X is covered by the grid X_δ with density proportional to $\mu_0(x)$, then x_s^+ and x_s^- can coincide with its nodes and E_s , D_s with some cells of this grid.

From (15), it is clear that to provide approximately the steepest descent on every step of the discretized version of procedure (13), one has to find

$$x_s^+ = \operatorname*{Argmin}_{x \in X_\delta \setminus A_{\delta s}} \gamma(x, \xi_s) \text{ and } x_s^- = \operatorname*{Argmax}_{x \in A_{\delta s}} \gamma(x, \xi_s), \tag{16}$$

where $A_{\delta s}$ is a discrete analogue of A_s. It is worthwhile to point out that for the discretized version of the iterative procedure, one can use a recursive formula for M_{s+1}^{-1} (see (17)) to simplify calculations. Complementing assumptions (a)–(f) by assumptions:

(g) for any design ξ with $\Phi[M(\xi)] \le Q < \infty$ and any C $\xi_0\{A : \gamma(x,\xi) = C\} = 0$,

(h) $\delta_s \to 0$, $\sum_s \delta_s \to \infty$

the weak convergency: $$\lim_{s \to \infty} \Phi[M(\xi_s)] = \min_{\xi \in \Xi(\xi_0)} \Phi[M(\xi)] \tag{17}$$

can be proven.

Result (17) is based on the fact that the fulfillment of the inequality:

$$\max_{x \in X^*} \gamma(x, \xi^*) \le \min_{x \in X \setminus X^*} \gamma(x, \xi^*)$$

is a necessary and sufficient condition for a design ξ^* to be optimal (X^* is a supporting set of ξ^*).

4. OPTIMAL DESIGNS WHEN CONTROLS BELONG TO A FUNCTIONAL SPACE

This case will be illuminated here by a rather specific example which nevertheless reflects the major difficulties.

Let $f(v) \in R^m$ and $x = \int_V f(v)h(v)dv$, where $h(v)$ can be controlled, $h(v) \in H$. If one manages to construct the mapping $X(H) \subset R^m$ of the set H, then all approaches discussed in the previous sections can be used without any alterations to find an optimal design ξ_x^* on X. The problem to be faced afterwards is the construction of an inverse mapping $X \to H$ to convert ξ_x^* to some design ξ_H^* defined on H. The latter problem is beyond the scope of this paper and its discussion can be find in Kozlov, 1981, Ermakov (ed), 1983, Ch.7. For the case discussed in section 2, the situation is slightly simpler because of the Equivalence Theorem (see, for instance, Ermakov (ed), 1983 Ch.2); thus only boundary points $\bar{X}(H)$ of $X(H)$ are needed for optimal design construction. Unfortunately, the numerical construction of $\bar{X}(H)$ happens to be sometimes a very difficult problem and it could be more efficient to work in the original space H.

To be more specific, let us assume that $v \in V \subset R^1$ and $0 \leq h(v) \leq 1$ and restrict ourself to the design problem in section 2. The most straightforward approach consists of discretization of V and approximation of $h(v)$ by same piecewise function. Under rather mild conditions, it is possible to prove that there exists optimal design which supporting points belong to the set $\bar{H} = \{h(v): h(v)[1-h(v)] = 0, v \in V\}$, see, for instance Fedorov, 1986. If V is discretized by a grid with elements Δ_j, then the simplest version of procedure (4), (5) (without "deleting" operation) converts to the following one:

a) $\xi_{s+1} = (1-\alpha_s)\xi_s + \alpha_s \xi(h_s)$;

b) steps for finding h_s :

- collect all Δ_j which negatively contribute to the sum

$$\varphi(\xi_s) = \sum_{j,j'} F_j^T \dot{\Phi} F_{j'},$$

where $F_j = \int_{\Delta_j} f(v) dv$ (usually $F_j \approx f(v_j)\Delta_j$).

- put $h_s(v) = 1$, $v \in \Delta_j$ if Δ_j was chosen in the previous stage, otherwise $h_s(v) = 0$,
- the fulfillment of the inequality $\varphi(\xi_s) < tr\, \dot{\Phi}(\xi_s) M(\xi_s)$ tests that h_s can be used for $\xi(h_s)$.

This iterative procedure guarantees that $\lim_{s \to \infty} \Phi[M(\xi_s)] = \Phi[M(\xi_\Delta^*)]$, where ξ_Δ^* is an optimal design for the discretized design problem and can be called a Δ-optimal design. When $a \leq v \leq b$ and functions $f(v)$ constitute a Tchebycheff system over the open interval (a,b), where a and b are possibly infinite, then the rather effective iterative procedure can be used for optimal design construction. The idea of this procedure is based on the following result (see, for instance, Fedorov, 1986).

Let $h(v) \in \bar{H}$ and let I be the number of separate nondegenerate intervals where $h(v)=1$ with the special convention that an interval whose closure contains point a or b, is counted as 1/2. For any point $x \in X, I^*$ stands for the least possible I. Then a necessary and sufficient condition that x belongs to the boundary of X is that $I^* \leq (m-1)/2$. Moreover, every boundary point corresponds to a unique $h(v)$ with $I(x)=I^*(x)$.

Let now $\bar{v}=(v_1,\ldots,v_{m-1})$, where $a \leq v_1 \leq \cdots \leq v_{m-1} \leq b$. According to the previous result, there exist optimal designs with all supporting points (in the operability region H) which have the following structures:

$$\bar{h}(v)=1, v \in (a,v_1); 0, v \in (v_1,v_2); 1, v \in (v_2,v_3); \cdots$$

and $\underline{h}(v)=1-\bar{h}(v)$.

That fact allows for modification of the iterative procedure (4), (5) (without deleting "bad" points) to the procedure with maximization in space, with dimension less than or equal to $(m-1)$, where m is a number of basic functions:
$$\xi_{s+1}=(1-\alpha_s)\xi_s+\alpha_s\xi(h_s)\ , \quad h_s = \underset{\gamma,\bar{v}}{\mathrm{Argmin}}\,\Phi[x_\gamma(\bar{v}),\xi_s],$$

where $a \leq v_1 \leq \ \ldots \ \leq v_{m-1} \leq b, x_1(\bar{v})=\int_a^b f(v)\,\bar{h}(v)dv$ and $x_2(\bar{v})=\int_a^b f(v)\underline{h}(v)dv$.

The design problem considered in this section comprises the major difficulties which can be met when X is a functional space. Other examples can be found in Mehra, 1976, Pazman, 1986. These authors use mainly the same ideas surveyed here. In concluding this section, it would be worthwhile to notice that parametrization of controls (a rather standard method in optimal control theory), e.g. linear approximation $\delta^T q(v)$ of $h(v)$ in our example, could be a useful tool allowing one to convert the original problem to a finite dimension design problem.

5. DISCRETE DESIGNS

To construct optimal discrete (or exact) designs, a number of exchange type algorithms can be used (for detailed information, see Cook and Nachtsheim, 1980; Johnson and Nachtsheim, 1983; Steinberg and Hunter, 1984).

The idea of the simplest algorithm (originated by Mitchell, 1974) can be formulated in the following way:

After the s-th step there is a design $\xi_{Ns}=\{x_{1s},\ldots,x_{Ns}\}$, where some supporting points can coincide. This design is complemented by k points:

$$x^+_{N+j,s} = \underset{x_j \in X}{\mathrm{Argmin}}\,\Phi[M(\xi_{N+j-1,s}+x)], j=\overline{1,k} \tag{18}$$

Then the same number of points:

$$x_{N+k-1,s} = \underset{x_l \in X_s}{\mathrm{Argmin}}\, \Phi[M(\xi_{N+k-1,s} - x_l)], l = \overline{1,k}, \tag{19}$$

are deleted and one arrives at the new design $\xi_{N,s+1}$ containing N observations. The notation $\xi_K + x\,(or - x)$ means that a point is included in (or excluded from) design ξ_K, X_s comprises all of the different supporting points of the design from the previous stage.

In practice, the excursion length k is usually rather modest (1-3) and there are no indications that an increase could be useful. Iterative procedure (18), (19) are computationally simple and often lead to very good results, especially when one faces discrete X, for example, $x_\alpha = \pm 1$.

In the iterative procedures (18), (19), the deletion and complementary steps are separated. If we unite them (Fedorov, 1972), then we arrive at the following iterative procedure (with excursion length 1):

$$\xi_{N,s+1} = \underset{x^+ \in X, x^- \in X_s}{\mathrm{Argmin}} \Phi[M(\xi_{Ns} + x^+ - x_j^-)], \tag{20}$$

where X_s is the supporting set of ξ_{Ns}. This procedure demands $N/2$ times more calculations at every step than (18), (19), but in most cases it gives better final results, see Johnson and Nachtsheim, 1983. The above minimization problem is equivalent to coordinate wise minimization of $\Phi[M]$:

$$\min_j \min_{x_j \in X} \Phi[M(x_{1s}, \ldots, x_{js}, \ldots x_{Ns}] \tag{21}$$

if one starts the numerical optimization in (21) with $x_j^0 = x_{js}$.

A similar choice of an initial point is appropriate in many optimization problems, but not in the optimal design of experiments when the objective function usually has a large number of local minima along the variation of x_j, and (21) will lead to the local minimum closest to x_{js}. The application of (20) helps to approach the global minimum by explicit forcing of x^+ to be away of x_j^-. Procedures (18), (19) or (20) become a practical tool when one manage to find a simple formula for calculation of increments for $\Phi[M]$ at every stage. For the majority of widely used criteria (D-criterion, linear criteria and so on) these formulas can be found in the above cited publications (see also section 2).

In spite of the rather long history of the numerical procedures discussed above, their convergence properties are not well known except for numerous empirical results. It is not a problem, for instance, to prove the convergence of (20) to some design better than an initial one but it has not yet been proven that the limit design has to be optimal.

REFERENCES

Atwood, C.L. (1973) Sequences Converging to D-Optimal Designs of Experiments, Ann.Stat., Vol. 1:342-352.

Box, G.E.P. and Hunter, W. (1965) Sequential Design of Experiments for Nonlinear Models. Proc. IBM Scientific Computing Symp. Statistics, 113-120.

Cook, R.D. and Nachtsheim, C.J. (1980) A Comparison of Algorithms for Constructing Exact D-Optimal Designs, Technometrics, Vol. 22:682-688.

Ermakov, S.M. (ed.) (1983) Mathematical Theory of the Design of Experiments (in Russian), p.386.

Fedorov, V.V. (1972) Theory of Optimal Experiments, Academic Press, N.Y., p. 292.

Fedorov, V.V. and Uspensky, A.B. (1975) Numerical Aspects of Design and Analysis of Experiments, Moscow State University, Moscow (in Russian), p. 167.

Fedorov, V.V. (1981) Active Regression Experiments in Mathematical Methods of Experimental Design, ed. Penenko, V.V., Nauka (in Russian), p.19-73.

Gaivoronsky, A. and Fedorov V. (1984) Design of Experiments under Constraints, WP-84-8, IIASA, Laxenburg, Austria, p. 11.

Gaivoronsky, A. (1985) Stochastic Optimization Techniques for Finding Optimal Sub-measures, WP-85-28, IIASA, Laxenburg, Austria, p.52.

Gribik, P.R. and Kortanek, K.O. (1977) Equivalence Theorems and Cutting Plane Algorithms for a Class of Experimental Design Problems, SIAM J. Appl. Math., Vol. 32:232-259.

Johnson, M.E. and Nachtsheim, C.J. (1983) Some Guidelines for Constructing Exact D-Optimal Designs on Convex Design Spaces, Technometrics, Vol. 25:271-277.

Karlin, S. and Studden, W.J. (1966) Tchebycheff Systems: with applications in Analysis and Statistics, Wiley and Sons, N.Y., p. 578.

Kozlov, V.P. (1981) Design of Regression Experiments in Functional Spaces, in Mathematical Methods of Experimental Design, ed. Penenko, V.V., Nauka (in Russian), p. 74-101.

Mehra, R.K. (1976) Synthesis of Optimal Inputs for Multiinput- Multioutput Systems with Process Noise, in "System Identification: Advances and Case Studies" ed. D.K. Mehra, D.G. Lainiotis, Academic Press, N.Y. pp. 211-250.

Mitchell, T.J. (1974) An Algorithm for the Construction of D-Optimal Experimental Designs, Technometrics, Vol. 16:203-210

Pazman, A. (1986) Foundations of Optimum Experimental Design, VEDA, D. Reidel Publishing Company, p.228.

Silvey, S.D. (1980) Optimal Design, Chapman and Hall, London, p.86.

Steinberg, D.M. and Hunter W.G. (1984) Experimental Design: Review and Comment, Technometrics, Vol. 26:71-97.

Sokolov, S.N. (1963) Continuous Design of Regression Experiments. Teor. Verejatuost.; Primen., Vol. 8:95-101 (a), 318-323 (b).

Wu, C.F. (1978) Some Iterative Procedures For Generating Nonsingular Optimal Designs, Commun. Statist. (Theor. Math.), Vol. A7(14):1399-1412.

Wu, C.F. and Wynn, H. (1978) The Convergence of General Step-Length Algorithms for Regular Optimum Design Criteria, Ann. Statist., Vol. 6:1273-1285.

Wynn, H. (1982) Optimum Submeasures with Applications to Finite Population Sampling, in "Statistical Decision Theory and Related Topics III", Vol. 2, Academic Press, N.Y., pp. 485-495.

ORDERING EXPERIMENTAL DESIGNS

Friedrich Pukelsheim

Institut für Mathematik, Universität Augsburg, Federal Republic of Germany

Abstract. We present an overview of certain two-stage orderings of experimental designs which are such that they reflect an increase in information. These orderings use group majorization, in addition to the Loewner ordering of nonnegative definite matrices. The groups act through congruence on the moment matrices and information matrices of the problem, and a table of known results and open problems depending on the particular group is presented. The examples of quadratic regression on the symmetrized unit interval and of linear regression over the unit cube are discussed in some detail.

Key words: Information functionals, p-means, invariance, group majorization, information increasing orderings, universal optimality, simultaneous optimality.

AMS 1980 subject classification: 62K05, 62K10, 06F20.

1 Introduction

Experimental design orderings which reflect an increase in information are useful in that they allow to discriminate between competing designs. For a detailed technical derivation the reader is referred to Giovagnoli et al. (1986). A survey of the present state of experimental design theory will be found in Atkinson (1986) and Pukelsheim (1986) and the references given there.

2 Maximizing information

As usual in experimental design theory we consider a classical linear model

$$Y(x) = x'\beta + \sigma e$$

assuming uncorrelated observations with unit variance. The vectors $x \in \mathbb{R}^k$ represent the *experimental conditions*, and in their totality are assumed to form a compact set $\mathcal{X} \subset \mathbb{R}^k$, the *experimental domain*. A *design* ξ then is a discrete probability distribution on the experimental domain $\mathcal{X}$, determining allocations and frequencies of the observations.

2.1 Moment matrices

The essential quantity associated with the design ξ is its $k \times k$ *moment matrix*

$$M(\xi) = \int xx'd\xi = \sum_{i=1}^{l} \xi(x_i)x_i x_i'.$$

The set of all moment matrices forms a convex compact subset of nonnegative definite matrices. Since in some problems it is desirable to distinguish between feasible and non-feasible moment matrices, we simply assume to start from a set $\mathcal{M}$ of moment matrices which is convex and compact. This covers the case which is often dealt with that the set $\mathcal{M}$ consists of *all* moment matrices, as well as allowing for the possibility of $\mathcal{M}$ being a genuine subset of moment matrices.

2.2 Information matrices

We shall assume that an s-dimensional parameter system $K'\beta$ is of interest, where the $k \times s$ matrix K has rank s. Here the moment matrix may degenerate, with its rank varying between s and k, depending on whether the nuisance parameters remain identifiable or not. For the parameter system $K'\beta$ identifiability obtains if and only if the range of M contains the range of K. Thus the $s \times s$ *information matrix* for $K'\beta$ is defined to be

$$C(M) = \begin{cases} (K'M^{-}K)^{-1} & \text{in case of identifiability} \\ 0 & \text{otherwise.} \end{cases}$$

We shall tacitly assume that the parameter system $K'\beta$ is identifiable under at least one moment matrix in the set $\mathcal{M}$, in order to deal with a non-void problem.

As an example consider quadratic regression

$$Y(x_t) = \beta_0 + \beta_1 t + \beta_2 t^2 + \sigma e = x_t'\beta + \sigma e.$$

We allow t to vary over the symmetrized unit interval, resulting in the experimental domain

$$\mathcal{X} = \left\{ x_t = \begin{pmatrix} 1 \\ t \\ t^2 \end{pmatrix} \mid t \in [-1, +1] \right\}.$$

Thus $\mathcal{X}$ essentially looks like a parabola in three-dimensional space. The parameter systems of interest here may be the set of all parameters, the subsets of any two parameters, or any single parameter, i. e.

$$\begin{pmatrix} \beta_0 \\ \beta_1 \\ \beta_2 \end{pmatrix}, \begin{pmatrix} \beta_0 \\ \beta_1 \end{pmatrix}, \begin{pmatrix} \beta_0 \\ \beta_2 \end{pmatrix}, \begin{pmatrix} \beta_1 \\ \beta_2 \end{pmatrix}, \beta_0, \beta_1, \beta_2.$$

2.3 Information functionals

The proper information matrix for the parameter system of interest actually is

$$\frac{n}{\sigma^2} C(M),$$

being directly proportional to sample size n and inversely proportional to the model variance σ^2. In a last step we need some real-valued functionals which appropriately preserve the properties of information matrices.

To this end we define an *information functional* ϕ to be a real-valued function on the set of nonnegative definite $s \times s$ matrices such that ϕ is

(a) nonnegative, since information ought to be a nonnegative quantity,

(b) positively homogeneous, whence we can dispose of the scalar factor n/σ^2,

(c) concave, because information cannot possibly be increased by interpolation, and

(d) increasing in the Loewner ordering, which actually is implied by (a) – (c).

This set of properties forms a minimal set of requirements for any specific application, while at the same time being sufficiently strong to build a general theory.

As an example we mention the *p-means*, with $p \in [-\infty, +1]$, i. e. the generalized means of order p of the eigenvalues of the information matrices. They are defined through

$$\begin{aligned} \phi_0(C) &= (\det C)^{1/s}, && \text{i. e. } p = 0, \\ \phi_p(C) &= (\text{trace} C^p/s)^{1/p}, && \text{for } 0 \neq p \leq 1, \\ \phi_{-\infty}(C) &= \lambda_{\min}(C), && \text{i. e. } p = -\infty. \end{aligned}$$

In classical terms ϕ_0 is D-optimality, ϕ_{-1} is A-optimality, and $\phi_{-\infty}$ is E-optimality.

3 Ordering information matrices

Group majorization appears to be the right tool to model increasing symmetry or increasing balance of a design. This is closely related to the invariance properties of the underlying problem. A comprehensive treatment is presented in Giovagnoli et al. (1986), and we here outline only such details as are necessary to sketch the development.

Let $\tilde{G}$ be a subgroup of the general linear group $GL(k)$, and assume that $\tilde{G}$ acts linearly on the experimental conditions x, i. e.

$$x \longrightarrow Qx, \text{ with } Q \in \tilde{G} \subset GL(k).$$

We give two simple examples. For quadratic regression over the symmetrized unit interval $[-1, +1]$ a natural candidate is the *sign-change group* which consists of the identity and of

$$x_t = \begin{pmatrix} 1 \\ t \\ t^2 \end{pmatrix} \longrightarrow \begin{pmatrix} 1 & 0 & 0 \\ 0 & -1 & 0 \\ 0 & 0 & 1 \end{pmatrix} x_t = \begin{pmatrix} 1 \\ -t \\ t^2 \end{pmatrix}.$$

Here the group consists of two transformations only. As a second example consider linear regression over the unit cube $[0,1]^k$. In this case the permutation group is appropriate to catch the apparent symmetry, according to

$$x = \begin{pmatrix} x_1 \\ \vdots \\ x_k \end{pmatrix} \longrightarrow \pi x = \begin{pmatrix} x_{\pi(1)} \\ \vdots \\ x_{\pi(k)} \end{pmatrix}.$$

3.1 Induced group actions

Since our problem formulation heavily depends on moment and information matrices it is important to recognize that the linear group action on the experimental conditions x translates into *congruence action* on matrices:

$$M(\xi) = \int xx'd\xi \longrightarrow \int Qxx'Q'd\xi = QM(\xi)Q',$$

$$C(M) \longrightarrow (K'(QMQ')^{-}K)^{-1} = \check{Q}C(M)\check{Q}'.$$

In order for this to work out we must verify the following assumptions:

- The experimental domain $\mathcal{X}$ must be invariant.
- The set $\mathcal{M}$ of feasible moment matrices must be invariant.
- The reduction C from moment matrices to information matrices must be equivariant.
- The information functionals ϕ to be considered must be invariant.

We mention in passing that the matrices $\check{Q}$ which act on the information matrices C form a subgroup $\check{G}$ of the $s \times s$ general linear group, and that the passage from $\tilde{G}$ to $\check{G}$ is a group homomorphism.

3.2 Information increasing orderings

Assume from now on that the problem is invariant under a group $\tilde{G}$ as just outlined. We shall call a moment matrix B *more centered than* another moment matrix A when

$$B = \sum \alpha_i Q_i A Q_i' \in \text{ convex hull of the orbit of } A.$$

This is the usual concept of group majorization; our terminology of "being more centered" is tailored to the design problem.

The strongest reasonable ordering of moment matrices and of information matrices is, of course, the Loewner ordering defined by $M \geq B$ when $M - B$ is nonnegative definite.

The superposition of group majorization and Loewner ordering produces the *information increasing ordering* which has been found to be appropriate for the design problem, as follows. A moment matrix M is called *at least as informative* as another moment matrix A, denoted by $M \gg A$, when M is larger in the Loewner ordering than some matrix B which is more centered than A. Formally:

$$M \geq B \in \text{ convex hull of the orbit of } A, \quad \text{for some } B.$$

The corresponding information increasing ordering for information matrices will also be denoted by $\gg$. That these information preorderings nicely agree with the various levels of our problem is shown by the following.

Theorem. *(Giovagnoli et al. (1986))*

$$\begin{aligned} &M \gg A \\ &\Longrightarrow C(M) \gg C(A) \\ &\Longrightarrow \phi(C(M)) \geq \phi(C(A)), \text{ for all invariant } \phi. \end{aligned}$$

3.3 Universal optimality vs. simultaneous optimality

The preceeding theorem suggests to discriminate between the notions of *universal optimality* whenever

$$C \gg D, \text{ for all competing } D,$$

and of *simultaneous optimality* whenever

$$\phi(C) \geq \phi(D), \text{ for all competing } D \text{ and for all invariant } \phi.$$

Frequently these notions will coincide according to the following.

Theorem. *(Giovagnoli et al. (1986)) If the underlying group is compact and the information matrix C is invariant then*

$$C \text{ is universally optimal} \iff C \text{ is simultaneously optimal.}$$

When the group fails to be compact or the matrix C is not invariant it seems that the notion of simultaneous optimality is of a greater bearing. The following table gives an overview of some known results and open problems.

group	ordering	invariant functionals
$\{I_s\}$	Loewner	all ϕ
Perm(s)	?	?
Orth(s)	upper weak majorization of ordered eigenvalues	symmetric functions of ordered eigenvalues
Unim(s)	?	determinant
reflection groups	?	?
?	?	p-means

As an outstanding result we mention that this provides a further justification for the most popular criterion of D-optimality as being the sole invariant information functional under the group of unimodular linear transformations (i. e. those with determinant ± 1). On the other hand it would be of interest to study finite reflection groups as they also arise in other aspects of multivariate analysis, or to find a group such that the invariant functionals are determined by the p-means.

4 Quadratic regression; regression over the unit cube

As mentioned above the model for quadratic regression over the symmetrized unit interval $[-1, +1]$ is

$$Y(x_t) = \beta_0 + \beta_1 t + \beta_2 t^2 + \sigma e.$$

A design ξ is invariant under the *sign-change group* if and only if ξ is symmetric about 0. This reduces the corresponding moment matrices to a two-parameter subset. If we augment this with an improvement in the Loewner ordering we obtain a reduction to the one-parameter family of *symmetric three-points designs* ξ_α given by

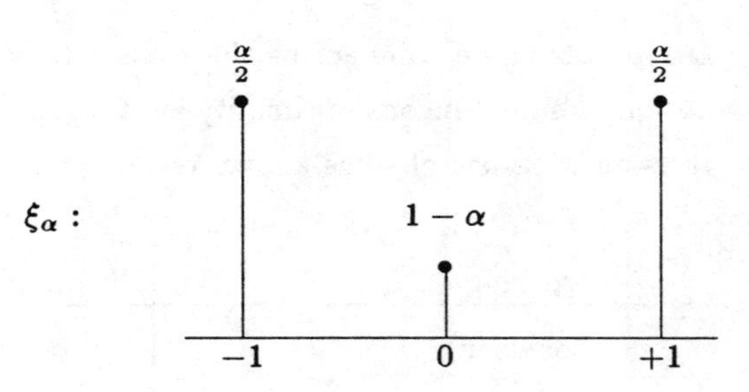

This approach yields the following results, cf. Preitschopf and Pukelsheim (1986): *(a) For every ξ there exists some α such that $\xi_\alpha \gg \xi$. (b) For every p there exists some $\alpha(p)$ such that $\xi_{\alpha(p)}$ is p-optimal.*

Another very instructive example is linear regression over the unit cube which has recently been resolved in a brilliant paper by Cheng (1986). With experimental conditions x varying over the k-dimensional unit cube $[0,1]^k$ the model

$$Y(x) = x'\beta + \sigma e$$

is invariant under the permutation group. Now an invariant design ξ has a moment matrix $M(\xi)$ which belongs to the two-parameter family of completely symmetric matrices (i. e. having identical on-diagonal elements and identical off-diagonal elements). At this stage the General Equivalence Theorem is invoked to obtain a further reduction to the one-parameter family of *uniform vertex designs* ξ_t, defined as follows.

A vertex x of the unit cube will be called a *c-vertex* if x has c components equal to unity and $n-c$ components equal to zero. The unique permutation invariant design which is supported by the c-vertices is the uniform distribution on the c-vertices and will be denoted by ξ_c. In addition we also need mixture designs as defined by

$$\xi_t = \big(1-(t-c)\big)\xi_c + (t-c)\xi_{c+1}, \text{ with } t \in (c, c+1),$$

which are the permutation invariant designs supported by the c-vertices and the $(c+1)$-vertices and hence are convex combinations of ξ_c and ξ_{c+1}. The parameterization chosen evidently is continuous. The family of uniform vertex designs now is given by ξ_t with t varying continuously between 0 and k.

Cheng (1986) proves the following result: *For all p there exists some $t(p)$ such that $\xi_{t(p)}$ is p-optimal.* Moreover he derives an explicit formula for $t(p)$. A monotone behaviour emerges, in that as p increases from $-\infty$ towards 1 one has that $t(p)$ increases from the integer part of $(k+1)/2$ towards k. More precisely $t(p)$ is constantly equal to an integer value c over closed intervals of p, and strictly increasing in-between. Almost all of the variation of $t(p)$ occurs for positive values of p, a qualitative feature which is also encountered in quadratic regression.

References

Atkinson, A. C. (1986). Recent developments in the design of experiments. *Internat. Statist. Rev.*, to appear.

Cheng, C.-S. (1986). An application of the Kiefer-Wolfowitz equivalence theorem. *University of California, Berkeley,* Preprint, 14 pp.

Giovagnoli, A., Pukelsheim, F., and Wynn, H. P. (1986). Group invariant orderings and experimental designs. *J. Statist. Plann. Inference*, to appear.

Preitschopf, F., and Pukelsheim, F. (1986). Optimal designs for quadratic regression. *J. Statist. Plann. Inference*, to appear.

Pukelsheim, F. (1986). Information increasing orderings in experimental design theory. *Internat. Statist. Rev.*, to appear.

Bernoulli, Vol. 2, pp. 167-173

OBSERVATION AND EXPERIMENTAL DESIGN FOR AUTOCORRELATED PROCESSES

H. P. Wynn
Department of Mathematics
The City University
Northampton Square
LONDON
EC1V OHB

1. INTRODUCTION

The papers by Kiefer and Wynn (1983, 1984) treat the study of experimental design for autocorrelated error processes by letting the mechanism that defines the design be the realization of a discrete state process: $\{X_t\}$. This makes the asymptotic properties easier to understand. The work followed that of Kiefer (1960) under the heading of "exact treatment design".

In this paper we rework the principal results in an elementary fashion for two treatments. This case also makes the comparison with sampling theory more transparent. In this way we hope to underline the unity of design and sampling under a general banner of "optimum data collection". Indeed it is within the context of autocorrelated models that the barriers between the different areas of this broader discipline have most been broken down.

2. THE MODEL

Let $\{Y_t\}(t = 1, \ldots, N)$ be an autoregressive process of order p, AR(p):

$$Y_t + \alpha_1 Y_{t-1} + \ldots + \alpha_p Y_{t-p} = \varepsilon_t,$$

where $\{\varepsilon_t\}$ is an uncorrelated zero mean process, with variance σ^2.

We shall consider, for ease of explanation, a two-state process, $\{X_t\}$ which defines the allocation of treatments A and B with parameters θ_A and θ_B. Thus we <u>observe</u>

$$Z_t = Y_t + \theta_A \quad \text{if} \quad X_t = 1, \quad Z_t = Y_t + \theta_B \quad \text{if} \quad X_t = 0$$

Let $\phi = \theta_A - \theta_B$. We shall consider $\hat{\phi}$, the best linear unbiased estimator (BLUE) for ϕ. The design problem is then to define $\{X_t\}$ to minimize $\text{var}(\hat{\phi})$.

Let Γ be the covariance matrix of the $\{Y_t\}$ process. Let $\Gamma^{-1} = \{m_{st}\}$ which we assume to exist.

Then define

$$d_{11} = \frac{1}{N} \sum_{s,t} m_{st} X_s X_t \ ,$$

$$d_{22} = \frac{1}{N} \sum_{s,t} m_{st}(1-X_s)(1-X_t) \ ,$$

$$d_{12} = \frac{1}{N} \sum_{s<t} m_{st} X_s (1-X_t) \ .$$

Using standard theory the BLUE of (θ_1, θ_2) has covariance matrix

$$\frac{1}{N(d_{11}d_{22}-d_{12}^2)} \begin{bmatrix} d_{11} & -\ d_{12} \\ -d_{12} & d_{22} \end{bmatrix} .$$

Then

$$\text{Var}(\hat{\phi}) = \frac{d_{11}+d_{22}+2d_{12}}{N(d_{11}d_{22}-d_{12}^2)} \ .$$

Now consider a situation in which for all allocations $\{X_t\}$, in some class, we have:

(i) $d_{11} + d_{12} \to$ a constant, independent of $\{X_t\}$,

(ii) $d_{11} = d_{22} + 0\left(\frac{1}{N}\right)$,

(iii) $d_{11} \to \overline{d}_{11}$ a constant, depending on $\{X_t\}$ $(d_{22} \to \overline{d}_{22},\ d_{12} \to \overline{d}_{12})$.

Then

$$N \, var(\hat{\phi}) \to 2(\overline{d}_{11} - \overline{d}_{12})^{-1}$$

By virtue of (i) we achieve min var $(\hat{\phi})$ within the class by achieving

$$\max \{\overline{d}_{11}\} .$$

This is a special case of the "universal optimality result" of Kiefer.

We can now specialize to an AR(P) in order to elucidate the conditions (i), (ii) and (iii). A general quadratic form $\frac{1}{N} z^T \Gamma^{-1} z$ in $\Gamma^{-1} = \{m_{st}\}$ can be written

$$\frac{\sigma^2}{N} z^T \Gamma^{-1} z = \frac{1}{N}(1 + \alpha_1{}^2 + \ldots + \alpha_p{}^2) \Sigma z_t$$

$$+ \frac{2}{N}(\alpha_1 + \alpha_1\alpha_2 + \ldots + \alpha_{p-1}\alpha_p) \Sigma z_t z_{t-1} + \ldots + \frac{2}{N} \alpha_p \sum_t z_t z_{t-p} +$$

$$\ldots + 0\left(\frac{1}{N}\right).$$

We can then capture the d_{11}, d_{22} and d_{12} by suitable adaptation. Notice the important fact that the behaviour only depends except for a $0\left(\frac{1}{N}\right)$ term on the neighbourhood structure up to lag p of the $\{X_t\}$ sequence.

Now let $\{X_t\}$ be stable in the sense that all of the following limits exist:

$$\pi_1(0) = \lim\left\{\frac{1}{N} \Sigma X_t\right\} ,$$

$$\pi_2(0) = \lim\left\{\frac{1}{N} \Sigma (1-X_t)\right\} ,$$

$$\pi_{11}(r) = \lim\left\{\frac{1}{N} \Sigma X_t X_{t-r}\right\} ,$$

$$\pi_{12}(r) = \lim\left\{\frac{1}{N} \Sigma X_t(1-X_{t-r})\right\} ,$$

$$\pi_{21}(r) = \lim\left\{\frac{1}{N} \Sigma (1-X_t)X_{t-r}\right\} ,$$

$$\pi_{22}(r) = \lim\left(\frac{1}{N}\Sigma\,(1-X_r)(1-X_{t-r})\right) \qquad (r = 1, \ldots, p).$$

Then we can write

$$\bar{d}_{11} = (1 + \alpha_1{}^2 + \ldots + \alpha_p{}^2)\pi_1(0) + 2\sum_{r=1}^{p}\pi_{11}(r)\,c_r\,,$$

$$\bar{d}_{22} = (1 + \alpha_1{}^2 + \ldots + \alpha_p{}^2)(1 - \pi_1(0)) + 2\sum_{r=1}^{p}\pi_{22}(r)\,c_r\,,$$

$$\bar{d}_{12} = \Sigma\pi_{12}(r)c_r + \Sigma\pi_{21}(r)\,c_r$$

where

$$c_r = \alpha_1 + \alpha_1\alpha_2 + \ldots + \alpha_{p-r}\alpha_r \qquad (r = 1, \ldots, p, \text{ with } \alpha_0 = 1).$$

Basic relationships exist between the $\pi(r)$'s:

$$\pi_1(0) = 1 - \pi_2(0)\,,$$

$$\pi_{12}(r) = \pi_{21}(r),$$

$$\pi_{12}(r) + \pi_{11}(r) = \pi_1(0)\,,$$

$$\pi_{21}(r) + \pi_{22}(r) = \pi_2(0) \qquad (r = 1, \ldots, p).$$

If, in addition we have a balanced sequence then $\pi_1(0) = \pi_2(0) = \frac{1}{2}$ and $\pi_{11}(r) = \pi_{22}(r)$ $(r = 1, \ldots, p)$.

We may now return to (i), (ii) and (iii). It follows that $\bar{d}_{11} + \bar{d}_{12}$ is a constant. Moreover if the $\{X_t\}$ sequence is balanced then $\bar{d}_{11} = \bar{d}_{22}$. Thus the optimality problem is to maximize

$$\sum_{r=1}^{p}\pi_{11}(r)c_r \qquad (1)$$

with respect to the choice of $\{X_t\}$, satisfying the stability conditions.

The work of Kiefer and Wynn (1984) points out the connection between sequences satisfying the stability conditions and binary processes. We may interpret

$$c(r) = \pi_{11}(r) - (\pi_1(0))^2 \quad (r = 1, \dots p)$$

as the sample autocovariance function of $\{X_t\}$. There is a 1 - 1 correspondence between such sequences and realizations (with probability one) of stationary binary sequences. Consider the balanced case, so that $\pi_1(0) = \frac{1}{2}$. The set of $c(r)$ or $\pi_{11}(r)$ $(r = 1, \dots p)$ is a closed convex polytope whose extreme points are given by very special periodic sequences. Moreover since (1) is a <u>linear programme</u> for fixed c_r one (or more) of these sequences must be the optimum. Computational methods of Martins de Carvalho and Clark (1983) and of Karakostas and Wynn (1986) give results up to $p = 5$. Thus theoretically and to some extent computationally the design problem is solved. Kiefer and Wynn (1984) contains a full discussion of the k-treatment case.

3. SAMPLING

The most closely related sampling model is the following. Let

$$Z_t = \theta + Y_t$$

and <u>observe</u> Z_t if $X_t = 2$ and <u>do not</u> observe Z_t when $X_t = 0$. Let Γ_s be the covariance matrix of the sample. Then consider various criteria, based on the observed values.

(i) min var (θ) where $\hat{\theta}$ is the BLUE of θ.

(2) min $E(\tau - \hat{\tau})^2$ where $\tau = \sum_{t=1}^{N} Z_t$ and $\hat{\tau}$ is the BLUE of τ in the sense of prediction.

(3) min var $(\hat{\theta}_{LS})$ where $\hat{\theta}_{LS} = \frac{1}{n} \sum_{s} Z_t$ and $n \leq N$ is the sample size.

Under the conditions of Section (2) we see that (1) and (2) lead to the same criterion asymptotically: maximize the sum of the elements of Γ_s^{-1}. Criterion (3) leads to minimizing the sum of the elements of Γ_s. We can compare these criteria with that from Section (2) which with obvious notation can be reexpressed as minimizing the sum of the elements of

$$\left[\Gamma_s - \Gamma_{s,\bar{s}}\, \Gamma_{\bar{s}}^{-1} \Gamma_{\bar{s},s}\right]^{-1} \tag{1}$$

where $\bar{s}$ is the complement of the sample.

A final class of criteria is obtained in the pure prediction problem when $\theta = 0$. In this case we may be interested again in $E(\tau - \hat{\tau})^2$ which is the sum of the elements of

$$\Gamma_{\bar{s}/s} = \Gamma_{\bar{s}} - \Gamma_{\bar{s},s}\, \Gamma_s^{-1}\, \Gamma_{s,\bar{s}}\ , \tag{2}$$

the conditional covariance matrix of the unsampled Z_t given the sample Z_t. This quantity is asymptotically independent of the choice of $\{X_t\}$ (for fixed n). A criterion under investigation is to minimize $\det(\Gamma_{\bar{s}|s})$ which is equivalent to maximizing $\det(\Gamma_s)$. We have referred elsewhere to this criterion as "maximum entropy sampling" and there are analogies with statistical mechanics. It is clear from consideration of the asymptotic form of Γ^{-1} that this criterion again only depends on the structure of the $\{X_t\}$ process up to lag p.

REFERENCES

Kiefer, J. (1960). "Optimum experimental designs V. with applications to systematic and rotatable designs". Proc. Fourth Berk. Symp., 2, 381-405.

Kiefer, J. and Wynn, H.P. (1983). "Autocorrelation - robust design of experiments". Scientific Inf. Data Analysis and Robustness. Academic Press, New York.

Kiefer, J. and Wynn, H.P. (1984). "Optimum and minimax exact treatment designs for one-dimensional autoregressive error processes". Ann. Statist. 12, 431-450.

Karakostas, K. and Wynn, H.P. (1976). "Optimum systematic sampling for autocorrelated superpopulations". J. Stat. Plan. Inf. submitted.

Martins de Carvallo, J.L. and Clark, J. M.-C. (1983). "Characterising the autocorrelation of binary sequences". IEEE. Trans. Inf. Th. 24, 502-508.

THE DESIGN OF EXPERIMENTS FOR MODEL SELECTION

A.M. Herzberg
Department of Mathematics, Imperial College of Science & Technology
London, U.K.

A.V. Tsukanov
Sevastopol Instrument Making Institute, Sevastopol, U.S.S.R.

1. INTRODUCTION

The problem of the selection of the optimal model has a large literature. For example, Mallows (1973), Akaike (1974), Woodruffe (1982), Shibata (1980, 1981), Allen (1974) and Vapnik (1982) were concerned with techniques for selecting an appropriate model to fit a given set of data; Andrews (1971) and Atkinson and Cox (1974) were concerned with the design of experiments for model selection.

Herzberg and Tsukanov (1985a) discussed the design of experiments for linear model selection with the jackknife criterion. In other papers, Herzberg and Tsukanov (1985b, 1986) gave a Monte-Carlo comparison of the C_p criterion with that of the jackknife under different measures and considered modifications of the usual jackknife procedure to include the possibility of the removal of different numbers of observations at a time, and selected observations.

In this paper, further considerations for the optimal design in the selection of models will be presented.

2. THE CRITERION FOR THE SELECTION OF MODELS

Let the true functional relationships be represented by

$$y_i = \eta(\mathbf{x}_i) + \varepsilon_i \qquad (i=1,\dots,N) \quad , \qquad (1)$$

where y_i is the ith observation of the dependent variable at the k-dimensional design point, $\mathbf{x}_i$, the independent and controlled variable, $\eta(.)$ is the true but unknown function, model, and ε_i is an independent random variable with mean 0 and constant variance σ^2. For ease in presentation and without loss of generality, suppose that the problem is to choose one of two models

$$\eta_j(\mathbf{x},\boldsymbol{\alpha}_j) \qquad (j=1,2) \quad , \tag{2}$$

where $\boldsymbol{\alpha}_j$ is a vector of unknown parameters to be determined by least squares.

Consider as a measure for the adequacy of the jth model, the jack-knife criterion

$$TJK_j = \frac{1}{N}\sum_{i=1}^{N}[y_i-\eta_j\{\mathbf{x}_i,\hat{\boldsymbol{\alpha}}(-i)\}]^2 \qquad (j=1,2) \quad , \tag{3}$$

where $\hat{\boldsymbol{\alpha}}_j(-i)$ is the least squares estimator of $\boldsymbol{\alpha}_j$ determined from the N-1 points consisting of all the design points except $\mathbf{x}_i$. The model $\eta_1(.)$ will be preferred to $\eta_2(.)$ if $TJK_1 < TJK_2$. This and other related criteria and measures for the discrimination among two or more models are given and elaborated on in the papers by Herzberg and Tsukanov.

Mallows (1973) suggested choosing the model for which

$$C_p = C_j/\hat{\sigma}^2 - N \tag{4}$$

is a minimum, where

$$C_j = RSS_j + \ell p_j\hat{\sigma}^2$$

and $\hat{\sigma}^2$ is an estimate of σ^2. Further, RSS_j is the residual sum of squares and p_j is the number of unknown parameters for the jth model. The constant ℓ may be changed; Mallows set $\ell = 2$.

In particular, Herzberg and Tsukanov (1986) considered modifications of the usual jackknife procedure to include the possibility of the removal of different numbers of observations at a time and selected observations.

3. THE DESIGN OF EXPERIMENTS FOR MODEL SELECTION

In order to determine the optimal design for model selection, the following method is used:

(i) a function r_{ij} specifying the goodness of decisions is determined, where r_{ij} is the price of the selection of the model from set S_i when the true model is in set S_j;

(ii) the function of average risk is obtained, i.e. $R=\sum_{i,j} r_{ij}p_{ij}$, where p_{ij} is the probability of the selection of the model from the set S_i when the true model is in the set S_j;

(iii) the p_{ij} are varied according to the criteria and the design used.

The value of R depends on the vector of unknown parameters, $\boldsymbol{\alpha}$. In this case, a minimax or Bayesian approach may be used.

It is always possible to transform the response function in such a way that

$$a \leqslant \eta(\mathbf{x}) \leqslant b \quad \text{and} \quad c \leqslant \mathbf{x}_i \leqslant d \qquad (i=1,\ldots,k) \ . \tag{5}$$

Consequently, a vector of unknown parameters $\boldsymbol{\alpha}$ is restricted to a finite field Ω. The function R can be investigated further for Ω and fixed variance of the error, σ^2.

4. A MONTE-CARLO EXAMPLE

Consider one-dimensional polynomial models. In order to compare the criteria and the design, it is necessary to choose a set of test models. The restrictions of (5) are used with $a = c = -1$ and $b = d = 1$. It is possible to use a Tchebycheff system of orthogonal polynomials as a network of models which apprimates the behaviour of risk in the region Ω. One set of such polynomials is

$$\eta_1 = x, \quad \eta_2 = -1+2x^2, \quad \eta_3 = 3x-4x^2, \quad \eta_4 = 1-8x^2+8x^4,$$
$$\eta_5 = -5x+20x^3-16x^5 \ . \tag{6}$$

Consider the following two designs with 12 points:

X_1: 12 equally spaced points ±1, ±0.82, ±0.64, ±0.45, ±0.27, ±0.09;

X_2: the D-optimal design for a polynomial of degree five, i.e. 12 points, two replicates at each of ±1, ±0.77, ±0.29.

Table 1 gives the result of a computer simulation for two designs. Observations were generated from the models given in (6) with normally distributed errors with zero mean and variance $\sigma^2 = 1$. The table gives the number of correct decisions out of 500 simulations. The computing was done on the computer complex CDC Cyber 174 and CDC 6500 of the Imperial College of Science and Technology, London.

Table 1: Frequency of correct decisions for C_p and *TJK* criteria for X_1 and X_2 and $\sigma^2=1$

		Design of true model				
Design	Criterion	1	2	3	4	5
X_1	C_p	325	282	267	271	334
	TJK	330	295	248	180	113
X_2	C_p	328	300	259	286	395
	TJK	352	311	255	270	330

If the matrix of the r_{ij}'s is known, then the function R can be approximated and the choice of the design and criterion made together.

REFERENCES

Akaike, H. (1974). A new look at statistical model identification. IEEE Trans. Automatic Control, 19, 716-723.

Allen, D.M. (1974). The relationship between variable selection and data augmentation and a method for prediction. Technometrics, 16, 125-127.

Andrews, D.F. (1971). Sequentially designed experiments for screening out bad models with F-tests. Biometrika, 58, 427-432.

Atkinson, A.C. and Cox, D.R. (1974). Planning experiments for discriminating between models (with discussion). J.R. Statist. Soc. B36, 321-348.

Herzberg, A.M. and Tsukanov, A.V. (1985a). The design of experiments for linear model selection with the jackknife criterion. Utilitas Mathematica, 28, 243-253.

Herzberg, A.M. and Tsukanov, A.V. (1985b). The Monte Carlo comparison of two criteria for the selection of models. J. Statist. Comput. Simul., 22, 113-126.

Herzberg, A.M. and Tsukanov, A.V. (1986). A note on modifications of the jackknife criterion for model selection. Utilitas Mathematica, 29, 209-216.

Mallows, C.L. (1973). Some comments on C_p. Technometrics, 15, 661-675.

Shibata, R. (1980). Asymptotically efficient selection of the order of the model for estimating parameters of a linear process. Ann. Statist. 8, 147-164.

Shibata, R. (1981). An optimal selection of regression variables. Biometrika, 68, 45-54.

Vapnik, V. (1982). Translated by S. Kotz. *Estimation of Dependences Based on Empirical Data*. Springer-Verlag, New York.

Woodruffe, M. (1982). On model selection and the arc sine laws. Ann. Statist. 10, 1182-1194.

THE DESIGN AND ANALYSIS OF FIELD TRIALS IN THE PRESENCE OF FERTILITY EFFECTS

C. Jennison
School of Mathematical Sciences
University of Bath
BATH BA2 7AY
United Kingdom

The recent interest in analyses of field trials incorporating adjustments for variations in fertility or other systematic effects can be traced back to the work of Papadakis (1937) who demonstrated how conventional treatment estimates can be improved by performing a second analysis using the average of the residuals of its neighbours as a covariate for each plot. Over thirty years later Atkinson (1969) investigated this unconventional use of a function of the response variables as a covariate and showed the resulting treatment estimates to be close to those obtained by fitting the first-order autoregressive model of Williams (1952).

The use of spatial models for field experiments has since developed in its own right with major contributions from the work of Besag (1974,1977) and Bartlett (1978). In general, spatial models define a covariance structure for the observations, possibly involving variance ratios to be estimated from the data, and both treatment estimates and estimates of standard error are obtained by the usual methods for general linear models. A convincing model for one-dimensional layouts has recently been developed by Besag and Kempton (1986) and consists of a fertility process with independent first differences plus superimposed independent error for each plot; Williams (1986) proposes a similar model based on the relationship between correlation and inter-plot distance determined by Patterson and Hunter (1983) for a set of 166 cereal variety trials.

The approach of Wilkinson, Eckert, Hancock and Mayo (1983) is more in the spirit of Papadakis although "adjustment" is by the yields rather than residuals of neighbouring plots. These authors propose a *smooth trend plus independent error* model

$$Y=D\tau+\xi+\eta,$$

where Y is the vector of yields, D the design matrix, τ the vector of treatment effects, ξ represents a trend term which varies smoothly within columns, and η denotes independent errors. Let plots be indexed along columns and suppose ξ is locally approximately linear within each column so $\xi_i - \frac{1}{2}(\xi_{i-1}+\xi_{i+1})=0$; estimating equations are formed from adjusted yields, $Y_i' = Y_i - \frac{1}{2}(Y_{i-1}+Y_{i+1})$ thereby removing almost completely the effect of trend, ξ.

In the "least squares smoothing" method of Green, Jennison and Seheult (1985) this same model is fitted by the penalty function approach well known in non-parametric regression. Values of τ, ξ and η are found by minimizing the penalty function

$$\lambda(\Delta\xi)^T\Delta\xi + \eta^T\eta,$$

where $\Delta\xi$ is the vector of second differences $\xi_{i-1}-2\xi_i+\xi_{i+1}$ and λ a tuning constant controlling the degree of smoothness of the fitted ξ. An appropriate value of λ must be chosen either by inspection of the fitted ξ and η or by an automatic method such as cross-validation - see Green (1985). A full decomposition of Y is obtained and the fitted trend and residuals η_i can be inspected for features of interest.

Note that minimizing the above penalty function is equivalent to solving the pair of simultaneous equations

$$\tau=(D^TD)^{-1}D^T(Y-\xi)$$

$$\xi=S(Y-D\tau)$$

where $S=(I+\lambda\Delta^T\Delta)^{-1}$. Thus, for given τ, ξ is obtained by applying the smoothing matrix S to $Y-D\tau$, and for given ξ, τ is the ordinary least squares estimate based on adjusted yields $Y-\xi$. The form of these equations suggests extensions in which τ is estimated robustly from $Y-\xi$ and ξ is obtained by applying a robust, non-linear smoother to $Y-D\tau$, a solution being found by iterating between the two equations. In a recent paper Papadakis (1984) describes modifications to his original method to deal with single abnormal observations and apparent discontinuities in fertility attributable to, say, changes in soil type or drainage pattern; both these problems can be handled by the simultaneous equation approach using a treatment estimate which downweights extreme values and a smoother that recognises jumps in fertility and does not smooth across them.

The use of blocks in the design and analysis of field trials deserves some comment. Patterson and Hunter (1983) discuss incomplete block designs for cereal variety trials and demonstrate the substantial reduction in variance of treatment estimates from complete block designs (typically 35% for large blocks) but they show that the further improvements obtained by fitting a full spatial model are rather small (5 or 10%). The blocks in these designs are physically contiguous and divisions between blocks have no direct physical meaning, rather the fitting of block effects allows a step function approximation to a smooth trend ξ. The inclusion of such artificial blocks is unnecessary in other methods of analysis (although they do appear in the model of Williams (1986), ostensibly as a means of curtailing long range correlations). Blocking by real physical criteria is of course desirable and a method which can detect from the data where blocks should be introduced is most useful; there is greater scope for detection of "regions" when the experimental layout is two-dimensional and in this case there are interesting parallels with the identification of objects and distinct areas in image analysis.

As the above discussion illustrates, recent research has led to a variety of analyses and the experimenter may be faced with a bewildering choice. Fortunately, different methods usually give very similar estimates of treatment effects - most give the least squares estimate for some assumed correlation structure of the observations and changes in this assumed structure tend to affect the estimates only slightly. Rather than find fault with particular methods we should recognise the potential of a selection of tools for data analysis: model based methods which provide both treatment estimates and estimates of standard errors, as long as we accept the assumptions of the model, and more exploratory methods which allow a full investigation of the data and have greater flexibility to adapt to features of the data as they are discovered.

I would now like to turn to the problem of design. Firstly, it should be pointed out that special designs are in no way essential for a "neighbour" analysis, in fact, these methods can be used to retrieve a satisfactory analysis from a poorly designed experiment; for example, fitting an appropriate spatial model can remove the bias that would otherwise be introduced by an improperly randomized or even a completely systematic design. Good design will of course improve efficiency and several recent papers discuss optimal designs for correlated observations, see for example Gill and Shukla (1985) and references therein. The general conclusion is that designs should be balanced, i.e. treatments should be neighbours and possibly also second neighbours of each other an equal number of times but no treatment should appear next to itself.

One aspect of the theory of optimal design that may need to adapt to new methods of analysis is the role of blocks. As mentioned previously, artificial blocking is no longer necessary for analysis and experiments with a small number of very large blocks may become more common: a typical variety trial can consist of three replicates of 50 varieties so if the replicates were physically separate we would have just three blocks of size 50.

To assess the importance of optimal design I performed calculations for an example with four replicates of 20 varieties, comparing the average standard error of treatment differences from a second order balanced design and a design with treatments allocated randomly within each replicate. Using a variety of autoregressive and moving average processes for the true model and both correct and slightly incorrect models in the analysis I found the balanced design to be always superior but often only marginally and at most by 1 or 2%. Other factors may be of greater practical importance. When correlations are high it is noticeable that the variance of treatment estimates for treatments appearing on end plots is considerably higher than average. The suggestion by Wilkinson et al

(1983) of adding extra plots at the end of each column in order to give an "adjusted" yield for each internal plot has led to some confusion - clearly these plots are in many ways no different from other plots and they must certainly be counted when discussing efficiency - but such additional plots could be used to ensure that no single treatment estimate is too variable. Alternatively, one or more treatments may be given an extra replicate in return for appearing on several end plots, thereby equalising as nearly as possible the variances of treatment estimates.

To conclude, there is presently a great deal of practical and theoretical interest in the analysis and design of field trials and we are seeing an influx of ideas from many different areas of statistics. Future work offers an exciting prospect as ideas not traditionally associated with field trials are developed and the areas of application are extended to the whole range of agricultural experiments.

ACKNOWLEDGEMENTS

My own work in this area has been in collaboration with Peter Green and Allan Seheult. I am particularly grateful to Julian Besag for stimulating my interest in this topic.

REFERENCES

Atkinson, A.C. (1969) The use of residuals as a concomitant variable. *Biometrika,* **56**, 33-41.

Bartlett, M.S. (1978) Nearest neighbour models in the analysis of field experiments (with Discussion). *J.R.Statist.Soc.*,B, **40**, 147-174.

Besag, J.E. (1974) Spatial interaction and the statistical analysis of lattice systems (with Discussion). *J.R.Statist.Soc.*,B, **36**, 192-236.

Besag, J.E. (1977) Errors-in-variables estimation for Gaussian lattice schemes. *J.R.Statist.Soc.*,B **39**, 73-78.

Besag, J.E. and Kempton, R.A. (1986) Statistical analysis of field experiments using neighbouring plots. *Biometrics,* **42**, 231-251.

Gill, P.S. and Shukla, G.K. (1985) Efficiency of nearest neighbour balanced block designs for correlated observations. *Biometrika,* **72**, 539-544.

Green, P.J.(1985) Linear models for field trials, smoothing and cross-validation. *Biometrika,* **72** 527-537.

Green, P.J., Jennison, C. and Seheult, A.H. (1985) Analysis of field experiments by least squares smoothing. *J.R.Statist.Soc.*,B, **47**, 299-315.

Papadakis, J.S. (1937) Méthode statistique pour des expériences sur champ. *Bull. Inst. Amél Plantes á Salonique,* **23**.

Papadakis, J.S. (1984) Advances in the analysis of field experiments. Πρακτικα τη Ακαδημιας Αθηνων, **59**, 326-342.

Patterson, H.D. and Hunter, E.A. (1983) The efficiency of incomplete block designs in National List and Recommended List cereal variety trials. *J.Agric.Sci.,* **101**, 427-433.

Wilkinson, G.N., Eckert, S.R., Hancock, T.W. and Mayo, O. (1983) Nearest neighbour (NN) analysis of field experiments (with Discussion). *J.R.Statist.Soc.*,B, **45**, 151-211.

Williams, E.R. (1986) A neighbour model for field experiments. *Biometrika,* **73**, 279-287.

Williams, R.M. (1952) Experimental designs for serially correlated observations. *Biometrika,* **39** 151-167.

ON THE EXISTENCE OF MULTIFACTOR DESIGNS WITH GIVEN MARGINALS

Krafft, Olaf
Institut für Statistik der RWTH Aachen
Federal Republic of Germany

Consider the following row-and column-design with p=3 treatment levels (index k), m=3 row-factor levels (index i) and n=3 column-factor levels (index j)

i \ j	1	2	3
1	3	1	1
2	3	3	2
3	2	1	2

When performing an analysis of variance for such a design, it turns out that only the row-frequencies r_{ik} and column-frequences c_{jk} of the treatment levels are relevant:

$$R_1 = (r_{ik}) = \begin{pmatrix} 2 & 0 & 1 \\ 0 & 1 & 2 \\ 1 & 2 & 0 \end{pmatrix}, \quad C_1 = (c_{jk}) = \begin{pmatrix} 0 & 1 & 2 \\ 2 & 0 & 1 \\ 1 & 2 & 0 \end{pmatrix}.$$

E.g. for investigations on design-optimality one is thus inclined to take these matrices as a basis. But changing R_1 and C_1 only slightly into

$$R_2 = \begin{pmatrix} 1 & 2 & 0 \\ 0 & 1 & 2 \\ 2 & 0 & 1 \end{pmatrix}, \quad C_2 = \begin{pmatrix} 1 & 0 & 2 \\ 2 & 1 & 0 \\ 0 & 2 & 1 \end{pmatrix},$$

one easily sees that a design corresponding to R_2 and C_2 does not exist.

Hence we have the problem: Given $R = (r_{ik}) \in \mathbb{N}_0^{m \times p}$, $C = (c_{jk}) \in \mathbb{N}_0^{n \times p}$. Which conditions on R and C guarantee the existence of a design corresponding to R and C ?

Using indicators x_{ijk} ($x_{ijk} = 1$ iff treatment level k is combined with row-and column levels i,j), this problem can formally be stated as of finding conditions for consistency of the system

$$
(\mathrm{I}) \qquad
\begin{aligned}
&\sum_{j=1}^{n} x_{ijk} = r_{ik} \;, \; (i,k) \in M \times P \;, \\
&\sum_{i=1}^{m} x_{ijk} = c_{jk} \;, \; (j,k) \in N \times P \;, \\
&\sum_{k=1}^{p} x_{ijk} = 1 \;, \; (i,j) \in M \times N \;, \\
&\qquad x_{ijk} \in \{0,1\}, (i,j,k) \in M \times N \times P \;,
\end{aligned}
$$

where $M = \{1,\ldots,m\}$, $N = \{1,\ldots,n\}$, $P = \{1,\ldots,p\}$.

In case p=2 the solution is known as Gale-Ryser theorem: Let $r_j^* = |\{i : r_{i1} \geqq j\}|$, c_j^* obtained from c_{j1} by arrangement in non-ascending order. Then (I) is consistent iff the vector c^* is majorized by the vector r^* in the Schur-ordering.

A generalization to the case $p \geqq 3$ is unknown: Neither an adequate generalization of the majorization ordering to matrices R and C has been found nor could the known proofs of the Gale-Ryser theorem (algorithm for explicit construction (Gale-Ryser), application of Hall's theorem on systems of distinct representatives (Higgins)) be carried over.

We have the following conjecture: (I) is consistent iff

$$
(1) \qquad
\begin{aligned}
&\sum_{k=1}^{p} r_{ik} = n, \; 1 \leqq i \leqq m \;, \; \sum_{k=1}^{p} c_{jk} = m, \; 1 \leqq j \leqq n \;, \text{ and} \\
&\sum_{E} r_{ik} + \sum_{F} c_{jk} \geqq \alpha(E,F) + \beta(E,F) \;\forall\, E \subset M \times P, \;\forall\, F \subset N \times P \;,
\end{aligned}
$$

where $\alpha(E,F) = |\{(i,j) \in M \times N : \forall\, k \in P \text{ are } (i,k) \in E \wedge (j,k) \in F\}|$
and $\beta(E,F) = |\{(i,j) \in M \times N : \forall\, k \in P \text{ is } (i,k) \in E \vee (j,k) \in F\}|$.

The conjecture is based on the following arguments:
Necessity of (1) is easily to prove. In case p=2 (1) is equivalent to the Gale-Ryser criterion. In case m=n=p=3 it has been proved by checking all possible cases. (E.g. in the example at the beginning (1) is violated for $E = \{(1,3),(3,2),(3,3)\}$, $F = \{(1,1),(1,2),(3,1)\}$.

Our idea to prove the sufficiency part is to treat a generalized version of (I): Consider x_{ijk} as the values of a mapping $f : M \times N \times P \to \{0,1\}$.

Let $D \subset M \times N$ and restrict all conditions in (I) to the set D to obtain a more general system (I'). A proof by induction on $|D|$ seems possible.

As a preliminary result we have the following generalization of the Gale-Ryser theorem for p=2:

Let $D \subset M \times N$, $B_i = \{j \in N : (i,j) \in D\}$, $A_j = \{i \in M : (i,j) \in D\}$, $r_i, c_j \in \mathbb{N}_0$.

Then the system

$$\sum_{j \in B_i} x_{ij} = r_i \ , \ i \in M \ ,$$

$$\sum_{i \in A_j} x_{ij} = c_j \ , \ j \in N \ ,$$

$$x_{ij} \in \{0,1\}, \ (i,j) \in D \ ,$$

is consistent if and only if $\sum_{i \in M} r_i = \sum_{j \in N} c_j$ and

$$\sum_{i \in S} (|B_i| - r_i) + \sum_{j \in T} c_j \geq \sum_{i \in S} |B_i T| \ \forall S \subset M \ , \ \forall T \subset N \ .$$

The proof runs by induction on $|D|$.

Work has been done with coauthor M. Schaefer. We thank N. Gaffke for helpful discussions.

References

Higgins, P.J. (1959). Disjoint transversals of subsets. Canad. J. Math., 11, 280-285.

Jurkat, W.B., and Ryser, H.J. (1968). Extremal configurations and decomposition theorems, Journ. Alg., 8, 194 - 222.

Krafft, O. (1978). Lineare statistische Modelle und optimale Versuchspläne, Göttingen: Vandenhoeck u. Ruprecht, 185 - .

Marshall, A.W. and Olkin, I. (1979). Inequalities: Theory of Majorization and Its Applications. New York: Academic, chap. 7 C.

Mirsky, L. (1971). Transversal Theory. New York: Academic, chap. 5.

Pukelsheim, F. and Titterington, D.M. (1985). On the construction of multifactor designs from given marginals. Preprint Nr. 75, Inst. f. Mathematik, Univ. Augsburg.

LOCAL ASYMPTOTIC NORMALITY IN GAUSSIAN MODEL OF VARIANCE COMPONENTS

Maljutov M.B.
Moscow Lomonosov University, Moscow, USSR

Let us consider a classical mixed ANOVA model

$$y = X\beta + \sum_{i=1}^{k} s_i U_i \phi_i . \tag{1}$$

Here $y = y^{(n)}$ is $(n \times 1)$-vector of measurements, $X = X^{(n)}$, $U_i = U_i^{(n)}$ are respectively $(n \times p)$ and $(n \times n_i)$-matrices of known parameters, ϕ_i is $(n_i \times 1)$ -normally distributed random vector $\phi_i \sim N(0, I_{n_i}), i = 1, \ldots, k$; I_n is an identity mattix, $\beta \in \mathbb{R}^p$ is an unknown parameter.

It is clear that

$$Ey = X\beta ,$$

$$V = Cov\, y = \sum_{i=1}^{k} s_i^2 G_i \quad , \quad G_i = U_i U_i^T ,$$

thus $d_i = s_i^2$ are called variance components.

The pioneer work where ANOVA methods were applied to testing hypotheses about variance components for a balanced model was Fisher (1918), later R.Fisher devoted some attention to those models in his famous book (1925). Important contributions to this theory were made later by F.Yates, A.Wald, C.Eisenhart, C.Sheffé, S.R.Searle, C.R.Rao, T.W.Anderson among many others.

Modern works on this subject may be classified into two main streams. In the first one (see Rao and Kleffe (1980)) invariant unbiased quadratic estimates (MINQUE and others) were examined. The second stream deals with maximum likelihood estimates (MLE) for $\theta = (d_1, \ldots, d_k, \beta^T)$ for the distribution $P_\theta^n(\cdot)$ of y in model (1). Asymptotics for MLE-s under conditions including $\sum_{i=1}^{k} tr\, G_i^{(n)} \le K < \infty$ which can be fulfilled only for hierarchical models (1) was heuristically discussed in Hartley and Rao (1967).

Asymptotic normality (AN) of MLE-s under more general conditions was proved in Miller (1977). There are many iterative methods of computing MLE (Goldstein (1986), Luanchi (1983), Maljutov (1983) and others. The one which was proposed in the latter references is suitable for defining each iterative step $\overset{s}{\theta}$ as an estimate not only in the Gaussian case but under rather general dependence of moments of y up to the 4-th degree on those of degree 1 and 2. The convergence in probability when $s \to \infty$ is proved, so is AN of any iteration provided a preliminary guess $\overset{o}{\theta}$ is consistent.

The asymptotic efficiency of MLE's does not follow from Cramer's classical theory because the measurements in (1) are essentially dependent. The first aim of the present report is to announce that local asymptotic minimaxity of MLE or the first iteration $\overset{1}{\theta}$ of Luanchi (1983), Maljutov (1983) (the modified Fisher score statistic) follows in a standard way (Ibragimov and Khasminsky (1981)) from the following decomposition of $L_n(u) = \ln[dP^{(n)}_{\theta + u\nu(n)} / dP^{(n)}_{\theta}]$, $u \in \mathbb{R}^{p+k}$ where $\nu(n) = (n_1, \ldots, n_k, \nu_{k+1}(n), \ldots, \nu_{k+p}(n))$, $\nu_i(n) \equiv \nu_0(n), i > k$, is such that for $1 \le i \le j \le k$

$$C_{ij} = \lim_{n \to \infty} (1/2) \operatorname{tr}(V^{-1} G_i V^{-1} G_j) n_i n_j, \quad B = \lim_{n \to \infty} \nu_0(n) \bar{X}^T \bar{V}^{-1} X \tag{2}$$

exist and some other conditions similar to those of Miller (1977) are fulfilled.

$$L_n(u) = u^T \lambda - (1/2) u^T \mathcal{I} u + \psi_n(u), \tag{3}$$

where $\lambda \sim N(0, \mathcal{I})$, $\mathcal{I} = \begin{pmatrix} C & O \\ O & B \end{pmatrix}$, $d_i > 0$, $i = 1, \ldots, k$ and for all $K > 0$

$$\sup_{\|u\| < K} |\psi_n(u)| \underset{n \to \infty}{\longrightarrow} 0 \tag{4}$$

$$\lim_{n \to \infty} P^{(n)}_{\theta} \left(\sup_{\|u\| > K} L_n(u) \ge 0 \right) = 0 \tag{5}$$

The condition (4) (respectively, (5)) is suitable for deriving AN of $\overset{1}{\theta}$ (of MLE).

The method of deriving (3) is standard. The first term of (3) is a principal part of the first term of Taylor's expansion, the second term is a principal part of the second degree term. The i -th component of λ, $i \le k$, is the limit of the normalized and centered expression $\| U_i^T V^{-1}(y - X\beta)\|^2$. Components λ_i and λ_j, $i \le k$, $j > k$, are uncorrelated because of symmetry of gaussian distribution. Uniform convergence of the residual $\phi_n(u) \to 0$ can be proved by the methods of Maljutov (1983).

Our second aim is to investigate designs optimizing some function of $\mathcal{J}$ in (3) which is simultaneously the normalized covariance matrix of MLE or of $\overset{1}{\theta}$. We begin with the simplest one-way mixed model

$$y_{ij} = \beta_0 + \beta_1 x_{ij} + \varphi_{(i)} + e_{ij}, \; i = 1,\dots,I, \; j = 1,\dots,m_i; \; n = \sum_{i=1}^{I} m_i .$$

Here $\beta_1 \in \mathbb{R}^1$, $x_{ij} \in \mathbb{R}^1$, $\beta_0 \in \mathbb{R}^1$ is a common mean, random vectors $\varphi = (\varphi_{(1)},\dots,\varphi_{(i)}) \sim N(0, a\mathbb{I}_{m_i})$ and $e = (e_{11},\dots,e_{1m_1}, e_{21},\dots,e_{Im_I}) \sim N(0, b\mathbb{I}_n)$ are independent. In matrix notations (1)

$$X = (\mathbb{1}_n \vdots x), \; UU^T = G = diag(G_1,\dots,G_I), \; G_i = \mathbb{1}_{m_i}\mathbb{1}_{m_i}^T, \; \mathbb{1}_m = \begin{pmatrix} 1 \\ \vdots \\ 1 \end{pmatrix}.$$

We consider the asymptotics of MLE-s when

$$I \to \infty, \; \min_{1 \le i \le I} m_i \to \infty, \; 0 < d \le \inf \frac{m_i}{m_j} \le \sup \frac{m_i}{m_j} \le D < \infty. \quad (6)$$

Using the obvious equalities

$$G_i^2 = m_i G_i, \; V^{-1} = b^{-1}\mathbb{I}_n - diag\left(\frac{a}{b(b + am_i)} G_i, i = 1,\dots,I\right),$$

$$V^{-2} = b^{-2}\mathbb{I}_n + diag\left[\left(-2\frac{a}{b(b+am_i)} + \frac{a^2 m_i}{b^2(b+am_i)^2}\right) G_i, i = 1,\dots,I\right),$$

. . . . . .

we obtain limiting equations

$$x^T V^{-1} x = b^{-1} \sum_{i=1}^{I} S_i^2 m_i + \sum_{i=1}^{I} a^{-1}(\bar{x}_i)^2 + o(1), \; \bar{x}_i = m_i^{-1} \sum_{j=1}^{m_i} x_{ij};$$

$$x^T V^{-1} \mathbb{1}_n = \sum_{i=1}^{I} \bar{x}_i (a^{-1} + o(1)), \; S_i^2 = m_i^{-1} \sum_{j=1}^{m_i} x_{ij}^2 - (\bar{x}_i)^2,$$

$$\mathbb{1}_n^T V^{-1} \mathbb{1}_n = I(a^{-1} + o(1)), \; tr V^{-2} = n(b^{-2} + o(1)),$$

$$tr V^{-2} G = I(2(ab)^{-1} + o(1)),$$

$$tr (V^{-1}G)^2 = I(a^{-2} + o(1)).$$

Thus when $x^T V^{-1} x = a^{-1} \sum_{i=1}^{I} (\bar{x}_i)^2$ (e.g. when $x_{ij} \equiv x_i$) we get that limiting distribution of the vector

$$([\sqrt{I}(\hat{\beta}_0-\beta_0),(\hat{\beta}_1-\beta_1),(\hat{a}-a)],\sqrt{n}(\hat{b}-b))$$

has the covariance matrix

$$J^{-1}=diag\left[a\begin{pmatrix}1 & \bar{x}\\ \bar{x} & \overline{x^2}\end{pmatrix},2a^2,2b^2\right],\ \bar{x}=\frac{\sum x_i}{I},\ \overline{x^2}=\frac{\sum x_i^2}{I}$$

which does not depend on mutual relationship between

$m_1,\dots,m_I\longrightarrow\infty$.

Designs optimizing a convex differentiable function $\Phi(J^{-1})$ by choosing $x_1,\dots,x_I$ may easily be found by standard methods known for homoscedastic independent measurements.

If $S^2\underset{n\to\infty}{\longrightarrow}D>0$, then the covariance matrix of the limiting distribution of the vector $(\sqrt{I}(\hat{\beta}_0-\beta_0),\sqrt{n}(\hat{\beta}_1-\beta_1),\sqrt{I}(\hat{a}-a),\sqrt{n}(\hat{b}-b))$ is $diag(a,D^{-1}b,2a^2,2b^2)$. Generalization to the general case of the mean $X\beta$ wher $X=\begin{pmatrix}f_1(x_1)\dots f_p(x_1)\\ \dots\dots\\ f_1(x_n)\dots f_p(x_n)\end{pmatrix}, f_i:X\to\mathbb{R}^1,$ is obvious. A survey of non-asymptotic designs for estimating multi-way variance components is in Anderson (1975).

REFERENCES

Anderson, R.L. (1975). Designs and Estimators for Variance Components. In: A Survey of Statistical Desig and Linear Models, Srivastava, J.N. (Ed.). N.Holland Publ. Co, pp. 1-29.

Fisher, R.A. (1918). The correlation between relatives.. Trans. R.Soc. Edinburgh, v. 52, 399-433.

Fisher, R.A. (1925). Statistical methods for research workers. Oliver & Boyd, London.

Goldstein, H. (1986). Multilevel mixed linear models analysis using iterative generalized least squares. Biometrika, v. 73, N 1, 43-56.

Hartley, H.O. and Rao, J.N.K. (1967). MLE for the mixed ANOVA model. Biometrika, v. 54, 93-108.

Ibragimov, I.A. and Khasminsky, R.Z. (1981). Asymptotic theory of estimation. Springer, N.Y.

Luanchi, M. (1983). Asymptotic investigation of iterati estimates (thesis). Moscow Lomonosov University, Depart. of Math. and Mech.

Maljutov, M.B. (1983). Lower bounds for average sample size ... Izv. vusov, Matematika, N 11, 19-41.

Maljutov, M.B. and Luanchi, M. (1985). Iterative quadratic estimates of mixed ANOVA models. Abstracts of III-th conference "Application of multivariate analysis..." part II, Tartu, pp. 49-51.

Miller, J.J. (1977). Asymptotics for MLE's in the mixed ANOVA model. Ann. Statist., v. 5, 746-762.

ASYMPTOTIC METHODS IN STATISTICS
(second order asymptotics, saddle point methods, etc.)

(Session 6)

Chairman: W. van Zwet

DIFFERENTIAL GEOMETRICAL METHOD IN ASYMPTOTICS OF STATISTICAL INFERENCE

Shun-ichi AMARI
University of Tokyo, Tokyo, 113 Japan

Abstract ----- Differential geometry provides a new powerful method of asymptotics of statistical inference. Geometrical concepts are explained intuitively in the framework of a curved exponential family without technical details. We show some fundamental results of higher-order asymptotics of estimation and testing obtained by the geometrical method. Further prospects of the geometrical method are given.

Why geometry? A typical statistical problem is to make some inference on the underlying probability distribution $p(x)$ based on N independent observations $x_1, \ldots, x_N$ therefrom. In many cases, statisticians do not directly treat the function space $F = \{ p(x) \}$ of all the possible distributions, but presume a parametric statistical model $M = \{ p(x, u) \}$, where u is an n-dimensional vector parameter. Then, a model M is regarded as an n-dimensional manifold imbedded in F, and it is assumed that the true distribution $p(x)$ is included in M or at least is close to M. Roughly speaking, a naive distribution $\hat{p}(x)$ is obtained in F from the observations as, for example, the empirical distribution or its smoothed version. Then, we infer based on this $\hat{p}$ on the true distribution which is supposed to belong to M. Hence, it is important to know the geometrical shape and the relative position of M inside F.

When the number N of observations is sufficiently large, $\hat{p}$ is very close to the true distribution $p(x, u)$, so that one may use linear approximation at p of M in F in evaluating inferential procedures. Hence, linear geometry is sufficient for the first order asymptotic theory. This is the reason why one can construct a first order asymptotic theory in a unified manner. In order to evaluate higher-order, for example second and third order, characteristics, linear approximation is insufficient. It is necessary to connect these tangent spaces or linear approximations of M obtained at various points, thus taking the non-linear effects into account. To this end, one needs to introduce invariant affine connections by which the curvatures are defined. However, this is not a trivial task. We introduce two dually coupled affine connections, the exponential or $\alpha = 1$ connection and the mixture or $\alpha = -1$ connection. Then it will be shown that the related exponential and mixture curvatures play a fundamental role in the higher order asymptotic theory. The notion of a more general fibre bundle is useful for studying non-parametric or semi-parametric models.

Curved exponential family. A curved exponential family is a very tractable statistical model by the following two reasons: It has a minimal sufficient statistic x, and the enveloping exponential family is flat with respect to the $\alpha = \pm 1$ connections. This makes the related geometrical theory very simple and transparent, and we can avoid technicality of differential geometry by using this model. Therefore, we use a curved exponential family to explain the results of the geometrical method. However, it should be noted that we can construct a differential geometrical theory for a general parametric model or even a non-parametric model by using proper differential geometric notions or their extentions.

An exponential family has the following probability density functions $q(x, \theta) = \sum_{i=1}^{m} \theta^i x_i - \psi(\theta)$

with respect to a suitable measure on the sample space, where $x = (x_i)$ and $\theta = (\theta^i)$, i = 1, --- , m, are m-dimensional vectors and $\psi(\theta)$ is the cumulant generating function. The family $S = \{q(x, \theta)\}$ is an m-dimensional manifold in F, where the natural or canonical parameter θ defines a coordinate system of S. A point θ implies a distribution $q(x, \theta)$ in S specified by θ. We say that this coordinate system is $\alpha = 1$ - affine (or e-affine) by regarding

θ as an $\alpha = 1$ (or exponentially) linear coordinate system of S. This is a definition of the e-linearity. (Obviously, we need to define the e-linearity in a general statistical model by introducing an e- or 1-affine connection.) A curve $\theta = \theta(t)$ parametrized by a scalar t is e-linear, when it is linear in t. More generally, a submanifold of S which is represented by a set of linear equations in θ is e-linear. An e-curvature of a submanifold can be defined in an ordinary way, when it is not e-linear. (This curvature is a tensor, but we do avoid technical descriptions).

There is another coordinate system $\eta = (\eta_i)$ called the expectation parameter or the expectation coordinate system, which is defined by

$$\eta_i = E[x_i],$$

where E denotes the expectation with respect to $q(x, \theta)$. This η is also an important coordinate system dually coupled with θ, and there is a one-to-one relation between them,

$$\theta = \theta(\eta), \quad \eta = \eta(\theta).$$

We may use this η to specify a distribution in S. The η is said to be m-affine or $\alpha = -1$-affine, and any submanifold of S which is defined by a set of linear equations in η is said to be m- or $\alpha = -1$-flat. We have thus defined the m- or $\alpha = -1$ flatness. When a submanifold is not m-flat, the m-curvature can be defined in a similar manner.

Let $x_1, \ldots, x_N$ be N independent vector observations. Their arithmetic mean $\bar{x} = (1/N) \sum_{t=1}^{N} x_t$

is a minimal sufficient statistic. This $\bar{x}$ defines a point (distribution) in S as follows. Let $\hat{\eta}$ be the point in S whose η-coordinates are put equal to the sufficient statistic

$$\hat{\eta} = \bar{x}.$$

We call the point $\hat{\eta}$ (or more precisely a distribution specified by $\hat{\eta}$) the observed point. Its θ-coordinates $\hat{\theta} = \theta(\hat{\eta})$ is the m.l.e. of θ.

A curved exponential family $M = \{p(x, u)\}$ is a submanifold of S parametrized by an n-dimensional parameter $u = (u^a)$, $(a = 1, ---, n)$, where $n < m$, such that

$$p(x, u) = q\{x, \theta(u)\}.$$

The submanifold M is represented by

$$\theta = \theta(u) \quad \text{or} \quad \eta = \eta(u)$$

in the respective coordinate systems. The e- and m-curvatures can be calculated by differentiating $\theta(u)$ and $\eta(u)$ twice with respect to u.

Inferential procedures. Estimation of the true parameter u is stated geometrically as follows : Given an observed point $\hat{\eta} = \bar{x} \in S$ which belongs to S but does not in general lie in M, find a point $\hat{u} \in M$ or point $\eta(\hat{u}) \in M$ which is closest to **u** in some sense. Let

$$\hat{u} = e(\hat{\eta})$$

be an estimator, which is a mapping from S to M, $e : \hat{\eta} \mapsto \hat{u}$. Let A(u) be the inverse image of this mapping, i.e.,

$$A(u) = e^{-1}(u) = \{\eta \in S \mid u = e(\eta)\}.$$

Then, A(u) forms an (m - n)-dimensional submanifold, and {A(u)} is a foliation of S. We call this A(u) the ancillary submanifold or estimating submanifold attached to u by the estimator e. The value of the estimator is u, when and only when the observed point $\hat{\eta}$ is in A(u).

When an estimator e is consistent, A(u) passes through the point $\eta(u) \in M$, because $\hat{\eta}$ tends to $\eta(u)$ as $N \to \infty$. Let us introduce a coordinate system $v = (v^{\kappa})$, which is (m - n)-dimensional, in each A(u) such that (u, v) is a coordinate system of S. Then, any point $\eta \in S$ is uniquely specified by (u, v) as

$$\eta = \eta(u, v),$$

where u shows that the point η belongs to A(u) and v shows its relative position in A(u). The origin v = 0 is put at the intersection of A(u) and M, so that v = 0 if and only if the η is in M.

The sufficient statistics $\bar{x}$ or equivalently the observed point $\hat{\eta}$ is decomposed into two statistics $(\hat{u}, \hat{v})$ by solving

$$\hat{\eta} = \eta(\hat{u}, \hat{v}).$$

As can be easily seen, the statistic $\hat{u}$ is an estimator including most of Fisher information in $\bar{x}$ and $\hat{v}$ is rather ancillary (asymptotically ancillary) including little information concerning the true u. We can obtain the Edgeworth expansion of the joint distribution of $(\hat{u}, \hat{v})$ up to the third order terms by using the geometric quantities related to the curvatures of M and A(u) and their angle of intersection. This elucidates how geometric quantities are related to the performances of estimators. Problems of conditional inference and ancillarity can also be understood from this geometric point of view.

Testing hypothesis $H_0 : u \in D$ against $H_1 : u \notin D$ can be analyzed in a similar way. Since observations give an observed point $\hat{\eta}$ in S, the critical region R of a test is set in S such that the hypothesis H_0 is rejected if, and only if, $\hat{\eta} \in R$. Now let us compose an ancillary family {A(u)} such that the critical region R is composed of some of these A(u)'s,

$$R = \bigcup_{u \in R_M} A(u).$$

Then, the decomposed statistics $(\hat{u}, \hat{v})$ with respect to this A = {A(u)} has the following meaning : The test statistic λ is a function of $\hat{u}$ only, $\lambda = \lambda(\hat{u})$. The statistic $\hat{v}$ has little information, but it can be used as a conditioning statistic. The characteristics of various tests can be analyzed by using the geometric shape of the related A(u) through the Edgeworth expansion of the distribution of $(\hat{u}, \hat{v})$. The characteristics of interval estimators are closely related to those of associated tests, and they can be analyzed in a similar manner.

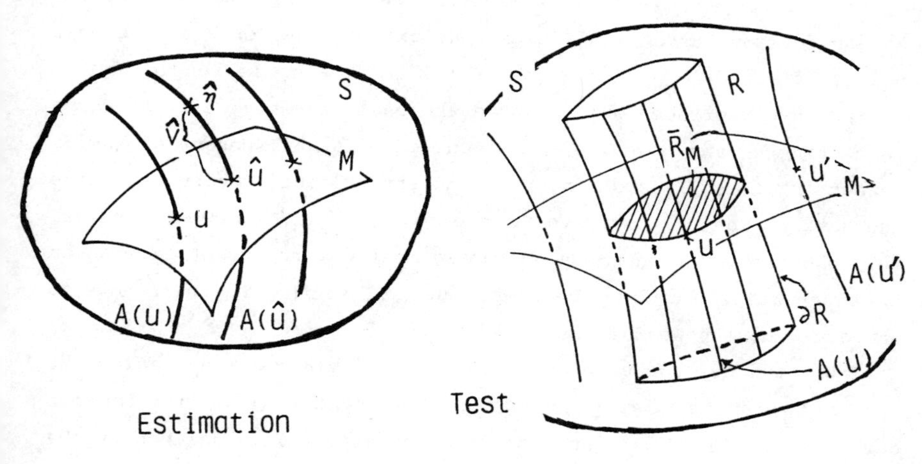

Asymptotic theory of estimation. Let $\hat{u}$ be a consistent estimator. Then, its mean square error is expanded as

$$NE[(\hat{u} - u)'(\hat{u} - u)] = A_1 + N^{-1}A_2 + O(N^{-2}).$$

Then, the first-order error matrix A_1 is given by

$$A_1 = (g - g_A)^{-1},$$

where g is the Fisher information matrix of M, g_A represents the square of the cosine of the angle between A(u) and M, and the angle is defined with respect to the Fisher information matrix of S. Therefore, as is well known, an estimator is efficient when the estimating manifolds A(u) are orthogonal to M, and $A_1 = g^{-1}$ holds. Let $\hat{u}^*$ be the one-step bias corrected version of an efficient estimator $\hat{u}$. Then, its A_2 term is decomposed into the sum of three non-negative terms as

$$A_2 = \Gamma_M^2 + (H_M^e)^2 + (H_A^m)^2.$$

Here, Γ_M^2 is the square of the mixrture connection of the coordinate system u, and $(H_M^e)^2$ is the square of the (α = 1)-curvature of M, both of which do not depend on the estimator. The third term $(H_A^m)^2$ is the square of the (α = -1)-curvature of the estimating manifold A(u), and it vanishes for the m.l.e., because the estimating manifolds of the m.l.e. are mixture-(α = -1)-flat. Hence, the m.l.e. is second-order efficient.

Asymptotic theory of tests. Before studying the power function of a test, we show a geometric result obtained from the Neyman-Pearson fundamental lemma: The critical region R of the most powerful test of $H_0 : u = u_0$ against $H_1 : u = u_1$ is bounded by an m-flat hypersurface which is orthogonal to the e-flat curve connecting u_0 and u_1. The e-flat curve forms an exponential family connecting $p(x, u_0)$ and $p(x, u_1)$, and the critical region R remains the same for any alternative $H_1 : u = u_1'$ if u_1' is on the curve. This shows the reason why a uniformly most powerful test exists for an exponential family. However, when M is curved, there are no uniformly most powerful tests.

We consider the simplest case of testing $H_0: u = u_0$ against $H_1 : u \neq u_0$ in a scalar parameter case where M = {p(x, u)} forms a curve in S. The power function $P_T(t)$ of a test T is defined by the probability of rejecting H_0 when the true distribution is at

$$u = u_0 + t/\sqrt{Ng},$$

where g is the Fisher information of M at u_0. It is expanded as

$$P_T(t) = P_{T1}(t) + P_{T2}(t)/\sqrt{N} + P_{T3}/N + O(N^{-3/2})$$

It is well known that there exist a number of first order uniformly most powerful tests in the sense that $P_{T1}(t)$ is most powerful at any t. They are, for example, the likelihood ratio test, Wald test, efficient score test, etc. Geometrically, a test is first-order efficient if and only if the boundary ∂R of R (or equivalently A(u)) is (asymptotically) orthogonal to M at their intersecting point. It is also known that a test is automatically second-order efficient in the sense that $P_{T2}(t)$ is most powerful at any t, whenever it is first-order efficient. However, there exist no third-order efficient (uniformly most powerful) tests, implying that a test can be good at a specific t_0 but not so good at other t's. Then, what are the third order characteristics of the above mentioned widely used tests? Geometry can answer this problem. The characteristics depend on the cosine of the asymptotic angle between ∂R and M, which plays a role of canceling the e-curvature (non-exponentiality) of M. We show the results.

Let us define the deficiency or third order power loss function $\Delta P_T(t)$ of an efficient test T by

$$\Delta P_T(t) = \lim_{N\to\infty} N\{P_3^*(t) - P_{T3}(t)\},$$

where $P_3^*(t)$ is the third order term of the test $T^*(t)$ which is most powerful at t (but not at other t'). Then $\Delta P_T(t)$ is obtained explicitly as

$$\Delta P_T(t) = a(t,\alpha)\,\{c - b(t,\alpha)\}^2\gamma^2,$$

where $a(t,\alpha)$ and $b(t,\alpha)$ are known functions depending on the level α, γ^2 is the square of the e-curvature of M (Efron's curvature) and c is a factor of compensating the e-curvature through the asymptotic angle between M and A(u) or ∂R.

The values of c are calculated for various tests. The results are as follows: c = 0 for the Wald test, c = 1/2 for the likelihood ratio test, c = 1 for the locally most powerful test, etc.

We show the universal deficiency curve for various tests. It should be noted that the results hold after both the level and bias of the corresponding test statistics are adjusted up to $O(N^{-1})$ like we do in the Bartlett adjustment. The adjustment procedures are given from the Edgeworth expansion of related $(\hat{u}, \hat{v})$. It is possible to extend the above results to a vector parameter case.

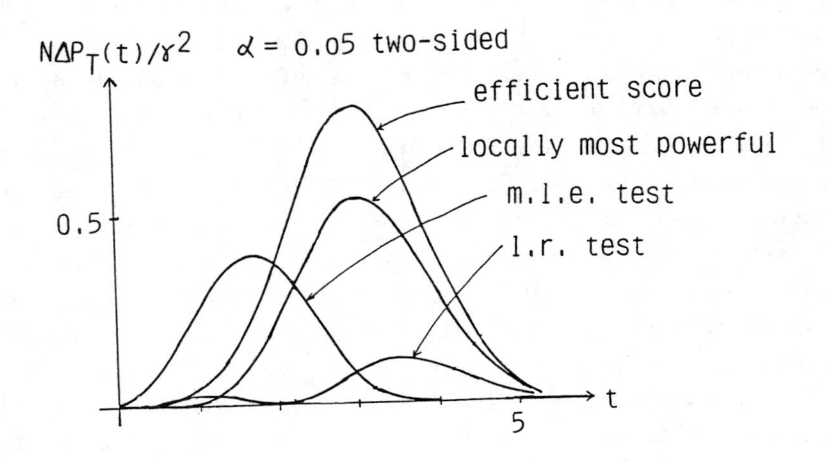

Estimation in the presence of an infinitely large number of nuisance parameters. In order to show that the geometrical method is applicable, not only to the above classical settings, but also to more modern ones such as non-parametric or semiparametric cases, we study this problem which is interesting in its own right. Let $M = \{p(x; u, \xi)\}$ be a statistical model with two parameters u and ξ. We estimate the structural parameter u from a sequence of independent observations $x_1, \ldots, x_N$, where x_i is subject to the distribution $p(x; u, \xi_i)$. Here, u is common but the incidental parameter ξ changes observation by observation and their values are unknown. Let us consider the following estimating equation

$$\sum_{i=1}^{N} y(x_i, u) = 0,$$

from which $\hat{u} = \hat{u}(x_1, \ldots, x_N)$ is calculated. We call $y(x, u)$ the estimating function. It is known that an estimator is consistent, when and only when $E[y(x, u)] = 0$ for any ξ, where E is the expectation with respect to $p(x; u, \xi)$. Let us define an asymptotic variance $AV[\tilde{\xi}]$ of a consistent estimator by

$$AV[\bar{\xi}] = \lim_{N\to\infty} NE[(\hat{u} - u)^2],$$

where $\bar{\xi} = (\xi_1, \xi_2, \ldots)$ is an infinite sequence of values of the incidental or nuisance parameter. An estimator $\hat{u}$ is said to be optimal, when its asymptotic variance $AV[\bar{\xi}]$ is not larger than that of any other estimators for any sequence $\bar{\xi}$. This definition of optimality is very strong so that there might not exist the optimal estimator. The geometrical method can solve the problem of obtaining a necessary and sufficient condition that guarantees the existence of the optimal estimator and obtaining the estimating function $y(x, u)$ of the optimal estimator when it exists.

To this end, we define a vector space $R(u, \xi)$ at each point (u, ξ) of M by the set of random variables

$$R(u, \xi) = \{r(x) \mid E[r] = 0, E[r^2] < \infty\}$$

where the expectation E is taken with respect to $p(x; u, \xi)$. The inner product is naturally introduced in R by the covariance, and R is a Hilbert space. The aggregate $\mathcal{R}$ of these R over all $(u, \xi) \in M$ forms a fibre bundle called the Hilbert bundle. We can define the e-parallel transport and the m-parallel transport in $\mathcal{R}$ by defining the respective affine connections which are dually coupled.

We can define the e-closed information subbundle R^I, which is a family of subspaces of $R(u, \xi)$. Formally, the R^I is the orthogonal complement of the subspace R^N composed of all the m-parallel transports of the ξ-score functions in the subspace R^T composed of all the m-parallel transports of ξ- and u-scores. A necessary and sufficient condition for the existence of a consistent estimator for any $\bar{\xi}$ is that $R^I \neq \{0\}$. When $R^I \neq \{0\}$, these exists a consistent estimator but it is not necessarily optimal. A necessary and sufficient condition for the existence of the optimal estimator is that the projection $u^I(x; u, \xi)$ of the u-score function to R^I is e-invariant, i.e., it dose not depend on ξ. When this is the case, the optimal estimating function $y(x, u)$ is given by this u^I.

<u>Time series analysis and theory of systems.</u> Let us consider a stationary Gaussian time series $\{x_t\}$, $t = 0, \pm 1, \pm 2, \cdots$. Let S be the set of all such stochastic processes, or the set of the

probability measures defining such processes. We can introduce to S the Fisher information metric and the α-connections in a simlar manner. The geometry of S is useful not only for analyzing mutual relations of various models of time serieses and their transformations, but also for identification, transformation and approximation of systems.

Information Theory and Statistics. The geometrical method opens a new field connecting information theory and statistics.

R e f e r e n c e s

Amari, S. [1982 a]. Differential geometry of curved exponential families---curvatures and information loss. Ann. Statist., 10, 357-385.

Amari, S. [1982 b]. Geometrical theory of asymptotic ancillarity and conditional inference. Biometrika, 69, 1-17.

Amari, S. [1983]. Comparisons of asymptotically efficient tests in terms of geometry of statistical structures. Bull. Inter. Statist. Inst., Proc. 44 Session, Book 2, 1190-1206.

Amari, S. [1985]. Differential Geometrical Methods in Statistics. Springer Lecture Notes in Statistics, Vol.28

Amari, S. [1986]. Differential Geometry of Statistics---Towards New Developments. To appear in Differential Geometry in Statistical Inference. IMS Monograph Series.

Amari, S. [1986]. Geometrical Theory on Manifold of Linear Systems. METR 86-1, Univ. Tokyo

Amari, S. and Kumon, M. [1985]. Optimal Estimation in the Presence of Infinitely Many Nuisance Parameter---Geometry of Estimating Functions, METR 85-2, Univ. Tokyo

Kumon, M. and Amari, S. [1983]. Geometrical theory of higher-order asymptotics of test, interval estimator and conditional inference. Pro. Roy. Soc., A 387, 429-458.

Kumon, M. and Amari, S. [1985]. Differential Geometry of Testing Hypothesis---A Higher Order Asymptotic Theory in Multi-Parameter Curved Exponential Family, METR 85-1, Univ. Tokyo

Bernoulli, Vol. 2, pp. 205-213

LIKELIHOOD, ANCILLARITY AND STRINGS

O.E. Barndorff-Nielsen
Department of Theoretical Statistics,
Institute of Mathematics, Aarhus University,
8000 Aarhus C, Denmark.

For a parametric model M with model function $p(x;\omega)$, d-dimensional parameter $\omega = (\omega^1,...,\omega^d)$, maximum likelihood estimator $\hat{\omega}$ and k-dimensional sufficient statistic t, let a be an <u>auxiliary (statistic)</u>, i.e. a is a statistic of dimension k-d such that $(\hat{\omega},a)$ is a smooth one-to-one transformation of t. For the purpose of inference on ω we may consider the log likelihood function l as a function of ω and $(\hat{\omega},a)$, i.e. $l = l(\omega;\hat{\omega},a)$. Further let $\partial_r = \partial/\partial\omega^r$, $\hat{\partial}_r = \partial/\partial\hat{\omega}^r$ and

$$(1) \qquad \chi_{r_1...r_p;s_1...s_q} = \chi_{r_1...r_p;s_1...s_q}(\omega;a) = \dagger\partial_{r_1}...\partial_{r_p}\hat{\partial}_{s_1}...\hat{\partial}_{s_q} l(\omega;\hat{\omega},a)$$

where $\dagger$ indicates the operation of substituting $\hat{\omega}$ by ω.

The discussion in the present paper evolves around the following formula for the conditional model function for $\hat{\omega}$ given a (Barndorff-Nielsen (1980,1983))

$$(2) \qquad p(\hat{\omega};\omega|a) \doteq p^*(\hat{\omega};\omega|a)$$

where p^* is defined by

$$(3) \qquad p^*(\hat{\omega};\omega|a) = c(\omega,a)|\hat{j}|^{\frac{1}{2}}e^{l(\omega)-l(\hat{\omega})}.$$

Let

$$(4) \qquad \bar{c} = \bar{c}(\omega,a) = \log\{(2\pi)^{d/2}c(\omega,a)\}.$$

This quantity $\bar{c}$ is often close to 0.

The interest of (2) is tied to the conditonality viewpoint, that is to cases where a is not only auxiliary, in the above sense, but is also distribution constant, either exactly or approximately. We refer to a statistic having both these properties as an <u>ancillary (statistic)</u>.

In the case of ordinary repeated sampling with sample size n, if the approximation (2) is correct to order $0(n^{-\nu/2})$ (typically, $\nu = 2,3$ or ∞) then to that order many developments traditionally requiring calculation of moments or cumulants can instead be carried out in terms of the <u>mixed log model derivatives</u> (1). For instance, if (2) holds to $0(n^{-3/2})$, confidence limits for a one-dimensional interest parameter, valid to $0(n^{-3/2})$ conditionally on a as well as unconditionally, can be thus determined. Bartlett adjustment affords another instance, to be discussed later. (Barndorff-Nielsen (1986a,b)). Yet another exemplification is provided by formula (5) below.

By Taylor expansion of the right hand side of (3) in $\hat{\omega}$ around ω one may derive an asymptotic expansion of $p^*(\hat{\omega};\omega|a)$ of the form

$$(5) \qquad p^*(\hat{\omega};\omega|a) = \varphi_d(\hat{\omega}-\omega;\jmath)\{1+R_1+R_2+\ldots\}$$

where $\varphi_d(\cdot;\lambda)$ denotes the d-dimensional normal probability density function with mean vector 0 and precision matrix λ, where $\jmath$ is the observed information matrix with elements

$$(6) \qquad \jmath_{rs} = -\chi_{rs} = \chi_{r;s} ,$$

and where R_ν is generally of order $0(n^{-\nu/2})$ under repeated sampling. In particular

$$R_1 = -\tfrac{1}{2}\Big\{ (\hat{\omega}-\omega)^t \, \not{j}^{rs} \overset{-1}{\Gamma}_{rst} + h^{rst}(\hat{\omega}-\omega;\not{j}) \overset{-\frac{1}{3}}{\Gamma}_{rst} \Big\} .$$

Here, and elsewhere, we employ Einstein's summation convention. Furthermore, $h^{rst}(\cdot;\not{j})$ is the tensorial Hermite polynomial of degree 3, corresponding to the precision $\not{j}$, and $\overset{-1}{\Gamma}$ and $\overset{-\frac{1}{3}}{\Gamma}$ are affine connections, in the sense of differential geometry, on the parameter space of the model M. For any real α the observed α-connection $\overset{\alpha}{\Gamma}$ is defined by (Barndorff-Nielsen (1986a))

$$\text{(7)} \qquad \overset{\alpha}{\Gamma}_{rst} = \frac{1+\alpha}{2} \not{l}_{rs;t} + \frac{1-\alpha}{2} \not{l}_{t;rs} .$$

These connections are 'observed analogues' of the Chentsov-Amari connections $\overset{\alpha}{\Gamma}$. (Chentsov (1972), Amari (1985).)

The expansion (5) has some similarity to but is distinct from the Edgeworth expansion for the distribution of $\hat{\omega}$. Thus (5) employs mixed log model derivatives instead of (approximate) cumulants of $\hat{\omega}$.

Note that (5) is valid as an asymptotic expansion irrespective of the accuracy with which $p^*(\hat{\omega};\omega|a)$ approximates $p(\hat{\omega};\omega|a)$ and irrespective of whether a is (approximately) distribution constant or not.

For fixed value of the auxiliary statistic a there is in general (locally, at least) a smooth one-to-one correspondence between $\hat{\omega}$ and the score vector $l_* = (l_1(\omega),\ldots,l_d(\omega))$. Hence, by the usual formula for transformation of probability densities, (2) can be transformed to a formula for the conditional distribution of l_*. Equivalently, one may transform to $\underset{\sim}{l}_* = l_* \not{j}^{-\frac{1}{2}}$ where $\not{j}^{-\frac{1}{2}}$ is the square root of the matrix $\not{j}$. Writing $l_;$ for the matrix with elements $l_{r;s}$ one finds

$$\text{(8)} \qquad P(\underset{\sim}{l}_*;\omega|a) = c(\omega,a)\{|\hat{j}||\not{j}|/|l_;|^2\}^{\frac{1}{2}} e^{l(\omega)-l(\hat{\omega})}$$

where on the right hand side $\hat{\omega}$ has to be expressed in terms of $\underset{\sim}{l}_*$ (and a).

The relation (2) is, in fact, exact for most transformation models as well as for a variety of other models (Barndorff-Nielsen (1983), Blæsild and Jensen (1984), Barndorff-Nielsen and Blæsild (1986a)). Outside these cases the best that can generally be achieved is an asymptotic (relative) error of order $O(n^{-3/2})$. In particular, if M is an exponential model of order k and if $d = k$ then (2) is valid to order $O(n^{-3/2})$.

Now suppose that (2) holds with error $O(n^{-3/2})$ and let M_0 be a submodel of M, having parametric dimension $d_0 < d$. Using the asymptotic normality of the score vector for M, under the hypothesis M_0, one may, by standardizing the part of the score vector orthogonal to M_0 so as to have variance matrix equal to the unit matrix I_{d-d_0} asymptotically, construct a supplementary auxiliary statistic of dimension $d-d_0$ so as to make (2) valid to order $O(n^{-1})$ under M_0. In fact, there is a considerable variety of approximately distribution constant statistics of dimension $d-d_0$ which could serve in the capacity of supplementary auxiliary and yielding accuracy $O(n^{-1})$. However, demanding accuracy $O(n^{-3/2})$ of (2) narrows down the choice significantly. More specifically and supposing, for simplicity, that $d_0 = d-1$, it can be shown that accuracy $O(n^{-3/2})$ of (2) as applied to M_0 may be achieved by taking as supplementary auxiliary

$$r^{\dagger} = r - \text{bias correction}$$

where

$$r = \pm\sqrt{2\{l(\hat{\omega})-l_0(\hat{\omega}_0)\}} \tag{9}$$

is the signed log likelihood ratio statistic for testing M_0 against M; and that this choice of a supplementary auxiliary is unique to the asymptotic order concerned. (Barndorff-Nielsen (1984,1986b).)

The statistic $r^\dagger$ is asymptotically $N(0,1)$ distributed to order $O(n^{-1})$. By introducing a variance adjustment it is possible to establish a statistic

$$r^* = r^\dagger/\text{s.d. adjustment} \tag{10}$$

which is asymptotically standard normal to order $O(n^{-3/2})$. This may be used for a refined test of M_0 versus M, as well as for the role of supplementary auxiliary. Moreover, with error $O(n^{-3/2})$,

$$r^* \doteq r - r^{-1} \log K , \tag{11}$$

where K is a certain explicitly given function of the observations, and the right hand side of (11) is often simpler to calculate than (10). In case M is a (k,k) exponential model while M_0 is a $(k,k-1)$ exponential model we have

$$K = |r| \begin{vmatrix} \frac{\partial\theta}{\partial\omega_0^1}(\hat{\omega}_0) \\ \vdots \\ \frac{\partial\theta}{\partial\omega_0^{d-1}}(\hat{\omega}_0) \\ \theta(\hat{\omega})-\theta(\hat{\omega}_0) \end{vmatrix}^{-1} \{|j_0(\hat{\omega}_0)|/|j(\hat{\omega})|\}^{\frac{1}{2}} .$$

For details and examples, see Barndorff-Nielsen (1986b).

Let

$$\bar{b} = d^{-1} E_\omega w \tag{12}$$

where w is the log likelihood ratio statistic for

testing a particular value of ω under M, i.e.

$$w = 2\{l(\hat{\omega})-l(\omega)\}. \tag{13}$$

The quantity $\bar{b}$ and suitable approximations thereof are termed <u>Bartlett adjustments</u> for the log likelihood ratio statistic. The Bartlett adjusted version $w' = w/\bar{b}$ of w is, in wide generality, asymptotically χ^2-distributed on d degrees of freedom, the degree of approximation to the limiting χ^2 distribution being $O(n^{-3/2})$, or even $O(n^{-2})$, as opposed to $O(n^{-1})$ for w itself.

The norming quantity $\bar{c}$ is related to $\bar{b}$ by the approximate relation

$$\bar{b} \doteq e^{-(2/d)\bar{c}}. \tag{14}$$

(Barndorff-Nielsen and Cox (1984).)

Decomposition of the norming quantity $\bar{c}$ (or, equivalently, of Bartlett adjustments) into invariant terms can be achieved by the use of <u>strings</u>, a differential geometric concept generalizing those of tensors, connections, and derivatives of scalars (functions). (Barndorff-Nielsen (1986c), Barndorff-Nielsen and Blæsild (1986b,c); see also McCullagh and Cox (1986) which provides and discusses the first example of such a decomposition.)

A <u>(p,q) string</u> of length $(\bar{m},\bar{n})$ $(\bar{m} \leq \infty,\ \bar{n} < \infty)$ is a sequence $\bar{M}$ of multiarrays M defined on a manifold M and satisfying the transformation law

$$M_{B_q C_m}^{A_p D_n} = \left\{ \sum_{\mu=1}^{m} \sum_{\nu=n}^{\bar{n}} M_{S_q T_\mu}^{R_p U_\nu} \omega^{T_\mu}{}_{/C_m} \psi^{D_n}{}_{/U_\nu} \right\} \psi^{A_p}{}_{/R_p} \omega^{S_q}{}_{/B_q} \tag{15}$$

for $m = 1,2,\ldots,\bar{m}$ and $n = 1,\ldots,\bar{n}$. Here ω and ψ are alternative coordinate systems on M with generic coordinates $\omega^r, \omega^s, \omega^t, \ldots$ and $\psi^a, \psi^b, \psi^c, \ldots$ respectively. Furthermore, $A_p = a_1 \ldots a_p$, $B_q = b_1 \ldots b_q$, etc., and

$$\omega^{T_\mu}_{/C_m} = C^{\Sigma}_m/\mu \; \omega^{t_1}_{/C_{m1}} \cdots \omega^{t_\mu}_{/C_{m\mu}},$$

etc., where the summation is over all partitions of the index set $C_m = c_1 \ldots c_m$ into μ blocks and where

$$\omega^{t}_{/c_1 \ldots c_m} = \frac{\partial^m \omega^t}{\partial \psi^{c_1} \ldots \partial \psi^{c_m}} .$$

The string $\bar{M}$ is said to be a <u>costring</u> if $\bar{n} = 0$ and a <u>contrastring</u> if $\bar{m} = 0$. These types of strings can be represented in terms of tensors and special, simple kinds of strings. In particular, any costring can be represented as the <u>intertwining</u> of a <u>connection string</u>, i.e. a (1,0) costring, and a sequence of tensors.

Mixed log model derivatives provide examples of strings, and so do moments and cumulants of log likelihood derivatives. Tensors determined by associated <u>intertwining</u> operations may be used to obtain invariant decompositions. For example, to order $0(n^{-3/2})$ we have

$$\bar{c} \doteq - \frac{d}{24} \Big\{ (3\not{\kappa}_{rstu} + 12\not{\kappa}_{rt;su}) \not{\imath}^{rs} \not{\imath}^{tu}$$

$$+ (3\not{\kappa}_{rst}\not{\kappa}_{uvw} + 2\not{\kappa}_{rtv}\not{\kappa}_{suw}) \not{\imath}^{rs}\not{\imath}^{tu}\not{\imath}^{vw} \Big\} ,$$

where the right hand side is a sum of four invariant (separately interpretable) terms, due to the fact that the quantities $\not{\kappa}$ are tensors. These tensors were obtained by means of intertwining, applied to the first few of the mixed log model derivatives.

For further examples and an extensive study of the mathematical properties of strings, see Barndorff-Nielsen and Blæsild (1986b,c).

REFERENCES

Amari, S.-I. (1985). Differential-Geometric Methods in Statistics. Lecture Notes in Statistics 28. Springer-Verlag, Heidelberg.

Barndorff-Nielsen, O.E. (1980). Conditionality resolutions. Biometrika 67, 293-310.

Barndorff-Nielsen, O.E. (1983). On a formula for the distribution of the maximum likelihood estimator. Biometrika 70, 343-365.

Barndorff-Nielsen, O.E. (1984). On conditionality resolution and the likelihood ratio for curved exponential models. Scand. J. Statist. 11, 157-170. Corrigendum: 1985, Scand. J. Statist. 12, 191.

Barndorff-Nielsen, O.E. (1986a). Likelihood and observed geometries. To appear in Ann. Statist.

Barndorff-Nielsen, O.E. (1986b). Inference on full or partial parameters, based on the standardized signed log likelihood ratio. Biometrika 73, 307-322.

Barndorff-Nielsen, O.E. (1986c). Strings, tensorial combinants, and Bartlett adjustments. Proc. Roy. Soc. London A 406, 127-137.

Barndorff-Nielsen, O.E. and Blæsild, P. (1986a). Combination of reproductive models. Research Report 107, Dept. Theor. Statist., Aarhus University. To appear in Ann. Statist.

Barndorff-Nielsen, O.E. and Blæsild, P. (1986b). Strings: Mathematical theory and statistical examples. Research Report 146, Dept. Theor. Statist., Aarhus University. To appear in Proc. Roy. Soc. London.

Barndorff-Nielsen, O.E. and Blæsild, P. (1986c). Strings: contravariant aspect. Research Report 152, Dept. Theor. Statist., Aarhus University.

Barndorff-Nielsen, O.E. and Cox, D.R. (1984). Bartlett adjustments to the likelihood ratio statistic and the distribution of the maximum likelihood estimator. J.R. Statist. Soc. B 46, 483-495.

Blæsild, P. and Jensen, J.L. (1985). Saddlepoint formulas for reproductive exponential models. Scand. J. Statist. 12, 193-202.

Chentsov, N.N. (1972). Statistical Decision Rules and Optimal Inference. (In Russian). Nauka, Moscow. English translation 1982. Translation of Mathematical Monographs Vol. 53. American Mathematical Society, Providence, Rhode Island.

McCullagh, P. and Cox, D.R. (1986). Invariants and likelihood ratio statistics. To appear in Ann. Statist.

ON ASYMPTOTICALLY COMPLETE CLASSES OF TESTS

Bernstein A.V.
Moscow, USSR

Let $X^{(n)}=(X_1,\dots,X_n)$ be the sample from the distribution P_ξ depending on the unknown parameter $\xi \in R^k$. Let P_ξ have the density $p(x,\xi)$ w.r.t. some σ-finite measure. Based on this sample the null hypothesis H_0: $\xi = 0$ is tested against the sequence of the contiguous alternatives $H_{1n}: \xi=\xi_n=\tau\lambda,\ \lambda\in\Lambda$, where Λ is the given subset of $R^k \setminus \{0\}$ and $\tau = n^{-1/2}$.

The decision rule in this asymptotic (as.) testing problem (t.p.) is determined by the test-sequence (t.s.) $\{\varphi_n\}$ where for each n the test $\varphi_n(X^{(n)})$ is based on the sample $X^{(n)}$ of size n. This test is characterized by their power function

$$\beta_n(\xi_n | \varphi_n) = \int \varphi_n(X^{(n)})\, dP_{\xi_n}^{(n)}$$

where $P_\xi^{(n)}$ denotes n-fold product measure of identical components P_ξ. Denote by $\alpha_n(\varphi_n) = \beta_n(0|\varphi_n)$ the level of the test φ_n.

Let $r \geq 0$ be the arbitrary given integer and Φ be an arbitrary class of the t.s. For instance Φ may be class Φ_0 of all t.s. or the class $\Phi_r(\alpha)$ of all t.s. with the level equal to the given α up to an error term $o(\tau^2)$.

<u>Definition</u>. The subclass Φ_r^* of the class Φ is as. complete of order r in the class Φ if for each t.s. $\{\varphi_n\}\in\Phi$ there exists the t.s. $\{\Psi_n\}\in\Phi_r^*$ such that for all $\lambda\in\Lambda$ and $n\to\infty$

$$\alpha_n(\Psi_n) \leq \alpha_n(\varphi_n) + o(\tau^2),$$

$$\beta_n(\tau\lambda|\Psi_n) \geq \beta_n(\tau\lambda|\varphi_n) + o(\tau^2).$$

If the class Φ_z^* consists of the single t.s. $\{\Psi_{n,z}^*\}$ then this t.s. is called as. optimal t.s. of order z in the class Φ.

Earlier the as. complete classes were known for the two special cases. In the first one $k=1$ and $\Lambda = \Lambda^+ = \{\lambda \in R^1 : \lambda > 0\}$. Le Cam (1960) and Chibisov (1974) showed that for $z=0$ and $z=1$ in the classes $\Phi_0(\alpha)$ and $\Phi_1(\alpha)$ there exists the as. optimal t.s. of order $z=0$ and $z=1$ respectively. Pfanzagl (1975) showed that for $z=2$ in the class $\Phi_2(\alpha)$ does not exist the single as. optimal t.s. of order $z=2$ and constructed the as. complete subclass $\Phi_2^*(\alpha)$ of order $z=2$. In the second case the dimension k is an arbitrary natural number but $z=0$. In this case Chibisov (1967) constructed the as. complete class Φ_0^* of order $z=0$ in the class Φ_0 of all t.s.

In this paper the general method for the description of the as. complete classes Φ_z^* of arbitrary order z in the class Φ_0 of all t.s. for arbitrary-dimensional parameter ξ is present.

Let us introduce the notations. Let

$$y(x) = \frac{\partial \ln p(x,\xi)}{\partial \xi}\bigg|_{\xi=0}$$

be the k -dimensional vector of the logarithmic derivatives, Σ be the covariance matrix of the random vector $y(X_1)$ under the null hypothesis H_0 and

$$Y_n = Y_n(X^{(n)}) = \tau \sum_{1\leq i\leq n} y(X_i)$$

be the normalized sum of such vectors.

Let $\nu = (\nu_1, \ldots, \nu_k)$ be a vector with integer -valued non-negative components,

$$\ell^{(\nu)}(x) = \left.\frac{\partial^{|\nu|} \ln p(x,\xi)}{\partial \xi^{|\nu|}}\right|_{\xi=0}$$

be logarithmic derivative of order ν, $m_\nu = E_0 \ell^{(\nu)}(X_1)$ be the expectation of the random variable $\ell^{(\nu)}(X_1)$ under H_0 and

$$W_n^{(\nu)} = W_n^{(\nu)}(X^{(n)}) = \tau \sum_{1 \le i \le n} [\ell^{(\nu)}(X_i) - m_\nu]$$

be the corresponding normalized sum.

Introduce the random vector $(Y_n; W_n)$ where the sub-vector W_n has the components $W_n^{(\nu)}$ with $2 \le |\nu| \le z+1$.

Theorem 1. The statistic (Y_n, W_n) is locally as. sufficient of order z for the family $\{P_{\tau\lambda}^{(n)}, \lambda \in R^k\}$.

In particular it means that the test $\Psi_n(X^{(n)})$ in any t.s. $\{\Psi_n\} \in \Phi_z^*$ may be chosen such that it will depend on the sample $X^{(n)}$ only through the value $(Y_n(X^{(n)}); W_n(X^{(n)}))$ of this statistic.

To formulate the other results we have to introduce the additional notations. Let $C \subset R^k$ be the convex set with the smooth boundary having the form

$$C = \{ y \in R^k : g(y) \le d \} \tag{1}$$

where $g(y)$ is a smooth convex function on R^k. Denote

$$n(z) = \frac{\nabla g(z)}{|\nabla g(z)|}$$

the unit normal vector to the set C in the point $z \in \partial C$ here $\nabla g(z) = \frac{\partial}{\partial z} g(z) \in R^k$ is the vector-gradient of g. Let

$$\delta^*(x^* | C) = \sup\{(x^*, y);\ y \in C\}$$

be the support function of the convex set C. In this notations the set C (1) may be rewritten in such form

$$C = \{ y \in R^k : (y, n(z)) \le \delta^*(n(z)|C) \text{ for all } z \in \partial C \}.$$

Consider the following auxiliary non-as. t.p. of the location parameter of the normal distribution. Let Y be the k -dimensional normal random vector with the parameters $(\Sigma\lambda, \Sigma)$ where Σ is the above-mentioned covariance matrix. Based on the single observation y on this vector Y we wish to test the hypothesis $\lambda = 0$ against the bounded alternative

$$\lambda \in \Lambda_n = \{\lambda \in \Lambda : 0 < |\lambda| \leq \ln \ln n\}.$$

From our results (Bernstein (1985)) it follows that the complete class $\mathcal{F}^*$ in this auxiliary t.p. consists of the non-randomized tests $\bar{\Psi}(\cdot)$ having the form

$$\bar{\Psi}(y) = \bar{\Psi}(y|C_n) = \begin{cases} 0 & \text{for } y \in C_n, \\ 1 & \text{for } y \bar{\in} C_n, \end{cases} \tag{2}$$

where their acceptance region C_n is the convex set of the form

$$C_n = C_n(d) = \{y \in R^k : g_n(y) \leq d\} \tag{3}$$

determined by the convex function

$$g_n(y) = a'y + \frac{\gamma}{2} y'Vy +$$

$$+ (1-\gamma) \int_{\Lambda_n} \frac{\exp(\lambda' y - \frac{1}{2}\lambda'\Sigma\lambda) - 1 - \lambda' y}{|\lambda|^2} \mu(d\lambda). \tag{4}$$

This function is determined by the vector $a \in R^k$, by the non-negative-definite matrix V of order k, by the number $\gamma \in [0,1]$ and by the probability measure $\mu(\cdot)$ on Λ_n.

Let us now turn to our as. t.p. Let the concrete test in above descripted auxiliary t.p. be chosen. In other words the convex set C_n is chosen, or, as is the same, the function g_n of the form (4) or, as is the same, the concrete value of the collection (a, V, γ, μ) is fixed. Then based on this choise we shall construct for arbitrary z the t.s. $\{\Psi_n^{(z)}\}$ in the following way. The corresponding test $\Psi_n^{(z)}(X^{(n)})$ is a non-randomized one and its acceptance region is the following

$$\{X^{(n)} : (Y_n, n(z)) \leq \delta^*(n(z)|C_n) + \qquad (5)$$

$$+ \sum_{1\leq j\leq z} \tau^j P_j(\{W_n^{(\nu)},\quad 2\leq |\nu| \leq j+1\}; z) \quad \text{for all } z \in \partial C_n\}.$$

Here $\delta^*(\cdot|C_n)$ is the support function of the chosen convex set C_n (3) in auxiliary t.p. and the functions P_j are the polynomials of the corresponding components $W_n^{(\nu)}$ and of their moments m_ν. These polynomials are completely and constructively determined by the chosen test $\overline{\Psi}$ (2)-(4) in the auxiliary t.p. More precisely their coefficients are the concrete constructively determined functions of the derivatives of the function (4) in the points $z \in \partial C_n$. For example

$$P_1 = -\sum_{|\nu|=2} \frac{g_n^{(\nu)}(z) W_n^{(\nu)}}{\nu! |\nabla g_n(z)|} - \sum_{|\nu|=3} \frac{g_n^{(\nu)}(z) m_\nu}{\nu! |\nabla g_n(z)|}.$$

Theorem 2. The class Φ_z^* of t.s. $\{\Psi_n^{(z)}\}$ constructed above is the as. complete subclass of order z in the class Φ_0 of all t.s.

Consider the particular case $z = 0$. The acceptance region of the test $\Psi_n^{(0)}(X^{(n)})$ has the form

$$\{X^{(n)} : (Y_n, n(z)) \leq \delta^*(n(z)|C_n) \text{ for all } z \in \partial C_n\} = \qquad (6)$$

$$= \{X^{(n)} : Y_n(X^{(n)}) \in C_n\}.$$

Thus the test $\overline{\Psi}$ (2) in the auxiliary t.p. and the corresponding test $\Psi_n^{(0)}(X^{(n)})$ in the original as. t.p. are related by the equality: $\Psi_n^{(0)}(X^{(n)}) = \overline{\Psi}(Y_n(X^{(n)}))$. Hence for the case $z = 0$ we have obtained one of the results of Chibisov (1867) with certain improvement concerning the determined structure of the convex set C_n (see (3),(4)).

Let $\Psi_n^{(z)}$ for $z \geq 1$ be the test with the acceptance region (5). The statistics $W_n^{(\nu)}$ with $|\nu| = z+1$ are situated only in the polynomial $P_z(W_n; z)$. Let $\overline{\Psi}_n^{(z)}(X^{(n)})$ be the test which is derived from the test $\Psi_n^{(z)}$ by replacing the statistics $W_n^{(\nu)}$ with $|\nu| = z+1$ in the polynomial $P_z(W_n; z)$ by the non-random quantities $(M_\nu \cdot \Sigma^{-1} \cdot z)$ where $M_\nu = E_0[\ell^{(\nu)}(X_1) \cdot y(X_1)] \in R^k$ is the conditional expectation $E_0(W_n^{(\nu)} | Y_n = z)$ up to an error term $o(1)$.

Theorem 3. As $n \to \infty$ the following is true

$$\sup_{|\lambda| \leq \ln \ln n} |\beta_n(\tau\lambda | \Psi_n^{(z)}) - \beta_n(\tau\lambda | \overline{\Psi}_n^{(z)})| = o(\tau^z).$$

Note that the test $\overline{\Psi}_n^{(z)}(X^{(n)})$ depends on the sample through the statistic $(Y_n; \{W_n^{(\nu)}, 2 \leq |\nu| \leq z\})$ only.

Corollary. The class Φ_z^* of t.s. $\{\overline{\Psi}_n^{(z)}\}$ constructed above is as. complete subclass of order z in the class Φ_0.

For example we may construct the corresponding as. complete subclass $\overline{\Phi}_1^*$ of order $z = 1$ in the class Φ_1. Let $\overline{\Psi}(y)$ be the concrete test of the form (2)-(4) in the auxiliary t.p., $\Psi_n^{(1)}(X^{(n)})$ be the corresponding test from the class Φ_1^* and $\overline{\Psi}_n^{(1)}(X^{(n)})$ be the corresponding already changed test from $\overline{\Phi}_1^*$. This test depends on the sample through the statistic Y_n only and has the acceptance region of the form

$$\{X^{(n)} : Y_n(X^{(n)}) \in C_{n,1}\}, \tag{7}$$

where the non-random convex set $C_{n,1} \subset R^k$ is constructively determined by the set C_n. In particular for all $x^* \in R^k$ it holds:

$$\delta^*(x^* | C_{n,1}) = \delta^*(x^* | C_n) + O(\tau).$$

The relation between the acceptance regions (6) and (7)

for the tests from the as. complete classes Φ_0^* and Φ_1^* is the multivariate analogue of the known result "the first order efficiency implies the second order efficiency" and it coincides with the above mentioned result in the one-dimensional case under one-sided alternatives.

From our results it follows in the one-dimensional case under the one-sided alternatives that the as. complete class $\overline{\Phi}_2^*(\alpha)$ coincides with the as. complete class of order $z=2$ which was obtained by Pfanzagl (1975) who investigated this special case directly.

In conclusion we present the as. approximation for the power function of the test $\Psi_n^{(z)}$ (5).

Theorem 4. The power function $\beta_n(\tau\lambda|\Psi_n^{(z)})$ is the following: as $n\to\infty$

$$\beta_n(\tau\lambda|\Psi_n^{(z)}) = 1-\int_{C_n}(1+\sum_{j=1}^{z}\tau^j R_j(y,\lambda))\times$$
$$\times\ \varphi_\Sigma(y-\Sigma\lambda)dy + \sum_{j=1}^{z}\tau^j\ \times \tag{8}$$
$$\times\sum_i\int_{\partial C_n}\frac{R_{ij}(z,\lambda)\,\varphi_\Sigma(z-\Sigma\lambda)}{|\nabla g_n(z)|^i}\,mes_{k-1}(dz)+o(\tau^z)$$

where $\varphi_\Sigma(\cdot)$ is the density function of the multivariate normal distribution in R^k with the parameters $(0;\Sigma)$. The functions $R_j(y,\lambda)$ are the polynomials of the Edgeworth type expansion for the random vector's Y_n distribution density under the distribution $P_{\tau\lambda}^{(n)}$ of the sample. Note that the sum of the two first terms in (8) is the as. approximation for the power function $\beta_n(\tau\lambda|\Psi_n^{(0)})$ up to an error term $o(\tau^z)$.

In the third term in (8) the sum over i has the finite number of summands. The integration in this term is fulfilled over the Lebesque measure on the $(k-1)$ -dimensional boundary of the set C_n . The functions $R_{ij}(z,\lambda)$ are completely determined by the derivatives of the function g_n at the points $z \in \partial C_n$.

For example when $z = 1$ there is only one term in the sum over i with $i = 1$ and the function R_{11} is the following

$$R_{11} = \sum_{|\nu|=2} \frac{g_n^{(\nu)}(z)(M_\nu \Sigma^{-1} z)}{\nu!} + \sum_{|\nu|=3} \frac{g_n^{(\nu)}(z) m_\nu}{\nu!} .$$

References

Bernstein, A.V. (1985). The specification of the theorems about the complete classes of the tests. Theory Probab. Appl. 30, 576-580.

Chibisov, D.M. (1967). A theorem on admissible tests and its application to an asymptotical problem of testing hypotheses. Theory Probab. Appl. 12, 96-111.

Chibisov, D.M. (1974). Asymptotic expansion for some asymptotically optimal tests. In: Proceedings of the Prague Symposium on Asymptotic Statistics, Vol. 2. (J. Hajek, ed.), 37-68.

Le Cam, L. (1960). Locally asymptotically normal families distributions. Univ. California Publ. Statist. 3, 37-98.

Pfanzagl, J. (1975). On asymptotically complete classes. In: Statistical Inference and related topics. Proceedings of the Summer Research Institute on Statistical Inference for Stochastic Processes, Vol. 2, 1-43.

TAIL PROBABILITY APPROXIMATIONS

Daniels, H.E.,
Statistical Laboratory,
University of Cambridge,
16 Mill Land,
Cambridge, CB2 1SB,
U.K.

The following is an abstract of the paper which will appear in International Statistical Review, April 1987.

This paper discusses two methods of approximating to the tail probability $P(\overline{X} \geq \overline{x})$ of the mean $\overline{x}$ of a sample of n observations from a distribution whose mean is E(X).

The standard method of finding an approximation whose relative error is controlled over the whole range of $\overline{x}$ is to use the ideas of Esscher [7] and Cramér [4] which involve an Edgeworth expansion of an exponentially shifted density recentred at the sample mean (Barndorff-Nielsen and Cox [1]). This approach gives an asymptotic expansion in powers of n^{-1} when $\overline{x} - E(X) = O(n^{-1/2})$. For large deviations it can be further reduced to an expansion whose terms decrease by a factor n^{-1}.

Recently Lugannani and Rice (1980) presented a new formula based on an extension of the saddlepoint technique due to Bleistein [2] [3] which makes proper allowance for the presence of a pole at the origin in the Fourier inversion integral. It yields an asymptotic expansion whose successive terms decrease in order by a factor n^{-1} over the whole range of $\overline{x}$, unlike those of the previous formula.

The object of the paper is to familiarise statisticians with the theory underlying the new formula and others related to it, and to show how the two apparently unrelated expansions for the tail probability arise from different approaches to the saddlepoint approximation.

References

1. Barndorff-Nielsen, O. and Cox, D.R. (1979). Edgeworth and saddlepoint approsimations with statistical applications.

2. Bleistein, N. (1966). Uniform expansions of integrals with stationary points near algebraic singularity. *Comm. Pure Appl. Math.* 19, 353-370.
3. Bleistein, N. and Handelsman, R.A. (1975). *Asymptotic expansions of integrals*. Holt, Rinehart and Winston, New York.
4. Cramer, H. (1938). Sur un nouveau theoreme - limite de la theorie des probabilites. *Actualites Scientifiques et Industrielles*, No. 736. Hermann et Cie, Paris.
5. Daniels, H.E. (1954). Saddlepoint approximations in statistics. *Ann. Math. Statist.* 25 No.4, 631-649.
6. Daniels, H.E. (1983). Saddlepoint approximations for estimating equations. *Biometrika*, 70, 89-96.
7. Esscher, F. (1932). On the probability function in the collective theory of risk. *Skand. Akt Tidsskr.* 175-195.
8. Olver, F.W.J. (1974). *Asymptotics and Special Functions*. Academic Press, New York.
9. Lugannani, R. and Rice, S.O. (1980). Saddlepoint approximations for the distribution of the sum of independent random variables. *Adv. Appl. Prob.* 12. 474-490.

ON SECOND ORDER ADMISSIBILITY IN SIMULTANEOUS ESTIMATION

B.Ya. Levit
Moscow, 129085, Godovikova 3-66

1. Introduction. In verifying admissibility of certain statistical estimators the basic role of the differential inequality

$$\sum_{i,j=1}^{s} \frac{\partial}{\partial\theta_i}\left(\omega^2(\theta)\, a_{ij}(\theta)\frac{\partial}{\partial\theta_j}\right)\varphi(\theta)\le 0, \quad \theta\in\Theta \qquad (1)$$

was recently recognized, with ω^2 a given positive function, (a_{ij}) a symmetric positively defined matrix, Θ a given open subset of R^s. See e.g. Brown (1981), Berger (1980). The point is whether there exists a solution to (1) such that $\varphi(\theta)\geqslant 0$, $\varphi(\theta)\neq$ const.

In simultaneous estimation to be discussed below (1) reduces to a simpler inequality.

$$\sum_{i=1}^{s} \frac{\partial}{\partial\theta_i}\left(\omega^2(\theta)\, a(\theta_i)\frac{\partial}{\partial\theta_i}\right)\varphi(\theta)\le 0, \quad \theta\in\Theta \qquad (2)$$

with $\omega(\theta)=\prod_{j=1}^{s}\Omega(\theta_j)$, $\Theta=\tilde{\Theta}\times\dots\tilde{\Theta}$, $\Omega(\cdot)$, $a(\cdot)$ being given smooth positive functions, $\tilde{\Theta}=(\tau_-,\tau_+)$, $-\infty\le\tau_-<\tau_+\le+\infty$. Setting $\mu_i(\theta)=\omega^2(\theta)a(\theta_i)\partial \ln\varphi/\partial\theta_i$ (2) is equivalent to

$$\sum_{i=1}^{s}\left(\frac{\partial\mu_i(\theta)}{\partial\theta_i}+\frac{\mu_i^2(\theta)}{\omega^2(\theta)\,a(\theta_i)}\right), \quad \theta\in\Theta, \qquad (3)$$

with $\mu=(\mu_1,\dots,\mu_s)\neq 0$.

For the one-dimensional case: $s=1$, the necessary and sufficient condition for the non-existence of a non-trivial solution to (1)-(3) which has been known for long (see e.g. Brown (1971)), is the divergence of the integral $\int(a\Omega^2)^{-1}d\tau$ at either of the points $\tau_\pm$. However for $s>1$ the problem in general is more involved even in the simplest case (2).

Berger (1980) followed by others gives a sufficient condition for the existence of the solution $\mu \not\equiv 0$ in a wider class of the inequalities (3). Brown (1981) presents more general necessary and/or sufficient conditions for the inequality (1), which however are not readily expressed in terms of the coefficients of (1).

The differential inequalities (1)-(3) arise as well in the context of the second order admissibility (s.o.a.) of statistical estimators (Levit (1985)). The asymptotic approach outlined in Sect. 2,3 has the advantage of reducing the admissibility questions to (1) in a more transparent and general fashion.

In Sect. 4 we show that under mild assumptions usually met in practice the divergence of the integral $\int (\tau^{s-1} a\, \Omega^{2s})^{-1} d\tau$ at the points $\tau_{\pm}$ is necessary for the non-existence of non-trivial solutions to (2)-(3). In Sect. 5 the result is applied to s.o.a. of the maximum likelihood estimators in some classical parametric families, including the normal and gamma with scale parameters, Poisson distribution, Pareto etc.

In Sect. 6 a sufficient condition for the non-existence of the solutions to (2),(3) is given and that is applied to example which generalizes an estimation problem proposed in Brown (1980,1981).

As a starting point we present an assertion in Sect.3 on the risk expansion of certain estimators, immediately reducing the problem of s.o.a. to the inequality (1). This is an extension of Theorem 3.1) in Levit (1982) and Theorem 5 in Levit (1983). The proof of the result will be given elsewhere.

2. <u>The statistical model</u>. Let $X_i = (x_{i1}, \dots, x_{is})$ be a sequence of i.i.d. random vectors with a pdf of the form $p(X, \theta) = \prod_{j=1}^{s} f(x_j, \theta_j)$, $\theta = (\theta_1, \dots, \theta_s) \in \Theta = \tilde{\Theta} \times \dots \times \tilde{\Theta}$, $\tilde{\Theta} = (\tau_-, \tau_+)$, $-\infty \le \tau_- < \tau_+ \le +\infty$. Denote by

$i(\tau)=-E_\tau \frac{\partial^2 \ln f(x,\tau)}{\partial \tau^2}$, $m(\tau)=\frac{\partial^3 \ln f(x,\tau)}{\partial \tau^3}$, $\tilde{m}(\tau)=m(\tau)/i(\tau)$, $I(\theta)=(\mathrm{diag}\ i(\theta_j))$ - the Fisher matrix related to the family p. Let θ_ε be an arbitrary estimator of θ, based on $X_1,\dots,X_n$ where $\varepsilon=n^{-1/2}$. From the very beginning we confine ourselves to the loss functions (l.f.) of the form

$$W_\varepsilon(\theta_\varepsilon,\theta)=W(\varepsilon^{-1}(\theta_\varepsilon-\theta),\theta)=w(\varepsilon^{-1}\mathcal{D}(\theta)(\theta_\varepsilon-\theta))$$

where either

$$w(y)=|y|^\alpha=\Big(\sum_{j=1}^{s} y_j^2\Big)^{\alpha/2}, \quad \mathcal{D}(\theta)=I^{1/2}(\theta), \tag{4}$$

or

$$w(y)=\sum_{j=1}^{s}|y_j|^\alpha, \quad \mathcal{D}(\theta)=(\mathrm{diag}\ d(\theta_j)), \tag{5}$$

$d(\cdot)$ being a given positive function and $\alpha>0$. Define the risk function $R_\varepsilon(\theta_\varepsilon,\theta)=E_\theta W_\varepsilon(\theta_\varepsilon,\theta)$.

Let $\varphi_\theta(y)$ be the normal $\mathcal{N}(0, I^{-1}(\theta))$ pdf,

$R(\theta)=\int\varphi_\theta(y)\,w(\mathcal{D}(\theta)y)dy$, $A(\theta)=\int\varphi_\theta(y)(yy^T-I^{-1}(\theta))w(\mathcal{D}(\theta)y)dy$.

Evaluating the last integral for the l.f. (4) one obtains in particular

$$A(\theta)=(\mathrm{diag}\ a(\theta_j))=C_{\alpha,s}\frac{\alpha}{s}(\mathrm{diag}\ i^{-1}(\theta_j)), \tag{6}$$

where $C_{\alpha,s}=2^{\alpha/2}\Gamma((\alpha+s)/2)/\Gamma(s/2)$ and for the l.f. (5)

$$A(\theta)=(\mathrm{diag}\ a(\theta_j))=C_{\alpha,1}(\mathrm{diag}\ d(\theta_j)\,i^{-1-\alpha/2}(\theta_j)). \tag{7}$$

We assume below that the pdf $f(x,\tau)$ satisfies the regularity assumptions of Levit (1982) on any compact $K\subset\tilde{\Theta}$, $d(\cdot)$ is locally Hölder, $d(\cdot)\in C^{\beta_0}(\tilde{\Theta})$ for some $\beta_0>0$, and for any $K\subset\tilde{\Theta}$ and $\alpha>0$ the maximum likelihood estimator (m.l.e.) $\hat{\theta}_{\varepsilon,1}$ of θ_1 satisfies the relation

$$\overline{\lim_{n\to\infty}}\ \sup_{\theta_1\in K} E_{\theta_1}|\varepsilon^{-1}(\hat{\theta}_{\varepsilon 1}-\theta_1)|^\alpha<\infty .$$

These assumptions imply in particular that $a(\tau)>0$ and $a(\cdot)\in C^{1,\beta}(\tilde{\Theta})$ for some β, $0<\beta\le\beta_0$.

3. Second order asymptotics. In the second order optimality theory one is interested in the class $\mathfrak{S}^{\beta}(\Theta)$ of estimators θ_ε admitting the risk expansion

$$R_\varepsilon(\theta_\varepsilon, \theta) = R(\theta) - \varepsilon^2 \rho(\theta) + o_K(\varepsilon^2), \quad (\varepsilon \to 0), \tag{8}$$

with a locally Hölder function $\rho \in C^{\beta}(\Theta)$, the subscript K meaning "uniformly on compacts $K \subset \Theta$ ". As we shall see shortly there exists a meaningfull relation between this problem and a second order linear elliptic differential operator

$$L\omega(\theta) = 2 \sum_{j=1}^{s} \frac{\partial}{\partial \theta_j}(a(\theta_j)\frac{\partial}{\partial \theta_j})\omega(\theta) + C(\theta)\omega(\theta) \tag{9}$$

with $C(\theta)$ to be defined later in remakr 2.

Provided the regularity conditions mentioned above hold one arrives at the following statements along the lines of Levit (1982,1983).

Theorem 1. An estimator θ_ε with risk expansion (8) exists if and only if there exists a positive solution (necessarily of the class $C^{2,\beta}(\Theta)$) $\omega(\theta)$ to the equation

$$L\omega(\theta) + \rho(\theta)\omega(\theta) = 0, \quad \theta \in \Theta. \tag{10}$$

Define the functions $g, z: \Theta \to R^s$

$$g_j(\theta) = \frac{\tilde{m}(\theta_j)}{6} \int_{R^s} \varphi_\theta(y)(i(\theta_j)y_j^4 - 3y_j^2)\omega(\mathcal{D}(\theta)y)dy,$$

$$z_j(\theta) = (\ln d(\theta_j))' \int_{R^s} \varphi_\theta(y)(i(\theta_j)y_j^4 - 4y_j^2 + i^{-1}(\theta_j))\omega(\mathcal{D}(\theta)y)dy.$$

For any arbitrary positive function $\omega(\cdot) \in C^{2,\beta}(\Theta)$ let

$$v(\theta) = I^{-1}(\theta)(\nabla \ln \omega^2(\theta) + A^{-1}(\theta)(z(\theta) + g(\theta)), \tag{11}$$

$$v_\varepsilon(\theta) = v(\theta)\chi_{\{|v(\theta)| < \varepsilon^{-1}\}}, \quad \theta_{\varepsilon,\omega} = \hat{\theta}_\varepsilon + \varepsilon^2 v_\varepsilon(\hat{\theta}_\varepsilon).$$

Theorem 2. Let $\omega(\cdot) \in C^{2,\beta}(\Theta)$, $\omega(\theta) > 0$.
Then

$$R_\varepsilon(\theta_{\varepsilon,\omega}, \theta) = R(\theta) + \varepsilon^2 \frac{L\omega(\theta)}{\omega(\theta)} + o_K(\varepsilon^2), \quad (\varepsilon \to 0).$$

Remark 1. Clearly $\theta_{\varepsilon,\omega}$ admits the expansion (8) provided ω satisfies (10). Theorems 1 and 2 imply that the class of estimators $\{\theta_{\varepsilon,\omega} \mid \omega \in C^{2,\beta}(\Theta), \omega > 0\}$ is second order complete in $\mathfrak{S}^{\beta}(\Theta)$.

Remark 2. The coefficient $C(\theta)$ in (9) may be duly defined as the ε^2 -term in the expansion (8) for $\theta_{\varepsilon,1}$.

Remark 3. For the l.f. (4) the $v(\theta)$ in (11) reduces to

$$v_j(\theta)=i^{-1}(\theta_j)\Big(\frac{\partial \ln \omega^2(\theta)}{\partial\theta_j} + \frac{3\alpha+2S-2}{2(S+2)}(\ln i(\theta_j))' + \frac{\alpha+S}{2(S+2)}\tilde{m}(\theta_j)\Big)$$

and for the l.f. (5)

$$v_j(\theta)=i^{-1}(\theta_j)\Big(\frac{\partial \ln \omega^2(\theta)}{\partial\theta_j} + (\ln d(\theta_j))' + \frac{\alpha+1}{6}\tilde{m}(\theta_j)\Big).$$

Therefore in this cases the m.l.e. $\hat{\theta}_\varepsilon$ coincides with $\theta_{\varepsilon,\omega}$ for $\omega^2(\theta)=\prod\limits_{j=1} \Omega^2(\theta_j)$ where for the l.f. (4)

$$\Omega^2(\tau)=i(\tau)^{-\frac{3\alpha+2S-2}{2(S+2)}} \exp\Big\{-\frac{\alpha+S}{2(S+2)}\int_{\tau_0}^{\tau}\tilde{m}(u)du\Big\} \tag{12}$$

and for the l.f. (5)

$$\Omega^2(\tau)= d(\tau)^{-1} \exp\Big\{-\frac{\alpha+1}{6}\int_{\tau_0}^{\tau}\tilde{m}(u)du\Big\}. \tag{13}$$

4. Some admissibility results.

Definition. An estimator $\theta_\varepsilon \in \mathfrak{S}^{\beta}(\Theta)$ is said to be second order admissible in $\mathfrak{S}^{\beta}(\Theta)$ (s.o.a.) if there do not exist estimators $\theta'_\varepsilon \in \mathfrak{S}^{\beta}(\Theta)$ with $\rho' \geq \rho$, $\rho' \not\equiv \rho$.

The next assertion is an immediate consequence from the theorems 1,2.

Assertion 1. The estimator $\theta_{\varepsilon,\omega}$ is s.o.a. iff there does not exist nontrivial positive solution to the inequality

$$L_\omega \varphi = \sum_{j=1}^{S} \frac{\partial}{\partial\theta_j}\Big(\omega^2(\theta) a(\theta_j)\frac{\partial}{\partial\theta_j}\Big)\varphi(\theta) \leq 0, \quad \theta \in \Theta.$$

Proof. Note that for any positive $\omega, \omega_1 \in C^2(\Theta)$

$$\frac{L\omega}{\omega} - \frac{L\omega_1}{\omega_1} = -\frac{2L_\omega\varphi}{\omega\omega_1} \quad \text{with} \quad \varphi = \frac{\omega_1}{\omega},$$

and that $L_\omega \varphi \leq 0$, $\varphi \not\equiv const$ implies $L_\omega \varphi^{1/2} \leq 0$ with strict inequality at least in a point. Then apply theorem 2 and the remark 1.

Let $W(\Theta)$ be the class of piecewise continuously differentiable functions φ vanishing outside compacts in Θ, with square integrable first order derivatives.

Theorem 3. The estimator $\theta_{\varepsilon,\omega}$ is s.o.a. iff $\inf\{F(\varphi) \mid \varphi \in W(\Theta),\ \varphi|_K \geq 1\} = 0$ for any compact $K \subset \Theta$ where

$$F(\varphi) = \int_\Theta \omega^2(\theta) \sum_{j=1}^{s} a(\theta_j) \left(\frac{\partial \varphi}{\partial \theta_j}\right)^2 d\theta \quad . \qquad (14)$$

The proof is a straight generalization of that of Theorem 4.1 in Levit)1985).

We present next a useful necessary condition for s.o.a. in the special case $\omega^2(\theta) = \Omega^2(\theta_1) \times \dots \times \Omega^2(\theta_s)$.

Theorem 4. Let for a given relatively open subset Q of $S_1 = \{\theta \mid |\theta| = 1\} \subset R^s$ the cone $C = \{\theta \mid \theta = \tau\vartheta, \tau > 0, \vartheta \in Q\}$ be contained in Θ. Suppose that both the functions $a(\tau)$ and $\Omega(\tau)$ are regularly varying at the points $0, \infty$ and that $\theta_{\varepsilon,\omega}$ is s.o.a. Then the integral $\int (\tau^{s-1} a(\tau) \Omega^{2s}(\tau))^{-1} d\tau$ diverges at the points 0 and ∞.

The proof is a straight generalization of that of Theorem 4.5 in Levit (1985). One has only to note that for any relatively open $Q \subset S_1$ not having any points in common with the coordinate axis the expression $a(\tau\vartheta_i)\omega^2(\tau\vartheta)/a(\tau)\Omega^{2s}(\tau)$ is bounded from zero uniformly in $\tau > 0$, $\vartheta \in Q$.

5. Some applications. It is clear from (12),(13) that the s.o.a. of the m.l.e. $\hat{\theta}_\varepsilon$ depends only on the behavior of the functions $i(\tau)$, $\tilde{m}(\tau)$ and $d(\tau)$ near the endpoints of $\tilde{\Theta}$ (but not on the particular family $f(x,\tau)$ at hand). However for the illustrative purposes we consider here two classical families, the first being the normal density

$$f_1(x,\tau)=(2\pi\tau^k)^{-1/2}\exp\{-x^2/(2\tau^k)\},\ -\infty<x<+\infty,\ \tau>0,\ k\neq 0,$$

the other beeing the Poisson distribution

$$f_2(x,\tau)=\tau^x e^{-\tau}/x! \qquad ,x=0,1,\ldots,\ \tau>0.$$

Let us confine to one of the loss functions

$$W_1 = \Big(\sum_{j=1}^{S} i(\theta_j)(\theta_{\varepsilon j}-\theta_j)^2\Big)^{\alpha/2}, \qquad (\alpha>0), \quad (15)$$

$$W_2 = \sum_{j=1}^{S} \theta_j^{\beta}\,|\theta_{\varepsilon j}-\theta_j|^{\alpha}, \qquad (\alpha>0, -\infty<\beta<+\infty). \quad (16)$$

We state first some admissibility results postponing till later a remark on their statistical meaning.

Assertion 2. The m.l.e. is s.o.a.

a) for f_1, W_1 iff $k=3$, $S=1,2$ (and then for any $\alpha>0$);
b) for f_1, W_2 iff $k=3, S=1$ or $k=3$, $S=2$, $\beta=-\alpha$;
c) for f_2, W_1 iff $S=1$, $\alpha=2$;
d) for f_2, W_2 iff $S=1, \alpha=2$ or $S=2$, $\alpha=2$, $\beta=0$.

Proof. a) From (6)(12) one easily obtains

$$a(\tau)=\text{const}\cdot\tau^2,\ \Omega^2(\tau)=\tau^{\gamma-1},\ \gamma=(\alpha+S)(3-k)/(2(S+2)).$$

According to Theorem 4 the necessary condition for s.o.a. is $\gamma=0$, in which case

$$a(\tau)=\text{const}\cdot\tau^2,\ \Omega^2(\tau)=\tau^{-1}. \qquad (17)$$

Substituting $\theta_i=\exp\{z_i\}$ in (14) reduces

$$F(\varphi)=\int_{R^{+S}} \Big(\prod_{i=1}^{S}\theta_i^{-1}\Big)\sum_{j=1}^{S}\theta_j^2\Big(\frac{\partial\varphi}{\partial\theta_j}\Big)^2 d\theta, \qquad \varphi\in W(R^{+S}),$$

to

$$\tilde F(\tilde\varphi)=\int_{R^S}\sum_{j=1}^{S}\Big(\frac{\partial\tilde\varphi}{\partial z_j}\Big)^2 dz, \qquad \tilde\varphi\in W(R^S).$$

Now it is well known that $\inf\{\tilde F(\tilde\varphi)\,|\,\tilde\varphi\in W(R^S), \tilde\varphi|_K\geq 1\}=0$ iff $S\leq 2$ (see e.g. Brown (1971) or Levit (1985)), from where the proof comes.

b) Here

$$a(\tau)=\text{const}\cdot\tau^{2+\alpha+\beta},\ \Omega^2(\tau)=\tau^{-\beta-(k+3)(\alpha+1)/6}$$

while any s.o.a. estimator $\theta_{\varepsilon,\omega}$ with $\omega^2(\theta)=\prod_{j=1}^{S}\Omega^2(\theta_j)$ is also necessarily s.o.a. for $S=1$. Thus one obtains two equations from Theorem 4:

$$(S-1)+(2+\alpha+\beta)-S(\beta+(k+3)(\alpha+1)/6)=1 \;,$$
$$(2+\alpha+\beta)-(\beta+(k+3)(\alpha+1)/6)=1 \;, \qquad (18)$$

resulting in $k=3$ and $(S-1)(\alpha+\beta)=0$ For $S=1$ the s.o.a. follows since the relation (18) is both necessary and sufficient for the case, as was mentioned in the introduction. For $\alpha+\beta=0$ we again find ourselves in case (17) considered above.

c) Here $a(\tau)=const\cdot\tau$, $\Omega^2(\tau)=\tau^{\frac{\alpha-2}{2(S+2)}}$. The necessary condition of theorem 4 which is also sufficient for s.o.a. provided $S=1$, results in the equation

$$S\left(1+\frac{\alpha-2}{2(S+2)}\right)=1 \;, \quad (S=1,2,\ldots,\; \alpha>0),$$

the only solution to which is $S=1$, $\alpha=2$.

d) Here $a(\tau)=const\cdot\tau^{(1+\beta+\alpha/2)}$, $\Omega^2(\tau)=\tau^{-\beta-(\alpha+1)/3}$. In the same manner as in the case b) one obtains

$$(S-1)+(1+\beta+\alpha/2)-S(\beta+(\alpha+1)/3)=1 \;,$$
$$(1+\beta+\alpha/2)-(\beta+(\alpha+1)/3)=1 \;,$$

resulting in $\alpha=2$, $(S-1)(\beta+\alpha/2-1)=0$. For $S=1$ the result follows immediately, while for $\beta+\alpha/2-1=0$ it again reduces to (17).

Remark 4. Related results on s.o.a. of the m.l.e. were obtained in Ghosh and Sinha (1981) for $S=1$, $\alpha=2$.

Remark 5. Note that $i(\tau)=const\cdot\tau^{-2}$ and $\tilde{m}(\tau)=(k+3)\tau^{-1}$ for $f=f_1$. Since in verifying s.o.a. of the m.l.e. only these functions matters, the assertion holds equally well for any parametric family with given behavior of $i(\cdot)$ and $\tilde{m}(\cdot)$. In particular the statements a), b) hold for the gamma family $f_3(x,\tau)=x^{\gamma-1}\exp\{-x/\tau^k\}/\tau^{\gamma k}\Gamma(\gamma)$ $(x>0, \tau>0, k\neq 0)$ independently of γ (cf. Berger (1980)); the Pareto family $f_4(x,\tau)=\lambda x^{-\lambda-1}$, $\lambda=\tau^{-k}$, $(x>1, \tau>0, k\neq 0)$ etc.

Remark 6. There is an abundant statistical literature on admissibility for the families f_i as well as for other distributions within the exponential family, mainly restricted to the case $\alpha=2$ in (15),(16); see the list of references below and further references therein.

Analyzing s.o.a. of the m.l.e. for different α exhibits the curiously sensitive way in which its admissibility depends on the particular choice of parametrization - as is the case with the families f_1, f_3, f_4 as well as on the loss function at hand, as in the case f_2. The situation thus presented seems to demonstrate that the whole affair of relying on the admissibility of estimators has in a way to be reconsidered.

But looking at it yet in another way forces one to admit the fruitfullness of relating the admissibility properties to the existence of corresponding non-trivial positive superharmonic functions and through this to some more involved mathematical fields of a running interest.

6. A sufficient condition for s.o.a. Turning back to Assertion 1, let $\Theta=R^S$, $\omega(\theta)=\Omega(\theta_1)\cdots\Omega(\theta_S)$, $A(\theta)=(\operatorname{diag} a(\theta_j))$ with both $\Omega(z)$ and $a(z)$ positive even functions, $a(\cdot)\in C^{1,\beta}(R^1)$, $\Omega(\cdot)\in C^{2,\beta}(R^1)$. Denote $h_-(z)=a(z)\Omega^2(z)(\int_0^z\Omega^2(u)du)^{S-1}$ and provided the last integral converges, $h_+(z)=a(z)\Omega^2(z)(\int_z^\infty\Omega^2(u)du)^{S-1}$.

Lemma 1. Let $\int_0^\infty h^{-1}(z)dz=\infty$ for $h=h_-$ or $h=h_+$. Then $\theta_{\varepsilon,\omega}$ is s.o.a.

Proof. Set $\varphi_R(\theta)=f(\max|\theta_j|)$, where

$$f(z)=\begin{cases}1, & z\le R_0,\\ C(R)\int_z^R h^{-1}(u)du, & R_0\le z\le R,\\ 0, & z\ge R\end{cases}$$

and $C(R)=\left(\int_{R_0}^R h^{-1}(u)\,du\right)^{-1}$. Note that $C(R)\to 0$ as $R\to\infty$ for any $R_0\ge 0$. For $h=h_-$ we have

$$F(\varphi_R)=\int_{R^S}\omega^2(\theta)\sum_{j=1}^S a(\theta_j)\Big(\frac{\partial\varphi_R}{\partial\theta_j}\Big)^2 d\theta = 2S\int_{0<\theta_j<\theta_S}$$

$$=2S\int_{R_0}^R a(\theta_S)\Omega^2(\theta_S)\int_{\substack{0<\theta_j<\theta_S\\ j=1,\dots,S-1}}\Big(\prod_{j=1}^{S-1}\Omega^2(\theta_j)\Big)\Big(\frac{\partial\varphi_R}{\partial\theta_S}\Big)^2 d\theta_1\cdots d\theta_{S-1}d\theta_S=$$

$$2sc^2(R)\int_{R_0}^{R} a(\theta_s)\Omega^2(\theta_s)\left(\int_0^{\theta_s}\Omega^2(u)du\right)^{s-1} h^{-2}(\theta_s)d\theta_s = 2sc(R)\to 0.$$

For $h=h_+$ replace $\{\theta_j<\theta_s, j=\overline{1,s-1}\}$ by $\{\theta_j>\theta_s, j=\overline{1,s-1}\}$ The assertion now follows from theorem 3.

Remark 7. Let $\Omega^2(r)$ be regularly varying at infinity with the exponent of regular variation $\rho\neq -1$. Then the sufficient condition of the Lemma 1 is equivalent to the divergence of the integral $\int^{\infty}(r^{s-1}a(r)\Omega^{2s}(r))^{-1}dr$ (cf. Theorem 4). This is readily verified using the well known relations (see e.g. Feller (1966), Ch. VII, 9, Theorem 1).

$$\lim_{r\to\infty} r\Omega^2(r)\Big/\int_0^r\Omega^2(u)du = |\rho+1|, \qquad \rho>-1,$$

$$\lim_{r\to\infty} r\Omega^2(r)\Big/\int_r^{\infty}\Omega^2(u)du = |\rho+1|, \qquad \rho<-1.$$

As for the last example we consider a slight generalization of the problem proposed in Brown (1980,1981). Let $\Theta=R^s$, $p(X,\theta)=(2\pi)^{-s/2}\exp\{-|X-\theta|^2/2\}$,

$$W(\theta_\varepsilon-\theta,\theta)=\sum_{j=1}^{s}(1+|\theta_j|)^{\beta}|\theta_{\varepsilon j}-\theta_j|^{\alpha}, \quad \alpha>0, -\infty<\beta<+\infty.$$

Here of course the m.l.e. is the sample mean.

Assertion 3. The m.l.e. is s.o.a. iff $\beta\geq 1$ or $\beta<1$, $s\leq(2-\beta)/(1-\beta)$.

Proof. With $i(r)\equiv 1$, $\tilde m(r)\equiv 0$ one has from (7),(13)

$$a(r)=\text{const}\cdot(1+|r|)^{\beta}, \quad \Omega^2(r)=(1+|r|)^{-\beta},$$

$$h_-(r)\asymp\begin{cases} r^{-(1-\beta)(s-1)}, & \beta<1,\\ (\ln r)^{s-1}, & \beta=1, \\ 1, & \beta>1.\end{cases} \quad (r\to\infty).$$

The assertion now follows from Lemma 1 since for $\beta\neq 1$ the condition of the Lemma is necessary and sufficient for s.o.a., according to Remark 7 (Note that the result does not depend on α).

REFERENCES

Berger J. (1980). Improving on inaddmissible estimators in continuous exponential families with applications to simultaneous estimation of gamma scale parameters. Ann. Statist. 8, 545-571.

Brown L.D. (1971). Admissible estimators, recurrent diffusions, and insoluble boundary value problems. Ann. Math. Statist. 42, 855-903.

Brown L.D. (1980). Examples of Berger's phenomenon in the estimation of independent normal means. Ann. Statist. 8, 572-585.

Brown L.D. (1981). The differential inequality of a statistical estimation problem. Unpublished manuscript.

Feller W. (1966). An introduction to probability theory and its applications. V. II. Wiley, New York.

Ghosh J.K., Sinha B.K. (1981). A necessary and sufficient condition for second order admissibility with applications to Berkson's bioassay problem. Ann. Statist. 9, 1334-1338.

Levit B.Ya. (1982). Minimax estimation and positive solutions of elliptic equations. Theor. Probab. Appl. 27, 525-546.

Levit B.Ya. Second order availability and positive solutions of the Schrödinger equation. (1983) Lect. Notes Math. 1021, 372-385.

Levit B.Ya. (1985). Second order asymptotic optimality and positive solutions of the Schrödinger equation. Theor. Probab. Appl. 30, 309-338.

Bounds for the asymptotic efficiency of estimators based on functional contractions; applications to the problem of estimation in the presence of random nuisance parameters

J.Pfanzagl
Mathematical Institute
University of Cologne
Weyertal 86
5000 Cologne 41
West Germany

INTRODUCTION

Let $(X, \mathcal{A})$ be a measurable space, and $\mathcal{P}$ a family of mutually absolutely continuous probability measures on $\mathcal{A}$. The problem is to estimate a functional $\kappa : \mathcal{P} \to \mathbb{R}$. Let $x_1, ..., x_n$ be independent realizations from $P \in \mathcal{P}$. Known to the experimenter are only the values $U(x_1), ..., U(x_n)$, where U is a measurable function from $(X, \mathcal{A})$ to some measurable space $(Y, \mathcal{B})$. What are the asymptotically optimal estimators for $\kappa(P)$, based on $U(x_1), ..., U(x_n)$?

In the first part of this paper we give an answer to this general problem. In the second part we illustrate the application of this general result to the problem of estimating a functional, based on observations governed by a probability measure depending on a random nuisance parameter.

PART 1

For any probability measure $P|\mathcal{A}$, let $P * U|\mathcal{B}$ denote the distribution of U under P (i.e. $P * U(B) = P(U^{-1}B)$, $B \in \mathcal{B}$). To be estimable on the basis of observations $U(x_\nu)$, $\nu = 1, ..., n$, the functional κ must necessarily satisfy a certain identifiability condition which allows us to consider $\kappa|\mathcal{P}$ as a functional on $\mathcal{P} * U$, say $\underline{\kappa}$. This identifiability condition is:

(1) For every $P, Q \in \mathcal{P}$, $\kappa(Q) = \kappa(P)$ implies $Q * U = P * U$.

To solve the general problem, we use certain results on tangent spaces.

The reader unfamiliar with this approach may consult Pfanzagl and Wefelmeyer (1982), p. 23 and (1985), p. 57.

Let $T(P, \mathcal{P})$ denote the tangent space of $\mathcal{P}$ at P, and $T(P*U, \mathcal{P}*U)$ the tangent space of $\mathcal{P}*U$ $(:= \{Q*U : Q \in \mathcal{P}\})$ at $P*U$. In most applications we have

$$P^U T(P, \mathcal{P}) = T(P*U, \mathcal{P}*U). \tag{2}$$

For a justification see Pfanzagl and Wefelmeyer (1982), p. 31. Here P^U denotes the conditional expectation operator, assigning to each function $f \in \mathcal{L}_2(X, \mathcal{A}, P)$ its conditional expectation $P^U f \in \mathcal{L}_2(Y, \mathcal{B}, P*U)$ with respect to P, given U.

Assume that κ is differentiable on $\mathcal{P}$, the canonical gradient being $\kappa^*(\cdot, P) \in T(P, \mathcal{P})$. Our problem is to determine the canonical gradient of $\underline{\kappa}$ on $\mathcal{P}*U$, say $\underline{\kappa}^*(\cdot, P*U) \in T(P*U, \mathcal{P}*U)$. It is straightforward to show that the identifiability condition (1) implies the following local counterpart, expressible in terms of the tangent space:

$$\text{For every } g \in T(P, \mathcal{P}),\ P^U g = 0 \text{ implies } g \perp \kappa^*(\cdot, P). \tag{3}$$

(For any two functions $f, g \in \mathcal{L}_2(X, \mathcal{A}, P)$, $f \perp g$ means that $P(fg) = 0$. To keep the formulas more transparent, we write $P(f)$ for $\int f(x)P(dx)$.)

Let $\kappa^\perp(\cdot, P*U)$ denote the orthogonal component of $P^U\kappa^*(\cdot, P)$ with respect to $P^U\{g \in T(P, \mathcal{P}) : g \perp \kappa^*(\cdot, P)\}$. With this notation, the canonical gradient of $\underline{\kappa}|\mathcal{P}*U$ at $P*U$ can be written as

$$\underline{\kappa}^*(y, P*U) = \frac{P(\kappa^*(\cdot, P)^2)}{P*U(\kappa^\perp(\cdot, P*U)^2)}\kappa^\perp(y, P*U), \quad y \in Y. \tag{4}$$

(It is condition (3) which guarantees that $P*U(\kappa^\perp(\cdot, P*U)^2) > 0$.)

Observe that $\{g \in T(P, \mathcal{P}) : g \perp \kappa^*(\cdot, P)\}$ has an intuitive interpretation. It is the set of all *directions* g in which the value of the functional remains unchanged. We call it the *level space* of the functional κ.

The knowledge of $\underline{\kappa}^*$ enables us to determine the asymptotic bound for the concentration of regular estimator-sequences for κ, based on the observations $U(x_\nu)$, $\nu = 1, ..., n$. For any asymptotically median unbiased estimator-sequence $\kappa^{(n)}$, $n \in \mathbb{N}$, the sequence of induced probability distributions, $P^n * n^{1/2}(\kappa^{(n)} - \kappa(P))$, $n \in \mathbb{N}$, cannot be more concentrated asymptotically than the normal distribution with mean 0 and variance $P*U(\underline{\kappa}^*(\cdot, P*U)^2)$. This expression will be called *minimal asymptotic variance* (of estimators based on U, in this case). See Pfanzagl and Wefelmeyer (1982), Section 9.2. What one would expect intuitively is, in fact, true: The minimal asymptotic variance of estimators based on $U(x_\nu)$, $\nu = 1, ..., n$, cannot be smaller than the minimal asymptotic variance of estimators based on x_ν, $\nu = 1, ..., n$: For all $P \in \mathcal{P}$,

$$P * U(\underline{\kappa}^*(\cdot, P * U)^2) \geq P(\kappa^*(\cdot, P)^2). \tag{5}$$

The canonical gradient is not only the key for computing the asymptotic variance bound. Estimators of the canonical gradient can also be used for the construction of asymptotically efficient estimator-sequences by a one-step-improvement procedure. (See Pfanzagl and Wefelmeyer (1982), Section 11.4, for the general background.)

PART 2

As a particular application we now discuss the problem of estimating a parameter from a sample $x_1, ..., x_n$, depending on a random nuisance parameter.

Let $\Theta \subset \mathbb{R}$ be an open set and H an arbitrary space, endowed with a σ-algebra $\mathcal{C}$. Let $\{P_{\vartheta,\eta} : \vartheta \in \Theta,\ \eta \in H\}$ be a family of mutually absolutely continuous probability measures on $\mathcal{A}$. Let Γ be a probability measure on $\mathcal{C}$, the distribution of the random nuisance parameter. The realization x_ν is obtained as the result of a two-stage experiment: First, the value of the nuisance parameter, η_ν, is obtained as a realization from Γ, then x_ν is obtained as a realization from P_{ϑ,η_ν}. The realizations for $\nu = 1, ..., n$ are mutually independent. The problem is to estimate ϑ.

To apply the results of Part 1, we consider (x_ν, η_ν) as the outcome of an experiment governed by the probability measure $Q_{\vartheta,\Gamma} \mid \mathcal{A} \times \mathcal{C}$, which is defined by

$$Q_{\vartheta,\Gamma}(B) := \int P_{\vartheta,\eta}(B_\eta)\Gamma(d\eta), \qquad B \in \mathcal{A} \times \mathcal{C}$$

(where B_η denotes the section of B at η).

If $x \to p(x, \vartheta, \eta)$ is a density of $P_{\vartheta,\eta}$ with respect to some dominating σ-finite measure μ, then $(x, \eta) \to p(x, \vartheta, \eta)$ is a density of $Q_{\vartheta,\Gamma}$ with respect to $\mu \times \Gamma$.

In the case of *unknown* random nuisance parameters it is only x_ν (and not the pair (x_ν, η_ν)) which is available for estimating $\kappa(Q_{\vartheta,\Gamma}) = \vartheta$. In other words, U is the projection $(x, \eta) \to x$. The induced measure $Q_{\vartheta,\Gamma} * U$ equals $P_{\vartheta,\Gamma}$, defined by

$$P_{\vartheta,\Gamma}(A) = \int P_{\vartheta,\eta}(A)\Gamma(d\eta), \qquad A \in \mathcal{A}.$$

Its μ-density is

$$x \to p(x, \vartheta, \Gamma) := \int p(x, \vartheta, \eta)\Gamma(d\eta). \tag{6}$$

The identifiability condition (1) reduces in this case to

$$P_{\vartheta',\Gamma'} = P_{\vartheta'',\Gamma''} \text{ implies } \vartheta' = \vartheta''.$$

Let $\mathcal{Q} := \{Q_{\vartheta,\Gamma} : \vartheta \in \Theta, \Gamma \in \mathcal{G}\}$, where $\mathcal{G}$ denotes the family of prior distributions $\Gamma|\mathcal{C}$. It is easy to see that (see e.g. Pfanzagl and Wefelmeyer (1982), p. 227 and p. 228) $T(Q_{\vartheta,\Gamma}, \mathcal{Q})$ contains all functions

$$(x,\eta) \to c\ell^\bullet(x,\vartheta,\eta) + k(\eta) \quad \text{with } c \in \mathbb{R} \text{ and } k \in T(\Gamma,\mathcal{G}),$$

and that

$$\kappa^*((x,\eta), Q_{\vartheta,\Gamma}) = \ell^\bullet(x,\vartheta,\eta)/Q_{\vartheta,\Gamma}(\ell^\bullet(\cdot,\vartheta,\cdot)^2),$$

where $\ell^\bullet(x,\vartheta,\eta) := \frac{\partial}{\partial\vartheta}\log p(x,\vartheta,\eta)$.

Finally, the level space is $\{(x,\eta) \to k(\eta) : k \in T(\Gamma,\mathcal{G})\}$.

According to the general prescription outlined in Part 1, we have to determine conditional expectations, given U. Thanks to the special nature of the function U as the projection into the first component, this becomes particularly simple: For any function $f : X \times H \to \mathbb{R}$ which is integrable with respect to $Q_{\vartheta,\Gamma}$, we have

$$(7) \qquad (Q^U_{\vartheta,\Gamma} f)(x) = \frac{\int f(x,\eta)p(x,\vartheta,\eta)\Gamma(d\eta)}{\int p(x,\vartheta,\eta)\Gamma(d\eta)}, \quad x \in X.$$

To obtain the canonical gradient, $\underline{\kappa}^*(\cdot, P * U)$, we determine the conditional expectations of the elements of the level space,

$$(8) \qquad K(\vartheta,\Gamma) := \{x \to \frac{\int k(\eta)p(x,\vartheta,\eta)\Gamma(d\eta)}{\int p(x,\vartheta,\eta)\Gamma(d\eta)} : k \in T(\Gamma,\mathcal{G})\},$$

and the conditional expectation of $\ell^\bullet(\cdot,\vartheta,\cdot)$, which is

$$(9) \qquad x \to \frac{\int \ell^\bullet(x,\vartheta,\eta)p(x,\vartheta,\eta)\Gamma(d\eta)}{\int p(x,\vartheta,\eta)\Gamma(d\eta)}.$$

Let $d(\cdot,\vartheta,\Gamma)$ denote the orthogonal component of this function with respect to $K(\vartheta,\Gamma)$. With this notation, the canonical gradient can be expressed as

$$(10) \qquad \underline{\kappa}^*(x, P_{\vartheta,\Gamma}) = d(x,\vartheta,\Gamma)/P_{\vartheta,\Gamma}(d(\cdot,\vartheta,\Gamma)^2).$$

From this we obtain the asymptotic variance bound

$$(11) \qquad 1/P_{\vartheta,\Gamma}(d(\cdot,\vartheta,\Gamma)^2).$$

The application of the general result of Part 1 to the problem of unknown random nuisance parameters has led us to a result which was obtained earlier (see Pfanzagl and Wefelmeyer (1982), Section 14.3) in a direct way.

Now we apply this result to a more special model, namely

$$p(x,\vartheta,\eta) = q(x,\vartheta)p_0(S(x,\vartheta),\vartheta,\eta). \tag{12}$$

The representation (12) is not unique, and it is convenient in applications to have a certain freedom in choosing q and p_0. For some purposes, it is, however, advantageous to use a certain canonical form of (12), in which $p_0(\cdot,\vartheta,\eta)$ is a density of $P_{\vartheta,\eta} * S(\cdot,\vartheta)$ (with respect to an appropriate σ-finite measure not depending on ϑ). Whenever a representation (12) exists, there also exists a *canonical* representation of this type. (The argument brought forward in connection with (14.3.11) in Pfanzagl and Wefelmeyer (1982), p. 232, remains true if the sufficient statistic S depends on ϑ.)

The level space $K(\vartheta,\Gamma)$ (see (8)) now consists of all functions

$$x \longrightarrow \frac{\int k(\eta)p_0(S(x,\vartheta),\vartheta,\eta)\Gamma(d\eta)}{\int p_0(S(x,\vartheta),\vartheta,\eta)\Gamma(d\eta)}, \quad k \in T(\Gamma,\mathcal{G}).$$

All these functions are contractions of $S(\cdot,\vartheta)$ (i.e. they depend on x through $S(x,\vartheta)$ only). Determining orthogonal components with respect to $K(\vartheta,\Gamma)$ becomes particularly simple if $K(\vartheta,\Gamma)$ is the class of *all* functions in $\mathcal{L}_2(X,\mathcal{A},P_{\vartheta,\Gamma})$ with expectation zero which are contractions of $S(\cdot,\vartheta)$. This is the case if (see Pfanzagl and Wefelmeyer (1985), p. 95, Proposition 3.2.5)

(*i*) the family $\mathcal{G}$ of prior distributions is *full* (i.e. if $T(\Gamma,\mathcal{G}) = \{k \in \mathcal{L}_2(H,\mathcal{C},\Gamma) : \Gamma(k) = 0\}$), and

(*ii*) the family $\{P_{\vartheta,\eta} * S(\cdot,\vartheta),\ \eta \in H\}$ is complete for every $\vartheta \in \Theta$.

Since many interesting applications are of this type, we consider it now in more detail.

Up to now we have not yet introduced explicitly the image space of $S(\cdot,\vartheta)$. Assume this is a measurable space $(Y,\mathcal{D})$. Then the class of all functions in $\mathcal{L}_2(X,\mathcal{A},P_{\vartheta,\Gamma})$ which are contractions of $S(\cdot,\vartheta)$ is

$$\{h \circ S(\cdot,\vartheta) : h \in \mathcal{L}_2(Y,\mathcal{D},P_{\vartheta,\Gamma} * S(\cdot,\vartheta))\}.$$

Hence our assumption about $K(\vartheta,\Gamma)$ may be written as

$$K(\vartheta,\Gamma) = \{h \circ S(\cdot,\vartheta) : h \in \mathcal{L}_2(Y,\mathcal{D},P_{\vartheta,\Gamma} * S(\cdot,\vartheta)), \\ P_{\vartheta,\Gamma} * S(\cdot,\vartheta)(h) = 0\}.$$

For any measurable function $f : X \to \mathbb{R}$ fulfilling $P_{\vartheta,\Gamma}(f) = 0$ its projection into $K(\vartheta,\Gamma)$ is its conditional expectation, given $S(\cdot,\vartheta)$, with respect to $P_{\vartheta,\Gamma}$. Since $S(\cdot,\vartheta)$ is sufficient for the family $\{P_{\vartheta,\eta} : \eta \in H\}$, it is sufficient for the family $\{P_{\vartheta,\Gamma} : \Gamma \in \mathcal{G}\}$. Hence the conditional expectation of f, given $S(\cdot,\vartheta)$, with respect to $P_{\vartheta,\Gamma}$ may be chosen independent of Γ (presuming that f is $P_{\vartheta,\Gamma}$-integrable). To emphasize the independence of Γ, the pertaining conditional expectation operator is denoted by $P_{\vartheta,\cdot}^{S(\cdot,\vartheta)}$. Hence the orthogonal complement of f with respect to $K(\vartheta,\Gamma)$ may be obtained as

$$x \to f(x) - (P_{\vartheta,\cdot}^{S(\cdot,\vartheta)} f)(S(x,\vartheta)). \tag{13}$$

The function for which we need this orthogonal complement is given by (9). For densities of the type (12) we have

$$\ell^\bullet(x,\vartheta,\eta) = \frac{q^\bullet(x,\vartheta)}{q(x,\vartheta)} + \frac{p_0^\bullet(S(x,\vartheta),\vartheta,\eta)}{p_0(S(x,\vartheta),\vartheta,\eta)} + \frac{p_0'(S(x,\vartheta),\vartheta,\eta)}{p_0(S(x,\vartheta),\vartheta,\eta)} S^\bullet(x,\vartheta)$$

(where $^\bullet$ means differentiation with respect to ϑ, and $p_0'(s,\vartheta,\eta) = \frac{\partial}{\partial s} p_0(s,\vartheta,\eta)$). Hence the function (9) becomes

$$x \to \frac{q^\bullet(x,\vartheta)}{q(x,\vartheta)} + c_1(S(x,\vartheta),\vartheta,\Gamma) + c_0(S(x,\vartheta),\vartheta,\Gamma)S^\bullet(x,\vartheta),$$

with

$$c_1(s,\vartheta,\Gamma) := \int p_0^\bullet(s,\vartheta,\eta)\Gamma(d\eta) / \int p_0(s,\vartheta,\eta)\Gamma(d\eta) \tag{14}$$

and

$$c_0(s,\vartheta,\Gamma) := \int p_0'(s,\vartheta,\eta)\Gamma(d\eta) / \int p_0(s,\vartheta,\eta)\Gamma(d\eta). \tag{15}$$

Let $a(\cdot,\vartheta)$ and $b(\cdot,\vartheta)$ denote the orthogonal component of $q^\bullet(\cdot,\vartheta)/q(\cdot,\vartheta)$ respectively $S^\bullet(\cdot,\vartheta)$ with respect to $K(\vartheta,\Gamma)$, determined according to (13). With these notations, we obtain

$$d(x,\vartheta,\Gamma) = a(x,\vartheta) + c_0(S(x,\vartheta),\vartheta,\Gamma)b(x,\vartheta). \tag{16}$$

From this, the canonical gradient can be obtained according to (10). The important point in (16) is that $d(\cdot,\vartheta,\Gamma)$ depends on the unknown prior distribution only through c_0.

Except for the case $a(\cdot,\vartheta) \equiv 0$, one can obtain an estimator for ϑ from the estimating equation

$$\sum_{\nu=1}^{n} a(x_\nu,\vartheta) = 0. \tag{17}$$

The advantage of this estimating equation lies in its independence of Γ. The asymptotic variance of the resulting estimators is

$$P_{\vartheta,\Gamma}(a(\cdot,\vartheta)^2)/(P_{\vartheta,\Gamma}(a^\bullet(\cdot,\vartheta)))^2.$$

Since $P_{\vartheta,\Gamma}(a^{\bullet}(\cdot,\vartheta)) = -P_{\vartheta,\Gamma}(a(\cdot,\vartheta)d(\cdot,\vartheta,\Gamma))$, this asymptotic variance is, in general, larger than the asymptotic variance bound (11). The asymptotic variance coincides with the asymptotic variance bound if $S^{\bullet}(\cdot,\vartheta)$ belongs to $K(\vartheta,\Gamma)$, for in this case $d(x,\vartheta,\Gamma) = a(x,\vartheta,\Gamma)$.

$S^{\bullet}(\cdot,\vartheta) \in K(\vartheta,\Gamma)$ is trivially fulfilled if $S(\cdot,\vartheta)$ is independent of ϑ. If (12) is in its canonical form, we have

$$P^S_{\vartheta,\cdot} \frac{q^{\bullet}(\cdot,\vartheta)}{q(\cdot,\vartheta)} = 0\ ,$$

hence $a(\cdot,\vartheta) = \frac{q^{\bullet}(\cdot,\vartheta)}{q(\cdot,\vartheta)}$, and the estimating equation (17) becomes

$$\sum_{\nu=1}^{n} \frac{q^{\bullet}(x_\nu,\vartheta)}{q(x_\nu,\vartheta)} = 0.$$

The estimators thus obtained are socalled *partial likelihood estimators*, and these are asymptotically optimal for densities of the type (12) with S not depending on ϑ, provided the conditions (i) and (ii) are fulfilled. This is the optimality result obtained in Pfanzagl and Wefelmeyer (1982), Section 14.3.

In the more general case of a sufficient statistic S depending on ϑ the problem remains to estimate the function $c_0(\cdot,\vartheta,\Gamma)$ in (16).

Let $c_0^{(n)}(\cdot,\vartheta;\ x_1,\dots,x_n)$, $n \in \mathbb{N}$, be such an estimator-sequence. We take any $n^{1/2}$-consistent estimator-sequence for ϑ, say $\vartheta^{(n)}$, $n \in \mathbb{N}$, and define

$$\begin{aligned}(18)\qquad & d^{(n)}(x;x_1,\dots,x_n) := a(x,\vartheta^{(n)}(x_1,\dots,x_n)) \\ & + c_0^{(n)}(S(x,\vartheta^{(n)}(x_1,\dots,x_n)),\ \vartheta^{(n)}(x_1,\dots,x_n);\ x_1,\dots,x_n) \\ & \quad b(x,\vartheta^{(n)}(x_1,\dots,x_n)).\end{aligned}$$

To obtain an asymptotically efficient estimator-sequence for ϑ, one needs an estimator-sequence for the canonical gradient $\underline{\kappa}^*(\cdot,P_{\vartheta,\Gamma})$. According to (8), this requires an estimator for $P_{\vartheta,\Gamma}(d(\cdot,\vartheta,\Gamma)^2)$, say $D^{(n)}$. Since x_ν, $\nu = 1,\dots,n$, are distributed according to $P_{\vartheta,\Gamma}$,

$$(19)\qquad D^{(n)}(x_1,\dots,x_n) := n^{-1}\sum_{\nu=1}^{n} d^{(n)}(x_\nu;x_1,\dots,x_n)^2$$

provides an example of such an estimator-sequence.

Then

$$(20)\qquad \underline{\kappa}^{(n)}(x;x_1,\dots,x_n) := \frac{d^{(n)}(x;x_1,\dots,x_n)}{D^{(n)}(x_1,\dots,x_n)}$$

is an estimator for $\underline{\kappa}^*(x, P_{\vartheta,\Gamma})$ which can be used to define an improved estimator,

$$\hat{\vartheta}^{(n)}(x_1, ..., x_n) := \vartheta^{(n)}(x_1, ..., x_n) + n^{-1} \sum_{\nu=1}^{n} \underline{\kappa}^{(n)}(x_\nu; x_1, ..., x_n). \tag{21}$$

Under suitable regularity conditions, the estimator-sequence $\hat{\vartheta}^{(n)}$, $n \in \mathbb{N}$, will be asymptotically optimal (in the sense of being regular and asymptotically normal with minimal asymptotic variance).

If interchanging of derivation with respect to s and integration with respect to Γ is permitted, the function c_0, as defined by (15), can be rewritten as

$$c_0(s, \vartheta, \Gamma) = p_0'(s, \vartheta, \Gamma)/p_0(s, \vartheta, \Gamma), \tag{22}$$

with $p_0(s, \vartheta, \Gamma) = \int p_0(s, \vartheta, \eta)\Gamma(d\eta)$.

If (12) is in its canonical form, $p_0(\cdot, \vartheta, \Gamma)$ is the density of $P_{\vartheta,\Gamma} * S(\cdot, \vartheta)$. Hence it appears natural to use $S(x_\nu, \vartheta)$, $\nu = 1, ..., n$, for estimating $p_0(\cdot, \vartheta, \Gamma)$. Of course, one cannot compute $S(x_\nu, \vartheta)$, since ϑ is unknown; but asymptotically one obtains the same result if one replaces ϑ by a preliminary estimator.

Bickel and Ritov (1985) suggested for this purpose, in connection with a regression model (where μ is the Lebesgue measure), the use of kernel estimators based on a logistic kernel. Following this suggestion, I tried to show (for the general case considered here) that estimators for $c_0(\cdot, \vartheta, \Gamma)$, based on kernel estimators of $p_0(\cdot, \vartheta, \Gamma)$, can be used in the improvement procedure (21). Simulation experiments revealed, however, that this does not work satisfactorily in practice with moderate sample sizes like 100, say. Hence it appears advisable to take advantage of the special structure of $p(\cdot, \vartheta, \Gamma)$ as a mixture (see (6)). Simulation experiments suggest that one can obtain satisfactory results if one uses $S(x_\nu, \vartheta^{(n)}(x_1, ..., x_n))$, $\nu = 1, ..., n$, to estimate first the prior distribution Γ, and replaces Γ by its estimator, say $\Gamma^{(n)}(x_1, ..., x_n)$, in (15) to obtain an estimator for c_0, and to estimate $P_{\vartheta,\Gamma}(d(\cdot, \vartheta, \Gamma)^2)$ by

$$\int d^{(n)}(x; x_1, ..., x_n)^2 p(x, \vartheta^{(n)}(x_1, ..., x_n), \Gamma^{(n)}(x_1, ..., x_n))\mu(dx).$$

Conditions under which such an estimation procedure gives rise to improved estimators which are asymptotically optimal are yet unknown. Available are results of simulation experiments which suggest that the following estimation procedure for Γ works satisfactorily.

(i) Approximate Γ by a discrete distribution, the support of which consists of $10 - 20$ points.

(ii) Use $S(x_\nu, \vartheta^{(n)}(x_1, ..., x_n))$, $\nu = 1, ..., n$, to estimate the probabilities of these supporting points, say by the EM-algorithm.

In a real application one will certainly use a more refined procedure for estimating Γ. If one takes the idea of a prior distribution serious, one would certainly expect that it has some smooth shape. Hence one would try different sets of supporting points and check whether the shape remains the same. One would interpolate in between supporting points with high probability and check whether the smooth shape remains unchanged, etc. Such a heuristic estimation procedure for Γ would certainly lead to good estimators, in spite of the fact that one can hardly think of a mathematical model for such a procedure which would lead to "proofs" for properties of the estimators thus obtained.

In order to establish the practicability of the estimation procedure outlined under (*i*)-(*ii*) we had to carry through a large number of simulation experiments for each of several examples. This, of course, excludes any of the refinements suggested above. Since the "unrefined" estimation procedure seems to be rather crude, we applied, in addition, the improvement procedure (21) with the true canonical gradient, $\underline{\kappa}^*(\cdot, P_{\vartheta,\Gamma})$, based on the knowledge of Γ, i.e. we computed, in addition,

$$\tilde{\vartheta}^{(n)}(x_1, ..., x_n) = \vartheta^{(n)}(x_1, ..., x_n) + n^{-1} \sum_{\nu=1}^{n} \underline{\kappa}^*(x_\nu, \vartheta^{(n)}(x_1, ..., x_n), \Gamma).$$

It seems plausible that no estimation procedure for Γ can lead to improved estimators asymptotically superior to $\tilde{\vartheta}^{(n)}$. The figures representing the results of Examples 1 and 2 show that, in general, these "fictitious" estimators $\tilde{\vartheta}^{(n)}$ are slightly superior to the improved estimators $\hat{\vartheta}^{(n)}$, but that this difference is not spectacular. Hence one cannot expect a remarkable gain from further refinements of the estimation procedure for Γ.

To illustrate that the estimation of $\underline{\kappa}^*(\cdot, P_{\vartheta,\Gamma})$ via *empirical Bayes estimates* for Γ really works, we consider two examples taken from literature. For each example, we present the asymptotic mean deviation from ϑ of the preliminary estimator and of the improved estimator, as well as 99% confidence intervals for the real mean deviation of the preliminary, the improved, and the "fictitious" estimator. The sample size is $n = 100$. Each of the confidence intervals is based on 10.000 simulations. In each case, the results are given for two different prior distributions.

EXAMPLE 1. Lindsay (1982), Kumon and Amari (1984).

$$p((x_1, x_2), \vartheta, \eta) = \vartheta\eta^2 \exp[-\eta(\vartheta x_1 + x_2)] ,$$
$$S((x_1, x_2), \vartheta) = \vartheta x_1 + x_2 ,$$
$$c_0(s, \Gamma) = \int_0^\infty \eta^3 \exp[-\eta s]\Gamma(d\eta) / \int_0^\infty \eta^2 \exp[-\eta s]\Gamma(d\eta) ,$$

$$d((x_1,x_2),\vartheta,\Gamma) = \frac{1}{2\vartheta}(x_2 - \vartheta x_1)c_0(S((x_1,x_2),\vartheta),\Gamma) \, .$$

Asymptotic variance bound:

$$(23) \qquad 12\vartheta^2 / \int_0^\infty c_0(s,\Gamma)^2 s^3 \int_0^\infty \eta^2 \exp[-\eta s]\Gamma(d\eta)ds.$$

Preliminary estimator: Median $\{x_{2\nu}/x_{1\nu} : \nu = 1,...,n\}$; asymptotic variance $4\vartheta^2$.

Simulations for mixing distributions δ_1 (with $\delta_1\{1\} = 1$) and Γ_3 (the Gamma distribution with shape parameter 3).

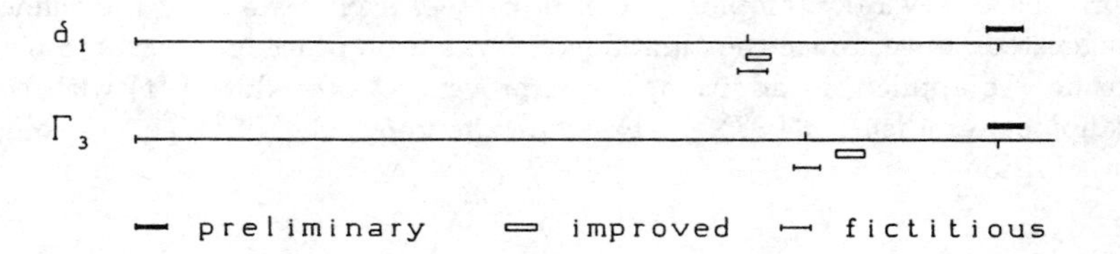

EXAMPLE 2. Kalbfleisch and Sprott (1970), Morton (1981), Kumon and Amari (1984).

$$p((x_1,...,x_q),\vartheta,\eta) = (2\pi\eta^2)^{-\frac{q}{2}} \exp[-\frac{1}{2\eta^2}\sum_{i=1}^{q}(x_i - \vartheta)^2] \, ,$$

$$S((x_1,...,x_q),\vartheta) = \sum_{i=1}^{q}(x_i - \vartheta)^2 \, ,$$

$$c_0(s,\Gamma) = \int_0^\infty \eta^{-2-q} \exp[-s/2\eta^2]\Gamma(d\eta) / \int_0^\infty \eta^{-q} \exp[-s/2\eta^2]\Gamma(d\eta) \, ,$$

$$d((x_1,...,x_q),\vartheta,\Gamma) = c_o(S((x_1,...,x_q),\vartheta),\Gamma)\sum_{i=1}^{q}(x_i - \vartheta) \, .$$

Asymptotic variance bound:

$$(24) \qquad 2^{\frac{q}{2}}\Gamma(\frac{q}{2}) / \int_0^\infty c_0(s,\Gamma)^2 s^{\frac{q}{2}} \int_0^\infty \eta^{-q} \exp[-s/2\eta^2]\Gamma(d\eta)ds \, .$$

Preliminary estimator: $(nq)^{-1}\sum_{\nu=1}^{n}\sum_{i=1}^{q} x_{i\nu}$; asymptotic variance $\frac{1}{q}\int_0^\infty \eta^2\Gamma(d\eta)$.

Simulations for $q = 2$. Mixing distribution δ_1 (with $\delta_1\{1\} = 1$) and G with density $\frac{8}{3\sqrt{\pi}}x^{-6}\exp[-x^{-2}]$, $x > 0$.

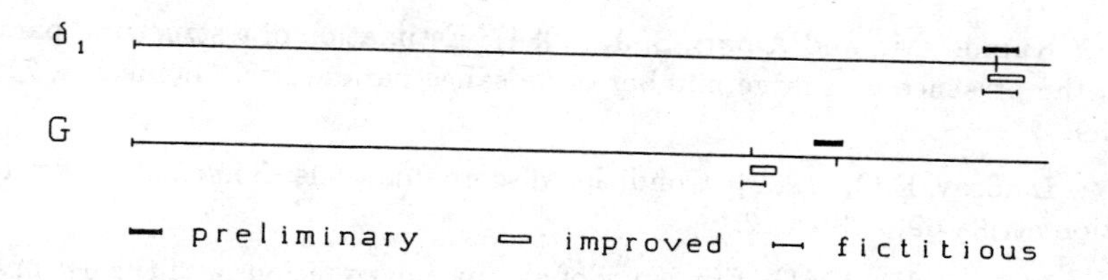

The estimator suggested in literature for Example 2 is based on estimators of the individual values of the nuisance parameters $\eta_1, ..., \eta_n$. If $\eta_1, ..., \eta_n$ are known, the optimal estimator for ϑ is

$$\sum_{\nu=1}^{n} \eta_\nu^{-2} \bar{x}_\nu / \sum_{\nu=1}^{n} \eta_\nu^{-2}, \qquad \text{with } \bar{x}_\nu := q^{-1} \sum_{i=1}^{q} x_{i\nu} .$$

In other words, ϑ is the solution of the estimating equation

$$\sum_{\nu=1}^{n} \eta_\nu^{-2} (\bar{x}_\nu - \vartheta) = 0 .$$

Replacing η_ν^2 by the estimator $q^{-1} \sum_{i=1}^{q} (x_{i\nu} - \vartheta)^2$, one arrives at the estimating equation

$$\sum_{\nu=1}^{n} \frac{\bar{x}_\nu - \vartheta}{q^{-1} \sum_{i=1}^{q} (x_{i\nu} - \vartheta)^2} = 0 .$$

The asymptotic variance of this estimator is $\frac{1}{q-2} / \int_0^\infty \eta^{-2} \Gamma(d\eta)$, an amount always larger than (24) (unless Γ is degenerate).

This example demonstrates that estimating the unknown nuisance parameters individually is not the best strategy. Estimating the distribution of the nuisance parameters leads to better estimators for ϑ.

ACKNOWLEDGMENT. The author wishes to thank Mr. W. Krimmel who helped with the numerical computations.

REFERENCES

Bickel, P.J., and Ritov, Y. (1985): Efficient estimation in the errors in variables model. Unpublished manuscript.

Kalbfleisch, J.D., and Sprott, D.A. (1970): Application of likelihood methods to models involving large numbers of parameters. J. Roy. Statist. Soc. Ser. B 32, 175 - 208.

Kumon, M., and Amari, S.-J. (1984): Estimation of a structural parameter in the presence of a large number of nuisance parameters. Biometrika 71, 445 - 459.

Lindsay, B.G. (1982): Conditional score functions: Some optimality results. Biometrika 69, 503 - 512.

Morton, R. (1981): Efficiency of estimating equations and the use of pivots. Biometrika 68, 227 - 233.

Pfanzagl, J., and Wefelmeyer, W. (1982): Contributions to a General Asymptotic Statistical Theory. Lecture Notes in Statistics 13. Springer-Verlag, New York.

Pfanzagl, J., and Wefelmeyer, W. (1985): Asymptotic Expansions for General Statistical Models. Lecture Notes in Statistics 31. Springer-Verlag, Berlin.

ON CHI-SQUARED GOODNESS-OF-FIT TESTS FOR LOCATION-SCALE MODELS

Drost F.C., Oosterhoff J.,
Free University, Amsterdam, Netherlands
Kallenberg W.C.M.,
Twente University of Technology, Enschede, Netherlands

Let $Y_1,\ldots,Y_n$ be i.i.d. real-valued random variables with distribution function (cdf) F^Y. Suppose one wants to test the composite null hypothesis

$$H_0: F^Y \in \mathcal{F}_0 = \{F_0(\tfrac{\cdot-\mu}{\sigma}) : \mu \in \mathbb{R},\ \sigma > 0\}$$

with $\theta = (\mu,\sigma)$ an unknown location-scale nuisance parameter and F_0 a given cdf with density f_0. Let $\hat{\theta}_n = (\hat{\mu}_n, \hat{\sigma}_n)$ be an equivariant estimator of θ based on the raw data $Y_1,\ldots,Y_n$, let $Z_j = (Y_j - \hat{\mu}_n)/\hat{\sigma}_n$, $j = 1,\ldots,n$, and let $\hat{F}_n$ denote the empirical cdf of $Z_1,\ldots,Z_n$. Introduce the cells $I_{ki} = (a_{k,i-1}, a_{k,i}]$, $i = 1,\ldots,k$, where $a_{k,0} = -\infty < < a_{k,1} < \ldots < a_{k,k-1} < a_{k,k} = \infty$. Denote the cell probabilities under H_0 by $p_{ki} = P_{F_0}(I_{ki}) = F_0(a_{k,i}) - F_0(a_{k,i-1})$, $i = 1,\ldots,k$, and the cell frequencies by

$$\hat{N}_{ni} = n(\hat{F}_n(a_{k,i}) - \hat{F}_n(a_{k,i-1})) = \#\{j: Z_j \in I_{ki},\ j = 1,\ldots,n\},\ i = 1,\ldots,k.$$

Chi-squared tests of H_0 can be based on the random (column) k-vector

$$V_k = ((\hat{N}_{ni} - np_{ki})/(np_{ki})^{\frac{1}{2}},\ i = 1,\ldots,k)$$

(this is the random cell approach). Well known test statistics are

(i) Watson-Roy statistic $WR_k = V_k' V_k$

(ii) Rao-Robson-Nikulin statistic $RR_k = V_k' \Sigma_k^{-1} V_k$

(iii) Dzhaparidze-Nikulin statistic $DN_k = V_k'[I_k - B_k(B_k'B_k)^{-1}B_k']V_k$.

Here Σ_k^{-1} is a generalized inverse of the asymptotic covariance matrix Σ_k of V_k under H_0, I_k is the k×k identity matrix and B_k is the k×2 matrix with i-th row

$$B_{ki} = p_{ki}^{-\frac{1}{2}} \nabla_{\mu,\sigma} \int_{I_{ki}} dF_0((y-\mu)/\sigma)\Big|_{\mu=0,\sigma=1},\ i = 1,\ldots,k.$$

We assume throughout that $\hat{\theta}_n$ admits the representation

$$n^{\frac{1}{2}}(\hat{\theta}_n - \theta) = n^{-\frac{1}{2}}\sigma\Sigma_{j=1}^{n} h((Y_j - \mu)/\sigma) + O_{P_\theta}(n^{-\frac{1}{4}})$$

where $h: \mathbb{R} \to \mathbb{R}^2$ is the influence function satisfying

$E_{F_0} h(Y_j) = 0$, $E_{F_0} [h(Y_j)h(Y_j)']$ is finite and nonsingular.
For m.l. estimators $\hat{\theta}_n^{ML}$ one has $h(y) = J^{-1} \nabla_{\mu,\sigma} \log \sigma^{-1} f_0((y-\mu)/\sigma) \big|_{\theta=(0,1)}$ where J is the Fisher information matrix of F_0 at $\theta = (0,1)$. In this special case, if $J - B_k'B_k$ is nonsingular,

$$\Sigma_k^{-1} = I_k + B_k(J - B_k'B_k)^{-1}B_k'.$$

For fixed k the null distributions of the statistics (i) - (iii) do not depend on θ and tend to χ^2 distributions (RR_k, DN_k) or to a linear combination of χ_1^2 distributions (WR_k) as $n \to \infty$.

The power of the three tests is studied for local alternative (contamination) hypotheses

$$H_{1n}: F^Y \in \mathcal{F}_{n1} = \{(1-\gamma_n)F_0(\tfrac{\cdot-\mu}{\sigma}) + \gamma_n F_1(\tfrac{\cdot-\mu}{\sigma}) : \mu \in \mathbb{R},\ \sigma > 0\}$$

where $n^{\frac{1}{2}}\gamma_n \to \gamma > 0$ and F_1 is a fixed alternative cdf.

We consider two questions:

(a) Let $k = k(n)$ depend on n. To optimize asymptotic local power, must $k(n)$ remain bounded or $k(n) \to \infty$ as $n \to \infty$?

(b) Are m.l.e.'s $\hat{\theta}_n^{ML}$ (under H_0) really the best choice for estimating θ or is a higher power possible with other estimators?

For an answer to question (a) we introduce the noncentralities

$$\delta_k^{WR} = \gamma^2(d_k - B_kH)'(d_k - B_kH)$$

$$\delta_k^{RR} = \gamma^2(d_k - B_kH)'\Sigma_k^{-1}(d_k - B_kH)$$

$$\delta_k^{DN} = \gamma^2(d_k - B_kH)'[I_k - B_k(B_k'B_k)^{-1}B_k'](d_k - B_kH)$$

where $d_k = ((\pi_{ki} - p_{ki})/p_{ki}^{\frac{1}{2}},\ i = 1,\dots,k)$, $\pi_{ki} = P_{F_1}(I_{ki})$ and $H = E_{F_1} h(Y_j)$.
We assume $\max_{1\le i\le k} p_{ki} \to 0$ and $k \min_{1\le i\le k} p_{ki} > \varepsilon > 0$ if $k \to \infty$.

<u>Theorem A</u>. Suppose $E_{F_1} h(Y_j)h(Y_j)'$ is finite and assume F_0 and F_1 satisfy some weak regularity conditions. Then for each of the level-α tests based on WR_k, RR_k or DN_k it holds that

$$\lim_{k\to\infty} \delta_k/k^{\frac{1}{2}} = \begin{cases} 0 \\ \infty \end{cases} \Rightarrow \text{asymptotic local power} = \begin{cases} \alpha \\ 1 \end{cases} \text{if } k(n) \to \infty;$$

hence small k is best if $\delta_k/k^{\frac{1}{2}} \to 0$ and $k \to \infty$ is best if $\delta_k/k^{\frac{1}{2}} \to \infty$.
Under stronger regularity conditions (f_1 is density of F_1)

$$\lim_{k\to\infty} \delta_k/k^{\frac{1}{2}} = \begin{cases} 0 \\ \infty \end{cases} \text{if } E_{F_0} |f_1(Y_j)/f_0(Y_j) - 1|^{\frac{4}{3}+\rho} \begin{cases} <\infty \text{ for a } \rho > 0. \\ =\infty \text{ for a } \rho < 0. \end{cases}$$

This extends a similar result for simple hypotheses in Kallenberg et al. (1985), cf. also Gvanceladze and Chibisov (1979). The proof is based on the asymptotic normality of the test statistics under H_0 and H_{1n} if $k = k(n) \to \infty$ slowly, see Drost (1986).

We now turn to question (b).

<u>Theorem B</u>. Let k be fixed, let $f_0(Y) > c > 0$ in neighborhoods of $a_{k,1}$ and $a_{k,k}$, let $\hat{\theta}_n$ be an estimator of θ and suppose the test statistics have a limit distribution under H_0. Then, if

$$n^{\frac{1}{2}} \| \hat{\theta}_n - \theta \| \to_{P_\theta} \infty \text{ under } H_{1n} \text{ as } n \to \infty, \qquad (*)$$

the power of the level-α tests based on WR_k and RR_k (if $r(\mathcal{Z}_k) = k-1$) tends to one under H_{1n} as $n \to \infty$.

Since by assumption $n^{\frac{1}{2}}(\hat{\theta}_n - \theta)$ is asymptotically normal under H_0, condition (*) expresses that $\hat{\theta}_n$ is highly non-robust w.r.t. local alternatives F_{n1} determined by F_1. This condition is satisfied e.g. for moment type estimators when F_1 is heavy-tailed.

Theorem B also holds for the tests of fit of Kolmogorov-Smirnov, Kuiper, Cramér-von Mises and Anderson-Darling, based on the empirical cdf $\hat{F}_n$ of $Z_1,\ldots,Z_n$. The condition $f_0(Y) > c > 0$ must now hold on a arbitrary open set. The situation w.r.t. DN_k is unsettled.

Our results are illustrated by some examples based on simulated data, employing various estimators $\hat{\theta}_n$. Estimators of location: sample mean $\hat{Y}$, trimmed sample mean $\bar{Y}_{.1}$ with 10% trimming on both sides, sample median M. Estimators of scale: sample standard deviation S, median absolute deviation Mad = median$\{|Y_j - M|, i = 1,\ldots,n\}/\Phi^{-1}(3/4)$ (the last estimator is appropriate in normal models). See Figure 1.

We expect that (i) for heavy tailed alternatives (first column of Fig.1) the power is highest for large k and for non-robust moment type estimators, (ii) for light or moderate tailed alternatives (second column) small k and estimators efficient under H_0 perform best, (iii) for skew alternatives (third column) intermediate behaviour. The numerical results agree reasonably well with expectations and show that the choice of $\hat{\theta}_n$ has often more effect than the choice of k.

References

Drost F.C. (1986). Generalized chi-square goodness-of-fit tests for location-scale models when the number of classes tends to infinity. Report 309, Dept. of Math. and Comp. Sc., Free University, Amsterdam.

Gvanceladze L.G., Chibisov D.M. (1979). On tests of fit based on grouped data. In: Contributions to Statistics, J. Hájek Memorial Volume. Prague: Academia, p. 79-89.

Kallenberg W.C.M., Oosterhoff, J. and Schriever B.F. (1985). The number of classes in chi-squared goodness-of-fit tests. J. Amer. Statist. Assoc., V.80, p. 959-968.

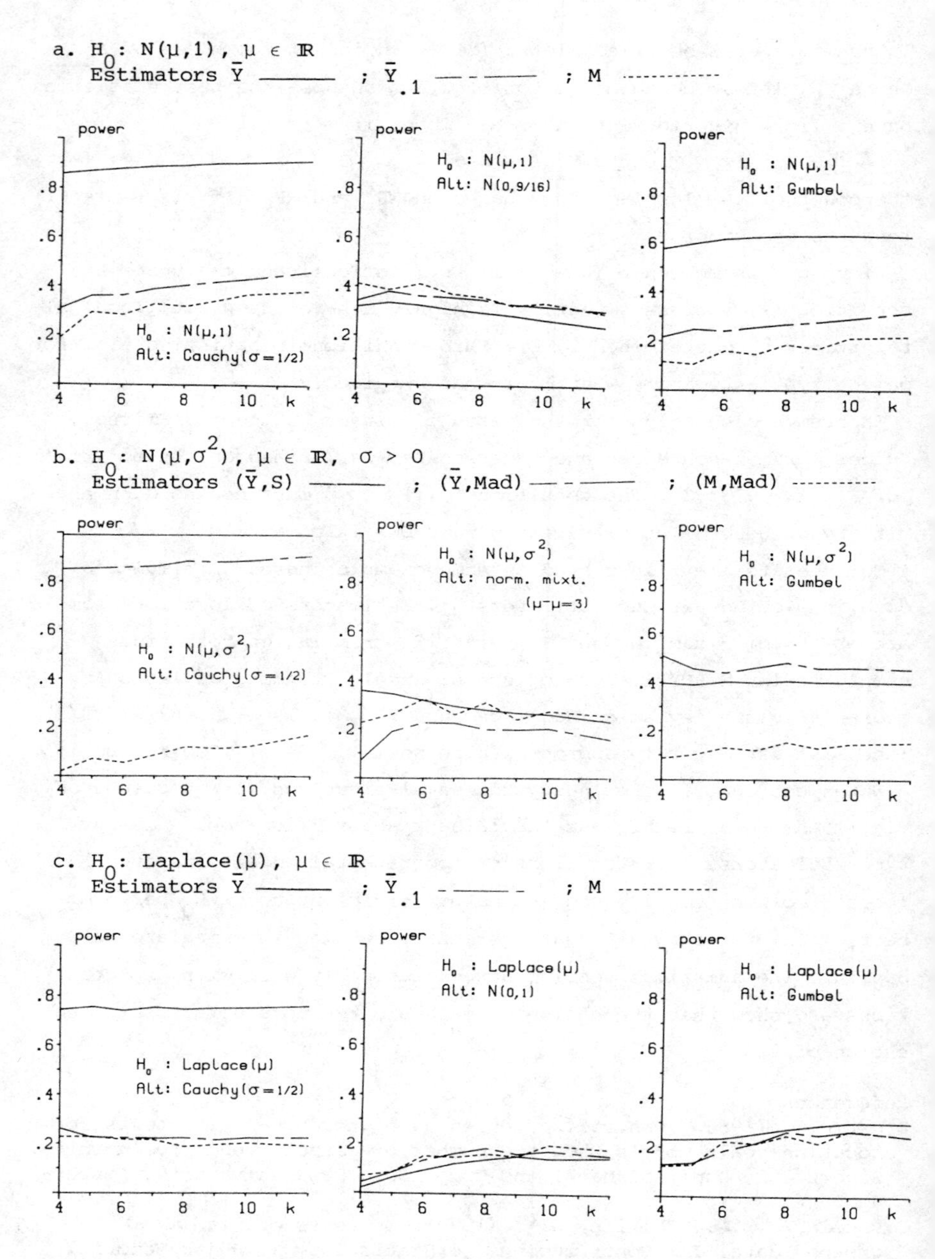

Figure 1. Simulated power of Rao-Robson-Nikulin test based on 10000 samples for 3 testing problems, employing various estimators $\hat{\theta}_n$ and equiprobable cells ($k = 4,\ldots,12$). Sample size $n = 50$, level $\alpha = .05$.

SOME PROBLEMS IN STATISTICS

F. Hampel
Seminar für Statistik
ETH-Zentrum, CH-8092 Zürich, Switzerland

This paper discusses several open or partly open problems in statistics and probability theory, in the hope of stimulating research.

1. Let us approach the central limit theorem (CLT) from a different angle. Let $X_1, X_2, \ldots$ be independent, identically distributed (i.i.d.) on R^1 with density $f(x)$, $EX_1 = 0$, $EX_1^2 = \sigma^2$ (and whatever regularity conditions will be needed). Let $p_n(t)$ be the density of $T = \sum_1^n X_i/n$. Define (for t in an interval around 0) $g_t(z) = c_t \exp(\alpha_t z) f(t+z)$ such that $\int g_t(z)dz = 1$ and $\int z g_t(z)dz = 0$.

Denote the density of the sum of n i.i.d. random variables with density g_t by $j_{n,t}(z)$. Then the log density derivative of T is given exactly by

$$p_n'/p_n(t) = n\int j_{n-1,t}(s)g_t(-s)f'/f(t-s)ds/\int j_{n-1,t}(s)g_t(-s)ds$$

(cf. Hampel 1973, Field and Hampel 1982). This formula, derived by elementary means, may be processed further in various ways.

Thus, by formal Taylor expansions, $\alpha_t = t/\sigma^2 + O(t^2)$ and $c_t = 1+t^2/(2\sigma^2) + O(t^4)$ whence $g_t(z) \approx (1+tz/\sigma^2)f(t+z)$ for small t. When it has been shown (perhaps also by elementary means?) that $j_{n,t}(s)/j_{n,t}(0)$ tends to a constant for $n \to \infty$ in a sufficiently large neighborhood of zero, then

$$\lim_{n\to\infty} p_n'(t)/(np_n(t)) = \int g_t(-s)f'/f(t-s)ds = -t/\sigma^2 + O(t^2) .$$

But this familiar form (the first, linear term of a Taylor expansion) is an unfamiliar variant of the central limit theorem (a "sublocal" CLT, though strictly speaking it is the only purely local one). In fact, normality means an exact straight line; and f'/f is both smoothed (essentially by g_t) and multiplied by n, so that a smaller and smaller interval carries almost all probability. The study of remainder terms and higher order terms may yield additional results about the CLT.

When the moment generating function for X_1 exists (as we have tacitly assumed), we obtain a "normal", i.e., nontrivial locally linear limit for all t, not only for t = 0 as in the classical CLT (cf.

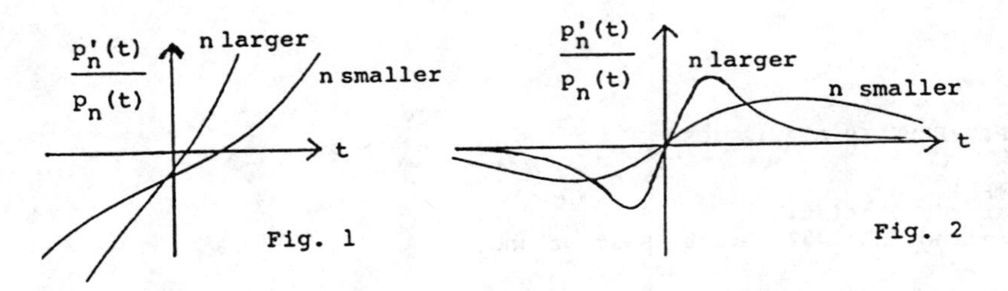

Fig.1). But when not all moments exist, multiplication by n is counteracted by smoothing towards zero away from the origin, so that $p_n'/p_n(t)$ gets actually steepened in a smaller and smaller region (Fig.2). In the boundary case of barely existing second moment, the locally linear central region barely collects all probability mass: this is the situation in the classical CLT. Under still weaker moment conditions, we may obtain convergence to stable laws with index $\alpha < 2$. Even if some of the exact details of this approach should prove to be difficult, the qualitative picture sketched still may provide much illumination and insight.

2. Many statistics can be viewed as functionals on the space of probability distributions, making study of their derivatives a useful tool. While in recent years the compact or Hadamard derivative has been popularized, it seems that for robust M-estimators (and other robust estimators?) the stronger Fréchet derivative is most appropriate. Clarke (1986/7) showed that even the Huber estimator of location (with discontinuous ψ'-function) is Fréchet differentiable. Perhaps it is possible to obtain results of the following kind: If the influence function of T at F is bounded with a bounded Lipschitz constant, and uniformly continuous as a function of F (in the weak topology for the F's), then T is Fréchet differentiable.

3. Consider for simplicity again the location parameter problem for i.i.d. observations. Given a parametric (location-scale) model for the data, and given the scale of the data, the distribution of an estimator is fixed for each location parameter and thus yields, in principle, confidence intervals for each n . By contrast, if nothing is known about the underlying distribution, the variance of the data is not sufficient for the variance of the estimator (except for the mean), and all data may be anywhere on one side of the parameter with probability > 5% for $n < 6$. For larger n , one may try to obtain non-parametric confidence intervals, for example by small sample asymptotic methods using conjugate distributions or "exponential tilting"

with the empirical cumulative distribution (cf. Field and Hampel 1978, p.15; cf. also recent work by C. Field and M. Tindley presented at the Joint Statistical Meetings 1986 in Chicago).

Given the exact or approximate distributions of a class of location estimators (perhaps under a robustness restriction) under a certain distribution for a fixed sample size, we may compare them by means of their variance, or (since the variance is very closely geared to the normal distribution, as already stressed by Fisher) by means of their Fisher-information, or perhaps also the length of confidence intervals on a fixed level. (The resulting optimal or "Pitman-type" estimators must not be confused with the "Pitman-like" estimators as used by Vernon Johns who considers the classical form and not the purpose of Pitman-estimators.)

The next step might be to study the stability or robustness of small sample standard errors, confidence intervals or Fisher-information under changes of the underlying distribution and perhaps find again optimal compromizes between size and stability. When a standard error etc. is rather stable, then it probably can be estimated rather accurately in a nonparametric setting, when the underlying distribution is unknown.

4. Large samples of supposedly independent data often show long-term (slowly decaying) serial correlations, which can most easily be modelled by means of self-similar processes (with hyperbolically decaying correlations). The effects of long-term serial correlations on confidence intervals may be quite serious: the true variance of the mean of 100 moderately correlated observations may easily be about ten to twenty times the nominal one under the independence assumption, and this ratio gets worse with increasing sample size. Cf. Graf et al. (1984) and Hampel et al (1986), Ch. 8.1; cf. also the talk by H.Künsch at this meeting. One of the (many) open questions is if and how robust randomization tests can be adapted to such serially correlated data.

5. There have been some investigations about the robustness of designs, as in the analysis of variance, against missing values, but hardly about the harder problem of robustness against outliers and gross errors. We may distinguish between "wild" (w) and "mean" (m), i.e., nasty outliers, and between discovery of a discrepancy (d) and correct identification and correction (c) of the outliers (cf. Hampel 1975 and Hampel et al. 1986, Ch. 8.4b). E.g., in the 3×3 design with main eff-

Table 1

0	0	0
0	-10	0
0	0	10

ects model of Table 1, it is not possible to identify the two mean outliers (without prior knowledge about the effects). Table 2 gives the maximum number of missing values or gaps (g) and outliers that can be safely handled in a k × k design with main effects model and unknown variance under favorable (f) and unfavorable (u) positioning of the deviant cells.

Table 2.	gd	gcf	gcu	wdf	wdu	wcf	wcu	mdu	mcu
2 × 2	4	1	1	-	-	-	-	-	-
3 × 3	9	4	2	5	3	3	1	2	1
4 × 4	16	9	3	12	8	8	2	3	1
5 × 5	25	16	4	21	15	15	3	4	2
6 × 6	36	25	5	32	24	24	4	5	2

R e f e r e n c e s

1. Clarke B.R. Nonsmooth analysis and Fréchet differentiability of M-functionals. - Probability Theory and Related Fields. (To appear) 1986/7.
2. Field C.A., Hampel F.R. Small sample asymptotic distributions of M-estimators of location. - Biometrika, 1982, Vol. 69, p. 29-46.
3. Graf H., Hampel F.R., Tacier J. The problem of unsuspected serial correlations. - In: Robust and Nonlinear Time Series Analysis. J. Franke, W. Härdle, R.D. Martin (eds.), Lecture Notes in Statistics. Vol. 26. - New York: Springer, 1984, p. 127-145.
4. Hampel F.R. Some small sample asymptotics. - In: Proceedings of the Prague Symposium on Asymptotic Statistics. Prague, 1973, p. 109-126.
5. Hampel F.R. Beyond location parameters: Robust concepts and methods (with discussion). - In: Proceedings of the 40th Session of the ISI, Vol. XLVI, Book 1, Warsaw, 1975, p. 375-391.
6. Hampel F.R., Ronchetti E.M., Rousseeuw P.J., Stahel W.A. Robust Statistics: The Approach Based on Influence Functions. - New York: Wiley, 1986, 502 p.

ADAPTIVE PROCEDURES FOR DETECTION OF CHANGE

Hušková M., Dept. of Statistics, Sokolovská 83,
186 00 Praha 8, Czechoslovakia

0. ABSTRACT

Sen (1980) developed the test procedures based on ranks for detection of small changes in the regression model occuring at an unknown time point. He proved that their powers attain asymptotic maximum for the same choice of the score-generating function as in e. g. the two-sample location problem. The aim of this paper is to construct adaptive test procedures for the mentioned testing problem when the distribution of errors is unknown.

1. INTRODUCTION

Let $Y_i = Y(t_i)$, $i = 1,\dots,n$, be indenpendent random variables taken at time points $t_1,\dots,t_n$, respectively, where $t_1 - t_2 - \dots - t_n$ (not all equal), Y_i have a distribution function F_i, $i = 1,\dots,n$ with $F_i(x) = F(x-\theta_o-\theta_i c_i)$ where $\theta_0,\theta_1,\dots,\theta_n$ are unknown parameters, $c_1,\dots,c_n$ are known regression constants, F has an absolutely continuous density f with finite nonzero Fisher´s information $I(f) = \int (f'(x))^2 (f(x))^{-1}\,dx$.

Consider the testing problem H_0: $\theta_1 = \theta_2 = \dots = \theta_n = \theta^*$ against $H^+_{1n}(\alpha,\beta)$: $\theta_1 = \theta_2 = \dots = \theta_\tau = \theta^* + \alpha C_n^{-1/2} \neq \theta_{\tau+1} = \dots = \theta_n = \theta^* + \beta C_n^{-1/2}$, $\alpha > \beta$, where $1 \leq \tau \leq n-1$ is an unknown time point, $C_n^2 = \sum_{i=1}^n (c_i - \bar{c}_n)^2$, $\bar{c}_n = \frac{1}{n}\sum_{i=1}^n c_i$.

Define the simple linear statistic

$$S_k(\varphi;\theta) = \sum_{i=1}^{k} (c_i - \bar{c}_k)\, a_k\, (\varphi; R_{ik}(\theta)) \quad k = 1,\dots,n,$$

where $R_{ik}(\theta)$ is the rank $Y_i - \theta c_i$ among $Y_1-\theta c_1,\dots$ $\dots, Y_k-\theta c_k$, $1 \leq i \leq k$, the scores $a_k(\varphi,i)$ are defined by $a_k(\varphi,i) = E\,\varphi(U_{ki})$, $i=1,\dots,k$, with $U_{k1} \leq \dots \leq U_{kk}$ being the ordered sample corresponding to the sample $U_1,\dots,U_k$ from the uniform (0,1)-distribution, the score-generating function φ is a nonconstant nondecreasing squared integrable defined on (0,1).

Sen (1980) proposed to establish the test for H_0 versus $H^+_{1n}(\alpha,\beta)$ on the statistic:

$$D^+_n(\varphi) = (A_n(\varphi)C_n)^{-1} \max_{1\leq k\leq n} S_k(\varphi, \hat{\theta}_n(\varphi)),$$

where

$$A^2_n(\varphi) = \frac{1}{n-1}\sum_{i=1}^{n}(a_n(\varphi,i) - a_n(\varphi))^2,\ a_n(\varphi) = \frac{1}{n}\sum_{i=1}^{n} a(\varphi,i)$$

and $\hat{\theta}_n(\varphi)$ is the usual R-estimator of θ. From the computational point of view a slightly modified procedure is more appealing. Namely, the procedure based on the statistic:

$$T^+_n(\varphi) = (A_n(\varphi)C_n)^{-1} \max_{1\leq k\leq n}\left\{ S_k(\varphi,\bar{\theta}_n) - \frac{A^2_k(\varphi)C^2_k}{A^2_n(\varphi)C^2_n}\, S_n(\varphi,\bar{\theta}_n)\right\},$$

where $\bar{\theta}_n$ is a preliminary shift equivariant estimator of θ with the property $C_n(\bar{\theta}_n-\theta) = O_p(1)$. Both procedures are asymptotically equivalent.

2. ADAPTIVE PROCEDURES

The test procedure based on $D^+_n(\varphi)$ $(T^+_n(\varphi))$ attains asymptotic maximum for $\varphi_f(u) = -f'(F^{-1}(u))(f(F^{-1}(u)))^{-1}$ $u \in (0,1)$ (Sen(1980)). Since in practice φ_f is rarely known it seems to be natural to use the procedure based on $D^+_n(\varphi)$ $(T^+_n(\varphi))$ with φ replaced by some suitable estimator $\hat{\varphi}_n$ of φ_f. Several types estimators of φ_f were developed (for references, see review papers Hogg (1974), Hogg and Lenth (1984) and Hušková (1984)). Most of them can be applied also in our problem of detection of small changes in the regression model. Here we shall exhibit some of them.

We shall assume that $\hat{\varphi}_n$ is an estimator of φ_f fulfilling:

a. $\hat{\varphi}_n$ is independent on $Y_1,\ldots,Y_n$;

b. $\int_0^1 (\hat{\varphi}_n(u) - \varphi_f(u))^2 \, du = o_p(1)$ under H_0

and either

c. $\hat{\varphi}_n$ is nondecreasing function on $(0,1)$ a.s.

or

c´. $\hat{\varphi}_n$ has the first and the second derivatives fulfilling

$$\sup_{u\in(0,1)} |\hat{\varphi}_n(u)| = O_p(n^{1/4-\varepsilon})$$

$$\sup_{u\in(0,1)} |\hat{\varphi}_n'(u)| = o_p(n^{1/2}), \quad \sup_{u\in(0,1)} |\hat{\varphi}_n''(u)| = o_p(n^{1/2}),$$

for some $\varepsilon > 0$.

Estimator $\hat{\varphi}_n$ satisfying a,b,c was proposed by van Eeden (1970) and estimator $\hat{\varphi}_n$ with a,b,c was suggested by Beran (1974) and further developed by Hušková (1984) and Hušková and Sen (1985). In the latter case, φ_f is approximated by a smooth function

$$\hat{\varphi}_n(u) = \sum_{i=1}^{M_n} \hat{d}_{in} \quad P_i(u) \quad u \in (0,1) ,$$

where $\{P_i(u),\ u \in (0,1)\}_{i=1}^{\infty}$ is a complete orthogonal system, $\hat{d}_{in}$ is an estimator of $\int_0^1 \varphi_f(u)\, P_i\,(u)\, du$ based on the asymptotic linearity of rank statistics, $M_n \to \infty$.

The main result is the following:

<u>Theorem.</u> Let assumptions formulated above be satisfied. Let the derivative of f be a.s. finite. Let

$$\lim_{n\to\infty} (n\,\theta)^{-1} \sum_{i=1}^{[n\theta]} c_1^s = \mu_s(\theta) \quad \text{for} \quad s = 1,2 \quad \text{exists and} \quad \mu_s(\theta)$$

be continuous for $\theta \in (0,1)$ and $\max_{1\le i\le n} |c_i| = O(1)$. If $\hat{\varphi}_n$ is an estimator of φ_f fulfilling either a,b,c or a,b,c´ then under $H_{1n}(\alpha,\beta)$:

$$\sup_{|(\theta-\theta^*)C_n|\le D} |T_n^+(\varphi_f,\theta) - T_n(\hat{\varphi}_n,\theta)| = o_p(1) \quad ,$$

where $D > 0$. The assertion remains true if we replace $T_n^+(\varphi,\theta)$ by $D^+(\varphi,\theta)$.

The resulting test procedure based on $T_n^+(\hat{\varphi}_n)(D_n^+(\hat{\varphi}_n))$ with asymptotic level α can be described as follows:

1. find a suitable location and scale invariant estimator $\bar{\theta}_n$ of θ,
2. compute

$$T_{kn}(\hat{\varphi}_n) = \frac{1}{A_n(\hat{\varphi}_n)C_n} \left(S_k(\hat{\varphi}_n,\bar{\theta}_n) - \frac{A_k^2(\hat{\varphi}_n)C_k^2}{A_n^2(\hat{\varphi}_n)C_n^2}\,,\ S_n(\hat{\varphi}_n,\bar{\theta}_n)\right)$$

$$1\le k\le n-1$$

$$\left(D_{kn}(\hat{\varphi}_n) = \frac{1}{A_n(\hat{\varphi}_n)C_n}\ S_k(\hat{\varphi}_n,\ \theta_n(\hat{\varphi}_n))\right);$$

3. if at least for one $1 \le k \le n-1$ $T_{kn}(\hat{\varphi}_n)(D_{kn}(\hat{\varphi}_n))$ exceeds Δ_α^+, where Δ_α^+ fulfills: $P(\sup_{0\ u\ 1} W_o(u) < \Delta_\alpha^+) = 1-\alpha$, we reject H_o: otherwise we accept H_o. Here $\{W^o(u);\ u \in (0,1)\}$ denotes the standard Brownian bridge.

REFERENCES

Beran, R. (1974). Asymptotically efficient adaptive rank estimates in location models. Ann. Statist. 2, 63-74.

Hogg, R. V.(1974). Adaptive robust procedures: a partial review and some suggestions for future application and theory. J. Amer. Statist. Assoc., 69, 909-923.

Hogg, R. V. and Lenth R. V. (1984). A review of some adaptive statistical techniques. Comm. in Statistics, 1551-1579.

van Eeden, C. (1970). Efficiency robust estimation of location. Ann. Math. Statist. 41, 172-181.

Hušková, M. (1984). Adaptive methods. Handbook of Statistics vol. 4,R. Krishnaiah and P. K. Sen (Ed.), 347-358.

Hušková, M. and Sen, P. K. (1985). On sequentially efficient rank statistics, Sequential Analysis, 4, 125-151.

Sen, P. K. (1980). Asymptotic theory of some tests for a possible change in the regression slope occuring at an unknown time point, Zeit. f. Wahr. 52, 203-218.

ON LOCAL AND NON-LOCAL MEASURES OF EFFICIENCY

W.C.M. Kallenberg
University of Twente, Faculty of Applied Mathematics, P.O.Box 217, Enschede, The Netherlands
(this research has been done partly together with
Jana Jurečková, Prague, partly with Teresa Ledwina, Wrocław)

Consider a sequence of estimators $\{T_n\} = \{T(\hat{P}_n)\}$ of the unknown parameter $\theta = T(P)$, where T is a real valued functional on a space of probability measures and where $\hat{P}_n$ denotes the empirical probability measure. To measure the performance of $\{T_n\}$ define

$$a_n(\varepsilon,T) = P\{|T_n - T(P)| > \varepsilon\}.$$

If $\{T_n\}$ is asymptotically normal, i.e.

$$T_n - T(P) \to N(0,\sigma^2(P))$$

as $n \to \infty$, then for each $c > 0$

$$(1)\quad \lim_{n\to\infty} a_n(n^{-1/2}c,T) = 2\Phi(-c\sigma^{-1}(P)),$$

and hence the well-known comparison based on asymptotic variances is equivalent to comparison based on the local ($n^{-1/2}c \to 0$) behaviour of a_n.

On the other hand non-local comparison is based on the inaccuracy rate

$$e(\varepsilon,\{T_n\}) = -\lim_{n\to\infty} n^{-1}\log a_n(\varepsilon,T).$$

The **non-local limit of the local measure** is obtained by sending $c \to \infty$ in (1) yielding

$$-\lim_{c\to\infty}\lim_{n\to\infty} c^{-2}\log a_n(n^{-1/2}c,T) = \{2\sigma^2(P)\}^{-1}.$$

Note that the first argument of a_n, i.e. $n^{-1/2}c$, tends to zero as $n \to \infty$, while the factor before the logarithm is kept fixed when taking the first limit ($n\to\infty$). More of less inverting the order of taking limits yields the **local limit of the non-local measure**:

$$\lim_{\varepsilon\to 0}\lim_{n\to\infty} (n\varepsilon^2)^{-1}\log a_n(\varepsilon,T),$$

where the first argument of a_n, i.e. ε, now is kept fixed and the factor before the logarithm tends to zero when taking the first limit $(n\to\infty)$.

QUESTION: **When do we have that the non-local limit of the local measure equals the local limit of the non-local measure, i.e. when do we have that**

$$(2) \quad \lim_{\varepsilon\to 0} \varepsilon^{-2} e(\varepsilon,\{T_n\}) = \{2\sigma^2(P)\}^{-1}?$$

So far no rigourous proof of (2) has been given.

To prove (2) rigourously we use the following approach:

(*) apply a very general large deviation theorem to obtain an expression for the involved inaccuracy rate $e(\varepsilon,\{T_n\})$;

(**) relate this complicated expression to the large deviation probability of a sum of i.i.d. r.v.'s in an expansion for $\varepsilon \to 0$; here we use suitable differentiability concepts for functionals;

(***) expand locally, in the usual way, the obtained large deviation probability of the sum of i.i.d. r.v.'s.

Remark 1. Consider the following example. Let $X_1, X_2, \ldots$ be i.i.d. r.v.'s normally $N(\theta,1)$ distributed. Consider the estimator

$$T_n = (1-H_n)Y_n + H_nX_1 \qquad n=2,3,\ldots,$$

where H_n, Y_n and X_1 are independent, $Y_n = (n-1)^{-1}\sum_{i=2}^n X_i$ and $P_\theta(H_n=1) = n^{-1} = 1-P_\theta(H_n=0)$. Then we have

$$(T_n-\theta)n^{1/2} \xrightarrow{D_\theta} N(0,1)$$

as $n \to \infty$, while $e(\varepsilon,\{T_n\}) = 0$ for each $\varepsilon > 0$; so (2) does not hold in this example. This example shows that although two estimators (Y_n and T_n) may be very close to each other locally, nevertheless their non-local behaviour may be very different.

In local theory one has the following approach. A statistical functional is approximated by a linear statistical functional leading, when applied to the empirical probability measure, to a sum of i.i.d. r.v.'s. Then it is shown that the remainder terms are small (in probability) and hence the distribution function of the statistic is approximated by the distribution function of a sum of i.i.d. r.v.'s, which can be handled. In large deviation theory such an approach fails as is seen in the above example, since the remainder terms have to be shown to be exponentially small. However the same **idea** as in local theory can be used here: the large deviation probability of the statistic (and not the statistic itself) is approximated locally by the large deviation probability of a sum of i.i.d. r.v.'s, cf. (**).

Elaboration of the three steps.

(*) Proposition 1. (Groeneboom et al.) **Under some regularity conditions we have**

$$-\lim_{n\to\infty} n^{-1}\log P\{|T_n - T(P)| > \varepsilon\} = K(\Omega_\varepsilon, P),$$

where

$$K(\Omega_\varepsilon,P) = \inf_{Q\in\Omega_\varepsilon} K(Q,P), \qquad \Omega_\varepsilon = \{Q : |T(Q) - T(P)| > \varepsilon\}$$

and

$$K(Q,P) = \begin{cases} E_Q \log(dQ/dP & \text{if } Q \ll P \text{ (Kullback-Leibler information number).} \\ \infty & \text{otherwise} \end{cases}$$

The expression $K(\Omega_\varepsilon,P)$ is in general rather complicated, but can be expanded locally ($\varepsilon \to 0$) for linear functionals, as is seen in the third step.

(***) Theorem 2. **Let** $g(\varepsilon) = \varepsilon + o(\varepsilon)$ **as** $\varepsilon \to 0$. **Suppose that**

$$\begin{aligned}(3)\quad &\inf\{K(Q,P) : |T(Q) - T(P)| \geq \varepsilon\} = \\ &\inf\{K(Q,P) : |\textstyle\int\psi dQ - \int\psi dP| \geq g(\varepsilon)\}\end{aligned}$$

with ψ **satisfying** $E_P \exp(r\psi) < \infty$, $E_P \exp(-r\psi) < \infty$ **for some** $r > 0$ **and** $E_P(\psi - E_P\psi)^2 > 0$. **Then**

$$(4)\quad K(\Omega_\varepsilon,P) = \{2E_P(\psi - E_P\psi)^2\}^{-1}\varepsilon^2 + o(\varepsilon^2) \text{ as } \varepsilon \to 0.$$

Note that (4) implies (2) since $\sigma^2(P) = E_P(\psi - E_P\psi)^2$. The right-hand side of (4) may be interpreted as the large deviation probability of $\sum_{i=1}^n \psi(X_i)$. What remains is the second step describing how we can approximate $T(P)$ by the linear functional $\int\psi dP$ to obtain condition (3). We therefore invoke differentiability of functionals.

(**) First we consider **Fréchet-type differentiability**, implying the existence of a bounded function ψ such that

$$(5) \qquad \lim_{\|Q-P\|\to 0} \frac{T(Q)-T(P)-\int \psi d(Q-P)}{\|Q-P\|} = 0$$

uniformly for $Q \in \{Q: Q<<P\}$. Here $\|\cdot\|$ denotes total variation.

Theorem 3. **If T is Fréchet-type differentiable with nonconstant function $\psi[P]$, then (4) holds true.**

When T is not Fréchet-type differentiable we may invoke a weaker differentiability concept: T is **Hadamard-type differentiable** if there exists a bounded function ψ such that for each compact subset C (5) holds true **uniformly** for $Q \in \{Q<<P, (Q-P)/\|Q-P\| \in C\}$.

Theorem 4. **If T is Hadamard-type differentiable and some regularity conditions are satisfied, then (4) holds true.**

In view of the nature of the problem (inverting the order of taking limits) uniformity is essential. The above differentiability concepts imply uniformity and are therefore very natural in this context.

Remark 2. There is an intimate relation between (2), and the equality of the local limit of Bahadur efficiency and the non-local limit of Pitman efficiency.

The theory can be applied e.g. on linear rank tests yielding the local behaviour of its Bahadur slope and in many other testing problems. In estimation theory the above theory is applied e.g. on L- and M-estimators.

More details can be found in Jurečková and Kallenberg (1987), and in Kallenberg and Ledwina (1987).

References

1. Groeneboom, P., Oosterhoff, J. and Ruymgaart, F.H. (1979). Large deviation theorems for empirical probability measures. Ann. Probability 7, 553-586.
2. Jurečková, J. and Kallenberg, W.C.M. (1987). On local inaccuracy rates and asymptotic variances. Statistics and Decisions, to appear.
3. Kallenberg, W.C.M. and Ledwina, T. (1987). On local and non-local measures of efficiency. The Annals of Statistics, to appear.

MAXIMAL DEVIATIONS OF GAUSSIAN PROCESSES AND EMPIRICAL DENSITY FUNCTIONS

V.D.Konakov
Central Economics Mathematical Institute of the USSR Academy of Sciences.Moscow.

Let X_1 and X_2, $t\in[0,T]$, be a pair of independent gaussian processes of the form $X_i(t)=\mu_i(t)+Y_i(t)$, where $\mu_i(t)\in C^2[0,T]$, $i=1,2$. $Y_i(t)$ are stationary gaussian processes with zero mean and covariance functions $r_i(t)$. Suppose that

$$r_i(t)=1-\frac{t^2}{2}+\frac{\lambda_{4.i}t^4}{4!}+\frac{\lambda_{6.i}t^6}{6!}+o(t^6), \quad i=1,2 \tag{1}$$

The random vectors

$$\left(X_i(0),X_i(t),X_i'(0),X_i'(t),X_i''(0),X_i''(t)\right),\ i=1,2 \tag{2}$$

have nondegenerate densities in R^6 for $t\in[0,T]$.
Definition. The process $V(t)=\mu(t)+X\cos t+Y\sin t$, $t\in[0,T]$, where (X,Y) is a two-dimensional standard gaussian vector, is called a cosine-process.
Denote $\|\varphi\|_2=\max\limits_{[0,T]}\{|\varphi(t)|,|\varphi'(t)|,|\varphi''(t)|\}$.

Theorem 1. Let the conditions (1) and (2) be satisfied. Then it is possible to find $\varepsilon>0$ depending only on a covariance structure of the processes $Y_i(t)$ and constants $C(T)$ and $0<\rho_T<1$ such that $\|\mu\|_2\le\varepsilon u$, $i=1,2$, implies

$$\left|P(\max_{[0,T]}|X_1(t)|\le u)-P(\max_{[0,T]}|X_2(t)|\le u)\right|\le C(T)Tu^2e^{-\frac{u^2}{1+\rho_T}} \tag{3}$$

If, in addition, $\int|r_i(t)|\,dt<\infty$ and $u_T\le T\exp(-\delta u_T^2)$ for some $\delta>0$, then one can choose C and $0<\rho<1$ independent of T, i.e.,

$$\left|P(\max_{[0,T]}|X_1(t)|\le u_T)-P(\max_{[0,T]}|X_2(t)|\le u_T)\right|\le CTu_T^2 e^{-\frac{u_T^2}{1+\rho}} \quad (4)$$

Inequality (3) holds true if one of the processes is cosine-process and $T<\pi$.

Let $k(x)$ be a function with finite support such that $k^{(i)}(x)\in L_2(-\infty,\infty)$, $i=1,2,3$ and $\int k^2(x)dx=\int (k'(x))^2dx=1$. Suppose $Y_T(t)=M_T(t)+\int k(t-s)\,dW(s)$ where $W(s)$ is a two-sided Wiener process. We consider a curve in (x,y) plane discribed by the equation $\tau(\varphi)=u_T-M_T(\varphi)$, $\varphi\in[0,T]$, in polar coordinates. The straight line perpendicular to the radius-vector $\tau(\varphi)$ at its end point divides the plane into two half-planes, let $S_{M,\varphi}$ be the half-plane which contains the origin. Denote $I_k=[(k-1)t_0,kt_0]$, $k=1,\ldots,N_1$, $N_1=[T/t_0]$, $t_0\in(0,\pi/4)$ and define

$$p(k+1)=P_\xi\{\bigcap_{I_{k+1}} S_{M,\varphi}\}+P_\xi\{\bigcap_{I_{k+1}} S_{-M,\varphi}\}$$

$$p(k,k+1)=P_\xi\{\bigcap_{I_k\cup I_{k+1}} S_{M,\varphi}\}+P_\xi\{\bigcap_{I_k\cup I_{k+1}} S_{-M,\varphi}\}$$

where $\xi=(X,Y)$ is a standard two-dimensional gaussia vector, $P_\xi(A)=P(\xi\in A)$.

Therem 2. Assume that $M_T(t)\in C^2[0,T]$ for any T. Then it is possible to find $\varepsilon>0$, $C<\infty$ and $0<\rho<1$ depending only on k and t_0 such that $\|M_T\|_2<\varepsilon u_T$ implies

$$P\{\max_{[0,T]}|Y_T(t)|\le u_T\}=\exp\{-\sum_{k=1}^{N_1-1}(p(k+1)-$$

$$-p(k,k+1))\}\cdot(1+L(T,u_T))$$

where $|L(T,u_T)|\le C\cdot T\cdot u_T^2\exp(-\frac{u_T^2}{1+\rho})$.

Consider $M_k(\varphi)=M_T((k-1)t_0+\varphi)$, $\varphi\in I$, $I=[0,t]$, $t<\frac{\pi}{2}$. We draw perpendiculars l_1 and l_2 at the end points of $\tau_k(0)$ and $\tau_k(t)$, $\tau_k(\varphi)=u_T-M_k(\varphi)$, $\varphi\in I$, directed outside the curvilinear sector which is enclosed by the curve $\tau_k(\varphi)$, $\varphi\in I$. Let Λ_k^+ be the set under the curve composed of ℓ_2, $\tau_k(\varphi)$, $\varphi\in I$, and ℓ_1 and let Λ_k^- correspond to the change of $\mu_k(\varphi)$ by $-\mu_k(\varphi)$. Define

$$R_T(I)=\sum_{k=1}^{N_1-1} P_\xi(\Lambda_k^+ \Delta \bigcap_I S_{\mu_k,\varphi})+P_\xi(\Lambda_k^- \Delta \bigcap_I S_{-\mu_k,\varphi})$$

$$R_T=\max(R_T(I_1),R_T(I_1\cup I_2)),\ I_1=[0,t_0],\ I_2=[t_0,2t_0]$$

Theorem 3. Let the conditions of theorem 2 be satisfied and $R_T\to 0$ as $T\to\infty$. Then it is possible to find constants $\varepsilon>0$, $C<\infty$ and $0<\rho<1$ such that $\|\mu_T\|_2<\varepsilon u_T$ implies

$$P\{\max|Y_T(t)|\le u_T\}=\exp\{-\frac{T}{2\pi}\int(e^{-\frac{(u_T-x)^2}{2}}+e^{-\frac{(u_T+x)^2}{2}})d\eta_T\}\times \tag{5}$$

$$\times(1+L(T,u_T))$$

where $\eta_T(x)=T^{-1}\cdot\lambda\{s\in[0,T],\mu_T(s)\le x\}$, λ is the Lebesgue measure on R^1.

$$|L(T,u_T)|\le C(R_T+\exp(-\frac{u_T^2(1-\varepsilon)^2}{2})+Tu_T^2\exp(-\frac{u_T^2}{1+\rho}))$$

Denote $u_T=\ell_T+x/\ell_T$, $\ell_T=\sqrt{2\ln(T/2\pi)}$, $b_T(s)=u_T\,\mu_T(s)$,

$$\omega(\delta,x(\cdot))=\sup_{|s'-s|\le\delta,\ s\in[0,T]}|x(s')-x(s)|,\ \Delta_T=\max(\Delta_T^+,\Delta_T^-)$$

Δ_T^+ is the maximal increment $\Delta\varphi$ for which inequality

$$\left(1-\frac{b_T(\varphi)}{u_T^2}\right)(1-\cos\Delta\varphi)\le\frac{b_T(\varphi+\Delta\varphi)-b_T(\varphi)}{u_T^2}$$

holds true for $\varphi\in[0,T]$, Δ_T^- corresponds to the substitution $-\mu_T(\varphi)$ for $\mu_T(\varphi)$.

Corollary 1. Assume that the conditions of theorem 2 are satisfied, $b_T(s)$ is uniformly bounded in T and $s\in[0,T]$, $\lim_{T\to\infty}\omega(\Delta_T,b_T(\cdot))=0$. Then there exist $\gamma>0$ and $C<\infty$ such that (5) holds true with $|L(T,u_T)|\le C(\omega(\Delta_T,b_T(\cdot))+T^{-\gamma})$.

A refinement of the result of Bickel and Rosenblatt (1973) easily followes from corollary 1, namely, assume that

a) $b_T(s)$ is uniformly bounded in T and $s \in [0,T]$

b) there exists a uniformly continuous function $b(s)$ on $[0,\infty)$ such that $\gamma_T = \sup_{[0,T]} |b_T(s) - b(s)| \to 0$ as $T \to \infty$

c) there exists a distribution function $\eta(x)$ such that

$\tilde{\gamma}_T = \sup_{[0,T]} |T^{-1} \cdot \lambda\{s \in [0,T], b_T(s) \le x\} - \eta(x)| \to 0$ as $T \to \infty$.

Corollary 2. Let the conditions of theorem 2 and a) - c) be satisfied. Then it is possible to find a constant C such that

$$P\{\max_{[0,T]} |Y_T(t)| \le \ell_T + x/\ell_T\} = \exp\{-e^{-x} \cdot \int (e^{-z} + e^{z}) d\eta(z)\} \times$$

$$\times (1 + L(T,u_T)), \ |L(T,u_T)| \le C(\omega(\Delta_T, b(\cdot)) + \gamma_T + \tilde{\gamma}_T + \frac{1}{\ln T})$$

Using an appropriate approximation sceme, the results stated above are applied to investigation of the maximal deviation of empirical density functions.

References

Bickel, P., Rosenblatt, M. (1973). On some global measures of the deviations of density function estimates. Ann. Statist. 1, 1071-1095.

CHI-SQUARED TEST STATISTICS BASED ON SUBSAMPLES

Mirvaliev M., Institute of Mathematics Academy of Sciences Uzbek SSR, Tashkent, USSR

Moore and Spruill (1975), hereafter refered to as MS, gave unified large-sample theory of general chi-squared statistics for tests of fit (see also Chibisov, 1971). Chase (1972) studied χ^2 statistics of Pearson-Fisher and Chernoff-Lehmann type when parameters are estimated independently of the sample. Murty and Gafarian (1970) obtained the limiting distributions of the same statistics when additional observations are required for estimation of parameters. We will consider the case in which the parameters are estimated by subsamples.

Let $Y_1, Y_2, \ldots$ be independent R^k - valued random variables with df $F(x|\theta,\eta)$, $\theta \in \Omega_1 \subset R^m$, where Ω_1 is an open set, η ranges over a neighborhood of a point η_o in R^p. Suppose one wants to test the composite null hypothesis H_o: $\eta = \eta_o$ against sequences of alternatives H_{1n}: $\eta = \eta_n = \eta_o + n^{-1/2}\gamma$ for fixed $\gamma \in R^p$.

In forming chi-square type statistics we partition R^k into $M = \prod_{i=1}^{k} M_i$ cells $I_1(\varphi),\ldots,I_M(\varphi)$. They are formed by the Cartesian products of the cells of the partition of the x_i - axis by functions of

$$-\infty = x_{i,o}(\varphi) < x_{i,1}(\varphi) < \ldots < x_{i,M_i-1}(\varphi) < x_{i,M_i}(\varphi) = \infty$$

for each $i=1,\ldots,k$ and φ ranges over an open set Ω_2 in R^r. Let $V_n(\theta,\eta,\varphi)$ - is M-vector with components

$$v_j(\theta,\eta,\varphi) = \frac{N_{nj}(\varphi) - np_j(\theta,\eta,\varphi)}{[np_j(\theta,\eta,\varphi)]^{1/2}}, \quad j=1,\ldots,M$$

where $N_{nj}(\varphi)$ is the number of $Y_1,\ldots,Y_n$ falling in the j-th cell and

$$p_j(\theta,\eta,\varphi) = \int_{I_j(\varphi)} dF(x|\theta,\eta).$$

We will assume that (see MS)

A5. Under (θ_o,η_n)

$$n^{1/2}(\theta_n - \theta_o) = n^{-1/2}\sum_{i=1}^{n} h(Y_i,\eta_n) + A\gamma + o_p(1)$$

for some m x p matrix A and measurable function $h(x,\eta)$ from $R^k \times R^r$ to R^m satisfying the following condition

$$E[h(Y,\eta_n)|(\theta_o,\eta_n)]=0,\ E[h(Y,\eta_n)h(Y,\eta_n)'|(\theta_o,\eta_n)] = L(\eta_n)$$

where $L(\eta_n)$ is a nnd m x m matrix converging to the finite nnd matrix $L = E[h(Y)h(Y)']$ as $n \to \infty$.

Let us suppose that m(n) elements of the sample $Y_1,\dots,Y_n$ are chosen such that assumptions of i. and i.d. are saved. For our χ^2 test statistics the inknown parameters are estimated, without loss generality, by subsample $Y_1,\dots,Y_{m(n)}$. Here we give only limiting alternative distributions of statistics, but it is suffisient to suppose $\gamma = 0$ to obtain null case results.

Denote

$$B = \left(p_i^{-1/2} \frac{\partial p_i}{\partial \theta_j} \right),\quad B_{12} = \left(p_i^{-1/2} \frac{\partial p_i}{\partial \eta_l} \right)$$

for each $i = 1,\dots,M,\ j = 1,\dots,m,\ l = 1,\dots,p$ and $q = (p_1^{1/2},\dots,p_M^{1/2})'$.

Theorem 1. Suppose that as $n \to \infty$, $\frac{m(n)}{n} \to \tau$, $0<\tau\le 1$. If assumptions A1 - A3, A6 and A5 with n replaced by m(n) hold, then under (θ_o,η_n)

$$V_n(\theta_{m(n)},\eta_n) \xrightarrow{\mathcal{L}} N(\mu_\tau, \Sigma_\tau),$$

where

$$\mu_\tau = (B_{12} - \tau^{-1/2} BA)\gamma,$$

$$\Sigma_\tau = I_M - qq' + \frac{1}{\tau}BLB' - BE[h(Y)W(Y)'] - E[W(Y)h(Y)']B'$$

W(Y) -is M-vector with j-th component $[\chi_j(x) - p_j]/p_j^{1/2}$

and $\chi_j(x)$ is indicator function of $I_j(\varphi_o)$.

The Moore-Penrose inverse Σ_τ^+ of the asymptotic covariance matrix Σ_τ of $V_n(\theta_{m(n)},\eta_n)$ under H_o is applied to the construction of test statistics having a chi-squared limiting distribution (see Rao and Mitra, 1971). The results are obtained for estimators satisfying the assumption A5. Here we formulate them for the special cases of m.l.e.'s based on grouped $\tilde{\theta}_{m(n)}$ and ungrouped $\hat{\theta}_{m(n)}$ data $Y_1,\ldots,Y_{m(n)}$.

Let

$$\tilde{\Sigma}_\tau^+ = I_M - qq' - \frac{1-2\tau}{1-\tau} B(B'B)^{-1}B'$$

and

$$\mu_1=(I_M - B(B'B)^{-1}B')B_{12}\gamma, \tilde{\mu}_\tau = \mu_1+(1-\tau^{-1/2})B(B'B)^{-1}B'B_{12}\gamma.$$

Theorem 2. Suppose that as $n \to \infty$, $\tau_n = m(n)/n \to \tau$. Let C1, C2 and C3 (see MS) with n replaced by m(n) hold.

If $0 < \tau \leq 1$, then under (θ_o,η_n)

$$T_n^{(1)} = \| V_n(\tilde{\theta}_{m(n)},\varphi_n) \|^2$$

has limiting distribution

$$\chi^2_{M-m-1}(\|\mu_1\|^2) + \frac{1-\tau}{\tau} \chi^2_m(\|\tilde{\mu}_\tau - \mu_1\|^2)$$

If $0 < \tau < 1$, and $\tilde{\Sigma}^+_{m(n)} = \tilde{\Sigma}^+_{\tau_n}(\tilde{\theta}_{m(n)},\varphi_n)$ then under (θ_o,η_n)

$$T_n^{(2)} = V_n(\tilde{\theta}_{m(n)},\varphi_n)' \tilde{\Sigma}^+_{m(n)} V_n(\tilde{\theta}_{m(n)},\varphi_n)$$

has limiting distribution $\chi^2_{M-1}(\|\tilde{\mu}_\tau\|^2)$.

Let J is the information matrix for $F(x|\theta)$ at θ_o, J_{12} is the $m \times p$ matrix with (i,j)-th entry

$$E\left[\left(\frac{\partial \log f}{\partial \theta_i}\right)\left(\frac{\partial \log f}{\partial \eta_j}\right)\right]$$

and $\lambda_{M-m},\ldots,\lambda_{M-1}$ are the m roots of the determinantal equation

$$|B'B - (1-\lambda)J| = 0.$$

Denote

$$\hat{\Sigma}_{\tau}^{+} = I_M - qq' - \frac{1-2\tau}{\tau} B(J + \frac{1-2\tau}{\tau} B'B)^{-1}B'$$

and

$$\mu_2 = (B_{12} - BJ^{-1}J_{12})\gamma, \hat{\mu}_\tau = \mu_2 + (1-\tau^{-1/2})BJ^{-1}J_{12}\gamma.$$

Theorem 3. Suppose that as $n \to \infty$, $\tau_n \to \tau$, $0 < \tau \leq 1$. Let C1, C2, C4, C6 and C5 with n replaced by m(n) hold.

1°. Then under (θ_o, η_n)

$$T_n^{(3)} = \| V_n(\hat{\theta}_{m(n)}, \varphi_n) \|^2$$

has limiting distribution

$$\chi^2_{M-m-1}(\|\mu_1\|^2) + \sum_{j=M-m}^{M-1} \varepsilon_j \chi^2_{1j}(\beta_j^2/\varepsilon_j)$$

where

$$\varepsilon_j = 1 + \frac{1-2\tau}{\tau}(1-\lambda_j), \sum_{j=M-m}^{M-1} \beta_j^2 = \| \hat{\mu}_\tau - \mu_1\|^2.$$

2 . If $\hat{\Sigma}_{m(n)}^{+} = \hat{\Sigma}_{\tau_n}^{+}(\hat{\theta}_{m(n)}, \varphi_n)$, then under (θ_o, η_n)

$$T_n^{(4)} = V_n(\hat{\theta}_{m(n)}, \varphi_n)' \hat{\Sigma}_{m(n)}^{+} V_n(\hat{\theta}_{m(n)}, \varphi_n)$$

has limiting distribution $\chi^2_{M-1}(\|\mu_1\|^2 + \sum_{j=M-m}^{M-1} \beta_j^2/\varepsilon_j)$.

Note that when $\tau = 1/2$ the limiting null distributions of statistics $T_n^{(i)}$, $i = 1,2,3,4$ mentioned above are χ^2_{M-1}. This case for continuous distributions earlier was investigated by Mirvaliev (1984).

REFERENCES

Chase,G.R.(1972). Chi-square test when parameters are estimated independently of the sample. J.Amer.Statist. Assoc. 67, 609-611.

Chibisov,D.M.(1971). Certain chi-square type tests for continuous distributions. Theor. Probability Appl. 16, 1-22.

Mirvaliev,M.(1984). On the Pearson's statistic for continuous distributions when parameters are estimated by subsamples. Doklady AN UzSSR. 11, 7-10.

Moore,D.S.,and Spruill,M.C.(1975). Unified large-sample theory of general chi-squared statistics for tests of fit. Ann.Statist. 3, 599-616.

Murty,V.K.,and Gafarian,A.V.(1970). Limiting distribution of some variations of the chi-square statistic. Ann.Math.Statist. 41, 188-194.

Rao,S.R.,and Mitra,S.K.(1971). Generalized inverse of matrices and its applications. Wiley. N.-Y.

ON HODGES-LEHMANN INDICES OF NONPARAMETRIC TESTS.

NIKITIN Ya.Yu.
Leningrad State University, Leningrad, USSR.

The Hodges-Lehmann asymptotical relative efficiency (ARE) was introduced by Hodges and Lehmann(1956). It is a very natural means for the comparison of tests but it is not so widely investigated as Pitman and Bahadur ARE's.

Let $X_1, X_2, \dots$ be a sequence of i.i.d. observations taking values in some measurable space and having there the distribution P_θ, $\theta \in \Theta$. Consider testing the hypothesis $H: \theta \in \Theta_0 \subset \Theta$ against the alternative $A: \theta \in \Theta_1 = \Theta \setminus \Theta_0$ with the aid of the sequence of statistics $\{T_n(X_1, \dots, X_n)\}$. Let $\beta_n(\alpha; \theta)$ be the probability of the second-kind error for T_n under the fixed level $\alpha \in (0,1)$ and the simple alternative $\theta \in \Theta_1$. If there exists such a function $d_T(\theta)$, $0 < d_T(\theta) < \infty$, that

$$\lim_{n\to\infty} n^{-1} \ln \beta_n(\alpha; \theta) = -\frac{1}{2} d_T(\theta), \quad (1)$$

then $d_T(\theta)$ is called the Hodges-Lehmann index of $\{T_n\}$. For each two sequences of this kind the Hodges-Lehmann ARE is defined as the ratio of their indices.

It follows from (1) that the computation of indices is based on the logarithmic asymptotics for probabilities of large deviations of $\{T_n\}$ not under the null-hypothesis as in Bahadur theory, but under the alternative. Bahadur(1967), Rao(1965) and Raghavachari(1982) note that this is a difficult problem. Indices of some parametric statistics were found by Brown(1971), Brown et al.(1984) and Raghavachari(1982), but their methods are not suitable for the nonparametric situation.

Let $K(\theta, \theta')$ be the Kullback-Leibler information corresponding to distributions P_θ and $P_{\theta'}$ and

$$K(\Theta_0, \theta) = \inf \{K(\theta_0, \theta) : \theta_0 \in \Theta_0\}.$$

The following result generalizes the well-known Stein lemma.

Theorem 1. For any sequence of statistics $\{T_n\}$

$$d_T(\theta) \leqslant 2K(\Theta_0, \theta) \qquad (2)$$

The sequence $\{T_n\}$ is called asymptotically optimal (AO) if the equality takes place in (2) and locally asymptotically optimal (LAO) if a weaker condition holds

$$d_T(\theta) \sim 2K(\Theta_0, \theta), \; \theta \to \partial\Theta_0 .$$

Our first main result is that two-sided statistics of Kolmogorov-Smirnov and ω^2-type for testing goodness-of-fit, symmetry, homogeneity and independence are A(We emphasize that the property of AO in the Pitman or Bahadur sense takes place for these statistics in exceptional cases and only locally (Nikitin(1984)).

The second main result consists in the fact that linear rank statistics are not AO. Their indices are locally equivalent to Bahadur's exact slopes under certai regularity conditions imposed on score functions and distributions of observations. As linear rank statistic are equivalent in Bahadur and Pitman senses (see,e.g. Kremer (1979)),it follows that the local ordering of su statistics doesn't depend on the type of ARE.

Consider for example the two-sample problem. Let $X_1, \ldots, X_m$ and $Y_1, \ldots, Y_n$ be two independent sample with continuous distribution functions F_1 and F_2. Suppose that

$$\lim_{m,n\to\infty} \frac{m}{m+n} = \rho_1 \in (0,1)$$

and let $\rho_2 = 1-\rho_1$. We are testing the hypothesis H_1: $F_1 \equiv F_2$ against the parametric alternative $A_1: F_1(x) = G(x;0)$, $F_2(x) = G(x;\theta)$, $\theta > 0$, where

$$G(x;\theta) \neq G(x;\theta'), \quad \theta \neq \theta'$$

and the distribution function $G(x;\theta)$ has the positive density $g(x;\theta)$.

Theorem 2. For any sequence $\{T_{m,n}\}$ for this problem

$$d_T(\theta) \leq -2\ln \int_{-\infty}^{\infty} g^{\rho_1}(x;0)g^{\rho_2}(x;\theta)dx \quad (3)$$

The property of AO means equality in (3).

Let $\hat{F}_m$, $\hat{G}_n$ and $\hat{H}_{m+n}$ be empirical d.f.'s based on the first, second and the pooled sample, q be a positive weight function with bounded derivative. Consider the statistics

$$\mathcal{D}_{m,n} = \sqrt{\frac{mn}{m+n}} \sup_x |\hat{F}_m(x) - \hat{G}_n(x)|,$$

$$\omega^2_{m,n,q} = \frac{mn}{m+n} \int_{-\infty}^{\infty} (\hat{F}_m(x) - \hat{G}_n(x))^2 q(\hat{H}_{m+n}(x)) d\hat{H}_{m+n}(x).$$

Theorem 3. Sequences of statistics $\{\omega^2_{m,n,q}\}$ and $\{\mathcal{D}_{m,n}\}$ are AO. The proof uses (3) and some inequalities for large deviations obtained with the aid of Sanov's theorem.

Now proceed to the linear rank statistic

$$S_{m,n} = (m+n)^{-1} \sum_{i=1}^{m} J(R_i/m+n+1),$$

where R_i is the rank of X_i in the pooled sample and the score function J satisfies the conditions

$$\int_0^1 J(u)du = 0, \quad \int_0^1 J^2(u)du = 1, \quad \sup |J'''(u)| < \infty.$$

We impose also some conditions on the family $\{G(x;\theta)\}$ in terms of the function $\Psi(x;\theta) = G(G^{-1}(x;0);\theta)$. One requires specifically the differentiability of Ψ two times in x and one time in θ.

Theorem 4. Under these conditions, as $\theta \to 0$

$$d_S(\theta) \sim \rho_1\rho_2 \left(\int_0^1 J'(u)\Psi'_\theta(u;0)du\right)^2 \theta^2.$$

The proof uses the Chernoff-Savage representation of linear rank statistics. Then using some suitable variant of Sanov's theorem from Groeneboom et al. (1979) we proceed to a variational problem with restrictions. The Euler-Lagrange equation with small parameters is investigated by methods of nonlinear functional analysis.

It's interesting to know when the index $d_S(\theta)$ obtain locally, as $\theta \to 0$, its upper bound from (3). This leads to a kind of characterization theorems similar to the case of Bahadur ARE discussed in Nikitin(1984). Suppose for instance that $G(x;\theta)=G(x-\theta)$ and G satisfies some additional regularity conditions. Note that $S_{m,n}$ coincides with the Wilcoxon statistic $W_{m,n}$ when $J(u)= =\sqrt{12}(u-1/2)$.

Theorem 5. The sequence $\{W_{m,n}\}$ is LAO iff G is the logistic d.f.

REFERENCES.

Bahadur R.R.(1967). Rates of convergence of estimates and test statistics. Ann.Math.Stat.38,303-324.

Brown L.D.(1971). Non-local asymptotic optimality of appropriate likelihood ratio tests. Ann.Math.Stat.42, 1206-1240.

Brown L.D., Ruymgaart F.H., Truax D.R.(1984). Hodges-Lehmann efficacies for likelihood ratio type tests in curved bivariate normal families. Stat.Neerl.38,2,21-36

Groeneboom P., Oosterhoff J., Ruymgaart F.H.(1979). Large deviation theorems for empirical probability measures. Ann.Prob.7,553-586.

Hodges J., Lehmann E.(1956). The efficiency of some nonparametric competitors of the t-test. Ann.Math.Stat. 26,324-335.

Kremer E.(1979). Lokale Bahadur-Effizienz linearer Rangtestes. Dissertation. Hamburg, 132 pp.

Nikitin Ya.Yu.(1984). Localasymptotic Bahadur optimality and characterization problems. Theory Prob.Appl.29,79-9

Raghavachari M.(1982). On the computation of the Hodges-Lehmann efficiency of test statistics. In: Festschrift for E.Lehmann. Wadsworth, USA, pp.367-378.

Rao C.R.(1965). Linear statistical inference and its applications. J.Wiley, 510 pp.

DIFFERENTIAL GEOMETRY AND STATISTICAL INFERENCE

SKOVGAARD, L.T., Statistical Research Unit, University of Copenhagen, Denmark.

Any smooth, finite-dimensional family of probability distributions can in a natural fashion be regarded as a Riemannian manifold, with the Fisher-information as Riemannian metric (4).
If the family is a regular, m-dimensional exponential family

$$P_{\Theta} = \{P_\theta | \theta \in \Theta \subseteq \mathbb{R}^m\} ,$$

dominated by a σ-finite measure λ on $\mathbb{R}$,
such that the minimal representation of the densities are given by

$$f_\theta(x) = {dP_\theta}/{d\lambda(x)} = \exp(\theta' q(x) - \psi(\theta)) ,$$

the Fisher-information can be written as

$$g_\theta = V_\theta q(x) = \ddot{\psi}(\theta)$$

It is useful to introduce a family of invariant connections $\overset{\alpha}{\nabla}$, indexed by a real-valued parameter α, and defined by its components, the Christoffel symbols

$$\overset{\alpha}{\Gamma}_{ijk} = 1/2(1-\alpha)T_{ijk} ,$$

where T is the skewness tensor $T = \dddot{\psi}$.

For α=0, we get the Riemannian connection $\overset{o}{\nabla}$, corresponding to the Fisher-information g, α=1 is the exponential connection, and α=-1 is the mixture connection (1). Further the geometries corresponding to α's with opposite signs are its others conjugate in the sense of Amari (1,6). For every α, we define α-geodesic curves γ through the requirements

$$\overset{\alpha}{\nabla}_{\dot{\gamma}} \dot{\gamma} = 0 \quad \text{on} \quad \gamma$$

For estimation in a curved subfamily (hypothesis) of the exponential family, we define the α-estimator to be the projection along an α-geodesic, orthogonally onto the hypothesis. All of these estimators will be consistent and first-order efficient, and when bias-corrected, the second-order term will consist of three parts: The naming curvature (depending on the parametrization), the "Efron excess" (imbedding curvature) and the estimator-dependent curvature. The last of these terms vanishes for the MLE (1,7).

The geometric quantities also enter in the interpretation of higher-order asymptotic properties of estimators and tests and are also useful in connection with topics such as conditional inference, ancillarity and sufficiency (1,3).

The multivariate normal distribution N_p is an important special case of a regular exponential family, with dimension $m=p+p(p+1)/2$. When using the holonomic basis for the set of vector fields and the identifications

$$\frac{\partial}{\partial\mu_i} \sim e_i \quad , \frac{\partial}{\partial\sigma_{ij}} \sim E_{ij}$$

where e_i is the i'th unit vector and E_{ij} is the matrix

$$E_{ij} = \begin{cases} 1_{(i,i)} & ; \quad i=j \\ 1_{(i,j)} + 1_{(j,i)} & ; \quad i \neq j \end{cases} ,$$

the components of the Riemannian metric can be written as

$$g(e_i,e_j)=e_i'\Sigma^{-1}e_j = \sigma^{ij}$$
$$g(e_i,E_{rs}) = 0$$
$$g(E_{ij},E_{rs}) = 1/2 \, tr(\Sigma^{-1}E_{ij}\Sigma^{-1}E_{rs})=\sigma^{is}\sigma^{jr}+\sigma^{ir}\sigma^{js}$$

The family of α-connections are given by

$$\overset{\alpha}{\nabla}_{e_i} e_j = (1-\alpha)/2 \, (e_i{}'e_j+e_je_i{}')$$

$$\overset{\alpha}{\nabla}_{e_i} E_{rs} = -(1+\alpha)/2 \, E_{rs} \, \Sigma^{-1} e_i$$

$$\overset{\alpha}{\nabla}_{E_{ij}} E_{rs} =-(1+\alpha)/2 \, (E_{ij}\Sigma^{-1}E_{rs} +E_{rs} \, \Sigma^{-1}E_{ij})$$

and the geodesic equations become (8)

$$\ddot{\mu} - (1+\alpha)\dot{\Sigma}\Sigma^{-1}\dot{\mu} = 0$$
$$\ddot{\Sigma} + (1-\alpha)\dot{\mu}\dot{\mu}' - (1+\alpha)\dot{\Sigma}\Sigma^{-1}\dot{\Sigma} = 0$$

For $\alpha=0$ (the Riemannian case), these equations have long since been solved for the univariate case (2,8), and recently for the multivariate case (5). In the general α-case, only the univariate equations have been solved. Here, the α-geodesics are coordinate curves for μ (i.e. curves with fixed μ) or parabolas in the (μ,σ^2)-coordinate system

$$\sigma^2 = a + b\mu + 1/2\ (\alpha-1)\mu^2$$

As an example of geodesic estimation, let us consider the univariate normal model, and the hypothesis of a fixed coefficient of variation $\varkappa>0$. The submodel thus is

$$M_\varkappa = \{N(\xi,\varkappa^2\xi^2) \mid \xi > 0\},$$

and the estimation geodesics orthogonal to the hypothesis are given by

$$a = \varkappa^2\xi^2 + 1/2\ (\alpha+1)\ \xi^2\ ,\ b = -\alpha\ \xi$$

If we start out with the traditional estimates $(\overline{X},s^2)$ in the full univariate model, we will end up with the geodesic α-estimate given as the positive solution to the equation

$$\xi^2(\varkappa^2 + 1/2(\alpha+1)) - \alpha\xi\overline{X} + (1/2(\alpha-1)\overline{X}^2 - s^2) = 0$$

Two values of α lead to simplification of this equation: The Riemannian estimate ($\alpha=0$) is

$$\sqrt{\frac{\overline{X}^2 + 2s^2}{1 + 2\varkappa^2}}$$

and for $\alpha=-(2\varkappa^2+1)$, we get the estimate

$$\frac{(1+\varkappa^2)\overline{X}^2 + s^2}{(1+2\varkappa^2)\overline{X}}$$

This latter value of α is interesting, since it corresponds to the dual geometry to that in which the hypothesis is itself a geodesic.

R e f e r e n c e s.

1. Amari S. Differential-Geometrical Methods in Statistics. - Springer, 1985.

2. Atkinson C. & Mitchell A.F.S. Rao's distance measure. - Sankhya 1981, vol. 43, p. 345-365.

3. Barndorff-Nielsen O.E., Cox D.R. & Reid N. The Role of Differential Geometry in Statistical Theory. - Int. Statist. Rev. 1986, vol. 54, p. 83-96.

4. Chentsov N.N. Statistical Decision Rules and Optimal Inference (in Russian). - Nauka, Moscow, 1972; translated in English, - AMS, Rhode Island, 1982.

5. Eriksen, P.S. (1986). Geodesics connected with the Fisher metric on the multivariate normal manifold. Lecture given at 2nd meeting between French and Danish Statisticians, Sandbjerg, Denmark, 1986.

6. Lauritzen S.L. Statistical manifolds. Technical Report 84-12, Institute of Electronic Systems, Aalborg University Centre, Denmark, 1984.

7. Madsen, L.T. The geometry of statistical models: A generalization of curvature. Research Report 79-1, Statistical Research Unit, Copenhagen, Denmark, 1979.

8. Skovgaard, L.T. A Riemannian geometry of the multivariate normal model. - Scand. J. Statist. 1984, vol. 11, p. 211-223.

LARGE SAMPLE PROPERTIES FOR GENERALIZATIONS OF THE TRIMMED MEAN

VERAVERBEKE, N., Limburgs Universitair Centrum, Diepenbeek, Belgium

1. Introduction. Let $X_1,\ldots,X_n$ be a sample of i.i.d. random variables with d.f. F and denote by $X_{(1)}\le\ldots\le X_{(n)}$ the order statistics. Let $0<\alpha,\beta<1/2$ and put $\gamma = 1-\beta-\alpha$. We deal with two different generalizations of the usual trimmed mean

$$\gamma^{-1}n^{-1}\sum_{i=[\alpha n]+1}^{n-[\beta n]} X_{(i)}$$

For this we consider a kernel $h(x_1,\ldots,x_m)$, symmetric in its m variables ($2\le m\le n$). The first generalization is given by

$$M_{n\alpha\beta} = \gamma^{-m}\binom{n}{m}^{-1}\sum_{i=[\alpha\binom{n}{m}]+1}^{\binom{n}{m}-[\beta\binom{n}{m}]} W_{n,i} \tag{1}$$

where $W_{n,1}\le\ldots\le W_{n,\binom{n}{m}}$ are the ordered values of $h(X_{i_1},\ldots,X_{i_m})$ $(1\le i_1<\ldots<i_m\le n)$. The second generalization is

$$\tilde{M}_{n\alpha\beta} = \gamma^{-m}\binom{n}{m}^{-1}\sum h(X_{(i_1)},\ldots,X_{(i_m)}) \tag{2}$$

where the sum is taken over all m-tuples $(i_1,\ldots,i_m)$ with $[\alpha n]+1\le i_1<\ldots<i_m\le n-[\beta n]$. The statistics (1) and (2) have been introduced in [1] and [2] respectively.

It is shown that both (1) and (2) can be represented as an ordinary U-statistic plus a remainder term. Conditions are given under which the remainder term tends to zero at a certain rate, either in probability or almost surely.

2. Asymptotic representations for $M_{n\alpha\beta}$. Denote the distribution function of the kernel h by $H_F(y) = P(h(X_1,\ldots,X_m)\le y)$ and introduce the quantiles $g_1 = H_F^{-1}(\alpha)$, $g_2 = H_F^{-1}(1-\beta)$.

Our analysis requires either condition (A) or the stronger condition (A') :

(A) : H_F has a density h_F which is continuous and positive at g_1 and g_2 and is bounded in some neighborhoods of g_1 and g_2.

(A'): (A) holds and h_F' exists and is bounded in some neighborhoods of g_1 and g_2.

Further define :

$$\theta(\alpha,1-\beta) = \gamma^{-m}\int_{g_2}^{g_1} y\, d\, H_F(y)$$

$$\begin{aligned}H(x_1,\ldots,x_m) = {}&\gamma^{-m} h(x_1,\ldots,x_m)\, I\,(g_1\le h(x_1,\ldots,x_m)\le g_2) - \theta(\alpha,1-\beta)\\ &+\gamma^{-m}g_1[I(h(x_1,\ldots,x_m)\le g_1)-\alpha]\\ &-\gamma^{-m}g_2[I(h(x_1,\ldots,x_m)\le g_2)-(1-\beta)].\end{aligned}$$

Theorem 1. Let $M_{n\alpha\beta}$ be given by (1). Then, as $n\to\infty$,

$$M_{n\alpha\beta} = \theta(\alpha,1-\beta) + \binom{n}{m}^{-1} \sum_{1\le i_1<\ldots<i_m\le n} H(X_{i_1},\ldots,X_{i_m}) + R_n$$

where

$R_n = o_p(n^{-1/2})$ under (A),

$R_n = O(n^{-3/4}(\log n)^{3/4})$ a.s. under (A').

3. Asymptotic representations for $\tilde{M}_{n\alpha\beta}$. Introduce the quantiles $\tilde{g}_1 = F^{-1}(\alpha)$, $\tilde{g}_2 = F^{-1}(1-\beta)$. Also introduce the following quantities

$$g(x;\alpha,\beta) = \gamma^{-m} I(\tilde{g}_1 \le x \le \tilde{g}_2) \int_{\tilde{g}_1}^{\tilde{g}_2} \ldots \int_{\tilde{g}_1}^{\tilde{g}_2} h(x,x_1,\ldots,x_{m-1}) dF(x_1)\ldots dF(x_{m-1})$$

$\mu(\alpha,1-\beta) = E\, g(X;\alpha,\beta)$ $\quad A = -g(\tilde{g}_1;\alpha,\beta)$ $\quad B = g(\tilde{g}_2;\alpha,\beta)$

$$\psi(x) = m([g(x;\alpha,\beta)] - \mu(\alpha,1-\beta)] + A[\alpha - I(x \le F^{-1}(\alpha))] + B[1-\beta-I(x \le F^{-1}(1-\beta))]).$$

Our analysis requires either the set of conditions $(\tilde{A})$-$(\tilde{C})$ or the stronger set of conditions $(\tilde{A}')$ - $(\tilde{C}')$:

$(\tilde{A})$: F has a density f which is continuous and positive at $\tilde{g}_1$ and $\tilde{g}_2$ and is bounded in some neighbourhoods of $\tilde{g}_1$ and $\tilde{g}_2$.

$(\tilde{B})$: For some $a < F^{-1}(\alpha)$ and $b > F^{-1}(1-\beta)$: $\sup_{a\le x_1,\ldots,x_m\le b} |h(x_1,\ldots,x_m)| = M_o < \infty$.

$(\tilde{C})$: The function $\int_{\tilde{g}_1}^{\tilde{g}_2} \ldots \int_{\tilde{g}_1}^{\tilde{g}_2} h(x,x_1,\ldots,x_{m-1}) dF(x_1)\ldots dF(x_{m-1})$ is continuous at $\tilde{g}_1$ and $\tilde{g}_2$

$(\tilde{A}')$: $(\tilde{A})$ holds and f' exists and is bounded in some neighborhoods of $\tilde{g}_1$ ang $\tilde{g}_2$.

$(\tilde{B}')$: $(\tilde{B})$ holds and for some neighborhoods U_1 of g_1 and U_2 of g_2

$$\sup_{x\in U_i} \sup_{a\le x_2,\ldots,x_{m-1}\le b} \int_{\tilde{g}_1}^{\tilde{g}_2} |h(x,dx_1,x_2,\ldots,x_{m-1})| = M_i < \infty, \ i = 1,2.$$

$(\tilde{C}')$:The function in $(\tilde{C})$ is Lipschitz continuous in neighborhoods of $\tilde{g}_1$ and $\tilde{g}_2$.

Theorem 2. Let $\tilde{M}_{n\alpha\beta}$ be given by (2). Then, as $n \to \infty$,

$$\tilde{M}_{n\alpha\beta} = \mu(\alpha,1-\beta) + n^{-1} \sum_{i=1}^{n} \psi(X_i) + \tilde{R}_n$$

where

$\tilde{R}_n = o_p(n^{-1/2})$ under $(\tilde{A})$ - $(\tilde{C})$,

$\tilde{R}_n = O(n^{-3/4}(\log n)^{3/4})$ a.s. under $(\tilde{A}')$ - $(\tilde{C}')$.

4. Applications. The representations in probability typically lead to central limit results, while the a.s. representations provide applications beyond asymptotic normality, such as L.I.L. and functional central limit results.

References

1. Serfling R.J. Generalized L-, M- and R-statistics. Ann. Statist., 1984, 12, 76-86.

2. Janssen P., Serfling R.J., Veraverbeke N. Asymptotic normality of U-statistics based on trimmed samples. J. Statist. Planning Inf., 1986, to appear.

MULTIVARIATE ANALYSIS
(large number of parameters . . .)
(Session 7)

Chairman: Y. Escoufier

DISCRIMINANT ANALYSIS FOR SPECIAL PARAMETER STRUCTURES

J. Läuter, Academy of Sciences of the G.D.R.,
Karl Weierstraß Institute of Mathematics, Mohrenstraße 39, Berlin 1086

1. Introduction

The discriminant analysis deals with the problem of assigning a multivariate observation to one of several populations. It yields diagnostic and prognostic decisions for applications in biological, social and technical fields.

Usually the considered populations are not known completely. In general only learning samples which correspond to the populations are available.

To separate two known p-dimensional populations with the mean vectors $\mu^{(1)}$, $\mu^{(2)}$ and the common covariance matrix Σ, Fisher has published fifty years ago the linear discriminant function $(\mu^{(1)} - \mu^{(2)})' \Sigma^{-1} y$. For the normal distribution with unknown parameters, the plug-in and the likelihood discrimination rules were investigated by Wald (1944), Sitgreaves (1952), Anderson (1958), Okamoto (1963), DasGupta (1965), Schaafsma (1972), McLachlan (1974/75) and others. The Bayes approach was introduced in the discriminant analysis with unknown parameters by Geisser (1964).

The discriminant rules considered by these authors are invariant under translations and under any regular transformations of the p variables. Thus DasGupta [1] treated the maximum likelihood discrimination rule for two normal populations $N(\mu^{(1)}, \Sigma)$, $N(\mu^{(2)}, \Sigma)$ which assigns an observation $y^{(0)}$ to the population

$$j=\begin{cases} 1 \text{ if } \frac{n^{(1)}}{n^{(1)}+1} x^{(1)'}G^{-1}x^{(1)} < \frac{n^{(2)}}{n^{(2)}+1} x^{(2)'}G^{-1}x^{(2)}, \\ 2 \text{ otherwise,} \end{cases}$$

where $x^{(j)}=y^{(0)}-y_{\cdot}^{(j)}$ $(j=1,2)$, $n^{(j)}$ are the sample sizes, $y_{\cdot}^{(j)}$ are the sample means and G is the matrix of sums of products,

$$G = (g_{hi}) = \sum_{j=1}^{2} \sum_{k=1}^{n^{(j)}} (y_k^{(j)}-y_{\cdot}^{(j)})(y_k^{(j)}-y_{\cdot}^{(j)})'.$$

DasGupta proved that this rule is a minimax and admissible rule. In his proof, the population 1 which $y^{(0)}$ belongs to and the Mahalanobis distance $\Delta^2=(\mu^{(1)}-\mu^{(2)})' \cdot\Sigma^{-1}(\mu^{(1)}-\mu^{(2)})$ are the only maximal invariants in the parameter space. Therefore additional suppositions on the structure of the parameters such as a factorial structure of Σ cannot lead to an improved invariant rule.

Non-invariant discrimination rules are usually considered only in connection with selection of variables. The practical experience states that the selection process results often in diminished error rates. This is interpreted in such a way that special parameter structures, in particular high correlations of the variables, multicollinearity and "overfitting", prevail in the applications. Therefore suitable non-invariant decision rules should be investigated more.

Though the procedures of selection of variables represent a certain aid for such structures, they are not the best and not the adequate way to attain a good discrimination. Instead of "annihilation" of information, an equalizing, a smoothing between the variables should be tried. We ourself have constructed special stable discriminators which have smaller error rates than the customary discriminant analysis including selection of variables.

Special parameter structures can be described by means of certain prior distributions of the parameters or by algebraical restrictions. Schaafsma and Steerneman [2] and Eben [3] have assumed a special ordering of the variables.

Basic assumptions:

Learning samples $y_k^{(j)} \sim N(\mu^{(j)}, \Sigma)$

$(j=1,2; k=1,\ldots,n^{(j)};\ n^{(j)} \geq 1;\ n=n^{(1)}+n^{(2)} \geq p+2)$,

observation to be assigned $y^{(0)} \sim N(\mu^{(l)}, \frac{1}{t}\Sigma)$ $(l=1,2;\ t>0)$. $n^{(j)}$, $y_{\bullet}^{(j)}$, $x^{(j)}$, G are defined as before. The parameters of the decision problem to be solved are $\mu^{(1)}$, $\mu^{(2)}$, Σ, l and t. Here l denotes the unknown population and t the precision of vector $y^{(0)}$ which is to be assigned. In general the parameters are unknown.

Basic transformations according to DasGupta [1] for a given $t_0 > 0$:

$$t^{(l)}=((n^{(l)}+t_0/n^{(l)})^{1/2} \qquad (l=1,2),$$

$$\mu=(\mu_i)=t^{(1)}(\mu^{(1)}-\mu^{(2)}),$$

$$y=(y_i)=-t^{(2)}x^{(1)}+t^{(1)}x^{(2)},\quad z=(z_i)=t^{(2)}x^{(1)}+t^{(1)}x^{(2)}.$$

2. Ridge Method in Discriminant Analysis

By using a normal-Wishart distribution as prior distribution, the ridge discrimination rule proposed by DiPillo [4][5] arises as the corresponding Bayes rule. In this method the customary discriminant analysis is modified in such a way that a matrix $G+\nu\Sigma_0$ is applied instead of G, where $\nu\Sigma_0$ is a given and fixed positive definite matrix. In practice a suitable diagonal matrix is mostly substituted for Σ_0.

Theorem 1 (A. Fuentes [6]): For fixed t_0, ν, Σ_0 ($t_0 >$ 0, $\nu \geq p$, Σ_0 symm., pos. def.), $t=t_0$ and the 0-1 loss function, the ridge rule which decides for the population j with the minimal value

$$\frac{n^{(j)}}{n^{(j)}+t_0}(y^{(0)}-y_{\bullet}^{(j)})'(G+\nu\Sigma_0)^{-1}(y^{(0)}-y_{\bullet}^{(j)}) \qquad (j=1,2)$$

is unbiased and admissible in the class of rules that are invariant under translations (i. e. that depend only on $x^{(1)}$, $x^{(2)}$, G).

3. Discriminators for One-Factor Structures

The considered one-factor structure is defined by

$$\Sigma = K + \omega(\mu^{(1)} - \mu^{(2)})(\mu^{(1)} - \mu^{(2)})',$$

where K is a positive definite diagonal matrix and ω is a positive scalar. This structure includes an algebraical relation between the mean values and the covariances which was also investigated by Sörbom [7] and in the program LISREL [8] .

In former papers (J. Läuter [9] [11]) we have proved that the customary discriminant analysis is not an admissible decision rule for this restricted structure. We have proposed a maximum likelihood approach for joint estimating the mean values and covariances [10] and a method based on variance components estimation by MINQUE [12] .

Theorem 2 (J. Läuter [12]): For fixed positive t_0, c_i, ε_i with $c_1^{-1}+\ldots+c_p^{-1} < 1$, $t=t_0$ and the 0-1 loss function, the rule which decides for the population j with the minimal value

$$\frac{n^{(j)}}{n^{(j)}+t_0}(y^{(0)}-y_{\cdot}^{(j)})'(C+E-G)^{-1}(y^{(0)}-y_{\cdot}^{(j)}) \quad (j=1,2)$$

is admissible in the class of rules that are invariant under translations. Here

$$C=\mathrm{Diag}(c_i(ay_i^2+bz_i^2+g_{ii})), \quad E=\mathrm{Diag}(\varepsilon_i),$$

$$a=\frac{t_0}{2t^{(1)}t^{(2)}(t^{(1)}t^{(2)}-1)}, \quad b=\frac{t_0}{2t^{(1)}t^{(2)}(t^{(1)}t^{(2)}+1)}.$$

Remark: Each rule of this kind corresponding to any t_0 is also admissible in the case of unknown precision t.

For practical purposes the limits $t_0 \to 0$, $E \to 0$, $c_i \to p$ or $c_i \to \infty$ are important:

<u>One-factor discrimination rule:</u>

$$j=\begin{cases}1 \text{ if } (y_{\bullet}^{(1)}-y_{\bullet}^{(2)})'(C-G)^{-1}(y^{(0)}-\frac{1}{2}(y_{\bullet}^{(1)}+y_{\bullet}^{(2)}))>0,\\ 2 \text{ otherwise,}\end{cases}$$

where $C=\mathrm{Diag}(p(G+H))$, $H=\frac{n^{(1)}\,n^{(2)}}{n^{(1)}+n^{(2)}}(y_{\bullet}^{(1)}-y_{\bullet}^{(2)})(y_{\bullet}^{(1)}-y_{\bullet}^{(2)})'$.

<u>Special one-factor discrimination rule:</u>

$$j=\begin{cases}1 \text{ if } (y_{\bullet}^{(1)}-y_{\bullet}^{(2)})'C^{-1}(y^{(0)}-\frac{1}{2}(y_{\bullet}^{(1)}+y_{\bullet}^{(2)}))>0,\\ 2 \text{ otherwise,}\end{cases}$$

where $C=\mathrm{Diag}(G+H)$.

These limit rules possess the property of being <u>limits of sequences of admissible rules.</u> But they themselves are not necessarily admissible.

4. Stable Discriminators for Multiple-Factor Structures

The derivation of stable discriminators for multiple-factor structures is performed under strong restrictions and the supposition that the parameters are known. The parameters $\mu=\mu^{(1)}-\mu^{(2)}$ and Σ are assumed to have the shape $\mu=M1$, $\Sigma=M\Omega M'$, where M is a pxf matrix of rank f which contains exactly one non-zero element in each row, Ω is a positive definite diagonal matrix, 1 is the vector of unities.

Thus the total set of p variables is splitted into f disjoint partitions each corresponding to a factor. All variables of a partition have the same discriminant power, the correlation between them is +1 or -1.

Then the optimal discrimination rule is

$$j=\begin{cases}1 \text{ if } \mu'\Sigma^{-}(y^{(0)}-\frac{1}{2}(\mu^{(1)}+\mu^{(2)}))>0,\\ 2 \text{ otherwise}\end{cases}$$

with Σ^{-} being a generalized inverse of Σ, namely

$$\Sigma^{-}=C^{-1}(\mathrm{Diag}(\Sigma C^{-1}\Sigma))^{-1}\,\mathrm{Diag}(\Sigma),$$

where C is an arbitrary fixed or random positive definite diagonal matrix. Σ^{-} is also of diagonal form. Substituting the estimates $y_{\bullet}^{(1)}$, $y_{\bullet}^{(2)}$, G for the true parameters, rule

$$j=\begin{cases}1 \text{ if } (y_{\bullet}^{(1)}-y_{\bullet}^{(2)})'T(y^{(0)}-\frac{1}{2}(y_{\bullet}^{(1)}+y_{\bullet}^{(2)}))>0,\\ 2 \text{ otherwise,}\end{cases}$$

$$T=C^{-1}(\mathrm{Diag}(GC^{-1}G))^{-1}\mathrm{Diag}(G)$$

arises. In our applications the <u>multi-factor rule</u> corresponding to $C=\mathrm{Diag}(G)$, $T=(\mathrm{Diag}(G(\mathrm{Diag}(G))^{-1}G))^{-1}$ is used. This rule does not need an algorithm of matrix inversion; only diagonal matrices must be inverted.

The elementary multivariate calculus treated in this section enables us to establish a method of selecting variables, namely by maximizing the simple measure tr(HT). Furthermore the found matrix T can be applied for determining a suitable matrix $\nu\Sigma_0$ of the <u>ridge method</u>: $\nu=p$, $\Sigma_0=T^{-1}/(n+p-3)$.

5. Simulation Results and a Medical Application

In the simulations described now 5 different situations (tasks a, b, c, d, e) of the parameters $\mu^{(1)}-\mu^{(2)}$ and Σ are considered. For all tasks, p=10, $\Delta^2=4$. The tasks a to d correspond to the one-factor structure of section 3, task e has a more complicated form.

<u>Task a:</u> 10 variables with the same parameters and the correlation coefficient $\rho=0,8$ between them (high correlation).

<u>Task b:</u> 10 highly correlated variables with different discriminant power: $t_0=0$, $\omega=10/41$, $\mu_i^2/\kappa_i=164i^2/385$ (i=1,...,10).

<u>Task c:</u> 10 uncorrelated variables with identical discriminant power.

<u>Task d:</u> 10 uncorrelated variables with different discriminant power: $t_0=0$, $\omega=0$, $\mu_i^2/\kappa_i=4i^2/385$ (i=1,...,10).

<u>Task e:</u> 4 uncorrelated blocks of variables consisting of 1, 2, 3, 4 variables, respectively. Each block has $\Delta^2=1$. Within a block, all variables have the same discriminant power and the correlation $\rho=0.8$ between them

In table 1 simulation results of the following discriminators are contained:

<u>Method 1:</u> Classical discriminant analysis according to Fisher/Anderson.

Method 2: Classical discriminant analysis with selection of variables by means of the universal computer program MVD [13].

Method 3: Discrimination rule for independent variables (using only the diagonal of G).

Method 4: One-factor rule of section 3.

Method 5: Special one-factor rule of section 3.

Method 6: Multi-factor rule of section 4.

For each performed test two learning samples of the populations $N(\mu^{(j)}, \Sigma)$ with sample size $n^{(1)}=n^{(2)}=20$ are generated. Then the arising discriminator is applied to two working samples of the same populations (t=1) which consist of 2000 vectors each. Every error rate in table 1 is the average of the discrimination error of 10 independent tests of this kind. For all tasks the infimum of the error rate is $\phi(-\frac{1}{2}\Delta) = 0.1587$ (optimum error with known parameters).

Discussion of the results: The classical discriminant analysis (method 1) possesses high average errors which additionally vary strongly. The standard process of selecting variables (method 2) is successful in case of highly correlated variables but not in the case

Table 1: Average error rate of 6 different discriminators (simulation results)

	Discrimination method					
	1	2	3	4	5	6
Task a	.220	.188	.164	.164	.164	.164
Task b	.220	.188	.161	.161	.161	.167
Task c	.220	.262	.202	.192	.193	.202
Task d	.220	.235	.204	.201	.201	.204
Task e	.220	.205	.190	.190	.185	.175

of null correlation. The methods 4 and 5 which have especially been developed for the one-factor covariance structure yield very good results in the cases a to d. The multi-factor method 6 shows an excellent result in case of task e, and it is altogether the best one, too. The attained minimum error for each task is missed by

0.010, at the most, if the multi-factor rule is employed.

In order to test the elementary strategy of selecting variables and the ridge strategy mentioned in section 4, data from the epidemiology of heart diseases are used. Starting from a sample of 1210 observation vectors of two populations with 30 variables, we form 10 independent learning samples consisting of 121 vectors each. Table 2 provides the average error rate of these 10 partial samples (n=121) for the discriminators 1 to 6 and the ridge method 7. The methods 6 and 7 are applied without and with the selection process. The error rate belonging to a learning sample is determined by using the observations of all other partial samples (1089 vectors) as a working sample. It can be seen that the selection of variables and the ridge method result in a slight diminution of the error rate.On the other hand, table 2 shows also that the created stable discrimination rules lose their importance if the sample sizes increase. In the case of fivefold sample size (n=605), the ridge method has the same errors as Fisher's usual discriminant analysis.

Table 2: Average error rate of 7 different discriminators; methods 6 and 7 without and with selection of variables (data from the epidemiology)

	Discrimination method								
	1	2	3	4	5	6	6 with sel.	7	7 with sel.
n=121	.317	.310	.308	.316	.311	.297	.289	.295	.286
n=605	.265	.273	.306	.312	.308	.285	.286	.265	.273

6. Analogous Considerations in Regression Analysis

The investigations in discriminant analysis suggest analogous considerations in regression analysis. Contrary to the comprehensive literature on regression under parameter restrictions connected with the names

Marquardt (1970), Toutenburg (1968), H. Läuter (1970), Kuks and Olman (1972), Bunke (1972), Krafft (1986) and others, our approach uses a special one-factor structure.

It is started from a (1+p)-dimensional normally distributed random variable $(y_0\ y')'$ where y_0 is 1x1 and y is px1 with the covariance matrix

$$\begin{pmatrix} \sigma_{00} & \sigma_0' \\ \sigma_0 & \Sigma \end{pmatrix}.$$

The special structure $\Sigma = K + \omega\sigma_0\sigma_0'$ (K pos. def. diag. matrix, $\omega > 0$) is assumed. Let $(y_{0.}\ y_{.}')'$ be the sample mean of n observations ($n \geq p+2$) and

$$\begin{pmatrix} g_{00} & g_0' \\ g_0 & G \end{pmatrix}$$

the corresponding matrix of sums of products. The following theorem enables the prediction of y_0 by y.

Theorem 3: For fixed c_i ($c_i > 0$; $i=0,\dots,p$; $c_0^{-1}+\dots+c_p^{-1} < 1$) and ε_i ($\varepsilon_i > 0$; $i=1,\dots,p$) and the quadratic loss $(\hat{y}_0 - y_0)^2$, the prediction

$$\hat{y}_0 - y_{0.} = g_0'((c_0-1)(C+E-G) - c_0 \frac{n}{n+1}(y-y_{.})(y-y_{.})')^{-1}(y-y_{.})$$

is admissible in the class of all predictions of $(y_0 - y_{0.})$ by $(y-y_{.})$, g_{00}, g_0, G. Here

$$C = \operatorname{Diag}_{i=1,\dots,p}\left(c_i\left(\frac{n}{n+1}(y_i - y_{i.})^2 + g_{ii}\right)\right),\quad E = \operatorname{Diag}(\varepsilon_i).$$

7. Conclusion Remark

We were forced to leave the least square method to attain stabilized decision rules and to take into account special parameter restrictions. The new rules have the best results in those very cases which are the most complicated ones for the solution of the normal equation.

References

[1] DasGupta, S. (1965): Optimum Classification Rules for Classification into Two Multivariate Normal Populations. Ann. Math. Statist. 36, 1174-1184.

[2] Schaafsma, W. and T.Steerneman (1981): Discriminant Analysis when the Number of Features is Unbounded. IEEE Transactions on Systems, Man and Cybernetics, 11, 2, 144-151.

[3] Eben, K. (1986): Discriminant Analysis in the Case of Variables Ordered According to Their Importance. Proc. Conf. DIANA 2, Liblice, ČSSR.

[4] DiPillo, P.-J. (1976): The Application of Bias to Discriminant Analysis. Comm. Statist.-Theor. Meth., A5(9), 843-854.

[5] DiPillo, P.-J. (1979): Biased Discriminant Analysis: Evaluation of the Optimum Probability of Misclassification. Comm. Statist.-Theor. Meth., A8(14), 1447-1457.

[6] Fuentes Rodriguez, A. (1986): Admissibility and Unbiasedness of the Ridge Classification Rules for Two Normal Populations with Equal Covariance Matrices. Submitted to Statistics.

[7] Sörbom, D. (1974): A General Method for Studying Differences in Factor Means and Factor Structure between Groups. Br. J. Math. Statist. Psychol., 27,229-239.

[8] Jöreskog, K.G. and D.Sörbom (1984): LISREL VI - Analysis of Linear Structural Relationships by a Method of Maximum Likelihood. Mooresville, In: Scientific Software, Inc.

[9] Läuter, J. (1982): Improvement of the Discriminant Analysis under Parameter Restrictions. Proc. Conf. DIANA, Liblice, ČSSR, 183-193.

[10] Läuter, J. (1984): Algorithms of Discriminant Analysis Using Parameter Restrictions for Diminishing the Error Rate. In: COMPSTAT 1984, Physica-Verlag, Vienna, 131-136.

[11] Läuter, J. (1985): Discriminant Analysis under Parameter Restrictions - Statistical and Computational Aspects. Statistics 16, 1, 125-137.

[12] Läuter, J. (1986): Discriminant Analysis in Special Models - Theoretical and Practical Results. Proc. Conf. DIANA 2, Liblice, ČSSR.

[13] Läuter, J. and K. Hermann (1984): Program System Multivariate Analysis of Variance and Discriminant Analysis. Report, Acad. of Sc. of the GDR, Inst of Math.

ASYMPTOTICS OF INCREASING DIMENSIONALITY IN CLASSIFICATION

Meshalkin L.D., Moscow, USSR

1. INTRODUCTION

A classification of vector $x \in R^p$ ($p>>1$) into one of two statistical populations F_i ($i=1,2$) given by learning samples and priory assumptions about distributions F_i is discussed. The problem is studied under the condition of sample information deficiency when n_i ($i=1,2$) the learning samples sizes are comparable with m - the number of parameters differentiating distributions. Typical data for medical applications are: $p=5\div10$, $n_1=15\div100$, $n_2=100\div1000$, Kullback distance between distributions $J=2\div4$. Follow A.N. Kolmogorov a sequence of classification problems is considered in which

$$p,\ m,\ n_i \to \infty\ ;\ p/n_i < C < \infty\ ;\ m/n_i \to u_i < \infty\ (i=1,2)\,. \qquad (1)$$

Idealized mathematical models and the asymptotics (1) give us the possibility to realize what is observed by mathematical simulation, to make recomendations for constructing classification procedures.

2. SUBSTITUTION OF MAXIMUM LIKELIHOOD ESTIMATES INSTEAD OF UNKNOWN PARAMETERS IN LIKELIHOOD RATIO. RESULTS OF 70-THS

As one knows the best test of classification of two known distributions F_i ($i=1,2$) is based on the likelihood ratio and has a form

$$\frac{f_2(x)}{f_1(x)} \gtrless C \Longrightarrow \begin{array}{l} F_2 \text{ is accepted} \\ F_1 \text{ is accepted} \end{array}$$

where f_i is a density of F_i $(i=1,2)$. When distributions F_i are known only with accuracy to unknown parameters Θ_i and learning samples are given, then it is quite natural to look at classificators

$$\frac{f_2(x,\hat{\Theta}_2)}{f_1(x,\hat{\Theta}_1)} \gtrless C \Longrightarrow \begin{matrix} F_2 \text{ is accepted} \\ F_1 \text{ is accepted} \end{matrix} \qquad (2)$$

where $\hat{\Theta}_i$ $(i=1,2)$ are maximum likehood estimates derived from priory information and learning samples. Asymptotics (1) make a surprise because in general case

$$\sup_x \left| \frac{f_2(x,\hat{O}_2)}{f_1(x,\hat{O}_1)} - \frac{f_2(x,O_2)}{f_1(x,O_1)} \right| \not\to 0 \ .$$

For example. In the classical Fisher's model of classification, where two normal distributions with common covmatrix are given $F_i = N(a_i, \Sigma)$ when Σ is known the minimax probability of misclassification

$$\alpha \longrightarrow \Phi(-J/2\sqrt{J+u_1+u_2}),$$

where $\Phi(t) = (2\pi)^{-1/2} \int_0^t \exp\{-v^2/2\} dv$,

$J = (a_2-a_1)' \Sigma^{-1} (a_2-a_1) < C < \infty$ - Machalanobis distance between distributions and in definition of u_i $m = p$. In the traditional asymptotics, where p is fixed and $n \to \infty$, u_i in (4) are equal to zero.

The general Fisher's model of discriminant analysis, in which Σ is unknown, was studied by Deev (1970). Under $u_1^{-1}+u_2^{-1} > 1$ he has shown that minimax error of new vector x classification

$$\alpha \to \Phi(-J(1-u_1u_2/(u_1+u_2))^{1/2}/2\sqrt{J+u_1+u_2}). \qquad (5)$$

As it follows from comparison (4) and (5), in the terms of α the price of $p(p+1)/2$ unknown common parameters is high enough.

Pikjalis (1976) has shown by mathematical simulation

that Deev's formulaes give good approximation even under medial sample size and have evident advantage before the traditional asymptotical expansion. This result has become a serious argument for futher investigation of classification problem in asymptotics (1).

Assume, that coordinates of x and Θ could be divided into k mutually disjoint sets (bloks) $x = (x^1,\ldots,x^k)$, $\Theta^j = (\Theta^1,\ldots,\Theta^k)$ where x^j has dimensionality p_j $(\Sigma p_j = p)$ and $\Theta^j - m_j$ $(\Sigma m_j = m)$ and for all x,Θ $f(x,\Theta) = \prod_j f^j(x^j,\Theta^j)$. Such distribution we would call independent-block one. Let us assume that in asymptotics (1)

$$p_j,\ m_j < C < \infty ; \qquad (6)$$

$$|\Theta_2^{(j)} - \Theta_1^{(j)}| < C/\sqrt{\min n_i} \qquad (7)$$

$$J_j = J < C < \infty , \qquad (8)$$

where $J_j = \sum_{t,s} (\Theta_2^{(t)} - \Theta_1^{(t)}) i_{ts}(\Theta_1)(\Theta_2^{(s)} - \Theta_1^{(s)})$ and summation is made over all t,s belonging to the j-th block,

$$\| i_{ts}(\Theta) \| = \| \int \frac{\partial \ln f(x,\Theta)}{\partial \Theta^{(t)}} \cdot \frac{\partial \ln f(x,\Theta)}{\partial \Theta^{(s)}} f(x,\Theta)\mu(dx) \| -$$

the Fisher's information matrix. Meshalkin and Serdobolskiy (1978) has shown that under (6)-(8) and some additional regularity type conditions for independent - block distributions the result (4) holds. Moreover if in the j-th block there are m_j - differentiating distributions parameters and l_j - unknown common parameters $(l_j < C < \infty)$ and the same estimates of common parameters are substituted into both densities then (4) holds too. In other words O(m) common parameters in case of independent-block distributions do not make worse the results of classification.

Let us remember some notions (Chow et al. (1968), Aivazian et al. (1985)).

The p-dimensional vector x has tree dependence structure if such replacement of its coordinates $\beta(1,\dots,p) = (\beta(1),\dots,\beta(p))$ exists that for every $\beta(i)$ one could find $j(\beta(i)) \in \{0, \beta(1),\dots,\beta(i-1)\}$ that for all z

$$P\{x^{(\beta(i))} < z \mid x^{(\beta(1))},\dots,x^{(\beta(i-1))}\} =$$

$$P\{x^{(\beta(i))} < z \mid x^{(j(\beta(i)))}\},$$

and j=0 corresponds to the fictitious coordinate $x^{(0)}=1$

The graph G = (V,E), where V(G) = {0,1,...,p} is the set of nodes of G and $E(G) = \bigcup_{1} (i,j(i))$ - is the set of nonoriented edges of G, is called the graph of dependence structure.

The remarkable peculiarity of normal distribution with tree dependence structure is following: its covmatrix Σ depends on no more then 2p-1 parameters. More exactly, in $\Sigma^{-1} = \| \sigma^{ij} \|$ besides of diagonal elements only elements with $(i,j) \in E(G)$ are different from zero.

Zarudskiy (1978, 1980) has found that under some additional assumption when x has known tree dependance structure then in the classical Fisher's model with unknown Σ the (4) holds too. Moreover the graph of dependence structure is renewed in the asymptotics (1) with accurasy to unsignificant edges.

3. VARIABLES SELECTION

It is shown Estes (1965), Raudys (1979) by mathematical simulation that a problem of selecting the best set of variables for classification is unexpectedly delicate. To understand the matter the asymptotics (1) has to be added by the assumption that r - number of selected variables (coordinates of x) is proportional to p (Meshalkin (1977)

$$r/p \to \rho > 0. \tag{9}$$

It is natural to start study with the simplest model of independent-block distributions in which every block consists from one variable (model of independent variables). When there are many approximately equivalent (on their input in discrimination) variables, in learning samples selected variables have larger deviations from expected values and therefore, the contrast between the results of classification of learning and examining samples is considerable. Serdobolskiy (1983) has found the limit classification error in asymptotics (1),(2) for model of independent-block distributions in addition to (6) - (9) assumptions that there exists the limit distribution of J_j the Kullback distance between distributions of j-th block; $k^{-1}\sum_{i:J_i<v/k} 1 \rightarrow F(u)$; $m_j = m/k$; $n_1 = n_2$ and the selection rule has a form $\hat{J}_j \gtrless C(n)$, where $\hat{J}_j$ is the estimate of Kullback distance between distributions of the j-th block.

In practice the variable selection is made by different heuristical stepwise procedures which are extremly difficult to study theoretically. Therefore it is very interesting to investigate the Fisher's model in details with additional assumption that the distributions have tree dependance structure. A matter is due to above mentioned pecularity of Σ^{-1} for differentiating between distributions only those variables are important which are selfimportant or which are not selfimportant but are directly related upon graph G with selfimportant variables.

4. BIASED ESTIMATION

One could expect from (3)-(5) that the use of unbiased estimates in discriminant function without simplifying additional assumption on Σ does not lead to minimal errors. This is supported by mathematical simulation (Di Pillo (1979), Shurygin (1985)). For Fisher's model a theoretical investigations have been done to find new classifiers (Serdobolskiy (1983b), Barsov (1985)). The used classifiers were of type:

$$w_{\Gamma}(x) = (x-(\hat{a}_1+\hat{a}_2)/2)'\Gamma(S)(\hat{a}_1-\hat{a}_2) \gtrless 0 \quad (9$$

where S - is unbiased estimate of Σ ,

$$\Gamma(S) = \int_{t>0}(I_p+tS)^{-1}db(t) \quad (10$$

and b(t) - a function of limited variation.

Numerical realization of this algorithm has shown its good efficiency both on artificial samples from normal populations and on known empirical collections.

Recent resultes of Girko (1985) for estimators of Stieltjes transformation of normalised spectral functions of covmatrix for nonnormal distributions promise that (9 (10) approach could be investigated for more general distributions then normal ones.

5. ASSIGNING OF NUMERICAL SCORES

For practice a problem of comparison in asymptotics (1 of different methods of assigning of numerical scores to values of qualitative variables is important. Let us consider the simplest model of independent discrete variables and assume that for every variable j its values could be ordered and could be regarded as result of quantification of continuous variables with corresponding distributions

$$F_{1j}(t) = F_j(t+\delta_j),\ F_{2j}(t) = F_j(t-\delta_j),\ \delta_j = o(1)$$

where F_j are known. Let us look at the procedure in which the boundaries of quantification are estimated as the first step , then δ_j and after the logarithms of ratio (2 are scored as linear function from δ_j. It gives the limit minimax probability of misclassification as (4) with m = p But if in the same model for every variable j with k_j possible values they are scored independently then in formula (4) $m = \sum_j (k_j-1)$ (Meshalkin (1977)).

6. CONCLUSION

It is necessary to remark that the report was restricted by very idealized models with strong parametrical assumptions. This helps to evaluate different methods and to understand acting mechanisms. But it is hard to predict what could happen with conclusions when basic idealised assumptions are fulfiled only approximately. And what should it mean "approximately" in asymptotics (1)? In the statistics the complexity of approximating functions usually depends on the number of observations. In application to the classification problem this is done in the traditional asymptotics in Vapnik (1982). But in the asymptotics (1) there no results in this direction.

REFERENCES

Aivazyan, S.A., Yenyukov, I.S., and Meshalkin L.D. (1985). Applied Statistics. Study of relationships. Chapter 4. Moscow: Finansy i statistika (in Russian).

Barsov, D.A. (1985). Minimization of classification error by use of biased discriminant functions. In: Statistics, Probability, Economics. Moscow: Nauka, pp. 376-379 (in Russian).

Chow, C.K., and Lie, C.N. (1968). Approximating discrete probability with dependence trees. IEEE Transactions on Information theory IT-14, pp. 462-467.

Deev, A.D. (1970). Representation of statistics of discriminant analysis and asymptotic expansions in dimensionalities comparable with sample size. Reports of Academy of Sciences of the USSR, v. 195, n. 4, pp. 759-762 (in Russian).

Di Pillo, P.J. (1979). Biased discriminant analysis: evaluation of the optimum probability of misclassification. Commun. Statist.-theor. meth., A8(14) pp. 1447-1457.

Estes, S.E. (1965). Measurment selection for discriminants used in patern classification. Ph. D. Dissertation, Stanford University. Stanford, California.

Girko, V.L. (1985). Contend with dimensionality in multivariaye statistical analysis. In: Application of multivariate statistical analysis in economics and quality control. Tartu, v. 1, pp. 43-52 (in Russian).

Haff, L.R. (1979). Estimation of inverse covariance matrix random mixtures of the inverse covariance Wishart matrix and identity. Ann. Statist., v. 7, n. 6, pp. 1264-1276.

Meshalkin, L.D. (1976). Theory of statistical analysis of chronicaly progressing diseases. Ph. D. Dissertation, Moscow State University (in Russian).

Meshalkin, L.D., and Serdobolskiy, V.I. (1978). Errors in classifying multivariate observations. Theory of Probabilities and its Applications, v. 23, n. 4, pp. 772-781 (in Russian).

Pikelis, V.S. (1976). Comparison of methods of computing the expected classification errors. Automatika and Remote Control, n. 5, pp. 59-63 (in Russian).

Raudys, S. (1979). Classification errors when features are selected. Statistical Problems of Control, Issue 38, Vilnus: Inst. of Math. and Cyb. Press, pp. 9-26 (in Russian).

Serdobolskiy, V.I. (1983a). On minimal error probability in discriminant analysis. Reports of Academy of Sciences of the USSR, v. 270, n. 5, pp. 1066-1070 (in Russian).

Serdobolskiy, V.I. (1983b). The influence of selecting components of random variable on classification. Mathematics, Izvestija VUZov, USSR, n. 9, pp. 46-55 (in Russian).

Shurygin, A.M. (1985). Ways of improving linear discrimination in normal case. In: Statistics, Probability, Economics. Moscow: Nauka, pp. 379-382 (in Russian).

Vapnik, V.N. (1982). Estimation of Dependences based on empirical data. Springer, N.Y.

Zarudskiy, V.I. (1978). Classification of normal vectors with a simple dependence structure in multidimensional space. In: Applied multivariate statistical analysis. Moscow: Nauka, pp. 37-51 (in Russian).

Zarudskiy, V.I. (1980). Determination of some connections graphs for normal vectors in large dimensionality space. In: Algorithmic and Programic Supply of Applied Statistical Analysis. Moscow: Nauka, pp. 189-208 (in Russian).

EXTENSIONS AND ASYMPTOTIC STUDIES OF MULTIVARIATE ANALYSES

POUSSE Alain
U.A.CNRS n° 1204, Faculté des Sciences, Pau, France.

1. INTRODUCTION.

The aim of the multivariate factor methods, such as Principal Component Analysis (P.C.A.), canonical analysis (C.A.), discriminant analysis (D.A.), was initially to study statistical variables. The joint law of these variables was supposed to be gaussian, and within this framework good properties were obtained. In the France of the sixties, these methods met with new interest with the possibilities offered by computers, and their studies have developed along two lines : a) with the aim of using them in a descriptive objective, giving as much importance to the statistic units (the "individuals") as to the variables, b) the ambition to get free from all hypothesis on the variable law. This has led (see Cailliez and Pages (1975)) to a geometric representation and to a very common form of exploratory practice. The purpose was to get representations in a plane, or in a small dimension space, of the individuals and of the variables. That is obtained through representing each individual or, symmetrically, each variable in a finite dimension Euclidian space, and through projecting them upon an optimal plane, with respect to choosen criteria.

From there on things were sufficiently clear to be able to, on the one hand, extend these methods and on the other hand, reintroduce various statistical structures. Thus was made clearer if necessary the model as considered according to the applications, and thus to return to a possible inference about the parameters of this model, which would have been impossible with an only descriptive and geometric approach.

2. EXTENSIONS IN EXPLORATORY FACTOR ANALYSES.

2.1. Principal component Analysis.

2.1.1. Geometric starting point in finite dimension. $X=(X_1,\cdots,X_p)$ is a $\mathbb{R}^p$-valued centered statistical variable observed on n individuals $\omega_1,\cdots,\omega_n$ each individual ω_i having a weight p_i ($\Sigma p_i=1$). Any real variable Z observed on these individuals is represented in R^n by the vector $Z(\omega_1),\cdots, Z(\omega_n)$. $\mathbb{R}^n$ has usually $\mathrm{diag}(p_i)_{i=1,\cdots,n}$ as metric, so that the inner product of two centered variables is their empirical covariance. Similarly the individuals are represented in $\mathbb{R}^p$ by means of $X_1,\cdots,X_p$, and the Euclidian space of the individuals is $E=(\mathbb{R}^p,M)$. The P.C.A. of X is associated with the operator U : $f\in F \to Uf = E(fX) = \Sigma p_i f(\omega_i)X(\omega_i) \in E$. Its adjoint is U^*: $u\in E \to U^*u = \langle u,X\rangle_M$ and the operator $V = U\circ U^*$: $u \to E[\langle X,u\rangle_M X]$ is the empirical covariance operator of X. The Schmidt (singular value) decomposition $U = \sum\sqrt{\lambda_i}\, f_i \otimes u_i$ gives the principal values λ_i, reduced components f_i and vectors u_i. The optimal representations of individuals (resp.variables) are obtained by projection onto the subspace spanned by the u_i's (resp.the f_i's). Problems of choice of metrics on E or F are dealt with either on the lefthand side (when the metric is depending of the variables) or on the righthand side (when the metric take into consideration a possible connection between individuals).

2.1.2. Extensions of infinite dimension.

α) on the individual level. The study of the stability by sampling, for example, requires P.C.A. to extend to random variables, with an infinite number of individuals (this takes us back to Hotelling). If X is a r.v. from $(\Omega,\mathcal{A},P)$ into $\mathbb{R}^p$, let's say of the second order, the Hilbert space $L^2(\Omega,\mathcal{A},P)$ is an extension of F, the operators and the analysis being analogous as above.

β) as regards variables. We may want to consider an infinite number of variables (for example to study time series). An individual is then represented by a trajectory which, if $X = (X_t)_{t\in T}$ is of the second order (i.e. $E[\int_T (X_t)^2 d\mu(t)]<\infty$), belongs to $E=L^2(T,\mathcal{E},\mu)$. The previous operators are now $\forall f \in L^2(\Omega,\mathcal{A},P)$, $\forall t\in T$ $[U(f)](t) = E(fX_t)$ and $\forall u\in E$ $U^*(u) = \int_T u(t)X_t d\mu(t)$. The P.C.A. elements are again obtained by the Schmidt decomposition of U.

If the trajectories are regular, a Sobolev space H(T) can replace $L^2(T,\mathcal{E},\mu)$. The approximation of this type of analysis is dealt with by Besse (1979). One can find a study of descriptive methods for processes in Saporta (1981).

2.1.3. Hilbert variables and complex variables. The above statement leads us to consider various types of spaces in the diagram, which are Hilbert spaces. This leads to developing the P.C.A. of a random variable with values in a separable Hilbert space (Kleffe (1973), Dauxois and Pousse(1976), Boudou (1979)). Since P.C.A. depends only on the law of variables (in fact only on the covariances), one can consider the P. C.A. of a probability (or a measure, see Qannari (1983)) on H with a second order moment. All the previous cases will be particular cases (Dauxois and Pousse (1976)).

Some data may naturally be considered as complex variable values (modulus : intensity , argument : direction). Time series are also often treated as a family of complex variables. The P.C.A. of complex Hilbertian variables has thus been studied (cf.Chakak (1983)). It is not the equivalent of the P.C.A. of real variables, but it may be quite easily put to use on a practical scale. A comparison can be drawn in Deville (1981) between the complex P.C.A. and the spectral analysis of stationary time series.

2.1.4. P.C.A. under constraints. The main constraints used in P.C.A. are linear ones (orthogonality to a subspace). One can consider them in all the above extensions, and further (to belong to a convex cone : see Téchené (1980)). A more general study of many optimality properties of the P.C.A. and equivalent criteria leading to a P.C.A. were considered, after Rao (1964), in Boudou (1979), Croquette (1980), Dauxois, Pousse and Téchené (1979), Ramasy (1984), Téchené (1980). Similarly, Sabatier (1984) worked on the choice of metrics, when using instrumental variables.

2.2. Canonical analysis.

Associated with the variables X (into $\mathbb{R}^p$) and Y (into $\mathbb{R}^q$) we can here consider the operators (as in P.C.A.) U_1 and U_2. One has : $U_i \circ U_i^* = V_i$ (i=1,2) and $U_1 \circ U_2^* = V_{12} = E(Y \otimes X) = V_{21}^*$ (with $x \otimes y : u \to \langle u,x \rangle y$).

The elements of this analysis can be obtained through the eigenvalues and eigenvectors of $R = V_1^{-1} V_{12} V_2^{-1} V_{21}$ (see Cailliez and Pages (1975)). The previous extensions are still to be taken into account, with in addition :

2.2.1. The study of the non-linearity. Usually, the (linear) C.A. of $X = (X_1, \cdots, X_p)$ and $Y = (Y_1, \cdots, Y_q)$ is the C.A. of the two subspaces $F_X = sp(X_1, \cdots, X_p)$ and $F_Y = sp(Y_1, \cdots, Y_q)$ of $L^2(\Omega, \mathcal{A}, P)$; it is the classical procedure of searching, at each step, a normalized $f = \sum a_i X_i$ which is the nearest to F_Y. The non linear C.A. of X and Y is obtained by replacing F_X (resp.F_Y) by $L^2(\mathfrak{B})$ (resp. $L^2(\mathfrak{C})$), where $\mathfrak{B}$ (resp. $\mathfrak{C}$) is the σ-field spanned by X (resp.Y). This is equivalent to the search of a normalized $f = h(X) \in L^2$ which is the nearest to $L^2(\mathfrak{C})$, where h is a measurable, but not necessarily linear, function (see Lancaster (1969)). This analysis amounts to the spectral decomposition of $E^{\mathfrak{B}} \cdot E^{\mathfrak{C}}$. An approximation can be easily made by a correspondence analysis (see Dauxois and Pousse (1976), and § 3.2.).

Depending on the form of f sought, a whole family of semi-linear canonical analyses can be developed between linear and non-linear C.A.. It is always a C.A. of two subspaces. The semi-linear P.C.A. has been studied and used for the descriptive treatment of a qualitative process (see Boumaza (1980))

2.2.2. The extensions to k subspaces. It is an usual extension in the linear case , which can also be considered (see Dauxois, Fine and Pousse (1979), Dauxois and Pousse (1975)) in the non-linear case (more particularly so in the case of multiple correspondence analysis , a type of analysis commonly used (Baccini et al. (1986)) as an exploratory and descriptive method associated with several qualitative variables).

3. ASYMPTOTIC STUDIES.

Two types of problems will be considered : some due to the sampling, others resulting from the different types of approximation. In a practical situation the analysis in infinite dimension as mentioned above can only be obtained from an approximation of finite dimension. For instance, the time is discretized in analysing a random process (the process is considered at times t_1, ...,t_k which are fixed or random). On the other hand, one may deal with a sample of size n of observations which can be either independent and identically distributed or not.

3.1. Approximation.

The next proposition (Dauxois and Pousse (1977), Arconte (1980)) is the main tool used in deriving approximations. Let T be a compact selfadjoint and positive operator on a separable Hilbert space H, with the spectral decomposition $T = \sum \lambda_i e_i \otimes e_i$. Let us denote $I' = \{ i \in I; \lambda_i \text{ simple} \}$, $I_j = \{ i \in I; \lambda_i = \lambda$ ($k_j = \text{card } I_j$ is the multiplicity of λ_j). Let L be a subset of I such that

$\{\lambda_i\}_{i\in L}$ is the strictly decreasing sequence of all eigenvalues of T, and let P_j be the orthogonal projector on the eigenspace $E_j=sp\{e_k;k\in I_j\}$ associated with λ_j : $P_j = \sum_{k\in I_j} e_k \otimes e_k$ and $T = \sum_{k\in L} \lambda_k P_k$.

Let $\sigma^*(H)$ be the cone of all compact selfadjoint positive operators on H and, for any n of $\mathbb{N}$, let T_n be a $\sigma^*(H)$-valued r.v.. Denote λ_j^n, e_j^n, E_j^n, P_j^n the elements associated with T_n as λ_j, e_j, E_j, P_j with T.

<u>Proposition.</u> If T_n converges almost surely (a.s.) to T in $\mathcal{L}(H)$, then :

$\forall j \in I$ $\quad \lambda_j^n$ converges a.s. to λ_j

$\pi_j^n = \sum_{k\in I_j} P_k^n$ converges a.s. to P_j

$\forall j \in I'$ $\quad e_j^n$ converges a.s. to e_j.

These results will be used when deriving approximations, either by discretisation, or, in the next section, by sampling.

<u>3.1.1. Approximation of a non linear canonical analysis by a correspondence analysis.</u> We consider the non linear C.A. of a pair (X,Y) of real r.v.. Let, for any n of $\mathbb{N}$, $(A_i^n)_{i\in I}$ (resp. $(B_j^n)_{j\in J}$) be a finite partition of $\mathbb{R}$ in intervals, which spans the sub σ-field $\mathcal{D}_n$ (resp. $\mathcal{E}_n$) of $\mathcal{B}_{\mathbb{R}}$, the Borel σ-field of $\mathbb{R}$. For any n, the non linear C.A. of the restriction to $\mathcal{D}_n \otimes \mathcal{E}_n$ of the law of (X,Y) is obtained by the spectral analysis of $E^{\mathcal{B}_n} \cdot E^{\mathcal{C}_n}$, where $\mathcal{B}_n$ is $X^{-1}(\mathcal{D}_n)$ and $\mathcal{C}_n = Y^{-1}(\mathcal{E}_n)$, and this C.A. is a correspondence analysis. If limsup $\nu(A_i^n)=0$, where ν is the Lebesgue measure of $\mathbb{R}$, one shows that $\mathcal{B}_n$ strongly converges to $\mathcal{B}$ (i.e. $\forall f\in L^2 \lim E^{\mathcal{B}_n}(f) = E^{\mathcal{B}}(f)$), and similarly for $\mathcal{C}_n$ to $\mathcal{C}$. One can prove (see Dauxois and Pousse (1976)) that, if $E^{\mathcal{B}} \cdot E^{\mathcal{C}}$ is compact, $E^{\mathcal{B}_n} \cdot E^{\mathcal{C}_n} \cdot E^{\mathcal{B}_n}$ uniformly converges to $E^{\mathcal{B}} \cdot E^{\mathcal{C}} \cdot E^{\mathcal{B}}$. Thus, from the above proposition, the elements of correspondence analysis converge to those of the non linear C.A. of (X,Y).

In the case of multiple C.A., one gets in the same way an approximation by a multiple correspondence analysis.

<u>3.1.2. Discretization upon time for the analysis of random functions.</u>

$[0,T] \subset \mathbb{R}$ is divided into $t_1,\cdots,t_p$, either at random or in an organised way. One can also show the convergence (the demonstration depends on the method and on the type of the discretization ; see Besse (1979), Dossou-Gbete (1980)). Other forms of approximations are studied in Martin (1980).

3.2. Convergence and asymptotic laws when sampling.

Under the same assumptions given in § 3.1. for the operators T_n and T, one has the following general result (Dauxois et al. (1982),Arconte (1980)).

Theorem. If $\sqrt{n}(T_n - T)$ converges in law to a r.v. U in the separable Banach space $\sigma_\infty(H)$ (the space of the compact operators on H), then :

a) the joint law of the family $\{\sqrt{n}(\pi_j^n - P_j)\}_{j\in I}$ converges, as $n\to\infty$, to the joint law of $\{\varphi_j(U)\}_{j\in I}$, where $\varphi_j(U) = -(S_jUP_j + P_jUS_j)$ and $S_j = \sum_{k\in I-I_j} e_k \otimes e_k/(\lambda_k-\lambda_j) = \sum_{i\in L-\{j\}} P_i/(\lambda_i-\lambda_j)$.

b) the joint law of the family $\{[\sqrt{n}(\lambda_k^n-\lambda_j)]_{k\in I_j}\}_{j\in L}$ converges, as $n\to\infty$, to the joint law of $\{\Delta_j(P_jUP_j)\}_{j\in L}$, where Δ_j maps to a selfadjoint of finite rank $k_j=\text{card}(I_j)$ operator the vector of its decreasing eigenvalues.

c) the joint law of $\{\sqrt{n}(e_j^n - e_j)\}_{j\in I'}$ converges, as $n\to\infty$, to the joint law of $\{S_jUe_j\}_{j\in I'}$.

All these results are obtained by using the Rubin theorem (Billingsley (1968)). One can note that all those results remain valid whether the eigenvalues are simple or multiple. The results hold whatever U. We do not use asymptotic developments. One can use another approach, which comes to extending the perturbation theory (Kato (1966)) to the $V'=V+\varepsilon(U)U$ case (se Fine (1981) and Fine (1986)). If U is a gaussian centered Hilbertian r.v., then $S_jUP_j+P_jUS_j$, P_jUP_j and S_jUe_j also are gaussian centered.

3.3.Applications.

3.3.1. Principal component analysis. From an i.i.d. sample $(X_i)_{i=1..n}$ o the r.v. X which has a 4th moment in the Hilbert space H (often $\mathbb{R}^p$), the hy potheses of the above theorem are verified for $T_n=V_n$, the sample covarian ce operator (Central limit theorem).Thus one can make explicit the asymp totic laws of all elements, for any law of X. Those results are then given i Dauxois et al.(1982) and Romain (1979) (centered X) or Arconte (1980) (uncentered X). Results concerning inferential applications (confidence sets, tests) if X has a gaussian or elliptic law can also be found there (see als Anderson(1963), Davis(1977), Muirhead(1978), Tyler(1983), Waternaux(1976)

The case of reduced P.C.A. is more complex, because they are many reductions for a random function. This asymptotic study can be found in Fang and Krishnaiah(1982), Fine and Romain(1984), Fujikoshi (1980), Konishi(1979)).

3.2.2. Linear canonical analysis. Let X and Y be two centered r.v. which admit a moment of second order in $\mathbb{R}^p$ and $\mathbb{R}^q$ respectively. The linear C.A. of (X,Y) can be obtained (§ 2.2.) from $R=V_1^{-1}V_{12}V_2^{-1}V_{21}$, where $V_1 = E(X \otimes X)$, $V_{12} = E(Y \otimes X)$, $V_{21} = E(X \otimes Y)$ and $V_2 = E(Y \otimes Y)$. Without loss of generality one can consider $V_1 = I_1$ and $V_2 = I_2$ (i.e. the components of X or Y are reduced and orthogonal : the principal components of X or of Y are often used). Thus $R = V_{12}V_{21} = \sum \rho_i e_i \otimes e_i$, where the ρ_i's are the canonical coefficients, the e_i's (resp. e'_i's) the factors associated with X (resp. Y : $\rho_i e'_i = V_{21} e_i$) and the f_i's (resp. g_i's) the canonical variables ($\langle X,e_i\rangle = \rho_i f_i$, $\langle Y,e'_i\rangle = \rho_i g_i$).

From a sample of size n, the C.A. is obtained through $(V_1^n)^{-1}V_{12}^n(V_2^n)^{-1}V_{21}^n$. Denote $R_n = (V_1^n)^{-1/2}V_{12}^n(V_2^n)^{-1}V_{21}^n(V_1^n)^{-1/2}$, a positive selfadjoint operator. R_n has the empirical canonical coefficients ρ_i^n for eigenvalues, with the associated eigenvectors h_i^n. The factors are $e_i^n = (V_1^n)^{-1/2} h_i^n$. One can prove (see Arconte(1980)) that R_n converges a.s. to R in $\sigma_2(\mathbb{R}^p)$ and that $\sqrt{n}(R_n-R)$ converges in law to a gaussian centered r.v. T, and we can apply the above theorem. One makes completely explicit the results for the ρ_i^n and h_i^n in either the gaussian or elliptic cases. One must go back to the e_i^n to obtain the asymptotic laws of the factors. The complete statement of these results and various inferential applications are given in Arconte (1980). In the case of simple eigenvalues, part of those results were already known through very different approaches (cf. Muirhead and Waternaux (1980)). If X and Y are not centered, one uses $X_i-\bar{X}$ and $Y_i-\bar{Y}$ and shows that the asymptotic laws remain the same (see Boutayeb (1983)).

3.3.3. Discriminant analysis. The discriminant analysis implies a variable X with values in $\mathbb{R}^p$ and a nominal variable Y with q modalities. Taking into account the indicator variables $(Y^i)_{i=1,\cdots,q}$ of the modalities of Y, the D.A. on a sample of size n appears as a canonical analysis in which the X variable is centered (one works with the $X_i-\bar{X}$) when the indicator variables are not. So, one considers an operator R'_n instead of R_n, and one shows that $\sqrt{n}(R'_n-R)$ has the same asymptotic law as $\sqrt{n}(R_n-R)$. Therefore, one can use the results obtained for the C.A., but the explicit forms of the covariance operators are more difficult to obtain (see Boutayeb (1983)). One must notice that the discriminant analysis is also a particular P.C.A. (it is the P.C. A. of the centers of gravity of the groups of individuals taking the same modality of Y). However, since the sampling does not apply to the centers of gravity, the asymptotic results for the P.C.A. cannot be used.

3.3.4. Correspondence analysis. We now consider two nominal variables X and Y with p and q modalities. The correspondence analysis can be shown as being simultaneously a non linear canonical analysis (C.A. of $L^2(\mathcal{B})$ and $L^2(\mathcal{C})$, where $\mathcal{B}$ -resp. $\mathcal{C}$- is the finite σ-field spanned by X-resp.Y-) and a linear non centered canonical analysis (of the 2 groups of indicator variables). One starts (Ghomari (1983)) with the asymptotic results for the non centered linear C.A. and one makes them explicit in the case of the correspondence analysis. This way, one furthers results obtained in O'Neil(1978).

3.3.5. Complex P.C.A. The P.C.A. of a variable with values in a complex Hilbert space is not equivalent to a P.C.A. of real variables. Furthermore, the complex multivariate gaussian law is more constraining than the real one and we cannot use the central limit theorem. We needed an isometric map between complex Hilbert spaces and orthogonal sums of real Hilbert spaces. Once obtained the asymptotic law of $U_n = \sqrt{n}(V_n - V)$, one can use again the above mentionned general theorem after having shown that it remained valid in the complex case. Refer to Chakak (1983) for the definition of the P.C.A. of a complex hilbertian variable, for its properties, approximations, computation possibilities, and for its asymptotic results (see also Krishnaiah and Lee (1977)).

3.3.6. Multivariate "errors-on-variables" regression models. A possible application of the above results in a different frame would be about the linear model : $\forall i=1,\cdots,n \quad Y_i = y_i + e_i$ and $y_i \in E_q$, where : $y_i \in \mathbb{R}^p$ is unknown , the e_i are independent r.v. with the same law, and the covariance matrix Γ of which is either known or of the form $\sigma^2\Gamma^*$ with known Γ^*. E_q is an affine subspace of $\mathbb{R}^p$ with dimension q.

The estimator $\hat{E}_q$ is spanned by $\hat{a}_1,\cdots,\hat{a}_q$, the eigenvectors of $\Gamma^{-1/2}V_n\Gamma^{-1/2}$ associated with the q largest eigenvalues $\hat{\lambda}_i$, where $V_n=(1/n)\sum_{i=1,\cdots,n}(Y_i-\bar{Y})\otimes(Y_i-\bar{Y})$.

$$\hat{\sigma}^2 = (1/(p-q))\sum_{i=q+1}^{p} \hat{\lambda}_i \text{ is a convergent estimator of } \sigma^2 .$$

Under the usual hypotheses (in Malinvaud (1981)), but without the assumption that the errors are gaussian, Gedler(1986) gets, through obtaining in several stages an appropriate T_n and T, the asymptotic laws of $\sqrt{n}\,(\hat{\sigma}^2-\sigma^2)$ and $\sqrt{n}\,(\hat{a}_i-a_i)$. This work extends results of Malinvaud (1981),Gleser (1981).

REFERENCES.

ANDERSON,T.W.(1963). Asymptotic theory for Principal Component Analysis. The Annals of Math. Statistics, vol.34, n°1.
ARCONTE,A.(1980).Etude asymptotique de l'analyse canonique.Th.3°C. PAU.
BACCINI,A. and al.(1986).Comparaison et évaluation des approches française et britannique de l'analyse de données complexes [Lancaster(Gr.-Britain) -Toulouse (France)]. Rapport C.N.R.S., Univ. de TOULOUSE III (France).
BESSE,P.(1979). Etude descriptive d'un processus. Approximation et interpolation. Thèse de 3° cycle, Univ. de TOULOUSE III (France).
BILLINGSLEY, P.(1968).Convergence of Probability measures.Wiley, N.York.
BOUDOU,A.(1979).Différents types d'A.C.P. de fonctions aléatoires hilbertiennes; étude de certaines contraintes.Pub.Labo.Stat.,Univ.TOULOUSE III.
BOUMAZA,R.(1980).Et.descript.d'une f.aléat.qualitative,Th.3°C TOULOUSE III
BOUTAYEB, A.(1983). Etude asympt.de l'anal.discriminante. Thèse 3°C. PAU.
CAILLIEZ,F.et PAGES,J.P.(1975).Intr.à l'analyse des données. SMASH.Paris.
CHAKAK, A.(1983).Analyse en composantes principales de fonctions aléatoires complexes.Stabilité et étude asymptotique.Thèse 3° cycle,Univ.de PAU.
CROQUETTE,A.(1980).Quelques résultats synth.en analyse des données multidim. : optimalité et métriques à effets relationnels. Th. 3°C.,TOULOUSE III.
DAUXOIS,J., FINE,J. et POUSSE,A.(1979). Echantillonnage en segmentation, étude de la convergence. Stat. et Analyse des données, vol.3.
DAUXOIS,J. et POUSSE,A.(1975). Une extension de l'analyse canonique. Quelques applications. Ann. Inst. Henri Poincaré B, vol.21 n°4, p.355-379.
DAUXOIS,J. et POUSSE,A.(1976).Les analyses factorielles en Calcul des Probabilités et en Statistique;essai d'étude synthétique.Thèse,Univ.TOULOUSE III
DAUXOIS,J. et POUSSE,A.(1977). Some convergence problems in factor analyses. Recent dev. in Stat., Barra et al.(Ed.), North Holland Pub. Co.
DAUXOIS,J.,POUSSE,A. et ROMAIN,Y.(1982).Asymptotic theory for the Principal Component Analysis of a vector random function : some applications to Statistical Inference.Journal of multivariate analysis,vol.12 n°1,p.136-154
DAUXOIS,J., POUSSE,A. et TECHENE,J.J.(1979). Quelques propriétés d'optimisation sous contraintes. Applic. à l'analyse factorielle d'une fonction aléatoire. Publ. du Labo. de Stat. Univ. de TOULOUSE III.
DAVIS, A.N.(1977). Asympt. th.for P.C.A.: nonnormal case, Aust.J.St., 19.
DEVILLE, J.C.(1981).Décomp. des processus aléatoires : comparaison entre les méthodes factorielles et l'anal.spectrale.Stat.et An.des données,6 n°3.
DOSSOU-GBETE, S.(1980).Approx.de l'A.C.P. semi-lin. d'une fonction aléat. par l'A.C.P. semi-lin. d'une f.a. qualitative. Labo. de Stat., TOULOUSE III.
FANG,C. and KRISHNAIAH,P.R.(1982). Asymptotic distributions of functions of the eigenvalues of some random matrices for nonnormal populations. Journal of Multivariate Analysis vol.12 n°1, p.39-63.
FINE,J.(1981). Analyses en composantes principales réduites et théorie des

perturbations. Thèse de 3° cycle, Univ. de TOULOUSE III.
FINE, J. (1986). On the validity of the perturbation methods in asymptotic theory. Statistics (à paraitre).
FINE,J. et ROMAIN,Y.(1984).Reduced Princ.Comp.Analysis,Statistics, 15 n°4.
FUJIKOSHI,Y.(1980). Asymptotic expansions for the distributions of the sample roots under nonnormality. Biometrika, vol.67.
GEDLER, G. (1986). Modèles de régression sur variables entachées d'erreurs. Etude asymptotique. Thèse de 3° cycle, Univ. de PAU.
GHOMARI,A.(1983).A.C.d'une table de contingence.Et.asympt.,Th.3°C.PAU.
GLESER,J.L.(1981). Estimation in a multivariate "errors-in-variables" regression model; large sample results. Ann.of Stat., vol.9, n°1, p. 24-44.
KATO,T.(1966).Perturbation theory for linear operators.Springer,New York.
KLEFFE,J.(1973).Principal component of random variables with values in a separable Hilbert space. Math. Oper. forch. v. Statist., vol.4, p.391-406.
KONISHI,S.(1979). Asymptotic expansions for the dist. of statistics based on the sample correlation matrix in Princ. Comp.Anal., Hiroshima Math.J.n°9.
KRISHNAIAH,P.R. and LEE,J.C.(1977).Inference of the eigenvalues of the covariance matrix of real and complex multivariate normal populations, Multivariate Anal. IV, P.R. Krishnaiah (Ed.), North Holland Publishing Co.
LANCASTER,H.(1969). The Chi-squared Distributions. Wiley, New-York.
MALINVAUD, E.(1981). Méthodes statistiques de l'économétrie. Dunod, Paris.
MARTIN, J.F.(1980). Le codage flou et ses appl. en Stat., Thèse 3°C. PAU.
MUIRHEAD, R.J.(1978). Latent roots and matrix variates : a review of some asymptotic results. Annals of Statistics, vol.6 n°1.
MUIRHEAD, R.J. and WATERNAUX,C.M.(1980).Asymptotic distr.in can. correlation anal.and other multiv.proc.for nonnormal populations. Biometr.,66.
O'NEILL,M.E.(1978). Asymptotic distributions of the canonical correlation from contingency tables. Australian J. of Statistics, vol.20 n°1,p.75-82.
QANNARI,E.M.(1983).Anal.factor.de mesures;Appl.,Thèse 3°C.TOULOUSE II
RAMASY,A.(1984). Propr. extrémales des valeurs singulières d'opér. compacts. Applications stat. à la réduction des variables. Thèse 3° cycle,PAU
RAO, C.R.(1964). The use and interpretation of principal component analysi in applied research. Sankhya ser.A, n°26, p. 329-358.
ROMAIN,Y.(1979).Et.asym.de l'A.C.P.d'une fonct.aléat.,Th.3°C TOULOUSE I
SABATIER,R.(1984). Quelques généralisations de l'anal.en comp.principale de variables instrumentales. Statistique et Analyse des données, vol.9 n°3
SAPORTA,G.(1981).Méth.explor.d'anal.de données temporelles, Th.PARIS V
TECHENE,J.J.(1980). Réductions optimales d'opérateurs. Application aux analyses factorielles. Thèse de 3° cycle, Univ. de PAU.
TYLER,D.E.(1983). The asympt. distr. of P.C. roots under local alternative to multiple roots /A class of asymptotic tests···Annals of Math. Stat.,11 n
WATERNAUX, C.M.(1976). Asymptotic distribution of the sample roots for a nonnormal population. Biometrika vol.63.

Bernoulli, Vol. 2, pp. 317-326

ESTIMATION OF SYMMETRIC FUNCTIONS OF PARAMETERS AND ESTIMATION OF COVARIANCE MATRIX

Kei Takeuchi and Akimichi Takemura
Faculty of Economics, University of Tokyo
Bunkyo, Tokyo, 113 JAPAN

1. GENERAL METHODOLOGY.

In most of the problems of multivariate analysis, number of parameters to be estimated is very large. It is especially true when unknown covariance matrix of many variables is involved. C.Stein (1956) first noted that when the dimensionality of the parameter is large, componentwise application of one dimensional optimal estimation procedures may not be admissible and uniform improvement may be possible. Actually the improvement can be substantial when the dimensionality of the parameter is large. Stein's procedure has been extensively investigated and extended into various cases.

The purpose of this paper is to look at the problem of estimation of multi-dimensional parameter from slightly different viewpoint and apply the idea to some estimation problems related to the covariance matrix.

Suppose that we are to estimate p real parameters $\theta_1,\ldots,\theta_p$ based on sample $X=(X_1,\ldots,X_n)$. Let $\hat{\theta}_1,\ldots,\hat{\theta}_p$ be componentwise optimal estimators of $\theta_1,\ldots,\theta_p$. It has been now well established that in many cases when p is not smaller than some integer, the set of unbiased estimators is jointly inadmissible in terms of squared error loss.

This phenomenon can be interpreted in the following manner. Suppose that the problem is symmetric about the parameters, i.e., the problem is invariant under permutations of the parameters. More precisely for any possible set of parameters $(\theta_1,\ldots,\theta_p)$ and for any permutation $(i_1,\ldots,i_p)$ of $(1,\ldots,p)$, $(\theta_{i_1},\ldots,\theta_{i_p})$ is also a possible parameter point.

Let $\theta_{(1)} \leq \ldots \leq \theta_{(p)}$ be the ordered set of values of p parameters and let $r(i)$ be the rank of the i-th parameter θ_i in the set of the parameter values. Now the problem of joint estimation of p parameters can be decomposed into two parts, i.e., estimation of the set of the parameter values and estimation of the rank of each parameter in the parameter set. Let $\hat{\theta}_{(1)} < \ldots < \hat{\theta}_{(p)}$ be ordered set of estimators $\hat{\theta}_1,\ldots,\hat{\theta}_p$ and let $r^*_{(i)}$ be the rank of $\hat{\theta}_i$ in the set. We assume that the probability that any pair of two $\hat{\theta}_i$'s having identical values is zero, hence we can define the rank $r^*_{(i)}$ uniquely. Now it is natural to estimate the rank $r_{(i)}$ of the parameter by the rank of the estimator $r^*_{(i)}$.

However for estimation of the ordered set it is obvious that the maximum of the estimators $\hat{\theta}_{(p)}$ tends to be larger than the maximum of the parameters $\theta_{(p)}$. Similarly $\hat{\theta}_{(1)}$ tends to be smaller than $\theta_{(1)}$. Hence the estimated range $\hat{\theta}_{(p)} - \hat{\theta}_{(1)}$ tends to be larger than the parameter range $\theta_{(p)} - \theta_{(1)}$ and the difference can be substantial especially when the latter is small. Thus it is intuitively justified to modify the estimators $\hat{\theta}_1,\ldots,\hat{\theta}_p$ so that the range be shrunken. James-Stein estimator of the mean vector of the multivariate normal distribution is clearly of such type.

Estimation of the set of p values $\theta_{(1)} \leq \ldots \leq \theta_{(p)}$ is equivalent to estimating p functionally independent symmetric functions $\psi_1=\psi_1(\theta_1,\ldots,\theta_p)$,..., $\psi_p=\psi_p(\theta_1,\ldots,\theta_p)$ defining one to one correspondence. Then the estimators of ψ's derived from $\hat{\theta}_1, \ldots,\hat{\theta}_p$, i.e., $\hat{\psi}_1= \psi_1(\hat{\theta}_1,\ldots,\hat{\theta}_p)$,..., $\hat{\psi}_p= \psi_p(\hat{\theta}_1,\ldots,\hat{\theta}_p)$ may not be good estimators of $\psi_1,\ldots,\psi_p$. If we have intuitively better estimators

$\hat{\psi}_1^*,\ldots,\hat{\psi}_p^*$, then we can get another set of estimators $\hat{\theta}_{(1)}^* \leq \ldots \leq \hat{\theta}_{(p)}^*$ by solving the equations $\hat{\psi}_1^* = \psi_1(\hat{\theta}_1^*,\ldots,\hat{\theta}_p^*)$ $,\ldots,$ $\hat{\psi}_p^* = \psi_p(\hat{\theta}_1^*,\ldots,\hat{\theta}_p^*)$.

Clearly one such simple transformation is the ordered set of values themselves $\psi_1 = \theta_{(1)},\ldots,\psi_p = \theta_{(p)}$.

Another possible transformation is to take p first moments of θ_i's,

$$\psi_j = \frac{1}{p}\sum_{i=1}^{p} \theta_i^{\,j}, \qquad j=1,\ldots,p.$$

For illustration consider estimation of independent normal means. Let X_i, $i=1,\ldots,p$ be independent normal random variables with mean θ_i and variance 1. Then the unbiased estimators of ψ_j's are given as follows.

$$\hat{\psi}_1^* = \frac{1}{p}\sum X_i ,$$

$$\hat{\psi}_2^* = \frac{1}{p}\sum X_i^{\,2} - 1 ,$$

$$\ldots,$$

$$\hat{\psi}_p^* = \frac{1}{p}\sum X_i^{\,p} - \binom{p}{2}\hat{\psi}_{p-2}^* - 3\binom{p}{4}\hat{\psi}_{p-4}^* - 3\cdot 5\binom{p}{6}\hat{\psi}_{p-6}^* \cdots$$

Modified estimates $\hat{\theta}_1^*,\ldots,\hat{\theta}_p^*$ can be obtained by solving

$$\sum_{i=1}^{p} \hat{\theta}_i^{*j} = p\,\hat{\psi}_j^* , \qquad j=1,\ldots,p. \tag{1}$$

A necessary condition for having real-valued solution of these equations is that $\hat{\psi}_{2k}^{*1/2k}$ is monotone nondecreasing in integer k and that $\hat{\psi}_{2j}^* \geq \hat{\psi}_j^{*2}$ for odd j. Since some necessary conditions may not be satisfied, we may proceed as follows.

Suppose that the system of equations (1) does not have real-valued solutions for $\hat{\theta}_1^*,\ldots,\hat{\theta}_p^*$. Let k be such an integer that the partial system of equations $\sum_i \hat{\theta}_i^{*j} = p\,\hat{\psi}_j^*$ for $j=1,\ldots,k$ has real valued solutions but when the k+1-st equation is added the system no longer has real valued solutions. Then we may determine $\hat{\theta}_i^*$, $i=1,\ldots,p$, so that

$$\left|\sum_{i=1}^{p} \hat{\theta}_i^{*k+1} - p\,\hat{\psi}_{k+1}^*\right|$$

is minimized under the restriction $\sum_{i=1}^{p} \hat{\theta}_i^{*j} = p\,\hat{\psi}_j^*$, $j=1,\ldots,k$.

The first step of this procedure goes as follows. Let $\bar{X} = (1/p)\sum X_i$. If $(1/p)\sum_i X_i^2 < \bar{X}^2 + 1$, or equivalently if $\sum(X_i-\bar{X})^2 < p$, then $\hat{\theta}_j^* = \bar{X}$, $j=1,\ldots,p$. Otherwise we proceed to the next step. Ignoring the third and higher order moment conditions we obtain the following Stein type estimator.

$$\hat{\theta}_i^* = \bar{X} \qquad \text{if} \qquad \sum(X_i-\bar{X})^2 < p$$

$$= \left(1 - \frac{p}{\sum(X_j-\bar{X})^2}\right)^{1/2} (X_i-\bar{X}) + \bar{X}, \qquad \text{otherwise.}$$

Therefore if we consider only first two moment conditions, either all estimated parameter values collapse to the mean or they are proportionally shrunken toward the mean. Now if we consider higher order moment conditions, collapsing of estimated parameter values to several clusters occur. Let k be as above. Under the restriction $\sum \hat{\theta}_i^{*j} = p\,\hat{\psi}_j^*$, $j=1,\ldots,k$, we have to maximize or minimize $\sum \hat{\theta}_i^{*j+1}$. Let $\lambda_1,\ldots,\lambda_k$ be the Lagrange multipliers for the restrictions. Differentiating the Lagrange form with respect to $\hat{\theta}_i^*$ we obtain

$$(k+1)\,\hat{\theta}_i^{*k} - k\,\lambda_k \hat{\theta}_i^{*k-1} - \ldots - \lambda_1 = 0, \qquad i=1,\ldots,p.$$

Therefore $\hat{\theta}_i^*$'s are the roots of a k-th degree polynomial (k<p). This implies that there are no more than k distinct values among $\hat{\theta}_1^*,\ldots,\hat{\theta}_p^*$. If some of the natural estimates, say $\hat{\theta}_i = X_i$, $i\in I$, are close to each other, then it may be reasonable to replace $\hat{\theta}_i$, $i\in I$,

by some middle value $\bar{\theta}_I$. This kind of clustering can be realized by our procedure. A concrete example of this clustering will be given in the next section.

The basic idea can be extended into more general situations. Suppose that over the parameter space $\Xi = \{\theta\}$ a group of one to one transformations $G = \{g\}$ is defined. Then the parameter space is partitioned into the orbits of transformations. Let Ξ_0 be the set of the representative elements of the orbits. Then the parameter point $\theta \in \Xi$ can be represented by a pair $\theta = (\eta, g)$, where $\eta \in \Xi_0$, $g \in G$, and $\theta = g\eta$. Now let $\hat{\theta} = \{\hat{\eta}, \hat{g}\}$ be some natural estimator of θ. Assume that the distribution of $\hat{\eta}$ depends only on η and independent of g. Then we may consider estimation of η independently of g and depending on $\hat{\eta}$ alone. Let $\psi = \psi(\eta)$ define one to one transformation between η and ψ. If we have some reasonable estimator of the form $\hat{\psi}^* = \psi^*(\hat{\eta})$, where ψ^* is not necessarily identical with ψ, then a modified estimator of η can be obtained by $\hat{\eta}^* = \psi^{-1}(\hat{\psi}^*)$ and of θ by $\hat{\theta}^* = \{\hat{\eta}^*, \hat{g}\}$.

We apply these ideas to the estimation of covariance matrix in the next section.

2. ESTIMATION OF COVARIANCE MATRIX.

In this section we consider estimation of covariance matrix Σ of a multivariate normal population. Let $G = O(p)$ be the group of $p \times p$ orthogonal matrices acting on Σ as $g \Sigma g'$. This leads to the decomposition $\Sigma = P \Lambda P'$ where $\Lambda = \mathrm{diag}(\lambda_1, \dots, \lambda_p)$ is a diagonal matrix consisting of ordered characteristic roots of Σ and $P \in O(p)$ is an orthogonal matrix. Then instead of estimating Σ, we may estimate Λ and P separately. Let $\hat{\Sigma} = W/n$ be the usual estimator of Σ where W is a matrix distributed according to the Wishart distribution $W(n, \Sigma)$. Let $\hat{\Sigma} = \hat{P} \hat{\Lambda} \hat{P}'$ and let $\hat{\lambda}_1 < \dots < \hat{\lambda}_p$ be the characteristic roots of $\hat{\Sigma}$. It is easy to see that $\hat{\lambda}_p$ tends to be larger than λ_p and $\hat{\lambda}_1$ tends to be smaller than λ_1

Therefore better estimators $\hat{\lambda}_i^*$ may be obtained by shrinking $\hat{\lambda}_i$ toward the average. Takemura(1984b) discussed minimax estimation of this type. For a survey of other estimators of covariance matrix see Lin and Perlman(1985).

Based on the ideas given in the previous section we define a set of p independent symmetric polynomials $\psi_j = \psi_j(\lambda_1,\ldots,\lambda_p)$ and obtain unbiased estimators $\hat{\psi}_j^*$ of ψ_j. Then we solve for $\hat{\lambda}_i^*$ by $\hat{\psi}_j^* = \psi_j(\hat{\lambda}_1^*,\ldots,\hat{\lambda}_p^*)$. If real valued solutions do not exist for these equations similar procedures as was discussed above must be applied. A natural choice of ψ_j's will be the zonal polynomials. Zonal polynomials are symmetric homogeneous polynomials of the characteristic roots and eigenfunctions of the expectation operator. See Takemura(1984a) and Takeuchi and Takemura(1985) for this aspect of zonal polynomials. Let Z_κ be the zonal polynomial of degree k indexed by a partition κ of positive integer k. Then

$$E[\ Z_\kappa(\hat{\lambda}_1,\ldots,\hat{\lambda}_p)\ ] = (c_{n,\kappa}/n^k)\ Z_\kappa(\lambda_1,\ldots,\lambda_p)\ ,$$

where $c_{n,\kappa}$ is the eigenvalue associated with Z_κ. Therefore unbiased estimator of $Z_\kappa(\lambda_1,\ldots,\lambda_p)$ is simply given by $(n^k/c_{n,\kappa})\ Z_\kappa(\hat{\lambda}_1,\ldots,\hat{\lambda}_p)$. As discussed above the modified roots $\hat{\lambda}_1^*,\ldots,\hat{\lambda}_p^*$ are obtained by solving

$$Z_\kappa(\hat{\lambda}_1^*,\ldots,\hat{\lambda}_p^*) = (n^k/c_{n,\kappa})\ Z_\kappa(\hat{\lambda}_1,\ldots,\hat{\lambda}_p)\ .$$

There are still many zonal polynomials to choose. Simple ones which we investigate here are top order zonal polynomials $Z_{(k)}$ and last order zonal polynomials $Z_{(1^k)}$. Another reason to choose these zonal polynomials is that their eigenvalues are minimum and maximum, respectively, among zonal polynomials. This implies that sample values of these zonal polynomials tend to have large biases.

The last order zonal polynomials (with $\kappa=(1^k)$) are simply the elementary symmetric functions e_k of the characteristic roots. For the elementary symmetric functions the eigenvalues $c_{n,(1^k)}$ are given as

$$c_{n,(1^k)} = n(n-1)\cdots(n-k+1) .$$

Note that $c_{n,(1^k)}/n^k=(1-1/n)\cdots(1-(k-1)/n)\leq 1$. This implies that the elementary symmetric functions of the population roots are under-estimated by the elementary symmetric functions of the sample roots. Our procedure then is to correct these biases in p elementary symmetric functions of the sample roots and then solve them for p modified roots $\hat{\lambda}_1^*,\ldots,\hat{\lambda}_p^*$.

Top order zonal polynomials are somewhat more complicated. But its coefficients in terms of moments of the roots $\psi_j= \lambda_1^j+\ldots+ \lambda_p^j$ are explicitly known. See Section 4.6 of Takemura(1984a). See Parkhurst and James(1974) for extensive table of coefficients of zonal polynomials. First several top order zonal polynomials are

$$Z_{(1)}= \psi_1 ,$$

$$Z_{(2)}= \psi_1^2 + 2 \psi_2 ,$$

$$Z_{(3)}= \psi_1^3 + 6 \psi_1\psi_2 + 8 \psi_3.$$

$(\psi_1,\ldots,\psi_p)$ and $(Z_{(1)},\ldots,Z_{(p)})$ are in one to one correspondence. Hence $(\lambda_1,\ldots,\lambda_p)$ and $(Z_{(1)},\ldots,Z_{(p)})$ are in one to one correspondence as well. The eigenvalue $c_{n,(k)}$ is known to be

$$c_{n,(k)}= n(n+2)\cdots(n+2(k-1)) .$$

Hence top order zonal polynomials of the population roots are over-estimated by top order zonal polynomials of the sample roots. Correcting these biases we obtain another modification of the sample roots.

For concreteness consider the second order condition. Let $\bar{\lambda} = (1/p)\Sigma\lambda_i$ and $e_2 = \hat{\lambda}_1\hat{\lambda}_2+ \hat{\lambda}_1\hat{\lambda}_3+\ldots+ \hat{\lambda}_{p-1}\hat{\lambda}_p$. When we modify the sample roots by adjusting the elementary symmetric functions, we replace e_2 by $n/(n-1)e_2$. Note that $2e_2 = p(p-1)\bar{\lambda}^2 - \Sigma(\hat{\lambda}_i - \bar{\lambda})^2 \leq$ $p(p-1)\bar{\lambda}^2$. Therefore if $n/(n-1)e_2 > p(p-1)\bar{\lambda}^2/2$ or equivalently if

$$\Sigma\ (\hat{\lambda}_i/\bar{\lambda} -1)^2 < p(p-1)/n$$

then $\hat{\lambda}_i^* = \bar{\lambda}$, $i=1,\ldots,p$. Similarly when top order zonal polynomials are used, it can be shown that $\hat{\lambda}_i^* = \bar{\lambda}$, $i=1,\ldots,p$, if

$$\Sigma\ (\hat{\lambda}_i/\bar{\lambda} -1)^2 < p(p+2)/n$$

This calculation suggests that the shrinking of the roots are more drastic when top order zonal polynomials are used.

For higher order moment conditions it becomes more and more difficult to explicitly describe the pattern of the shrinking of the roots because it involves obtaining roots of higher degree polynomials. Here we illustrate the shrinking and the clustering of the roots with the following numerical example. Consider the cases $(\hat{\lambda}_1,\hat{\lambda}_2,\hat{\lambda}_3,\hat{\lambda}_4)=(1,2,3,4)$, $(\hat{\lambda}_1,\hat{\lambda}_2,\hat{\lambda}_3,\hat{\lambda}_4)=(1,2,4,8)$. The following Table 1 shows the modified roots $(\hat{\lambda}_1^*,\hat{\lambda}_2^*,\hat{\lambda}_3^*,\hat{\lambda}_4^*)$ for various degrees of freedom. We only show those degrees of freedom, where the pattern of clustering changes.

TABLE 1

Modification based on ele sym functions					Modification based on top order zonal polynomials				
DF	$\hat{\lambda}_1^*$	$\hat{\lambda}_2^*$	$\hat{\lambda}_3^*$	$\hat{\lambda}_4^*$	DF	$\hat{\lambda}_1^*$	$\hat{\lambda}_2^*$	$\hat{\lambda}_3^*$	$\hat{\lambda}_4^*$
∞	1.0	2.0	3.0	4.0	∞	1.0	2.0	3.0	4.0
		...					...		
42	1.13	2.23	3.26	3.38	212	1.06	1.98	3.47	3.49
41	1.14	2.24	3.31	3.31	211	1.06	1.98	3.48	3.48
		...					...		
27	1.19	2.81	3.00	3.00	73	1.04	2.88	3.04	3.04
26	1.22	2.93	2.93	2.93	72	1.04	2.99	2.99	2.99
		...					...		
16	2.00	2.67	2.67	2.67	31	2.16	2.61	2.61	2.61
15	2.5	2.5	2.5	2.5	30	2.5	2.5	2.5	2.5

Modification based on ele sym functions

DF	$\hat{\lambda}_1^*$	$\hat{\lambda}_2^*$	$\hat{\lambda}_3^*$	$\hat{\lambda}_4^*$
∞	1.0	2.0	4.0	8.0
		...		
11	1.87	2.28	4.10	6.74
10	2.18	2.18	4.03	6.61
		...		
8	2.75	2.75	3.22	6.29
7	3.08	3.08	3.08	5.77
6	3.50	3.50	3.50	4.50
5	3.75	3.75	3.75	3.75

Modification based on top order zonal polynomials

DF	$\hat{\lambda}_1^*$	$\hat{\lambda}_2^*$	$\hat{\lambda}_3^*$	$\hat{\lambda}_4^*$
∞	1.0	2.0	4.0	8.0
		...		
516	1.41	1.45	4.27	7.87
515	1.43	1.43	4.27	7.87
		...		
42	1.53	1.53	5.79	6.15
41	1.37	1.71	5.96	5.96
		...		
21	0.86	4.19	4.98	4.98
20	0.91	4.70	4.70	4.70
		...		
12	3.12	3.96	3.96	3.96
11	3.75	3.75	3.75	3.75

Apparently the modification based on the top order zonal polynomials shrink the differences of the roots too much and somewhat erratically, while shrinking by the elementary symmetric functions seems to be just appropriate. We carried out Monte Carlo experiments to confirm this observation. We generated 1000 Wishart matrices for the following four types of population covariance matrices: Σ=diag(1,1,1,1), diag(1,1,1,5), diag(1,2,3,4), and diag(1,2,4,8). Table 2 shows the averages and standard deviations of sample roots and roots modified by the above two methods for the degrees of freedom 20 and 50.

TABLE 2

Degrees of Freedom = 20

Pop roots		1.0	1.0	1.0	1.0	1.0	1.0	1.0	5.0
Sample	AVE	0.46	0.76	1.13	1.65	0.53	0.89	1.40	5.17
roots	STD	(.14)	(.17)	(.23)	(.33)	(.16)	(.21)	(.31)	(1.6)
Ele sym	AVE	0.73	0.96	1.04	1.28	0.77	0.99	1.26	4.96
func	STD	(.26)	(.22)	(.23)	(.34)	(.26)	(.25)	(.35)	(1.6)
Top ord	AVE	0.93	1.01	1.02	1.03	0.51	0.78	2.93	3.76
zonal	STD	(.22)	(.17)	(.18)	(.18)	(.30)	(.53)	(.72)	(1.4)

Pop roots		1.0	2.0	3.0	4.0	1.0	2.0	4.0	8.0
Sample	AVE	0.73	1.62	2.85	4.81	0.75	1.77	3.88	8.59
roots	STD	(.24)	(.43)	(.68)	(1.2)	(.25)	(.50)	(1.1)	(2.4)
Ele sym	AVE	1.02	2.08	2.83	4.09	0.95	2.16	3.98	7.90
func	STD	(.46)	(.66)	(.81)	(1.2)	(.42)	(.77)	(1.3)	(2.6)
Top ord	AVE	1.45	2.64	2.95	2.97	0.76	3.21	5.37	5.64
zonal	STD	(.76)	(.68)	(.75)	(.77)	(.72)	(1.5)	(1.5)	(1.9)

Degrees of Freedom = 50

Pop roots		1.0	1.0	1.0	1.0	1.0	1.0	1.0	5.0
Sample	AVE	0.64	0.86	1.09	1.40	0.69	0.95	1.27	5.07
roots	STD	(.11)	(.12)	(.14)	(.19)	(.12)	(.14)	(.19)	(1.0)
Ele sym	AVE	0.81	0.97	1.03	1.18	0.85	0.99	1.15	5.00
func	STD	(.17)	(.14)	(.14)	(.21)	(.17)	(.16)	(.20)	(1.0)
Top ord	AVE	0.95	1.00	1.01	1.03	0.64	0.64	2.10	4.60
zonal	STD	(.15)	(.11)	(.12)	(.13)	(.16)	(.16)	(.26)	(.97)

Pop roots		1.0	2.0	3.0	4.0	1.0	2.0	4.0	8.0
Sample	AVE	0.89	1.81	2.90	4.38	0.90	1.89	3.94	8.24
roots	STD	(.18)	(.33)	(.48)	(.75)	(.19)	(.36)	(.76)	(1.6)
Ele sym	AVE	0.99	2.04	2.92	4.03	0.99	2.01	4.00	7.97
func	STD	(.24)	(.43)	(.58)	(.82)	(.23)	(.44)	(.88)	(1.7)
Top ord	AVE	1.08	2.39	3.13	3.33	1.09	1.80	5.16	6.92
zonal	STD	(.34)	(.66)	(.53)	(.68)	(.36)	(.85)	(.83)	(1.7)

From our simulation results, modification by elementary symmetric functions seems to be very useful.

REFERENCES

Lin, S.P. and Perlman M.D. (1985). A Monte Carlo comparison of four estimators of a covariance matrix. Multivariate Analysis-VI, 411-429.

Parkhurst A,M. and James, A.T. (1974). Zonal polynomials of order 1 through 12. Selected tables in mathematical statistics, Vol. 2, IMS.

Stein, C. (1956). Inadmissibility of the usual estimator for the mean of a multivariate normal distribution, Proc. Third Berkeley Symp. on Math. Statist. Probab., 1, 197-206.

Takemura, A. (1984a). Zonal polynomials. IMS Lecture Notes-Monograph Series, Vol.4, IMS.

Takemura, A. (1984b). An orthogonally invariant minimax estimator of the covariance matrix of a multivariate normal population, Tsukuba J. Math., 8. 367-76.

Takeuchi, K. and Takemura, A. (1985). Eigenfunctions of association algebra of pairings and zonal polynomials. Discussion Paper 85-F-5, Faculty of Economics, Univ. of Tokyo.

INTRODUCTION IN GENERAL STATISTICAL ANALYSIS

V.L.Girko
Dept. of Cybernetics
Kiev T.G.Shevchenko State University
ul. Vladimirskaja 64
Kiev 252017
U.S.S.R.

The general statistical analysis of observations (G-analysis) is a mathematical theory studying some complex systems S, such that the number m_n of parameters of their mathematical models can increase together with the growth of the number n of observations over the system S. The purposes of this theory consist in finding by the observations of the system S such its mathematical models (G-estimates) which would approach the system S in a sense with a given rate at the minimal number of observations and under the general assumptions on the observations: the existence of the distribution densities of observed random vectors and matrices is not needed; the existence of several first moments of their components is only required, the numbers m_n and n satisfy the G - condition: $\lim_{n\to\infty} f(m_n, n) < \infty$, where $f(x,y)$ is some positive function increasing along x and decreasing along y. In most cases the function $f(x,y)$ is equal to xy^{-1}. In this case G-condition is also called the Kolmogorov-Deev condition.

In the general statistical analysis two conditions (postulates) are assummed.

1) the dimension (a number of parameters) of estimated characteristics of this system won't change with the

increase of the number m_n of parameters of system S mathematical models.

2) the dimension m_n of mathematical models can increase with the growth of number n of observations over the system S and, on the contrary, depends on m_n and cannot growt arbitrarily fast with the increase of m_n.

Suppose that an absolutely integrable Borel function $f(x)$ is given which has the partial derivatives of the second order, and a consistent estimate of the value $f(a)$ to be found by the independent observations $x_1, ,\dots, x_n$ over the random vector η distributed according to the normal law $N(a, R_{m_n})$.

Let us consider the functions $u(t,z) = Ef(z+a+\nu t^{1/2} n^{-1/2})$, where $t>0$ is a real - valued parameter, $z \in R^{m_n}$, ν is a random vector distributed according to the normal law $N(0, R_{m_n})$. These functions satisfy the equation

$$(\partial/\partial t)u(t,z) = Au(t,z), \quad u(1,z) = E f(z+\hat{a}),$$

$$A = (2n)^{-1} \sum_{i,j=1}^{m_n} r_{ij}(\partial^2/\partial z_i \partial z_j), \; u(z,0) = f(z+a), 0 \le t \le 1,$$

where r_{ij} are the entries of the matrix R_{m_n}, $\hat{a} = n^{-1} \sum_{k=1}^{n} x_k$.

Let us take the Fourier transform of both sides of this equation. Then for the function $\varphi(t,s) = \int \exp(i(s,z)) \times u(t,z)\,dz$, $s \in R^{m_n}$, $dz = \prod_{i=1}^{m_n} dz_i$ we have $(\partial/\partial t)\varphi(t,s) = -(2n)^{-1}(R_{m_n}s,s)\varphi(t,s)$. From this equation we have $\varphi(0,s) = \exp\{(2n)^{-1}(R_{m_n}s,s)\}\varphi(1,s)$. As the G-estimate of the function $\varphi(0,s)$ we take the function $G_n(s) = \exp\{(2n)^{-1}(R_{m_n}s,s)\} \times \hat{\varphi}(1,s)$ where $\hat{\varphi}(1,s) = \int \exp(i(s,z)) f(z+\hat{a})\,dz$.

Theorem 1. Let

$$\lim_{n\to\infty} \sum_{i,j=1}^{m_n} r_{ij} \int \left|\frac{\partial}{\partial z_i} f(z)\right| dz \int \left|\frac{\partial}{\partial z_j} f(z)\right| dz \, n^{-1} = 0,$$

$$\lim_{n\to\infty} \iiint \Big[\int_0^1 \int \Big(\sum_{k=1}^{m_n} (R_{m_n}^{1/2} x)_k \frac{\partial}{\partial z_k}\Big)^2 \Big| f(z + R_{m_n}^{1/2} y \frac{\sqrt{n-1}}{n} +$$

$$+tR_{m_n}^{1/2}\frac{x}{n})\Big|\,dz\,dt\Big]^2\exp\Big\{-\tfrac{1}{2}(y,y)-\tfrac{1}{2}(x,x)\Big\}(2\pi)^{-m_n}dy\,dx\,n^{-3}=0.$$

Then for any $0<c<\infty$

$$\lim_{n\to\infty}\ \sup_{s:\,(2n)^{-1}(R_{m_n}s,s)<c} E\,|G_n(s)-\varphi(0,s)|^2=0.$$

Using the estimate $G_n(s)$ for the Fourier transforms of the functions $f(a+z)$ we introduce the regularized estimate of the function $f(a+z)$

$$G(\varepsilon)=(2\pi)^{-m_n/2}\int\exp(-\tfrac{1}{2}(z,z))\,f(\varepsilon(I-\varepsilon^{-2}n^{-1}R)^{1/2}z+\hat{a})\,dz,\ \varepsilon>0.$$

Theorem 2. Let

$$\lim_{n\to\infty}E\sum_{i,j=1}^{m_n}r_{ij}\int\Big|\frac{\partial}{\partial u_i}f(\varepsilon A_\varepsilon z+a+u+\nu\frac{\sqrt{n-1}}{n})\Big|_{u=0}e^{-\frac{1}{2}(z,z)}dz\,\times$$

$$\times\int\Big|\frac{\partial}{\partial u_j}f(\varepsilon A_\varepsilon z+u+a+\nu\frac{\sqrt{n-1}}{n})\Big|_{u=0}e^{-\frac{1}{2}(z,z)}dz\,n^{-1}(2\pi)^{-m_n/2}=0,$$

$$\lim_{n\to\infty}\iint\Big[\int_0^1\int\Big|\Big(\sum_{k=1}^{m_n}(R_{m_n}^{1/2}x)_k\frac{\partial}{\partial u_k}\Big)f(\varepsilon A_\varepsilon z+u+a+R_{m_n}^{1/2}\times$$

$$\times y\frac{\sqrt{n-1}}{n}+tR_{m_n}^{1/2}\frac{x}{n})\Big|e^{-\frac{1}{2}(z,z)}dz\,dt\Big]^2\exp\Big\{-\tfrac{1}{2}(y,y)-$$

$$-\tfrac{1}{2}(x,x)\Big\}(2\pi)^{-2m_n}dy\,dx\,n^{-3}=0$$

where $A_\varepsilon=(I-\varepsilon^{-2}n^{-1}R_{m_n})^{1/2}$,

$$\lim_{n\to\infty}n^{-1}\lambda_i(R_{m_n})=0,\quad i=\overline{1,m_n},$$

$\lambda_i(R_{m_n})$ are the eigenvalues of R_{m_n}

Then for any $\varepsilon>0$

$$\operatorname{plim}_{n\to\infty}\left[G(\varepsilon)-E\, f(\varepsilon \eta + a)\right]=0$$

where η is a normal standard vector in R_{m_n}.

By virtue of G-analysis techniques the estimates $G_1 \div G_{11}$ for the following quantities are found: generalized variance, Stieltjes transform of the covariance matrix normalized spectral function, inverse covariance matrix, traces of powers of covariance matrices, smoothed normalized spectral functions of the covariance matrix, Kolmogorov-Wiener filter solution, Mahalanobis distance, regularized Mahalanobis distance, Anderson-Fisher statistic, regularized Anderson-Fisher statistic, nonlinear discriminant function obtained from observations of random vectors with different covariance matrices. Under some conditions these estimates are proved to be asymptotically normal as $\overline{\lim}_{n\to\infty} m_n n^{-1} < 1$. In particular if $\lim_{m\to\infty} m(n_1+n_2-2)^{-1} = c < \infty$, $\overline{\lim}_{m\to\infty}\left[n_1 n_2^{-1} + n_2 n_1^{-1}\right] < \infty$, $\lambda_i \le c < \infty$ where λ_i are the eigenvalues of the covariance matrix R_m, then

$$\left[G_8 - b'(I\varepsilon + R_m)^{-1} b\right]\sqrt{n_1+n_2-2}\; a_m(\varepsilon) \Rightarrow N(0,1),$$

where $G_8 = \left(\left(I\varepsilon + \varepsilon\hat{\theta}_m^{-1}\hat{R}\right)^{-1}(\hat{x}_1-\hat{x}_2), (\hat{x}_1-\hat{x}_2)\right) - (n_1^{-1}+n_2^{-1}) \times \operatorname{Tr} \varepsilon \hat{\theta}_m^{-1}\hat{R}\left(I\varepsilon + \varepsilon\hat{\theta}_m^{-1}\hat{R}\right)^{-1}$, $\hat{R} = (n_1+n_2-1)^{-1}\left[\sum_{i=1}^{n_1}(x_i-\hat{x}_1)\times(x_i-\hat{x}_1) + \sum_{i=1}^{n_2}(y_i-\hat{x}_2)(y_i-\hat{x}_2)'\right]$, x_i, y_i – are independent observations respectively over random m-dimensional vectors ξ and η distributed by the normal laws $N(a_1, R)$, $N(a_2, R)$, $b = a_1 - a_2$, $\hat{\theta}_m$ is a nonnegative solution of the equation $1 - K_m + K_m \theta m^{-1} \operatorname{Tr}(I\theta + \hat{R})^{-1} = \theta\varepsilon^{-1}$, $\varepsilon > 0$, $K_m = \frac{m}{n_1+n_2-2}$, $a_m(\varepsilon)$ - is a sequence of nonrandom bounded constants,

$$\hat{x}_1 = n_1^{-1}\sum_{i=1}^{n_1} x_i, \qquad \hat{x}_2 = n_2^{-1}\sum_{i=1}^{n_2} y_i .$$

ESTIMATION AND TESTING OF HYPOTHESES IN MULTIVARIATE GENERAL GAUSS-MARKOFF MODEL

Wiktor Oktaba

Institute of Applied Mathematics, Academy of Agriculture, Lublin, Poland

ABSTRACT

The unified theory of linear estimation (Rao, 1971) for a general univariate Gauss-Markoff model is used to the transformed Multivariate general Gauss-Markoff model (MGM) $(\underline{U}, \underline{XB}, \underline{\Sigma} \otimes \sigma^2\underline{V})$ to obtain estimators of parameters and test functions of hypotheses. $\underline{U}$ is a random matrix $n \times p$ with covariance matrix $\underline{\Sigma} \otimes \sigma^2\underline{V}$, $\underline{X}$ - a known design matrix $n \times m$, $\underline{B}$ - a matrix $m \times p$ of parameters, $\underline{XB}$ - the expected value of the matrix $\underline{U}$, $\underline{\Sigma} = \underline{\Sigma}'$ - a fixed and nonsingular matrix, $\underline{V}$ - a known non-negative definite matrix $n \times n$, σ^2 - an unknown positive scalar. $\otimes$ - Kronecker product of matrices. Let us state that $\underline{T}=\underline{V}+\underline{XMX}'$, where $\underline{M}=\underline{M}'$ is any arbitrary matrix such that $R(\underline{X}) \subset R(\underline{T})$. The symbol $R(\underline{A})$ is used to denote the vector space spanned on the columns of a matrix $\underline{A}$. Some results on estimating and testing are collected in 3.1 and 3.2. Moreover the estimator of a set of estimable parametric functions, $\underline{L}\hat{\underline{B}}\underline{A}$ and some matrix covariances in MGM are given in 4.2 by using the th. 4.1, $\underline{L}$ and $\underline{A}$ are $a \times m$ and $p \times b$ fixed matrices of the full rows and full columns ranks respectively. The unbiased estimators of the covariance matrix $\sigma^2\underline{\Sigma}$ and of σ^2 are presented in th. 5.1.

1. NOTATION

$\underline{\Sigma} = [\sigma_{ij}]$ is matrix of the order p; $i,j=1,\ldots,p$. $r(\underline{A})$ is the rank of matrix $\underline{A}$, $tr(\underline{A})$ denotes the trace of matrix $\underline{A}$. We are interested in testing the estimable hypothesis $\underline{LBA}=\underline{0}$ i.e. such that

$$(1.1) \qquad R(\underline{L}') \subset R(\underline{X}'),$$

where

$$(1.2) \quad r(\underline{L})=a \leqslant r(\underline{X}), \quad r(\underline{A})=b \leqslant p, \quad r(\underline{A}) \leqslant n-r(\underline{X}),$$

Moreover we want to estimate the matrices $\underline{B}$, $\underline{\Sigma}$, $\underline{LBA}$ and the scalar σ^2.

Let

$$(1.3) \qquad \underline{U} = [\underline{Y}_1,\dots,\underline{Y}_p]$$

where $\underline{Y}_1,\dots,\underline{Y}_p$ are column vectors $n\times 1$ of the matrix $\underline{U}$ and let

$$(1.4) \qquad \underline{B} = [\underline{\beta}_1,\dots,\underline{\beta}_p] ,$$

where

$$(1.5) \qquad \underline{\beta}_i' = [\beta_{1i},\dots,\beta_{mi}] , \quad i=1,\dots,p,$$

$$(1.6) \qquad \underline{Y}' = [\underline{Y}_1',\dots,\underline{Y}_p'] ,$$

$\underline{Y}$ being the vector $p\times 1$ with $\underline{Y}_1,\dots,\underline{Y}_p$ as its columns. $\underline{Y}$ is the development of the matrix $\underline{U}$. The covariance matrix of $\underline{U}$ is

$$(1.7) \qquad \mathrm{Cov}(\underline{U}) = \mathrm{Cov}(\underline{Y}).$$

The ć-inverse matrix of $\underline{A}$ is $\underline{A}^-$ if $\underline{A}\underline{A}^-\underline{A} = \underline{A}$. $(V \vdots X)$ is a partitioned matrix.

2. TRANSFORMATION OF MGM INTO THE MODEL WITH INDEPENDENT COLUMNS AND THE SAME COVARIANCE MATRIX $\sigma^2 V$ FOR EACH COLUMN

Using the transformation

$$(2.1) \qquad \underline{U}^* = \underline{U}\underline{K} ,$$

we get instead of $(\underline{U}, \underline{X}\underline{B}, \underline{\Sigma} \otimes \sigma^2\underline{V})$ the following model

$$(2.2) \qquad (\underline{U}^*, \underline{X}\underline{B}^*, \underline{I}_p \otimes \sigma^2\underline{V}),$$

where

$$(2.3) \quad \underline{K} = \underline{\Sigma}^{-1/2} = \{k_{ij}\} = [\underline{k}_1,\dots,\underline{k}_p] , \quad i,j=1,\dots,p,$$

$$(2.4) \qquad \underline{B}^* = \underline{B}\underline{K} , \quad \underline{\beta}_i^* = \underline{B}\underline{k}_i , \quad i=1,\dots,p,$$

where $\underline{\beta}_i^*$ is the vector $m\times 1$.

Let us observe that MGM can be presented in the following form containing p variates:

$$(2.5) \quad (\underline{Y}_i, \underline{X}\underline{\beta}_i, \sigma_{ii}\sigma^2\underline{V}), \quad \mathrm{Cov}(\underline{Y}_i, \underline{Y}_j) = \sigma_{ij}\sigma^2\underline{V}.$$

Under transformation (2.1) the model (2.5) is of the form

(2.6) $(\underline{Y}_i^*, \underline{X}\beta_i^*, \sigma^2\underline{V})$, $i=1,\ldots,p$,

where

(2.7) $\underline{Y}^* = \underline{U}\underline{k}_i$, $\underline{k}_i = (k_{i1},\ldots,k_{ip})'$, $i=1,\ldots,p$

and

(2.8) $$\underline{K}\,\Sigma\,\underline{K} = \underline{I}_p$$

(2.9) $\operatorname{Cov}(\underline{Y}_i^*, \underline{Y}_j^*) = \underline{O}$; $i\neq j$, $i,j=1,\ldots,p$.

3. ESTIMATION OF PARAMETER MATRIX B AND OF SCALAR σ^2 AND TESTING HYPOTHESES IN MGM

Theorem 3.1. The estimator of $\underline{B}$ is (Rao, 1971)

(3.1) $\hat{\underline{B}} = \underline{C}_3\underline{U} = \underline{C}_2'\underline{U} = (\underline{X}'\underline{T}^-\underline{X})^-\underline{X}'\underline{T}^-\underline{U}$.

BLUE of an estimable parametric function $\underline{\lambda}'\underline{\beta}_i^*$ is

(3.2) $$\underline{\lambda}'\hat{\underline{\beta}}_i^* = \underline{\lambda}'\hat{\underline{B}}\underline{k}_i$$

with dispersion matrix

(3.3) $$V(\hat{\underline{\beta}}_i^*) = \sigma^2\underline{C}_4$$

and dispersion matrix of $\underline{\lambda}'\hat{\underline{\beta}}_i^*$

(3.4) $$V(\underline{\lambda}'\hat{\underline{\beta}}_i^*) = \sigma^2\underline{\lambda}'\underline{C}_4\underline{\lambda}.$$

The covariance matrix between estimators of estimable parametric functions $\underline{\lambda}_1'\hat{\underline{\beta}}_i^*$ and $\underline{\lambda}_2'\hat{\underline{\beta}}_i^*$ is

(3.5) $\operatorname{Cov}(\underline{\lambda}_1'\hat{\underline{\beta}}_i^*, \underline{\lambda}_2'\hat{\underline{\beta}}_i^*) = \sigma^2\underline{\lambda}_1'\underline{C}_4\underline{\lambda}_2 = \sigma^2\underline{\lambda}_2'\underline{C}_4\underline{\lambda}_1$.

An unbiased estimator of σ^2 is

(3.6) $\hat{\sigma}_i^2 = \nu_e^{-1}(\underline{k}_i'\underline{U}\underline{C}_1\underline{U}\underline{k}_i)$, $i=1,\ldots,p$,

where

(3.7) $$\begin{cases}\underline{C}_1=\underline{T}^- - \underline{T}^-\underline{X}(\underline{X}'\underline{T}^-\underline{X})^-\underline{X}'\underline{T}^-, & \underline{C}_4=(\underline{X}'\underline{T}^-\underline{X})^- - \underline{M}\\ \underline{C}_3 = \underline{C}_2' = (\underline{X}'\underline{T}^-\underline{X})^-\underline{X}'\underline{T}^- \end{cases}$$

and

(3.8) $$\nu_e = r(\underline{V}\,\vdots\,\underline{X}) - r(\underline{X}).$$

Theorem 3.2. Let $\underline{L}\hat{\underline{\beta}}_i^*$ be the vector of BLUE's of a set of estimable parametric functions $\underline{L}\underline{\beta}_i^*$. If

(3.9) $\underline{Y}_i^* \sim N_n(\underline{X}\underline{\beta}_i^*, \sigma^2\underline{V})$, $\operatorname{Cov}(\underline{Y}_i^*,\underline{Y}_j^*)=\underline{O}$; $i\neq j$, $i,j=1,\ldots,p$,

then

1° $\underline{L}\hat{\underline{\beta}}_i^*$ and $\underline{Y}_i^{*\prime}\underline{C}_1\underline{Y}_i^*$ are independently distributed with

(3.10) $\underline{L}\underline{\beta}_i^* = \underline{L}\underline{C}_3\underline{U}\underline{k}_i \sim N(\underline{L}\underline{\beta}_i^*, \sigma^2\underline{L})$, $\operatorname{Cov}(\underline{L}\hat{\underline{\beta}}_i^*, \underline{L}\hat{\underline{\beta}}_j^*)=\underline{O}$, $i\neq j$;

and

(3.11) $\operatorname{tr}\left[\underline{U}'\underline{C}_1\underline{U}(\Sigma\,\sigma^2)^{-1}\right] \sim \chi^2_{p\cdot\nu_e}$, $\underline{L} = \underline{L}\underline{C}_4\underline{L}'$.

2° Let

(3.12) $$\underline{Ł}\beta_i^* = \underline{\varphi}_i$$

be a null consistent hypothesis with known wektors $\underline{\varphi}_i$.
The null hypothesis is consistent iff

(3.13) $$\underline{L}\underline{L}^- \underline{m}_i = \underline{m}_i \ , \quad \underline{m}_i = \underline{Ł}\hat{\beta}_i^* - \underline{\varphi}_i \ .$$

If the hypothesis is consistent then

(3.14) $$F_i = \nu_1^{-1}(\underline{m}_i' \underline{L}^- \underline{m}_i) : \hat{\sigma}_i^2$$

has a central F distribution on $\nu_1 = r(\underline{L})$ and ν_e degrees of freedom when the hypothesis is true and otherwise non-central.

4. ESTIMATOR OF A SET OF ESTIMABLE PARAMETRIC FUNCTIONS AND ITS COVARIANCE MATRIX

Theorem 4.1. In MGM for the estimable hypotheses

(4.1) $$\underline{Ł}_i \underline{B}\underline{A} = \underline{O} \quad , \ i=1,2$$

we have

(4.2) $$\underline{Ł}_1 \underline{C}_2' \underline{V} \underline{C}_2 \underline{Ł}_2 = \underline{Ł}_1 \underline{C}_4 \underline{Ł}_2' \ ,$$

where $\underline{Ł}_1$, $\underline{Ł}_2$ and $\underline{A}$ are fixed matrices.

Theorem 4.2. In MGM BLUE of estimable ŁBA is

(4.3) $$\underline{Ł}\hat{\underline{B}}\underline{A} = \underline{Ł}\underline{C}_3 \underline{U}\underline{A} \ .$$

The covariance matrices are:

(4.4) $$\mathrm{Cov}(\hat{\underline{B}}) = \underline{\Sigma} \otimes \underline{C}_2' \underline{V} \underline{C}_2 \sigma^2 \ ,$$

(4.5) $$\mathrm{Cov}(\underline{Ł}\hat{\underline{B}}\underline{A}) = \underline{A}' \underline{\Sigma} \underline{A} \otimes \underline{Ł}\underline{C}_4 \underline{Ł}' \sigma^2 \ ,$$

(4.6) $$\mathrm{Cov}(\underline{Ł}_i \hat{\underline{B}}\underline{A}, \underline{Ł}_j \hat{\underline{B}}\underline{A}) = \underline{A}' \underline{\Sigma} \underline{A} \otimes \underline{Ł}_i \underline{C}_4 \underline{Ł}_j' \sigma^2 \ ,$$

where $\underline{Ł}$, $\underline{A}$, $\underline{Ł}_i$, $\underline{Ł}_j$ are matrices: $a \times m$, $p \times b$, $1 \times m$, and $1 \times m$ respectively.

5. ESTIMATION OF MATRIX $\Sigma\sigma^2$ AND OF SCALAR σ^2 IN MGM

Theorem 5.1. The unbiased estimators of the matrix $\underline{\Sigma}\sigma^2$ and of the scalar σ^2 are:

(5.1) $$\widehat{\sigma^2 \underline{\Sigma}} = \nu_e^{-1} \underline{U}' \underline{C}_1 \underline{U} = \nu_e^{-1} (\underline{U} - \underline{X}\hat{\underline{B}})' \underline{T}^- (\underline{U} - \underline{X}\hat{\underline{B}})$$

and

(5.2) $$\hat{\sigma}^2 = (p\nu_e)^{-1} \mathrm{tr}(\underline{U}' \underline{C}_1 \underline{U} \underline{\Sigma}^{-1}).$$

REFERENCES

Rao, C.R.,(1971): Unified theory of linear estimation, Sankhyā 33A, 371-394

SYMMETRY GROUPS AND INVARIANT STATISTICAL TESTS FOR FAMILIES OF MULTIVARIATE GAUSSIAN DISTRIBUTIONS

E.A. Pukhal'skii
Moscow Physical and Technical Institute, Moscow, USSR

A general scheme is suggested to find invariant criteria for testing hypotheses about parameters of multivariate Gaussian distributions.

Let $\mathcal{X}$ be a finite-dimensional real Euclidean space, the sample $x \in \mathcal{X}$ and let

$$\mathcal{P} = \{N(\mu, \varphi(\lambda)) : (\mu, \lambda) \in \Omega\}, \quad \Omega \subset \mathcal{I} \times \mathcal{A}^+, \qquad (1)$$

be a family of nonsingular Gaussian distributions $N(\mu, \Lambda)$ on $\mathcal{X}$ with $N(0,E) \in \mathcal{P}$. Here (Pukhal'skii(1981,1983)) $\mathcal{A}$ is an associative algebra with involution, $\varphi: \mathcal{A} \to \mathcal{R}$ is an isomorphism of $\mathcal{A}$ onto the operator algebra $\mathcal{R}$ generated by all covariance operators Λ of the family, $\mathcal{I}$ is the minimal $\mathcal{R}$-invariant subspace in $\mathcal{X}$ containing all mean vectors μ of the family, Ω is a subset in the direct product of the set $\mathcal{I}$ and the set $\mathcal{A}^+$ of all positive elements of $\mathcal{A}$, and E is the unit operator on $\mathcal{X}$. Let $\mathfrak{J}(\Omega)$ be the group of all affine maps $g: \mathcal{X} \to \mathcal{X}$ such that $P \in \mathcal{P}$ iff $Pg^{-1} \in \mathcal{P}$ and $\mathcal{G}(\Omega)$ be the subgroup of all affine g such that $P = Pg^{-1}\ \forall P \in \mathcal{P}$. Let $\overline{\mathcal{P}}$ be a subfamily in $\mathcal{P}$ with corresponding $\overline{\mathcal{A}} \subset \mathcal{A}$, $\overline{\mathcal{I}} \subset \mathcal{I}$ and $\overline{\Omega} \subset \Omega$. A criteria $\phi: \mathcal{X} \to [0,1]$ for testing the hypothesis $H_0: P \in \overline{\mathcal{P}}$ against the alternatives $P \in \mathcal{P} \setminus \overline{\mathcal{P}}$ satisfying the conditions $\phi(g(x)) = \phi(x)\ \forall x \in \mathcal{X}$, $g \in G$ with a transformation group G, $\mathcal{G}(\Omega) \subset G \subset \mathfrak{J}(\Omega) \cap \mathfrak{J}(\overline{\Omega})$, has the form $\phi(x) = \tilde{\phi}(Q(x))$ where $Q: \mathcal{X} \to \mathcal{V}$ is a maximal invariant of G, $\mathcal{V}$ is the range of values of Q and $\tilde{\phi}: \mathcal{V} \to [0,1]$ is a test with the range of definition $\mathcal{V}$.

Let $\psi: \mathcal{A} \to \mathcal{L}(\mathcal{E})$ be a homomorphism of $\mathcal{A}$ into the algebra $\mathcal{L}(\mathcal{E})$ of all linear operators on an Euclidean space $\mathcal{E}$.

We denote $\mathcal{H}(\psi)=\{A\in \mathcal{L}(\mathcal{E}): A\psi(a)=\psi(a)A \ \forall a\in \mathcal{A}\}$ the commu tant to $\psi(\mathcal{A})$ and $\mathcal{U}(\psi)=\{U\in \mathcal{H}(\psi): UU^*=E\}$ the group of al unitary elements of $\mathcal{H}(\psi)$. We use the following diagram o isomorphisms (van der Waerden(1979))

$$\mathcal{H}(\psi)\xrightarrow{\pi'} \prod_{\alpha\in\Delta} M^{\delta_\alpha}_{m_\alpha} \quad ; \text{ (i) } \Pi Ax= \Pi x\cdot\pi'(A)^*, \quad \forall x\in\mathcal{E},\ A\in\mathcal{H}(\psi);$$

$$\mathcal{E} \xrightarrow{\Pi} \prod_{\alpha\in\Delta} M^{\delta_\alpha}_{d_\alpha,m_\alpha}; \text{ (ii) } \Pi\varphi(a)x=\pi(a)\cdot\Pi x, \quad \forall x\in\mathcal{E},\ a\in\mathcal{A}; \tag{2}$$

$$\mathcal{A} \xrightarrow{\pi} \prod_{\alpha\in\Delta} M^{\delta_\alpha}_{d_\alpha} \quad ; \pi(a)\equiv\{\pi(a)_\alpha;\ \alpha\in\Delta\},\ \pi(a)_\alpha\in M^{\delta_\alpha}_{d_\alpha}.$$

Here $M^r_{s,t}$ denotes the set of all s×t-matrices over the re al field if r=1, over the complex field if r=2 and over the division algebra of quaternions if r=4 respectively, $M^r_s=M^r_{s,s}$, Δ is a finite index set and $\prod_{\alpha\in\Delta}$ denotes the direct product. $M^r_{s,t}$ is equipped with the canonical Eucl: dean metric, the sum of squares of matrix elements. Let $(a,b)_{\mathcal{A}}=(\pi(a),\pi(b))$ be the canonical inner product in $\mathcal{A}$ We denote e_α the principal idempotents, $\alpha\in\Delta$, $\pi(e_\alpha)_{\alpha'}=\{1_{d_\alpha}$ if $\alpha=\alpha'$ and 0 if $\alpha\neq\alpha'\}$, and e the unit element of $\mathcal{A}$. We let $\mathcal{Z}(\mathcal{M})$ denote the volume of a manifold $\mathcal{M}$. Let $U^r_{s,t}=\{u\in M^r_{s,t}: uu=1_s\}$, $U^r_s=U^r_{s,s}$ and

$$\mathcal{Z}(U^r_{s,t})=2^s(2/\pi)^{rs(s-1)/4}\pi^{rst/2} / \prod_{i=0}^{s-1}\Gamma(r(t-i)/2).$$

For $w\in\mathcal{A}^+$ let q(w) be the operator $a\mapsto waw$ on $\mathcal{S}\mathcal{A}=\{a\in$ S $a=a^*\}$. Let $N_\psi(w)=\det\psi(w)$, $\gamma(w)=[\det q(w)]^{-1/2}$ and

$$K_\psi=\prod_{\alpha\in\Delta} 2^{-d_\alpha-\delta_\alpha d_\alpha(d_\alpha-1)/2}\mathcal{Z}(U^{\delta_\alpha}_{d_\alpha,m_\alpha}).$$

Let $x\in\mathcal{X}$, $Z(x)=\mathrm{Proj}_{\mathcal{Z}}\,x$, $Y(x)=x-Z(x)$ and $S(x)\in\mathcal{A}$:(a, $S(x))_{\mathcal{A}}=(\varphi(a)x,x)_{\mathcal{X}}$ $\forall a\in\mathcal{A}$. Let $\mathcal{Y}=\mathcal{X}\ominus\mathcal{Z}$, $\varphi_1(a)=\varphi(a)|_{\mathcal{Z}}$ and $\varphi_0(a)=\varphi(a)|_{\mathcal{Y}}$. Let n_α, p_α and $n_\alpha-p_\alpha$ are the values of m_α in (2) for $\mathcal{E}=\mathcal{X}$ and $\psi=\varphi$, $\mathcal{E}=\mathcal{Z}$ and $\psi=\varphi_1$ and $\mathcal{E}=\mathcal{Y}$ and $\psi=\varphi_0$ respectively.

Lemma 1. (Z,S(Y)) is a maximal invariant of $\mathcal{G}(\Omega)$. I: $P(dx)=f(Z(x),S(Y(x)))dx$ and $n_\alpha-p_\alpha\geq d_\alpha$ $\forall\alpha\in\Delta$ then

$$P_{Z,S(Y)}(dz,dw)=K_{\varphi_0} f(z,w) N_{\varphi_0}(w)^{1/2} \gamma(w)dzdw, \quad z\in \mathcal{Z},\ w\in \mathcal{A}^+ . \tag{3}$$

If $P(dx)=f_1(x)dx$ then (3) holds with

$$f(z,w)=\int_{\mathcal{U}(\varphi_0)} f_1(z+\varphi_0(w^{1/2})Uy_0)dU/\ell(\mathcal{U}(\varphi_0)), \tag{4}$$

where $y_0\in \mathcal{Y}$: $S(y_0)=e$.

Let (see(2)) $B_\delta=\{\tau\in\{0,1\}^\Delta :\ \delta_\alpha\neq 2$ imply $\tau_\alpha=0\ \forall\alpha\in\Delta\}$, and $\mathcal{G}_{\delta,d,n,p}$ be the set of all permutations ℓ on Δ such that $(\delta,d,n,p)_{\ell(\alpha)}=(\delta,d,n,p)_\alpha\ \forall\alpha\in\Delta$. Letting $x_\alpha=(\Pi x)_\alpha$ we can define the operators C_τ and W_ℓ on $\mathcal{E}$ by $(C_\tau x)_\alpha=\{{}^t x_\alpha^*$ if $\tau_\alpha=1$ and x_α if $\tau_\alpha=0\}$ and $(W_\ell x)_\alpha=x_{\ell^{-1}(\alpha)}$. Let $\mathcal{GA}$ be the group of all invertible elements of the algebra $\mathcal{A}$.

Lemma 2. $\mathcal{J}(\Omega)\subset\mathcal{J}(\mathcal{Z}\times\mathcal{A}^+)$. The group $\mathcal{J}(\mathcal{Z}\times\mathcal{A}^+)$ has the form

$$(Z,Y)\mapsto(h+\varphi_1(a)U_1C_\tau^1W_\ell^1 Z,\ \varphi_0(a)U_0C_\tau^0W_\ell^0 Y): \quad h\in\mathcal{Z},\ a\in\mathcal{GA},\ U_i\in\mathcal{U}(\varphi_i),\ \tau\in B_\delta,\ \ell\in\mathcal{G}_{\delta,d,n,p} . \tag{5}$$

Suppose $\Omega=\mathcal{Z}\times\mathcal{A}^+$ and $\bar\Omega=\bar{\mathcal{Z}}\times\bar{\mathcal{A}}^+$. Let $\bar{\mathcal{T}}$ be a subgroup in $\mathcal{G}\bar{\mathcal{A}}$ such that any $w\in\bar{\mathcal{A}}^+$ is uniquely representable in the form $w=tt^*$ with $t\in\bar{\mathcal{T}}$. Let G be the subgroup in (5) given by $h\in\bar{\mathcal{Z}}$, $a\in\bar{\mathcal{T}}$, $U_1=E$, $\tau=0$ and $\ell=id$. We define the statistics $\bar T(\bar Y)\in\bar{\mathcal{T}}$: $\bar T(\bar Y)\bar T(\bar Y)^*=\bar S(\bar Y)$, $V=\varphi_1(\bar T(\bar Y)^{-1})(Z-\bar Z)$ and $R=[\bar T(\bar Y)]^{-1}[\bar c^{1/2}S(Y)\bar c^{1/2}+\bar S(Z-\bar Z)][\bar T(\bar Y)^*]^{-1}$, where $\bar c=\Sigma_{\alpha\in\bar\Delta}\|\bar e_\alpha\|^2_{\mathcal{A}}\bar d_\alpha^{-1}\bar e_\alpha$. Let $\bar\eta(t)dt$ be the left-invariant measure on $\bar{\mathcal{T}}$ with $\bar\eta(e)=1$ and $\bar J_0=\det[d(tt^*)/dt|_{t=e}]$. Suppose $n_\alpha-p_\alpha\geqslant d_\alpha\ \forall\alpha\in\Delta$.

Theorem. (V,R) is a maximal invariant of G. If $\bar P(dx)=f(\bar Z(x),\bar S(\bar Y(x)))dx$ then (V,R) is stochastically independent with $(\bar Z,\bar S(\bar Y))$ and

$$\bar P_{V,R}(dv,dw)=K_{\bar\varphi_0}^{-1}K_{\varphi_0}[N_{\varphi_0}(\bar c)\,\bar\gamma(\bar c)]^{-1/2}N_{\varphi_0}(w-\bar S(v))^{1/2}\times \gamma(w-\bar S(v))dvdw,\quad v\in\mathcal{Z}\ominus\bar{\mathcal{Z}},\ w\in e+\mathcal{SA}\ominus\mathcal{S}\bar{\mathcal{A}}:\ \bar S(v)<w.$$

If $P(dx)=f(Z(x),S(Y(x)))dx$ then

$$dP_{V,R}/d\bar{P}_{V,R}(v,w)=K_{\bar{\varphi}_0}\bar{J}_0 \cdot \int_{\bar{\mathfrak{I}}}\int_{\bar{\mathcal{C}}} f(h+\varphi_1(t)v,\ \bar{c}^{-1/2}t(w-\bar{S}(v))t^*\bar{c}^{-1/2})N_{\bar{\varphi}_0}(tt^*)^{1/2}\bar{\eta}(t)dhdt \ . \qquad (6)$$

If $P(dx)=f_1(x)dx$ then (6) holds with $f(z,w)$ given by (4)

Let $\mathcal{U}\mathcal{A}=\{u\in\mathcal{G}\mathcal{A}:\ uu^*=e\}$ be the group of all unitary elements of $\mathcal{A}$. Let $\Xi_{\alpha i}(w)$ be the i-th maximum eigenval of $\pi(w)_\alpha$ and $\Xi(w)=\{(\Xi_{\alpha 1}(w),\dots,\Xi_{\alpha d_\alpha}(w)):\alpha\in\Delta\}$ be t spectrum of $w\in\mathcal{A}^+$.

Lemma 3. Ξ is a maximal invariant of the group of transformations $w\mapsto uwu^*$: $u\in\mathcal{U}\mathcal{A}$ of $\mathcal{A}^+$. If $P(dw)=f(\Xi(w))dw$, $w\in\mathcal{A}^+$, then

$$P_\Xi(d\xi)=K_{\mathcal{A}}f(\xi)V(\xi)d\xi,\quad \xi_{\alpha 1}>\dots>\xi_{\alpha d_\alpha}>0,$$

where

$$V(\xi)=\prod_{\alpha\in\Delta}\prod_{i<i'}(\xi_{\alpha i}-\xi_{\alpha i'})^{\delta_\alpha},\quad K_{\mathcal{A}}=\prod_{\alpha\in\Delta}\mathcal{E}(U_{d_\alpha}^{\delta_\alpha})/\mathcal{E}(U_1^{\delta_\alpha})^{d_\alpha}.$$

In particular, Theorem and Lemmas 1 and 3 give the wel -known statistics, e.g., B-statistics, matrices of sampl correlations, Hotelling's canonical correlations, etc. A the same time the methods developed allow us to find inv riant statistical tests when we deal with multivariate G ussian models with compound symmetry (Sysoev and Shaikin (1981)).

Pukhal'skii, E.A. (1981). Minimal sufficient statistics for the normal models with an algebraic structure. Teo Veroyatnost. i Primenen. 26, 574-583.
Pukhal'skii, E.A. (1983). Two classes of statistical tes for invariance of distributions with respect to groups of linear transformations. Dokl. Akad. Nauk SSSR 270, 551-554.
Sysoev, L.P., and Shaikin, M.E. (1981). Optimal paramete estimates in regression models with a special covarian structure and their application to two-factor experiments. Avtomat. i Telemekh. 6, 44-56.
van der Waerden, B.L. (1979). Algebra. "Nauka", Moscow.

PRESCRIBED CONDITIONAL INTERACTION MODELS FOR BINARY CONTINGENCY TABLES

T. Rudas
Institute of Sociology, Eotvos University, Pesti B. u. 1.
H-1052 Budapest, Hungary

INTRODUCTION

Log-linear models are known to prescribe that certain conditional interactions among the subsets of a set of categorical variables are zero. In this paper such models are defined and investigated for binary contingency tables that prescribe the values of certain conditional interactions arbitrarily (not necessarily zero). This class of models contains, as a special case, the class of log-linear models. Every prescribed conditional interaction model is equivalent to an exponential family. A version of the iterative proportional fitting procedure is shown to converge to the maximum likelihood estimates, if these exist, for which a sufficient condition is given.

CONDITIONAL INTERACTIONS

Let Z be a set of binary variables. If V is a subset of Z consisting of v variables then the marginal table generated by V is T(V), a cell (i.e. an index) in T(V) is i_V, where i_V is a series of 1s and 2s of lenght v. If W is the complement of V then $i_Z=(i_V,i_W)$. The interaction in the marginal table T(V) is defined as

$$I(V) = \Sigma\, s(i_V)\log(P(i_V)$$

where $s(i_V)$ is 1 if the number of 2s in i_V is even and -1 if odd. This interaction is the logarithm of a generalized odds ratio and is said to be of order v-1.

It was proved by Good (1963) that supposing that all the qth and higher order interactions are zero is equivalent to supposing that in a log-linear (LL) model the LL interactions pertaining to these and higher order marginal tables are zero.

The conditional interaction of V given $W=i_W$ is defined as

$$I(V|W=i_W) = \sum s(i_V)\log(P(i_V,i_W)$$

Lemma 1. Let V be a subset of U, the complement of U is T Let the conditional interaction $I(V|W=i_W)$ be given for al i_W. Then $I(U|T=i_T)$ is given for all i_T. More exactly in this case

$$I(U|T=i_T) = \sum s(i_V,i_U)I(V|W=i_W)$$

where the summation goes for the indices of those variabl that are in U but not in V and $s(i_V,i_U)$ is 1 if the numbe of 2s among the indices of the variables in U but not in V is even and -1 if this is odd (i.e. $s(i_V,i_U)$ depends only on those indices in i_U which are not in i_V).

PRESCRIBED CONDITIONAL AND PARTIAL INTERACTION MODELS

Let $G=(V_1,V_2,\ldots,V_n)$ be an ascending class of subsets of Z. Let i_{Wj} denote an index for the complement of V_j, W_j $(j=1,\ldots,n)$. Let the positive numbers $a(i_{Wj})$ be given for each j and i_{Wj}. Then the set of all probability distributions over T(Z) with the property

$$I(V_j|W_j=i_{Wj}) = a(i_{Wj})$$

for all j and i_{Wj} is the prescribed conditional interacti (PCIN) model defined by G and the $a(i_{Wj})$ numbers. Let G' denote the descending class which is the complement of G with respect to the class of all subsets of Z.

Theorem 1. Let $a(i_{Wj})$ be zero for all j and i_{Wj}. Then the PCIN model defined by G and the $a(i_{Wj})$ numbers is equivalent to the LL model defined by G'.

In the multivariate normal theory the conditional covariance of two variables does not depend on the values of the fixed variables. Therefore conditional association are partial associations at the same time. This is clearly not the case with categorical variables. We shall say that the partial interaction of variables V is a if

$$I(V|W=i_W) = a \qquad (1)$$

i.e. when the conditional interaction of variables V is independent of the indices of the variables in W.

A prescribed partial interaction (PPIN) model is defined by an ascending class of subsets of Z, $G=(V_1,\ldots,V_n)$ and a set of positive constants, $(a_1,\ldots,a_n)$ and contains those probability distributions over T(Z) for which (1) holds with $V=V_j$, $W=W_j$, $i_W=i_{W_j}$ and $a=a_j$ $(j=1,\ldots,n)$.

As a PPIN model is a special PCIN model, it follows from Th 1. that LL models are exactly those PPIN models in which the partial interactions are equal to zero. This shows that LL models are in a certain sense best characterized not by the sets of variables that are LL interactions (G'), but by the sets of those variables that have zero partial interactions (G). This was also suggested by results on connections between and common generalizations of LL and covariance selection models, cf. Kiiveri, Speed (1982), Lauritzen, Wermuth (1984), Wermuth (1976).

MAXIMUM LIKELIHOOD ESTIMATION OF PCIN MODELS

Theorem 2. For any system of $a(i_{W_j})$ numbers the PCIN model, if not empty, is an exponential family. More exactly if a probability distribution Q is contained in it then the PCIN model consists of those and only those distributions that are of the following form:

$$P(i_Z) = Q(i_Z)\exp(\sum f_k((i_Z)_k))$$

where the summation goes for the U_k elements of G' and f_k is a real function on the $T(U_k)$ marginal table and $(i_Z)_k$ is the coordinate projection of the i_Z index from T(Z) to $T(U_k)$.

Now consider the exponential family generated by a given distribution, say, Q and functions over certain marginals, say, G'. Q and G' define a linear family as well. It is known (see Cencov (1972)) that these families, if are not disjoint, have one common element and this distribution is the ML projection of all distributions from the linear family into the exponential family and the MDI projection of all distributions from the exponential family into the linear family.

Minimum discrimination information projection of a given distribution into a linear family can be obtained by the well known iterative proportional fitting procedure (see Csiszar (1975), Ireland, Kullback (1968)).

Corollary 1. Maximum likelihood estimates under a PCIN model and a given sample can be obtained by adjusting the G' marginals of an arbitrary distribution in the PCIN model to the corresponding observed marginals using the IPFP.

The MDI projection of a distribution on a linear family exists if and only if there is a distribution in that family which is absolutely continuous with respect to the given distribution (see Csiszar (1975)).

Corollary 2. If in the sample all cells have positive frequencies then there exists a unique ML estimate.

REFERENCES

Cencov, N. N. (1972) Statistical Decision Rules and Optimal Decisions. Nauka, Moscow (Russian).
Csiszar, I. (1975) I-divergence geometry of probability distributions and minimization problems. Ann. Probab. 3, 146-158.
Good, I. J. (1963) Maximum entropy for hypothesis formulation especially for multidimensional contingency tables. Ann. Math. Statist. 34, 911-934.
Ireland, C. T., Kullback, S. (1968) Contingency tables with given marginals. Biometrika 55, 179-188.
Kiiveri, H. T., Speed, T. P. (1982) Structural analysis of multivariate data: a review. in S. Leinhardt (ed.) Sociological Methodology 1982. Jossey-Bass, San Francisco.
Lauritzen, S. L., Wermuth, N. (1984) Mixed interaction models. Preprint, Aalborg Universitetscenter.
Wermuth, N. (1976) Analogies between multiplicative models in contingency tables and covariance selection. Biometrics 32, 95-108.

QUADRATIC INVARIANT ESTIMATORS WITH MAXIMALLY BOUNDED MEAN SQUARE ERROR

ŠTULAJTER František

Comenius University, Bratislava, Czechoslovakia

Let us consider the usual linear regression model $Y = X\beta + \varepsilon$, where ε is a nx1 Gaussian random vector with $E[\varepsilon] = 0$ and $E[\varepsilon\varepsilon'] = \sum_{i=1}^{p} \nu_i V_i$. We assume that $V_1, \dots, V_p$ are known symmetric nxn matrices with the property $(V_i, V_j) = tr(V_i V_j) = \delta_{ij}$. Let f be any px1 vector and let $\gamma(\nu) = f'\nu$ be a function of an unknown variance-covariance components vector $\nu = (\nu_1, \dots, \nu_p)'$. Let us consider invariant quadratic estimators $\tilde{\gamma} = (B, Z_{M'}) = tr(BZ_{M'})$ of γ, where B is any matrix from the set $\mathcal{H}$ of all symmetric nxn matrices, $Z_{M'} = MYY'M'$ and M is any matrix from the set $\mathcal{M} = \{M: M^2 = M, MX = 0\}$.

It is well known that in many cases there does not exist any unbiased invariant estimator for the function $\gamma = f'\nu$ and some other criterion than unbiasedness must be used. Our criterion is based on the following inequality which holds for the mean square error (MSE) of any arbitrary invariant quadratic estimator .

Lemma 1.

$$MSE_\nu[\tilde{\gamma}] \leq \|\nu\|^2 \left(\left\| \frac{1}{\sqrt{3p}} \sum_{j=1}^{p} f_j V_j - \sqrt{3p}\, M'BM \right\|^2 + \|f\|^2 - \left\| \frac{1}{\sqrt{3p}} \sum_{j=1}^{p} f_j V_j \right\|^2 \right).$$

Proof: It is well known that for any invariant quadratic estimator $\tilde{\gamma}$ we have $MSE_\nu[\tilde{\gamma}] = D_\nu[\tilde{\gamma}] +$

$+ (E_\nu[\tilde{\gamma}] - f'\nu)^2 = 2 \sum_{i,j=1}^{p} \nu_i \nu_j (MV_i MBV_j M', B) +$

$+ \{ \sum_{j=1}^{p} \nu_j [(MV_j M', B) - f_j] \}^2$. Thus we have, using Schwarz inequalities:

$$MSE_\nu[\tilde{\gamma}] \leqq 2 \sum_{i,j=1}^{p} |\nu_i| |\nu_j| |(MV_i M'B, MBV_j M')| +$$

$$+ \sum_{j=1}^{p} \nu_j^2 \sum_{j=1}^{p} [(MV_j M', B) - f_j]^2 \leqq$$

$$\leqq 2 (\sum_{j=1}^{p} |\nu_j| \|V_j\| \|M'BM\|)^2 +$$

$$+ \sum_{j=1}^{p} \nu_j^2 \sum_{j=1}^{p} [(M'BM, V_j) - f_j]^2 \leqq$$

$$\leqq \|\nu\|^2 [3p \|M'BM\|^2 - 2 (M'BM, \sum_{j=1}^{p} f_j V_j) + \|f\|^2] \leqq$$

$$\leqq \|\nu\|^2 [\| \frac{1}{\sqrt{3p}} \sum_{j=1}^{p} f_j V_j - \sqrt{3p} M'BM \|^2 + \|f\|^2 -$$

$$- \| \frac{1}{\sqrt{3p}} \sum_{j=1}^{p} f_j V_j \|^2].$$

<u>Definition:</u> An estimator $\hat{\gamma} = (B, Z_{M'})$ is called the maximally bounded mean square error invariant estimator (MBMSEIE) for a function $\gamma = f'\nu$ if the symmetric matrix B and the matrix $M \in \mathcal{M}$ are such that

$$\| \sum_{j=1}^{p} f_j V_j - 3p M'BM \|^2 = \min_{A \in \mathcal{H},\, N \in \mathcal{M}} \| \sum_{j=1}^{p} f_j V_j -$$

$$- 3p N'AN \|^2 .$$

<u>Theorem:</u> The MBMSEIE $\hat{\gamma}$ for the function $\gamma = f'\nu$ is given by $\hat{\gamma} = \frac{1}{3p} \sum_{j=1}^{p} f_j (V_j, Z_{M_o})$, where $M_o = I - X(X'X) X'$ and $MSE_\nu[\hat{\gamma}] \leqq \|\nu\|^2 [\|f\|^2 - \frac{1}{3p} (Gf, f)]$ for every ν, where G is a pxp matrix with $G_{ij} = tr (M_o V_i M_o V_j)$; $i,j=1,2,\ldots,p$.

Proof: Let M be any matrix from the set $\mathcal{M}$ and let $\mathcal{H}_M$ denotes the subspace $\mathcal{H}_M = \{M'BM;\ B \in \mathcal{H}\}$ of $\mathcal{H}$. Then it can be easily proved that for every $M \in \mathcal{M}$ $\mathcal{H}_M$ is a subspace of $\mathcal{H}_{M_o}$ and $\mathcal{H}_{M_o} = \{A \in \mathcal{H} : AX = 0\}$.

Using these properties we have for any $A \in \mathcal{H}$ and any $D \in \mathcal{H}_M$ the following inequalities: $\|A - D\|^2 \geqq$ $= \|A - \mathcal{P}_M A\|^2 \geqq \|A - \mathcal{P}_{M_o} A\|^2 = \|A - M_o A M_o\|^2$,

where $\mathcal{P}_M$ denotes the orthogonal projector on the subspace $\mathcal{H}_M$. Using these results we get that $\|\sum_{j=1}^{p} f_j V_j -$ $- 3p\, M'BM\|^2$ will be minimal iff $3p\, M'BM = \mathcal{P}_{M_o} \sum_{j=1}^{p} f_j V_j =$ $= M_o \sum_{j=1}^{p} f_j V_j M_o$, hence iff $M = M_o$ and

$B = \frac{1}{3p} \sum_{j=1}^{p} f_j V_j$. Thus we get

$$\min_{M \in \mathcal{M},\ B \in \mathcal{H}} \left\| \frac{1}{\sqrt{3p}} \sum_{j=1}^{p} f_j V_j - \sqrt{3p}\, M'BM \right\|^2 =$$

$= \frac{1}{3p} \left[\|f\|^2 - (Gf, f)\right]$, from which the inequality for the MS $E_v[\hat{\gamma}]$ easily follows.

R e f e r e n c e s

1. Kleffe, J., (1977) Optimal estimation of variance components - a survey. Sankhyá 39, Series B, 211 - 244.

TIME SERIES

(long range dependence, non-linear processes, estimation of spectra)
(Session 8)

Chairman: H. Tong

NON-GAUSSIAN SEQUENCES AND DECONVOLUTION

ROSENBLATT, M., Department of Mathematics, University of California, San Diego, La Jolla, California, 92093, U.S.A.

Consider a linear process X_t generated by the independent identically distributed input real-valued sequence ξ_k with $E\xi_k \equiv o$, $E\xi_k^2 \equiv 1$ convolved with a linear filter a_k

$$X_t = \sum_k a_k \xi_{t-k} \quad .$$

This model has been of some interest in geoexploration [1,5,8]. Given observations on the sequence X_t only, a principal interest is the estimation of the coefficients a_k and the input sequence ξ_k by deconvolving X_t. Most of the work on deconvolution has focused on Gaussian processes X_t [5]. Because of the nonidentifiability problem in the case of Gaussian processes, to arrive at a uniquely determined solution, it has been conventional to assume that the transform of the filter sequence

$$a(z) = \sum a_k z^k$$

is minimum phase. If $a(z)$ is a polynomial this amounts to saying that its roots are outside the unit disc in the complex plane. The spectral density of the process $\{X_t\}$ is

$$f(\lambda) = |a(e^{-i\lambda})|^2/2\pi \quad .$$

$f(\lambda)$ is specified by the second order moments of $\{X_t\}$ and doesn't contain the phase information of $a(e^{-i\lambda})$. If $\{X_t\}$ is Gaussian, its structure is determined by knowledge of $f(\lambda)$ alone and observations on $\{X_t\}$ can only determine $a(e^{-i\lambda})$ up to its modulus $|a(e^{-i\lambda})|$. This is the reason for making the minimum phase assumption in the case of a Gaussian process $\{X_t\}$. In recent years there has been interest in nonGaussian linear processes $\{X_t\}$ and the associated problem of deconvolution (see [1,6,8]). It is curious that in the nonGaussian case one can obtain most of the phase information of $a(e^{-i\lambda})$, something impossible in the case of a Gaussian process. The following lemma suggests the degree to which this is possible:

Lemma. Let X_t be a nonGaussian linear process with the ξ_t's having all their moments finite. Assume that $\sum |ja_j| < \infty$ and $a(e^{-i\lambda}) \neq o$ for all λ. The function $a(e^{-i\lambda})$ can then be determined in terms of observations only on $\{X_t\}$ up to an indeterminate integer m in a factor $e^{im\lambda}$ that corresponds to a shift of index of X_t and an indeterminate sign of $a(1) = \sum a_k$. This result is still valid if there is a finite moment of order $k > 2$ with corresponding cumulant $\gamma_k \neq 0$.

It is clear that if one can estimate $a(e^{-i\lambda})$ consistently and $a(e^{-i\lambda}) \neq o$ for all λ, that one can deconvolve the process X_t to obtain the sequence ξ_t. The lemma cited above gives conditions under which one can deconvolve to obtain the ξ_t sequence except for an indeterminacy in index and sign. The idea in estimating $a(e^{-i\lambda})$ is as follows. Estimate the modulus $|a(e^{-i\lambda})|$ of $a(e^{-i\lambda})$ by using a spectral density estimate of $f(\lambda)$. The phase

$$h(\lambda) = \arg \{a(e^{-i\lambda})a(1)/|a(1)|\}$$

of $a(e^{-i\lambda})$ is determined by making use of the relation

$$h(\lambda_1) + \cdots + h(\lambda_{k-1}) - h(\lambda_1 + \cdots + \lambda_{k-1})$$
$$= \arg \left[\left\{\frac{a(1)}{|a(1)|}\right\}^k \lambda_k^{-1} b_k(\lambda_1,\ldots,\lambda_{k-1})\right]$$

where $b_k(\lambda_1,\ldots,\lambda_{k-1})$ is the kth order cumulant spectral density of the process $\{X_t\}$. This relation follows from the fact that $b_k(\lambda_1,\ldots,\lambda_{k-1})$ has the form

$$b_k(\lambda_1,\ldots,\lambda_{k-1})$$

$$= \frac{\gamma_k}{(2\pi)^{k-1}} a(e^{-i\lambda_1}) \cdots a(e^{-i\lambda_{k-1}}) a(e^{i(\lambda_1+\cdots+\lambda_{k-1})}) \quad .$$

In terms of an estimate $\hat{b}_k$ of b_k one can obtain an estimate $\hat{h}(\lambda)$ of $h(\lambda)$. Let $\hat{f}(\lambda)$ be an estimate of $f(\lambda)$. Then

$$\hat{a}(e^{-i\lambda}) = \{2\pi\ \hat{f}(\lambda)\}^{1/2}\ e^{i\hat{h}(\lambda)}$$

is a plausible estimate of $a(e^{-i\lambda})$. One can deconvolve X_t by using

$$\hat{\xi}_t = \hat{a}^{-1} * X_t \quad .$$

A discussion of the form of these estimates and their asymptotic behavior can be found in [2,7]. Some suggestions as to alternative forms of computation in the estimation of $a(e^{-i\lambda})$ are given in [4].

In the lemma cited above it was assumed that $a(e^{-i\lambda}) \neq 0$ for all λ. In theory and practice it ofteh happens that $a(e^{-i\lambda})$ occasionally may be zero or close to zero. In [3] it is indicated that deconvolution can still be effected if the zero set is small enough and behavior near zero appropriate.

To illustrate the deconvolution procedure consider the moving average

$$x_t = \xi_t - 2.333\xi_{t-1} + 0.667\xi_{t-2}$$

where the ξ_t's are independent exponential random variables with mean zero. The roots of the polynomial $a(z) = 1 - 2.333z + .667z^2$ are 3 and 1/2 and so the process X_t is not minimum phase. The first line of our figure is the moving average process generated from $t = 10$ to $t = 298$. Actually we generated 640 successive values of x_t by Monte Carlo simulation. The second line is the independent exponential sequence ξ_t that generated x_t for the same t range. Line 3 is the estimated $\hat{\xi}_t$ as generated by our procedure. Line 4 is the difference $\xi_t - \hat{\xi}_t$. The last line is an x_t deconvolution attempt using the minimum phase assumption, that is, a Wiener-Levinson deconvolution.

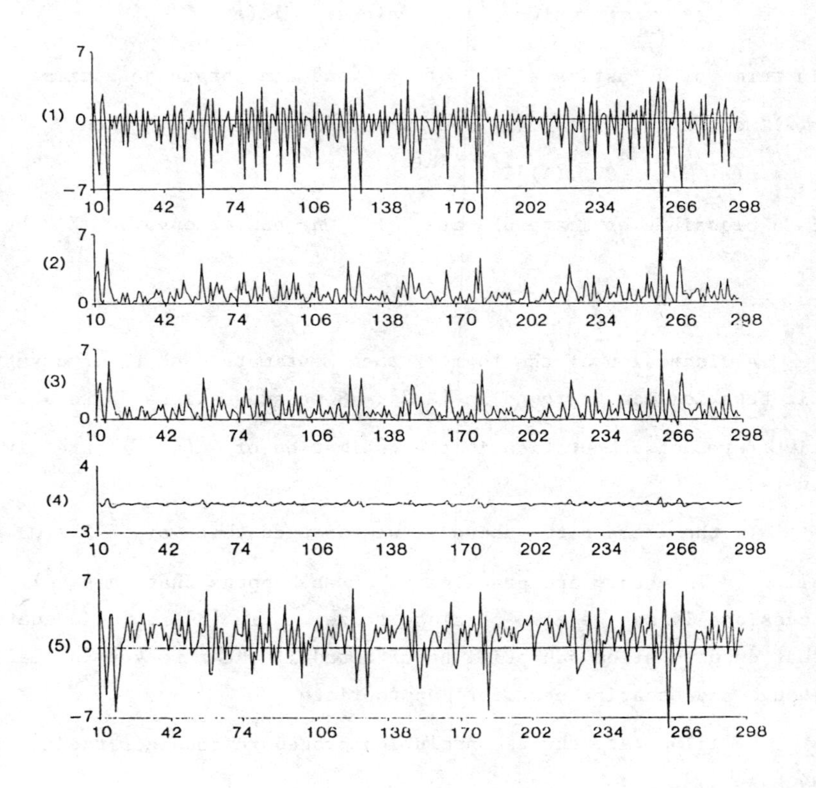

References

1. Donoho, D., On minimum entropy deconvolution, In: Applied Time Series Analysis II, Academic Press,]98], pp. 565-608.
2. Lii, K.S. and Rosenblatt, M., Deconvolution and estimation of transfer function phase and coefficients for nonGaussian linear processes, Ann. Statistics, 10 (1982), 1195-1208.
3. Lii, K.S. and Rosenblatt, M., Deconvolution of nonGaussian linear processes with vanishing spectral values, Proc. Nat'l Acad. Sci., U.S.A., 86 (]986), 119-200.
4. Matusoka T. and Ulrhch, T. Phase estimation using the bispectrum, Proc. IEEE, 72 (1984), 1403-1411.

5. Peacock, K. and Treitel, S., Predictive deconvolution: theory and practice, Geophysics, 34 (1969), 155.
6. Rosenblatt, M., Linear processes and bispectra, J. Appl. Probab., 17 (1980), 265-270.
7. Rosenblatt, M., *Stationary Sequences and Random Fields*, Boston: Birkhaüser, 1986, 258p.
8. Wiggins, R., Minimum entropy deconvolution, Geoexpl., 16 (1978), 21-35.

Bernoulli, Vol. 2, pp. 355-367

NON-LINEAR TIME SERIES MODELS OF REGULARLY SAMPLED DATA: A REVIEW

H. Tong (University of Kent at Canterbury, U.K.)

1. INTRODUCTION

1.1 Linearity, Gaussinity and Non-linear Oscillations

As far as time series modelling is concerned, the impact of works like Box and Jenkins (1970) was such that the field in the 1970's was pre-dominantly linear and, implicitly or explicitly, Gaussian. However, towards the late 1970's and early 1980's it became increasingly clear that *linearity* and *Gaussianity*, although powerful and useful concepts they are in many ways and will remain so indefinitely, are unrealistic assumptions beyond the first approximations in many practical situations.

For example, sustained animal population cycles, Oster et al (1978), sustained oscillations of water level in a rock channel, Whittle (1954), sustained solar cycles, etc., all point to the necessity of non-linear oscillations theory in modern time series modelling. The modelling of this aspect of a dynamical system, in part, led to the class of threshold models of Tong (1983), who started it in 1977.

The most primitive form of a threshold model is the so-called *self-exciting threshold autoregressive model* (SETAR) which, for a time series $\{X_t : t = 0, \pm 1, \pm 2, \ldots\}$, is typified by the very simple example

$$X_t = \begin{cases} a\, X_{t-1} + \text{white noise if } X_{t-1} \le \text{threshold} \\ b\, X_{t-1} + \text{white noise if } X_{t-1} > \text{threshold} \end{cases}$$

where a and b are constants, which need not be equal.

A succinct form of the general threshold model is (for $t = 0, \pm 1, \pm 2, \ldots$)

$$\underset{\sim}{X}_t = \underset{\sim}{B}(J_t)\, \underset{\sim}{X}_t + \underset{\sim}{A}(J_t)\, \underset{\sim}{X}_{t-1} + \underset{\sim}{H}(J_t)\, \underset{\sim}{e}_t + \underset{\sim}{C}(J_t), \qquad (1.1)$$

where $\{\underset{\sim}{X}_t\}$ is a k-dimensional time series, J_t for each t is an observable (indicator) random variable taking integer values 1,2,.., say, and, for each $j \in \{1, 2, \ldots\}$, $\underset{\sim}{A}(j)$, $\underset{\sim}{B}(j)$ and $\underset{\sim}{H}(j)$ are $k \times k$ (non-random) matrix coefficients, $\underset{\sim}{C}(j)$ is a $k \times 1$ vector of constants, and $\underset{\sim}{e}_t$ is a sequence of i.i.d. k-dimensional random vectors with zero mean and a covariance matrix Σ say, and $\underset{\sim}{e}_t$ independent of $\underset{\sim}{X}_s$, $s < t$. For each t, the value taken by $\underset{\sim}{J}_t$ indicates the *regime* the system $(\underset{\sim}{X}_t)$ is currently in and the switching from one regime to the next may be related to the crossing of a threshold by an exogenous or endogenous variable.

The special case with

$$J_t = j \text{ iff } \underset{\sim}{X}_{t-1} \in R_j, \qquad (1.2)$$

where $R_1, R_2, \ldots, R_\ell$ form a partition of $\mathbb{R}^k$, defines a *self-exciting threshold model*, which is a system of piece-wise linear (stochastic) difference equations. They are analogous to a system of piece-wise linear differential equations, which were studied extensively by the Russian school of non-linear oscillations under the leadership of academician A.A. Andronov in the 1930's. At a different level, Chen (1971) has given a plausible approach to piece-wise linearization of a non-linear system in continuous time by essentially focusing the modelling on neighbourhoods of singularities over the state space.

The basic idea is the introduction of regimes via thresholds. Let us agree to call this the *threshold principle*, under which we may group the following models: threshold moving average models, Wecker (1977), Jolliffe et al (1985); threshold ARMA models, W.Y. Wang et al (1984); piecewise polynomial AR models, Ozaki (1982), Pemberton (1986); Markov chain driven piecewise linear AR models, Tong (1983, p.62) and smooth threshold models, Chan et al (1986a).

About the same time, inspired by physical consideration in non-linear oscillations, especially energy flow and amplitude-frequency relations, Ozaki has developed a suggestion, attributed to H. Akaike in 1978, into the class of exponential autoregressive models, EXPAR. It should be mentioned that a similar idea was earlier developed in Jones (1976, esp.p.7 and p.158). Ozaki (1982) summarizes the development of EXPAR. Ozaki(1985a) and Tong et al (1980b) are also relevant. Typically, an EXPAR model (of order k) for a univariate time series $\{X_t : t = 0, \underline{+}1, \underline{+1}, \ldots\}$ takes the form

$$X_t = \sum_{j=1}^{k} A_{t,j} X_{t-j} + e_t, \quad A_{t,j} = a_j + b_j \exp(-c^2 X_{t-1}^2), \qquad (1.3)$$

where $\{e_t\}$ is a sequence of i.i.d. random variables with zero mean and constant variance and e_t independent of X_s, $s < t$, a_j's, b_j's and c are constants. The *physical/plausible* idea seems to be that the 'impulse response', $A_{t,j}$, at time t is controlled by the rise and fall of X_{t-1}^2, interpreted as 'energy at time t_. In discussing EXPAR models, Tong et al (1980a, p.278 and pp.283-285) raised the issue of a precise mathematical formulation of the amplitude-frequency notion. The general question remains open.

1.2 Algebra, Control Engineering and Bilinearity

From an algebraic viewpoint, a linear system/model is 'additive' in the sense that the principle of superposition applies. The 'obvious' extra algebraic operation to include in a non-linear system/model would be 'multiplication', which may destroy the principle of superpositon. By denoting the inputs and outputs by X_t and Y_t respectively, the formal functional power series for a single input/single output system takes the form

$$Y_t = a_i X_{t-i} + \sum_{i=0}^{\infty} \sum_{j=0}^{\infty} a_{ij} X_{t-i} X_{t-j} + \ldots \qquad (1.4)$$

Brillinger (1969) has shown how a *small* perturbation of a linear system may lead to the above non-linear system. However, Kalman

(1968) concentrated on the sub-collection of multilinear systems, which includes of course the bilinear systems as special cases. The general idea is that the system (with say r input/ output variables) is linear in each of the variables, holding the others constant. The fundamental paper by Brockett (1976) has provided much impetus to the development of bilinear systems (i.e. r=2) in the control engineering literature in the West. In his later paper (1977) he discussed the limitations of these systems from the point of view of non- linear oscillations. Borrowing the above idea, Granger & Anderson (1978) introduced the class of bilinear time series models by essentially replacing the deterministic input of the above bilinear system by a sequence of *unobservable* i.i.d. random variables. A typical example of a bilinear time series model is, in its simplest but non-trivial form,

$$X_t = a X_{t-1} + b e_{t-1} + c e_{t-j} X_{t-1} + e_t , \qquad (1.5)$$

where $\{e_t\}$ is a sequence of i.i.d. random variables with zero mean and finite variance σ_e^2 and e_t is independent of X_s, $s < t$. Here, j is an integer. The formal difference between a bilinear time series model and an autoregressive/moving average (ARMA) model is the bilinear term eX . Subba Rao et al(1984) gives a comprehensive account of this class of models. Feigin et al (1985), Quinn (1982), Tong (1981) and Wang et al (1983) may be added to its list of references, especially in connection with a deeper probabilistic analysis of the rather particular case with j = 0 = b.

1.3 Time Irreversibility, Long-range Dependence

Time irreversibility of time series data, in that the probability structure is not preserved upon reversal of the time direction, is probably the most obvious manifestation of non-linearity/non-Gaussianity, see, e.g., Brillinger et al (1967), Lawrance et al (1985) and Tong (1983). Another important notion which might be relevant to non-linear time series analysis is long-range dependence, in which the autocovariance function decays only algebraically fast , e.g. Cox (1981).

1.4 Discrete Time vs. Continuous Time

Quite often the given regularly sampled time series data are obtained from an originally time-continuous record (i.e. an analogue record), e.g. an EEG record. Equally often it is found more convenient to formulate the problem in continuous time (e.g. linear or non-linear differential equations) although available data are almost invariably in discrete time, e.g. some econometric models of economic time series data. The question arises as to the general inference about the underlying continuous-time models (e.g. in the form of a stochastic differential equation) based on discrete-time models fitted to the regularly sampled data. This area is fraught with potential pitfalls even in the simplest linear ARMA cases. An illustration is given by Chan et al (1986d). The case with non-linear models seems much more treacherous despite the apparent optimism of Ozaki (1985b) and merits careful scrutiny. Bartlett(1946) and Robinson (1977a) represent earlier references.

1.5 Instantaneous Non-linear Transformation

Given a time series $\{X_t : t = 0, \pm 1, \pm 2, \ldots\}$, we may often prefer to analyse for various reasons the derived series $\{Y_t : t = 0, \pm 1, \pm 2, \ldots\}$, where $Y_t = f(X_t)$, for each t, f being an instantaneous non-linear transformation. These include the stabilization of variances, the decoupling of multiplicative effects, the induction of finite moments of all orders, etc. Such a non-linear transformation represents non-linearity of possibly the simplest type. A popular approach is via the well-known Box-Cox transformations. Tong (1983) has given examples.

1.6 General Comments on More General Models

The aforementioned models are, in a sense, 'first-generation' finite parametric models and it would be tempting to obtain a more general purpose finite parametric model encompassing these. Priestley (1980) is an attempt in this direction, in which he develops a *state-dependent model* (SDM). The general idea is to allow the coefficients of an ARMA model (c.f. a and b of (1.5)) to become dependent on finite past X's and finite past e's which are grouped under a 'state' vector. It seems too early to assess the full impact of SDM. Before then the following questions may be raised.

(i) Exactly how general are the SDM's? If, for instance, the functional forms of the ARMA coefficients are assumed 'smooth' such that a truncated functional power series obtains for X_t in terms of $\{e_s : s \leq t\}$, then might the difficulties encountered by Haggan et al (1984) in 'identifying' SETAR models within the SDM's perhaps be expected because threshold models are typically 'discontinuous'?

(ii) Does the use of the adjective 'state-dependent' conform to the emphasis usually placed on the minimality of a state-vector? See, e.g. Akaike (1974). The question raised seems to be a fundamental one because it relates to the uniqueness of the model, without which it would seem difficult to justify using SDM's for discriminating among different classes of non-linear time series models.

(iii) Might the greater flexibility of the SDM's well be gained at the expense of the greater intractability of the probabilistic structure (e.g. conditions for ergodicity/ stationarity, distributions and moments of the X's, etc.) and statistical inference (e.g. sampling properties of the parameter estimates) etc. of the models?

Similar questions may be raised with Tjøstheim's *doubly*

stochastic model (1986).

2. PROBABILISTIC ASPECTS

2.1 Which time series model is non-linear?

As a working definition, we agree to classify as *essentially linear* a time series model for which, given infinite past observations, the linear least-squares predictor is also the least-squares predictor. A model which is not essentially linear is classified as non-linear. See also Hannan (1979,pp.84-5). In each case we have assumed the existence of the predictor.

Now, according to the definition the following bilinear model is classified as non-linear:

$$X_t = (a + b\, e_{t-j})X_{t-1} + e_t , \quad (j \geq 1) \qquad (2.1)$$

but setting $j=0$ changes the model to an essentially linear one. In each case, we have assumed that $\{e_t\}$ is a sequence of i.i.d. random variables with e_t independent of X_s, $s<t$. By a similar token, the class of *random coefficient autoregressive models*, Andel (1976), Nicholls et al (1982), and the related class of *non-linear autoregressive models in exponential variables* of Lawrance & Lewis (1985) are classified as essentially linear.

2.2 Innovations

We restrict our discussion to strictly stationary time series $\{X_t\}$. Let $\hat{X}_t$ denote the least-squares predictor (assumed to exist) of X_t given $\{X_s, s<t\}$. We denote the difference $X_t - \hat{X}_t$ by ε_t and call $\{\varepsilon_t\}$ the *innovation sequence* . For non-linear autoregressive (NLAR) models of the form

$$X_t = f(X_{t-1}, X_{t-2}, \ldots, X_{t-k}) + e_t , \qquad (2.2)$$

it is obvious that

$$\varepsilon_t = e_t, \text{ all } t.$$

For bilinear models, the situation is slightly more complicated, which may be best illustrated by reference to the special case (2.1). For $j \geq 1$, $\varepsilon_t = e_t$ all t and least-squares predictor is non-linear in X. For $j = 0$, $\varepsilon_t = b\, e_t X_{t-1}$ all t, and least-squares predictor is linear in X. In effect, the bilinear model (2.1) with $j=0$ is essentially the linear autoregressive model

$$X_t = a X_{t-1} + \varepsilon_t , \qquad (2.3)$$

on using the general equation

$$X_t = \hat{X}_t + \varepsilon_t , \qquad (2.4)$$

where $\{\varepsilon_t\}$ is a sequence of uncorrelated random variables each with zero mean and variance $b^2(\text{Var } e_t)$ (Var X_t). Here and elsewhere

$$X_t = \hat{X}_t \tag{2.5}$$

describes the bone or the skeleton of the stochastic model (2.4). We refer to model (2.5) as the *skeleton* of model (2.4) and denote it by M_o. Model (2.4) is sometimes denoted by M_ε. Let $\text{Var}(.|M)$ denote the variance under model M.

Now, it would seem intuitively reasonable to expect

$$\text{Var}(X_t|M_\varepsilon) \to \text{Var}(X_t|M_o) \quad \text{as} \quad \text{Var } \varepsilon_t \to 0. \tag{2.6}$$

This vaguely resembles the requirement of stability of the model M_ε in the sense that the energy of the excited system (i.e.$\text{Var}(X_t | M_\varepsilon)$) should be reduced to the energy of the unexcited system (i.e. $\text{Var}(X_t|M_o)$) upon the cessation of the excitation. Some examples are given in the *expanded version of this review*.

2.3 Stability, Ergodicity and Stationary Distributions

The relationship between the skeleton process M_o and the stochastic process M_ε goes deeper than is apparent in§ 2.2. Chan et al (1985a) and Chan (1986) have obtained an explicit link between the stability of M_o and the ergodicity of M_ε.

Let $\{\underset{\sim}{X}_t\}$ satisfy the equation

$$\underset{\sim}{X}_{t+1} = T(\underset{\sim}{X}_t) + \underset{\sim}{e}_{t+1}, \quad t \geq 0, \quad T : \mathbb{R}^m \to \mathbb{R}^m, \tag{2.7}$$

where $\underset{\sim}{X}_t$ takes values in $\mathbb{R}^m$. Let T be continuous and homogeneous (i.e. $T(c\underset{\sim}{x}) = c\, T(\underset{\sim}{x})$, $\forall\, c > 0$, $x \in \mathbb{R}^m$). Let the origin, $\underset{\sim}{0}$, be a fixed point of T. In the case of $\{\underset{\sim}{e}_t\}$ being i.i.d. with an absolutely continuous marginal distribution possessing a positive p.d.f. f(.) over $\mathbb{R}^m$, we assume that $\int ||t|| f(t)\mu_m(dt) < \infty$, μ_m being the Lebesgue measure on Borel sets of $\mathbb{R}^m$. If $\underset{\sim}{e}_t$ is a vector with all components being zero except for the first, and if these components are i.i.d. each having an absolutely continuous distribution, the p.d.f., f(.), of which is positive everywhere in $\mathbb{R}$, we assume that $\int |t| f(t)dt < \infty$ and that for $\underset{\sim}{x} = (\underset{\sim}{x}_1, \underset{\sim}{x}_2, \ldots, \underset{\sim}{x}_m)'$, $T(\underset{\sim}{x}) = (h(\underset{\sim}{x}), x_1, \ldots, x_{m-1})'$.

Theorem: Under the above convention for $\{\underset{\sim}{e}_t\}$, the existence of a continuous Lyapunov function, V, in a neighbourhood of the origin (i.e. $V(\underset{\sim}{0}) = 0$; $V(\underset{\sim}{x}) > 0$ when $\underset{\sim}{x} \neq \underset{\sim}{0}$; $V(T(\underset{\sim}{x})) - V(\underset{\sim}{x}) < 0$ when $\underset{\sim}{x} \neq \underset{\sim}{0}$; $V(\underset{\sim}{x})$ is continuous in $\underset{\sim}{x}$; $V(\underset{\sim}{x}) \to \infty$ when $||\underset{\sim}{x}|| \to \infty$) implies the geometric ergodicity of $\{\underset{\sim}{X}_t\}$ given by (2.7).

The above theorem enables us to investigate the ergodicity of fairly general classes of non-linear time series models simply by appealing to the stability of the corresponding skeleton processes, which may have been established by the applied mathematicians/engineers. Chan et al (1985a) and Chan (1986) have given several examples which include SETAR and EXPAR. However, the technique does not seem to apply to the general bilinear models.

A direct probabilistic approach to the study of ergodicity/strict stationarity of specific non-linear time series models has been studied by many authors including, e.g. Jones (1976,1978) -

non-linear autoregressive models mostly of the first order, Chan et al(1985b), Petruccelli et al (1984) and Tong (1983) - SETAR models mostly of order one, Pemberton (1986), Tjøstheim (1984a, 1984b) - EXPAR Models of order one. Bhaskara Rao et al (1983), Feigin et al (1985), Guegan (1981), Hannan (1982), Pham et al (1981), Quinn (1982), Subba Rao et al (1984) and Tong (1981) - various specific cases of of bilinear models. One of the basic tools is based on an extension of Foster's approach, e.g. Tweedie (1975), to cover the case of Markov chains on a general state space. The case of non-linear models of higher orders is insufficiently studied although the approach of Chan et al (1985a) seems helpful.

Explicit evaluation of stationary marginals and moments (if existing) is rarely available except in very special cases; some of these are given by Andel et al (1984, 1986) and Chan et al(1986c), all for SETAR models. Granger et al (1978) and Tong (1981) have proved the *non-existence* of moments of all orders for simple bilinear models and Wang et al (1983) has discussed the marginal p.d.f. of a simple bilinear model. Numerical procedures have been developed by Jones (1978) and Pemberton (1986) for evaluating marginal densities of general non-linear autoregressive models of order one. All in all, the problems here are characteristically complicated and solutions are incomplete.

3. STATISTICAL ASPECTS

3.1 Graphical Methods

Graphical methods should form a very important part of identification of non-linear time series models and this point should be stressed more often in the literature. These include among others visual inspection of sample autocorrelations for e.g. possible trend deletion, of marginal histograms and two-dimensional histograms of data for e.g. departure from Gaussianity, etc. Tong (1983) has also emphasised the importance of non-parametric lag regressions. Sampling properties of these estimates are given by Robinson (1983) and Yakowitz (1985). *Discrete-time phase diagrams* consisting of lines joining each point (X_t, X_s) to the next (X_{t+1}, X_{s+1}) (c.f. phase diagrams in dynamics) are used extensively in Tong (1983). Priestley's SDM identification technique involves use of graphics.

Post-model examination is more crucial in non-linear modelling because the view taken here is that *each particular non-linear model is best suited for each particular type of non-linearity*. Post-model examination should include not just the usual examination of the fitted residuals but also other 'goodness-of-fit', e.g. observed (non-parametric) lag regressions vs. the fitted (parametric) lag regressions, the observed low-dimensional marginal distributions vs. the fitted, the observed auto and bi-spectra vs. the fitted, etc. Lim (1981) has conducted a comparative study along these lines on the SETAR models, the bilinear models and the EXPAR models fitted to the Canadian lynx data and Wolf's sunspot numbers.

3.2 Estimation, Tests for Linearity, Prediction, and Applications.

There seems to be essentially two principal methods used for the estimtion of parameters of finite parametric non-linear time series models, namely the maximum likelihood method and the conditional least-squares method.

Subba Rao et al (1984) adopted the former for the bilinear models. The parameter estimates so obtained are consistent (op.cit). However, asymptotic distribution of the estimates is not available. One difficulty may be due to the lack of moment properties of bilinear models mentioned in §2.3.

The method of conditional least-squares (CLS) seems to be popular among users of SETAR models, EXPAR models and, more generally, non-linear autoregressive models appropriately parameterized. The method is essentially one of minimizing the sum of squares $\Sigma\ \varepsilon_t^2$. The paper by Klimko et al (1978) provides most of the theoretical framework for sampling properties of the estimates. Specifically, strong consistencey, asymptotic normality and rates of convergence of parameter estimates of SETAR models are discussed, Chan et al (1986b), Petruccelli (1986) and Chan (1986). Results here bear resemblance, in many ways, to those for linear autoregressive models. Now, for practical purposes, especially for higher order SETAR models, equation (1.2) is often specialised to

$$J_t = j \text{ iff the d-th component of } \underset{\sim}{X}_{t-1} \in R_j, \qquad (3.1)$$

where $R_1, R_2, \ldots, R_\ell$ form a partition of $\mathbb{R}$. The parameter d is called a *delay parameter* . Some sampling properties of CLS estimate of d are given in Chan (1986). Pemberton (1986), Tjøstheim (1984a, 1984b) have discussed some but not all corresponding results for the EXPAR case.

Tjøstheim (1984a,b) has adapted the arguments of Klimko et al (1978) and made the results of CLS estimation more readily accessible to several classes of time series models, e.g. bilinear models, random coefficient autoregressive models and EXPAR. His results concerning SETAR models should be reinforced by those summarised in Chan et al (1986b). Robinson (1977b) and Hinich et al (1980) have discussed the estimation of a simple non-linear moving average model which includes the product $e_t e_{t-1}$. Aase (1983) has discussed recursive estimation, based on Kalman's filter, of non-linear autoregression. Some of his results might be relevant for Priestley's SDM. The general approach based on *stochastic regression models,* Lai et al (1983), begins to show promising results for the estimation problem of non-linear time series models, S.R. Wang et al (1984).

Tests for linearity are reviewed in W.S. Chan et al (1986). Prediction and applications are reviewed in the expanded version of this review.

4. SOME UNSOLVED PROBLEMS NOT MENTIONED EARLIER

(1) Probabilistic and statistical aspects of non-linear models of higher order are inadequately understood. Often even ergodicity/stationarity is difficult to establish although Chan et

al (1985a) helps in some cases.

(2) Most of the non-linear models here have a *constant* variance conditional on past X's. This is clearly a limitation by reference to the concept of 'instantaneous variance per unit time' in the theory of Markov processes with continuous state space. (See, e.g., Cox et al (1965, p.214)

(3) The connection between (deterministic) non-linear oscillations and (stochastic) non-linear time series models is undoubtedly strengthened via Chan et al (op.cit). However, their theorem leads to sufficient conditions for stability and ergodicity only. What about necessary conditions? Chan (1986) gives some initial results.

(4) The white noise term is often assumed to have a density of infinite support. What would happen if it has only finite support? This consideration might be related to a proper formulation of 'stochastic limit cycles' (c.f. § 3.7 of Tong (1983)), which remains open.

(5) A deeper theoretical study of $E[X_t | X_{t+j}]$'s, for $j=\pm 1, \pm 2, \ldots$, should pay dividends in terms of insights into time irreversibility and model discrimination.

(6) In linear prediction theory, the celebrated result of Kolmogorov (1941) implies that, under mild conditions, the logarithm of the m.s.e. of prediction is bounded below by $\int \ell n\, f(\omega) d\omega$, where f(.) is the spectral density of $\{X_t\}$. In a sense, this lower bound gives us an idea about the maximum amount of (linear) information in the data. It guards us against exaggerated claims of 'superiority' of our prediction. Is there an analogous guard with the non-linear case which may be simply expressed in terms of characteristics of the observables, $\{X_t\}$? Kanter (1979), Rosenblatt (1979) and Shepp et al (1980) may be relevant.

(7) The expanded version of this review has reviewed non-linear prediction, and pointed out the omission of references to prediction interval as an outstanding area of research.

ACKNOWLEDGEMENTS

I am particularly grateful to Professor Sir David Cox, FRS, for many valuable comments and suggestions. I also thank Drs. K.S. Chan, Ian T. Jolliffe and Byron J.T. Morgan for comments, and the Royal Society of London for financial support.

REFERENCES

Aase, K. K. (1983) Recursive estimation in non-linear time series models of the autoregressive type. J.R.Statist.Soc., B, 45, 228-237

Akaike, H. (1974) Stochastic theory of minimal realizations, IEEE

Trans. Auto Control , AC-19, 667-674.
Andel, J. (1976) Autoregressive series with random parameters. Math.Op.u.Statist., 7, 735-741.
Andel, J. and Barton, T. (1986) A note on threshold AR(1) Model with Cauchy innovations J.Time Series Anal, 7, 1-5.
Andel, J., Netuka, I., Svara, K. (1984) On threshold autoregressive processes. Kybernetika, 20, 89-106.
Andronov, A.A., Khaikin, S.C. (1937) Theory of oscillations (in Russian), Moscow. English translation by S.Lefschetz, Princeton Univ. Press, 1949.
Bartlett, M.S. (1946) The theoretical specification and sampling properties of autocorrelated time series. J.Res.Statist.Soc.Suppl. 8, 27-41.
Bhaskara Rao, M., Subba Rao, T. and Walker, A.M. (1983) On the existence of some bilinear time series models. J.Time Series Analysis, 4, 95-119.
Box, G.E.P. and Jenkins, G.M. (1970) Time Series Analysis. Forecasting and Control, San Franciso: Holden-Day.
Brillinger, D.R. (1969) The identification of polynomial systems by means of higher order spectra. Appl. & Meth.Random Data Anal, Southampton, U.K., July 1969, pp.M-1 - M.21.
Brillinger, D.R. and Rosenblatt, M. (1967) Asymptotic theory of estimates of k-th order spectra" Spectral Analysis of Time Series. ed. B.Harris, 189-232, New York: Wiley.
Brockett, R.W. (1976) Volterra series and geometric control theory. Automatica, 12, 167-176.
Brockett, R.W. (1977) Convergence of Volterra Series on infinite intervals and bilinear approximations. In: Non-linear Systems and Applications. Ed.V.Lakshmikanthan, 39-46.
Chan, K.S. (1986) Topics in nonlinear time series analysis. Unpublished Ph.D. thesis, Princeton University, USA.
Chan, K.S. and Tong, H. (1984) A note on sub-system stability and system stability. J.Eng.Maths. (China), 1, 43-51.
Chan, K.S. and Tong, H. (1985a) On the use of the deterministic Lyapunov function for the ergodicity of stochastic difference equations. Adv.Appl.Prob., 17, 666-678.
Chan, K.S. and Tong, H. (1986a) On estimating thresholds in autoregressive models. J.Time Series Anal, 7, 179-190.
Chan, K.S. and Tong, H. (1986b) A survey of the statistical analysis of univariate threshold autoregressive models. To appear in: Advances in Statistical Analysis and Statistical Computing: Theory & Applications, Ed. R.S. Mariano, JAI Press Inc., U.S.A.
Chan, K.S. and Tong, H. (1986c) A note on certain integral equations associated with non-linear time series analysis. Prob.Th. & Related Fields, 73, 153-158.
Chan, K.S. and Tong, H. (1986d) A note on embedding a discrete parameter ARMA model in a continuous parameter ARMA model. To appear: J.Time Series Analysis.
Chan, K.S., Petruccelli, J.D., Tong, H. and Woolford, S.W. (1985b) A multiple-threshold AR(1) model. J.Appl.Prob., 22, 267-279.
Chan, W.S. and Tong, H. (1986) On tests for non-linearity in time series analysis. To appear: J.Forecasting.
Chen, C.F. (1971) Hurwitz' stability criterion and Fuller's

aperiodicity criterion in non-linear systems analysis. Int.J. Electronics, 31, 609-619.
Cox, D.R. (1981) Statistical analysis of time series: some recent developments. Scand.J.Statist., 8, 93-115.
Cox, D.R. and Miller, H.D. (1965) The Theory of Stochastic Processes. London: Methuen (now in 2nd ed. by J. Wiley)
Feigin, P.D. and Tweedie, R.L. (1985) Random coefficient autoregressive processes: a Markov chain analysis of stationarity and finiteness of moments. J.Time Series Anal, 6, 1-14.
Granger, C.W.J. and Andersen, A.P. (1978) An introduction to bilinear time series models. Vandenhoek and Ruprecht, Gottingen.
Guegan, D. (1981) Etudes d'un Modele nonlineaire, le modele superdiagonal d'ordre. C.R.Acad.Sci., Paris, 293, Series 1, 95-98.
Haggan, V., Heravi, S.M. and Priestley, M.B. (1984) A study of application of state-dependent models in non-linear time series analysis. J.Time Series Anal, 5, 69-102.
Hannan, E.J. (1979) The statistical theory of linear systems. In: Developments in Statistics, 2, Ch.2, 83-121, New York: Academic Press.
Hannan, E.J. (1982) A note on bilinear time series models. Stochastic Processes and their Appl., 12, 221-224.
Hinich, M.J. (1982) Testing for Gaussianity and linearity of a stationary time series. J.Time Series Anal, 3, 169-176.
Hinich, M.J. and Patterson, D.M. (1980) Identification of the coefficients in a non-linear time series of the quadratic type. J.Econ., 30, 269-288.
Jolliffe, I.T. and Kumar, K. (1985) Discussion of the paper by Lawrance and Lewis. J.R.Statist.Soc.,B, 47, 190-191.
Jones, D.A. (1976) Non-linear autoregressive processes. Unpublished Ph.D. Thesis, University of London, U.K.
Jones, D.A. (1978) Non-linear autoregressive processes. Proc.R.Soc.,London, A, 360, 71-95.
Kalman, R.E. (1980) Pattern recognition properties of multi-linear response functions, Parts I and II. Control and Cybernetics, 9, 5-31. (Original paper in Russian in 1968.)
Kanter, M. (1979) Lower bounds for nonlinear prediction error in moving average processes. Ann.Prob., 7, 128-138.
Klimko, L.A. and Nelson, P.I. (1978) On conditional least squares estimation for stochastic processes. Ann.Statist., 6, 629-643.
Kolmogorov, A.N. (1941) Stationary sequences in Hilbert space. Bull.Math.Univ., Moscow, 2, (6), 1-40.
Lai, T.L. and Wei, C.Z. (1983) Asymptotic properties of general autoregressive models and strong consistency of least squares estimates of their parameters. J.Multi.Anal, 13, 1-23.
Lawrance, A.J. and Lewis, P.A.W. (1985) Modelling and residual analysis of non-linear autoregressive time series in exponential variables (with Discussion). J.R.Statist.Soc., B, 47, 165-202.
Lim, K.S. (1981) On threshold time series modelling. Unpublished Ph.D. thesis., Univ. of Manchester, U.K.
Nicholls, D.F. and Quinn, B.G. (1982) Random coefficient autoregressive models: an introduction. Lecture Notes in

Statistics, Vol.No.11, New York: Springer-Verlag.
Oster, G. and Ipaktchi, A. (1978) Population Cycles. Theor. Chem.Periodicity. In: Chem.& Biol. (Eds.H.Eyring and D.Henderson New York: Academic Press, 111-132.
Ozaki, T. (1982) The statistical analysis of perturbed limit cycle processes using nonlinear time series models. J.Time Series Anal, 3, 29-41.
Ozaki, T. (1985a) Nonlinear time series models and dynamical systems. In: Handbook of Statistics, Vol.5, Eds. E.J. Hannan and P.R.Krishnaiah, North-Holland.
Ozaki, T. (1985b) Statistical identification of storage models with application to stochastic hydrology. Water Resources Bull., 21, 663-675.
Pemberton, J. (1986) Contributions to the theory of non-linear time series models. Unpublished Ph.D. Thesis, Univ. of Manchester, U.K.
Petruccelli, J. (1986) On the consistency of least squares estimators for a threshold AR(1) model. J.Time Series Anal, 7, 269-278.
Petruccelli, J.D. and Woolford, S.W. (1984) A threshold AR(1) mode. J.Appl.Prob., 21, 270-286.
Pham, Tuan Dinh and Tran, L.T. (1981) On first-order bilinear time series models" J.Appl.Prob., 18, 617-627.
Priestley, M.B. (1980) State-dependent models: a general approach to nonlinear time series analysis. J.Time Series Anal, 1, 47-71.
Quinn, B.G. (1982) A note on the existence of strictly stationary solutions to bilinear equations. J.Time Series Anal, 3, 249-252.
Robinson, P.M. (1977a) The construction and estimation of continuous time models and discrete approximations in econometrics. J.Econometrics, 6, 173-197.
Robinson, P.M. (1977b) The estimation of a nonlinear moving average model. Stoch. Proc. & Applic., 5, 81-90.
Robinson, P.M. (1983) Nonparametric estimators for time series. J.Time Series Anal, 4, 185-207.
Rosenblatt, M. (1979) Linearity and nonlinearity in time series - prediction. Bull.Int.Statist.Inst., 42.
Shepp, L.A., Slepian, D. and Wyner, A.D. (1980) On prediction of moving-average processes. Bell System Tech.J., 59, 367-415.
Subba Rao, T. and Gabr, M.M. (1984) An introduction to bispectral analysis and bilinear time series models. Lecture Notes in Statistics, Vol.No.24, New York: Springer-Verlag.
Tjøstheim, D. (1984a) Estimation in nonlinear time series models I: stationary series. Research Report, Dept. of Mathematics, Univ. of Bergen, Norway.
Tjøstheim, D. (1984b) Estimation in nonlinear time series models II: some non-stationary series. Research Report, Dept. of Mathematics, Univ. of Bergen, Norway.
Tjøstheim, D. (1986) Some doubly stochastic time series models. J.Time Series Anal, 7, 51-72.
Tong, H. (1981) "A note on a Markov bilinear stochastic

process in discrete time. J.Time Series Anal, 2, 279-284.
Tong, H. (1983) Threshold Models in Non-linear Time Series Analysis. In: Lecture Notes in Statistics, Vol.No.21, New York: Springer-Verlag.
Tong, H. and Lim, K.S. (1980a) Threshold autoregression, limit cycles and cyclical data (with Discussion). J.Roy. Statist.Soc. B, 42, 245-292.
Tong, H. and Pemberton, J. (1980b) On stability and limit cycles of non-linear autoregression in discrete time. Cahiers du CERO (Bruxelles), 22, 137-147.
Tweedie, R.L. (1975) Sufficient conditions for ergodicity and recurrence of Markov chain on a general state space. Stoch.Proc.Appl., 3, 385-403.
Wang, S.R. and An, H.Z. (1984) On the statistical analysis of general linear models. (In Chinese) Research Report, Inst. of Appl.Maths., Academia Sinica, Beijing, China.
Wang, S.R., An, H.Z. and Tong, H. (1983) On the distribution of a simple stationary bilinear process. J.Time Series Anal, 4, 209-216.
Wang, W.Y., Du, J.G. and Xiang, J.T. (1984) Threshold Autoregressive Moving Average Model. (In Chinese) Comp.Maths., 4, 41-419.
Wecker, W.E. (1977) Asymmetric time series. ASA Proc.Bus. & Econ.Sect., pp.417-422. Also (March 1981) JASA.
Whittle, P. (1954) The statistical analysis of a seich record. Sears Foundation J.Marine Res., 13, 76-199.
Yakowitz, S.J. (1985) Nonparametric density estimation, prediction and regression for Markov sequences. J.Amer.Statist. Assoc., 80, 215-221.

ROBUST SPECTRAL ESTIMATION

Zhurbenko I.G.
Department of Mathematics and Mechanics, Moscow State University, Moscow 119899 USSR

In recent decade time series analysis was one of the most rapidly developing fields of probability theory and mathematical statistics. Arising from fundamental theoretical results of the theory of stationary stochastic processes based on the works of A.N.Kolmogorov, A. Ya.Hinchin, N.Wiener, G.Cramer, G.Doob, Yu.A.Rozanov, A.M.Yaglom time series analysis became one of the most significant tendencies in the theory and applications of mathematical statistics to which contributions were made by J.Tukey, M.Bartlett, E.Parzen, U.Grenander, M. Rosenblatt, E.Hannan, T.Anderson, D.Brillinger, V.V. Kotelnikov, V.V.Pesarenko, K.O.Dzhaparidze, N.I.Yadrenko, and many others. Various investigations in physics, economics, engineering, linquistics, biology, and sociology involve samples stochastically stationary or different from stationary ones by a distinguishable trend, seasonal components, periodical constituents, etc. In the study of such phenomena one can distinguish two main approaches. The first deals with stochastic relationship in time, it is called time series analysis in the time domain; the second deals with the frequency characteristics of series, this one is called spectral analysis of time series. The first approach is based on the study of correlation functions of observations and is connected, to a greater extent, with parametric, multivariate methods of investigation. The second approach is very demonstrative, it is based on spectral, asymptotic, functional techniques that in many usages are concerned with the physical nature of phenomena.

One of the first statistical studies of frequency characteristics was started by Schuster (1898) who introduced the definition of periodogram and used it to reveal hidden periodicities. He also pointed out statistical nonstability that fails the periodogram as a spectral density estimate. The idea of averaging the periodogram with a spectral window to have a consistent spectral density estimate was first suggested by Einstein (1914), but this early work of his went unnoticed in the statistical world where Bartlett (1948) and Daniel (1946) are usually given priority. The development of methods of the statistical estimation of spectra was stimulated by the results of A.N.Kolmogorov, M.D.Milliontcshikov, A.M.Obuchov, A.S.Monin, A.M.Yaglom (see Kolmogorov (1985)) concerning gas and liquid turbulence where formulations of many physical laws are of spectral form. Spectral analysis of time series experienced a long depression in its development as a result of, in particular, much computational work necessary for realization of discrete Fourier transform algorithms. The appearance of fast Fourier transform (Cooley and Tukey (1965)) changed the situation. This algorithm brought about reduction of operations of discrete Fourier transform from the usual CN^2 operations to $CN\ln N$ operations, so that it became convenient to calculate a covariance function estimate through double Fourier transform from an initial sample. All this with enlarged computer possibilities resulted in a tremendous popularity of the spectral approach to problems of stochastic analysis of time series. Of importance was the problem of discretization of a stationary stochastic process with continuous time solved by V.V.Kotelnikov and K.Shannon (see Yaglom (1962)) independently of each other. It turned out that any process with continuous time having spectrum in a bounded band $-\frac{\pi}{\Delta}\leq\omega\leq\frac{\pi}{\Delta}$ could be expressed as a linear combination of its discrete values on a lattice $K\Delta$, thus it is only the process with bounded spectrum that is entirely defined by its values $X(K\Delta)$. When estimating a spectrum in an unbounded band by the values of the process at moments $K\Delta$ one has the superimposed frequencies effect. In detail spectrograph estimates were first studied in Grenander and Rosenblatt (1957). Statistical estimates of higher-order spectra or spectral cumulants (Shiryaev (1960) as it is customary to call them were studied in detail by D.Brillinger and M.Rosenblatt.

The spectral cumulants of a stochastic process $X(t)$, $-\infty<t<+\infty$ with $E|X(t)|^n \leqslant C < \infty$ are defined by its moments

$$M^{(n)}(t_1, \dots, t_n) = E\{X(t_1) \cdot \dots \cdot X(t_n)\}$$

and cumulants

$$S^{(n)}(t_1, \dots, t_n) = \frac{i^{-n} \partial^n}{\partial u_1 \dots \partial u_n} \ln E \exp\{i(u_1 X(t_1) + \dots + u_n X(t_n))\} \Big/_{u_1 = \dots = u_n = 0},$$

$X(t) \in \Phi^{(n)}$ if for all $1 \leqslant K \leqslant n$ there is a measure $M^{(K)}(\Delta)$ of bounded variation in the Euclidean space R^K such that for all $t_1, \dots, t_K$

$$M^{(K)}(t_1, \dots, t_K) = \int_{R^K} \exp\{i(t_1\lambda_1 + \dots + t_K\lambda_K)\} M^K(d\lambda_1 \dots d\lambda_K).$$

The measure $M^K(d\lambda)$ is called the spectral moment of the stochastic process $X(t)$. The density of this measure can be defined only as a generalized function. The measure $F^{(n)}(d\lambda)$ is called a spectral cumulant if for all $t_1, \dots, t_n$

$$S^{(n)}(t_1, \dots, t_n) = \int_{R^n} \exp\{i(t, \lambda)\} F^{(n)}(d\lambda)$$

The measure $F^{(n)}(d\lambda)$ exists and is of bounded variation if $X(t) \in \Phi^{(n)}$. For the stationary process $X(t)$ the cumulants $S^{(K)}(t_1, \dots, t_K)$ are shift invariant

$$S^{(K)}(t_1 + \tau, \dots t_K + \tau) = S^{(K)}(t_1, \dots, t_K) \qquad \text{(I)}$$

and the spectral measures $F^{(K)}(d\lambda)$, $M^{(K)}(d\lambda)$ are concentrated on the manifold $\lambda_1 + \dots + \lambda_K = 0$. $X(t) \in S^{(K)}$

if (I) holds for all $K \leq n, t_1, \dots t_K, -\infty < \tau < +\infty$. The class $\Delta^{(K)}$ suggested by A.N.Kolmogorov and introduced in Shiryaev (1960) is defined as a subclass $\varphi^{(n)} \cap$ for which the measures $F^{(K)}(d\lambda), 1 \leq K \leq n$ are absolutely continuous with respect to the Lebesque measure in the planes $\lambda_1 + \dots + \lambda_K = 0$ and there is a cumulant spectral density of the K-th order defined by

$$S^{(K)}(t_1, \dots, t) = \int_{R^K} \exp\left(i \sum_{p=1}^{K} \lambda_p t_p\right) \delta(\lambda_1 + \dots + \lambda_K) f_K(\lambda_1, \dots, \lambda_K) \, d\lambda_1 \dots d\lambda_K$$

where $\delta(x)$ is the Dirac delta-function, the equalities holding for all $t_1 \dots t_K$.

Under wide mixing conditions of $X(t)$ the cumulant spectral density $f_n(\lambda_1, \dots, \lambda_n)$ is a continuous bounded function of its variables on the manifold $\lambda_1 + \dots + \lambda_n = 0$ (see Zhurbenko (1986)). The mixed derivative is bounded

$$\frac{\partial^s f_n(\lambda_1, \dots, \lambda_n)}{\partial \lambda_1^{s_1} \dots \partial \lambda_n^{s_n}} \leq C, \quad s_1 + \dots + s_n = s, \quad s_K = 0 \tag{2}$$

if there are the first pn moments $(p > 2)$ of $X(t)$, and the Rosenblatt mixing coefficient

$$\alpha(\tau) = \sup_{t,\, A \in \mathfrak{A}_{-\infty}^{t},\, B \in \mathfrak{A}_{t+\tau}^{\infty}} |P(AB) - P(A)P(B)|$$

($\mathfrak{A}_a^b$ is the σ-algebra generated by $X(t), t \in [a, b]$) satisfies the condition

$$\sum_{\tau=0}^{\infty} \tau^{n+s-1} \alpha^{\frac{p-2}{p}}(\tau) < \infty$$

The moment spectral density $g_n(\lambda_1, \dots, \lambda_n) = g(\bar\lambda)$ of the measure $M^{(n)}(d\lambda)$ which is formally defined from the relation

$$M^{(n)}(t_1, \dots, t_n) = \int g(\bar\lambda) \exp\{i(\bar t, \bar\lambda)\} \delta(|\bar\lambda|) \, d\bar\lambda$$

$$(\bar t = (t_1, \dots, t_n), \; \bar\lambda = (\lambda_1, \dots, \lambda_n), \; |\bar\lambda| = \lambda_1 + \dots + \lambda_n)$$

is a generalized function on the manifold $\lambda_1 + \dots + \lambda_n = 0$

and is connected with the cumulant spectral density $f_n(\lambda_1,\dots,\lambda_n) = f(\bar\lambda)$ by

$$f(\bar\lambda)\,\delta(|\bar\lambda|) = \sum (-1)^{K-1}(K-1)!\,g(\bar\lambda_1)\delta(|\bar\lambda_1|)\dots g(\bar\lambda_p)\delta(|\bar\lambda_p|),$$

$$g(\bar\lambda)\,\delta(|\bar\lambda|) = \sum f(\bar\lambda_1)\delta(|\bar\lambda_1|)\dots f(\bar\lambda_p)\delta(|\bar\lambda_p|) \qquad (3)$$

where the summation is done over all the unordered partitions of the set $\bar\lambda = (\lambda_1,\dots,\lambda_n)$ into unordered disjoint collections $\bar\lambda_K = (\lambda_{i_1},\dots,\lambda_{i_K})$, $K = 1,\dots,p$. The second-order cumulant spectral density that for $n=2$ coincides with the moment density $f_2(\lambda_1,\lambda_2) = g_2(\lambda_1,\lambda_2)$ on the line $\lambda_1 = -\lambda_2$ is called just spectral density of the second order. For discrete time in the above formulas R^n means a n-dimensional cube $-\pi \le \lambda_i \le \pi$, $i = 1,\dots,n$, the Dirac δ-function is to be replaced by the Dirac comb

$$\delta^*(x) = \sum_{K=-\infty}^{\infty} \delta(x + 2\pi K)$$

The main element of statistical estimates of various spectral characteristics of the stationary process $X(t)$, $t = \dots -1, 0, 1, \dots$ which are constructed from the sample $X(1),\dots,X(N)$ is the finite Fourier transform

$$d_N(\lambda) = \sum_{t=1}^{N} exp\{it\lambda\} X(t)$$

The construction of spectral estimates is based on the periodogram

$$I_N(\bar\lambda) = I_N(\lambda_1,\dots,\lambda_n) = (2\pi)^{-n+1} N^{-1} \prod_{j=1}^{n} d_N(\lambda_j), \quad \sum_{j=1}^{N} \lambda_j = 0 \,(mod\; 2\pi)$$

Under the conditions of smoothness of cumulant densities the periodogram $I_N(\bar\lambda)$ is an asymptotically non-biased estimate of the moment spectral density $g(\bar\lambda)$ at the points $\bar\lambda$ of the plane $|\bar\lambda| = 0\,(mod\; 2\pi)$ that do not belong to any of its subplanes $|\lambda_p| = 0\,(mod\; 2\pi)$. According to (3) for these and only these points $f(\bar\lambda) = g(\bar\lambda)$ and the function $f(\bar\lambda)$ exists as the usual function of a real variable, for which, for example, the inequality (2) holds under the mixing conditions. According to

(3) for the other points of the hyperplane $\bar{\lambda}=0 \pmod{2\pi}$ the function $g(\bar{\lambda})$ can be considered only as a generalized function and $E(I_N(\bar{\lambda}))_{N\to\infty} \to \infty$ (see Zhurbenko (1985)). The periodogram $I_N(\lambda)$ is not a consistent estimate of the spectral density $f(\bar{\lambda})$ even for $n=2$, to have consistent estimates two methods given below are usually applied.

The first of them dates back to Einstein (1914) and Bartlett (1948), it employs an asymptotical lack of correlation of the values of the periodogram on the lattice $\bar{X}_K = \frac{2\pi\bar{K}}{N}$, $\bar{K} \in Z^n \subset R^n$, Z being a set of integers. Averaging the values of the periodogram with the weight function $\varphi_N(x)$ (a spectral window) concentrated around the point 0 gives, through the central limit theorem, a consistent estimate of the spectral density $f(\bar{\lambda})$. Growth of concentration of $\varphi_N(x)$ due to the smoothness condition on $f(\bar{\lambda})$ reduces the bias but enlarges the variance of

$$\hat{f}_N(\lambda) = \sum_K \varphi_N(\bar{x}_K - \bar{\lambda}) I_N(\bar{x}_K) \tag{4}$$

Lessening concentration of $\varphi_N(x)$ at zero has an inverse effect on the bias and variance. In (4) summation is done over the points $\bar{x}_K$ in the neighbourhood of $\bar{\lambda}$, which belong to the hyperplane $|\bar{x}| = 0 \pmod{2\pi}$ and are absent in its subplanes of the form $|\bar{x}_p| = 0 \pmod{2\pi}$ where $\bar{x}_p$ is the projection of $\bar{x}$ onto a frame of reference of smaller dimension. For $n=2$ the optimal spectral window is obtained to minimize the mean square error

$$\inf_{\varphi_N(x)} \sup_{X(t)\in\mathcal{X}(\lambda,\alpha)} E(\hat{f}_N(\lambda) - f(\lambda))^2 \sim CN^{-\frac{2\alpha}{1+2\alpha}} \tag{5}$$

where $\mathcal{X}(\lambda,\alpha)$ is a class of stationary processes with spectral density $f(x)$ satisfying the Hölder condition with index α at point λ, the constant C can explicitly be written out (see Zhurbenko (1986)). The asymptotic equality (5) is attained on the function $\varphi_N(x) = \frac{2\pi}{N} A_N \varphi(A_N x)$ where $A_N = C(\alpha) N^{\frac{1}{1+2\alpha}}$, $C(\alpha)$ is defined by α,

$$\varphi(x) = \begin{cases} \frac{\alpha+1}{2\alpha}(1-|x|^{\alpha}), & |x| \leq 1 \\ 0 & |x| > 1 \end{cases}$$

According to the results of Grenander and Rosenblatt (1957) the suggested minimum (5) coincides with the minimum for all possible estimates $\hat{f}_N(\lambda)$ of arbitrary quadratic forms in N observations of the process $X(t)$, which in view of the dimension theory exhausts the class of all admissible second-order spectral estimates. Estimates of the form (4) are traditionally called spectrograph estimates.

The second method suggested by J.Tukey and A.N.Kolmogorov (see Zhurbenko (1985, 1986)) consists in averaging the values of the periodogram $I_M(\lambda)$ calculated with time shift in observation. The estimate becomes asymptotically consistent due to the mixing conditions of series of observations and by applying, on this account, the values of the shifted periodograms of the limit theorem to the sequence. In the general form the spectral estimate obtained with a time shift operator is defined as

$$\bar{f}_N(\lambda) = \frac{1}{T} \sum_{K=0}^{T-1} |W_M^{L_K}(\lambda)|^2 \tag{6}$$

where $|W_M^Q(\lambda)|^2$ is a smoothed periodogram

$$W_M^Q(\lambda) = \sum_{t=-\infty}^{\infty} a_M(t-Q) e^{it\lambda} X(t),$$

the integer-valued L, M, T, N are connected by the relations: $L \ll M \ll N$, $LT \sim N$, $N = L(T-1)+M+1$ the non-negative symmetric data window $a_M(t)$, $t = \ldots = -1, 0, 1, \ldots$ equals zero outside $[0, M]$ and satisfies the standardization $2\pi \sum a_M^2(t) = 1$. Note that the estimate of the form (6) is not, generally speaking, a spectrograph estimate.

As far as the asymptotics of the mean square error is concerned the class of spectrograph estimates are optimal within the class of all quadratic forms in observations of the process. At the same time mean square error is not the only requirement to spectral estimates in practice. In many applications where estimates of spectra of different nature are employed one often has to deal with deeply indented spectra, a considerable trend, "floating" of frequencies, strong non-stationary noises, etc. The frequent result of using spectrograph estimates in these situations is an undesirable distortion in estimation of some interesting peculiarities in a spectrum, a wrong estimate of the rate of the decay of the spectrum relative to the frequency axis, and so on. These undesirable effects in real spectrum estimation are considerably less due to spectral estimates with weak dependence on remote frequencies. With spectrograph estimates such dependence falls off at the rate no greater than CN^{-2} and the "memory" of a strong jitter remains over rather a long stretch of frequencies. The influence of remote frequencies is considerably reduced by smoothing a sample sequence at the ends (for detail see Hannan (1970), Anderson (1971), Brillinger (1975)). By this procedure the influence of remote frequencies becomes considerably less but for a given sample length the variance as well as mean square error becomes larger.

One can benefit much by a strong tapering of observations on time segments at their ends, then by averaging the modified periodograms calculated for different time intervals of a series of observations.

Application of a polynomial smoothing and then a time-shift operator produces estimates whose dependence on remote frequencies is of order $exp(-bN^{\varepsilon}), b>0, \varepsilon>0$ and mean square error is nearly optimal (see Zhurbenko

(1986)).

Such properties as a particularly weak dependence on remote frequencies and shift in time being combined in this estimate give a possibility to check a series for stationarity in a given frequency-band. The estimate is very good for a computer, it is economical of the total number of operations and takes up little room in the computer storage (Zhurbenko (1986), Brillinger (1975), Otnes and Enochson (1978), Priestley (1981), Handbook of Statistics (1983)). It can be used in spectral estimation with real time, it analyses all changes in the spectrum making it possible to study in detail real processes arising in various applications.

REFERENCES

Anderson, T. (1971). The statistical analysis of time series. J.Wiley, N.-Y., 704 p.

Bartlett, M.S. (1948). Smoothing periodograms from time series with continuous spectra. Nature, 161, 686-687.

Brillinger, D.R. (1975) Time series. Data analysis and theory. J.Wiley, N.-Y.

Cooley,J.M. and Tukey, J.W. (1965). An algorithm for the machine calculation of complex Fourier series. Mathematics of Computation, 19, 297-301.

Daniel, F.J. (1946). Discussion of paper of M.S.Bartlett. Journ. of Royal Stat. Society Suppl., 8, 27.

Einstein, A. (1914). Méthode pour la détermination des valeures statistiques d'observation concentrant des soumises a des fluctuations irrégulieres. Arch. Sci. Phys. et Nature, 37, 254-256.

Grenander, U. and Rosenblatt, M. (1957). Statistical analysis of stationary time series. J.Wiley, N.-Y.

Handbook of Statistics 3. (1983). Brillinger, D.R. and Krishnaiah P.R. (Ed.). North-Holland, Amsterdam.

Hannan, E.J. (1970). Multiple time series. J.Wiley, N.-Y.

Kolmogorov,A.N. (1985) Collected Works. Mathematics and Mechanics. Nauka, Moscow (in Russian).

Otnes, R.K. and Enochson, L. (1978). Applied time series analysis. J.Wiley, N.-Y.

Priestley, M.B. (1981). Spectral analysis and time series, I, II. Academic Press, N.-Y.

Schuster, A. (1898). On the investigation of hidden periodicities with application to a supposed 26 period of meteorological phenomena. Terr. Magn., 3, 13-41.

Shiryaev, A.N. (1960) Some problems in the spectral theory of higher order moments, I. Theor. Probab. Appl.,

5, 265-286.
Yaglom, A.M. (1962). Introduction to the theory of stationary random functions. Prentice Hall, Englewood Cliff.
Zhurbenko, I.G. (1985). Statistical estimation of higher spectra. Teor. veroyatn. i primen., v. 30, № 1, 66-77 (in Russian).
Zhurbenko, I.G. (1986). The spectral analysis of time series. North Holland, Amsterdam.

ON MARGINAL DISTRIBUTIONS OF THRESHOLD MODELS

Jiří Anděl
Charles University, Sokolovská 83, 186 00 Prague 8, Czechoslovakia
Alexandr Fuchs
Institute of Hygiene and Epidemiology, Šrobárova 48, 100 42 Prague 10, Czechoslovakia

1. INTRODUCTION

Let $\{Y_t\}$ be a sequence of uncorrelated random variables with vanishing mean and the same finite variance σ^2. Such a sequence is called a white noise or an innovation process. The autoregressive (AR) process $\{X_t\}$ is the linear process satisfying

$$X_t = a_1 X_{t-1} + \ldots + a_n X_{t-n} + Y_t \quad .$$

The moving average (MA) process $\{X_t\}$ is defined by

$$X_t = b_o Y_t + \ldots + b_m Y_{t-m} \quad .$$

In the last years some non-linear generalizations of these models have been investigated. We mention here briefly only so called threshold models. A more detailed review of non-linear models is given by Anděl et al. (1984).

Let $r_1, \ldots, r_h$ be given real numbers (called thresholds) such that $r_1 < r_2 < \ldots < r_h$ $(h \geq 1)$. Define $r_o = -\infty$, $r_{h+1} = \infty$, $B_o = (r_o, r_1)$, $B_j = [r_j, r_{j+1})$ for $j=1,\ldots,h$. Let a_{ij} be given autoregressive coefficients and let

$$X_t = \sum_{i=1}^{n_i} a_{ij} X_{t-i} + Y_t \quad \text{if} \quad X_{t-1} \in B_j \quad .$$

Then $\{X_t\}$ is called a threshold AR process. Further information about threshold models can be found in Tong (1983).

Wecker (1981) defined an asymmetric MA model by

$$X_t = Y_t + b_1 Y_{t-1}^+ + \ldots + b_m Y_{t-m}^+ + c_1 Y_{t-1}^- + \ldots + c_m Y_{t-m}^- ,$$

where $Y_k^+ = \max(0, Y_k)$, $Y_k^- = \min(0, Y_k)$.

Fuchs (1986) generalized this model in the following way. Let χ_{B_i} be the characteristic function of the set B_i and let b_{ik} be some constants. Define

$$\beta_k(y) = \sum_{i=0}^{h} b_{ik} \chi_{B_i}(y) .$$

Then

$$X_t = Y_t + \sum_{k=1}^{n} \beta_k(Y_{t-k}) Y_{t-k}$$

is called the threshold asymmetric MA process and

$$X_t = Y_t + \sum_{k=1}^{n} \beta_k(Y_{t-d}) Y_{t-k} \qquad \text{(d fixed)}$$

is called the lagged asymmetric MA process. Fuchs (1986) analyzed multidimensional generalizations of these models.

2. MARGINAL DISTRIBUTIONS OF THRESHOLD AR MODELS

We shall consider the following special model. Let

$$X_t = \begin{cases} \alpha X_{t-1} + Y_t & \text{for } X_{t-1} < 0 , \\ -\alpha X_{t-1} + Y_t & \text{for } X_{t-1} \geq 0 , \end{cases} \tag{2.1}$$

where $\alpha \in (0,1)$ is a given parameter.

Theorem 2.1. Let $\{Y_t\}$ be i.i.d. N(0,1) variables. Then there exists a unique stationary density $f(x)$ of the process $\{X_t\}$ defined by (2.1), and this stationary density is given by the formula

$$f(x) = \{ 2(1-\alpha^2)/\pi \}^{1/2} \exp\{ -\tfrac{1}{2}(1-\alpha^2)x^2 \} \Phi(-\alpha x) . \tag{2.2}$$

Proof. See Anděl et al. (1984) .

It can be calculated that the mean μ_0 and the variance σ_0^2 corresponding to the density $f(x)$ are

$$\mu_0 = -(2/\pi)^{1/2}\alpha(1-\alpha^2)^{-1/2},$$

$$\sigma_0^2 = (1-\alpha^2)^{-1}(1-2\alpha^2/\pi),$$

respectively.

The second case concerns the Cauchy innovations. Denote $C(a,b)$ the Cauchy distribution with the density

$$\pi^{-1} b\{b^2 + (x-a)^2\}^{-1}, \qquad b > 0 .$$

Let X_0 be a random variable with a density $h(x)$. Let $Y_1, Y_2, \ldots$ be i.i.d. $C(0,1)$ variables independent of X_0. Assume that X_t for $t \geq 1$ are generated by (2.1). If X_t $(t \geq 1)$ have also density $h(x)$, we say, that $h(x)$ is the stationary density of the threshold AR process (2.1) with Cauchy innovations.

Theorem 2.2. There exists a unique stationary density $h(x)$ of the process (2.1) with Cauchy innovations, and it is given by

$$\begin{aligned} h(x) = 2\pi^{-2}A[&-\{4A^2x^2 + (1-A^2+x^2)^2\}^{-1}x \ln\{A^{-2}(1+x^2)\} - \\ &- \{4A^2x^2 + (1-A^2+x^2)^2\}^{-1}(A^2-1+x^2) \operatorname{arctg} x + \\ &+ (2A)^{-1}\{(1+A)^2+x^2\}^{-1}(1+A)\pi], \end{aligned} \tag{2.3}$$

where $A = (1-\alpha)^{-1}\alpha$.

Proof. See Anděl and Bartoň (1986).

3. MARGINAL DISTRIBUTION OF THRESHOLD MA PROCESSES

In this part of the paper we shall deal with the model

$$X_t = Y_t + \beta_i Y_{t-1} \qquad \text{for} \quad Y_{t-1} \in B_i . \tag{3.1}$$

Let $\{Y_t\}$ be i.i.d. variables with a density $g(y)$. Let $p_i(x|y)$ be a conditional density of X_t, given $Y_{t-1} = y \in B_i$. Then the marginal density of X_t is

$$q(x) = \sum_{i=0}^{h} \int_{r_i}^{r_{i+1}} p_i(x|y)g(y)\, dy . \qquad (3.2)$$

Theorem 3.1. Let $\{Y_t\}$ be i.i.d. $N(0, \sigma^2)$ variables. Then the marginal density of X_t is

$$q(x) = \sum_{i=0}^{h} (2\pi)^{-1/2} w_{i+1}^{-1} \exp\left(-\frac{1}{2} x^2/w_{i+1}^2\right) \times \qquad (3.3)$$

$$\times \{\Phi(\sigma^{-2} r_{i+1} w_{i+1} - w_{i+1}^{-1} \beta_{i+1} x) - \Phi(\sigma^{-2} r_i w_{i+1} - w_{i+1}^{-1} \beta_{i+1} x)\},$$

where $w_i = \sigma(1+\beta_i^2)^{1/2}$ and Φ is the distribution function of $N(0,1)$.

Proof. in this case $p_i(x|y)$ is the density of $N(\beta_i y, \sigma^2)$ and $g(y)$ is the density of $N(0, \sigma^2)$. Inserting into (3.2) one gets after some computation the formula (3.3).

If we consider the model (3.1) with only one threshold $r_1 = 0$ (i.e. $h=1$) and with $\beta_1 = -\beta_2 = \beta$, then (3.3) can be simplified to

$$q(x) = 2^{1/2}\{\pi\sigma^2(1+\beta^2)\}^{-1/2} \exp[-x^2/\{2\sigma^2(1+\beta^2)\}] \times$$
$$\times \Phi[-\{\sigma^2(1+\beta^2)\}^{-1/2}\beta x] .$$

It is interesting to notice that $q(x)$ is the same as $f(x)$ in (2.2), if $\sigma^2 = 1$ and $\alpha = (1+\beta^2)^{-1/2}\beta$.

A similar result can be derived also for the model (3.1) with Cauchy innovations. To simplify the formulas, we introduce only the result for the model with one threshold

Theorem 3.2. Let $\{Y_t\}$ be i.i.d. $C(0,b)$ variables. If X_t is given by (3.1), where $h = 1$, $r_1 = 0$, $\beta_1 = -\beta_2 = \beta$, then the marginal density of X_t is

$$q(x) = 2b^{-1}\pi^{-2}\beta\Big[-[4\beta^2(x/b)^2+\{1-\beta^2+(x/b)^2\}^2]^{-1}(x/b)\times$$
$$\times \ln \beta^{-2}\{1+(x/b)^2\} -$$
$$-[4\beta^2(x/b)^2+\{1-\beta^2+(x/b)^2\}^2]^{-1}\{\beta^2-1+(x/b)^2\} \operatorname{arctg}(x/b) +$$
$$+[2\beta\{(1+\beta)^2+(x/b)^2\}]^{-1}(1+\beta)\pi\Big] . \qquad (3.4)$$

Proof. It is clear that $p_i(x|y)$ is the density of $C(\beta y,b)$ for $y<0$ and of $C(-\beta y,b)$ for $y \geq 0$. The function $g(y)$ is the density of $C(0,b)$. Inserting into (3.2) we obtain after tedious calculations the formula (3.4).

If $b = 1$ and $\beta = A$, i.e. for $\beta = (1-\alpha)^{-1}\alpha$, the density $q(x)$ from (3.4) is the same as the density $h(x)$ from (2.3). This is a direct proof that the function $h(x)$ is positive. Anděl and Bartoň (1986) proved by a different way only that $h(x)$ is nonnegative.

It seems to be surprising that even in the special cases the marginal distribution of threshold MA processes have the same form as those of threshold AR processes.

REFERENCES

Anděl J., Netuka I., Zvára K. (1984). On Threshold Autoregressive Processes. Kybernetika 20, 89-106.
Anděl J., Bartoň T. (1986). On a Threshold AR(1) Model with Cauchy Innovations. J.Time Series Anal. 7, 1-5.
Fuchs A. (1986). Threshold Moving Average Model. (Submitted).
Tong H. (1983). Threshold Models in Non-Linear Time Series Analysis. Lecture Notes in Statistics, Vol. 21, New York.
Wecker W.E. (1981). Asymmetric Time Series. (1981). J. Amer. Statist. Assoc. 76, 16-21.

Bernoulli, Vol. 2, pp. 385-388

ON THE BOUNDARY OF THE CENTRAL LIMIT THEOREM FOR STATIONARY ρ-MIXING SEQUENCES

Richard C. Bradley
Mathematics Dept., Indiana University, Bloomington, Indiana, USA

For strictly stationary random sequences satisfying the strong mixing condition, it is almost completely known what combinations of "moment assumption" and "mixing rate" will imply the central limit theorem. The "borderline" for the CLT under strong mixing was specified quite sharply by the CLT's of Ibragimov (1962)(see also [6]) together with the counterexamples (satisfying barely weaker conditions) of Davydov (1973). More recently, under the ρ-mixing condition the "borderline" for the CLT has also been specified quite sharply by the results of Ibragimov (1975), Peligrad (1987), and the author (1987a,b). The purpose of this note is to summarize this recent progress under ρ-mixing.

Notations: "Log" means natural logarithm, and $\log^+ x := \max\{0, \log x\}$. Also, "<<" means "$O(\cdot)$". If $(X_k, k = \ldots, -1,0,1,\ldots)$ is a strictly stationary random sequence, then for each $n = 1,2,\ldots$, $S_n := X_1 + \ldots + X_n$, $\sigma_n^2 := \text{Var } S_n$ (if it exists), and $\rho(n) := \{\sup|\text{Corr}(f,g)| : f \in L^2(X_k, k \le 0), g \in L^2(X_k, k \ge n)\}$. Here $L^2(\ldots)$ denotes the set of real-valued square-integrable functions which are measurable with respect to the σ-field generated by $(\ldots)$. The sequence (X_k) is "ρ-mixing" [8] if $\rho(n) \to 0$ as $n \to \infty$.

Our usual assumptions are as follows:

(1) The random sequence (X_k) is strictly stationary, $EX_0 = 0$, $EX_0^2 < \infty$, $\sigma_n^2 \to \infty$ as $n \to \infty$, and (X_k) is ρ-mixing.

CENTRAL LIMIT THEOREMS

A. (Ibragimov (1975)) If (X_k) satisfies (1) and $E|X_0|^{2+\delta} < \infty$ for some $\delta > 0$, then $S_n/\sigma_n \to N(0,1)$ in distribution as $n \to \infty$.

B. (Ibragimov (1975)) If (X_k) satisfies (1) and $\Sigma_{n=1}^{\infty} \rho(2^n) < \infty$, then $S_n/\sigma_n \to N(0,1)$.

C. (Peligrad (1987)) Suppose (X_k) satisfies (1), and for some $0 < \varepsilon < 1$, some $c > 0$, one has that $EX_0^2(\log^+|X_0|)^{2c/(1-\varepsilon)} < \infty$ and $\rho(n) \le c(\log n)^{-1}$ for all n sufficiently large. Then $S_n/\sigma_n \to N(0,1)$.

D. (Peligrad (1987)) Suppose (X_k) satisfies (1), and for some $\alpha > 0$ and $\beta > 0$ such that $\alpha + \beta > 1$ one has that $EX_0^2\exp((\log^+|X_0|)^\alpha) < \infty$ and $\rho(n) \ll (\log n)^{-\beta}$ as $n \to \infty$. Then $S_n/\sigma_n \to \mathbf{N}(0,1)$.

E. (Peligrad (1987)) (Essentially the general case): Suppose (X_k) satisfies (1) and for some $d > 3$, some non-decreasing function $g : [0,\infty) \to [0,\infty)$ one has that $g(n^{1/2}) \gg \exp(d\cdot\Sigma_{i=1}^{[\log n]}\rho(2^i))$ and $EX_0^2 g(|X_0|) < \infty$. Then $S_n/\sigma_n \to N(0,1)$.

F. (Bradley (1987b), Khinchin-Levy-Feller for i.i.d. case): Suppose (X_k) is strictly stationary and non-degenerate, $H(C) := EX_0^2 I(|X_0| \le C)$ satisfies $\mathrm{Lim}_{C\to\infty}H(2C)/H(C) = 1$, $EX_0 = 0$, $\rho(1) < 1$, and $\Sigma_{n=1}^\infty\rho(2^n) < \infty$. Then $S_n/a_n \to N(0,1)$ for some constants $a_n \to \infty$.

These results are all essentially sharp. Their assumptions are barely stronger than the properties satisfied by counterexamples discussed below.

All of these CLT's have been extended to the weak invariance principle with little or no strengthening of the assumptions; see Ibragimov (1975), Peligrad (1982), and Shao (1986a,b).

The "natural" mixing rate $\Sigma_{n=1}^\infty\rho(2^n) < \infty$ (see B. and F. above) was introduced in Ibragimov and Rozanov (1970), where it was shown (for second-order stationary processes) that it implies a continuous spectral density.

COUNTEREXAMPLES

Our counterexamples (X_k) will have the following property:

(2) There exists a sequence $n(1) < n(2) < n(3) < \ldots$ of positive integers such that as $k \to \infty$, $S_{n(k)}/\sigma_{n(k)}$ converges in distribution to the law whose distribution function is

$$F(x) := e^{-1}I_{[0,\infty)}(x) + \sum_{J=1}^{\infty}(1/J!)e^{-1}(2\pi J)^{-1/2}\int_{-\infty}^{x} e^{-u^2/(2J)}\,du .$$

The following counterexamples are taken from Bradley (1987a).

a. If $q : [0,\infty) \to [0,\infty)$ is a function satisfying mild regularity assumptions (specified in [1]), such that $\forall\delta > 0$, $q(x) = o(x^{2+\delta})$ as $x \to \infty$, then there exists a random sequence (X_k) satisfying (1) and (2) such that $Eq(|X_0|) < \infty$.

b. If $(\tau(n), n = 1,2,\ldots)$ is a sequence of positive numbers such that $\tau(1) \geq \tau(2) \geq \tau(3) \geq \ldots \to 0$, $\Sigma_{n=1}^{\infty} \tau(2^n) = \infty$, and $\lim_{n\to\infty} \tau(2n)/\tau(n) = 1$, then there exists (X_k) satisfying (1) and (2) such that $\forall n \geq 1$, $\rho(n) \leq \tau(n)$.

c. Given $\alpha > 0$, there exists (X_k) satisfying (1) and (2) such that $EX_0^2(\log^+|X_0|)^{\alpha} < \infty$ and $\rho(n) << (\log n)^{-1}$ as $n \to \infty$.

d. Given $\alpha > 0$, $\beta > 0$ such that $\alpha + \beta \leq 1$, there exists (X_k) satisfying (1) and (2) such that

$$EX_0^2 \exp((\log^+|X_0|)^{\alpha}) < \infty \text{ and } \rho(n) << (\log n)^{-\beta} \text{ as } n \to \infty.$$

e. (Essentially the general case): Suppose that $q : [0,\infty) \to [0,\infty)$ and $(\tau(n), n = 1,2,\ldots)$ are as in a. and b. above. Suppose there exists $d > 0$ such that $q([n\cdot\exp[-d\ \Sigma_{k=1}^{n} k^{-1}\tau(k)]]^{1/2}) = o(n)$ as $n \to \infty$. Then there exists (X_k) satisfying (1) and (2) such that $Eq(|X_0|) < \infty$ and $\rho(n) << \tau(n)$.

Peligrad (1987) noted that the properties in e. are only slightly weaker than the assumptions in E.

REFERENCES

1. Bradley, R. C. (1987a). The central limit question under ρ-mixing. Rocky Mountain J. Math. (To appear).
2. Bradley, R. C. (1987b). A central limit theorem for stationary ρ-mixing sequences with infinite variance. Ann. Probab. (To appear).
3. Davydov, Yu. A. (1973). Mixing conditions for Markov chains. Theory Probab. Appl. 18, 312-328.
4. Ibragimov, I. A. (1962). Some limit theorems for stationary processes. Theory Probab. Appl. 7, 349-382.
5. Ibragimov, I. A. (1975). A note on the central limit theorem for dependent random variables. Theory Probab. Appl. 20, 135-141.
6. Ibragimov, I. A., and Linnik, Yu. V. (1971). Independent and Stationary Sequences of Random Variables. Wolters-Noordhoff, Groningen.

7. Ibragimov, I. A., and Rozanov, Yu. A. (1970). Gaussian Random Processes. Izd-vo Nauka, Moscow. (In Russian.) English Translation: Springer-Verlag, Berlin, 1978.

8. Kolmogorov, A. N., and Rozanov, Yu. A. (1960). On strong mixing conditions for stationary Gaussian processes. Theory Probab. Appl. 5, 204-208.

9. Peligrad, M. (1982). Invariance principles for mixing sequences of random variables. Ann. Probab. 10, 968-981.

10. Peligrad, M. (1987). On the central limit theorem for ρ-mixing sequences of random variables. Ann. Probab. (To appear).

11. Shao, Q. (1986a). On the invariance principle for ρ-mixing sequences of random variables. Preprint, Hangzhou University.

12. Shao, Q. (1986b). An invariance principle for stationary ρ-mixing sequences with infinite variance. Preprint, Hangzhou University.

This work was partially supported by NSF grant DMS 86-00399.

Bernoulli, Vol. 2, pp. 389-392

DETECTION OF PARAMETER CHANGES AT UNKNOWN TIMES IN LINEAR REGRESSION MODELS

Jandhyala, V. K. and MacNeill, I. B
University of Western Ontario, London, Canada

1. INTRODUCTION

Chernoff and Zacks (1964) introduced a Bayesian-type approch to deriving test statistics for the one-sided detection of parameter changes at unknown times in the mean of a sequence of independent normal variables. Subsequently, this approach was adapted by Gardner (1969) for the two-sided detection of parameter changes. This paper presents the derivation of a "Bayesian-type" statistic for the detection of two-sided parameter changes at unknown times in general linear regression models. Distributional results are derived when the underlying regression model is harmonic.

2. DETECTING TWO-SIDED CHANGES IN REGRESSION PARAMETERS

In this section, derivations of two-sided change detection statistics for regression models are discussed. Consider the standard linear regression model $\underset{\sim}{Y} = X\underset{\sim}{\beta} + \underset{\sim}{\varepsilon}$, where $\underset{\sim}{\varepsilon}' = (\varepsilon_1, \ldots, \varepsilon_n) \sim N(0, \sigma^2 I)$, $\underset{\sim}{\beta}' = (\beta_0, \ldots, \beta_{p-1})$ is the vector of regression parameters, $\underset{\sim}{Y}' = (Y_1, Y_2, \ldots, Y_n)$ is the vector of observations of the dependent variable and X is the design matrix. The problem of interest is to detect two-sided changes in $\underset{\sim}{\beta}$ where the points of change are unknown. The result stated here is limited to the case of detecting at most one change (AMOC) with a uniform prior on the change point and in which one is interested in detecting two-sided changes in the parameter β_i at the unknown change point. More general results are derived in Jandhyala and MacNeill (1986).

If one lets $\underset{\sim}{\delta}' = (\delta_0, \delta_1, \ldots, \delta_{p-1})$ be the amount of change in the parameter $\underset{\sim}{\beta}$ at the unknown change point, then the hypotheses to be tested are: H_0: $\underset{\sim}{\delta} = 0$, versus H_a: $\underset{\sim}{\delta} \neq 0$. The two-sided change detection statistic is stated in the following theorem.

Theorem 1 The "Bayesian-type" two-sided change detection statistic for testing the null hypothesis against the alternative with a uniform prior on the unknown change point is given by:

$$U_{ni} = \frac{1}{(n-1)\hat{\sigma}^2} \sum_{k=1}^{n-1} \left\{ \underset{\sim}{R}' \underset{\sim}{X}_{(k)i} \right\}^2 \tag{1}$$

where $\underset{\sim}{R}' = \underset{\sim}{Y}'\{I-X(X'X)^{-1}X'\}$ is the vector of residuals, $\underset{\sim}{X}_{(k)i}$ is the ith column vector of X with the first k elements replaced by zeros and $\hat{\sigma}$ is a suitable consistent estimator of σ^2.

Let $\{B_{f_i}(t),\ t \in [0, 1]\}$, $i=0, 1, 2, \ldots, p-1$ be the limit process for sequences of stochastic processes defined by partial sums of linear functions of regression residuals where the linear regressor functions are denoted by $f_i(\cdot)$. Then the asymptotic form of U_{ni} given by (1) is as follows:

Theorem 2 If one is testing for two-sided change in β_i alone, then, $n^{-1}U_{ni}$ converges in distribution to

$$\int_0^1 B_{f_i}^2(t)\ dt. \tag{2}$$

3. DISTRIBUTIONS OF STOCHASTIC INTEGRALS.

The characteristic function of the stochastic integral in (2) is

$$\Phi(s) = \prod_{n=1}^{\infty} (1-2is\lambda_n)^{-1/2} \tag{3}$$

where λ_n is the n th eigenvalue of the Karhunen expansion. These eigenvalues satisfy the Fredholm integral equation,

$$\int_0^1 K_f(s, t)\Phi(s)ds = \lambda\Phi(t), \tag{4}$$

where $K_f(s, t)$ is the covariance kernel of the limit process, $\{B_f(t),\ t \in [0, 1]\}$. In this connection MacNeill (1978a) obtained a solution of the Fredholm-integral equation for the case of detecting a two-sided change in the intercept parameter of a general pth order polynomial regression model. In this section a general method of solving Fredholm-integral equations is outlined. This method is then adapted to solve the Fredholm-integral equation for the case of detecting two-sided changes in the intercept parameter of a harmonic regression model.

According to the method we propose, the Fredholm integral equation is differentiated twice to transform it into a differential equation with appropriate boundary conditions. This differential equation is then solved assuming a series solution. The solution is obtained using an argument that the determinant of a certain matrix is zero. Jandhyala and MacNeill (1986) give details and examples of this methodology.

Consider the case of fitting an harmonic polynomial of degree p to a set of data. That is,

$$Y_{nj} = \beta_0 + \sum_{i=1}^{p} \{\beta_i \cos 2\pi i(j/n) + \beta_{p+i} \sin 2\pi i(j/n)\} + \varepsilon_i. \tag{5}$$

The asymptotic form of the two-sided change detection statistic for detecting at most one change in the intercept β_0 with uniform prior on the change point is that of $\int \beta_p^2(t)\, dt$ where $\{B_p(t),\ t \in [0, 1]\}$ is the limit process for the sequences of residual partial sums of (5). The distribution of the stochastic integral may be obtained by solving the Fredholm integral equation (4) where $K_f(s, t) = K_p(s, t)$, the covaraiance kernel of $B_p(t)$. MacNeill (1978b) showed this covariance kernel to be

$$K_p(s, t) = min(s, t) - st - \sum_{j=1}^{p} (2\pi^2 j^2)^{-1} \{(1-\cos 2\pi jt)(1-\cos 2\pi js) + \sin 2\pi js \cdot \sin 2\pi jt\}. \tag{6}$$

The sequence of eigenvalues $\{\lambda_{pn}\}_{n=1}^{\infty}$ satisfying the Fredholm equation are then identified by the following theorem which may be proved by applying the previously outlined methodology.

Theorem 3 The sequence of eigenvalues $\{\lambda_{pn}\}_{n=1}^{\infty}$ satisfying the Fredholm equation (4) corresponding to (6) are

$$\lambda_{pn} = \frac{1}{4\pi^2 n^2}, \quad n=p+1, p+2, \ldots$$

and those satisfying the equation

$$\tan\left(\frac{1}{2\sqrt{\lambda_{pn}}}\right) = \frac{1}{2\sqrt{\lambda_{pn}}} \left\{ \left(\sum_{j=0}^{p} \frac{1}{1-4\pi^2 j^2 \lambda_{pn}} \right)^{-1} \right\}, \quad n=1, 2, \ldots.$$

This sequence of eigenvalues then determines the characteristic function $\Phi_p(s)$ given by (3). Selected quantiles of the distribution of $\int \beta_p^2(t)\, dt$ appearing in Table 1 were obtained numerically by inverting the characteristic function; see MacNeill (1978a) for details.

Table 1. Selected Quantiles for $\Omega_p(a)=P[\int B_p^2(t)dt\leq a]$

Proba-bilities	P 0	1	2	3	4	5	6
.90	.34730	.12837	.07680	.05450	.04215	.03433	.02893
.95	.46136	.16919	.10096	.07158	.05534	.04507	.03799
.99	.74345	.26990	.16049	.11364	.08782.	07150	.06272
Mean	.16667	.06534	.04002	.02876	.02242	.01837	.01556
Variance	.02222	.00138	.00048	.00024	.00014	.00009	.00006

The mean of $\int \beta_p^2(t)dt$ is $\mu_p = \sum_{j=p+1}^{\infty} (\pi j)^{-2}$.

REFERENCES

Chernoff, H. and Zacks, S. (1964). Estimating the current mean of a normal distribution which is subject to changes in time. Ann. Math. Statist. 35, 999-1018.

Gardner, L. A. (1969). On detecting changes in the mean of normal variates. Ann. Math. Statist. 40, 116-126.

Jandhyala, V. K. and MacNeill, I. B. (1986). Detection of parameter changes at unknown times in linear regression models. Technical Report No.: TR-86-23, U. W. O.

MacNeill, I. B. (1978a). Properties of sequences of partial sums of polynomial regression residuals with applications to tests for change in regression at unknown times. Ann. Statist. 6, 422-433.

MacNeill, I. B. (1978b). Limit processes for sequences of partial sums of regression residuals. Ann. Prob. 6, 695-698.

Bernoulli, Vol. 2, pp. 393-396

SOME ASPECTS OF DIRECTIONALITY IN TIME SERIES ANALYSIS

A.J. Lawrance
Department of Statistics, University of Birmingham, Birmingham B15 2TT, England

1. BACKGROUND TO DIRECTIONALITY

This paper gives a short informal account of directionality in the time series area. A time series $\{X_t\}$ is said to be *directional* when, for all t (=0,±1,±2,...) and for each r (=0,1,2,...) the joint distribution of $(X_t, X_{t+1}, \ldots, X_{t+r})$ does *not* equal the joint distribution of $(X_{t+r}, X_{t+r-1}, \ldots, X_t)$. When the two joint distributions are equal the time series $\{X_t\}$ is said to be *reversible*. Directionality is an aspect of dependence; it is usually only considered for stationary series. There is no directionality in a random series. Some directional features in actual time series plots will be easy to identify by eye; for instance, when a series is more prone to descending behaviour than to ascending behaviour; this is typical of daily riverflow series.

Directionality is not a widely discussed apsect of time series. Over the last twenty years there have been many advances in the linear modelling of time series, and particularly in Gaussian linear processes with autoregresive and moving average structures (arma processes); the Gaussian assumption then implies reversibility. Moreover, Weiss (1975) proved for arma proceses, but with some moving-average exceptions, that the only reversible processes are Gaussian. Thus, the need to consider directionality did not arise; directional analysis of data series was also ignored. Time series analysis is now well into the non

Gaussian and non-linear era; the classes of models being developed, such as nonlinear autoregresive, threshold, random coefficient and bilinear are all directional. The understanding and analysis of directionality thus assumes new importance; it may be useful in selecting an appropriate class of model, and in particular one which can produce simulated sequences in sympathy with observed sequences.

The rest of the paper will briefly touch on a selection of topics; a fuller account is being planned.

2. DIRECTIONALITY AND STATIONARITY

For the stationary time series $\{X_t\}$ suppose that $f_{X_t,X_{t+1}}(x,y)$ is the joint probability density function of X_t and X_{t+1}; if the process is reversible, then

$$f_{X_t,X_{t+1}}(x,y) = f_{X_{t+1},X_t}(x,y). \tag{2.1}$$

Integrating out the variable y, gives

$$f_{X_t}(x) = f_{X_{t+1}}(x). \tag{2.2}$$

Thus, the minimum or first-order form of reversibility implies that the time series is stationary as far as its marginal distribution of X_t is concerned. This result could be extended to show that reversibility implies stationarity. It is, however, obvious that stationarity does not imply reversibility.

First-order Markov time series models, linear or not, are reversible if the joint distribution of (X_t,X_{t+1}) is reversible; McKenzie (1985) has proved that a similar result holds for first-order moving average models.

3. REVERSED VERSIONS OF MODELS

In the study of directional properties of time series models it is sometimes useful to know the model which results from reversing the direction of time; for instance, maximised likelihoods of two parametrized models could then be compared to infer the direction of time. Consider reversing time in the first-order linear AR(1) model in which

$$X_t = \rho X_{t-1} + \epsilon_t \qquad |\rho| < 1 \qquad (3.1)$$

$$= \epsilon_t + \rho\epsilon_{t-1} + \rho^2\epsilon_{t-2} + \cdots \qquad (3.2)$$

This series form, only possible because $|\rho|<1$, shows that the model has been based on past innovations; if the reversed model is required, it should be based on future innovations. Thus, the reversed AR(1) has series representation

$$X_t^R = \epsilon_t' + \rho\epsilon_{t+1}' + \rho^2\epsilon_{t+2}' + \cdots \qquad (3.3)$$

which yields

$$X_t^R = \rho X_{t+1}^R + \epsilon_t'. \qquad (3.4)$$

Usually, interest is in generating models based on past behaviour, so (3.4) should be written

$$X_{t+1}^R = \rho^{-1}(X_t^R - \epsilon_t'). \qquad (3.5)$$

This form has been studied, Rosenblatt (1986); it is Markovian but in general has a nonlinear structure, ϵ_t' being nonlinearly dependent on X_t^R. Using the backward shift operator B, there is the alternative representation

$$(B-\rho)X_{t+1}^R = \epsilon_t'. \qquad (3.6)$$

Notice here that the root of the autoregresive polynomial, ρ, is inside the unit circle. The usual condition of roots outside the unit circle can thus be interpreted as a directionality condition; it has also been called the 'minimum phase' condition. Standard invertibility conditions can also be given directional interpretations.

4. STATISTICAL ANALYSIS OF DIRECTIONALITY

This topic is not at all well developed. One problem is to know what general aspects of directionality to emphasize in a preliminary analysis. For tractability, some use of the pairwise distributions of (X_t, X_{t+r}) seems necessary. Estimation of $pr(X_t > X_{t+r})$, which is ½ for reversible processes, could be considered; other suggestions include using the skewness of $X_t - X_{t+r}$ for $r=1,2,\ldots$ and examining the cross-correlation function $Corr(X_t^2, X_{t\pm r})$ for symmetry.

Some other, as yet more speculative possibilities, are model based. Suppose the dependency in $\{X_t\}$ is non-linear autoregressive; then any linear aspect could be removed by least squares leaving a residual sequence for study of its nonlinear directionality. Residuals and the corresponding reversed residuals can also offer new views of partial correlations and introduce their directional versions; reversed residuals arise from models in which the direction of time has been reversed.

More theoretical insight might be obtained by comparison of a directional model with a corresponding reversible one. In the case of a nonGaussian first-order linear autoregressive one, a possible form is given by

$$-\rho_1 X_{t-1} + (1+\rho_1\rho_2)X_t - \rho_2 X_{t+1} = \epsilon_t$$

where ρ_1 and ρ_2 are parameters; this has the two-sided series representation

$$X_t = (1-\rho_1\rho_2)^{-1}\left[\sum_{r=1}^{\infty}\rho_1^r\epsilon_{t-r} + \epsilon_t + \sum_{r=1}^{\infty}\rho_2^r\epsilon_{t+r}\right].$$

Little appears to be known about such models; two null hypotheses of directional interest are $\rho_1=\rho_2$ and $\rho_2=0$. Nonlinear forms of directional modelling and analysis also need investigation.

In particular situations there may be special features of the data, not just its time series structure, which could usefully be exploited to describe and quantify directionality.

REFERENCES

McKenzie, E. (1985). The distribution structure of finite moving-average processes. Preprint, Mathematics Department, Strathclyde University, Glasgow, Scotland.

Rosenblatt, M. (1986) Prediction for some nonGaussian autoregressive schemes. Advances in Applied Mathematics, 7, 182-198.

Weiss, G. (1975). Time reversibililty of linear stochastic processes. J. Appl. Prob., 12, 831-836.

THE ALGORITHM OF MAXIMUM MUTUAL INFORMATION FOR MODEL FITTING AND SPECTRUM ESTIMATION

Xie Zhongjie, Department of Probability and Statistics, Peking University, Beijing China

1. INTRODUCTION

In this paper, we discussed the model as $y_t = x_t + n_t$, where n_t is an MA(q) noise, $n_t = \Theta(U)\varepsilon_t$, $\Theta(z) = \Sigma_{k=0}^{q} \theta_k z^k \neq 0$, $|z| \leq 1$, $\Theta_0 > 0$. x_t is a stationary signal series, independent of n_t. Put

$$K = \left\{ \zeta_t : \begin{array}{l} \zeta_t \text{ stationary, independent of } n_t\text{, the correlation} \\ \text{function of } \eta_t = \zeta_t + n_t \text{ satisfies } R_\eta(k) = R_y(k),\ k = \overline{0,p} \end{array} \right\}$$

Suppose that p+1 covariance values $\{R_y(k);\ k=\overline{0,p}\}$ of y_t are given, we want to select a ξ_t such that

A. $\xi_t \in K$,

B. For any $k > 0$, the mutual information between $\underline{\xi}(k+p) = (\xi_1, \xi_2, \cdots, \xi_{k+p})$ and $\eta(k+p) = (\eta_1, \eta_2, \cdots, \eta_{k+p})$, i.e.

$$I(\underline{\xi}(k+p), \underline{\eta}(k+p)) = \int_x \int_y p(\underline{x},\underline{y}) \log \frac{p(\underline{x},\underline{y})}{p(\underline{x})p(\underline{y})} \, d\underline{x}d\underline{y} \tag{1}$$

is the maximum in K. Such ξ_t will be called as the optimum fitting for $\{R_y(k)\}_0^p$ under the Criterion of Maximum Mutual Information (CMMI).

2. SOME THEORETICAL RESULTS

Let $(\theta_0^2, \phi_1, \cdots, \phi_p)$ be the solution of the Yule-Walker Eqt. determined by $\{\gamma_0, \gamma_1, \cdots, \gamma_p\}$, where $\gamma_k = R_y(k)$, and let $\{\beta_k\}_0^{p+q}$ be p+q+1 real numbers determined by

$$(\Sigma_{k=0}^{p} \phi_k z^k)(\Sigma_{\ell=0}^{q} \theta_\ell z^\ell) = \Sigma_{\mu=0}^{p+q} \beta_\mu z^\mu \tag{2}$$

Put $d_\mu = \theta_0^2 \cdot \delta_{\mu,0} - \Sigma_{k=0}^{p+q-\mu} \beta_k \beta_{k+\mu}$, $0 \leq \mu \leq p+q$ (3)

$$b_\tau = \begin{cases} d_\tau, & 0 \leq \tau \leq p+q \\ 0, & \tau > p+q \end{cases}, \qquad B_m = \begin{pmatrix} b_0, & \cdots, & b_{m-1} \\ \vdots & & \vdots \\ b_{m-1}, & \cdots, & b_0 \end{pmatrix},$$

then the following theorems hold.

THEOREM 1. Suppose that y_t, n_t are Gaussian stationary series, y_t is nonsingular (see Rozanov (1976)). If $b_0>0$ and $\mathbb{B}_m \geq 0$, $\forall$ integer $m>0$, then under the CMMI, the optimum fitting ξ_t for $\{\gamma_k\}_0^p$ exists which is an ARMA(p,p+q) Gaussian series and (1) becomes

$$I_{max}(\underline{\xi}(k+p), \underline{\eta}(k+p))=\log(\det \mathbb{R}_{k+p-1}/\det \Gamma_{p+k-1})^{\frac{1}{2}} \quad (5)$$

where $\Gamma_{k+p-1}=\begin{pmatrix}\rho_0, & \cdots, & \rho_{k+p-1}\\ \vdots & & \vdots\\ \rho_{k+p-1}, & \cdots, & \rho_0\end{pmatrix}$ is the covariance matrix of n_t,

$$\mathbb{R}_{k+p-1}=\begin{pmatrix}\gamma_0 & \cdots, & \gamma_{k+p-1}\\ \vdots & & \vdots\\ \gamma_{k+p-1}, & \cdots, & \gamma_0\end{pmatrix} \quad \text{and} \quad \gamma_{p+s}=-\Sigma_{j=1}^{p}\phi_j\gamma_{p+s-j}, \quad s=\overline{1,k-1}.$$

Conversely, in order to find an optimum Gaussian ARMA series which makes (5) true, the necessary conditions are $b_0>0$ and $\mathbb{B}_m\geq 0$, $\forall$ $m>0$.

THEOREM 2. Suppose that y_t, n_t are Gaussian stationary series, y_t is nonsingular and $b_0>0$, $\mathbb{B}_m\geq 0$, $\forall$ $m>0$. Let ξ_t be a Gaussian stationary series which satisfies:

A. $\xi_t \varepsilon K$

B. ξ_t is the optimum solution which keeps (5) true, then ξ_t must be an ARMA(p,p+q) series and the parameters of the model are uniquely determined by $\{\gamma_k\}_0^p$ of y_t and $\{\theta_k\}_0^q$ of n_t.

THEOREM 3. Suppose that y_t is a regular (see Rozanov (1976)) Gaussian stationary series, the $\{c_k\}$ of the Wold's decomposition of $y_t=\Sigma_{k=0}^{\infty} c_k\varepsilon_{t-k}$ satisfy the following conditions:

A. $\Sigma_0^{\infty}|c_k|<+\infty$

B. $\Sigma_0^{\infty} c_k z^k \neq 0, \quad |z|\leq 1.$

Let x_t be a Gaussian stationary series, the spectral density function $f_x(\lambda)$ of x_t satisfies $f_x(\lambda)\geq\delta>0$. Suppose that y_1, y_2, ---, y_N are observations of y_t, put

$$\hat{\gamma}_j^{(N)}=N^{-1}\Sigma_{k=1}^{N-j} y_k y_{k+j}, \quad j=\overline{0,p}(N); \quad p(N)=o(N^{1/3}/(\ln\ln N)^{3/4}).$$

Then the optimum ARMA(p(N), p(N)+q) series $\xi_t^{(N)}$ under the CMMI for $\{\hat{\gamma}_j^{(N)}\}_0^{p(N)}$ almost sure exists when N sufficiently large and

$$\sup_{\lambda}|f_x(\lambda)-f_{\xi^{(N)}}(\lambda)|=o(1). \quad \text{a.s.}$$

3. THE ALGORITHM OF MMI

Suppose that $\{y_i\}_1^N$ are observations of y_t, $\{\rho_k\}_0^q$ the covariance values of the MA(q) noise, then the practical fitting procedures under the CMMI are following:

A. Determine a possitive integer number $p(N)=INT[LN^{0.3}]$, $1\leqq L\leqq 3$.

B. Let $(\hat{\phi}_1, \hat{\phi}_2, \cdots, \hat{\phi}_{p(N)})$ be the solution of the Yule-Walker Eqt. for $\{\hat{\gamma}_k^{(N)}\}_0^{p(N)}$. Put $\hat{\gamma}_k^{(N)}=-\Sigma_{\ell=1}^{p(N)} \hat{\phi}_\ell \hat{\gamma}_{k-\ell}^{(N)}$, $p(N)<k\leqq 2p(N)+q$,

C. Let $R_s^{(N)}=\hat{\gamma}_s^{(N)}$, $q<s\leqq 2p(N)+q$; $R_s^{(N)}= \hat{\gamma}_s^{(N)}-\varepsilon_N\rho_s$, $0\leqq s\leqq q$, where $\varepsilon_N\varepsilon(0,1]$ such that $b_0>0$, $B_m>0$ ($\forall$ m>0).

D. Suppose that $\{h_k\}_0^{p(N)+q}$ are solutions of the following Eqt.:

$$\Sigma_{k=0}^{p(n)+q-\tau} h_k h_{k+\tau}=d_\tau, \quad \tau=0,1,\cdots, p(N)+q.$$

Put $g_s=|\Sigma_{\ell=0}^{p(N)+q} h_\ell \exp\{-i\ell\frac{2s\pi}{T}\}|^{-2}$, $s=0,1,\cdots,T-1$, where $T=2^M>>p(N)+q$

and calculate $R_i(k)=T^{-1}\Sigma_{s=0}^{T-1} g_s\exp\{ik \frac{2\pi s}{T}\}$, $k=0,1,\cdots, p(N)+q$ by FFT.

E. Let $(\hat{\theta}_0, \hat{\theta}_1, \cdots \hat{\theta}_{p(N)+q})$ be the solution of the Yule-Walker Eqt. for $\{R_i(k)\}_0^{p(N)+q}$, then the optimum fitting ξ_t for x_t under the CMMI is the following ARMA (p(N), p(N)+q) model:

$$\Sigma_{k=0}^{p(N)} \hat{\phi}_k \xi_{t-k}=\Sigma_{\ell=0}^{p(N)+q} \hat{\theta}_\ell e_{t-\ell}, \quad t=0,\pm 1,\cdots$$

and the strong consistant estimate for $f_x(\lambda)$ is

$$f_\xi^{(N)}(\lambda)=(2x)^{-1}|\Sigma_{k=0}^{p(N)+q} \hat{\theta}_k e^{-ik\lambda}/\Sigma_{\ell=0}^{p(N)} \hat{\phi}_\ell e^{-i\ell\lambda}|^2, \quad -\pi\leqq\lambda\leqq\pi.$$

REFERENCES

Rozanov, Yu., A.(1967). Stationary Random Processes. Holden Day. San Francisco, pp. 52-55.

Xie Z. (1984). Principle of Maximum Mutual Information for Observed Data with Additive Noise. In: China-Japan Symposium on Statistics. Peking University Press, Beijing, pp. 356-359.

BOUNDARY CROSSING PROBLEMS AND SEQUENTIAL ANALYSIS
(Session 11)

Chairman: D.O. Siegmund

OPTIMAL SEQUENTIAL TESTS FOR RELATIVE ENTROPY COST FUNCTIONS

H.R. Lerche
Institute for Applied Mathematics
University of Heidelberg
Im Neuenheimer Feld 294
6900 Heidelberg
Federal Republic of Germany

Introduction

Sequential tests with parabolic boundaries, like the repeated significance test (RST), are currently in wide use in the empirical sciences. In recent years the operating characteristics of these tests were studied theoretically by many authors. A survey of these results provides Siegmund's recent book "Sequential Analysis". This lecture will center around a related but also somewhat different topic, the optimality properties of tests with parabolic boundaries.

Most of the results presented here, can be found in my research report (Lerche (1986a)), some others are completely new. Here I shall take a somewhat different view than earlier (Lerche (1985)). There I stressed the connections to boundary crossing and especially to the tangent approximation, from which I derived the results at first heuristically. Here I shall emphasize a certain decision-theoretic Bayes approach.

1. The repeated significance test as an adapted version of Wald's sequential probability ratio test

Let $X_1, X_2, \ldots$ be independent random variables which are identically distributed according to the measures P or Q. Let $L_n(X_1,\ldots,X_n) = \prod_{i=1}^{n} \frac{dQ}{dP}(X_i)$ denote the likelihood ratio after n observations. Let A,B be given constants with

$0<A<1<B$. Wald's sequential probability ratio test (SPRT) stops sampling at the first $n\geq 1$ at which $L_n \notin (A,B)$ (we call this stopping time T). It rejects P, as the measure which underlies the observations, if $L_T \geq B$ and Q if $L_T \leq A$.

A classical example for the SPRT is this: P and Q are the measures of normal random walks with drift $-\theta$ and θ, where θ is positive and known. Let $S_n = \sum_{i=1}^{n} X_i$. Then $L_n(X_1,\dots,X_n) = \exp(2\theta S_n)$ and the SPRT stops at the first n at which $\theta S_n \notin (\frac{1}{2}\log A, \frac{1}{2}\log B)$. The adapted version of Wald's SPRT one obtains as follows: If θ is unknown, one estimates it by $\hat{\theta}_n = \frac{|S_n|}{n+r}$ with $r>0$ and gets as modified stopping time

$$\hat{T} = \min\{n\geq 1 \mid \hat{\theta}_n\ S_n \notin (\tfrac{1}{2}\log A,\ \tfrac{1}{2}\log B)\}$$

which, if $A=B^{-1}$, is the stopping time of the repeated significance test

$$\hat{T} = \min\{n\geq 1 \mid \frac{S_n^2}{n+r} \geq \tfrac{1}{2}\log B\}.$$

2. Optimal properties of the repeated significance test

Let P_θ denote the measure of Brownian motion W(t) with drift θ. We consider the problem of deciding the sign of the drift θ, $H_o:\theta<0$ versus $H_1:\theta>0$. At first we discuss the case that $|\theta|$ is known and state a classical optimal property of the SPRT. Let (T,δ) denote a decision procedure. It consists of a stopping rule T and a final decision rule δ, according to which after stopping the sign is decided. Let $c>0$ denote the cost-factor for sampling and assume $\theta>0$. The Bayes risk of (T,δ) for the symmetric prior is given by

$$(1)\quad R(T,\delta) = \tfrac{1}{2}(P_{-\theta}\{(T,\delta) \text{ rejects } H_o\}+cE_{-\theta}T) + \tfrac{1}{2}(P_\theta\{(T,\delta) \text{ rejects } H_1\}+cE_\theta T) .$$

<u>Theorem 1</u>: *Let* B(c) *denote the position of the minimum of the function* $g(x) = \frac{e^{-2\theta x}}{1+e^{-2\theta x}} + \frac{cx}{\theta}\frac{1-e^{-2\theta x}}{1+e^{-2\theta x}}$. *Wald's SPRT with constants* -B(c) *and* B(c) *minimizes the Bayes risk* (1)

This is a classical result and can be found for instance in Shiryayev's book.

Let $G_{x,t}$ denote the posterior distribution with respect to the symmetric prior after observing $(W(s),s)$ up to the space-time point (x,t). Let $\Gamma(c) = \frac{e^{-2\theta B(c)}}{1+e^{-2\theta B(c)}}$. An equivalent representation of the optimal SPRT is this. It stops at

$$T = \inf\{t>0 \mid G_{W(t),t}\{-\theta\} \wedge G_{W(t),t}\{\theta\} \leq \Gamma(c)\}$$

and decides according to

$$\delta = 1 \text{ or } 0 \text{ if } G_{W(T),T}\{-\theta\} \lesseqgtr G_{W(T),T}\{\theta\} .$$

In the sequel we call this type of stopping rule *simple Bayes rule*. It stops at the first time at which the posterior probability of the hypothesis or alternative is too small.

We turn now to the case that $|\theta|$ is unknown. The parameter sets of H_o and H_1 are given by $\Theta_o=\{\theta<0\}$ and $\Theta_1=\{\theta>0\}$. The sampling cost is taken to be $c\theta^2$ where c is a positive constant. On the parameter space $\Theta_o \cup \Theta_1$ we put the normal prior $G(d\theta)=\varphi(\sqrt{r}\theta)\sqrt{r}d\theta$ with $\varphi(y) = \frac{1}{\sqrt{2\pi}} e^{-y^2/2}$. The Bayes risk for a decision procedure (T,δ) is given by

$$\text{(2)} \quad R(T,\delta) = \int_{-\infty}^{0} (P_\theta\{(T,\delta) \text{ rejects } H_o\}+c\theta^2 E_\theta T)G(d\theta) + \int_{0}^{\infty} (P_\theta\{(T,\delta) \text{ rejects } H_1\}+c\theta^2 E_\theta T)G(d\theta).$$

The objective is to find a decision procedure (T^*,δ^*) which minimizes the risk (2).

Let $G_{x,t}$ denote the posterior of θ given that the process $(W(s),s)$ has reached (x,t); $G_{x,t}=N(\frac{x}{t+r}, \frac{1}{t+r})$ where $N(\rho,\sigma^2)$ denotes the normal distribution with mean ρ and variance σ^2. For $\lambda>0$ the simple Bayes rule is defined as

$$T_\lambda = \inf\{t>0 \mid \min_{i=0,1} G_{W(t),t}(\Theta_i) \leq \Phi(-\sqrt{\lambda})\}$$

where Φ denotes the standard normal distribution function. It can also be expressed as

$$T_\lambda = \inf\{t>0 \,\|\, |W(t)| \geq \sqrt{\lambda(t+r)}\}$$

which defines a repeated significance test. The following result states that one of these rules is optimal for the risk (2).

Theorem 2: Let $c>0$. *Let* $\lambda(c)$ *denote the solution of the equation* $\varphi(\sqrt{\lambda})/\sqrt{\lambda} = 2c$ *and let* $T^* = T_{\lambda(c)}$. *Let* $\delta^*=1_{\{W(T^*)>0\}}$. *Then the procedure* (T^*,δ^*) *minimizes the Bayes risk* (2).

The meaning of the assumption about the observation cost becomes apparent by the following consideration. Let us consider the two testing problems

1) H_o : $\theta = -\theta_1$ versus H_1 : $\theta = \theta_1$ and
2) H_o : $\theta = -\theta_2$ versus H_1 : $\theta = \theta_2$ with

$\theta_i>0$, $i=1,2$. Let t_i, $i=1,2$, denote the sample sizes. Then the level-α Neyman-Pearson tests for both problems have the same power if and only if $\theta_1^2 t_1=\theta_2^2 t_2$. (This follows from the form of the power function of a Neyman-Pearson test of level α: $\Phi(-c_\alpha + \sqrt{t}(\theta_i+\theta))$. We note also that the cost-factor is proportional to the relative entropy

$$E_\theta \log \frac{dP_{\theta,1}}{dP_{-\theta,1}} = 2\theta^2.$$

We add some further remarks to the relevance of the theorem.

1) For fixed cost $c>0$, simple Bayes rules turned out as optimal or approximately optimal if there is an indifference zone in the parameter space, but they are not optimal if there is none. This shows the work of Wald-Wolfowitz (1948), Schwarz (1962), Lorden (1967), Chernoff (1972) et al..

2) To let the cost depend on the underlying parameter in a monotone way, is a natural assumption in connection with clinical trials, although then $|\theta|$ seems to be the more natural cost than θ^2 (see e.g. Anscombe (1963)). But, since the RST is a Bayes test for the cost $c\theta^2$, it is especially sensitive to big drifts.

Proof of Theorem 2: Let $Q = \int P_\theta G(d\theta)$ where $G(d\theta) = \varphi(\sqrt{r}\theta)\sqrt{r}d\theta$. A well known argument yields $R(T,\delta^*) \le R(T,\delta)$ for every δ and every stopping time T. It can be shown that

$$R(T,\delta^*) = \int f\left(\frac{W(T)^2}{T+r}\right)dQ \text{ holds}$$

with $f(x) = \Phi(-\sqrt{x}) + cx$.

This can be seen for the sampling term as follows:

$$\begin{aligned}\int \theta^2 E_\theta T\ G(d\theta) &= \int \theta E_\theta W(T)\ G(d\theta) \\ &= \int W(T)\left(\int \theta N\left(\frac{W(T)}{T+r}, \frac{1}{T+r}\right)(d\theta)\right)dQ \\ &= \int \frac{W(T)^2}{T+r}\ dQ\end{aligned}$$

by Wald's lemma, Bayes formula and Fubini's theorem.

Since $\lambda(c)$ is the position of the minimum of f

$$R(T,\delta^*) = \int f\left(\frac{W(T)^2}{T+r}\right)dQ \ge f(\lambda(c)).$$

Let $T^* = \inf\{t>0 \mid \frac{W(T)^2}{t+r} \ge \lambda(c)\}$. $R(T^*,\delta^*) = f(\lambda(c))$ if $Q(T^*<\infty) = 1$. But this holds since $P_\theta(T^*<\infty) = 1$ for all $\theta \neq 0$.

q.e.d.

A final remark to the preceding section. A similar consideration for the SPRT as the preceding one shows that

$$T^* = \inf\{t>0 \mid \theta|W(t)| \ge B(c)\}$$

is its natural form. (The Bayes risk for the symmetric prior is given by $R(T,\delta^*) = \int h(\theta|W(T)|)dQ$ where

$$h(x) = \frac{e^{-2\theta x}}{1+e^{-2\theta x}} + cx\,\frac{1-e^{-2\theta x}}{1+e^{-2\theta x}}.)$$

3. *Detecting an effect θ≠0 (Brownian motion case)*

We consider now a new problem, detecting an effect $\theta \neq 0$. We show that nearly parabolic boundaries are optimal if $|\theta|$ is unknown. At first we discuss the case that $|\theta|$ is known. Let $\gamma>0$ and $\theta>0$ be given. We set up the Bayes risk as

$$(3)\quad R(T) = \gamma P_o(T<\infty) + (1-\gamma)c\theta^2 E_\theta T$$

and ask for its minimizing stopping rule T*. Since

$$E_\theta \log \frac{dP_{\theta,T}}{dP_{o,T}} = E_\theta(\theta W(T) - \tfrac{1}{2}\theta^2 T)$$
$$= \tfrac{1}{2}\theta^2 E_\theta T \ ,$$

we again have a simple representation of the risk:

$$R(T) = \int h(\log \frac{dP_{\theta,T}}{dP_{o,T}})dP_\theta \quad \text{where}$$

$h(x) = \gamma e^{-x} + 2(1-\gamma)cx$. Let $b(c)$ denote the location of the minimum of h. Then $e^{-b(c)} = \frac{2(1-\gamma)c}{\gamma}$ and the optimal stopping rule is equal to

$$T^* = \inf\{t>0 | \frac{dP_{\theta,t}}{dP_{o,t}} \geq e^{b(c)}\}$$
$$= \inf\{t>0 | F_{W(t),t}\{0\} \leq \frac{2c}{1+2c}\} \ .$$

The second representation shows that T^* is a simple Bayes rule. Here $F_{W(t),t}$ denotes the posterior with respect to the prior $\gamma\delta_o + (1-\gamma)\delta_\theta$.

The analogous simple Bayes rule is also (nearly) optimal if $|\theta|$ is unknown. For this case we assume as prior

$$F = \gamma\delta_o + (1-\gamma)N(0,r^{-1}).$$

Here the Bayes risk is equal to

$$(4) \quad R(T) = \gamma P_o(T<\infty) + (1-\gamma)c\int \theta^2 E\ T\ \varphi(\sqrt{r}\theta)\sqrt{r}d\theta \ .$$

For c sufficiently small the minimizing stopping rule T^* will be a test of power one. The posterior with respect to F is given by

$$F_{x,t} = \gamma(x,t)\delta_o + (1-\gamma(x,t))N(\frac{x}{t+r}, \frac{1}{t+r}) \quad \text{with}$$
$$\gamma(x,t) = F_{x,t}\{0\} = \frac{\gamma}{\gamma + (1-\gamma)L_t(x)} \quad \text{where}$$
$$L_t(x) = \int \frac{dP_{\theta,t}}{dP_{o,t}} \varphi(\sqrt{r}\theta)\sqrt{r}d\theta = \sqrt{\frac{r}{t+r}}\ e^{x^2/2(t+r)} \ .$$

We introduce the relevant simple Bayes rule

$$\tilde{T} = \inf\ \{t>0 | F_{W(t),t}\{0\} \leq \frac{2c}{1+2c}\} \ .$$

With $b(c) = \frac{\gamma}{2(1-\gamma)c}$ it can be expressed as

$$\tilde{T} = \inf\{t>0 | L_t(W(t)) \geq b(c)\}$$

$$= \inf\{t>0 \mid |W(t)| \geq \sqrt{(t+r)(\log(\frac{t+r}{r}) + 2\log b)}\}.$$

Theorem 3: Let R(T) *be defined by* (4). *Then* $R(\tilde{T})-R(T^*)=o(c)$ *as* $c\to 0$.

For details see Lerche (1986b). We note that the asymptotic approach "$c\to 0$" is due to Chernoff (1959).

4. Optimality and overshoot

We consider again the problem of deciding the sign of a drift but now for the normal random walk with unit variance instead of Brownian motion. Here one additional difficulty arises. The overshoot of the process over the boundary has to be taken into account. We use the notation of Section 1, denoting by P_θ the random walk with drift θ and by S_n its value at stage $n\in\mathbb{N}$. Since S_n can be considered as the observation of Brownian motion at time n, we can think of having the same setup as in Section 2 (Bayes risk (2)) but restricted to positive integer-valued stopping times.

By the preceding remark it is obvious that the random walk has to stop somewhat earlier than the Brownian motion. The counterpart to Theorem 2 is the following.

Theorem 4: Let T_c *denote the stopping time*

$$T_c = \min\{n\geq 1 \mid \frac{S_n^2}{n+r} \geq \lambda(c)-2a(\hat{\theta}_n)\}$$

where $\hat{\theta}_n = \frac{S_n}{n+r}$, $e^{-a(\theta)} = \lim_{a\to\infty} E_\theta e^{-(Z_{\tau_a}-a)}$ *with* $Z_n = \theta(S_n-n\theta/2)$ *and* $\tau_a = \min\{n\geq 1 \mid Z_n\geq a\}$ *and* $\lambda(c)$ *is defined as in Theorem 2.*

Then $R(T_c,\delta^*)-R(T^*,\delta^*) = o(c)$ *as* $c\to 0$.

To explain this result somewhat, we mention a related (exact) result due to Lorden (1977). We consider now the risk (3) but for the normal random walk. Lorden's result is the following.

Theorem 5: Let $c>0$ *be fixed. Let*

$$T^* = \min \{n\geq 1 | \theta(S_n - n\theta/2) \geq b(c) - a(\theta)\}$$

where $a(\theta)$ *is defined in Theorem 4. Then* T^* *minimizes the risk* (3).

The linkage between Theorem 4 and 5 is the following: If the process S_n is very near to the stopping boundary the setup of Theorem 4 reduces to that of Theorem 5. This transition can be done with a modification of a result of Lai-Siegmund (1977) on nonlinear renewal theory.

References

Anscombe, F.J. (1963) Sequential medical trials. J. Amer. Statist. Assoc. 58, 365-383

Chernoff, H. (1959) Sequential design of experiments. Ann. Math. Statist. 30, 755-770

Chernoff, H. (1972) *Sequential Analysis and Optimal Design*. Regional conference series in applied mathematics of SIAM, Philadelphia

Lai, T.L., Siegmund D. (1977) A nonlinear renewal theory with applications to sequential analysis I. Ann. Statist. 5, 946-954

Lerche, H.R. (1985) On the optimality of sequential tests with parabolic boundaries, *Proceedings of the Berkeley Conference in Honor of Jerzey Neyman and Jack Kiefer*, Vol. II, L. LeCam, R. Olshen, eds., Wadsworth, Belmont, 298-316

Lerche, H.R. (1986a) *Boundary Crossing of Brownian Motion*. Springer Lecture Notes in Statistics No. 40

Lerche, H.R. (1986b) The shape of Bayes tests of power one Ann. Statist. 14

Lorden, G. (1967) Integrated risk of asymptotically Bayes sequential tests. Ann. Math. Statist. 38, 1399-1422

Lorden, G. (1977) Nearly-optimal sequential tests for finitely many parameter values. Ann. Statist. 5, 1-21

Schwarz, G. (1962) Asymptotic shapes of Bayes sequential testing regions. Ann. Math. Statist. 33, 224-236

Shiryayev, A.N. (1978) *Optimal Stopping Rules*. Springer Verlag, Berlin

Siegmund, D. (1985) *Sequential Analysis*. Springer-Verlag, Heidelberg

Wald, A. and Wolfowitz, J. (1948) Optimum character of the sequential probability ratio test, Ann. Math. Statist. 19, 326-339

ASYMPTOTIC EXPANSIONS IN SOME PROBLEMS OF SEQUENTIAL TESTING

V.I.Lotov,
Institute of Mathematics, Siberian Branch of Academy of Sciences of USSR, Novosibirsk, 630090, USSR
A.A.Novikov,
Steklov Mathematical Institute of Academy of Sciences of USSR, Vavilov 42, Moscow, GSP-1 117966, USSR

§1. Introduction. Let X_t be a process with independent homogeneous increments, $t \in R^+ = [0,\infty)$ or $t \in Z^+ = 0,1,\ldots$. Suppose the distribution of X_t depends on a parameter $\theta \in \Theta$ and the probability measures $P_\theta(\cdot)$ on σ-algebras $F_t = \sigma(X_s, s \leq t)$ are equivalent. We denote by $\frac{dP_\theta}{dP_\varphi}(t)$ the density of $P_\theta(\cdot)$ with respect to $P_\varphi(\cdot)$ on the σ-algebra F_t.

We consider the problem of sequential testing of hypotheses $H_1: \theta = \theta_1$ versus $H_2: \theta = \theta_2$.

In the first part of this report results (obtained by Lotov) about asymptotic expansions of characteristics of the sequential probability ratio test (SPRT) are presented. The stopping time for this test (Wald's test) is

$$\tau = \inf(t: S_t \equiv \log \frac{dP_{\theta_2}}{dP_{\theta_1}}(t) \notin (-a_1, a_2)), \ a_1>0, \ a_2>0.$$

We reject H_1 iff $S_\tau \geq a_2$. As is well known this test minimizes the average sample number for all tests with given bounds on the probabilities of errors.

In the second part of the report results (obtained by Novikov jointly with Dragalin) about the asymptotic properties of the so-called double sequential probability ratio test (2-SPRT) are given, which also possess some optimal properties. The stopping time for this test

(Lorden's test) is

$\tau(\varphi)=\min(\tau_1(\varphi),\tau_2(\varphi))$, $\tau_i(\varphi)=\inf(t:\log \frac{dP_\varphi}{dP_{\theta_i}}(t)\geq a_i)$, $a_i>0$,

and the decision rule is $D(\varphi)= 1 + I\{\log\frac{dP_\varphi}{dP_{\theta_1}}(\tau(\varphi)) \geq a_1\}$, where $\{D=i\}=\{\text{reject } H_i\}$, $i = 1,2$, and $I\{\cdot\}$ is an indicator function.

§2. Asymptotic expansions for the SPRT. We consider here only the case $t\in Z^+$. Denote

$$\alpha_1= P_{\theta_1}(S_\tau \geq a_2),\ \alpha_2= P_{\theta_2}(S_\tau \leq -a_1),$$

These are the error probabilities of the SPRT. The usage of this test requires knowledge of the dependence on a_1,a_2 of the error probabilities and average sample number (the direct problem). The problem of determining the levels a_1, a_2 given α_1,α_2 is also of interest (the inverse problem). The algorithms for calculation of exact values of α_i and $E_\theta\tau$ are known in some special cases, but they are too cumbersome and do not allow to solve these problems in the general case. At the same time several approximation formulae for α_i and $E_\theta\tau$ are known and Wald's approximation

$$(1)\quad \exp(a_1)\sim(1-\alpha_1)/\alpha_2,\quad \exp(a_2)\sim(1-\alpha_2)/\alpha_1,$$

is one of them. It is based on the neglection of the excess of the value S_τ over the bounds of the interval $(-a_1,a_2)$. Attempts to improve (1) which were done in a number of papers are connected with various excess approximation methods and with evaluation of the influence of the excesses.

We present here the asymptotic solution of these direct and inverse problems as $a_1,a_2\to\infty$ or $\alpha_1\alpha_2\to 0$ respectively. The so-called factorization methods in the works of Borovkov (see, f.e. Borovkov (1962a)), proved them to be powerful in the studies of asymptotic properties of the distribution in the problems with one straight-line boundary. A closely related approach was later suggested by Lotov to solve the problems with two straight-line boundaries. In this way the complete asymptotic expansions for

$P(S_\tau \in A, \tau = n)$, $P(S_n \in B, \tau > n)$ as $a_i \to \infty$, $C\sqrt{n} \leq a_i = o(n)$,

in the case of general random walk were obtained (Lotov (1979), (1982)). Factorizations method is used here to study the SPRT.

Denote $r(\lambda) = 1 - E_{\theta_1} \exp(\lambda S_1)$. The size of the domain of regularity of the function $r(\lambda)$ and its zeros play an important role in the following expansions. As is known $|r(\lambda)| < \infty, r(\lambda) \neq 0$ for $Re\lambda \in [0,1]$ except $r(0) = r(1) = 0$. Suppose that $|r(\lambda)| < \infty$ for $-\gamma \leq Re\lambda \leq 1+\beta$ for some positive numbers γ, β. This condition holds, as a rule, in applications. It is possible for the function $r(\lambda)$ to have some complex zeros in the set $Re\lambda \in [-\gamma, 0) \cup (1, 1+\beta]$. If such zeros exist, denote them by

$-\mu_1, \ldots, -\mu_{2s}$ $(0 < Re\mu_j < \gamma, \mu_{2j-1} = \overline{\mu_{2j}}, j=1,\ldots,s)$,

and $1 + \lambda_1, \ldots, 1 + \lambda_{2r}$ $(0 < Re\lambda_j < \beta, \lambda_{2j-1} = \lambda_{2j}$, $j = 1,\ldots,r)$, $\mu_0 = \lambda_0 = 0$. Assume, for simplicity, that all zeros are prime and the distribution of S_1 is absolutely continuous with respect to Lebesgue measure.

Under these conditions the asymptotic expansions for

$\alpha_i = \alpha_i(a_1, a_2)$, $E_{\theta_i}\tau$ $(i = 1, 2)$ as $a_1, a_2 \to \infty$

and for $\exp(a_1)$, $\exp(a_2)$ as $\alpha_1, \alpha_2 \to 0$ are obtained. The remainder terms are determined by the bounds β and γ of the regularity strip. For example the remainder term in asymptotic expansion for $\alpha_1(a_1, a_2)$ is of the order $o(\exp(-(1+\beta)a_2) + o(\exp\{-(1+\gamma)a_1 - a_2\})$. Moreover this technique permits us to prove the inequalities which govern the accuracy of approximation. We present here one of our results. Denote

$$\eta_{\pm} = \inf(n: S_n \gtrless 0), \ \chi_{\pm} = S_{\eta_{\pm}},$$

$$r_{\pm}(\lambda) = 1 - E_{\theta_1}[\exp(\lambda\chi_{\pm}); \eta_{\pm} < \infty],$$

$$\beta_{ij} = -\frac{r_+(-\mu_i)}{r'_+(1+\lambda_j)(1+\lambda_j+\mu_i)}, \quad \gamma_{ji} = \frac{r_-(1+\lambda_j)}{r'_-(-\mu_i)(1+\lambda_j+\mu_i)},$$

$i = 0,\ldots,2s$, $j = 0,\ldots, 2r$.

Theorem 1. For $\alpha_1 \to 0, \alpha_2 \to 0$

$$\exp(a_1) = \alpha_2^{-1} \Big[\sum_{i=0}^{2s} \gamma_{0i} \Big(\frac{\alpha_2}{\gamma_{00}}\Big)^{\mu_i} - \sum_{i=0}^{2s} \gamma_{0i} \Big(\frac{\alpha_2}{\gamma_{00}}\Big)^{\mu_i} \sum_{j=0}^{2r} \beta_{ij} \Big(\frac{\alpha_1}{\beta_{00}}\Big)^{1+\lambda_j+\mu_i} + \ldots + o(\alpha_1^{\beta+1}) + o(\alpha_2^{\gamma}) \Big].$$

A similar expansion is true for $\exp(a_2)$. The algorithm of calculation of all the following terms in these expansions is established. The method is applicable for th lattice valued random variables S_n as well, and permits us to obtain also asymptotic expansions for $P_\theta(S_\tau \geqslant a_2)$, $P_\theta(S_\tau \leqslant -a_1)$, $E_\theta \tau$ for $\theta \in \Theta$.

§3. Asymptotic expansions for 2-SPRT.

1. The 2-SPRT was introduced by Lorden (1976) who showed that in case $t \in Z^+$ (that is a discrete time case)

$$\inf E_\varphi \tilde{\tau} = E_\varphi \tau(\varphi) + o(1),$$

where the infimum is taken over all tests $(\tilde{\tau},\tilde{D})$ which satisfy the equalities $P_{\theta_i}(\tilde{D} = i) = \alpha_i$ and $\max(\alpha_1,\alpha_2) \to 0$ (that means $(\tau(\varphi), D(\varphi))$ is an asymptotic solution of the so-called modified Kiefer-Weiss problem in which the "indifference" zone $\theta \in (\theta_1,\theta_2)$ is supposed). For the case $t \in R^+$ (that is a continuous time case) a similar result is obtained by Dragalin, Novikov (1987) (see Theorem 3).

Though for the 2-SPRT the values of $E_{\theta_i} \tau(\varphi)$ is greater then $E_{\theta_i} \tau$ for Wald's test (about 10-25% in tipical cases), the advantage of the 2-SPRT is that its stopping time is bounded by a constant (in case of an exponential family $P_\theta(\cdot)$). Moreover with properly chosen $\varphi = \varphi(a_1,a_2)$ the 2-SPRT is asymptotically optimal up to order O(1) in the general Kiefer-Weiss problem (see Dragalin, Novikov (1987) and Theorem 4 below).

For the practical usage of the test it is desirable to know the dependence of the error probabilities $P_{\theta_i}(D(\varphi)=i)$ and average sample $E_\varphi \tau(\varphi)$ on the levels a_1, a_2. Here we present asymptotic expansions of characteristics of the 2-SPRT for both the modified and general Kiefer-Weiss problems (and for both $t \in Z^+$ and $t \in R^+$) for the case when the probability measures $P_\theta(\cdot)$ belong an exponential family. These results generalize some results in this direction obtained by Lorden (1976) and by Huffman (1983).

2.We suppose below that measures $P_\theta(\cdot)$, $t \in R^+$ or $t \in Z^+$, belong to an exponential family, that is

$$\log \frac{dP_\varphi}{dP_\theta}(t) = (\varphi - \theta)X_t - (b(\varphi) - b(\theta))t \quad \text{for } \varphi, \theta \in \Theta,$$

where Θ is an open interval in R^1 and $b(\theta)$ is a convex function such that

$$E_\theta X_1 = b'(\theta),\ \sigma^2(\theta) = E_\theta(X_1 - b'(\theta))^2 = b''(\theta).$$

Denote $C_i(\varphi) = (\varphi - \theta_i)/I(\theta_i, \varphi)$, $i = 1, 2$,

$$t(\varphi) = \left(\frac{a_2}{I(\theta_2,\varphi)} C_1(\varphi) - \frac{a_1}{I(\theta_1,\varphi)} C_2(\varphi)\right) (C_1(\varphi) - C_2(\varphi))^{-1},$$

$$S(\varphi) = (a_1/I(\theta_1,\varphi) - t(\varphi))/C_1(\varphi),$$

and the Kullback-Leibler information numbers

$$I(\theta,\varphi) = E_\varphi \log \frac{dP_\varphi}{dP_\theta}(1) = (\varphi - \theta)b'(\varphi) - b(\varphi) + b(\theta).$$

Using these notations we can rewrite the definition of the stopping time for the 2-SPRT as follows: $\tau(\varphi) = \min(\tau_1(\varphi), \tau_2(\varphi))$,

$$\tau_i(\varphi) = \inf(t: (\operatorname{sgn} C_i(\varphi))(X_t - b'(\varphi)t) \geq$$
$$\geq (\operatorname{sgn} C_i(\varphi))(S(\varphi) + (t(\varphi) - t)/C_i(\varphi)).$$

Note that $C_2(\varphi) < 0 < C_1(\varphi)$. It is easy to see that the boundary continuation region for the process $\tilde{X}_t = X_t - b'(\varphi)t$ is triangular with vertex $(t(\varphi), S(\varphi))$.

Denote $\alpha_3 = E_\varphi(X_1)^3$, $\varkappa_i(a_i) = \log\frac{dP_\varphi}{dP_{\theta_i}}(\tau(\varphi)) - a_i$,

$\gamma_i(\varphi) = \lim\limits_{a_i\to\infty} E_\varphi \exp(-\varkappa_i(a_i))$, $\rho_i(\varphi) = \lim\limits_{a_i\to\infty} E_\varphi \varkappa_i(a_i))$,

where $\varkappa_i(a_i)$ $(i = 1, 2)$ are the excesses over the boundaries. The existence of these limits is a consequence of the general results of Borovkov (1962b) for the case $t\in Z^+$ and Mogulskii (1976) for the case $t\in R^+$ (for the case $t\in Z^+$ see also books Woodroofe (1982), Siegmund (1985)).

3. Here we consider the 2-SPRT with fixed parameter and under the assumption that levels a_i increase in a such way that

$$\frac{a_2}{I(\theta_2,\varphi)} = \frac{a_1}{I(\theta_1,\varphi)} + r\sigma(\varphi)(C_1(\varphi) - C_2(\varphi)\left(\frac{a_1}{I(\theta_1,\varphi)}\right)^{1/2}, \tag{2}$$

where r is some finite constant.

Let $\Phi(x)$ and $f(x)$ denote the standard normal distribution and its density function, respectively.

<u>Theorem 2</u>. Let relation (2) hold and $a_1\to\infty$. Then

1) $P_{\theta_i}(D(\varphi) = i) = \gamma_i(\varphi)\Phi((-1)^{i+1}r)\exp(-a_i)(1+o(1))$,
   $i = 1, 2$;

2) $E_\varphi\tau(\varphi) = \dfrac{a_1}{I(\theta_1,\varphi)} + A_1\left(\dfrac{a_1}{I(\theta_1,\varphi)}\right)^{1/2} + A_2 + o(1)$,

where the constants

$$A_1 = \sigma(\varphi)(C_1(\varphi) - C_2(\varphi))(r\Phi(-r) - f(r),$$

$$A_2 = \frac{r}{2}(C_1(\varphi) - C_2(\varphi))f(r)\left[C_1(\varphi)\sigma(\varphi) - \frac{\alpha_3}{3\sigma^3(\varphi)}\right] +$$

$$+ \Phi(r)\rho_1(\varphi)/I(\theta_1,\varphi) + \Phi(-r)\rho_2(\varphi)/I(\theta_2,\varphi).$$

The next theorem gives an asymptotic solution of the inverse problem.

Theorem 3. Let $\alpha_1 \to 0$ and

$$\frac{|\log \alpha_2|}{I(\theta_2,\varphi)} = \frac{|\log \alpha_1|}{I(\theta_1,\varphi)} + r\sigma(\varphi)(C_1(\varphi)-C_2(\varphi))\left(\frac{|\log \alpha_1|}{I(\theta_1,\varphi)}\right)^{1/2},$$

where r is some finite constant and the levels of the 2-SPRT are determined by

$$a_i = |\log \alpha_i| + \log[\gamma_i(\varphi)\Phi((-1)^{i+1} r)], \quad i=1,2.$$

Then

1) $P_{\theta_i}(D(\varphi)=i) = \alpha_i(1+o(1)), \quad i=1,2;$

2) $E_\varphi \tau(\varphi) = \dfrac{|\log \alpha_1|}{I(\theta_1,\varphi)} + \sigma(\varphi) A_1 \left(\dfrac{|\log \alpha_1|}{I(\theta_1,\varphi)}\right)^{1/2} + A_2' + o(1),$

where

$$A_2' = A_2 + \Phi(-r)\frac{\log[\gamma_2(\varphi)\Phi(-r)]}{I_2(\theta_2,\varphi)} + \Phi(r)\frac{\log[\gamma_1(\varphi)\Phi(r)]}{I_1(\theta_1,\varphi)};$$

3) $\inf E_\varphi \tilde{\tau} = E_\varphi \tau(\varphi) + o(1),$

where the infimum is taken over all tests $(\tilde{\tau},\tilde{D})$ for which

$$P_{\theta_i}(\tilde{D}=i) \le \alpha_i(1+o(1)), \quad i=1,2.$$

4. Sketch of the proofs of Theorems 2 and 3. First note that

$$(3) \quad P_{\theta_1}(D(\varphi)=1) = P_{\theta_1}(\tau(\varphi)\le t(\varphi)) - P_{\theta_1}(\tau_2(\varphi)\le \tau_1(\varphi)\le t(\varphi)).$$

By equivalence of the measures $P_{\theta_1}(\cdot)$ and $P_\varphi(\cdot)$ and the definition of $\tau_1(\varphi)$ and $\varkappa_1(a_1)$ we have

$$P_{\theta_1}(\tau_1(\varphi)\le t(\varphi)) = \exp(-a_1)E_\varphi I\{\tau_1^* \le r\}\exp\{-\varkappa_1(a_1)\},$$

where $\tau_1^* = (\tau_1(\varphi) - a_1/I(\theta_1,\varphi))(|\varphi-\theta|^2\sigma^2(\varphi)a_1/I^3(\theta_1,\varphi))^{-1/2}$.

In case $t \in Z^+$, it is well known that τ_1^* and $\varkappa_1(a_1)$ are asymtotically independent as $a_1 \to \infty$ and τ_1^* is asymtotically normally distributed N(0,1). In case $t \in R^+$ these results can be proved by a similar way. Thus we have

$$P_{\theta_1}(\tau_1(\varphi)\le t(\varphi)) = \gamma_1(\varphi)\Phi(r)\exp(-a_1)(1+o(1)).$$

To estimate the second term in (3) note that

$$P_{\theta_1}(\tau_2(\varphi) < \tau_1(\varphi) \leq t(\varphi)) < \exp(-a_1) P_\varphi(\tau_2(\varphi) < \tau_1(\varphi) \leq t(\varphi)).$$

This inequality results from neglecting the excess over the boundary of the process X_t at the moment $\tau_1(\varphi)$. Using the properties of independence and homogeneity of X_t (which imply the strong Markov property) it is shown that this term has order $o(\exp(-a_1))$. Similarly the asymptotic expansion of $P_{\theta_2}(D(\varphi)=2)$ is deduced.

The proof of statement 2) is based on the following representation for the stopping time of the 2-SPRT:

$$\tau(\varphi) = t(\varphi) - \max_{i=1,2} [C_i(\varphi)(\tilde{X}_{\tau(\varphi)} - S(\varphi)] + \delta, \tag{4}$$

where

$$\delta = I\{\tau_1(\varphi) < \tau_2(\varphi)\} \varkappa_1(a_1)/I(\theta_1,\varphi) + I\{\tau_2(\varphi) \leq \tau_1(\varphi)\} \varkappa_2(a_2)/I(\theta_2,\varphi).$$

By condition (2) $S(\varphi) = -r\sigma(\varphi)(a_1/I(\theta_1,\varphi))^{1/2}$ and after some elementary calculations we obtain from (4) that

$$E_\varphi \tau(\varphi) = t(\varphi) - \frac{C_1(\varphi) - C_2(\varphi)}{2} E_\varphi |\tilde{X}_{\tau(\varphi)} - S(\varphi)| +$$

$$+ \frac{C_1(\varphi) + C_2(\varphi)}{2} r\sigma(\varphi) \left(\frac{a_1}{I(\theta_1,\varphi)}\right)^{1/2} + E_\varphi \delta.$$

So the main difficulty is to estimate the term $E_\varphi |\tilde{X}_{\tau(\varphi)} - S(\varphi)|$. Using the submartingale property of the process $|\tilde{X}_t - S(\varphi)|$ it is shown that

$$0 \leq E_\varphi |\tilde{X}_{t(\varphi)} - S(\varphi)| - E_\varphi |\tilde{X}_{\tau(\varphi)} - S(\varphi)| = o(1).$$

The asymptotic expansion of $E_\varphi |\tilde{X}_{t(\varphi)} - S(\varphi)|$ can be calculated with the help of asymptotic expansions associated with Central limit theorem (see f.e. Petrov (1972)).

Statement 3) of Theorem 3 is proved by a method which is close to that of Lorden (1976).

5. The technique discribed above also allows us to obtai asymptotic expansions in the general Kiefer-Weiss problem.

Let φ^* satisfy the relation

$$\frac{|\log \alpha_1|}{I(\theta_1,\varphi^*)} = \frac{|\log \alpha_2|}{I(\theta_2,\varphi^*)}$$

(this value φ^* is unique) . Denote

$$\tilde{\varphi} = \varphi^* + r(\sigma^2(\varphi)\, t(\varphi^*))^{-1/2}.$$

In the paper of Huffman (1983) (in which the result of Eisenberg (1982) was improved) it was shown that for the 2-SPRT with $\varphi = \tilde{\varphi}$ and $a_i = 1/\alpha_i$, $i=1,2$, $|\log\alpha_1| \asymp |\log\alpha_2|$,

$$\inf \sup_{\theta\in\Theta} E_\theta \tilde{\tau} = E_{\tilde{\varphi}}\tau(\tilde{\varphi}) + o(|\log\alpha_1|^{1/2}),$$

where the infimum is taken over all tests $(\tilde{\tau},\tilde{D})$ for which

$$P_\theta(\tilde{D} = i) \leqslant \alpha_i \qquad (i = 1, 2).$$

The next theorem contains further improvement of this result.

Theorem 4. Let $\alpha_1 \to 0$ and

$$\lim_{\alpha_1\to 0} |\log\alpha_1|/|\log\alpha_2| = C, \quad 0 < C < \infty,$$

where C is some constant. Assume

$$a_i = |\log\alpha_i| + \log[\gamma_i(\tilde{\varphi})\Phi((-1)^{i+1} r)], \quad i=1,2.$$

Then

1) $P_{\theta_i}(D(\tilde{\varphi}) = i) = \alpha_i(1+o(1)), \quad i=1,2;$

2) $E_{\tilde{\varphi}}\tilde{\tau}(\tilde{\varphi}) = \dfrac{|\log\alpha_1|}{I(\theta_1,\varphi^*)} - \sigma(\varphi^*)(C_1(\varphi^*) - C_2(\varphi^*)) f(r)\left(\dfrac{|\log\alpha_1|}{I(\theta_1,\varphi)}\right)^{1/2} + O(1);$

3) $\sup_{\theta\in\Theta} E_\theta\tau(\tilde{\varphi}) = E_{\tilde{\varphi}}\tau(\tilde{\varphi}) + O(1);$

4) $\inf \sup_{\theta\in\Theta} E_\theta\tilde{\tau} = E_{\tilde{\varphi}}\tau(\tilde{\varphi}) + O(1),$

where the infimum is taken over all tests $(\tilde{\tau},\tilde{D})$ for which

$$P_{\theta_i}(\tilde{D} = i) \leq \alpha_i(1 + o(1)), \quad i = 1, 2.$$

Statements 1) and 2) of this theorem are a consequence of Theorem 2 and 3. Statements 3) and 4) are proved by a method which is close to that of Huffman (1983).

REFERENCES

Borovkov, A.A. (1962a). New limit theorems in the boundary problems. Siberian Math. J., 3, N 5, 645-694.

Borovkov, A.A. (1962b). Some theorems on a nonlattice random walk on an interval. Theory Probab. and its Appl., 7, N 2, 170-184.

Dragalin, V.P. and Novikov, A.A. (1987) An asymptotic solution of the Kiefer-Weiss problem for processes with independent homogeneous increments. Theory Probab. and its Appl., 26, N 3.

Eisenberg, B. (1982). The efficient solution of the Kiefer-Weiss problem. Commun.Statist.-Sequential Analysis. 1, N 1, 81-88.

Huffman, M.D. (1983). An efficient approximate solution to the Kiefer-Weiss problem. Ann. Statist., 11, N 1, 306-316.

Lorden, G. (1976). 2-SPRT's and the modified Kiefer-Weiss problem of minimizing an expected sample size. Ann. Statist., 4, N 1, 281-291.

Lotov, V.I. (1982). Asymptotical analysis of distributions in two-sided boundary problems. I - II. Theory Probab. and its Appl., 24, N 3, 475-485, N3 , 873-879.

Lotov, V.I. (1982). On asymptotic of distributions related to a first passage problem from an interval of a nonlattice random walk. Proceed. of Inst.Math. of Siberian Academy of Sci., 1, 18-25.

Mogulskii A.A. On a value of a first excess for processes with independent homogeneous increments. Theory Probab. and its Appl., 21, N 3, 470-480.

Siegmund D. (1985). Sequential Analysis. Tests and Confidence Intervals. Springer-Verlag, Berlin - New-York.

Woodroofe M. (1982). Nonlinear Renewal Theory and Sequential Analysis. SIAM, Filadelfia.

ASYMPTOTIC METHODS FOR BOUNDARY CROSSINGS OF VECTOR PROCESSES

Karl Breitung
Seminar für angewandte Stochastik der Universität München,
Akademiestr.1/IV, D-8000 München 40, FR of Germany

For simplicity we consider only two-dimensional processes. The following results can be obtained also for n-dimensional processes (with obvious modifications). Let be given a two-dimensional stationary Gaussian process $\underline{x}(t) = (x_1(t), x_2(t))$ with $E(x_i(t)) = 0$ and $var(x_i(t)) = 1$ $(i=1,2)$ and with independent components. Assume further that the sample paths are continuously differentiable, that the joint distribution of $\underline{x}(0)$ and $\underline{x}(t)$ is always non-singular for $t \neq 0$ and that the covariance functions $r_i(t) = cov(x_i(0)x_i(t))$ admit the expansions :

$$r_i(t) = 1 + r_i''(0)t^2/2 + r_i^{(4)}(0)/4! + o(t^4) \text{ for } t \to 0.$$

$$r_i(t)\log(t) \to 0 \text{ for } t \to \infty.$$

Further is given a domain $D \ R^2$ by a twice continuously differentiable function $g: R^2 \to R$ by $D = \{\underline{x}; g(\underline{x}) > 0\}$ with boundary curve $G = \{\underline{x}; g(\underline{x}) = 0\}$. We assume that : 1) The origin is in D, 2) The distance of G to the origin is 1 and that there is exactly one point $\underline{x}_o$ on G with $|\underline{x}_o| = 1$ and for all other points $\underline{x}$ on G $|\underline{x}| > 1$, 3) G is a closed regular curve.

To simplify the calculations, we assume that $\underline{x}_o = (0,1)$. By multiplication with $u > 1$ we get new domains $D(u) = \{\underline{x}; g(u^{-1}\underline{x}) > 0\}$ with boundary curves $G(u) = \{\underline{x}; g(u^{-1}\underline{x}) = 0\}$.

Now, we consider the mean number of crossings of the process $\underline{x}(t)$ through G(u) during the time from 0 to T. Under some regularity conditions (/1/,/7/) this mean number $E(C_u)$ is given by :

$$E(C_u) = T \int_{G(u)} E(|z'(0)|; \underline{x}) p(\underline{x}) ds_u(\underline{x})$$

with $z'(0) = \underline{n}^T(\underline{x}(0))\underline{x}'(0)$, where $\underline{n}(\underline{x})$ is the normal vector to G(u) at $\underline{x}$ and $\underline{x}'(t)$ is the derivative process to $\underline{x}(t)$ at t, $E(Z;\underline{x})$ is the conditional expectation of Z under the condition $\underline{x}(0) = \underline{x}$, $p(\underline{x})$ is the normal probability density of $\underline{x}(0)$ and $ds_u(\underline{x})$ denotes curve integration over the curve G(u). This gives then, making the following substitution $\underline{x} \to y = u^{-1}\underline{x}$:

$$E(C_u) = Tu \int_G E(|z'(0)|; u\underline{y}) p(u\underline{y}) ds_1(\underline{y})$$

For this integral we obtain the following asymptotic approximation (/3/,/5/) for $u \to \infty$:

$$E(C_u) \sim T\phi(u)(2/\pi)^{1/2}(-r_2''(0)/(1-\kappa))^{1/2}$$

with $\phi(u)$ the standard normal density and κ the curvature of G at $\underline{x}_o$.

For the two-dimensional product density $\lambda_u(t)$ of the point process of crossings through G(u) a formula is given in /1/, which is with the transformation $\underline{x} \to u^{-1}\underline{x}$:

$$\lambda_u(t) = u^2 \int_{G^2} E(|z'(0)z'(t)|; u\underline{y}_1, u\underline{y}_2) p_t(u\underline{y}_1, u\underline{y}_2) ds_1(\underline{y}_1) ds_1(\underline{y}_2)$$

with $E(Z;\underline{x},\underline{y})$ denoting the conditional expectation of Z under the condition $\underline{x}(0)=\underline{x}$ and $\underline{x}(t)=\underline{y}$ and $p_t(\underline{x},\underline{y})$ the joint normal density of $\underline{x}(0)$ and $\underline{x}(t)$.

The factorial moment $E(C_u(C_u-1))$ is given by :

$$E(C_u(C_u-1)) = 2\int_0^T (T-t)\lambda_u(t)dt$$

Now, we replace $\lambda_u(t)$ by a simpler function $\bar{\lambda}_u(t)$, which is obtained by taking instead of the conditional expectation of the product the product of the conditional expectations, i.e. due to the symmetry :

$$\bar{\lambda}_u(t) = u^2 \int_{G^2} (E(z'(0); u\underline{y}_1, u\underline{y}_2))^2 p_t(u\underline{y}_1, u\underline{y}_2) ds_1(\underline{y}_1) ds_1(\underline{y}_2)$$

By estimating the difference it can be shown that asymptotically :

$$\int_0^T (T-t)\lambda_u(t)dt \sim \int_0^T (T-t)\bar{\lambda}_u(t)dt \text{ for } u \to \infty .$$

$\bar{\lambda}_u(t)$ is now replaced by by simple approximations using methods of asymptotic analysis (/2/). We choose a parametrization of G in such a way that near the point $\underline{x}_o=(0,1)$ on G the first coordinate is the parameter of the curve G. Then the integral can be written as a two-dimensional integral over a rectangle $I^2 \subset R^2$ with functions f_o and f_1 of the parameters τ_1 and τ_2 :

$$\bar{\lambda}_u(t) = u^2 \int_{I^2} f_o(\tau_1, \tau_2) \exp(u^2 f_1(\tau_1, \tau_2)) d\tau_1 d\tau_2$$

The exponential function appears, since $p_t(\underline{x},\underline{y})$ is a normal density. The function f_1 has its unique global maximum at the point (0,0), since there the normal density on the curve achieves its global maximum.

The results of /2/, chapter 82. show that then asymptotically for $u \to \infty$:

$$\chi_u(t) \sim 2\pi f_o(0,0)(\det(\underline{\underline{H}}(0,0)))^{-1/2} \exp(u^2 f_1(0,0))$$

with $\underline{\underline{H}}(0,0)$ the Hessian of f_1 at $(0,0)$. The values are :

$$f_o(0,0) = (2\pi)^{-2}((1-r_1^2)(1-r_2^2))^{-1/2}(ur_2')^2(1+r_2)^{-2}$$

$$f_1(0,0) = -(1+r_2)^{-1}$$

$$\det(\underline{\underline{H}}(0,0)) = ((1-r_1^2)-\kappa(1+r_2)^{-1})^2 - r_1^2(1-r_1^2)^{-2}$$

(We write r_i instead of $r_i(0)$.)
This yields further, using the relations $1-r_2^2 \sim -r_2''(0)t^2$ and $r_2' \sim r_2''(0)t$, that for t near 0 we have the following asymptotic approximation for $\chi_u(t)$:

$$\chi_u(t) \sim (4\pi)^{-1}(-r_2''(0))^{3/2}(1-\kappa)^{-1/2}(t+o(t))u^2\exp(-u^2(1+r_2)^{-1})$$

To obtain the asymptotic form of the factorial moment, we use the Laplace method for the asymptotic evaluation of univariate integrals (/2/,chapter 5). The exponential function has an unique global maximum at 0. Using equation 5.2.18 in /2/ we obtain :

$$2\int_0^T (T-t)\chi_u(t)dt \sim T\Phi(u)(2/\pi)^{1/2}(-r_2''(0)/(1-\kappa))^{1/2} \sim E(C_u)$$

Therefore :

$$E(C_u(C_u-1)) \sim E(C_u) \text{ for } u \to \infty .$$

With similar arguments as in /6/, chapter 12.2, the asymptotic Poisson character of the point process of outcrossings out of D(u) can be shown now. We have for p_i=P(i crossings in the time from 0 to T) :

$$E(C_u)-E(C_u(C_u-1))/2 \leq p_1+p_2 \leq E(C_u)$$

Since $p_1 \leq 2P(g(u^{-1}\underline{x}(0)<0)$ and for this probability we have (/3/,/4/):

$$P(g(u^{-1}\underline{x}(0))<0) \sim \Phi(-u)(1-\kappa)^{-1/2} = o(E(C_u)) \text{ for } u \to \infty.$$

this yields with the obvious inequality $p_2 \leq E(C_u)/2$:

$$p_2 \sim E(C_u)/2$$

This means that asymptotically for $u \to \infty$ the probability of one outcrossing out of D(u) in the time from 0 to T is equal to the expected number of outcrossings, which is $E(C_u)/2$. Using the asymptotic independence of crossings in distant intervals, which is obtained from the decrease of the covariance functions, then the asymptotic Poisson form of the point process of crossings can be shown.

The outlined method can be generalized for the case that G has a finite number of points $\underline{x}_1,..,\underline{x}_k$ with $|\underline{x}_i|=1$ and for all other points $\underline{x}$ on G $|\underline{x}|>1$. A different case is treated in /8/, where it is assumed that whole parts of the curve (or surface) have the same minimal distance to the origin.

References

1. Belyaev Yu. K. On the number of exits across a boundary of a region by a vector process.-Theor Prob. Appl.,1968,13,p.32o-324.
2. Bleistein N. and Handelsman R.A.. Asymptotic expansions of integrals.- New York: Holt, Rinehart and Winston, 1975.
3. Breitung K.. Asymptotic approximations for multinormal domain and surface integrals.- In Proceedings of IFIP 11, Springer Lecture Notes in Control and Information science 59- New York: Springer, 1984.
4. Breitung K.. Asymptotic approximations for multinormal integrals. - Journal of the engineering mechanics division ASCE, 1984, 11o/3, p.357-366.
5. Breitung K.. Asymptotic approximations for the crossing rates of stationary Gaussian vector processes.- Techn. Report. 1984:1, Dept. of Math. Stat., Univ. of Lund, Lund, Sweden.
6. Cramer H. and Leadbetter M.R.. Sattionary and related stochastic processes.- New York: Wiley, 1967.
7. Lindgren G.. Model processes in non-linear prediction with applications to detection and alarm.- Ann. Probab., 198o,p.775-792.
8. Lindgren, G.! Extremal ranks and transformations of variables for extremes of functions of multivariate Gaussian processes.- Stochastic processes appl., 17, 1984, p. 285-312.

FIRST PASSAGE DENSITIES OF GAUSSIAN AND POINT PROCESSES TO GENERAL BOUNDARIES WITH SPECIAL REFERENCE TO KOLMOGOROV-SMIRNOV TESTS WHEN PARAMETERS ARE ESTIMATED.

J. Durbin
London School of Economics & Political Science, London, England.

Suppose that $y(\tau)$ is a continuous Gaussian process with mean zero and covariance function $\rho(\sigma,\tau)$ for $0 \leq \sigma \leq \tau$ and that we wish to compute the first passage density of $y(\tau)$ at time t to a boundary $y=a(\tau)$. We assume that $a(\tau)$ is differentiable and that the variance of the increment $y(t)-y(s)$ satisfies the condition

(1) $\lim\limits_{s\uparrow t} (t-s)^{-1}[V\ y(t)-y(s)] = \lambda_t$ where $0<\lambda_t<\infty$ for $0<t\leq T$.

We assume also that $\rho(\sigma,\tau)$ is positive definite and continuously differentiable with respect to σ and τ and that λ_t is continuous in t. Let

(2) $b(t) = \lim\limits_{s\uparrow t}(t-s)^{-1}[E\ I(s,y)(a(s)-y(s))|y(t)=a(t)]$

where $I(s,y)$ is an indicator function defined to equal one if the sample path of $y(\tau)$ does not cross the boundary prior to time s and to equal zero otherwise. It was shown in Durbin (1985) that the first passage density from below at time t is

(3) $p(t) = b(t)\ f(t)$

where $f(t)$ is the density of $y(t)$ on the boundary.

Although the form (3) has an appealing structure, it cannot normally be used for direct calculation of $p(t)$ since the expression (2) for $b(t)$ is generally computationally intractable. In this paper a rapidly converging series expression is given for $b(t)$ the terms of which are computable. The partial sums of this series provide a sequence of approximations to $p(t)$ of increasing accuracy. A similar series expression and an analogous sequence of approximations are given for the first passage density of a point process to a general boundary.

Let b (t) be the expression obtained by omitting the indicator function from (2), i.e.

(4) $b_1(t) = \lim_{s\uparrow t} (t-s)^{-1} E[a(s)-y(s) \mid y(t) = a(t)]$.

It is known from my (1985) paper that on substituting $b_1(t)$ for $b(t)$ in (3) we obtain a first approximation $p_1(t)$ to $p(t)$ which is good when the boundary is remote. Let $\overline{I}(s,y) = 1-I(s,y)$. Then $\overline{I}(s,y)$ is an indicator function which is 1 if the boundary is crossed prior to s and is 0 otherwise. We obtain the required series for $p(t)$ by repeated substitution of $1-\overline{I}(s,y)$ for $I(s,y)$ starting with (2). Let us denote

$$\lim_{s_k\uparrow t_k,\ldots,s_1\uparrow t_1,\ s\uparrow t} \prod_{j=1}^{k} \frac{(a(s_j)-y(s_j))}{t_j - s_j} \frac{(a(s)-y(s))}{t-s} \text{ by } \prod_{j=1}^{k} \frac{d(a-y)}{dt_j} \frac{d(a-y)}{dt},$$

let $f(t_k,\ldots,t_1|t)$ be the joint probability density of $y(t_k),\ldots, y(t_1)$ at $a(t_k),\ldots, a(t_1)$ given $y(t) = a(t)$ and let

$$\text{(5) } b_i(t) = \int_o^t \int_o^{t_1} \ldots \int_o^{t_{i-2}} E[\frac{d(a-y)}{dt_{i-1}} \ldots \frac{d(a-y)}{dt_1} \frac{d(a-y)}{dt} | y(t_{i-1})=a(t_{i-1}),\ldots, y(t_1=a(t_1),y(t)=a(t)] \; f(t_{i-1},\ldots,t_1|t)dt_{i-1}\ldots dt_1$$

for i=3, 4,... with the same definition for $b_2(t)$ except that there is just a single integral with respect to t_1 from 0 to t. The series expression for $p(t)$ is then

$$\text{(6)} \quad p(t) = \sum_{i=1}^{\infty} (-1)^{i-1} b_i(t) f(t).$$

The $j^{\underline{th}}$ approximation to $p(t)$ is

$$\text{(7)} \quad p_j(t) = \sum_{i=1}^{j} (-1)^{i-1} b_i(t) f(t).$$

While the first approximation $b_1(t)$ is the same as in my (1985) paper the higher-order approximations are slightly different. The terms in the series are computable by numerical integration since the expectation in the integrand of (5) can be obtained from elementary properties of the normal distribution while $f(t_{i-1},\ldots t_1|t)$ is just a conditional normal density. Details of the proof of (6) will be given in a later paper.

The motivation for this work has been the calculation of asymptotic significance points for Kolmogorov-Smirnov test statistics when parameters have been estimated. In this application $y(\tau)$ is the empirical process with covariance function

$$\rho(\sigma,\tau) = \sigma(1-\tau)\, g(\sigma)'\mathcal{J}^{-1}g(\tau) \quad , 0\leq \sigma \leq \tau \leq 1,$$

where $\mathcal{J}$ is the information matrix per observation and $g(\tau)$ is a vector related to the score function. The process is far from Markovian but the conditions for applicability of the above results are satisfied.

We now seek analogous results for point processes. Suppose that $y(\tau)$ is an orderly process with unit jumps for $\tau \geq 0$ with $y(0)=0$ and that we wish to calculate the first passage density at time $\tau=t$ to a boundary $a(\tau)$. The problem differs slightly according to whether crossings from below or above the boundary are under consideration. For simplicity we confine ourselves to crossings from above. Strictly speaking, $a(\tau)$ need only be defined at possible crossing points but for simplicity let us assume it is continuous and increasing in τ; without loss of generality we may suppose that it is linear between adjacent crossing points. Let $a(t) = n$ and define the time points $t_1,\ldots, t_n$ by the relations $a(t_j) = n-j$, $j=1,\ldots n$. Since the boundary is linear between possible crossing points it has slope $(t_i-t_{i+1})^{-1}$ for $t_{i+1}<\tau \leq t_i$, $i=1,\ldots,n-1$ and slope $(t-t_1)^{-1}$ for $t_1<\tau\leq t$.

We first investigate whether a formula analogous to (3) exists for point processes. We start with the counterpart to (2) for crossings from above, that is, we define

(8) $b(t) = \lim_{s\uparrow t} (t-s)^{-1}E[I(s,y)(y(s)-a(s))|y(t) = a(t)]$.

The expected value of the random variable $I(s,y)$ as $s\uparrow t$ given $y(t) = a(t)$ is found to be $p(t)/f(t)$ where $p(t)$ is the first passage density and $f(t)$ is the density of $y(t)$ on the boundary. Hence $b(t) = (t-t_1)^{-1}p(t)/f(t)$ so we obtain finally

(9) $$p(t) = b(t)f(t)(t-t_1).$$

This is the analogue for point processes of the expression (3) for continuous Gaussian processes. It differs from (3) by including an extra factor $(t-t_1)$, representing the reciprocal of the slope of the boundary at time t. Because of its tautological structure, (9) is of no direct use for computing first passage densities. However, we can use the same techniques as for the Gaussian case to derive an exact series expression for p(t) and hence obtain a sequence of approximations of increasing accuracy.

To obtain a series expression for p(t) let

$$(10)\quad b_{i+1} = \sum_{0<j_1<\ldots<j_i\leq n-1} E\left[\frac{d(y-a)}{dt_{j_i}} \cdots \frac{d(y-a)}{dt_{j_1}} \frac{d(y-a)}{dt} \,\middle|\, y(t_{j_i}) = a(t_{j_i}), \ldots y(t_{j_1}) = a(t_{j_1}), y(t)=a(t)\right] f(t_{j_i}, \ldots t_{j_1} | t) \prod_{k=1}^{i} (t_{j_k} - t_{j_k - 1})$$

for $i=1,\ldots,n-1$ where $d(y-a)/dt_h$ and $f(t_h,\ldots t)$ are defined by analogy with the continuous case. Let

$$(13)\quad p_j(t)=[b_1(t)-b_2(t)+\ldots+(-1)^{j-1}b_j(t)]f(t)(t-t_1), j=1,\ldots,n.$$

Then $p(t) = p_n(t)$ exactly and $p_1(t)$, $p_2(t),\ldots$ are approximations to p(t) of increasing accuracy. These results are obtained as in the continuous case by repreated substitution of $1-\bar{I}(s,y)$ for the indicator function $I(s,y)$. Details of the proof will be given elsewhere.

No assumption of Markov-like behaviour has been made in obtaining these results. The motivation for the work was to obtain good approximations to percentage points of Kolmogorov-Smirnov statistics in the finite sample case when parameters have been estimated. It is hoped that some numerical results will be given in a later paper.

Durbin, J. (1985). The first-passage density of a continuous Gaussian process to a general boundary. J. Appl. Prob., 22, 99-122.

CONVERSE RESULTS FOR EXISTENCE OF MOMENTS FOR STOPPED RANDOM WALKS

Allan Gut
Uppsala University
Uppsala, Sweden

1. INTRODUCTION

The classical limit theorems, such as the law of large numbers, the central limit theorem, moment convergence and the law of the iterated logarithm are concerned with the asymptotic behaviour of sums of i.i.d. random variables and can thus also be viewed as results for the asymptotic behaviour of random walks.

Frequently, however, one is interested in the (asymptotic) behaviour of random walks evaluated at *randomly* selected timepoints. For renewal processes it is of interest to study the process evaluated at the time points generated by the counting process, for random walks one may investigate the behaviour at ladder epochs, first passage times etc.

This naturally leads to research concerning the limiting behaviour of stopped random walks, where Anscombe's theorem seems to be the first result.

In Gut (1986), Chapter I all the limit theorems for random walks mentioned above are presented for stopped random walks. In this context, and motivated by some applications in Gut (1986), Chapter III, it turned out that possible converses to the Burkholder-Gundy-Davis inequalities for martingales would be of interest. This led to some research, Gut and Janson (1986), which is presented here.

2. AN EXAMPLE

Let $\{S_n, n \geq 0\}$ be a random walk with i.i.d. increments $\{X_k, k \geq 1\}$ and set $S_0 = 0$.

In the context of renewal theory for random walks on the real line the following problem is of interest. Assume that $EX_1 > 0$ and define the first passage time process $\{\nu(t),\ t \geq 0\}$ by

(2.1) $\nu(t) = \min\{n: S_n > t\}$.

It is then possible to prove (see Gut (1974), Theorem 2.1) that, for $r \geq 1$,

(2.2) $E(S_{\nu(t)})^r < \infty \iff E(X^+)^r < \infty$

(2.3) $E(\nu(t))^r < \infty \iff E(X^-)^r < \infty$.

Moreover, it turns out that one major step in the proof is to truncate and to show that

(2.4) $E|X_1|^r < \infty$ and $E(S_{\nu(t)})^r < \infty \Rightarrow E(\nu(t))^r < \infty$

for $r > 1$.

This leads to the more general problem of finding relations between existence of moments for stopped sums and the stopping time and, in particular, to converses to some Burkholder-Gundy-Davis inequalities.

3. A BURKHOLDER-GUNDY-DAVIS INEQUALITY

Let $\{S_n, n \geq 0\}$ be a random walk whose i.i.d. increments have mean 0 and a finite moment of order $r \geq 1$. By using the celebrated inequalities by Marcinkiewicz and Zygmund one can show that there exists a numerical constant depending on r only such that

(3.1) $E|S_n|^r \leq B_r \cdot n^{\frac{r}{2}\vee 1} \cdot E|X_1|^r \ (r \geq 1)$.

Now, by using some martingale inequalities due to Burkholder, Gundy and Davis it is possible to extend (3.1) to stopped random walks as follows.

Let N be a stopping time with respect to an increasing sequence of sub-σ-algebras $\{F_n, n \geq 0\}$, where $F_0 = \{\emptyset, \Omega\}$. Further, assume that X_n is F_n-measurable and independent of F_{n-1} for all n. (The typical case is $F_n = \sigma\{X_1,\ldots X_n\}$). Then

(3.2) $E|S_N|^r \leq B_r \cdot EN^{\frac{r}{2}\vee 1} \cdot E|X_1|^r \ (r \geq 1)$, where B_r is as before.

By applying the triangle inequality to (3.2) it follows immediately that

(3.3) $E|S_N|^r \leq B_r' \cdot EN^r \cdot E|X_1|^r \quad (r \geq 1)$

for the case $EX_1 \neq 0$.

The following theorem is a consequence of (3.2) and (3.3).

Theorem 3.1. Let $r \geq 1$ and suppose that $E|X_1|^r < \infty$. Then

(i) $EN^r < \infty \Rightarrow E|S_N|^r < \infty$

(ii) $EX_1 = 0$ and $E(N^{\frac{r}{2}\vee 1}) < \infty \Rightarrow E|S_N|^r < \infty$. □

The theorem thus provides conditions which ensure that the stopped random walk has a finite moment of order r.

4. CONVERSES

The most general converse to (i) would be that $E|X_1|^r < \infty$ and $EN^r < \infty$ whenever $E|S_N|^r < \infty$. However, in view of (2.2) and (2.3) this is not possible in general. But, the following holds.

Theorem 4.1. Let $r \geq 1$ and suppose that $P(X_1 \geq 0) = 1$ and $P(X_1 > 0) > 0$. Then

$$ES_N^r < \infty \Longrightarrow EX_1^r < \infty \quad \text{and} \quad EN^r < \infty. \qquad \square$$

The following weaker converse to Theorem 3.1(i) holds for random walks with non-zero mean.

Theorem 4.2. Let $r \geq 1$ and suppose that $EX_1 \neq 0$. Then

$$E|S_N|^r < \infty \text{ and } E|X_1|^r < \infty \Longrightarrow EN^r < \infty. \qquad \square$$

The case $EX = 0$ is more complicated as is seen by the following example.

Example 4.1. Consider a symmetric simple random walk, that is, suppose that $P(X_1 = 1) = P(X_1 = -1) = \frac{1}{2}$. Also, let $N_+ = \min\{n: S_n = +1\}$. Here X_1 and S_{N_+} obviously have moments of all orders and yet it is well known that N_+ has no moment of order $\geq 1/2$. □

In Theorem 3.2 we made additional assumptions on the moments of X_1 which, together with the assumption that $E|S_N|^r < \infty$, implied that $EN^r < \infty$. Our next results show under what additional assumptions on the moments of N we can infer that $E|X_1|^r < \infty$.

<u>Theorem 4.3.</u> Let $r \geq 1$. Then

$$E|S_N|^r < \infty \quad \text{and} \quad EN^r < \infty \implies E|X_1|^r < \infty . \qquad \square$$

Note that no assumption was made about the existence of EX_1. However, if we assume that $EX_1 = 0$, the assumption on the moments of N can be weakened as follows.

<u>Theorem 4.4.</u> Let $r > 1$ and suppose that $EX_1 = 0$. Then

$$E|S_N|^r < \infty \quad \text{and} \quad EN < \infty \implies E|X_1|^r < \infty . \qquad \square$$

We do not know the minimal condition on N here; it seems possible that $EN^{\frac{1}{2}} < \infty$ would suffice (at least if $EX_1^2 < \infty$).

5. FURTHER RESULTS

<u>A</u>. We first observe that (2.2) and (2.3) do not follow immediately from any result in the previous section. However, the important step (2.4) follows from Theorem 4.2 ((2.3) then follows by truncation). For details we refer to Gut (1986), Section III.3 (or Gut (1974)).

<u>B</u>. Next we note that (2.2) and (2.3), in fact, are one-sided statements in the sense that only moments of X_1^+ or X_1^- are involved. In Gut and Janson (1986) we prove results like those of Section 4 for the positive and negative tails of X_1 and S_N separately.

C. To prove moment convergence in the strong law and the central limit theorem for $S_{\nu(t)}$ or $\nu(t)$ defined in Section 2 it is natural to investigate uniform integrability for the relevant families of random variables (properly normalized). The last part of Gut and Janson (1986) is devoted to extensions of the results in Section 4 to uniform integrability, where now a family of indices $\{N(t), t \geq 0\}$, with assumptions like those for N above, is considered.

References

1. Gut, A. (1974). On the moments and limit distributions of some first passage times. Ann. Probab. 2, 227-308.

2. Gut, A. (1986). Stopped random walks. Limit theorems and applications. Book manuscript, 290 p.

3. Gut, A., Janson, S. (1986). Converse results for existence of moments and uniform integrability for stopped random walks. Ann. Probability 14.

MATHEMATICAL PROGRAMMING IN SEQUENTIAL TESTING THEORY

Müller-Funk, U.
Institut für Mathematische Stochastik, Freiburg, BRD

To formulate the duality underlying our approach in abstract terms, let (S_i, T_i) be dual pairs of linear spaces $(i = 1,2)$, i.e. there are nondegenerate bilinear mappings $S_i \times T_i \ni (\sigma_i, \tau_i) \to \sigma_i \cdot \tau_i \in \mathbb{R}^1$. They induce a pairing $\sigma \cdot \tau = (\sigma_1, \sigma_2) \cdot (\tau_1, \tau_2) = \sigma_1 \cdot \tau_1 + \sigma_2 \cdot \tau_2$ on $S = S_1 \times S_2$ and $T = T_1 \times T_2$. We shall equip S_i and T_i with the corresponding weak topologies. With that understanding each of these spaces becomes locally convex as well as Hausdorff and S_i has T_i as its topological dual and vice versa. The same properties are enjoyed by the dual spaces S and T. Let $K \subset S_1$ be a closed convex cone with vertix at zero and denote its dual cone by K^*. R is some set and $g: R \to \mathbb{R}^1$, $r_i : R \to S_i$ $(i = 1,2)$ are functions satisfying the following assumptions

(A1) $\forall\, x,y \in R\ \forall\, 0 < \eta < 1 \quad \exists\, z \in R$ such that

$$g(z) \geq (1-\eta) g(x) + \eta g(y) ,$$
$$r_1(z) \leq (1-\eta) r_1(x) + \eta r_1(y) ,$$
$$r_2(z) = (1-\eta) r_2(x) + \eta r_2(y) .$$

(A2) $r_1(x) \in K \quad \forall\, x \in R$.

(A3) Define for any $\tau = (\tau_1, \tau_2)$

$$v(\tau) = \begin{cases} \sup\{g(x) - r_1(x)\ \tau_1 - r_2(x)\ \tau_2 : x \in R\} & , \tau_1 \in K^* \\ \infty & , \tau_1 \notin K^* \end{cases}$$

and assume that $v(\tau^o) < \infty$ for some $\tau^o \in K^* \times T_2$.

For any fixed $\sigma = (\sigma_1, \sigma_2) \in K \times S_2$ we look at the

Primal program $(\mathbb{P}_\sigma)$: *Dual program* $(\mathbb{D}_\sigma)$

$$g(x) \overset{!}{=} \sup , \qquad h(\tau) = v(\tau) + \sigma \cdot \tau \overset{!}{=} \inf ,$$

$$r_1(x) \leqq \sigma_1 \quad , \qquad\qquad \tau_1 \geqq 0 \quad ,$$
$$r_2(x) = \sigma_2 \quad , \quad , \qquad \tau_2 \,\varepsilon\, T_2 \quad .$$
$$x \,\varepsilon\, R \quad .$$

The value functions (in dependence upon the side-conditions σ) are labelled $\bar{g}$ resp. $\underline{h}$. Both programs are mutally bounded, i.e. for any x_o

$$-\infty < g(x_o) \leqq \bar{g}(x) \leqq \underline{h}(\sigma) \leqq v(\tau^o) + \sigma \cdot \tau^o < \infty, \ \sigma \,\varepsilon\, K \times S_1.$$

A moment's consideration shows that (because of (A1)) $\bar{g}$ and $\underline{h}$ are proper concave functions (where, as usual, we extend both to all of S by putting them $-\infty$ outside $K \times S_2$). The rationale of the result to come extends an argument due to Kennedy (1982). The whole approach is a variant of the well-known "perturbation method". The subgradient of $\bar{g}$ at σ is denoted by $\partial\bar{g}(\sigma)$ etc.. Note that $\underline{h}$ is the concave-conjugate function associated with $-v$. Accordingly, $\partial(-v)$ and $\partial\underline{h}$ are inverse to each other.

Theorem. Suppose that (A1)-(A3) hold true.

a) (*Weak duality*) $\bar{g}|\mathrm{int}(K \times S_2) = \underline{h}|\mathrm{int}(K \times S_2)$.

b) (*Strong duality*) If $\sigma \,\varepsilon\, \mathrm{int}(K \times S_2)$ and if $x^* \,\varepsilon\, R$ is an optimal solution to $(\mathbb{P}_\sigma)$, then $g(x^*) = h(\tau^*) \ \forall \tau^* \,\varepsilon\, \partial\bar{g}(\sigma)$. Furthermore,

i) x^* solves the Lagrangean problem
$$g(x) - r_1(x) \cdot \tau_1^* - r_2(x) \cdot \tau_2^* \overset{!}{=} \sup,$$
ii) $(\sigma_1 - r_1(x)) \cdot \tau_1^* = 0.$

Conversely, if $x^* \,\varepsilon\, R$ satisfies i),ii) then x^* is optimal for $(\mathbb{P}_\sigma)$ provided that it is feasible, i.e. that $r_1(x^*) \leqq \sigma_1$, $r_2(x^*) = \sigma_2$.

Proof. The proof becomes notationally less unsightly if we only deal with "$\leqq$" contraints, i.e. if we can delete all $S_2, T_2, \ldots$ components in which case we shall simply write $\sigma = \sigma_1$ etc.. The mathematics involved are virtually the same for both formulations. To show a) we fix $\tau \geqq 0$ As

$$v(\tau) = \sup\{\sup\{g(x)-r(x)\cdot\tau\}: r(x) \leqq \delta\} : \delta \geqq 0\}$$
$$\geqq \sup\{\sup\{g(x): r(x) \leqq \delta\} - \delta\cdot\tau : \delta \geqq 0\} = \sup\{\bar{g}(\delta) - \delta\cdot\tau : \delta \geqq 0\},$$
$$v(\tau) \leqq \sup\{g(r(x)) - r(x)\cdot\tau\} : x \in R\} \leqq \sup\{\bar{g}(\delta) - \delta\cdot\tau\} : \delta \geqq 0\},$$

we realize that $h(\tau) = \sigma\cdot\tau - \inf\{\delta\cdot\tau - \bar{g}(\delta) : \delta \geqq 0\} = \sigma\cdot\tau - \bar{g}^*(\tau)$, where $\bar{g}^*$ is the concave-conjugate of $\bar{g}$. This identity remains valid for all $\tau \notin K$ as well because $h(\tau) = \infty$ and $\bar{g}(\tau) = -\infty$ in this case. The Fenchel-Moreau Theorem now entails that $\underline{h} = \text{cl}(\bar{g})$ form which the assertion follows. To verify b) we suppose that $\bar{g}(\sigma) = \underline{h}(\sigma)$, $\partial\bar{g}(\sigma) = \partial\underline{h}(\sigma) \neq \phi$, and $\exists\, x^*$ such that $g(x^*) = \bar{g}(\sigma)$, $r(x^*) \leqq \sigma$. By definition,

$$\underline{h}(\delta) \leqq h(\sigma) + (\delta-\sigma)\cdot\tau^* \;\forall\, \delta \in S, \quad \tau^* \in \partial\bar{g}(\sigma).$$

Accordingly, we obtain for all $x \in R$ and $\delta = r(x)$

$$g(x^*) = \bar{g}(\sigma) = \underline{h}(\sigma) \geqq h(\delta) - (\delta-\sigma)\cdot\tau^* = h(r(x)) - (r(x)-\sigma)\cdot\tau^*$$
$$\geqq \bar{g}(r(x)) - (r(x)-\sigma)\cdot\tau^* \geqq g(x) - r(x)-\sigma)\cdot\tau^* \quad .$$

In particular, this holds true with $x = x^*$ implying that $(r(x^*)-\sigma)\cdot\tau^* = 0$. Consequently,

$$g(x^*)-r(x^*)\cdot\tau^* = g(x^*)-\sigma\cdot\tau^* \geqq g(x)-(r(x)-\sigma)\cdot\tau^*-\sigma\cdot\tau^*$$
$$= g(x)-r(x)\cdot\tau^* \Rightarrow v(\tau^*) = g(x^*)-r(x^*)\cdot\tau^* = g(x^*)-\sigma\cdot\tau^*$$

which yields the result.

Most questions in sequential analysis can be regarded as constrained optimization problems and we want to apply the above theorem to testing problems reformulated in that way. To this end we need randomized procedures. Let $\underline{\underline{A}} = (A_n, n \in \bar{\mathbb{N}})$ be a filtration, $A_\infty = \sigma\{A_n, 0 \leqq n < \infty\}$, $P|A_\infty$ a probability measure (p.m.), and $\underline{\underline{X}} = (X_1, X_2, \ldots)$ a sequence of $\mathbb{R}^1$-valued random variables (r.v.). (It is assumed that all quantities are defined on the same abstract set.) Put $\underline{\underline{C}} = (C_n, n \in \bar{\mathbb{N}})$ where $C_n = \sigma(X_1, \ldots, X_n)$ and where C_o denotes the trivial field. We suppose that A_o supports a uniformly distributed randomization variable U which is independent from $\underline{\underline{X}}$ (under P). A (sequential) test $(N, \underline{\underline{\phi}})$ consists of a Markov time N w.r. to $\underline{\underline{A}}$ and a $\underline{\underline{C}}$-adapted sequence $\underline{\underline{\phi}} = (\phi_n, n \in \bar{\mathbb{N}})$ of ordinary test functions. A stopping rule is a $\mathcal{C}$-adapted sequence $\underline{\underline{\xi}} = (\xi_n, n \in \bar{\mathbb{N}})$ such that P-a.s. $\xi_n \geq 0$ and $\Sigma\{\xi_n : n \in \bar{\mathbb{N}}\} \equiv 1$. With every such $\underline{\underline{\xi}}$ we associate a Markov time $N = N(\underline{\underline{\xi}}, U)$ w.r. to $\underline{\underline{A}}$ by setting

$$N = n \quad \Leftrightarrow \Sigma\{\xi_k : 0 \leq k < n\} < U \leq \Sigma\{\xi_k : 0 \leq k \leq n\}, \ 1 \leq n < \infty,$$

and analogously for $n = 0, \infty$. Throughout this note we shall suppose that the following condition (R) holds true (cf. /1/, p.9).

(R) $\forall n \in \mathbb{N} \ \forall A \in A_n \ \exists \ \xi_A \in \mathcal{C}_n$ such that $P(A|\mathcal{C}_\infty) = P(A|\mathcal{C}_n) = \xi_A$ p-a.s..

Note that this condition is tantamount to the requirement that for alle Markov times M (always w.r. to $\underline{\underline{A}}$) there is a version of

$$P(N = n|\mathcal{C}_\infty) = P(N = n|\mathcal{C}_n) = \xi_{\{N = n\}} = \xi_n$$

that is a stopping rule. The time $N(\underline{\underline{\xi}}, U)$ contructed above surely leads back to $\underline{\underline{\xi}}$. Generally speaking, however, there are numerous other Markov times inducing the same stopping rule. That non-uniqueness is of no importance to us because the expectation to any stopped $\underline{\underline{\mathcal{C}}}$-adapted sequence $\underline{\underline{W}}$ of nonnegative integrable r.v. only depends on the stopping rule, i.e.

$$E(W_N) = E(\Sigma W_n 1(N = n)) = \Sigma E(W_n P(N = n \ \mathcal{C}_n)) = E(\Sigma W_n \xi_n) \ .$$

Accordingly, every test $(N, \underline{\underline{\phi}})$ can equivalently be expressed in the form $(\underline{\underline{\xi}}, \underline{\underline{\phi}})$ and we shall switch freely between both formalizations.

Mathematical programs corresponding to sequential problems take on various forms. For the sake of definitivenes we shall look at the following program which comprises the Kiefer-Weiss problem as well as tests of power one. Let $\underline{\underline{W}}$ be as before, $\underline{\underline{Y}}(j)$ be other sequences of the same kind, and $0 < \alpha_j < 1$, where $1 \leq j \leq d$. Consider the optimization problem

$$g(\underline{\underline{\xi}}, \underline{\underline{\phi}}) = g(N, \underline{\underline{\phi}}) = E(-\Sigma W_n \xi_n) \qquad \stackrel{!}{=} \sup \qquad ,$$

$$E(\phi_N Y_N(j)) \qquad = E(\Sigma \phi_n Y_n(j) \xi_n) \qquad \leq \ \alpha_j \qquad (1 \leq j \leq c),$$

$$E((1-\phi_N)Y_N(j)) = E(\Sigma(1-\phi_n)Y_n(j)\xi_n) \leq \alpha_j \quad (c < j \leq d),$$

where the optimization process extends over the class R of all tests $(N,\underline{\phi})$ for which $E(W_N) < \infty$ and $P(N<\infty) = 1$. Note that the side-conditions are linear in $\underline{\phi}$ (resp. $\underline{\xi}$) if $\underline{\xi}$ (resp. $\underline{\phi}$) is kept fixed but are neither convex nor concave as functions of $(\underline{\phi},\underline{\xi})$. The above problem, however, fits into the framework of the foregoing duality theorem. To verify (A1) we associate with every $(\underline{\xi}(k),\underline{\phi}(k)) \in R$, $k=1,2$, and $0<\gamma<1$ the pseudo convex combination $(\underline{\xi},\underline{\phi}) \in R$, where

$$\xi_n = (1-\gamma)\xi_n(1)+\gamma\xi_n(2), \phi_n = ((1-\gamma)\phi_n(1)\xi_n(1)+\gamma\phi_n(2)\xi_n(2))/\xi_n.$$

Putting $S=S_1=T=T_1=\mathbb{R}^d$ and $K=\mathbb{R}^d_+$, we realize that (A2) is automatically satisfied and that (A3) is the crucial requirement. Here, the Lagrangean problem turns out to be an optimal stopping problem. In fact,

$$v(\tau) = \sup\{E(-W_N-(\sum_1^c \tau_j Y_N(j)) \wedge (\sum_{c+1}^d \tau_j Y_N(j))) : N \in R\}.$$

The general formulation of a stopping problem requires a $\underline{C}$-adapted reward sequence $\underline{V}$ and the related class N of P-a.s. finite Markov times N (w.r. to $\underline{A}$) for which $E(V_N^-) < \infty$. Let $\underline{T}$ denote the smallest regular supermartingale dominating $\underline{V}$. ($\underline{T}$ is again $\underline{C}$-adapted because of (R).) Put

$$M_1 = \inf\{m \geq 0 : V_m = T_m\},\ M_{k+1} = \inf\{m > M_k : V_m = T_m\},\ k \geq 1\ (\inf \emptyset = \infty).$$

The optimality check to follow generalizes a well-known result.

<u>Theorem</u>. Assume that $v = \sup\{E(V_N) : N \in N\} < \infty$ and let N be given. N is an optimal solution to the stopping problem if there is a partition $\underline{D} = (D_k, k \geq 1)$, $D_k \in A_{M_k}$, such that $N = \Sigma\{I(D_k)M_k : k \geq 1\}$ and such that $(T_{N \wedge n}, n \in \mathbb{N})$ is uniformly integrable.

Sketch of proof. The necessity part is obvious. To see that the optimality of N implies the RHS we first note that a standard argument yields

$$n \in \mathbb{N} : N \geq n \Rightarrow E(V_N | A_n) = T_n \quad \text{P-a.s.} \qquad (\dagger)$$

Now, the regularity of $\underline{\underline{T}}$ entails that

$$v = E(V_N) \leq E(T_N) \leq E(T_0) = v \Rightarrow T_N = V_N \quad P\text{-a.s.} \; .$$

The last identity is tantamount to the stated representation for N. Relation (†) implies

$$E(V_N | A_n) = T_{N\wedge n} \quad P\text{-a.s.} \Rightarrow (T_{N\wedge n}, n\in\mathbb{N}) \text{ is uniformly integrable.}$$

□

The foregoing theorems combined yield a method by means of which we can derive necessary and sufficient conditions (i.e. sequential fundamental lemmata) for testing problems like the one formulated above provided that we can settle the following points:

(1) The sequential weak compactness theorem or a direct argument ensures the existence of an optimal solution to the testing problem ($\mathbb{P}_\sigma$) and satisfies a Slater condition.

(2) If $N(\tau)$ is the class of Markov times related to the reward sequence $-W_n - (\Sigma\{\tau_j Y_n(j) : 1 \leq j \leq c\}) \wedge (\Sigma\{\tau_j Y_n(j) : c < j \leq d\})$, then $N(\tau) = R$ and $v(\tau) < \infty$ for all τ of interest.

Both (1) and (2) can be answered in the affirmative for the above and various other problems under broad assumptions. Computing subgradients and their inverses we can also characterize those values of σ for which the optimal stopping rule takes on the form $M_1(\tau)$; cf. /3/ for details

R e f e r e n c e s .

1. Heckendorff, H.. Grundlagen der sequentiellen Statistik Teubner, 1982.
2. Kennedy, P.D.. On a constraint optimal stopping problem. J. Appl. Prob. 19, 631-641.
3. Müller-Funk, U.. A weak compactness theorem and Lagrangean multipliers in sequential testing. Unpublished manuscript.

EXTREME VALUES AND APPLICATIONS
(strenght of materials)
(Session 12)

Chairman: R.L. Smith

THEORY OF EXTREMES AND ITS APPLICATIONS TO MECHANICS OF SOLIDS AND STRUCTURES

BOLOTIN V.V.
Institute of Mechanical Engineering, Academy of Sciences
Moscow, USSR

1. INTRODUCTION

The theory of extremes includes a number of related topics from the theories of probability, random processes and random fields:

- extremes of independent and indentically distributed random values;
- extremes of random sequences;
- local extremes of continuous random functions of continuous time;
- global extremes of those functions at a given time segment;
- critical points of random fields, i.e. continuous and double differentiable functions of two and more continuous variables.

In addition, certain problems of the theory of crossings are in a close connection with the theory of extremes:

- upcrossings of continuous scalar processes above a given level;
- excursions of continuous vector processes out of a given region;
- distributions of times up to the first crossing or excursion;
- excursions of random fields above a given level.

Properties of rare events, and, therefore, asymptotic behaviour of extremes are of special interest. The following non-mathematical applications of the theory of extremes are significant:

- theory of brittle fracture of solids;
- theory of damage accumulation in solids under long-acting or cyclic loading; theory of initiation and propagation of fatigue cracks;
- mechanics of fracture of composite materials;
- statistics of rare natural events (earthquakes, hurricanes, etc.);
- analysis of extremal cases in demography, biometry, econometry, ecology, etc.;
- statistical theory of reliability;
- study of loads and actions on structures and machines such as large-scale turbulence, wave motion, external vibration, etc.;
- theory of safety and service life of structures.

2. CLASSICAL THEORY OF EXTREMES

Let consider a sequence $x_1,\ldots,x_n$ of independent and identically distributed random values with cumulative distribution function (CDF) $F(x)$. Calculation of CDF $F_n(x)$ of $\max\{x_1,\ldots,x_n\}$ and CDF $F_1(x)$ of $\min\{x_1,\ldots,x_n\}$ is elementary. The central point is asymptotic behaviour of $F_n(x)$ and $F_1(x)$ at $n \to \infty$. After apropriate normalization, CDFs of maxima take the form depending on behaviour of the "tail" of $F(x)$:

$$\begin{aligned}
&\text{Type I:} \quad F_n(u)=\exp\left[-\exp(-u)\right], \; -\infty < u < \infty \;; \\
&\text{Type II:} \quad F_n(u)=\exp(-u^{-\alpha}), \; u > 0; \qquad (2.1)\\
&\text{Type III:} \quad F_n(u)=\exp\left[-(-u)^{\alpha}\right], \; u < 0.
\end{aligned}$$

Here $\alpha > 0$. Asymptotic CDFs of minima are

$$\begin{aligned}
&\text{Type I:} \quad F_1(u)=1-\exp\left[-\exp(u)\right], \; -\infty < u < \infty \;; \\
&\text{Type II:} \quad F_1(u)=1-\exp\left[-(-u)^{-\alpha}\right], \; u < 0 \qquad (2.2)\\
&\text{Type III:} \quad F_1(u)=1-\exp(-u^{\alpha}), \; u > 0.
\end{aligned}$$

Most frequently, CDFs of types I and II for maxima, and CDF of type II for minima are used in applications. They are called double exponentional, Fisher-Tippett's and Gnedenko-Weibull's distributions respectively.

It is imposible to present in this paper a complete survey of literature. Earlier publications are not included into the list of references, and those which are included are noted with an asterisc*. Most of latters contain a representative bibliography.

Properties of extremes had been mentioned primary by N. Bernoulli (1709) in connection with the problem of long-livers. Up-to-date theory of extremes takes the origin from Mises (1923), Fréchet (1927), Fisher and Tippett(1928). Classification of asymptotic distributions is referred usually to Mises (1936), and comprehensive theory to Smirnov (1933) and Gnedenko (1941, 1943). Systematic presentation can be found in books by Gumbel (1958*), Galambos (1978*), and Leadbetter, Lindgren and Rootzén (1983*).

3.PROBABILISTIC THEORY OF BRITTLE FRACTURE OF SOLIDS

It is the simplest and the most common application of the theory of extremes. A simplest model is presented in Figure 1a. A chain is composed of n consequently jointed links and subjected to the load s. The ultimate load r of each link is a random value. CDF F(r) is the same for all the links, and ruptures of links are independent events. The chain is ruptured if even one of links - the weakest - is ruptured. Let the number of links is large, and the "tail" of F(r) at sufficiently small $r-r_o > 0$ may be approximated as $(r-r_o)^{\alpha}/r_c^{\alpha}$ for $u=(r-r_o)/r_c$.

Similar consideration are valid for a stochastic model of ideal brittle solid (Figure 1b). Let a body is composed of a large number of structural elements with random ultimate stress r distributed identically and homogeneously. Stress field in the body is homogeneous and depends on the single parameter s. The body is fractured if even one element is fractured. Asymptotic aproximation for CDF of ultimate stress s is

$$F(s;M)=1-\exp\left[-\frac{M}{M}\left(\frac{s-r_o}{r_c}\right)^{\alpha}\right]. \tag{3.1}$$

At $s \geqslant r_o$ we assume $F(s;M)=0$. Here r_o is minimal, and r_c is characteristic ultimate stress, α is shape parameter of the distribution, M is a measure (say, the volume) of the stressed body, M_o is a characteristic value of this measure (say, the unity of volume).

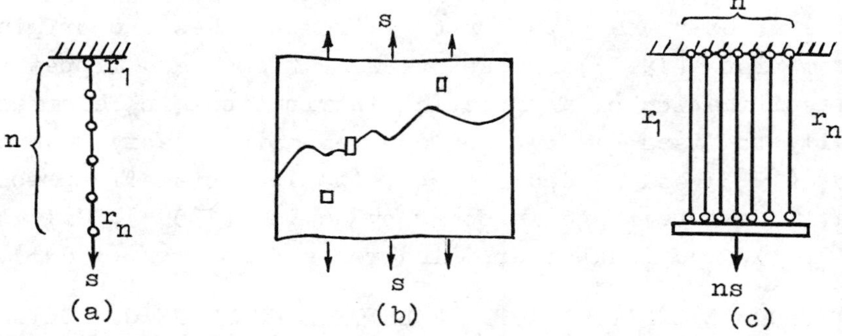

Figure 1. Stochastic models of fracture: (a) - chain of brittle links; (b) - model of ideal brittle body; (c) - set of ideal plastic elements.

Equation (3.1) presents the distribution introduced by Weibull (1939). It is illustrated in Figure 2a where the graphs are presented of $F_r(r)$ and $F(s;M)$. The size effect of brittle fracture follows from Equation (3.1) as well from the formula for mean ultimate stress (Figure 2b)

$$E\left[s(M)\right] = r_o + r_c\left(\frac{M_o}{M}\right)^{1/\alpha}\Gamma\left(1+\frac{1}{\alpha}\right). \tag{3.2}$$

Stochastic model of brittle fracture is in fact too far going idealization. All structural materials are in a certain degree ductile, and therefore local mechanical damage does not necessary results into global fracture of a body. An alternative to the model of brittle fracture is presented in Figure 1c. It is a set of parallel elements which can elongate unboundly if a certain yield stress is reached. Let yield stresses in elements are independent and identically distributed random values. The limit stat

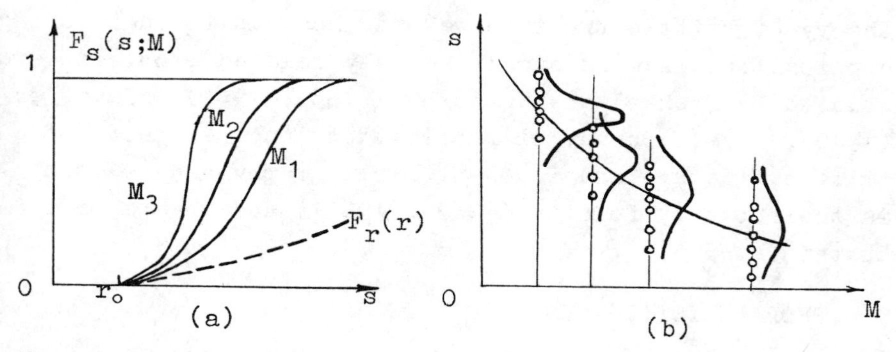

Figure 2. Size effect of brittle fracture: (a) - CDF of ultimate stress $F_s(s;M)$ for various sizes of specimens $M_1 < M_2 < M_3$; (b) - mean ultimate stress s versus the size of specimen M.

of the system takes place when the last element begins to elongate plastically. The ultimate load ns is equal to the sum of yield loads r in all elements. If n is large, the ultimate parameter s is distributed asymptotically normal.

Properties of real materials vary in a broad way between the two extreme cases. Many stochastic models of fracture were suggested taking into account material properties, type of loading, environmental conditions, etc. Some of models are discussed later.

Engineers had began to be interested in stochastic models of fracture before the theory of extremes was completed. The weakest link model was introduced by textile engineers who studied strength of fibers and threads using simple probabilistic considerations. In this context papers by Chaplin (1880, 1882) and Peirce (1926) should be mentioned. Griffith (1920) whose contributions are well known in connection with the theory of cracks in brittle solids discussed the influence of stochastically distributed microcracks on the ultimate stress. Aleksandrov and Zhurkov (1933) presented experimental results concerning the strength of various materials and, particularly, of thin fibers. Special emphasis was given to the size effect of brittle fracture. In papers by Weibull (1939, 1940) the theory of brittle fracture was advanced up to engineering applications. Almost simultaneously related problems were studied by Frenkel and Kontorova (1939, 1941), Afanasyev (1940), et al. Survey of publications on the theory of brittle fracture and revelent bibliography can be found in the books by Bolotin (1969*, 1981*) and the paper by Epstein (1985*).

4.PROBABILISTIC THEORY OF FATIGUE

Delayed damage and fracture of solids under cyclic (e. g. sinusoidal) loading is called fatigue. Under cyclic stresses which maxima are lower than the ultimate values, microcracks are accumulating in the body which are gradually growing, aggregating and at a certain time transform into a macroscopic crack (Figure 3a). Final rupture occurs as further propagation of a macrocrack under the same cyclic loading. Hence, fatigue fracture consist of the two stages. The first stage is microdamage accumulation, and the second is the macrocrack growth. The first stage take a significant and frequently a major part of the service life, and we consider this stage more closely.

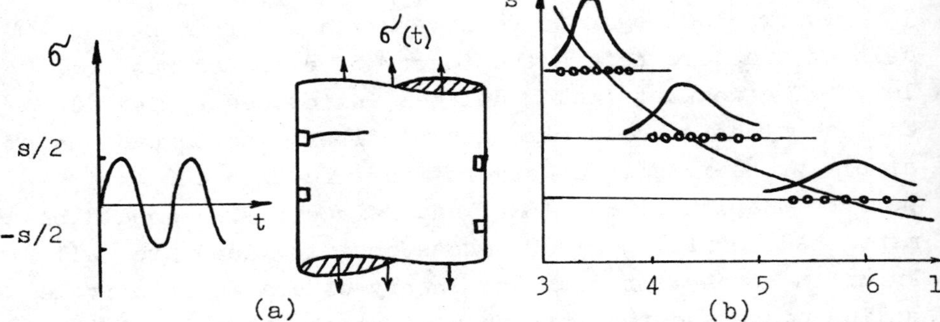

Figure 3. Stochastic models of fatigue: (a) - initiation and growth of a macrocrack under cyclic loading;(b) - relationship between the parameter of cyclic loading s and the cycle number N up to initiation of the first macrocrack.

The sources of macrocracks are weakest and/or most stre

ssed elements of material structure, e.g. grains or inclusions. Formation of such sources is a local event, and the weakest link model is applicable here. Let the ultimate stress amplitude s for an element of the material and the cycle number N up to rupture of this element are in relation

$$N=N_o\left(\frac{r_c}{s-r_o}\right)^m . \qquad (4.1)$$

Here $N_o > 0$, $r_c > 0$, $r_o \geq 0$. Strength of the element is a random value r with CDF $F_r(r)$. Other parameters entering into Equation (4.1) are assumed deterministic. Let the macrocrack initiation takes place when even of one of the elements is ruptured. Then the asymptotic approximation for CDF of the corresponding stress amplitude s at the given N is

$$F_s(s;M,N)=1-\exp\left[-\frac{M}{M_o}\left(\frac{s-r_o}{r_c}\right)^{\alpha}\left(\frac{N}{N_o}\right)^{\beta}\right] \qquad (4.2)$$

where $\beta = \alpha/m$, and $s \geq r_o$. Equation (4.2) presents the Weibull distribution for the ultimate amplitude. If the latter is fixed, the right-hand side in Equation (4.2)may be interpreted as CDF of the cycle number N for macrocrack initiation. This CDF is Weibull's too.

This model is recognized widely. It is used often by experimentators in treatment of fatigue test datas. There is a number of generalizations, e.g. for nonhomogeneous stress-state, nonstationary cyclic loading, etc.

Stochastic models of fatigue were discussed by Weibull (1940), Afanasyev (1940), Freudenthal (1946), Freudenthal and Gumbel (1953), Serensen (1954), Bolotin (1958), et al. Survey may be found in books by Bolotin (1961*, 1981*, 1984*) and Kogaev (1977*).

5. MECHANICS OF FRACTURE OF COMPOSITES

The basic type of industrial composites is a unidirectional fiber-reinforced composite. Fibers used for most composites are brittle, and their static strength may be described with Weibull distribution. Length of the stressed fiber segment λ and its chacteristic value λ_0 enter into Equation (3.1) instead of M and M_o:

$$F_r(r)=1-\exp\left[-\frac{\lambda}{\lambda_0}\left(\frac{r-r_o}{r_c}\right)^{\alpha}\right]. \tag{5.1}$$

If a fiber is ruptured, the transfer zone appears (Figure 4a). Its length λ_e is determined for the case of elastic matrix as

$$\lambda_e = f(v_f)\,\rho\left(\frac{E_f}{G_m}\right)^{1/2}. \tag{5.2}$$

Here ρ is fiber radius, E_f is fiber Young modulus, G_m is shear modulus of matrix, $f(v_f)$ a function of the fiber volume ratio v_f.

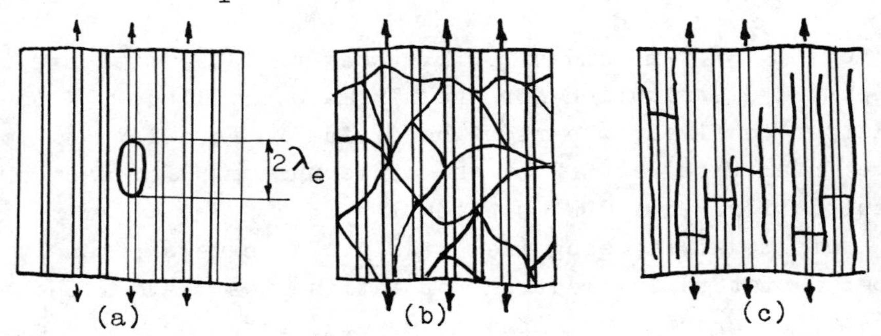

Figure 4. Patterns of damage and fracture of unidirectional fiber composite: (a) - single rupture of a fiber; (b) - dispersed damage accumulation leading to the loss of integrity; (c) - formation of a "brush-like" crack.

Thus, single ruptures of fibers do not mean failure of composite. Fracture toughness of composites is comparatively high even if fibers are very brittle. Fracture of composites in tention along fibers usually occurs in one of the two patterns. The first pattern is the loss of integrity, i.e. the global avalanche fracture at comparatively high density of local ruptures (Figure 4b). In the first approximation the ultimate load for the composite follows Weibull distribution with the form parameter αn_*. Here α is the form parameter for fiber strength, and n_* is the critical number of neighbouring local ruptures. More detailed analysis as well as experimental datas indicate that n_* varies from 6 to 10. The second pattern is fracture due to the growth of a macrocrack. Opposite to cracks in common structural materials which are usually modelled as mathematical cuts, transverse cracks in unidirectional fiber composites are of "brush-like" character due to splitting of fibers. When a crack is growing, the effective length required for the load transfer from ruptured fibers to the remaining part of composite is increasing too. Since the shear strength of matrix is low, a large part of fibers on the front of the crack becomes naked (Figure 4c). Due to the size effect strength of peripheral fibers diminishes. The critical state of the composite takes place when with probability of order of unity the size of the damage region grows under constant and even decreasing load. It means the final failure.

One of the first papers on stochastic models of fracture of composites was published by Rosen (1965). There is a number of various approaches to probabilistic modelling in the mechanics of composites. Papers by Kopyev and Ovchinskiy (1976, 1983), Tamuzs (1979, 1985), Harlow and Phoenix (1978, 1979), Smith (1980, 1984), Batdorf (1982) may be presented as examples. A model of dispersed damage accumulation in composites, and the model of "brush-like" fracture were proposed by Bolotin (1976*, 1981*). Earlier publications were surveyed by Argon (1974*). Some later papers are cited by Bolotin (1984*).

6.THEORY OF EXTREMES AND CROSSINGS OF RANDOM FUNCTIONS

We consider here continuous scalar functions x(t) of continuous scalar variable (usually time t). Assume all functions are double differentiable in the required probabilistic definition. Typical problems are calculation of CDFs of local maxima x_{max} and local minima x_{min}, of global maxima $x_{max}(T)$ and global minima $x_{min}(T)$ at a given segment $[0,T]$. More complicated problems arize in application, e.g. calculation of the joint distribution of two and more sequential extrema.

All these problems are in close connection with the theory of crossings. Asymptotic estimates of probabilities of rare crossings are of special interest. From euristical view-point, if a process is well-mixed, probability of the first upcrossing of a sufficiently high level may be estimated using the assumption that a set of upcrossings forms a Poissonian stream of events. Hence the approximate formula for the probability of non-crossing of the level $x^*=\text{const}$ at the time segment $[0,t]$ is

$$P\left\{x(\tau)\leqslant x^*;\tau\in[0,t]\right\}\sim \exp\left[-\int_0^T \nu(\tau;x^*)d\tau\right].\quad(6.1)$$

Here $\nu(\tau;x^*)$ is mathematical expectation of the number of upcrossings in the time unit, and it is assumed that with probability one $x(0)<x^*$. Some rigorous results leading to explicit formulae were obtained with the use of the theory of diffusion Markov processes.

The first publications on extremes of random processes were concerned with the theory of communications. The earliest are the paper by Rice (1939,1944,1945) dealing with distributions of maxima of continuous stationary normal processes. Systematic studies in the theory of extremes were done by Cramer (1966), Volkonskiy and Rozanov (1959-1972), Belyaev (1966-1972), Leadbetter (1963-1982), Lindgren (1970-1982), Nosko (1969-1982). Limit theorems for the number of crossings were primaly obtained by Volkonskiy and Rozanov (1959-1961). Generalization to vector processes was done by Belyaev (1968). Extremal properties of continuous random functions of two and more variables were studied by Longuet-Higgins (1957-1962) and Belyaev (1967). Earlier contributions to the theory of extremes and crossings of random processes are surveyed by Cramer and Leadbetter (1967*). The paper by Belyaev (1969*) contains some complementary results and generalizations. Recent developments are presented by Leadbetter, Lindgren and Rootzén (1983*). Applications and numerical methods are discussed by Tikhonov (1970*).

7. APPLICATIONS TO THE THEORY OF STRUCTURAL SAFETY AND RELIABILITY

Theory of structural safety and reliability is based on probabilistic models. Loads and actions are considered as random functions of coordinates and/or time. Properties of materials, components and joints are characterized with a set of random values and functions. Responce of a structure to loads and actions is a random process. Due to ageing, wear, corrosion, etc. material properties are varying in time, often also randomly. Reliability and safety regulations require that systems' parameters have to remain in the prescribed limits during all the planned lifetime, or during the time segment between two inspections. Probability of violation of those requirements has to be sufficiently small. Satety of large engineering systems must be garantied with probability $1-10^{-6}$, $1-10^{-7}$ and even higher.

The simplest model is presented in Figure 5a. Here $s(t)$ is load parameter, and $r(t)$ is strength parameter. Curves in Figure 5a represent the samples of random processes. The limit state is the first upcrossing by the process $s(t)$ of the level $r(t)$. Probability

$$R(t)=P\left\{\max\left[s(\tau)-r(\tau)\right]<0;\tau\in\left[0,t\right]\right\}, \qquad (7.1)$$

is called reliability function and considered as the fun-

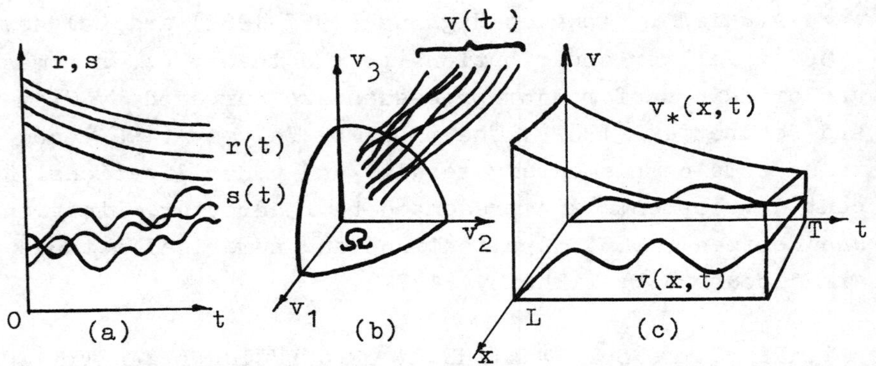

Figure 5. Statement of the problems of structural reliability: (a) - simplest model; (b) - generalization involving vector random processes; (c) - example of application of the theory of excursions of random fields.

damental measure of structural reliability. In most general cases we introduce the vector random process v(t) in a quality space V. The region Ω with the boundary Γ corresponds to admissible states (Figure 5b). Instead of Equation (7.1) we use the reliability function

$$R(t)=P\left\{ v(\tau)\in\Omega \;;\; \tau\in[0,t]\right\}. \tag{7.2}$$

Reliability analysis of continuous models of engineering systems involves applications of the theory of random fields. In Figure 5c an example of one-dimentional continuous system is shown. It is a segment of a pipeline with the length L and the planned lifetime T (for example, L=100km, T=40 years). Let structural reliability depends on relationship between the function v(x,t) describing stress-strain state of cross-section of the pipeline, and the function $v_*(x,t)$ for corresponding carrying capacity. Failure is the first excursion of the field v(x,t) above the level $v_*(x,t)$ in the rectangle $[0,L]\ \ [0,T]$. Equation (7.2) results into

$$R(t)=P\left\{ \max\left[v(x,\tau)-v_*(x,\tau)\right]<0;\ x\in[0,L]\ ;\ \tau\in[0,t]\right\}.$$

Most of problems of structural safety and reliability cannot be reduced to standard problems of the theory of crossings. An example presents simplest problem of fatigue crack growth under one-parameter random loading (Figure 6a). The crack length growth due to one positive range of nominal stress $\sigma(t)$ is

$$\frac{l(N)-l(N-1)}{\rho}=\frac{\left[K_{max}(N)-K_{min}(N)-K_{th}\right]^{m}}{\left(1-K^{2}_{max}/K^{2}_{c}\right)^{\alpha}} \qquad (7.3)$$

Here N is the number of range, ρ is characteristic size of material structure, K_c is fracture toughness, and K_{th} is damage threshold, m and α are positive numbers. Stress intensity factor $K=Y\sigma(\pi l)^{1/2}$ enters into the right-hand side of Equation (7.3) characterizing both the level of loading and the crack size. The constant Y is

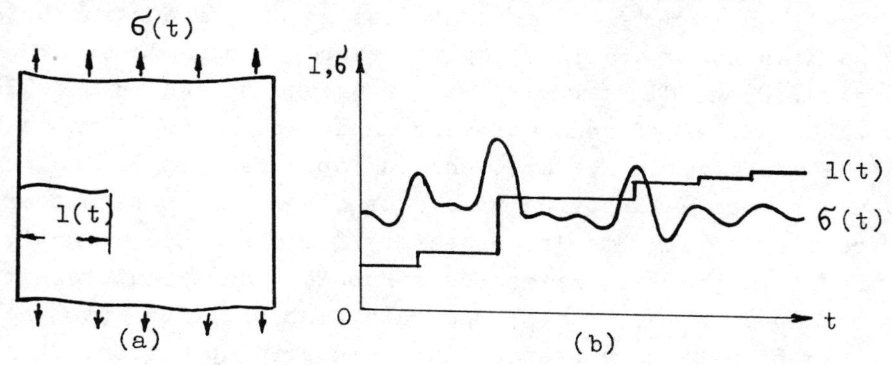

Figure 6. Fatigue crack growth under random loading: (a) - fatigue crack; (b) - samples of functions $\sigma(t)$ and l(t).

of order of unity. A crack becomes unstable when the maximum of stress intensity factor attains the critical magnitude K_c. If $\sigma(t)$ is a random process, l(t) is a random process too. Properties of the latter depend on the joint distribution of sequential extremes of $\sigma(t)$ (see Figure 6b). The problem is complicated with necessity to transform the variable t into the cycle number N, and vice versa. A solution was obtained by Bolotin (1980*) using approximation of the finite-difference equation (7.3) with a differential equation, and special assumptions on the properties of the loading process $\sigma(t)$. In most general cases, the Monte-Carlo method presents a way to access the reliability function

$$R(t)=P\left\{\max\left[Y\sigma(t)\sqrt{\pi l(\tau)}-K_c(\tau)\right]<0;\ \tau\in[0,t]\right\}. \qquad (7.4)$$

Fracture toughness K_c is assumed in Equation (7.4) varying in time due to ageing, corrosion and other types of deterioration.

Generally, most of practical problem of structural safety and reliability are very complicate, and the Monte-Carlo method is the only practical way to obtain numerical results. For example, in aseismic design of large structures, such as power stations, up to several thousands of generalized coordinates are used. Ground motion is modelled with samples of nonstationary random processes with maximal accelerations, durations and spectral properties which are random values. The volume of computations becomes exceedingly large to assess safety factors close to unity. Hence, a problem arizes to extrapolate numerical results obtained in the area of moderate probabilities into the area of very rare events. In engineering practice, functional scales are used corresponding to asymptotic distributions of extremes. For maximal stresses and accelerations the double exponentional distribution, and for load-carrying capacities of materials and structures Weibull-Gnedenko distribution are used.

First papers on application of the theory of random processes to fatigue life prediction were published by Fung (1954) and Crandall (1958). Bolotin (1959) evaluated safety factors against seismic actions using the theory of crossings. Since a number of contributions were done, e.g.

by Dimentberg (1963), Nikolayenko (1963), Caughey (1963). Shinozuka (1964), Shukaylo (1966), Freudenthal and Shinozuka (1965). Survey of earlier publications was presented by Bolotin (1969). Among the later publications papers by Sheputis (1967), Lin (1967), Cornell (1968), Weidenhammer (1968), Vanmarcke (1969), Crandall (1970), and Ditlevsen (1971) should be mentioned. Methods of the theory of random fields were applied to problems of reliability of continuous systems by Bolotin (1969), Belyaev (1970), and Volokhovsky (1973). References can be found in books by Bolotin(1969*,1981*,1984*), Svetlitsky(1976*), Heinrich and Hennig (1977*), Vanmarcke (1983*), Madsen, Krenk and Lind (1986*), as well as in proceedings edited by Moan and Shinozuka (1981*), Augusti, Borri and Vannucchi (1983*), Eggwertz and Lind (1985*).

References

Argon,A.(1974).Statistical aspects of fracture. In:Composite Materials/Eds. Broutman,L.J.,Krock,R.H.,5, Academic Press, New York.157-193.

Augusti,G.,Borri,A. and Vannucchi,G.,Eds.(1983). Applications of statistics and probability in soil and structural engineering.Proc. of ICASP-4.Florence,1983.Pitagora,Bologna.

Belyaev Yu.K.(1969).New results and generalizations of problems of crossings.-In Russian edition of book by Cramer,H. and Leadbetter,M.R.(1967).341-379.

Bolotin,V.V.(1969).Statistical methods in structural mechanics.Holden-Day,San Francisco.(Translated from Russian edition of 1965).

Bolotin,V.V.(1976).Statistical theory of damage accumulation in composites and size effect of reliability.-Mech.Polymers,2,p.247-255 (in Russian).

Bolotin,V.V.(1980).Time distribution up to fracture under random loading.-J.Appl.Math.Techn.Phys.5,149-158(in Rus).

Bolotin,V.V.(1981).Wahrscheinlichkeitsmethoden zur Berechning von Konstruktionen.Verlag für Bauwesen,Berlin.

Bolotin,V.V.(1981a).A unified model of fracture of composites under long-actings loads.-Mech. Composite Mater. 3,405-420(in Russian).

Bolotin,V.V.(1984).Service life prediction for mashines and structures.Mashinostroyenie,Moscow (in Russian).

Bolotin,V.V.(1984a).Random vibrations of elastic systems. Martinus Nijhoff, the Hague et al.(translated from Russian edition of 1979).

Cramer,H.,and Leadbetter,M.R.(1967).Stationary and related stochastic processes.Willey,New York.Russian translation:Mir, Moscow(1969).

Eggwertz,S.,and Lind,N.C.,Eds.(1985).Probabilistic methods in the mechanics of solids and structures.IUTAM Symposium,Stockholm,1984.Springer-Verlag, Heidelberg et al.

Epstein,B.(1985).Applications of extreme value theory to problems of material behaviour.-In: Probabilistic methods in the mechanics of solids and structures.Springer-Verlag,Heidelberg et al.3-9.

Galambos,J.(1978).The asymptotic theory of extreme order statistics.Wiley,New York.

Gumbel,E.J.(1958).Statistics of extremes.Columbia University Press,New York.

Heinrich,W.and Hennig,K.(1977).Zufallsschwingungen mechanischer Systeme.Akademie-Verlag,Berlin.

Kogayev,V.P.(1977).Evaluation of strength under stresses varying in time.Mashinostroyenie,Moscow (in Russian).

Leadbetter,M.R.,Lindgren,G.and Rootzén,H.(1983).Extremes and related properties of random sequences and processes.Springer-Verlag,New York et al.

Madsen,H.O.,Krenk,S.and Lind,N.C.(1986).Methods of structural safety.Prentice-Hall,Englewood Cliffs.

Moan,T.and Shinozuka,M.Eds.(1981).Structural safety and reliability.Proc.of ICOSSAR'81,Trondheim 1981.Elsevier, Amsterdam et al.

Svetlitskiy,V.A.(1976).Random vibrations of mechanical systems.Mashinostroyenie,Moscow (in Russian).
Tikhonov,V.I.(1970).Excursions of random processes.Nauka, Moscow (in Russian).
Vanmarcke,E.(1983).Random fields.Analysis and synthesys. MIT Press,Cambridge.

EXTREMES, LOADS, AND STRENGTHS

Rootzén, H.,
Institute of Mathematical Statistics, University of Copenhagen,
Universitetsparken 5, DK-2100 Copenhagen Ø, Denmark

A metal bar or a glass fibre breaks when the load exceeds its inherent strength. Stochastic extreme value theory provides models both for predicting maximum loads and for the strength of the material. The talk this note is based on was about the "Weibull theory" where the observed statistical variation of strengths of brittle materials is explained via extreme value theory and about extremes of load processes of the "filtered Poisson process" type. For a general discussion of the connections between structural engineering and extreme values we refer to the paper by Bolotin in these proceedings, and to [10] for a survey of the last few years developments in stochastic extreme value theory. For reasons of space we here concentrate on the first part, and refer to [11] for the second part of the talk.

STRENGTH OF BRITTLE MATERIALS

The starting point is the empirical fact that strengths of pieces of material manufactured under similar conditions show a practically important stochastic variation from piece to piece. This is connected with extreme values through the so-called *weakest link principle*, which says that for some materials, such as glass fibres or iron bars, the strength of a piece is determined by the strength of its weakest part. Weibull's argument ([12], cf. also [3,7]), which were quite informal, involved a "microscopic",unobservable, model for strengths, which then motivates an observable "macroscopic" model.

Here mathematically formalized versions of the two models will be introduced. We will show that the two are in fact equivalent, and discuss ways of testing the assumptions on observed strength data. A longstanding point of debate is which distribution is appropriate for material strengths. Of course the Weibull distribution is the main contender, but e.g. in [2] strengths of glass fibres is described by the product of two different Weibull distribution functions, to take into account surface and interior defects, and many other distributions, such as normal, lognormal, and gamma have been suggested [3,7]. An important feature of the statistical procedure proposed here is that it makes it possible to investigate homogeneity and weakest link behaviour without involving any assumptions about distributional forms.

We only consider the simplest case, of a specimen subjected to uniaxial tension. In the microscopic theory it is noted that observed strengths are much lower than the strength of the bonds between molecules, and the discrepancy is explained by the presence of small *flaws* or *microcracks* ([5]). The flaws are assumed to be "randomly" distributed in a homogeneous material,

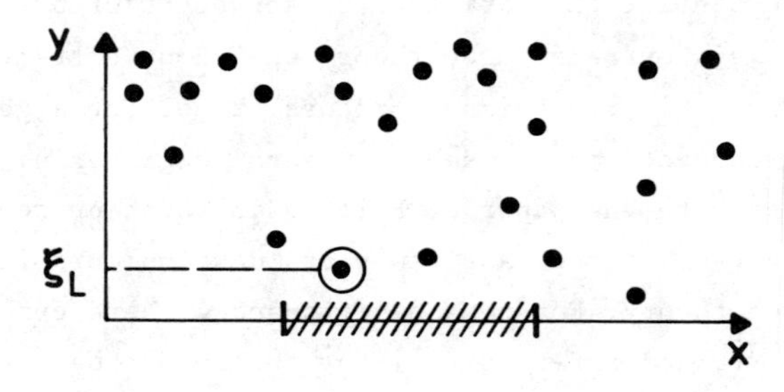

Figure 1 Plot of strengths x_i of microcracks against their location y_i.

with the material breaking when the local stress at any of the flaws exceeds its inherent strength. In Figure 1 this is illustrated by plotting the strength y_i of the i-th microcrack against its location x_i, measured along the specimen. The microscopic model for the strength of homogeneous brittle materials is that $\{(x_i,y_i)\}_{i=1}^{\infty}$ are the points of a Poisson process, N, in the first quadrant of the plane, with intensity measure $dx \times m(dy)$, and with m an arbitrary locally finite measure on $[0,\infty)$. The strength ξ_L of a piece of the specimen corresponding to the interval L is then

(1) $\qquad \xi_L = \min\{y_i; x_i \in L\},$

cf. Figure 1. Thus the survival function (s.f.) $S_L(x)$ of the piece L of length ℓ is given by

(2) $\qquad S_L(x) = P(\xi_L > x)$

$= P(N(L \times [0,x]) = 0)$

$= \exp\{-\ell\, m([0,x])\}$,

since $E(N(L \times [0,x])) = \ell\, m([0,x])$.

The mathematical formalization of the macroscopic model is as follows. We again consider a piece of material L with length ℓ and strength ξ_L and assume that it can be subdivided, at least hypothetically, into smaller pieces $L_1,\ldots,L_n$ of arbitrary lengths $\ell_1,\ldots,\ell_n$, and with definite (random) strengths $\xi_{L_1},\ldots,\xi_{L_n}$, respectively. We say that the material is stochastically[1]

(i) *brittle* if $\xi_L = \min(\xi_{L_1},\ldots,\xi_{L_n})$,

(ii) *homogeneous* if the marginal distribution of $\xi_{L_1},\ldots,\xi_{L_n}$ depends only on $\ell_1,\ldots,\ell_n$,

(iii) *disconnected* if $\xi_{L_1},\ldots,\xi_{L_n}$ are independent for all disjoint divisions $L_1,\ldots,L_n$ of L.

Of these properties, (ii) and (iii) are of purely statistical character, while (i) depends on the mechanism involved in a failure. The properties all have definite physical meanings. It follows at once from (ii) that the s.f. $S_L(x) = P(\xi_L > x)$ only depends on the length ℓ of L, i.e. $S_L(x) = S_\ell(x)$, and (i) and (ii) are then seen to imply that

(3) $$S_\ell(x) = S(x)^\ell, \quad x,\ell > 0,$$

with $S(x) = S_1(x)$.

Both models of course involve idealizations of reality. E.g. it may not be meaningful to assume that very short pieces have a definite (measurable) strength as is done in the macroscopic model, and microcracks have a physical extension which is not taken into account in the microscopic model. Nevertheless, on the scale of interest the models might still be quite accurate. It is then interesting to note that *the microscopic and macroscopic models are mathematically equivalent*. Thus, e.g. if one believes that a material shows the behaviour specified by (i)-(iii) then necessarily the physical mechanism behind failures must be the one given by the microscopic model.

We briefly outline a proof of this result. One half is immediate: clearly a material which satisfies (1) with $\{(x_i,y_i)\}$ the points of a Poisson process with intensity $dx \times m(dy)$ also satisfies the assumptions (i)-(iii) of the macroscopic model. For the converse, suppose the material satisfies (i)-(iii) with s.f. $S_\ell(x)$ given by (3). It is straightforward to see that this determines all joint distributions of strengths (of intervals). Now, by the first part of the proof the microscopic model specified by

$$m([0,x]) = -\log S(x), \quad x > 0,$$

satisfies (i)-(iii), and according to (2) also (3) holds. Hence, as was to be shown, this microscopic model leads to precisely the same distributions as the macroscopic model we started out with. The further more difficult problem of how to recover the Poisson process of microcracks from (hypothetical) measurements of strengths of all pieces will be treated elsewhere.

In practical situations one often needs not only the assumptions (i)-(iii) but also a parametric model for $S(x)$. One way to obtain this is to add an *ad hoc* notion which is that the material is

(iv) *Size-stable* if each S_ℓ is a location-scale transformation of S, i.e. if there are $\alpha_\ell > 0$, β_ℓ such that $S_\ell(x) = S(\alpha_\ell^{-1}(x-\beta_\ell))$, $\ell > 0$.

It follows from (3) and (iv) that $S(x)$ is min-stable, and hence is one of the three extreme value s.f.'s for minima (see [9], p.271 - 273). A final assumption is that strengths are

(v) *Non-negative* if $\xi_L \geqq 0$ for all L and if values arbitrarily close to zero are possible.

This further assumption makes the type III extreme value (or Weibull) s.f. the only possibility, so that then

$$S_\ell(x) = \begin{cases} 0 & x < 0 \\ \exp\{-\ell(x/\sigma)^\alpha\} & x \geqq 0, \end{cases} \tag{4}$$

where $\alpha, \sigma > 0$ are material parameters.

An alternative argument to (iv),(v) is to assume that $S(x)$ decreases as a power at $x = 0$, i.e. $S(x) \sim 1-(x/\sigma)^\alpha$ as $x \to 0$. Together with (3) and the standard criterion for the domain of attraction for minima for the Weibull distribution this again leads to (4). Further, in the engineering literature (4) is often advocated for directly, for its mathematical simplicity and flexibility.

We now turn to the problem of empirically testing the models. As far as I know, direct tests of the microscopic model is beyond present capabilities, and we will hence discuss how the assumptions of the macroscopic model can be checked using measurements of strengths of specimens of varying sizes. The independence assumption (iii) can be checked by applying any of the standard independence tests to a series of strength measurements. It seems less obvious how to test (i) and (ii) separately. Instead we investigate them together by embedding (3) into the larger model

(5) $\quad S_\ell(x) = S(x)^{\ell^\beta}$,

where $\beta > 0$ and $S(x)$ are free "parameters", and then test for $\beta = 1$. Writing $z = \log \ell$ and $h(x) = S'(x)/S(x)$ (assuming that $S(x)$ is differentiable) (5) takes the form $S_\ell(x) = \exp\{-e^{\beta z}\int_0^x h(t)dt\}$ and is recognized to be of the Cox-model type. We can hence use standard methods for the Cox model to estimatet β, test for $\beta = 1$, combine observations of strengths of pieces of different lengths to one estimate of the "underlying s.f." $S(x)$, and compare it with the Weibull s.f. resulting from (iv),(v).([8] is a general reference on the Cox model).

A. Deis in his masters thesis [4] makes a detailed study of these testing problems, and applies them to a number of data sets. Here we will as examples show the (still somewhat preliminary) results from two of his sets.

<u>Example 1</u> The first data are from Bader & Priest [1], and consist of strength measurements on about 60 carbon fibres of each of four different lengths $\ell_1 = 1$, $\ell_2 = 10$, $\ell_3 = 20$, and $\ell_4 = 50$ mm's. As of now we unfortunately have only had access to the ordered values and have hence not yet tested for independence. In Figure 2 the empirical s.f.'s for each of the four samples is plotted. The scales are choosen sc that if the Weibull model (4) holds then they should, except for random fluctuations, yield parallel straight lines of slope α, and with the i-th and j-th line a vertical distance $\log(\ell_j/\ell_i)$ apart. The plot roughly agrees with this expected behaviour. Nevertheless, the Cox estimator for β in the model (5) is

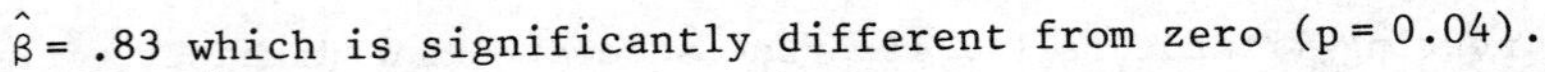

$\hat{\beta} = .83$ which is significantly different from zero ($p = 0.04$).

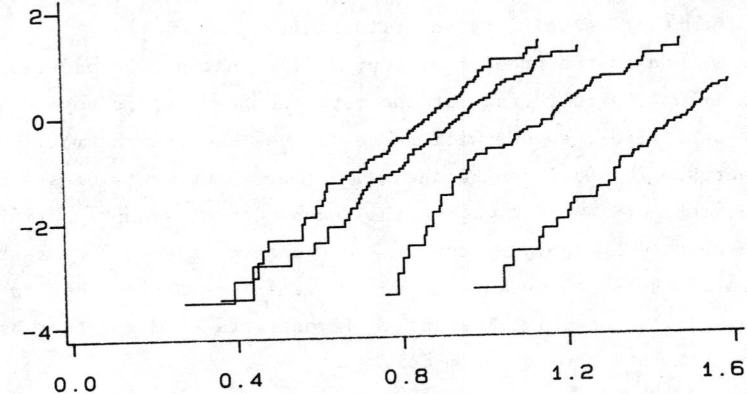

Figure 2 Plots of log {- log S(x)} against log x for glass fibres of four different lengths, in order from left to right 50 mm's, 20 mm's, 10 mm's, and 1 mm's.

Figure 3 contains a plot of the estimated survival function S in (5), using the method of Breslow to combine all four samples. There is a rather clear deviation from the straight line which would result if S(x) were the Weibull s.f. (4). We have not yet performed any formal tests of this deviation.

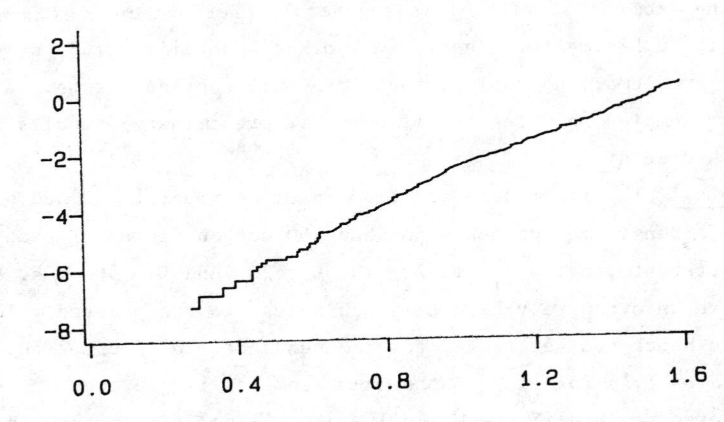

Figure 3 Plot of log (- log (underlying survival function) against log (strength), estimated using data from all four samples.

Example 2 The data here have been provided by L. Nilsson and

S. Uvell, Umeå University. The part we will discuss contains two strata (I and II) of measurements of strengths of optical fibres, which differ in experimental conditions (rate of increase of tension). Both contain 40 measurements on each of the two lengths $\ell_1 = 40$ and $\ell_2 = 80$ cm's. Correllations and partial autocorrellations were small in three of the four samples and formal tests did not indicate deviations from independence. Figure 4 contains plots of the survival functions for each set.

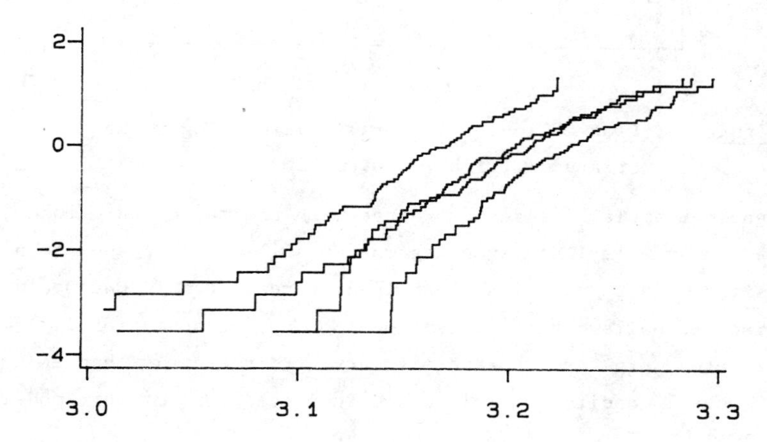

Figure 4 Plot of $\log\{-\log S(x)\}$ against $\log x$. The first and third line from the left are lengths 80 and 40 cm's from Stratum I, and the second and fourth are the lengths 80 and 40 cm's from Stratum II.

The Cox estimates for the model (5) are $\hat{\beta}=1.04$ and $\hat{\beta}=0.76$. With the size of the random variation taken into account, both estimates agree well with the hypothesis that $\beta=1$. Figure 5 shows Breslow's estimate for the underlying s.f.'s for the two strata. They are fairly linear, as predicted by the Weibull model (4).

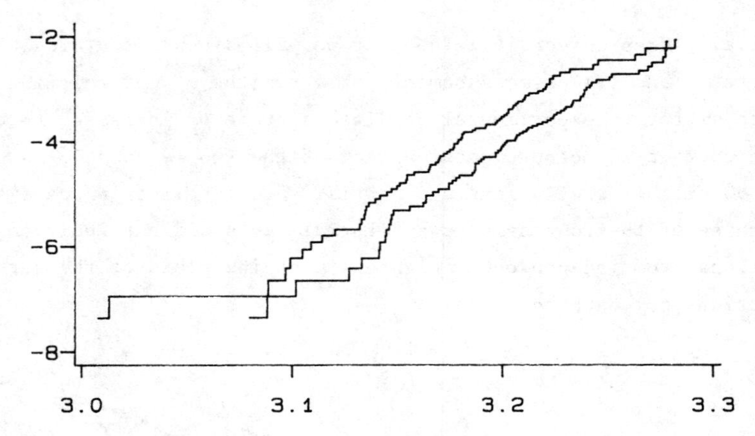

Figure 5 Plot of log {- log S(x)} against log x. Left Stratum I, right Stratum II. □

Hence Example 2 is seen to agree with the macroscopic model, while for Example 1 not even assumptions (i) and (ii) seem to be satisfied, since β is significantly different from 1. One would be inclined to believe that it is the homogeneity assumption (ii) which is violated, e.g. due to randomly varying diameters or changes in experimental conditions. This would then lead to a mixture model

$$(6) \qquad S_\ell(x) = \int S(x;\alpha)^\ell dF(\alpha),$$

where $F(\alpha)$ represents, say, variations in diameter or composition of the material.

Methods for analysing such models (sometimes called frailty models) are being developed, see [6].

If one tries to force the model (5) onto data which really come from (6), i.e. tries to find β such that

$$S_\ell(x) = \int S(x,\alpha)^\ell dF(\alpha) \approx S_1(x)^{\ell^\beta} = (\int S(x;\alpha)dF(\alpha))^{\ell^\beta}$$

then, as an easy consequence of Jensens's inequality, this leads to β - values less than 1. A further unfortunate consequence is that using (5) then leads to an overestimate of the strengths of large specimens.

In many situations one would (as for Examples 1 and 2) be rather convinced that (i) holds, and then the main practical use of the test for $\beta = 1$ is as a means to find inhomogeneities in the material or experimental setup, as mentioned above. This agrees with the experiences of Nilsson and Uvell. Initially their experiments showed very marked departures from (3) and (4), but after they had eliminated a number of causes for inhomogeneity in the material and experimental conditions, they were consistently able to obtain data agreeing with the model (4).

Acknowledgement I would like to thank Richard Smith for some valuable suggestions, in particular on the connection with the Cox model, and Bader & Priest and Nilsson & Uvell for the permission to use their data.

References

[1] Bader, M.G. and Priest, A.M. (1982). In: Progress in Science and Engineering of Composites, Hayashi, T., Kawata, K., and Umekawa, S. (Eds.) Japanese Society for Composite Materials, Tokyo, pp. 1129.

[2] Bolotin, V.V. (1969). Statistical methods in structural mechanics, Holden-Day, San Francisco, pp. 64-69.

[3] Chaplin, W.S. (1882). On the relative tensile strengths of long and short bars. Proc. of the Engineer's club, Philadelphia. 3, 15-28.

[4] Deis, A. (1986). Masters thesis in preparation. Inst. of Mathematical Statistics, University of Copenhagen.

[5] Griffith, A.A. (1921). The phenomena of rupture and flow in solids. Phil. Trans. Roy. Soc. London, Ser. A. 221, 163-198.

[6] Hougaard, P. (1984). Frailty models derived from the stable distributions. Report, Inst. of Mathematical Statistics, University of Copenhagen. 7, 1-23.

[7] Kontorova, T. and Frenkel, J. (1941). Statistical theory of brittle strength of real crystals. J. Techn. Physics. 11,173.

[10] Lawless, J.F. (1982). Statistical methods for lifetime data. Wiley, New York.

[8] Leadbetter, M.R., Lindgren, G., and Rootzén, H. (1983). Extremes and related properties of stationary sequences and processes. Springer, New York, pp. 271-273.

[9] Leadbetter, M.R., and Rootzén, H. (1986). Extremes of stochastic processes. To appear as special invited paper in Ann. Probab.

[11] Rootzén, H. (1986). Sample paths and extremes of filtered Poisson processes. Report, Inst. of Mathematical Statistics, University of Copenhagen.

[12] Weibull, W. (1939). A statistical theory of the strength of materials. Proc. Roy. Swedish Inst. Engineering Res. 151.

STATISTICAL MODELS FOR COMPOSITE MATERIALS

Richard L. Smith
University of Surrey
Guildford GU2 5XH
England

Paper presented at the First World Congress of the Bernoulli Society, Tashkent, USSR, September 1986.

Summary

Statistical models are constructed for the failure of composite materials consisting of strong parallel fibres embedded in a ductile matrix. Particular emphasis is given to the stress concentration factors, which govern the load transfer among the elements. The various methods of obtaining approximations to the probability of failure are reviewed, with emphasis on methods using extreme value theory. The last part of the paper is concerned with the verification of these models from experimental data, and with the new theoretical problems posed by this.

1. Introduction

The rapid growth of materials technology, particularly over the last twenty years, has led to the development of composite materials with properties quite different from those of classical materials. An important class of such materials consists of those in which parallel fibres of a strong but brittle material, such as carbon or glass, are embedded within a ductile matrix such as epoxy. Such materials have been found to have very high tensile strength while being much lighter than traditional materials such as steel or aluminium. The present paper is concerned with statistical models which attempt to explain the high tensile strength of these composites in terms of the strength of the individual fibres and the stress-distribution properties of the matrix. The development of such models was foreshadowed by earlier work on classical bundles, or bundles of fibres with equal load-sharing among the fibres. We first present a very brief review of the thoery of classical bundles, before going on to consider models which are more specific to composite materials. In the latter part of the paper, we shall discuss the verification of these models from experimental data.

2. Classical Bundles

Consider a bundle of fibres in tension. It is assumed that the fibres share the load equally, and moreover that they continue to share the load equally after some of the fibres have failed. Thus, if the number of fibres in the original bundle is n, the number of failed fibres is k, and the total load on the system is nL (i.e. load L per fibre), then the load on each fibre is $nL/(n-k)$. The strengths of the individual fibres, i.e. the largest loads they can support prior to failure, are assumed to be independent random variables with a common distribution function F. The model is appropriate for a system of parallel fibres in which there is no physical interaction among the fibres.

Daniels (1945) studied this model in detail and proved asymptotic normality (as $n \to \infty$) for the strength of the bundle. The problem is asymptotically equivalent to calculating the probability that Brownian motion crosses a sharply curved boundary, a subject with other areas of application (e.g. Daniels 1974) and one still of active research interest, motivated by the attempt to improve on the original normal approximation (Barbour 1981, Smith 1982a, Daniels and Skyrme 1985 and as yet unpublished work by P. Groeneboom). For a comparison of the various approximations available for classical bundles, see McCartney and Smith (1983). Phoenix and Taylor

(1973) extended the model to cover certain types of inhomogeneity among the fibres, such as random slack, but still within the basic framework of equal load-sharing.

In the 1950's B.D. Coleman, developed an alternative, dynamic, model with the emphasis on failure time rather than static strength. Some modern developments of this theory are due to Phoenix (1978, 1979) and Tierney (1981).

3. Local load-sharing and the chain-of-bundles model

Composite materials differ from classical bundles because of the matrix material, whose effect is that stress concentrations around a broken fibre are confined to a small region near the break. To analyse the statistical properties of these materials, we need a model for these local stress concentrations. Most work to date has been concerned with the chain-of-bundles model which, although oversimplified, appears to capture the most important properties of the materials.

The basic model (see Figure 1) represents the material as a system of m short bundles in series, each bundle containing n fibres. The lengths of the short bundles are taken to be equal to the ineffective length denoted δ, over which a failed fibre is inoperative. In typical composite materials, this ineffective length is of the order of 5-10 fibre diameters. The model allows us to treat the constituent short bundles as if they were independent, and thus to reduce the problem of calculating failure probabilities to the study of a single short bundle.

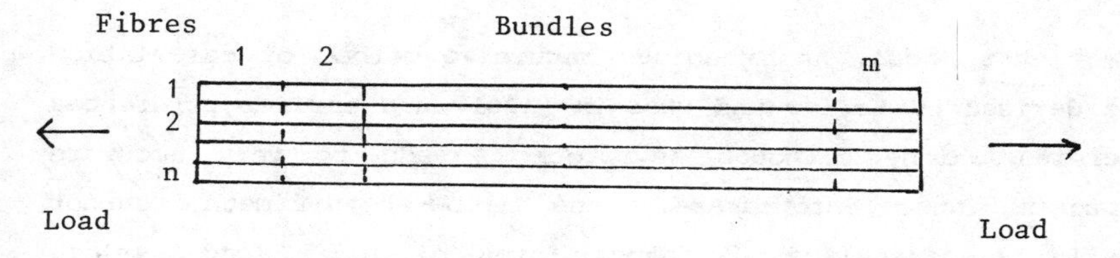

Figure 1: Chain-of-bundles model

Within each short bundle, the stress concentrations are also localised, in the sense that the stress concentrations resulting from a broken fibre are confined to fibres near the break, rather than spread over the whole bundle as with classical bundles. Theoretical calculations of stress concentration factors were made by, for example, Hedgepeth and Van Dyke (1967), and early work on the chain-of-bundles model (Rosen 1964, Zweben and Rosen 1970) was based on rough approximations for the probability of failure. The model is too complex however, for such rough approximations to be adequate, and much of the more recent work has been concerned with developing more precise approximations from a rigorous mathematical viewpoint.

A particularly simple version of this model, though one which is still complicated enough to pose challenging mathematical problems, is a one-dimensional array in which the load on a failed fibre is divided equally between its two nearest unfailed neighbours. Figure 2 illustrates a possible configuration under this model and associated stress concentration factors.

Figure 2: Example of stress concentration factors in the simple one-dimensional model

For this model, an ingenious recursive method of calculation was devised by Harlow and Phoenix (1981, and earlier references therein) which, although asymptotic, leads to very accurate approximations. Unfortunately, the Harlow-Phoenix method cannot easily be generalized to other forms of local load-sharing, though some progress in this direction was made by Pitt and

Phoenix (1982, 1983), and recent work by Kuo and Phoenix (1987), which is based on a different form of recursion, may point the way to such a generalisation. An alternative method of approximation, using extreme value theory, was devised by Smith (1980, 1982b). Although less accurate than the Harlow-Phoenix method this method is much more readily extended to cover alternative models, in particular models of three-dimensional composites. This was done by Smith et al. (1983). An extension in a different direction, to cover dynamic models and time to failure, was made by Tierney (1982).

Other recent approaches to failure-probability calculations are due to Batdorf (1982) and Tamuzs et al. (1982). These give results which are similar to the extreme-value approximations, though there are differences in the details of the calculations.

In the next section, we give a very bare outline of the use of extreme value theory to derive approximations for bundle failure probabilities. For full details, the reader is referred to the above-mentioned papers.

4. Approximations based on extreme value theory

Consider the model depicted in Figure 2. Suppose the distribution function for the strengths of individual fibres is denoted F, and let $F_n^{[k]}(x)$ denote the probability that there are at least k consecutive failed fibres in the bundle under applied load x per fibre.

The case k=1 is easy to handle, for then

$$F_n^{[1]}(x) = 1 - (1-F(x))^n. \qquad (1)$$

Limit laws for $F_n^{[1]}$ may be obtained from classical extreme value theory (Galambos 1978, Leadbetter, Lindgren and Rootzen 1983). For example, if $F(x) \sim cx^{\rho}$ as $x \downarrow 0$ $(c>0,\ \rho>0)$ then, defining $a_n^{[1]} = (cn)^{-1/\rho}$, we have

$$F_n^{[1]}(a_n^{[1]}x) \to 1-\exp(-x^{\rho}) \quad (0<x<\infty) \qquad (2)$$

as $n\to\infty$, which may also be interpreted as an approximation

$$F_n^{[1]}(x) \approx 1-\exp\{-(x/a_n^{[1]})^{\rho}\}. \quad (3)$$

This is the well-known Weibull distribution, the scaling constant $a_n^{[1]}$ having a useful interpretation as the <u>size effect</u>.

Consider now k=2. If we consider a pair of fibres in isolation, then the probability that both fail under an applied load x is $2F(x)F(K_1x)-F(x)^2$ where K_1 (here 1.5) is the stress concentration factor. This reduces to $(2K_1^{\rho}-1)c^2x^{2\rho}$ as $x\downarrow 0$. If we now define $a_n^{[2]} = [(2K_1^{\rho}-1)\, nc^2]^{1/2\rho}$, then arguments similar to (3) lead to

$$F_n^{[2]}(x) \approx 1-\exp\{-(x/a_n^{[2]})^{2\rho}\}.$$

This argument may be generalised to any k, and leads to

$$F_n^{[k]}(x) \approx 1-\exp\{-(x/a_n^{[k]})^{k\rho}\}, \quad (4)$$

where $a_n^{[k]} = [d_k nc^k]^{1/k\rho}$, and d_k is a constant depending on ρ and the stress concentration factors. This same approximation holds, with suitable definition of d_k, for three dimensional composites as well as the simple model depicted in Figure 2.

The quality of this approximation depends on choosing the right value of k. For k too large, the two sides of (4) will not be close to each other. For k too small, $F_n^{[k]}$ will not be a good approximation to the distribution function of bundle strength.

Theoretical and numerical work both suggest that for practical calculation, there is a critical value of k, say k^*, for which (4) is a very good approximation to the bundle failure probability. This k^* also has a physical interpretation as the number of failures required before the flaw in the material becomes sufficiently large to be unstable. When this happens, the stress concentrations become so large that the probability of further fibre failures, and hence failure of the material, is close to 1.

In this discussion we have only considered a single bundle in the chain-of-bundles model. Because the strength of the whole system is the strength of its weakest constituent bundle, extreme value theory may also be applied to the whole system, leading to the approximation

$$F_{m,n}^{[k]}(x) \approx 1-\exp\{-(x/a_{mn}^{[k]})^{k\rho}\} \quad (5)$$

for the probability of k consecutive failures in a chain of m bundles of n fibres. In other words, the chain-of-bundles model behaves like a single bundle of mn fibres.

One consequence of (4) and (5), with k set equal to the critical k^*, is that the Weibull shape parameter ρ, for a single fibre, is replaced by $k^*\rho$ for a bundle. Thus, if a Weibull distribution is fitted separately to bundle data and fibre data, the ratio of the two shape parameters gives an estimate of k^*.

5. Experimental verification

So far, experimental verification of these theories has been extremely limited. Nevertheless, there is a rapidly growing body of experimental work on the statistical properties of composites, including current work at the University of Surrey. Batdorf and Ghaffarian (1982) examined some limited experimental data, coming to rather negative conclusions about the suitability of the model, but this may be due to the limited amount of data they examined. Research in the Soviet Union includes the paper of Mikelsons and Gutans (1984), who applied the theory of Tamuzs et al. (1982) to boron-aluminium plastics. The following discussion is based primarily on Watson and Smith (1985), who analysed data obtained by Bader and Priest (1982) at Surrey.

Bader and Priest analysed carbon fibres made up in four different ways, (a) as single fibres (b) as dry bundles (no matrix), (c) as impregnated bundles (matrix of epoxy resin), (d) as a carbon-glass hybrid (carbon fibres embedded within a glass-epoxy composite). For each type of arrangement they also performed experiments at four different lengths. This was valuable, because it enabled us to consider whether the relationship between bundle strengths at different lengths was consistent with the weakest-link law, which is implied by the chain-of-bundles model. Combined with the Weibull distribution, this implies a relation of the form

$$F(x;\ell) = 1-\exp\{-\ell(x/x_1)^{\rho}\} \qquad (6)$$

for the distribution function of a specimen of length ℓ.

The distribution (6) was fitted, by numerical maximum likelihood, to each of the four data sets, with results given in Table 1. For the present discussion, interest lies in the relationship between the single-fibre parameters and the parameters for the impregnated and hybrid bundles.

TABLE 1

Equation (6) fitted to four kinds of assembly (standard errors in parenthesis)

	Single fibres	Dry bundles	Impregnated bundles	Hybrid bundles
x_1(GPa)	4.77	2.16	3.55	4.31
	(0.08)	(0.03)	(0.06)	(0.02)
ρ	5.6	14.7	18.7	26.4
	(0.2)	(0.8)	(1.2)	(1.1)

One object of interest here is the ratio between Weibull shape parameters of single fibres and bundles. For example, 18.7/5.6 = 3.3 which implies, by the theory outlined in Section 4, that the critical flaw size for impregnated bundles is between 3 and 4. Comparison of the corresponding values of x_1 (Watson and Smith, 1985) shows good consistency with the theory. For this comparison, we took δ=0.04mm and used a rather simple approximation for the value of d_k. For hybrid bundles, we find 26.4/5.6 = 4.7, implying a critical flaw size between 4 and 5, but in this case the bundle turns out to be stronger (by about 25%) than predicted by the theory. This implies a definite "hybrid effect" which is not explained by any theory developed so far.

Another aspect of the data analysis is whether model (6) is an adequate fit to the data collected for different lengths of specimen. There appears to be a serious question about this; Watson and Smith found that the ad hoc adjustment

$$F(x;\ell) = 1-\exp\{-\ell^{\alpha}(x/x_1)^{\rho}\}, \tag{7}$$

with $0<\alpha\leq 1$, fitted better than (6). The value of α was 0.90 for single fibres (not significantly different from 1), but 0.58 for impregnated bundles and 0.48 for hybrid bundles (both highly significant). It is difficult, however, to find a plausible physical explanation why (7) should fit better than (6); one was proposed by Watson and Smith, but it is by no means conclusive. The distribution (7) was also proposed by Gutans and Tamuzs (1984), but also apparently without physical explanation.

More recent work has focussed on the stress concentrations in hybrid composites. Bader and Pitkethly (1986) have reported more detailed experimental work on glass-carbon hybrids, and have suggested how the change of stress concentration factors at the edge of the carbon bundle could explain at least part of the hybrid effect. Future work may well be focussed on the chain-of-bundles model itself; the assumption that the short bundles are independent is obviously not very realistic, and it is possible that a model introducing some dependence among the short bundles would explain the failure of the weakest-link relation as well as casting further insight on the hybrid effect.

6. Conclusions

The chain-of-bundles model was proposed around twenty years ago to describe the statistical properties of composite materials. For many years, adequate mathematical analysis of the model was lacking, but this problem has now largely been solved through the development of suitable approximations to the probability of failure. Comparison with experimental data has confirmed the model in some respects but has highlighted weaknesses in others. Future work is likely to concentrate on more precise modelling of the stress concentrations and on a relaxation of the chain-of-bundles model itself.

References

Bader, M.G. and Pitkethly, M. (1986), Probabilistic aspects of the strength and modes of failure of hybrid fibre composites. To appear in the Proceedings of the Symposiums on Mechanical Characterisation of Fibre Composite Materials, Aalborg University, Denmark, June 1986.

Bader, M.G. and Priest, A.M. (1982), Statistical aspects of fibre and bundle strength in hybrid composites. In Progress in Science and Japanese Society for Composite Materials, Tokyo, p.1129.

Barbour, A.D. (1981), Brownian motion and a sharply curved boundary. Adv. Appl. Prob. 13, 736-750.

Batdorf, S.B. (1982), Tensile strength of unidirectionally reinforced composites I, J. Reinf. Plast. Compos. 1, 153-164.

Batdorf, S.B. and Ghaffarian, R. (1982), Tensile strength of unidirectionally reinforced composites II. J. Reinf. Plast. Compos. 1, 165-176.

Daniels, H.E. (1945), The statistical theory of the strength of bundles of threads I. Proc. R. Soc. A 183, 404-435.

Daniels, H.E. (1974), The maximum size of a closed epidemic. Adv. Appl. Prob. 6, 607-621.

Daniels, H.E. and Skyrme, T.H.R. (1985), The Maximum of a random walk whose mean path has a maximum. Adv. Appl. Prob. 17, 85-99.

Galambos, J. (1978), The Asymptotic Theory of Extreme Order Statistics. Wiley, New York.

Gutans, J. and Tamuzs, V. (1984), Scale effect of the Weibull distribution of fibre strength. Mech. of Compos. Mater., 20, 1107-1109.

Harlow, D.G. and Phoenix, S.L. (1981), Probability distributions for the strength of composite materials, I and II. Int. J. Fracture 17, 347-372 and 601-630.

Hedgepeth, J.M. and Van Dyke, P. (1967), Local stress concentrations in imperfect filamentary composite materials J. Compos. Mater. 1, 294-309.

Kuo, C.C and Phoenix, S.L. (1987), A limit theorem for size effects in the strength and lifetime distributions of a fibrous composite. To appear in Adv. Appl. Prob.

Leadbetter, M.R., Lindgren, G. and Rootzen, H. (1983), Extremes and Related Properties of Random Sequences and Processes. Springer Verlag, New York.

McCartney, L.N. and Smith, R.L. (1983), Statistical theory of the strength of fibre bundles. J. Appl. Mechanics 50, 601-608.

Mikelsons, M. and Gutans, J. (1984), Failure of boron-aluminium plastics under static and cyclic tension. Mech. of Compos. Mater., 20 52-59 (in Russian).

Phoenix, S.L. (1978), The asymptotic time to failure of a mechanical system of parallel members. SIAM J. Appl. Math 34, 294-309.

Phoenix, S.L. (1979), The asymptotic distribution for the time to failure of a fibre bundle. Adv. Appl. Prob. 11, 153-187.

Phoenix, S.L. and Taylor, H.M. (1973), The asymptotic strength distribution of a general fibre bundle. Adv. Appl. Prob. 5, 200-216.

Pitt, R.E. and Phoenix, S.L. (1982, 1983), Probability distributions for the strength of composite materials, III and IV. Int. J. Fracture 20, 291-311 and 22, 243-276.

Rosen, B.W. (1964), Tensile failure of fibrous composites. AIAA J. 2, 1985-1991.

Smith, R.L. (1980), A probability model for fibrous composites with local load sharing. Proc. R. Soc. A 372, 539-553.

Smith, R.L. (1982a), The asymptotic distribution of the strength of a series-parallel system with equal load-sharing. Ann. Prob. 10, 137-171.

Smith, R.L. (1982b), A note on a probability model for fibrous composites. Proc. R. Soc. A. 382, 179-182.

Smith, R.L., Phoenix, S.L., Greenfield, M., Henstenburg, R.B. and Pitt, R.E. (1983), Lower-tail approximations for the probability of failure in fibrous systems with hexagonal geometry. Proc. R. Soc. A. 388, 353-391.

Tamuzs, V., Azarova, M., Bondarenko, V., Gutans, J., Korabelnikov, J., Pikshe, P. and Silujanov, O. (1982). Fracture of unidirectional carbon fibre plastics and realization of fibre strength properties in them. Mech. of Compos. Mater., 18 34-41. (In Russian).

Tierney, L.J. (1981), The asymptotic time of failure of a bundle of fibres with random slacks. Adv. Appl. Prob. 13, 548-566.

Tierney, L.J. (1982), Asymptotic bounds on the time to fatigue failure of bundles of fibres under local load-sharing. Adv. Appl. Prob. 14, 95-121.

Watson, A.S. and Smith, R.L. (1985), An examination of statistical theories for fibrous materials in the light of experimental data. J. Mater. Sci. 20, 3260-3270.

Zweben, C. and Rosen, B.W. (1970), A statistical theory of material strength with application to composite materials. J. Mech. Phys. Solids, 18, 189-206.

THE DISTRIBUTION OF BUNDLE STRENGTH UNDER GENERAL ASSUMPTIONS

Daniels, H.E.,
Statistical Laboratory,
University of Cambridge,
16 Mill Lane,
Cambridge CB2 1SB,
U.K.

1. In the "classical" bundle of n parallel fibres stretched between two clamps, the breaking loads of the individual fibres are assumed to be independently distributed with a common distribution function and load is shared equally between surviving fibres (Daniels [2]). If the individual breaking loads are $\ell_{(1)} < \ell_{(2)} < \dots < \ell_{(n)}$, the bundle breaks at total load $L^* = \max_k \, (n-k+1)\ell_{(k)}$.

Phoenix and Taylor [6] showed that by reformulating the problem in terms of extension (strain) x rather than load ℓ the assumptions of the classical model could be relaxed to include more realistic models. Assuming Hooke's law, we have $\ell = \beta x$ before the fibre breaks, and for convenience we can take $\beta = 1$. Let ξ be the breaking extension of the fibre, with distribution function $F(\xi)$. Then the load on the fibre at extension x is $\ell(x) = xH(\xi-x)$ where $H(.)$ is the Heaviside function, and

$$E\ell(x) = x[1-F(x)] = \mu(x),$$

$$\mathrm{cov}\; \ell(x_1)\ell(x_2) = x_1x_2F(x_1)[1-F(x_2)] = c(x_1,x_2), \quad x_1 \le x_2.$$

Assuming the ξ's are independent, the total bundle load at extension x, $L_n(x) = \Sigma_1^n \ell_k(x) = x\Sigma_1^n H(\xi_k-x)$, has mean $n\mu(x)$ and convariance function $nc(x_1,x_2)$, $x_1 \le x_2$.

We require an approximation to $P(L^* \le L)$ where $L^* = \max L_n(x)$, $0 \le x < \infty$, which it attains at $x = x^*$.

Daniels (1945) showed that L^* is asymptotically normal with mean $n\mu(\tilde{x})$ and variance $n\tilde{x}^2F(\tilde{x})[1-F(\tilde{x})]$. Phoenix and Taylor [6] proved that the error was $(O(n^{-1/6})$. This approximation ignores the variation of x^* about $\tilde{x}$ which is $O(n^{-1/3})$.

When n is large, $L_n(x)$ approximates to a Gaussian process with covariance function $nc(x_1,x_2)$. But with the time transformation $t = F(x)$, $L_n(x)/x - n[1-F(x)]$ approximates to a rescaled Brownian Bridge on (0,1). This makes it possible to allow for the variation of x^* and improve the approximation error to $0(n^{-1/3})$ by increasing the mean by an amount $0(n^{1/3})$ (Daniels [3], Barbour [1], Smith [7], Daniels and Skyrme [4]). It is found that L^* is approximately normal with mean $n\mu(\tilde{x}) + \lambda n\ \ [\mu(\tilde{x})]\ [-\mu''(\tilde{x})]^{-1/3}$ and variance as before, where $\lambda = 0.995...$ is a certain constant, and $-\mu''(\tilde{x}) > 0$.

2. A generalization of the model which can be treated in a similar way is to allow the fibre cross-section to vary independently of its breaking stree and to assume the load is shared by the fibres in proportion to their cross-section. If the kth fibre has cross-section β_k it carries the load $\ell_k = \beta_k xH(\xi_k - x)$. Let $E(\beta_k) = \beta$, $\text{var}\ \beta_k = \sigma^2$, and $v = \sigma/\beta$. Then $E\ \ell_k(x) = \beta x[1-F(x)] = \beta\mu(x)$ as before, but

$$c(x_1,x_2) = \beta^2 x_1 x_2 [F(x_1) + V^2]\ [1-F(x_2)],\ x_1 \le x_2.$$

Rearranging this as

$$c(x_1,x_2) = \beta^2(1+V^2)^2 \left\{ \frac{F(x_1)+V^2}{1+V^2} \right\} \left\{ 1 - \frac{F(x_2)+V^2}{1+V^2} \right\}$$

the transformation $t = [F(x_1)+V^2]/[1+V^2]$ is again seen to reduce the problem to consideration of a Brownian Bridge on (0,1). It is found as before that L^* is approximately normal with

$$E(L^*) \sim \beta\{n\mu(\tilde{x}) + \lambda n^{1/3}(1+V^2)^{1/3}[\mu(\tilde{x})]\ [-\mu''(\tilde{x})]^{-1/3}\}$$

$$\text{var}\ L^* \sim \beta^2 n\tilde{x}^2 [F(\tilde{x})+V^2][1-F(\tilde{x})].$$

3. This seems to be the only generalization allowing a transformation to a Brownian Bridge. But the theory only requires the process to behave <u>locally</u> like a Brownian Bridge in a neighbourhood $x-\tilde{x} = 0(n^{-1/3})$ of $\tilde{x}$. Using this idea, a much wider class of models can be included. The idea of considering local Brownian behaviour was suggest by Durbin's [5] use of it for a boundary crossing problem, but our approach is rather different.

Consider a rescaled Brownian Bridge on (x_0, x_0-a), with covariance

$$c(x_1,x_2) = A.(x_1-x_0)[1-(x_2-x_0)/a,\ x_1 \le x_2.$$

Expand it near $\tilde{x}$, $x_0 < \tilde{x} < x_0+a$, within a neighbourhood $x-\tilde{x} = 0(n^{-1/3})$, so that

$$c(x_1,x_2) = A\{(\tilde{x}-x_0)[1-(\tilde{x}-x_0)/a]+[1-(\tilde{x}-x_0)/a](x_1-\tilde{x}) - [(\tilde{x}-x_0)/a](x_2-\tilde{x}) + O(n^{-2/3}),$$

the bilinear term being $O(n^{-2/3})$. For a given model, match this up with the linear expansion

$$c(x_1,x_2) = c(\tilde{x},\tilde{x}) + c'(\tilde{x},\tilde{x})(x_1-\tilde{x}) + \dot{c}(\tilde{x},\tilde{x})(x_2-\tilde{x}) + O(n^{-2/3}),$$

where $c'(x_1,x_2) = \partial c/\partial x_1$, $\dot{c}(x_1,x_2) = \partial c/\partial x_2$. Then, writing $c,c',\dot{c}$ for brevity, $c = A.(x-x_0)[1-(\tilde{x}-x_0)/a]$, $c' = A.[1-(\tilde{x}-x_0/a]$, $\dot{c} = -A(\tilde{x}-x_0)/a$. The process will be locally a Brownian Bridge if $c' > 0$, $\dot{c} < 0$, or Brownian motion if $\dot{c} = 0$. We find that $A = c'+|\dot{c}|$, $x_0 = \tilde{x}-c/c'$, $a = c[1/c+1/|\dot{c}|$.

As an example, consider the classical bundle for which $c(x_1,x_2) = x_1x_2F(x_1)[1-F(x_2)]$, $\mu(x) = x[1-F(x)]$ and $\mu'(\tilde{x}) = 0$. It gives $c' = \tilde{x}[1-F(\tilde{x})]$, $\dot{c} = 0$ and therefore behaves locally like Brownian motion with $c(x_1,x_2) = \tilde{x}[1-F(\tilde{x})]\{x_1-\tilde{x}[1-F(\tilde{x})]\} + O(n^{-}\ )$. Proceeding on this basis we arrive at the same approximation as before for $P(L^* \leq L)$ without having to transform the process.

The method can be used for the models incorporating random slack and plastic yield, proposed by Phoenix and Taylor [6] which are not amenable to transformation.

References

1. Barbour A.D. A note on the maximum size of a closed epidemic. - J. Roy. Statist. Soc., Ser. A, 1975, vol. 37, p. 137-171.
2. Daniels H.E. The statistical theory of the strength of bundles of threads. - Proc. Roy. Soc. London, Ser. A, 1945, vol. 1831, p. 405-435.
3. Daniels H.E. The maximum size of a closed epidemic. - Adv. Appl. Prob., 1974, vol. 6, p. 607-621.
4. Daniels, H.E. and Skyrme, T.H.R. The maximum of a random walk whose mean path has a maximum. Adv. Appl. Prob. 1985, vol. 17, p. 85-99.
5. Durbin, J. The first passage density of a continuous Gaussian process to a general boundary. J. Appl. Prob. 1985, vol. 22, p. 99-122.
6. Phoenix S.L. and Taylor H.M. The asymptotic strength of a general fibre bundle. - Adv. Appl. Prob., 1973, vol. 5, p. 200-216.
7. Smith R.L. The asymptotic distribution of the strength of a series-parallel system with equal load sharing. - Ann. Prob., 1982, vol. 10, p. 137-171.

AN ESTIMATE OF THE RATE OF CONVERGENCE IN THE LAW OF LARGE NUMBERS FOR SUMS OF ORDER STATISTICS AND THEIR APPLICATIONS.

Gafurov M.U., Khamdamov I.M., Institute of mathematics
AN UzSSR, Tashkent, USSR.

1. Let $X_1, \dots, X_n$ be independent, identically distributed random variables (r.v.) and let $X_n^{(1)}, \dots, X_n^{(n)}$ be corresponding order statistics, $X_n^{(1)} \geqslant \dots \geqslant X_n^{(n)}$.

Many papers are devoted to the investigation of distribution $S_n^{(r)} = \sum_{j=r+1}^{n} X_n^{(j)}$, $0 \leqslant r \leqslant n$ both for fixed and growing with n values of r, where the validness of the law of large numbers, of the law iterated logarithms, central limit theorem and similar questions were studied.

In the present paper we continue these investigations for a fixed $r \geqslant 0$. The main attention was paid to necessary and sufficient conditions for the convergence of large deviation probability series, to theorems on super large deviations and their refinements. Some applications of these theorems to the boundary value problems for $S_n^{(r)}$ are discussed too.

2. Reults.

Theorem 2.1. Let $r \geqslant 1$ and for some constant $C > 0$

$$P\{S_{n-k}^{(r)} > -C\} \geqslant q > 0$$

Then for all $x > -C$

$$P\{\max_{r \leq k \leqslant n} S_k^{(r)} \geqslant x\} \leqslant \frac{1}{q} P\{S_n^{(r)} \geqslant x - (r+1)C\}$$

This result have a important applications in proof of convergence of large deviation probability series, and in theorems on super large deviations too.

Given a non-negative monotone increasing function $H(x)$ and a non-negative function $\varphi(x)$, consider the following series

$$P^{(r)}(\varphi, H, \varepsilon) = \sum_{n>r} \varphi(n) P\{S_n^{(r)} \geq \varepsilon H(n)\},$$

$$M^{(r)}(\varphi, H, \varepsilon) = \sum_{n>r} \varphi(n) P\{\max_{r<k\leq n} S_k^{(r)} \geq \varepsilon H(n)\},$$

$$S^{(r)}(\varphi, H, \varepsilon) = \sum_{n>r} \varphi(n) P\{\sup_{k>n} \frac{S_k^{(r)}}{H(k)} \geq \varepsilon\}$$

Let $M = \{\varphi(x);\ \varphi(x)\geq 0,\ x\in[1,\infty), x^{1-\delta_1}\varphi(x)\uparrow$ and $e^{-\delta_2 x}\varphi(x)\downarrow$ for $x \geq x_0 > 1$ and for some $\delta_1, \delta_2 > 0\}$

Theorem 2.2. Suppose that $\varphi\in M$, $H(x)\uparrow\uparrow$, $x\in[1,\infty)$ and $H(x)$ satisfies condition

$$\limsup_{x\to\infty} \frac{H(cx)}{H(x)} < \infty \quad \forall c>1.$$

If $\liminf_{n\to\infty} P\{S_n^{(r)} \geq -\varepsilon H(n)\} > 0$ then

$$P^{(r)}(\varphi, H, \varepsilon) < \infty\ \forall\varepsilon>0 \Leftrightarrow M^{(r)}(\varphi, H, \varepsilon) < \infty\ \forall\varepsilon>0 \Leftrightarrow S^{(r)}(\varphi, H, \varepsilon) < \infty\ \forall\varepsilon>0$$

Theorem 2.3. Let α and β -are some real numbers, $\alpha<\beta$. Suppose that $EX_1=0$ for $\alpha\geq 1$ and $x^{+\alpha}F(-x)\to 0$, $P(X_1>x)\sim x^{-\beta}L(x)$ as $x\to\infty$, where $L(x)$ is slowly changing function.

If take place one of the following conditions

1) $x^{\alpha} \geq cn$, $\alpha\in(0,1)$;

2) $x^{\alpha}/\ln^{1-\alpha} x \geq cn$, $\alpha\in[1,2)$;

3) $x/\ln x \geq c\sqrt{n}$, $\alpha=2$, $EX_1^2=\infty$;

4) $x/\sqrt{\ln x} \geq c\sqrt{n}$, $\alpha=2$, $EX_1^2=1$;

then as $n \to \infty$

$$P\{S_n^{(r)} \geq x\} \sim P\{X_n^{(r+1)} \geq x\} \sim \frac{1}{(r+1)!}\left[n P(X_1 \geq x)\right]^{r+1}$$

3. Applications.

Now we show two applications for results, which takes above. Let $H(x)$ is positive strong increasing function in $[1,\infty)$ and let

$$\chi_r(\varepsilon) = \sup\{n \geq r;\ S_n^{(r)} \geq \varepsilon H(n)\},\quad \eta_r(\varepsilon) = \sup_{n \geq r}\left[S_n^{(r)} - \varepsilon H(n)\right]^+$$

Theorem 3.1. Let α and β are some real numbers, $\alpha < \beta$. Suppose that $EX_1 = 0$ for $\alpha \geq 1$ and $x^\alpha F(-x) \to 0$, $P(X_1 > x) \sim x^{-\beta} L(x)$ as $x \to \infty$. Futhermore, let take plase one of the following conditions

1) $H(n)/n^{1/\alpha} \uparrow$, $\alpha \in (0,1)$;

2) $H(n)/[n^{1/\alpha} \ln^{1-1/\alpha} H(n)] \uparrow$, $\alpha \in [1,2)$;

3) $H(n)/[n^{1/2} \ln H(n)] \uparrow$, $\alpha = 2$, $EX_1^2 = \infty$;

4) $C_0(r,\beta)/\varepsilon \leq H(n)/[n \ln H(n)] \uparrow$, $\alpha = 2$, $EX_1^2 = 1$.

Then

$$P\{\chi_r(\varepsilon) \geq n\} \asymp [n P(X_1 \geq \varepsilon H(n))]^{r+1},$$

$$P\{\eta_r(\varepsilon) \geq \varepsilon H(n)\} \asymp [n P(X_1 \geq \varepsilon H(n))]^{r+1}.$$

for all $\varepsilon > 0$, as $n \to \infty$

Theorem 3.2. Let $EX_1 = 0$, $EX_1^2 < \infty$ and

$$P(X_1 > x) \sim x^{-\beta} L(x), \ \beta > 2$$

Then for any $a > 0$

$$\left(S^{(r)}_{[n\cdot]}/n \,\middle|\, S^{(r)}_n > na \right) \overset{\mathcal{D}}{\Longrightarrow} J_{\alpha,\beta}\, I\{U \leq \cdot\}$$

where $P\{J_{\alpha,\beta} > x\} = (x/a)^{-\beta(r+1)}$ for $x \geq a$ and $P(U \leq t) = t^{r+1}$ for $0 \leq t \leq 1$.

REFERENCES.

Hatori H.,Maejima M.,Mori T. Convergence Rates in the Law of Large Numbers when Extreme Terms are Excluded. Z.Wahrsceinlichkeitstheorie Verw.Geb. 1979,B47,H1,p1-12

Gafurov M.U.,Slastnicov A.D.,Ischanov R. Convergence of large deviation probability series for sums without extemal terms DAN Uzb.SSR,1984,11,p.5-7

The index of the outstanding observation among n independent ones

L. de Haan, Erasmus University Rotterdam
I. Weissman, Technion, Haifa

INTRODUCTION

Let $X_1, X_2, \ldots$ be a sequence of independent random variables with distribution functions $F_1, F_2, \ldots$ respectively and let $M_n :=$ $\max(X_1, \ldots, X_n)$ for $n = 1, 2, \ldots$ Suppose there exist a non-degenerate distribution function G and sequences $a_n > 0$ and b_n $(n = 1, 2, \ldots)$ such that

$$(1) \qquad P\{\frac{M_n - b_n}{a_n} < x\} \to G(x) \text{ vaguely } (n\to\infty).$$

Mejzler (1956) characterized the limit distributions G under the regularity conditions

$$(2) \qquad |\log a_n| + |b_n| \to \infty,\ a_{n+1}/a_n \to 1,\ (b_{n+1} - b_n)/a_n \to 0\ (n\to\infty),$$

implied by

$$\min_{1\leq k\leq n} F_k(a_n x + b_n) \to 1 \ (n\to\infty) \text{ for all } x > {}_*x(G)$$ [1].

We call this class of distributions M.

Proposition (Mejzler 1956, Mejzler and Weissman 1969).
Any non-degenerate distributions function $G \in \mathcal{M}$ satisfies[1])

$$- \log G(x) \text{ is convex if } x^* = x^*(G) = \infty;$$
$$- \log G(x^* - e^{-x}) \text{ is convex if } x^* < \infty.$$

Conversely any distribution function satisfying this convexity property is in the class $\mathcal{M}$.
Now assume that (1) and (2) hold and $F_1, F_2, \ldots$ are continuous distribution functions. Let L_n be the index of the largest observation among $X_1, X_2, \ldots, X_n$. Clearly, if the X's are i.i.d., then L_n/n is asymptotically uniform [0, 1]. We investigate limiting distributions for L_n. It is proved e.g. that L_n/n has a non-degenerate limit distribution if

$$\text{(3)} \qquad \lim_{n\to\infty} \frac{a_n}{a_{[nt]}} = t^\rho, \quad \lim_{n\to\infty} \frac{b_n - b_{[nt]}}{a_{[nt]}} = -c \, \frac{t^\rho - 1}{\rho} \text{ for } 0 < t \leq 1$$

holds. The converse is true e.g. if $a_n \equiv 1$ in (1). We were unable to prove or disprove the conjecture that under (1) and (2) the distribution of L_n/n converges to a non-degenerate limit distribution if and only if (3) holds. Some examples related to this question have been studied.

1) For any distribution function P we define
$${}_*x(P) = \inf\{x \mid P(x) > 0\}$$
$$x^*(P) = \sup\{x \mid P(x) < 1\}.$$

RAIN FLOW CYCLE DISTRIBUTIONS FOR FATIGUE LIFE PREDICTION UNDER GAUSSIAN LOAD PROCESSES

Georg Lindgren & Igor Rychlik
Dept of Mathematical Statistics, Box 118, S-221 00 Lund, Sweden

Statistical wave analysis has found important application in the study of metal fatigue. When a piece of metal is subjected to a periodically varying load, microscopic inhomogeneities can develop into open cracks, leading to fatigue failure after a random time. The fatigue life depends on the amplitudes of the applied "load cycles" S_k, which are functions of the sequence of maxima and minima in the load process. A commonly used damage rule, due to Palmgren and Miner, postulates that the total damage caused by a stress history $\{S_k\}$ is

$$D(t) = \sum_{k=1}^{k(t)} 1/N_{S_k},$$

where the sum is extended over all cycles completed at time t, and N_S is the cycle life obtained from fatigue test with constant amplitude S. Fatigue failure occurs when $D(t)$ exceeds one. In most situations, N_S is large, and then, by ergodicity, the fatigue life is

$$T = \{\int N_s^{-1} f_S(s)ds\}^{-1}, \qquad (1)$$

where f_S is the density of the ergodic (long run) distribution of the cycle amplitudes S_k.

One important definition of cycle amplitudes, regarded to give the best predictions of fatigue life, is the Rain Flow Cycle count method (RFC). It is designed to catch both slow and rapid variations in the load process by forming cycles from pairs of high maxima and low minima even if they are separated by intermediate extremes.

A major point in the present paper is the following simple formulation of the RFC count, first presented in Rychlik (1986b). Let the load process $\xi(t)$ have a local maximum at t_k with height $u = \xi(t_k)$, and let t_k^- and t_k^+ be the times of the last and first down- and upcrossing of the level u before and after t_k. With

$$S_k^- = \xi(t_k)-\min\{\xi(t);t_k^-<t<t_k\},$$
$$S_k^+ = \xi(t_k)-\min\{\xi(t);t_k<t<t_k^+\},$$

the RFC-amplitude at t_k, is

$$S_k = \min(S_k^-, S_k^+).$$

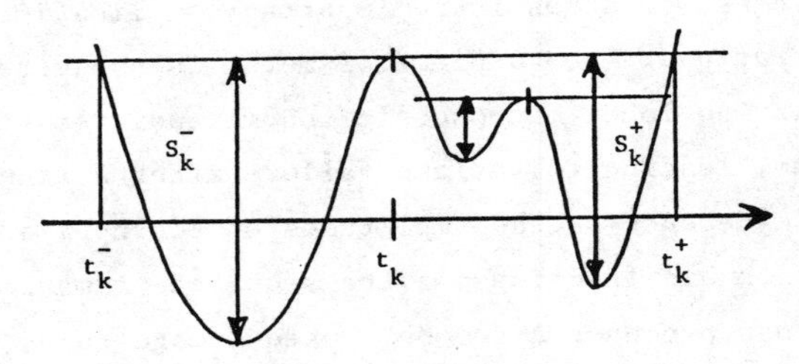

Figure 1 Definition of RFC-amplitude

Obvioulsy $S_k > h$, if and only if $\xi(t)$ reaches the level $u-h$ on both sides of t_k before it exceeds u again, time going both ways. Thus, the distribution of S_k is the solution to a first passage problem in the sequence of local maxima and minima.

Denote by $M_i = M_i(t_k)$, $i=0,\pm1,\ldots$, the sequence of local extrema of the function $\xi(t)$ indexed so that $M_0 = M_0(t_k)=\xi(t_k)$, and $M_{-1}(t_k)$, $M_1(t_k)$ are the surrounding local minima. The ergodic (long run) distribution of the RFC-amplitudes can be defined in terms of the ergodic distribution of the $\{M_i\}$-process. In particular, the conditional distribution of the RFC-amplitude S_k, given $M_0=u$, is

$$F_{S_k|M_0}(h|u) = 1 - P(M_i \text{ crosses } u-h \text{ before it crosses } u \text{ as } i \to \pm\infty \mid M_0=u) = \tag{2}$$
$$= 1 - \sum_{i,j=1}^{\infty} P(M_{-i}\le u-h, M_j\le u-h, u-h<M_n\le u \text{ for } -i<n<j \mid M_0=u).$$

In many cases the dependence between the extremes of a Gaussian load process can be approximated by an n-step Markov chain, and then the probabilities in (2) can be evaluated as follows.

First consider a one-step Markov chain of extremes $\{M_i\}$. Denote by $p(s_0)$, $u-h\le s_0\le u$, the conditional probability that M_i, $i>0$, crosses $u-h$ before it exceeds u, given $M_0=s_0$. Since $\{M_i\}$, $i<0$, and $\{M_j\}$, $j>0$, are conditionally independent, given $M_0=s_0$, the conditional distribution of the RFC-amplitude is

$$F_{S_k|M_0}(h|u) = 1 - p(u)^2.$$

To find $p(u)$, denote by

$$f_{M_2,M_1|M_0}(s_2,s_1|s_0), \text{ etc.},$$

the conditional densities of M_2, M_1 given $M_0=s_0$, etc, and write

$$f(s_2;s_0) = \int_{u-h}^{u} f_{M_2,M_1|M_0}(s_2,s_1|s_0)ds_1.$$

Then $p(s_0)$ is the unique solution of the integral equation

$$p(s_0) = P(M_1<u-h|M_0=s_0) + \int_{u-h}^{u} p(s_2)f(s_2;s_0)ds_2.$$

A two-step Markov chain of extremes can be treated similarly. If $p(s_0,s_1)$, $u-h\leq s_0,s_1\leq u$, is the conditional probability that M_i, $i>1$, crosses $u-h$ before it crosses u, given $M_0=s_0$ and $M_1=s_1$, the conditional distribution of the RFC-amplitude, given $M_0=u$, is

$$F_{S_k|M_0}(h|u) = 1 - P(M_{-1}\leq u-h,\ M_1\leq u-h|M_0=u) -$$

$$- \int_{u-h}^{u} P(M_{-1}\leq u-h|M_0=u,M_1=s_1)p(u,s_1)\cdot f_{M_1|M_0}(s_1|u)ds_1 -$$

$$- \int_{u-h}^{u}\int_{u-h}^{u} p(u,s_{-1})p(u,s_1)\cdot f_{M_{-1},M_1|M_0}(s_{-1},s_1|u)ds_{-1}ds_1.$$

Here, p is the unique solution of the integral equation

$$p(s_0,s_1) = P(M_3\leq u-h,\ u-h<M_2\leq u|M_0=s_0,M_1=s_1) +$$

$$+ \int_{u-h}^{u}\int_{u-h}^{u} p(s_2,s_3)\cdot f_{M_3,M_2|M_1,M_0}(s_3,s_2|s_1,s_0)ds_2ds_3.$$

The unconditional distribution of the RFC-amplitude is obtained by integrating the conditional density. The mean value of the RFC-amplitude is $E(S) = (2\pi\lambda_0)^{\frac{1}{2}}\alpha$, as derived by Rychlik (1986b), who also approximated the transition probabilities by means of a regression technique in a Slepian model, see Rychlik (1986a). Here $\alpha=\lambda_2/(\lambda_0\lambda_4)^{\frac{1}{2}}$ is a spectral width parameter, $\lambda_0=\mathrm{Var}(\xi(0))$,

$\lambda_2=\mathrm{Var}(\xi'(0))$, $\lambda_4=\mathrm{Var}(\xi''(0))$.

Figure 2 shows simulated RFC-distributions (Wirshing & Shehata (1977)), compared with the theoretical distribution, based on a one-step Markov chain approximation, for low frequency Gaussian white noise with $\alpha=0.745$. Figure 3 shows theoretical RFC-distributions, based on one-step and two-step Markov approximations.

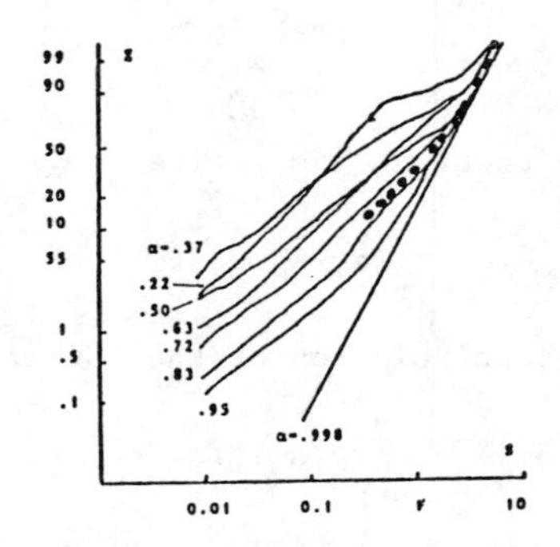

Figure 2 Simulated RFC-distributions for different α ———: (from Wirshing & Shehata (1977)); theoretical S-distribution ......, based on a one-step Markov assumption, plotted on Weibull probability paper.

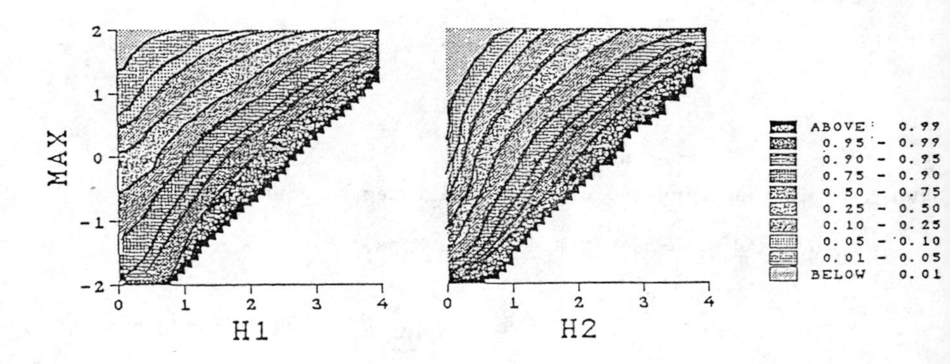

Figure 3 Isolines of approximative conditional distributions of RFC-amplitude given maximum, $F_{S|M}(h,u)$, $0\leq h\leq 4$ and $-2\leq u\leq 2$, based on one- and two-step Markov assumptions, H_1 and H_2, respectively.

In order to use (1) to predict fatigue life one also needs a model for the variation of the strength function N_s between units. We therefore assume there exists a standard strength function $\tilde{N}_s$, which gives the fatigue life if no random variability between units is present. A commonly used functional form for $\tilde{N}_s$ is

$$\tilde{N}_s = \begin{cases} as^{-k} & \text{for} \quad s \geq \tilde{s}_\infty, \\ \infty & \text{for} \quad s < \tilde{s}_\infty, \end{cases} \tag{3}$$

where $\tilde{s}_\infty$ is the fatigue limit.

In fatigue life experiments, where each unit is tested on only one stress level s, one sometimes finds the same variability in each set of data obtained from single stress levels. Further, in very long tests, some, but not all, units survive the whole test period. This has been taken as an indication that the fatigue limit $\tilde{s}_\infty$ is random.

A realistic model for the individual quality, which describes these phenomena, is obtained by introducing a multiplicative random fatigue strength and a random fatigue limit, taking

$$N_s = \nu \tilde{N}_{s\eta}, \tag{4}$$

where ν and η are random variables. In a log S – log N diagram, N_s defined by (4) is obtained from $\tilde{N}_s$ by shifting it by $-\log \eta$ and $\log \nu$, respectively. It is standard technique in fatigue life analysis to allow such random vertical and horizontal shifts in the S-N-curve.

Further, to describe the fact that the dispersion in the fatigue life distribution can be different for different stress levels, one can introduce a leveldependent strength variation, allowing different units to have different sensitivity on different levels. This can be achieved in the model

$$N_s = \nu (s\eta/s_0)^{-\gamma} \cdot \tilde{N}_{s\eta}, \tag{5}$$

where the random variable γ changes the form of the strength function, with $s = s_0/\eta$ as a fixed point. If N_s is defined by (5), the effect of γ is simply to change the slope k in the log S – log N diagram.

If the strength function is given by (3) with fatigue limit $\tilde{s}_\infty=0$, the fatigue life is the inverse of the k'th moment of the load cycle amplitude. In the following table we have compared the moments (evaluated numerically) of the RFC-amplitude with the corresponding moments of a Rayleigh variable with the same mean. Calculations are based on a one-step Markov sequence of extremes in a Gaussian low frequency white noise process.

Table 1 Moments of order k for RFC- and Rayleigh distribution

k	k'th RFC-moment	k'th Rayleigh-moment
1	1.86	1.87
2	5.53	4.47
3	21.56	12.56
4	103.69	39.95

REFERENCES

Rychlik, I. (1986a). Regression approximations of wave-length and amplitude distributions. To appear in Advances in Applied Probability.

Rychlik, I. (1986b). Rain Flow Cycle distribution for a stationary Gaussian load process. To appear in SIAM J. Appl. Math.

Wirshing, P.H., & Shehata, A.M. (1977). Fatigue under wide band random stresses using the Rain-Flow method. Journal of Engineering Materials and Technology. Trans. ASME, July 1977. 205-211.

HIGH-LEVEL EXCURSIONS OF GAUSSIAN FIELDS: A GEOMETRICAL APPROACH BASED ON CONVEXITY

Nosko V.P.
Moscow State University, Moscow, USSR

Let $X(\underline{t})$, $\underline{t} = (t_1,\ldots,t_d) \in R^d$, be a zero-mean, homogeneous, ergodic Gaussian random field. We assume that X has almost surely continuous partial derivatives of up to second order with finite variances. We write $\dot{\underline{X}}(\underline{t}) = (X_1(\underline{t}),\ldots,X_d(\underline{t}))$, $\ddot{\underline{X}}(\underline{t}) = (X_{11}(\underline{t}),\ldots,X_{1d}(\underline{t}), X_{22}(\underline{t}),\ldots,X_{2d}(\underline{t}),\ldots,X_{d-1,d-1}(\underline{t}), X_{d-1,d}(\underline{t}), X_{dd}(\underline{t}))$, where $X_i(\underline{t}) = \partial X(\underline{t})/\partial t_i$, $X_{ij}(\underline{t}) = \partial^2 X(\underline{t})/\partial t_i \partial t_j$, $1 \le i,j \le d$. We assume that the joint distributions of values $X(\underline{t})$, $\dot{\underline{X}}(\underline{t})$, $\ddot{\underline{X}}(\underline{t})$ at any n points $\underline{t} = \underline{t}^{(1)},\ldots,\underline{t}^{(n)} \in R^d$, $\underline{t}^{(i)} \ne \underline{t}^{(j)}$, $i \ne j$, $1 \le i,j \le d$, $n = 1, 2,\ldots$, are non-degenerate.

By excursions of X above level u are meant the parts of sample function of X, corresponding to connected components of the set $A_u(X) = \{\underline{t} \in R^d, X(\underline{t}) \ge u\}$. We write $F_u^X(\underline{0})$ for the excursion of X above u, corresponding to that component of $A_u(X)$ containing the origin. We are interested in the asymptotic (when $u \uparrow \infty$) distributions of various characteristics of "arbitrary" ("randomly chosen", "typical") excursion of X above level u.

Wilson and Adler (1982) considered the excursion $F_u^{\tilde{X}}(\underline{0})$ of the horizontal-window conditioned field $\tilde{X}(\underline{t}) = (X(\underline{t}) \mid \underline{0} \in S(u))$, where $S(u) = \{\underline{t} \in R^d, X(\underline{t}) = u, X_1(\underline{t}) = \ldots = X_{d-1}(\underline{t}) = 0, X_d(\underline{t}) > 0, \ddot{\underline{X}}(\underline{t}) < 0\}$, and regarded this excursion as the typical excursion of X above u. (Here $\ddot{\underline{X}}(\underline{t}) < 0$ denotes the negatively definiteness of the matrix $(X_{ij}(\underline{t}))$.) The distribution $\tilde{P}_u$ of the field $\tilde{X}$ is the Palm distribution of X relative to the point proces generated by random point set $S(u)$. Wilson and Adler obtained the asymptotic distribution of the height h_u of the excursion $F_u^{\tilde{X}}(\underline{0})$ above u for all $d \geq 2$:

$$\lim_{u \uparrow \infty} \tilde{P}_u(uh_u > z) = \exp(-z/m_o),\ z \geq 0,\ m_o = \mathrm{Var}\, X(\underline{0}).$$

Nosko (1985) considered (and regarded as typical excursion of X above u) the excursion $F_u^{\tilde{X}}(\underline{0})$ of the conditional field $\tilde{X}(\underline{t}) = (X(\underline{t}) \mid \underline{0} \in S(u))$, where

$$S(u) = S^{[u,\infty),+} = \{\underline{t} \in R^d,\ X(\underline{t}) \geq u,\ \dot{\underline{X}}(\underline{t}) = \underline{0},\ \ddot{\underline{X}}(\underline{t}) < 0\},$$

and obtained, for all $d \geq 2$, the asymptotic distributions of the $(d+1)$-dimensional volume V_u of the set $K_u(\tilde{X})$, bounded "above" by the surface of $F_u^{\tilde{X}}(\underline{0})$ and "below" by hyperplane $t_{d+1} = u$, and of the d-dimensional volume S_u of the base $B_u(\tilde{X})$ of $K_u(\tilde{X})$:

$$\lim_{u \uparrow \infty} \tilde{P}_u \{\gamma (u^d S_u)^{2/d} > z\} =$$

$$\lim_{u \uparrow \infty} \tilde{P}_u \{\varkappa (u^{d+1} V_u)^{2/(d+2)} > z\} = \exp(-z/m_o),\ z \geq 0.$$

In this relationship, $\tilde{P}_u$ is the Palm distribution of the underlying field X relative to the embedded random point process generated by the random point set $S^{[u,\infty),+}$. The positive constant factors γ and $\varkappa$ depend on dimension d and the matrix M of second-order spectral moments of the underlying field X. Note that our approach needs no functional limit theorems for deriving those distributions.

Now we introduce a new approach which allows us to obtain more full results. For any $u > 0$, let $K_u = u(K_u(\tilde{X}) \oplus \{\underline{u}_o\})$, where $K_u(\tilde{X}) \oplus \{\underline{u}_o\}$ is the translate of the set $K_u(\tilde{X})$ by the translation $\underline{u}_o = (0,\dots,0,-u) \in R^{d+1}$, $uC = \{u\underline{c},\ \underline{c} \in C\}$ is a positive homotetic of a set $C \subset R^{d+1}$. Let

$$K = \{(t_1,\dots,t_d,t_{d+1}) : 0 \leq t_{d+1} \leq Z - (2m_o)^{-1} \underline{t} M \underline{t}'\},$$

where Z is the random variable having the exponential

distribution $P\{Z > z\} = \exp(-z/m_o)$, $z \geq 0$, and $\underline{t} = (t_1, \ldots, t_d)$.

Theorem. Let X be a Gaussian random field as described above, whose covariance function $r(\underline{t})$ has finite fourth partial derivatives at $\underline{t} = \underline{0}$ and

$$\left| \frac{\partial^4 r(\underline{t})}{\partial t_1^{s_1} \ldots \partial t_d^{s_d}} - \frac{\partial^4 r(\underline{0})}{\partial t_1^{s_1} \ldots \partial t_d^{s_d}} \right| \leq C \left| \log |\underline{t}| \right|^{-(1+a)}$$

$s_i = 0, 1, \ldots, 4$, $i = 1, \ldots, d$, $\sum_i s_i = 4$,
for some finite $C > 0$, some $a > 0$ and all $\underline{t}$ with $|\underline{t}|$ small enough. Then as $u \uparrow \infty$ the volume and the surface area of the random set K_u (of its base $B_u = K_u \cap \{t_{d+1} = 0\}$, respectively) have the same asymptotic distributions as the volume and the surface area of the random set K (of its base $B = K \cap \{t_{d+1} = 0\}$, respectively).

The proof of the theorem is based on constructing a random set K_u^* with the following properties:

(1) K_u^* and K_u are identical on the set Q_u, the probability measure of which tends to one as $u \uparrow \infty$;

(2) K_u^* is a random element in the space $C(\mathcal{K}')$ of nonempty convex compact sets in R^{d+1}, the topology of which is defined by Hausdorff metric, see Matheron (1975);

(3) for the random element
$K_u^{**} = \{(t_1, \ldots, t_d, t_{d+1}) \colon 0 \leq t_{d+1} \leq u(\tilde{X}(\underline{0}) - u) - (2m_o)^{-1} \underline{t} M \underline{t}'\}$
in $C(\mathcal{K}')$, the Hausdorff distance $\rho(K_u^*, K_u^{**})$ converges to zero in probability as $u \uparrow \infty$.

It is a simple matter to check that

$$\lim_{u \uparrow \infty} \tilde{P}_u \{u(\tilde{X}(\underline{0}) - u) > z\} = \exp(-z/m_o), \quad z \geq 0.$$

From this it follows that K_u^{**} converges to K in distribution. Then by the property (3) above, K_u^* also converges to K in distribution. Since the volume and the surface area are the Minkowski functionals on $C(\mathcal{K}')$, they are continuous ones. Hence the distributions of the volume and the surface area of the random body K_u^* converge weakly to those ones of the random body K. Combining this with the property (1) then establishes the part of the theorem being concerned with the random set K_u. The proof of that part of the theorem being concerned with the base B_u of K_u is analogous.

The possibility of obtaining the explicit form for the asymptotic distributions of the surface area of the random set K_u and its base B_u is limited by the possibility of obtaining the explicit expressions for the surface area of the body K and its base B with the value $Z = z$ being fixed. For simplicity, consider only the case $d = 2$ and suppose that underlying random field X is isotropic as well as homogeneous. Then the body K is the segment of the paraboloid of revolution. Writing L_u for the perimeter of the base $B_u(\tilde{X})$ of the set $K_u(\tilde{X})$, it is easily to obtain (denoting $m_{11} = \partial^2 r(\underline{0})/\partial t_1^2$)

$$\lim_{u\uparrow\infty} \tilde{P}_u \left\{ \frac{m_{11}}{8\pi^2} (uL_u)^2 > z \right\} = \exp(-z/m_o), z \geq 0.$$

References

Matheron, G.(1975). Random sets and integral geometry. Wiley, New York.

Nosko V.P.(1985). The limiting distributions for characteristics of Gaussian random field high-level excursions. In: Fourth International Vilnius Conference on Probability Theory and Mathematical Statistics. Abstracts of Commun. V.II. Institute of Math. and Kybern. Press. Vilnius, pp. 269-271 (in Russian).

Wilson R.J., Adler R.J.(1982). The structure of Gaussian field near a level crossing. Adv.Appl.Probab. 14, 543-565.

EPIDEMIOLOGY
(mainly observational studies)
(Session 13)

Chairman: N.T.J. Bailey

Bernoulli, Vol. 2, pp. 507-516

EPIDEMIC PREDICTION AND PUBLIC HEALTH CONTROL, WITH SPECIAL REFERENCE TO INFLUENZA AND AIDS

N.T.J. Bailey and J. Estreicher,
Division of Medical Informatics, Hôpital Cantonal Universitaire,
CH - 1211 Geneva 4, Switzerland

1. INTRODUCTION

This paper is primarily concerned with the use of mathematical modelling and computer simulation to improve understanding of the population dynamics of epidemic outbreaks of infectious diseases at community level, in order to provide public health authorities with the predicted consequences of different available control strategies, thus helping to select optimal policies, special attention being paid to influenza and AIDS.

Most modelling efforts in the 19th century involved fitting only empirical curves to epidemic incidence data. For obvious statistical reasons it proved impossible to predict the future course of an epidemic by extrapolating from the initial stages. The first big advance came with Hamer's (1906) model in which the rate of spread of infection depended on the product of the numbers of susceptible and infectious persons, equivalent to the Law of Mass Action in chemical kinetics. The most influential development was later made by Kermack & McKendrick (1927 and later) in their studies of the basic nonlinear differential equations. This approach had already appeared in the "*a priori* pathometry" of Ross & Hudson (1916; 1917a,b), while the first probabilistic version is in En'ko (1889). For extensive reviews and references, see Anderson (1982), Bailey (1975, 1985, 1986b) and Dietz & Schenzle (1985).

The number of references dealing with the mathematical theory of infectious diseases in now over 1000. Unfortunately, few publications deal directly with support to ongoing clinical or public health decision-making. The theoretical knowledge required for improving practical action already exists, but is largely ignored by those who are best in a position to benefit from applying it. Starting from first principles, the present paper shows how the difficulties of implementation can be overcome.

2. BASIC APPROACH TO INFECTIOUS DISEASE MODELLING

It is clear from the research literature that quite simple kinds of dynamic modelling are potentially very powerful in understanding and predicting epidemic behaviour, provided they incorporate suitable tested assumptions about the underlying biological processes.

The most successful research has nearly always employed some kind of compartmental modelling, in which the boxes represent different epidemiological states, e.g. susceptible, infected, healthy carrier, recognised case, etc. The progress of any individual can be described theoretically in terms of movement from one box to another. This establishes a basic flow chart which can easily be used as an interface between biomathematicians and medical investigators. The next stage is to turn this qualitative model into a quantitative one by ascribing rates of flow between the boxes, involving rates of infection or recovery, lengths of latent or incubation periods, etc.

We then have an agreed basis for collaborative work. The flow chart is conceived in terms ordinarily used by clinicians, epidemiologists, immunologists, public health authorities etc., and can easily be modified to suit individual circumstances. Practically any infectious disease can be handled, and different degrees of realism and complexity can be investigated. The compartmental structure lends itself to mathematical analysis, describing the dynamic behaviour of the system in terms of differential equations for continuous time, or difference equations for discrete time. A wide range of powerful mathematical and computer methods are available, but the results can always be interpreted in the plain language concepts of the flow chart, and hence can be used directly in clinical or public health contexts. In addition, extensions can be made to the analysis of finer structural levels involving e.g. crucial aspects of clinical practice or immunological research, or wider systems interacting with the primary biological system, e.g. demography, social structure, socio-economic activity etc.

3.MODELS, DATA AND VALIDATION

Given the structure of any complex system, numerical values of all the parameters involved, and the set of initial conditions, the modelling approach enables the subsequent behaviour of the system to be calculated - usually by some form of computer simulation. This can be extremely valuable, even when the formulation, based on expert knowledge and experience, is essentially hypothetical. See, for example, the work of Cvjetanović *et al*. in Ch. 2 of Anderson (1982). Computer analysis allows one to calculate the approximate consequences of the chosen assumptions, thus helping to build up a more rational framework for administrative decision-making.

Good statistical practice demands that models be validated by being fitted to appropriate data. Parameters should be efficiently estimated and the goodness-of-fit tested for statistically significant gaps between theory and observation. But this is not enough. A statistically satisfactory model must also be shown to have adequate predictive power in operational situations if it is to be effectively incorporated in decision-making practice.

Consider, for example, the simplest kind of viral infections, like smallpox, influenza or measles, that spread from person to person and confer life-long immunity on the survivors of a single attack. We imagine three classes of individuals composed of (a) x susceptibles, (b) y infected & infectious persons, and (c) z recovered & immune. We ignore an initial latent period for the newly infected, and suppose there are no deaths, putting $x + y + z = n$, a constant.

We assume a form of homogeneous mixing in which each individual makes an average of β "effective contacts" in unit time, where "effective" means a contact sufficient to transmit infection when the contact is between an infectious person and a susceptible. Thus in Δt, the y infecteds contact $\beta y \Delta t$ other persons, of whom a proportion x/n are susceptible. The rate of new infections is thus $\beta xy/n$. We also assume a rate γ at which circulating infecteds are recognised as cases and removed from circulation, the overall rate being γy. Hence we easily obtain the following equations, which are of Kermack & McKendrick type:

$$dx/dt = -\beta xy/n, \ dy/dt = \beta xy/n - \gamma y, \ dz/dt = \gamma y. \qquad (1)$$

As initial conditions we could take :

$$x(0) = n - b, \ y(0) = b, \ z(0) = 0. \qquad (2)$$

Equations (1) and (2) represent one of the simplest formulations that has, along with various extensions and modifications, been exhaustively studied in the literature.

An important aspect of this very elementary model is that we cannot in general observe either $x(t)$ or $y(t)$. What we actually perceive is the rate γy at which circulating infecteds develop symptoms and are withdrawn from circulation. The observed data thus relate to the epidemic curve $w = \gamma y = dz/dt$. We can estimate both β and γ if we have enough data. It should be noted that β is almost never directly observable, while γ can sometimes be found from independent epidemiological data on the incubation periods observed for cases derived from uniquely identifiable contacts.

Many difficulties in epidemic modelling result from the hidden character of the underlying pathological processes. In some cases serological surveys may reveal the proportions of individuals in the susceptible or infectious groups. More realistic assumptions are easily incorporated where needed, e.g. latent period, healthy carriers, loss of immunity etc. However, we must not forget that every additional modification is liable to introduce more hidden activities and thus greater difficulties in identifying parameters, unless we can obtain extra sources of information from clinical observations, laboratory tests, serological surveys etc.

4. STATISTICAL PROBLEMS: ESTIMATION, FITTING AND SIGNIFICANCE TESTING

For modelling at the community level it is reasonable to start with deterministic formulations like (1), but epidemic curves showing morbidity incidence or serological surveys will still require statistical anlysis. As a first approximation we use a routine computerised optimistion algorithm for deriving least squares estimates in order to obtain the best fit between theoretically computed and observed epidemic curves. Baroyan & Rvachev (1967 and later) adopted this approach to influenza in the USSR, while Scherrer *et al.* (1985) recently developed new and improved methods for both solving large sets of nonlinear differential equations and carrying out the associated nonlinear least squares estimation of parameters. Practically useful results can be obtained, but care is needed in interpreting standard errors arising from the usual calculations. If the underlying dynamics are essentially independent of epidemic incidence level, then we may have a homoscedastic curvilinear regression situation for which the usual computed standard errors are valid.

Another situation arises if the morbidity data (i.e. new cases per day, week, or month) involve moderately large numbers showing approximately Poisson variation. We can then use the usual formula for χ^2 and minimise this to obtain quasi-maximum-likelihood estimates. This requires the residuals to be independently distributed, and we should examine them for the presence of any strong serial correlations.

There is also an important problem involving the initial conditions, as these will generally be unknown as well as the usual parameters. Thus with the model in (1), we have the additional unknown $\underline{b}$ in (2), assuming $\underline{n}$ is known. With influenza it turns out that we can estimate β, γ and $\underline{b}$ from the data. But when there are many different epidemiological states we are liable to have a considerable increase in the number of unknown parameters if we include all the initial values as well. The following trick may be useful.

Suppose that a single infected arises by genetic mutation and that all other groups apart from susceptibles are zero. Since $\underline{n}$ is large, $\underline{x}(\underline{t}) \sim \underline{n}$ for small $\underline{t}$, and all nonlinear equations can be linearised. For small $\underline{t}$ the latter can be regarded as equations for stochastic means, and will have solutions that are linear in exponential terms of the form $e^{\theta t}$. We then suppose that for larger $\underline{t}$, all transient terms with $\theta < 0$ can be neglected, and we treat what is left as deterministic quantities. In many models this greatly reduces the number of unknown initial values, following the process from a sufficiently large value of $\underline{t}$.

Finally, we note the importance of having good initial values of the parameters for starting computer optimisations. It has already been pointed out by Bailey (1975, pp. 86-7, 124-5) that, for a completed epidemic based on the model in (1) above we can obtain initial estimates β_o and γ_o from two equations giving the observed total size of epidemic and height of epidemic curve. Similarly (Bailey, 1986a), it is easy to see from (1) and (2) that the early stages of the epidemic are given by $\underline{w} = \underline{b}\,\gamma \exp\{(\beta - \gamma)\underline{t}\}$; and that the closing stages are given by $w = B\gamma \exp[\{\beta(1-i)-\gamma\}t]$, where $\underline{B}$ is some constant and $\underline{i}$ is the relative intensity of the epidemic, i.e. the proportion of susceptibles eventually infected. Fitting straights lines to logarithms of the data again gives two linear equations for estimating β_o and γ_o. We may also have an independent estimate of γ.

5. APPLICATIONS

5.1 Influenza in the USSR

The well-known work of Baroyan, Rvachev *et al*. (1967 and later) on modelling and predicting the spread of influenza is relatively complicated, involving integro-differential equations and partial differentials. Remarkably good results were obtained for epidemics within the USSR, and more recently, encouraging applications to the prediction of the global spread of influenza from Hongkong in 1968 were reported by Rvachev & Longini (1985) and Longini *et al*. (1986). The investigations of Scherrer *et al*. (1985) were applied to parts of Soviet data on the 1965 influenza outbreak. The simplified model in (1) above was used, together with a modified form of (2) to allow for the possibility that only a proportion α of the initial susceptibles were, for various social or medical reasons, effectively susceptible. The initial conditions were then :

$$x(0) = \alpha(n-b),\ y(0) = b,\ z(0) = (1-\alpha)(n-b). \tag{3}$$

Inspection of the data from the initial focus of Leningrad suggested that there was a fairly constant nonepidemic background. It was easy to incorporate a corresponding parameter d, the complete set being (α, β, γ, d, b). In general the numbers were large enough for a deterministic model to be appropriate. But inspection of residuals indicated a substantial source of variation independant of epidemic level. The formally computed standard errors then corresponded to the quasi-maximum-likelihood estimation relevant to this variation.

Parameters were then recalculated using a discrete-time analogue based on units of one day corresponding to the observed data. This was simpler and much faster than the continuous-time theory, and possibly more realistic, giving

$$\alpha = 0.189 \pm 0.020,\quad \beta = 3.98 \pm 0.34,\quad \gamma = 0.372 \pm 0.102, \quad d = 4956 \pm 779,\quad b = 5272 \pm 1305. \tag{4}$$

More detailed study of the residuals showed a marked weekly cycle, already well established in the Soviet literature. Removal of this trend reduced the residual variance by about 47%. It might therefore be reasonable to multiply all the standard errors in (4) by a factor $(0.53)^{0.5} = 0.73$.

Scherrer et al. (1985) also applied the model to predicting epidemic outbreaks in other Soviet cities. This could only be done conveniently where sufficiently unambiguous migration-rates could be read from the literature, e.g. Baroyan et al.(1973), and where a major connecting route could be identified to trigger off secondary outbreaks. This simplification worked surprisingly well. The epidemiological parameters α, β and γ for Leningrad were assumed equally valid for other cities, and pure predictions were made on the basis quoted migration-rates. Predicted results were very similar to the previously published Soviet figures in all the cities examined, namely Moscow, Riga, Gorkii, Kiev and Minsk, thus both confirming the Soviet claims and justifying the simpler model we have proposed.

5.2 AIDS in San Francisco

We shall now see to what extent these principles can be applied to the current spread of AIDS, with special reference to the situation in San Francisco. We investigated a spectrum of possible models, making different assumptions about latent, infectious and quiescent periods, as well as distinguishing between an AIDS and a non-AIDS path. Some of these models no longer fit the most recent clinical, serological and virological findings ; others fail to achieve any optimisation convergence. An early example is given by Bailey (1986c), but the considerably modified version described below appears to be more realistic.

There is no doubt about the extreme gravity of the global AIDS situation, and extensive information is available in a large and growing literature. Excellent reviews of present knowledge of the natural history of the disease can be found in Curran *et al.* (1985), Francis *et.al.* (1985), Koch (1985), Wong-Staal & Gallo (1985), and De Gruttola *et al.* (1986). The economic impact revealed by Hardy *et al.* (1986) shows a staggering estimate of direct costs of around $ 150'000 per patient for hospitalization expenses, and indirect costs due to losses from disability and premature death of more than three times that amount. Again, Curran *et al.* (1985) estimated the number of years of potential life lost (YPLL) up to age 65 in never-married men aged 25-44 years. In San Francisco and Manhattan, New York, the total loss in YPLL terms for AIDS was at least equal to the combined loss from accidents, cancer and homicide /suicide!

In order to have a relatively simple, but sufficiently realistic model - at least for provisional public health applications - we decided to explore the following model. The broad compartmental scheme is illustrated by the diagram :

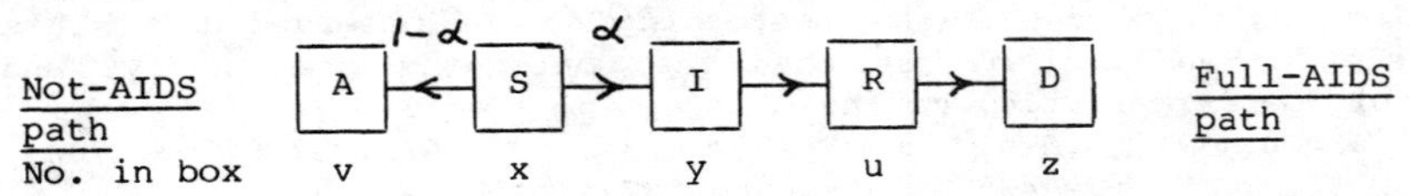

The group of susceptible individuals is indicated by S. For those who become infected with the AIDS virus LAV/HTLV-III, we distinguish between the path to the right leading to full AIDS, and the path to the left involving not-full AIDS (i.e. ARC, lesser AIDS, healthy carriers etc.). Group I involves circulating infectives (infected & infectious) who will develop AIDS. Group R contains those who have been recognised as cases, and D is the total of AIDS deaths. Box A, on the other hand, represents relatively healthy carriers and others who are more or less asymptomatic. S individuals have not yet had contact with the virus and are seronegative for antibodies to the virus. All other groups are seropositive. Following the suggestion of Wong-Staal & Gallo (1985), we assume that all A also carry the virus in their peripheral blood and are infectious. They thus constitute the major source of infection. We assume that known cases in R are not infecting others to any great extent, but this restriction could easily be relaxed.

Three kinds of main data are available from San Francisco, where we assume that the total size of the gay community is N = 100 000. The monthly incidence of both morbidity and mortality are directly available from the Bureau of Communicable Disease Control in San Francisco, while information on the incidence of infection, as judged from the rise in prevalence of seropositivity in a certain cohort of gay men, can be found in Echenberg *et al.*(1985) and Jaffe *et al.* (1985). Inspection of the seropositivity data indicates peak incidence around mid-1981, with prevalence tending to level off at about 70 %. We decided to divide the population into two parts: a proportion κ of homosexuals who contributed little to the epidemic, and $1-\kappa$ who were involved in a relatively high average rate of promiscuous contacts, with $n = (1-\kappa)N$. Inspection of the morbidity incidence data showed a somewhat similar curve (peak uncertain) shifted to the right by about 5 years, the latter being an indication of average incubation period.

Let us write α for the proportion of those infected on the full AIDS path. The effective contact-rate is β, as in section 2. Following a suggestion of Klaus Dietz (personal communication) we can allow for a possible gamma distribution of incubation period. Group $\underline{I}$ is accordingly divided into $\underline{m}$ identical parts, each with a transition rate $\underline{m}\gamma$ and virtual negative exponential sojourn time with parameter $\underline{m}\gamma$. The average incubation period thus remains at γ^{-1}. The number of infecteds in the $\underline{i}$'th subgroup is $\underline{y_i}$; let the total number of infectious seropositives be $\underline{P} = \underline{v} + \sum_1^m \underline{y_i}$. The epidemic curve of morbidity incidence is thus $\underline{w} = \underline{m}\,\gamma\,\underline{y_m}$. The general equations describing the system dynamics are therefore

$$dx/dt = -\beta xP/n;\quad dv/dt = (1-\alpha)\beta xP/n;\quad w = m\gamma y_m;$$
$$dy_1/dt = \alpha\beta xP/n - m\gamma y_1;\quad dy_i/dt = m\gamma(y_{i-1}-y_i),\ i = 2,\dots,m. \tag{5}$$

Since we are neglecting the possible infectiousness of the cases in $\underline{R}$, the mortality process can be decoupled from the rest of the system. In fact the average death-rate of AIDS cases is about 0.6 per year, with average survival time about 1.7 years.

We first assumed a constant incubation period, which means that the epidemic curve exactly follows the incidence curve of infection for full AIDS, and differs from the incidence of seropositivity only by a multiplying constant α. Moreover, the epidemic of infection is largely completed before any appreciable number of cases appears. The latter can thus be ignored in this context. Optimisation allows us to estimate α, β and κ, and location parameters for the two incidence curves. The incubation period can then be read from the distance between these two curves. We find

$$\alpha = 4.22 \pm 0.49\ \%,\quad \beta = 0.091 \pm 0.007 \text{ per month},$$
$$\kappa = 32.0 \pm 3.7\ \%,\quad \gamma^{-1} = 57 \text{ months}. \tag{6}$$

These results suggest that about 68 % of the male homosexual population are at major risk from AIDS, though only 4 % of these will develop the disease: a total of about 2900 in San Francisco. The average incubation period is around 5 years, and simulations predict a flat peak around Oct. 1985. Finally, the somewhat provisional quasi-goodness-of-fit χ^2 with 54 degrees of freedom was 68.2 (a formally non-significant result).

Although these results look encouraging, there are several difficulties. The constant incubation period may be an oversimplification. The relatively small value of α does not fit with recent reports by Brodt et al. (1986) and Metroka et al. (1986) suggesting that an ominously high fraction of seropositives may eventually develop full AIDS. It is possible, however, that the samples studied were drawn from higher than average risk groups: this aspect urgently needs further investigation. Even more telling is that the San Francisco morbidity incidence has continued to rise over the first 8 months of 1986 with no peak or flattening out in sight.

Further investigation of the full model in (5) shows unsatisfactory values of χ^2 over the range $\underline{m}$ = 1 to 9, the 5% point being passed only with $\underline{m}$ = 10. Although in principle more realistic, this is not yet satisfactory in practice.

So far, the constant incubation period model is the only one to yield a satisfactory χ^2, yet α may be too small and short term predictions are not validated. A possible explanation is that what we have labelled a "not-full AIDS" group does in fact lead to AIDS cases after a fairly lengthy period, insufficient time having elapsed previously for this morbidity to be observed. This suggests investigating a modified model in which the incubation period is in two parts: the first being long without much variation (large $\underline{m}$), and the second being much shorter and possibly negative exponential in distribution. Whether this will require much higher values of α remains to be seen, but full investigation of this frightening possibility is an urgent necessity.

6. PUBLIC HEALTH CONTROL

Our primary objective is to help public health authorities select optimal control policies for the suppression of epidemic disease through better understanding of the population dynamics involved. Possible interventions include improved and earlier diagnosis and treatment; preventive measures involving immunisation and special protection for high risk groups; improvements in health education leading to a reduction in the risk of infection, especially in relation to sexual behaviour; greater support for many kinds of pure and applied research, etc. Most, if not all, of such interventions involve changes in parametric values, and sometimes in system structure. If we have good validated models we can begin to make better quantitative predictions as to the likely consequences of any proposed intervention strategy, and thus facilitate an optimal choice.

While close integration of theory and practice is essential to building up a strong surveillance and control technology, philosophical conviction though necessary is not sufficient. Specific administrative and managerial action is required to set up the machinery required for actual implementation to be successfully achieved. Close collaboration is needed between modellers, biomathematicians, biostatisticians, clinicians, immunologists, epidemiologists and public health authorities responsible for actual decision-making. The development of such an integrated multidisciplinary approach clearly falls in the general area of applied systems analysis and operational research, but special steps are needed to ensure success in the context of the public health control of infectious diseases. Not only are many diseases highly complex in structure, but may involve a high degree of system resilience that may either resist change or involve sudden switching to new and unexpected stable equilibria.

A promising approach is that of Holling (1978), called "adaptive environmental assessment and management". This is a modern and updated version of operational research, operating on a workshop basis through a kind of "institute-without-walls". It has already been strongly advocated by Bailey (1986a,c) in relation to the control of both schistosomiasis and AIDS. There must be close integration of the investigating team and the policy makers. Exact details depend on circumstances. But the "adaptive" aspect entails an initial building up of good will in a sympathetic atmosphere, so as to facilitate a broad understanding of both the complex disease system and the decision-making process. Such an approach would ensure that biomathematical and modelling methodology was clearly concentrated on the task of resolving immediate medical, social and economic problems, especially those involved in the highly dangerous spread of AIDS to heterosexual groups.

ACKNOWLEDGEMENTS

This paper was supported in part by the Swiss National Foundation for Scientific Research, Grant No. 3.948-0.84. It is also a pleasure to recognise our debt to Dr. D. Echenberg, Director of the Bureau of Communicable Disease Control, San Francisco, for kindly supplying data on the local AIDS epidemic and for making many useful criticisms of our modelling efforts.

REFERENCES

ANDERSON, R.M. (Ed.) (1982) Population Dynamics of Infectious Diseases. London: Chapman & Hall.

BAILEY, N.T.J. (1975). The Mathematical Theory of Infectious Diseases. London: Griffin.

BAILEY, N.T.J. (1985). The role of statistics in controlling and eradicating infectious diseases. Statistician, 34, 3-17.

BAILEY, N.T.J. (1986a). The case for mathematical modelling of schistosomiasis. Parasitology Today, 2, 158-163.

BAILEY, N.T.J. (1986b). Macro-modelling and prediction of epidemic spread at community level. Math. modelling, 7.

BAILEY, N.T.J. (1986c). Use of simulation models to help control AIDS. Proc. 5th World Congress on Medical Informatics (MEDINFO 86).

BAROYAN, O.V. & RVACHEV, L.A. (1967). Deterministic epidemic models for a territory with a transport network. Kibernetika, 3, 67-74. (In Russian.)

BAROYAN, O.V., RVACHEV, L.A. et al. (1973). Mathematical and computer modelling of influenza epidemics in the USSR. Vestnik Akad. Med. Nauk, 28(5), 26-30. (In Russian.)

BAROYAN, O.V., RVACHEV, L.A. & IVANNIKOV, Yu. G. (1977). Modelling and Prediction of Influenza Epidemics in the USSR. Moscow:CVR. (In Russian.)

BRODT, H.R. et al. (1986). Spontanverlauf der LAV/HTLV-III-Infektion Dtsch. med. Wschr., 111, 1175-80.

CURRAN, J.W. et al. (1985). The epidemiology of AIDS: current status and future prospects. Science, 29, 1352-7.

De GRUTTOLA, V. et al. (1986). AIDS: has the problem been adequately assessed? Rev. Inf. Dis., 8, 295-305.

DIETZ, K. & SCHENZLE, D. (1985). Mathematical models for infectious disease statistics. In A Celebration of Statistics (Editions A.C. Atkinson & S.E. Fienberg), 167-204. New York : Springer.

ECHENBERG, D. et al. (1985). Update: Acquired Immunodeficiency Syndrome in the San Francisco Cohort Study, 1978-85. MMWR, 34, 573-5

EN'KO, P.D. (1889). The epidemic course of some infectious diseases. Vrac, 10, 1008-10, 1039-43, 1061-3. (In Russian.)

FRANCIS, D.P. et al. (1985). The natural history of infection with the lymphadenopathy-associated virus human T-lymphotropic virus type III. Ann. Int. Med., 103, 719-22.
HAMER, W. (1906). Epidemic disease in England. Lancet, 1, 733-9.
HARDY, A.M. et al. (1986). The economic impact of the first 10,000 cases of acquired immunodeficiency disease in the United States. JAMA, 255, 209-11.
HOLLING, C.S. (Ed.) (1978). Adaptive Environmental Assessment and Management. New York & Chichester, U.K.: Wiley.
JAFFE, H.W. et al. (1985). The acquired immunodeficiency syndrome in a cohort of homosexual men. Ann. Int. Med., 103, 210-14.
KERMACK, W.O. & McKENDRICK, A.G. (1927). Contributions to the theory of epidemics (Part I). Proc. Roy. Soc., A 115, 700-21.
KOCH, M.G. (1985) AIDS - Our Future ? Stockholm : Swedish Carnegie Institute. (In Swedish.)
LONGINI, I.M. Jr., FINE, P.E.M. & THACKER, S.B. (1986). Predicting the global spread of new infectious agents. Amer. J. Epidemiol., 123, 383-91.
METROKA, C.E. et al. (1986). A four year prospective study of clinical and immunological parameters of patients with generalized lymphadenopathy. (Poster 196, International Conference on AIDS, Paris, France.)
ROSS, R. (1916) ; ROSS, R. & HUDSON, H.P. (1917a, b.). An application of the theory of probabilities to the study a priori pathometry, Parts I, II & III. Proc. Roy. Soc., A92, 204-30 ; A93, 212-25 ; 225-40.
RVACHEV, L.A. & LONGINI, I.M. Jr. (1985). A mathematical model for the global spread of influenza. Math. Biosc., 75, 3-22.
SCHERRER, J.-R., WANNER, G., BAILEY, N.T.J. & ESTREICHER, J. (1985). Continuous-time modelling in biology. (Report to the Swiss National Foundation for Scientific Research, Grant No. 3.381-0.82)
WONG-STAAL, F. & GALLO, R.C. (1985). Human T-lymphotropic retroviruses. Nature, 317, 395-403.

MATHEMATICAL MODELS FOR CHRONIC DISEASE EPIDEMIOLOGY

Kenneth G. Manton
Duke University
Center for Demographic Studies
2117 Campus Drive
Durham, North Carolina 27707

1. INTRODUCTION

As populations age the mix of major public health problems shifts from acute infectious disease processes, for which the young are vulnerable, to chronic, degenerative disease processes to which, because of intrinsic biological aging processes, the elderly are especially vulnerable. The nature of the chronic disease processes is so different from that of infectious diseases it is necessary to develop both new mathematical and statistical models.

The need for new models results from several factors. First, there is the different nature of the outcome. Chronic disease processes are not well described as an event localized in time. Rather, it is necessary to describe chronic disease by the parameters of the underlying process which has a lengthy duration and a progressive, time directed trajectory. Because chronic diseases are processes with lengthy (relative to life span) duration, it is necessary to describe the interaction of concurrent chronic disease processes (especially at advanced ages where the prevalence of multiple chronic conditions is high) and between the disease process and the underlying, multidimensional physiological aging process. Because we will not observe all of the physiological parameters of the aging (or disease) processes, the models will also have to represent the effects of unobserved variables. Finally, the progressive nature of chronic disease processes requires developing measures of the increasing impact of the disease process on overall morbidity and functioning. Since many chronic disease processes are integral parts of aging, degenerative changes, we also have to identify several discrete stages of the process (e.g., latent, clinically manifest, manifest alteration of functioning) and represent them as qualitative, or discrete changes in chronic disease trajectories.

Second, because of the inter-linkage of multiple chronic disease and aging processes, the mode of assessing the influence of various variables, and techniques for controlling their change over time, must be different than for acute diseases. Specifically, because the models will be multivariate, dynamic and with feedback between process components, assessing the impact of a variable must be done by examining the system consequences of a change. In epidemiological studies the effect of a particular "risk" factor is often assessed by examining the size and significance of a logistic or Cox regression coefficient. In biological terms, since changing a physiological parameter can have multiple effects that evolve over time, a more realistic way of assessing impact is to conduct a simulation study of the influence of the risk factor on the evolution of the process.

Finally, because of the complexity of the processes, it is necessary to use multiple data sources to estimate parameters for multiple process components. This requires resolving problems of inference when using multiple data sources. The advantages of using multiple data sources to construct a model of the multiple component processes are manifold. Such a model will integrate, in a formal mathematical statement, the ensemble of available theoretical and empirical insights. Thus, it can be determined if new evidence, or hypotheses are consistent with the full range of available data and theory. Also, because the components of the process are formally represented specific substantively important questions about the mechanisms of the process can be evaluated. For example, suppose that changing the direction or rate of the age trajectory of a risk factor influences the rate of progression of a disease, its clinical incidence and survival. Yet, if the model does not contain a process component describing the evolution of risk factors, the question is moot--even if the longitudinal data on risk factor change is available. The effect of a given risk factor will be more accurately described by a fully elaborated process model because, instead of relying on unverifiable assumptions about the behavior of the process, the model components can be manipulated to investigate the implications of different assumptions. For example, in a model where there is a component process describing the evolution of risk factors among survivors the effects of diffusion on the trajectory of individual risk factor values on survival can be assessed. Without such a component the effect of the stochastic propagation of risk factor values on the age course of survival can not be studied.

Below we present a multivariate stochastic process model of chronic disease and aging, describe how components of that process represent biological mechanisms, and provide illustrations using epidemiological data.

2. A MATHEMATICAL MODEL OF CHRONIC DISEASE EVOLUTION AND AGING

To develop a model of chronic disease and aging we need to describe the individual level processes and how individual level processes are averaged in the population. Consider an individual whose change on multiple physiological variables is a function of his past physiological state, uncertainty in that state, and the probability of dying. This generates a "random walk" described by two equations. The first is a linear stochastic differential equation describing state changes as time directed movement in a J-dimensional space:

$$dz_t = A(t)\ z_t^*\ dt + D'(t)\ dw_t \tag{1}$$

where $z_t^* = (1,\ z_t')'$ and the J by (J+1) matrix A(t) describes deterministic change as a function of current position; w_t is a J-dimensional Wiener process with independent increments in nonoverlapping time intervals, with initial conditions $w_0=0$, and D'(t) is a bounded matrix of scale factors applied to dw_t to describe the randomness in the trajectory of change. The second equation describes the conditional mortality hazard for persons as a quadratic function of their position in the space:

$$\mu(t,z_t) = z_t^{*\prime}\ Q(t)\ z_t^* , \tag{2}$$

where Q(t) is a square symmetric nonnegative definite matrix of bounded time dependent coefficients.

This type of random walk produces a generalized Kolmogorov-Fokker-Planck (KFP) equation for the population:

$$\frac{\partial}{\partial t} f_t(z) = - \sum_{j=1}^{J} \frac{\partial}{\partial z_j} [u_{jt}(z)\ f_t(z)] + \tfrac{1}{2} \sum_{i=1}^{J} \sum_{j=1}^{J} \frac{\partial^2}{\partial z_i\ \partial z_j} [\sigma_{ij}(t)\ f_t(z)] \tag{3}$$

$$- [\mu(t,z) - \mu(t)]\ f_t(z)\ ,$$

where $f_t(z)$ is the multivariate density of z; u_{jt} is the jth element of u_t ($u_t = A(t)\ z^*_t$); and $\sigma_{ij}(t)$ is the (i,j)th element of the innovation covariance matrix $\Sigma(t)$ where $\Sigma(t) = D'(t)\ D(t)$.

The first term in (3) refers to deterministic changes in the distribution, the next term to diffusion, and the final term to systematic mortality loss from the multivariate density function. Woodbury and Manton (1977) show how the KFP equation can be used to evaluate multivariate human aging and mortality processes from longitudinal studies. Specifically, for multivariate Gaussian stochastic processes with linear dynamics and a quadratic force of mortality, the evolution of the multivariate risk factor distribution can be described by ordinary non-linear differential equations and the parameters of those differential equation systems can be estimated from epidemiological studies with continuous follow-up of mortality and measurement of risk factors at fixed times.

Using an exact discrete approximation, Woodbury and Manton (1983) showed that the likelihood function for I persons with risk factor measurements $\{x_{it}\}$ at fixed times $t \varepsilon \{0,1,...,T_i\}$, could be written:

$$L = \prod_{i=1}^{I} \{f_0\ (x_{i0})\} \{ \prod_{t=1}^{T_i} f_t(x_{it} | \hat{x}_{it}, \Sigma_t)\} \{ \prod_{t=0}^{T_i - 1} \exp[-h(x_{it})]\} \{1-\exp[-h(x_{iT_i})]\}^{\delta_i} \tag{4}$$

where term 1 represents initial conditions, term 2 represents the change in risk factor values for survivors, term 3 represents the probability of survival over the interval $[0, T_i]$, and term 4 represents the probability of dying in the interval (T_i, T_i+1), where $\delta_i = 1$ if death occurred in the interval and $\delta_i = 0$ otherwise. Term 2 of the likelihood function is parameterized by a linear autoregressive process, where $x_{it} \sim N\ (\hat{x}_{it}, \Sigma_t)$ and;

$$x_{it} = \hat{x}_{it} + e_{it};\ \text{where}\ E(e_{it}) = 0\ \text{and}\ E(e_{it}\ e'_{it}) = \Sigma_t, \tag{5a}$$

and

$$\hat{x}_{it} = x_{it-1} + A_t \, x^*_{it-1} \tag{5b}$$

$$= A_{0t} + (I + A_{1t}) \, x_{it-1} \, , \tag{5c}$$

where $x^*_{it} = (1, x'_{it})'$; A_t is partitioned at the first column so $A_t = [A_{0t} \vdots A_{1t}]$; I+A is the matrix of regression coefficients, and Σ_t is the diffusion covariance matrix.

It is statistically advantageous to make Σ_t and A_t parametric functions of age, i.e., let a_{it} be the age of person i at time t, and $a^*_{it} = (1, a'_{it})'$ and $x^{**}_{it} = (a^{*\prime}_{it}, x'_{it})'$. If Σ_t is replaced by Σ and A_t with A in (5b), then

$$\hat{x}_{it} = x_{it-1} + A \, x^{**}_{it-1} \tag{5d}$$

$$= A_0 \, a^*_{it-1} + (I+A_1) \, x_{it-1} \tag{5e}$$

where the J by (J+2) matrix A is now partitioned at the second column to yield $A = [A_0 \vdots A_1]$. For a single cohort, the first term in (5e) simplifies to A_{0t}, though the time trajectory of this vector is restricted to a linear form.

The hazard function appearing in (4) is quadratic in x^*_{it}:

$$h_t(x_{it}) = x^{*\prime}_{it} \, Q_t \, x^*_{it} \, . \tag{6a}$$

One could parameterize (6a) so that age is the only temporal effect:

$$h_t(x_{it}) = x^{**\prime}_{it} \, Q \, x^{**}_{it} \, , \tag{6b}$$

where the order of Q is (J+2) by (J+2). Alternatively, Q_t in (6a) can be replaced by an exponential function of age:

$$h_t(x_{it}) = x^{*\prime}_{it} \, [Q \exp(\beta \, a_{it})] \, x^*_{it} \, . \tag{6c}$$

All coefficients in "Q_t" in (6c) are age dependent. In (6b) only the constant and linear terms of the quadratic form in x_{it} depend on age. The parameter β may be modelled as $\beta_{t,t-a_{it}}$ to reflect changes over calendar periods or birth cohorts.

The exponential term makes (6c) a generalized Gompertz function that represents the effect of unobserved factors correlated with age and ensures that the effects of the individual physiological variables are not confounded with age. All three forms of (6) can be decomposed into cause specific terms to reflect competing risks (Yashin et al.,

1986).

Under the assumption that $x_{i0} \sim N(m_0, V_0)$ and $S_0=1$, we obtain

$$V_{t+} = (I + V_t B_t)^{-1} V_t \tag{7}$$

$$m_{t+} = m_t - V_{t+} (b_{1t} + B_t m_t) \tag{8}$$

$$h_t = \{2 h_t [\tfrac{1}{2} (m_t + m_{t+})] - \tfrac{1}{2} [h_t (m_t) + h_t (m_{t+})]\} + \tfrac{1}{2} \ln|I + V_t B_t| \tag{9}$$

where b_{1t} and B_t derive from the partitioning of Q_t:

$$2 Q_t = \left[\begin{array}{c|c} 2 b_{0t} & b'_{1t} \\ \hline b_{1t} & B_t \end{array}\right] .$$

These parameters can be used to simulate the mortality and risk factor processes in a cohort using equations describing the change in a.) survival, b.) the risk factor means and c.) the covariance matrix:

$$S_{t+1} = S_t \exp[-h_t] \tag{10}$$

$$m_{t+1} = A_{0(t+1)} + (I + A_{1(t+1)}) m_{t+} \tag{11}$$

$$V_{t+1} = \Sigma_{t+1} + [I + A_{1(t+1)}] V_{t+} [I + A'_{1(t+1)}] , \tag{12}$$

where $x_{i(t+1)} \sim N(m_{t+1}, V_{t+1})$. In (10) the variance and mean of the risk factor distribution affect survival through the relations in (9). The changes in the risk factor means and variances involve diffusion, mortality, and autoregressive effects (see Eqs. 7, 8, 11, 12).

3. EVALUATION OF CHRONIC DISEASE PROCESSES: DIFFERENT CONCEPTS

The mathematics help us understand the mechanisms of physiological aging and chronic disease progression and mortality in greater detail than is possible with existing hazard models. In the model chronic disease mortality is described as the exceedance of a probabilistically determined mortality threshold by a multivariate physiological stochastic process. In most hazard analysis models this physiological "threshold" does not exist. Instead, the mortality probability or hazard is a fixed function of covariates observed at specific points in time. There is no possibility to evaluate the underlying processes driving physiological aging changes with that type of model.

In the stochastic process model the matrices $I+A_{1t}$ and Σ_t represent the deterministic and stochastic dynamics of the physiological variables. The balance of deterministic vs. stochastic changes in risk factors has important implications for the evolution of mortality in risk heterogeneous populations. In (9) and (10) the probability of survival is a function of V_t, the risk factor covariance matrix. As the variance of risk factors increases, mortality increases. Depending upon whether individual risks are fixed (i.e., Σ_t=0) or stochastic (i.e., $\Sigma_t = V_t$) there are different implications for the evolution of mortality. Fixed risks cause mortality rates to decline as high risk persons die out. A white noise process implies no tendency for mortality to decrease because there is no systematic selection. Because diffusion maintains the variance of the distribution, mortality rates are still higher than in a homogeneous population. Since epidemiological data show persistence of physiological risk factors, and that genetic factors are important determinants of risk, there will be a strong deterministic component in most biological applications.

In the model age was treated specially because of inability to measure all factors changing with age and relevant to disease risk. Thus, we included age functions in both the dynamic and hazard components of the process. This makes the process strongly nonstationary. It also illustrates the generality of the multivariate Gaussian form which holds when risk factors are made explicit functions of time (Liptser and Shiryayev, 1977).

The model identifies many elements involved in chronic disease risk and provides an appropriate formalization of the multivariate stochastic process. It also provides a strategy for combining data from different sources. Estimates of the coefficient vectors and matrices (i.e., m_t, V_t, A_t, Σ_t, and Q_t (or Q and β)) may be derived from different data sets or ancillary theory, and combined in the differential equations describing the evolution of the population. For example, in most epidemiological studies the range of ages is limited as is the amount of time the cohort was followed. The process structure allows those different ranges of experience to be linked in ways accounting for individual physiological dynamics and the effects of systematic mortality selection on the population mix of individual physiological trajectories. Alternatively, with the coefficient matrices estimated from selected longitudinally followed populations the mortality of general populations can be described by applying the change and hazard coefficients to national estimates of the distribution of risk factors (e.g., cross-sectional estimates of m_t and V_t from health surveys).

The model can also identify the physiological mechanisms underlying age, period, and cohort effects by identifying those effects with specific features of the process. Period effects can be represented by the change, at a specified date, of the process parameters. Cohort effects are represented by birth cohorts having different process coefficients. Age effects are resident in the age functions representing unobserved covariate processes.

The conditional Gaussian form of the process allows updating of parameter estimates for new observations. The equations for the evolution of the mean and variance of the risk factors can be shown to lead to the following differential equations, which are the Kalman filter updating equations generalized for mortality selection, i.e.,

$$\frac{d}{dt} \hat{x}_t = A(t)\, \hat{x}^*_t - W_t\, [b_1(t) + B(t)\, \hat{x}_t] \tag{13}$$

$$\frac{d}{dt} W_t = A_1(t)\, W_t + W_t\, A_1'(t) + \Sigma(t) - W_t\, B(t)\, W_t\ , \tag{14}$$

where $b_1(t)$ and $B(t)$ result from partitioning $Q(t)$ the same way as b_{1t} and B_t are derived from Q_t. For each interval $[t_n, t_n+1)$, the initial conditions are:

$$\hat{x}_{t_n} = x_{t_n} \quad \text{and} \quad W_{t_n} = 0.$$

This yields $z_t \sim N(\hat{x}_t, W_t)$ for $t_n \leq t < t_n + 1$, under the assumptions (1) and (2). This provides an alternative to the Cameron-Martin (1944) equation, which, because of the necessity of re-estimating from the endpoint of the process, offers no simple approach to updating of process parameters (Yashin et al., 1986a).

The equations permit us to examine the interaction of forces of mortality (i.e., dependent competing risks) by using the differential equations to "forecast" survival with altered quadratic hazard equations for specific disease processes. The implications for the survival of persons with different risk factor trajectories is illustrated in Table 1 for the mortality process driven by eight risk factors (Pulse Pressure (PP), Diastolic Blood Pressure (DBP), Quetelet Index (QI), Cholesterol (SC), Blood Sugar (BS), Hemoglobin (HB), Vital Capacity (VC), and Cigarettes per Day (Cig).

Table 1. Effects of Risk Factor and Hazard Interventions on Evolution of Mortality Process

AGE	LIFE EXPECTANCY	PP (mmHg)	DBP (mmHg)	QI (hg/m^2)	SC (mg/dl)	BS (mg/dl)	HB (dg/dl)	VC (cl/m^2)	Cig
Baseline									
30	44.5	45.8	79.6	261.9	215.2	79.4	142.1	139.3	13.2
50	26.4	47.8	83.4	277.0	241.1	83.9	149.6	130.0	12.6
70	11.5	62.8	82.9	267.0	223.2	98.4	150.8	99.5	4.7
90	3.1	76.4	80.6	250.8	205.7	111.6	151.6	62.7	0.0
Control of Age Increase in Systolic & Diastolic Blood Pressure									
30	45.6	45.8	79.6	261.9	215.2	79.4	142.1	139.3	13.2
50	27.5	45.9	79.6	276.9	241.5	84.2	149.8	130.0	12.8
70	12.9	45.9	79.6	270.3	223.8	96.7	151.0	99.3	5.0
90	4.1	46.1	78.1	256.4	205.1	108.4	151.8	61.6	0.0
Reduction of Variance of Systolic & Diastolic Blood Pressure									
30	49.6	45.8	79.6	261.9	215.2	79.4	142.1	139.3	13.2
50	28.2	47.8	83.4	277.0	241.1	83.9	149.6	129.9	12.6
70	12.6	63.0	83.1	267.0	223.3	98.4	150.8	99.5	4.7
90	3.3	78.1	82.2	251.9	206.5	111.6	151.7	62.5	0.0
Eliminating Force of Mortality of Cancer									
30	46.8	45.8	79.6	261.9	215.2	79.4	142.1	139.3	13.2
50	28.3	47.8	83.4	277.0	241.1	83.9	149.6	129.9	12.7
70	12.8	62.8	82.9	267.0	233.1	98.2	150.8	99.3	4.8
90	3.6	76.4	80.5	248.8	204.4	111.0	151.7	62.0	0.0

This table is based on the experience of 2,336 males followed for 20 years in the Framingham Heart Study. We present, for ages 30, 50, 70, and 90 the life expectancy (e_t) and risk factor means (m_t) under baseline conditions and under the three alterations of the process equations described in the table. Each of these scenarios affects not only survival but also the means and variances of the other risk factors. Eliminating the age increase in blood pressure increases survival most at advanced ages. Reducing the variance of blood pressure increases survival most at earlier ages. Altering these two risk factors allows persons with other risk factors elevated to live longer, alters the pattern of mortality (because different forces of mortality relate differently to each risk factor), and alters the dynamics of other risk factors. Eliminating cancer also changed the survival of persons with different risk factor trajectories and generates a positive dependency between conditions.

4. SUMMARY

We presented a stochastic process model of human aging and mortality that reflected the temporal interaction of risk factors, risk factors and diseases, and diseases. Because the process components that underlie these interactions are represented explicitly these effects could be studied. In risk assessment procedures where such structure is not represented such effects cannot be directly assessed. Thus, if interactions are significant those models will not accurately describe mortality patterns, forecast mortality changes, nor realistically represent the relation of risk factors to disease risk. Not only is the process model biologically more realistic but it helps integrate evidence from multiple data sources and formally identifies the implications of theoretical specification. To accurately describe chronic disease, whose long term evolution makes such interaction likely, such process models are necessary.

REFERENCES

Cameron, R.H., Martin, W.T. (1944). J. Math. Phys. 23, 195-209.
Liptser, R., Shiryayev, A. (1977). Statistics of random processes. Springer-Verlag, New York.
Woodbury, M.A., Manton, K.G. (1977). Theor Popul Biol. 11, 37-49.
Woodbury, M.A., Manton, K.G. (1983). Human Biol. 55, 417-441.
Yashin, A.I. et al. (1986). J. Math Biol. 24, 119-140.
Yashin, A.I. et al. (1986a). Advances in Mathematics and Computers. (In Press).

GLOBAL FORECAST AND CONTROL OF FAST-SPREADING EPIDEMIC PROCESS

Vasilyeva V.,Belova L.,Rvachev L.A.,Shashkov V.Gamaleya Institute,Moscow, Rvachev L.L., Moscow University, USSR; Donovan D.,Commonwealth Department of Health, Canberra, Australia; Fine P., University of London, UK; Gregg M., Center for Disease Control, Atlanta, Longini I., Emory University, Atlanta, Fraser D., Swarthmore College, Pennsylvania, USA.

The present communication contains the brief account of the unpublished monograph "Experiment on pandemic process modelling" of the same authors, written in I982-I984 and dedicated to mathematical modelling of the epidemic process, global in potential and characterized by quickness of person-to-person transmission and mass morbidity. We distinguish two cases: the non-quarantine epidemics when the quarantine is not introduced, and the quarantine epidemics.

In contemporary world fast-spreading non-quarantine epidemics do not take place, except for those which are caused by influenza strains, either new or absent for a long time. So the experiment was carried out when the model had made a forecast of the daily course of the last great influenza pandemic - the one of I968-I969 - in the system of 52 world cities proceeding only from matrix of world air communications and from morbidity for july, I968, in Hong Kong (the initial point of this epidemic for the Earth). The results of the forecast are shown in Table I, where each sign means an average daily computer forecast for a given city and for four given days: a point "." is a weak morbidity (less than IO

Table I. Computer forecast of 1968- 1969 influenza pandemic.

i		City	1968						1969								
			July	Aug	Sept	Oct	Nov	Dec	Jan	Feb	March	April	May	June	July	Aug	Sept
1		London*															
2		Paris															
3	o	Rome															
4		Berlin															
5	o	Madrid															
6		Warsaw															
7		Budapest															
8		Sofia															
9		Stockholm															
10		Hong Kong															
11		Tokyo*															
12	o	Peking															
13	o	Shanghai															
14		Singapore															
15		Manila															
16		Bangkok															
17		Jakarta															
18		Calcutta															
19		Bombay															
20		Delhi															
21		Madras															
22	o	Seoul															
23		Teheran															
24	o	Karachi															
25		Cairo															
26		Kinshasa															
27		Johannesburg															
28		Casablanca															
29		Mexico															
30		Bogota															
31		Havana															
32		Caracas															
33		Lima															
34		Santiago															
35		Buenos Aires*															
36		Rio de Janeiro															
37	o	São Paulo															
38		Honolulu*															
39		Sydney															
40		Melbourne															
41	o	Perth															
42		Wellington															
43		Montreal*															
44		New York*															
45		Los Angeles*															
46		Washington*															
47		Houston*															
48		Chicago*															
49		San Francisco*															
50		Atlanta*															
51		Lagos															
52		Capetown*															

x Empirical data are joint with the surrounding region.

persons a day for IOO OOO of population), a dash "-" is an average morbidity (IO-IOO persons a day per IOO OOO of population), a plus "+" is a heavy morbidity (more than IOO persons a day per IOO OOO of population) and an exclamation mark "¡" means that the calculated peak of the epidemics falls on four given days in a given city.

However the actual morbidity according to the W.H.O. data on a city or to some other medical sourses is indicated by an unbroken line (where there may be gaps of information). The parts where this line is splitted into two correspond to the actual peak of the city epidemics, while arrows indicate the direction to the peak if this direction is known but nothing is known about the peak inself. Evidently the data on morbidity for some cities is absent at all because according to the computer forecast Havana is the only city where there is no epidemics (the transport system taken into account in the model is insufficient to carry the infection to Havana).

As to quarantine epidemics,the modelling is considered for processes mainly of unknown or unrecognized character. Thus all the parameters are free,and the paramount problem is their determination proceeding only from urgently coming in statistic data on undoubtedly observable phenomena. In conformity with the character of the data time t is considered to be discrete with I day step. To determine parameters and initial conditions of the so called "black" model which is intended for the subsequent control of the epidemic process (in contrast to the "white" model intended to generate scenarios of different epidemics in different parts of the world) it is necessary and sufficient to introduce the following statistical data into the model:

I. The daily morbidity in a city.
2. The number of patients in the city.
3. The number of isolated healthy contacts in city.

4. Time l from the appearance of clinical manifestation of the illness in a person to his isolation averaged by the number of cases in a city for a day. The present model is intended for the infections when the latent period (the time between the moment when a person gets infected and the moment when he can infect other people) is not shorter than the incubation period (the time between the moment when a person gets infected and the moment when the first signs of the illness appear). Consequently, a person who is a source of infection for other people can infect these people only during his individual period l ,thus determining the role of the mean l in a city as a control parameter. (The case when the latent period is longer than the incubation period is of no interest because the process will not exist at all, all infected people being isolated just before they could infect others).

5. The rate of ill persons $g(\tau)$, of recovered $h(\tau)$ and of the deceased $\hat{h}(\tau)$ in some statistical ensemble A of persons for whom the day of infection $\tau = 0$ is known where τ - is individual time for persons from A.

It is worth noting that the "black" model in contrast to the "white" model is intended only for the period of the epidemic process when the total number of cases does not exceed 7-IO per cent of the city population.

The values of the parameters and the initial conditions established, the model can give the forecast of values of all dynamic variables of a given epidemic process in a given city for every next day. In case of the quarantine epidemics the model should give not one but several forecasts for a city by means of varying the planning intracity strategy, i.e the total combination of parameter values which can be influenced by anti-epidemic measures (control parameters). When the strategy is chosen the forecast for a city becomes singlevalued.

Among the dynamic variables of the forecast there will be value U(t) - the number of non-isolated latent persons in a city for every next day t. It allows to solve not only the intracity but also the intercity control task, i.e. the control of intercity and international quarantine. Consider the concept of this solution.

The fact that for epidemic cities the intracity strategy is chosen means that to liquidate the epidemics in these cities is the matter of time only. Naturally, the less is current value U(t) the less time from the moment of starting the strategy is needed to turn U into zero. So if the intracity strategy is worked out at least in outline the cases of carrying single infected persons through the transport system into single cities of the world will be quite harmless for single countries and the mankind as a whole. The city where this will happen will quickly modify the strategy to itself and due to very small U cope with the situation in several days. It is very important not to allow the carriage to become plural as regards the number of persons or the number of cities. The latter is perfectly impossible as for obvious reasons it may lead to the epidemic crisis in a region or in the whole world.

We denote by t_o the day of starting the computer control of anti-epidemic measures. Then we denote by Sb(t) the probability of carrying at least one infected person into a given non-epidemic city during the time from t_o to t, regardless of the epidemic city this person came from. The safety requirements mentioned above would be insured if Sb(t) could be majorated by some small value α for all non-epidemic cities and for $t-t_o$ which is on the order of 2-3 months. Indeed smallness α for a single non-epidemic city means by definition that if infected persons were still carried into a city it would be single cases (the probability of carrying of infected persons $\geqslant$ n is majorated by value $\alpha_1^n/(1-\alpha_1)$, where $\alpha_1 < \alpha$).

As to the probability of carrying the infection into n cities, its accurate calculation is rather cumbersome due to two factors: a) carrying the infected persons into different cities is not an independent event, because an infected person who has reached one non-epidemic city cannot reach the other; b) any non-epidemic city can realize its probability Sb(t), i.e. it can become an epidemic on any day t_1, $t_0 < t_1 < t$, which will change probabilities Sb(t) of the rest cities at $t > t_1$. Similar change may repeat due to the carriage of infection into another non-epidemic city any day t_2, $t_1 < t_2 < t$, from the third city any day $t_3 > t_2$ etc. However, the probability for a city to become epidemic is small owing to $Sb(t) < \alpha$; secondly, those cities which became epidemic after day t_0 will have as it's mentioned above, only single cases of non-isolated infected persons, whereas even without the quarantine introduction and even in the most powerful passenger flows of the world such as New-York - Washington the probability of leaving the city during a day for a single person is $\leqslant 10^{-3}$; thirdly, we naturally perform the correction of all Sb(t) day t_1 . When these reasons are taken as a whole factor "b" can be fully neglected in this estimation. Here remains factor "a". It means that the true probability of carrying infected persons into two non-epidemic cities simultaneously will be lower than if we simply multiplied their Sb . That is why the probability of carrying the infection into n cities simultaneously is practically zero as it is majorated by α^n. These simple ideas do not need comments.

Now let's consider in brief the intracity control parameters, to be more precise the generalized parameters, because each of them expresses the overall effect of the whole group of concrete intracity anti-epidemic measures (for example: closing the places of mass recreation, catering, educational institutions, changes in the city transport and supply, limitting the movement of people in the city, accommodation of workers in the indispensible antiprises and closing of others, etc.). The definitions revealing the essence both of generalized parameters of the intracity control and of some of their useful functions are as follows:

I. λ - a mean number of effective contacts of an inhabitant of a city per a day. We call an effective contact of two persons such a contact that if one of them were an average sick person and the other were an average susceptible person the infection would inevitably happen (consequently, not every effective contact means getting infected because effectively contacting people do not always form a pair "a sick person - a susceptible"). Note that λ is not a pure control parameter, it has a mixed character because its basis component is the infection contageousness which does not refer to the control.

2. α - a rate of the city population which plays the role of inflammable material in the epidemic outbreak. And again α is not a pure control parameter as its basis component is the natural collective immunity of the population which does not refer to the control.

3. ℓ - mean time from the illness manifectation to the isolation of a sick person defined above.

4. η - a mean number of isolated contacts falling on one sick person; $\hat{\lambda}$ - a probability of a latent person isolation for reason that his contact with a sick person is fixed. These two values are not independent but are different forms of expression of the same control parameter. The connection between η and $\hat{\lambda}$ is empirical.

5. $M = \alpha \lambda \ell (1-\hat{\lambda})$ - It is the most generalized parameter of the intracity control.

The aim of the intracity control is to obtain $M < I$.We are short of time to dwell upon how to do this; we only note that on the basis of observable data coming in every day and which we recalculated earlier, the model works out about 20 types of the intracity information every day and sends it to every epidemic city. It is necessary to perform this procedure every day in order to correct continuously all calculations (the control should have IOO per cent reliability) to take into account not only the evolution of dynamic variables of the epidemic process but the fluctuation of parameter values calculated by the model as well (in connection with spontaneous changing of the anti-epidemic measures and of the population reaction, in connection with errors and incompleteness of observation , accidental process fluctuations and possible mutation of an agent).

Note to Table I. The spot to the left of the city name means that the actual morbidity for this city is taken according to Longini et al (I986). In nine cases the time of the actual peak of the city epidemics may be determined inderectly:

a) For Rome, Berlin, Peking, Shanghai, São Paulo and Capetown we only know the time of the beginning and the end of the epidemics. But the experience shows that during influenza the epidemic waves are more or less symmetrical, that is why it is quite natural to assume that the epidemic peak is in the middle between its beginning and its end. These six peaks are not marked in Table I.

b) For Mexico, Bogota and Buenos Aires the peak times according to the literature are too long, about a month. It's natural to assume that real peaks are in the middle of these intervals. In Table I these peaks are marked.

References

Longini I.M., Fine P.E.M., Thacker S.B.(I986) Predicting the Global Spread of New Infectious Agents. American Journal of Epidemiology, v. I23 , 3,383-39I.

A STATISTICAL ANALYSIS OF THE SEASONALITY IN SUDDEN INFANT DEATH SYNDROME (SIDS) .

Hans Bay

National Board of Health ,Statistical Department ,Denmark

INTRODUCTION

SIDS is defined as the sudden death of infants between one week and one year which is unexpected by history and in which a thorough post-mortem examination fails to demonstrate an adequate cause of death.

he syndrome has a characteristisk age distribution with a clearly defined peak at two to four month.

The syndrome has a seasonal variation with a peak incidence during winter month.

THE DATA SET

145 cases of SIDS in eastern Denmark in a four year period 1981 to 1984.This study examines whether the seasonality is related to the time of death, to the time of birth or both.

In fig.1 the observations (excl. death in 1984) are plotted in a diagram with time of birth on the x-axis and time of death on the y-axis.

The 145 SIDS have been grouped by month of birth and by month of death. This grouping gives rise to a grouping by age.

e.g. a SIDS born in january and who dies in march, will be given age-group 3. Age group 3 covers real life time between the firts and third month. see fig. 2.

In table 1 the data are grouped by each of the three variables, separately.

THE MODEL

X_{ijk}= number of SIDS from age-group i,who die in the month of j and are born in the month of k.

(X_{ijk}) are independent poisson-variates

$$E(X_{ijk}) = N_k \alpha_i \beta_j \gamma_k$$

N_k = the number of infants born in the month of k

α_i = the parameter of age

β_j = the parameter of the month of death

γ_k = the parameter of the month of birth

log-likelihood : $\sum_{ijk} X_{ijk} (\ln(\alpha_i) + \ln(\beta_j) + \ln(\gamma_k)) -$

$$\sum_{ijk} N_k \alpha_i \beta_j \gamma_k$$

the sufficient statistics

$X_{i..}$ = the total number of SIDS in age-group i

$X_{.j.}$ = the total number of SIDS died in the month of j

$X_{..k}$ = the total number of SIDS born in the month of k

The ML-estimates are found by solving the follwing equations

$$\frac{X_{i..}}{\alpha_i} = \sum_{jk} N_k \beta_j \gamma_k$$

$$\frac{X_{.j.}}{\beta_j} = \sum_{ik} N_k \alpha_i \gamma_k$$

$$\frac{X_{..k}}{\gamma_k} = \sum_{ij} N_k \alpha_i \beta_j$$

This are done by iteration and the estimates are shown in fig. 3-5.

The test for a vanishing effect of the month of birth was not significant (P=89.C%). The test for a vanishing effect of the month of deathwas more significant (P=14.5%) . A simultaneously test was found to be highly significant (P=0.4%).

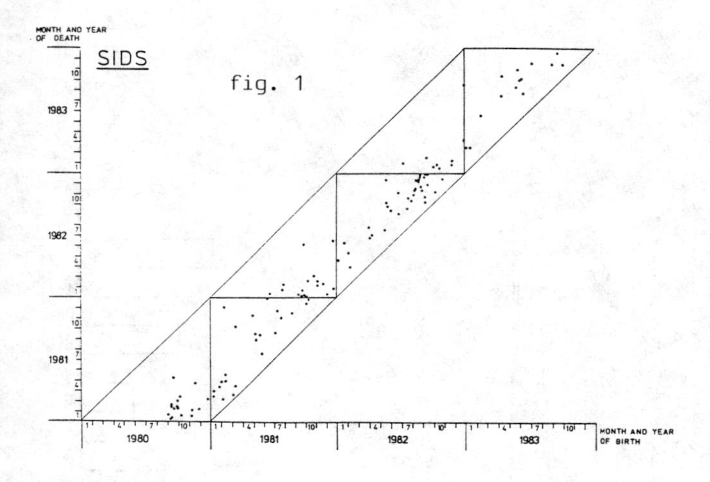

fig. 2

SIDS, grouped by month of birth and by month of death.

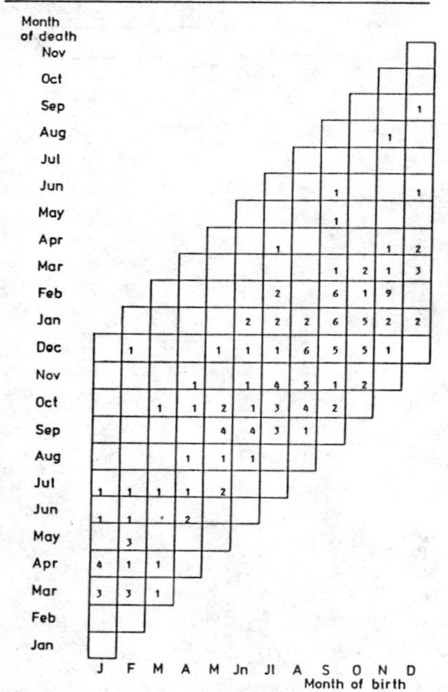

TABLE 1

NUMBER OF SIDS IN AGE-GROUPS AND GROUPED BY MONTH OF DEATH AND MONTH OF BIRTH

AGE-GROUP/ MONTH	NUMBER OF SIDS IN AGE-GROUPS	NUMBER OF SIDS IN EACH MONTH OF DEATH	NUMBER OF SIDS IN EACH MONTH OF BIRTH
1	1	21	9
2	12	18	10
3	24	14	4
4	43	10	6
5	28	4	10
6	17	6	10
7	7	6	16
8	7	4	18
9	1	13	23
10	4	14	15
11	1	14	15
12	0	21	9
TOTAL	145	145	145

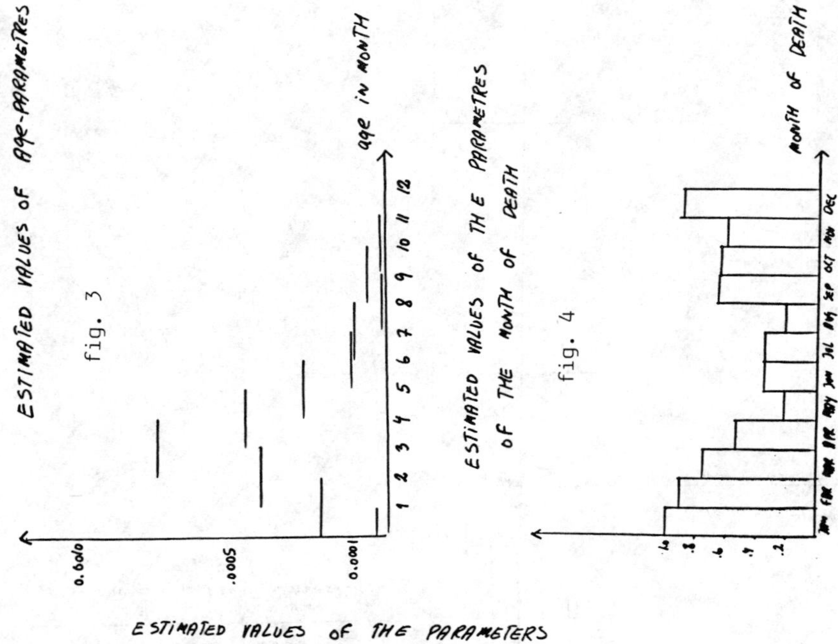

fig. 3

fig. 4

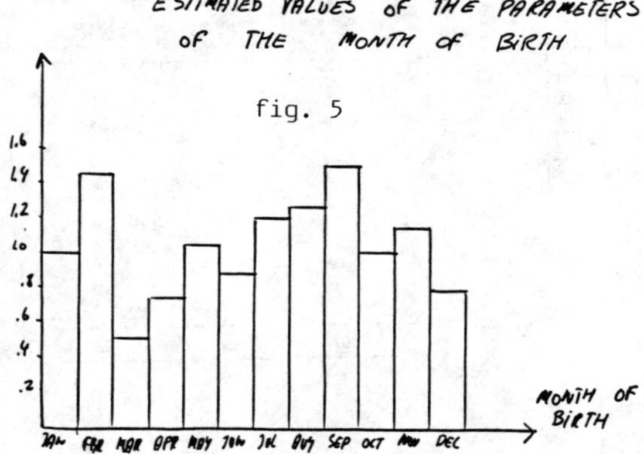

fig. 5

Considering the degrees of systematic variation in the estimated parameters of the month of birth and the month of death, the most reasonable conclusion on the basis of the estimates and of the tests is, that the age and the month of death independently influence the risk of SIDS.

Reference : K.Helweg-Larsen,Hans Bay and F.Mac. A statistical analysis of the seasonality in SIDS. International j.of Epidemiology,1985 vol 14, no.4

EPIDEMIOLOGICAL MODELS FOR SEXUALLY TRANSMITTED INFECTIONS

Dietz K.
Institute of Medical Biometry, Tübingen
Federal Republic of Germany

The classical models for sexually transmitted infections (Hethcote and Yorke, 1984) assume homogeneous mixing either between all males and females or between certain subgroups of males and females with heterogeneous contact rates. This implies that everybody is all the time at risk of acquiring an infection. These models ignore the fact that the formation of a pair of two susceptibles renders them in a sense temporarily immune to infection as long as the partners do not separate and have no contacts with other partners. The present paper takes into account the phenomenon of pair formation: Let $x(t)$ denote the proportion of singles at time t. If ρ denotes the rate of pairing, σ the rate of separation and μ the death rate, then x satisfies the following differential equation

$$\dot{x} = \mu+(\mu+\sigma)(1-x)-(\mu+\rho)x ,$$

if singles enter the population at rate μ. It is assumed that all rates are equal for male and females such that x refers either to single males or females. In equilibrium the proportion p of paired individuals is $p = 1-x = \rho/(2\mu+\sigma+\rho)$. The average number N of life time partners per individual is given by

$$N = (\rho/\mu)(2\mu+\sigma)/(2\mu+\sigma+\rho) = (\rho/\mu)(1-p) .$$

Let φ denote the rate of effective contact between the partners of a pair. The infection is supposed to terminate according to the rate γ. If an individual is susceptible again after an infection, then we have a generalization of the classical SIS-model with only two states for singles and four types of pairs. If an infection is followed by lifelong immunity then one has to distinguish three states for singles and nine types of pairs. The two models merge into a generalization of the classical "simple epidemic" (SI-model), if one lets γ tend to zero. Because of symmetry the proportion of pairs with a susceptible male and an infective female equals the proportion of pairs with a susceptible female and an infective male. Therefore the number of variables for pairs is reduced from four to three for the SIS-model and similarly from nine to six for the SIR-model. Let y and z denote the proportion of infectives and immunes, respectively, among singles. The proportion of paired individuals in pairs with one susceptible and one infective is denoted by q_{01}. Similarly, q_{11} denotes the proportion of paired individuals in pairs with two infective individuals. For the SIR-model one needs in addition the variables q_{02}, q_{12} and q_{22}, where the index "0" refers to a susceptible partner and the index "2" to an immune partner.

In the following all rates are expressed in units of the life expectancy $L = \mu^{-1}$: $r = \rho/\mu$; $s = \sigma/\mu$; $f = \varphi/\mu$; $g = \gamma/\mu$. Let $c = (1-p)/p$. Then for the SIS-model one can deduce the following differential equations:

$$\dot{y} = (1+s)c^{-1}((q_{01}/2)+q_{11})-(1+g+r)y ,$$

$$\dot{q}_{01} = 2rcy(1-y)+2gq_{11}-(2+s+g+f)q_{01} ,$$

$$\dot{q}_{11} = rcy^2+fq_{01}-(2+s+2g)q_{11} .$$

Similarly for the SIR-model one obtains

$$\dot{y} = (1+s)c^{-1}[(1/2)(q_{01}+q_{12})+q_{11}]-(1+r+g)y ,$$

$$\dot{z} = (1+s)c^{-1}[(1/2)(q_{02}+q_{12})+q_{22}]+gy-(1+r)z ,$$

$$\dot{q}_{01} = 2rc(1-y-z)y-(2+s+g+f)q_{01} ,$$

$$\dot{q}_{11} = rcy^2+fq_{01}-(2+s+2g)q_{11} ,$$

$$\dot{q}_{12} = 2rcyz+2gq_{11}-(2+s+g)q_{12} ,$$

$$\dot{q}_{22} = rcz^2+gq_{12}-(2+s)q_{22} ,$$

$$\dot{q}_{02} = 2rc(1-y-z)z+gq_{01}-(2+s)q_{02} .$$

It is assumed that the proportion of singles is equal to its equilibrium value. For both models one can explicitly derive a lower bound for the pairing rate r above which there exist positive equilibria:

$$r_{SIS} = \frac{(1+g)[f(s+2)+(s+2(g+1))(s+g+2)]}{fs-(1+g)(s+2(g+1))} ,$$

$$r_{SIR} = \frac{(1+g)(s+2(g+1))(s+2+g)(s+2+g+f)}{f[s(s+2+g)-2g(1+g)]-(1+g)(s+2+g)(s+2(g+1))} .$$

Both curves in the (r,s)-plane have a singularity which determines a lower bound for the separation rate s :

$$s^*_{SIS} = 2(1+g)/[f/(1+g)-1] ,$$

and

$$s^*_{SIR} = s^*_{SIS}+g .$$

For these minimum separation rates all individuals have to be paired ($p = 1$ for $r \to \infty$). The minimum number of life time partners is therefore $N^*_{SIS} = 2+s^*_{SIS}$ and $N^*_{SIR} = N^*_{SIS}+g$. These minima are subject to the condition $p = 1$. Depending on the values of f and g, there may be lower critical bounds for the number of life time partners for $p < 1$. From the formulas for r_{SIS} and r_{SIR} one can also easily deduce a lower boundary for the proportion of paired individuals: $p^* = (g+1)/f$. This lower boundary prevents the separation rate to become arbitrarily large: If f is finite then partnerships require in the average a minimum duration for transmission to be possible. This boundary is the same for both models. One obtains the classical models if one lets both f and s tend to infinity. Then nobody is paired and $r/(g+1)$ is the number of contacts during one infectious period.

In the classical theory a key parameter is the so-called basic reproduction rate R which is usually interpreted as the number of secondary cases which one case could infect if the total population were susceptible. For a positive endemic level R has to exceed one. The importance of this parameter results from its interpretation as a quantity which specifies the necessary elimination effort in the contact rate in order to make the zero equilibrium stable.

It turns out that for sexually transmitted diseases the elimination effort by chemotherapy is in general different from the elimination effort by reduction of the number of sexual partners. This raises the interesting problem to find out which control method is associated with the smallest elimination effort for given transmission conditions. For $f = \infty$ one can calculate the two elimination efforts explicitly for the SIS-model: $R_T = (N-2p)/[(1-p)(1+g)]$ and $R_N = N/(1+g+p(1-g))$, where R_T and R_N denote the elimination effort by chemotherapy and contact reduction, respectively. For $p = 0$, i.e. for the classical model $R_T = R_N$. But for $p > 0$ one finds that $R_T > R_N$. For p tending to one the ratio of the two quantities tends to infinity. If $f < \infty$, then one can specify domains for which $R_T < R_N$. If $p = 1$ then $R_T < R_N$ for $N < 2F(F+g)/(F(2+g)-1)$, with $F = f/(g+1)$. The detailed results for $p < 1$ and for the SIR-model will be published elsewhere.

The two models presented above can of course not claim to provide a realistic picture for the transmission of any particular sexually transmitted disease. But it is claimed that they offer a basic methodology which can be refined for special applications. It turns out that the endemic equilibrium is not only determined by the number of partners per unit of time but also by the duration of the partnership, by the duration of the search time for a new partner and by the contact rate during a partnership. Estimates on the basis of questionnaire surveys can thus be incorporated into the model in a meaningful way.

Of particular interest is the transient behavior of the systems given above. Compared to the classical models the rate of approach of equilibrium is much slower than for the classical models for similar elimination efforts. This finding has implications for the interpretation of data relating to the spread of human immunodeficiency virus infections out of the original risk groups into the heterosexual population.

Reference

Hethcote, H.W., and Yorke, J.A. (1984) Gonorrhea Transmission Dynamics and Control. - Lecture Notes in Biomathematics 56, Berlin: Springer-Verlag, 1984, 105p.

RESULTS AND PERSPECTIVES OF THE MATHEMATICAL FORECASTING OF INFLUENZA EPIDEMICS IN USSR

Ivannikov Yu.G.
All-Union Research Institute of Influenza, Leningrad, USSR

From 1970 in the All-Union Research Institute of influenza, Ministry of Health of the USSR, the All-Union information computerised system of the influenza epidemic forecasting is developing. This system is based on the L.A.Rvachev's mathematical model [I] and on an influenza surveillance in the USSR [2]. The epidemic modelling consists of two stages: the modelling of the local epidemic in a single city and the modelling of the epidemic spread over the territory of the country. The modelling of the local epidemic is carried out considering the number of the people of the given city, the probability to be ill for the patient on the first, second and so on day of illness, the number of the immune; succeptable and patients among the city population, the frequency of contacts among the city population.

The modelling of the epidemic spread over the terrytory of the country is carried out considering the transport communication between each pair of the cities including in the model.

The accuracy of coincidence of the local model with the real indices of the epidemic process in a single city is high, the error is I - 3 %.

The epidemic modelling in the scale of the country starts with the modelling of the local epidemic for the city ("initial" city) in which the current epidemic started for the first time in the country. Only daily morbidity in the "initial" city during beginning of the epidemic is used as initial data. By thia data the main parameters of the model expressing the frequency of contacts among the city population and the level of the community are determined. Then an assumption is made that these parameters are similar for those the whole terrytory of the country. The local models for all the other cities are made up with the same parameters. The start point of the epidemic in each city is determined considering the transport communications of the given city with other cities of the country. It is supposed that the source of infection is the patients who came into the city in question from the cities where the epidemic already started. Theire number is higher the greater are the transport communications with the cities - sources of the influenza patients and is the rise of the incidence of influenza disease in those cities.

From 1971 All-Union information computerised system of the influenza epidemic forecasting is used by All-Union Influenza Center for prediction of all influenza A epidemics developing on the territiry of the USSR [3,4]. The predictions of estimated numbers of sick men on each day of the epidemic were given 3 months ahead of epidemic for all main cities of the country (except "initial" city naturally).

The importance of computerised prognosis of influenza epidemics spread over the territory of the USSR is difficult to overestimate. In the cities for which the prognosis is given (and the range of the cities is practically unlimited) the public health care authorities reseive in advance the information about the supposed number of pati ents per day of the forthcoming epidemic period. Using this data it is possible to calculate the expected number of influenza complications, the number of deaths due to influenza, the excess incidence of somatic diseases, the excess deaths among the population and other indices. Taking all these indices and prognosis data as a guide it ispossible to distribute the drugs per the time exactly, the level of utilization of beds in hospitals of different profile (infectious and somatic), the demend of health services in medical staff and transport, the time of introducing into life different quaratine and other limiting measure.

The analysis of the prognosis results showed that its accurasy is quite sufficient for practical purposes. The evaluations of prediction's exactness of spread of epidemic morbidity are given in tables I - 3. Generaly the prediction of time of epidemic peak in some cities was satisfactory in 90% cases. The prediction of intensity of epidemic was satisfactory in 80% cases. The satisfactory prediction by both criteria was noted in 75% of cases.

Table I. Prediction's exactness of epidemic's peak

Epidemic (years)	Namber of cities	Punctually of prediction	Divergence of prediction and real peak of epidemic		
			≤3 days	≤7 gays	>7 days
I97I-I972	30	8	20	29	I
I972-I973	33	7	I6	30	3
I974-I975	58	2I	33	49	9
I976	35	5	I3	26	9
I977-I978	45	I6	23	42	3
I979-I980	29	6	I7	25	4
All	230	63	I22	20I	29
%	I00	27	53	87	I3

Table 2. Prediction's exactness of daily morbidity

Epidemic (years)	Number of cities	Prediction morbidity / real morbidity				
		≤ 0.7	0.7-I.0	I.0-I.5	I.5-2.0	≥ 2
I97I-I972	30	5	8	I5	I	I
I972-I973	33	0	0	I5	5	I3
I974-I975	58	2	I0	27	I7	2
I976	35	2	I6	7	3	7
I977-I978	45	9	I8	I8	0	0
I979-I980	29	3	II	I4	I	0
ALL	230	2I	63	96	27	23
%	I00	9	27	42	I2	I0

Perspectives of the influenza epidemic mathematical forecasting system development are related with further specification of the mathematical model, increasing of

Table 3. Prediction's exactness of epidemic's start

Epidemic (years)	Number of cities	Punctually of prediction	Divergence of prediction and real start of epidemic	
			≤I week	>I week
I972-I973	4I	26	38	3
I974-I975	I7	I2	I7	0
I976	38	I3	3I	7
I977-I978	45	28	42	3
I979-I980	32	I8	3I	I
All	I73	97	I59	I4
%	I00	56	92	8

importance of predition by including into it additional epidemiological characteristics, excess morbidity and necessary antiepidemic measures. The conditions of the improvement of influenza epidemic forecasting are also a developing of influenza surveillance and increase of the accuracy of the information about influenza and other acute respiratory diseases.

It is necessary to note that the considered method of mathematical prognosis of influenza epidemic because of its uneversal nature might be adapted for the territory of any country under the condition of epidemiological surveillance system which allows to registrate the current incidence of acute respiratiry diseases in cities. Certainly given method might be used for prognosis of repidly and widely spreading epidemics of other infections.

References

I. Baroyan O.V., Rvachev L.A., Ivannikov Yu.G. Modelling and forecasting of influenza epidemics for the territory of the USSA. M., I977. 546 p. (in Russian)
2. Ivannikov Yu.G., Ismagulov A.T. Epidemiology of influenza. Alma-Ata, Kazakhstan, I983, 204 p. (in Russian).
3. Ivannikov Yu.G., Shipulin E.S. The results mathematical modelling and prognosis of influenza epidemics in the scale of the country. The materials of the republic scientific conference on mechanism of transmission of infectious disease agents, Leningrad, I973, 57-59, (in Russian).
4. Smorodintsev A.A., Ivannikov Yu.G., Marynic I.G., Shipulin E.S. The epidemiological influenza situation in I97I-I972, J.Microbiol., M., I974, 4,40 (in Russian).

Bernoulli, Vol. 2, pp. 547-551

THE GENERALIZED DISCRETE-TIME EPIDEMIC MODEL WITH IMMUNITY

Longini, I.M.
Department of Statistics and Biometry
Emory University, Atlanta, Georgia 30322, U.S.A.

This paper deals with a discrete-time version of the "general epidemic" model as defined in chapter 6 in Bailey (1975) (i.e., an epidemic for those infectious diseases which confer immunity following infection for the period of interest). This model is based on that used by Soviet investigators over the last 20 years to analyze and predict influenza epidemics (see Baroyan, et al. (1977) and Rvachev and Longini (1985)).

MODEL FORMULATION AND PROBABILITY MODEL

We formulate the system on a discrete state space and index set using the following dynamic random variables:

- $U(t)$ - number of new infectives (latents) at time t, and $\underline{U}(t) = [U(0), U(1), \ldots, U(t)]$,
- $X(t)$ - number of susceptibles at time t ,
- $R(t)$ - number of immune individual at time t .

We also have the following parameters:

- λ - infectious contact rate,
- $g(\tau)$ - probability that an individual who was first infected τ time units ago is still infectious, where $g(\tau)>0$ for $0<\tau\leq\tau_2$ and $g(\tau)=0$ for $\tau=0$ and $\tau>\tau_2$,
- n - population size.

If a susceptible individual is infected at time t , then that individual becomes latent and then infectious for periods of time described by the distribution $\{g(\tau)\}$. Following that, the individual enters the immune state where he remains. We assume that individuals mix randomly and define the random function $P(t)=1-Q(t)$ as the probability that a susceptible individual is infected at time t, where

$$P(t) = 1-Q(t) = \lambda/n \sum_{\tau=1}^{\tau_2} g(\tau)U(t-\tau) \; , \; 0 \leq P(t) \leq 1 \; . \tag{1}$$

Using the above structure, we see for small populations, i.e., $n \leq 20$, that the probability mass function (pmf) of U(t) conditioned jointly on X(t-1) and $\underline{U}(t-1)$ is binomial, given by

$$\Pr[U(t)|X(t-1),\underline{U}(t-1)] = \binom{X(t)}{U(t)} P(t)^{U(t)} Q(t)^{X(t)-U(t)}, \; X(t) \geq U(t),$$

$$X(t) = X(t-1) - U(t-1) \; . \tag{2}$$

The initial conditions can be stated as

$$\Pr[X(0) = \alpha n] = 1 \text{ and } \Pr[U(0) = a] = 1 \; , \tag{3}$$

where αn and a are integers and $\alpha n + a = n$. Given an observed sequence $\{\tilde{U}(\ell)\}_{\ell=o}^{t}$, the likelihood function is given by

$$L_t(\alpha,\lambda) = \prod_{\ell=o}^{t} \binom{\tilde{x}(\ell)}{\tilde{U}(\ell)} \tilde{P}(\ell)^{\tilde{U}(\ell)} \tilde{Q}(\ell)^{\tilde{X}(\ell)-\tilde{U}(\ell)} \; . \tag{4}$$

The maximum likelihood (ML) estimators $\hat{\alpha}$ and $\hat{\lambda}$ are found using an iterative procedure such as Fisher's scoring method (see Saunders (1980) and Longini (1986)).

For larger populations, i.e. $n > 20$, the pmf (3) can be approximated by a Poisson distribution and the likelihood function becomes

$$L_t(\alpha,\lambda) = \prod_{\ell=o}^{t} \frac{[\tilde{P}(\ell)\tilde{X}(\ell)]^{\tilde{U}(\ell)}}{\tilde{U}(\ell)!} \exp\,[-\tilde{P}(\ell)\tilde{X}(\ell)] \; . \tag{5}$$

If α is known, then there is a closed form estimator for λ from equation (5) given by

$$\hat{\lambda} = n \sum_{\ell=o}^{t} \tilde{U}(\ell) \;/\; \sum_{\ell=o}^{t} \tilde{X}(\ell) \sum_{\tau=o}^{t_2} g(\tau)\, \tilde{U}(\ell-\tau) \; , \tag{6}$$

with approximate variance given by

$$\mathrm{Var}(\hat{\lambda}_t) \cong \hat{\lambda}_t^2 \;/\; E\,[\sum_{\ell=o}^{t} U(\ell)] = \hat{\lambda}_t^2 \;/\; \sum_{\ell=o}^{t} \hat{P}(\ell)X(\ell) \; . \tag{7}$$

Saunders (1980) has shown that $\hat{\lambda}_t$ from (6) is strongly consistent as $t \to \infty$.

DETERMINISTIC MODEL

The conditional expected value of U(t) (from (2)) is given by

$$E[U(t)|X(t-1), \underline{U}(t-1)] = x(t)\ P(t)\ , \tag{8}$$

which also holds for the Poisson approximation. This suggests the deterministic system of difference equations ($t=o,1,\ldots.$),

$$x(t+1) = x(t) - u(t)\ , \tag{9}$$

$$u(t) = x(t)\ P(t)\ ,\ x(0) = \alpha n\ ,\ u(\tau) = 0^{+}\ ,\ \tau\epsilon[-\tau_2, 0]\ ,$$

where $\{u(t)\}$ is substituted into (1) to obtain P(t). This system has been analyzed by Longini (1986) and Bykov and Kontratjev (1980). In order to state the results, we denote the infectious "contact number" as

$$\epsilon = \lambda \sum_{\tau=1}^{\tau_2} g(\tau)\ , \tag{10}$$

and the final infection attack rate as

$$\rho = r/\alpha n\ ,\ \text{where } r = \sum_{t=o}^{\infty} u(t)\ . \tag{11}$$

Then we have the following threshold theorem:

Theorem 1 (Longini (1986)): For the deterministic model specified by equations (9),

i) if $\alpha\epsilon \leq 1$, then $\rho \cong 0$ (no epidemic occurs),

ii) if $\alpha\epsilon > 1$, then $\rho > 0$ (an epidemic occurs), and ρ is bounded from below by ρ_1 which is the largest positive root of the equation $s = 1-\exp(-\alpha\epsilon s)$.

Theorem 1, which is proven in Longini (1986), gives a threshold result for the general discrete-time epidemic model with immunity that is analogous to the Kermack-McKendrick Threshold Theorem (see pg. 84 in Bailey (1975)) for the continuous-time epidemic model. It is well-known that the "initial infective replacement number" $\alpha\epsilon$ must exceed one for an epidemic to occur in a defined population.

An upper bound for ρ is given by corollary 1 below which was given by Bykov and Kontratjev (1980) for the special case where $\Sigma\ g(\tau) = 1$ and extended by Longini (1986) for a general distribution $\{g(\tau)\}$.

Corollary 1 (Bykov and Kontratjev (1980), Longini (1986)): For the deterministic model specified by equations (9), if $\alpha\epsilon > 1$, then ρ is bounded from above by ρ_2 which is given by

$$\rho_2 = \begin{cases} 2(1-1/\alpha\epsilon), & \text{if } 1 < \alpha\epsilon \leq 2, \\ 1, & \text{if } 2 < \alpha\epsilon . \end{cases}$$

We designate $\delta_{\alpha\epsilon} = (\rho_2/\rho_1) - 1$ as the relative error of the bounds on ρ for $\alpha\epsilon>1$. Then, we have the following corollary:

Corollary 2 (Bykov and Kontratjev (1980), Longini (1986)): The maximum value of $\delta_{\alpha\epsilon}$, for $1 < \alpha\epsilon$, is 0.255 at δ_2. Furthermore, $\delta_{\alpha\epsilon}$ is strictly increasing in $\alpha\epsilon$ in the interval (1,2) and strictly decreasing in $\alpha\epsilon$ in the interval $(2,\infty)$ with $\lim_{\alpha\epsilon\to\infty} \delta_{\alpha\epsilon} = 0$.

Bykov and Kontratjev (1980) gave corollary 2 for their model described above. Corollary 2 establishes that the bounds on ρ can be quite useful. The value of $\delta_{\alpha\epsilon}$ estimated from influenza epidemics in England and Wales and Greater London was found to range from 0.10 to 0.25 (see Longini (1986)).

An alternative to using the likelihood function (4) or (5) is to minimize the sum of squared differences between the observed sequence $\{\tilde{u}(\ell)\}_{\ell=o}^{t}$ and the expected sequence $\{u(\ell)\}_{\ell=o}^{t}$ generated by the deterministic model (9). The least squares estimators for α and λ can be found by numerically minimizing the function

$$\sum_{\ell=o}^{t} [\tilde{u}(\ell) - u(\ell)]^2 \, w_\ell , \tag{12}$$

over the parameter space, where the $\{w_\ell\}$ are appropriately selected weights in the case of weighted least squares.

Such a procedure has been carried out by Spicer (1979) (and Olesen) using data from influenza epidemics in England and Wales and Greater London for the years 1958-1973. They estimated $\alpha\epsilon$ with the weighted least squares estimates for $\alpha\epsilon$ ranging from 1.431 to 2.606. The results from theorem 1 and corollaries 1 and 2 have been used to compute bounds on ρ for their estimates (see Longini (1986)). Results for parameter estimation from influenza epidemics in small populations, i.e., households, can be found in Longini, et al. (1982, 1983).

References

Bailey N.T.J. (1975). The Mathematical Theory of Infectious Diseases. Charles Griffin, London.

Baroyan, O., Rvachev, L., Yvannikov, Y. (1977). Modeling and Forecasting of Influenza Epidemics for Territory of the U.S.S.R. (in Russian). Gamaleya Inst. Epidemiol. Microbiol, Moscow.

Bykov, S., Kontratjev, V. (1980). Some results of the study of a mathematical model of influenza epidemics (in Russian). Epidemiol. Prophylaxis Influenza: Trans. All-Union Res. Inst. Influenza, Leningrad, 122-129.

Longini, I., Koopman, J. (1982). Household and community transmission parameters from final distributions of infections in households. Biometrics 38, 115-126.

Longini, I., Koopman, J., Monto, A. (1983). Estimation procedures for transmission parameters from influenza epidemics: Use of serological data (in Russian). Voprosy Virusologii, No. 2, 176-181.

Longini, I. (1985). The generalized discrete-time epidemic model with immunity: A synthesis. Math. Biosci., 81, in print.

Rvachev, L., Longini, I. (1985). A mathematical model of the global spread of influenza. Math. Biosci., 75, 3-22.

Saunders, I. (1980). An approximate maximum likelihood estimator for chain binomial models. Austral. J. Statist., 22(3), 307 -316.

Spicer, C. (1979). The mathematical modelling of influenza epidemics. Brit. Med. Bull., 35, 23-28.

GEOLOGY AND GEOPHYSICS
(Session 14)

Chairman: D. Vere-Jones

STOCHASTIC MODEL OF MINERAL CRYSTALLIZATION PROCESS FROM MAGMATIC MELT

Rodionov D.A. Institute of Ore Deposits Geology, Mineralogy, Petrography and Geochemistry Academy of Sciences of USSR, Moscow, USSR

The most of igneous rocks petrographical researches include the problem of estimation of crystallization sequence of minerals forming the rock. Commonly, for the decision of this problem the structural correlation between grains is used. However frequently the personal factor acts on these observations and it leads to some subjectivity of the results. Becouse of it is very desirable to develope such criterium for the determination of crystallization mineral sequence, which would not depend on subjective perseption of investigator.

Let us imagine that whole volume of magmatic melt is divided into fixed small volumes, which are correspon - ding to the volumes of thinsections or samples in which the mineral contents are estimated. Let ξ be the melt content in that volume and let us consider ξ as random variate. Then accordingly to the mineral crystallization sequence we have the sequence of random variate which characterise the melt contents after crystallization of corresponding mineral, that is

$$\xi_0, \xi_1, \xi_2, \dots \xi_i, \dots \xi_{K-1}, \xi_K \qquad (1)$$

where ξ_0 is initial melt contents, ξ_i, ξ_{K-1}, ξ_K are the melt after the crystallization of the minerals with the numbers i, $K-1$, K correspondingly.

Let ε_i is random variate, which has the values in the interval from 0 to 1 t.i. $0 \leq \varepsilon_i \leq 1$. Besides $\varepsilon_i = \rho_i + \delta$, $M\varepsilon_i = \rho_i$, $D\varepsilon_i = D\delta \leq \alpha$. Then

$$\xi_1 = \xi_0 (1 - \varepsilon_1) \tag{2}$$

$$\xi_1 = \xi_1 (1 - \varepsilon_2) = \xi_0 (1 - \varepsilon_1)(1 - \varepsilon_2) \tag{3}$$

. . . .

$$\xi_\kappa = \xi_{\kappa-1}(1 - \varepsilon_\kappa) = \xi_0 \prod_{i=1}^{\kappa} (1 - \varepsilon_i) \tag{4}$$

Thus, we can consider the melt contents as the product of ξ_0 and κ random variates $(1 - \varepsilon_i)$. Let η_κ is random variate which is the model of the content of mineral with the number κ in their sequence. It is possible to consider this variate as the difference between the melt contents $\xi_{\kappa-1}$ before this mineral crystallization and the melt contents ξ_κ after its crystallization, that is

$$\eta_\kappa = \xi_{\kappa-1} - \xi_\kappa = \xi_0 \prod_{i=1}^{\kappa-1} (1 - \varepsilon_i) - \xi_0 \prod_{i=1}^{\kappa} (1 - \varepsilon_i) =$$

$$= \xi_0 \prod_{i=1}^{\kappa-1} (1 - \varepsilon_i) \left[1 - (1 - \varepsilon_\kappa)\right] = \xi_0 \varepsilon_\kappa \prod_{i=1}^{\kappa-1} (1 - \varepsilon_i) \tag{5}$$

Theorem 1. Let $\varepsilon_1, \varepsilon_2, \ldots, \varepsilon_i, \ldots, \varepsilon_\kappa, \varepsilon_{\kappa+1} \ldots$ and $\eta_1, \eta_2, \ldots, \eta_\kappa, \eta_{\kappa+1}$ are the sequences of positive random variates such that

$$\eta_\kappa = \xi_0 \varepsilon_\kappa \prod_{i=1}^{\kappa-1} (1 - \varepsilon_i) \tag{6}$$

and $0 \le \varepsilon_i \le 1$, $\varepsilon_\kappa > 0$. Then for all $\kappa = 1, 2, \ldots$ the next unequal of the variances will be truly:

$$D \ln \eta_\kappa \le D \ln \eta_{\kappa+1} \tag{7}$$

Let us consider new random variate $\ln \eta_\kappa$. It is possible to represent it as the sum:

$$\ln \eta_\kappa = \ln \xi_0 + \ln \varepsilon_\kappa + \sum_{i=1}^{\kappa-1} \ln (1 - \varepsilon_i) \tag{8}$$

If the variates ε_i are independent, the variance $D \ln \eta_\kappa$ of random variate $\ln \eta_\kappa$ is determined by the expression

$$D \ln \eta_\kappa = D \ln \xi_0 + D \ln \varepsilon_\kappa + \sum_{i=1}^{\kappa-1} D \ln (1-\varepsilon_\kappa) \tag{9}$$

and for the variate $D \ln \eta_{\kappa+1}$

$$D \ln \eta_{\kappa+1} = D \ln \xi_0 + D \ln \varepsilon_{\kappa+1} + D \ln (1-\varepsilon_\kappa) + \sum_{i=1}^{\kappa-1} D \ln (1-\varepsilon_i) \tag{10}$$

It is clearly that the expressions (7) and (8) are distinguished only by two items $D \ln \varepsilon_\kappa$ and $D \ln \varepsilon_{\kappa+1}$ + + $D \ln$ $(1-\varepsilon_\kappa)$. It is easy to show that $D \ln \varepsilon_\kappa \leq$ $\leq D \ln$ $(1-\varepsilon_\kappa)$ if to decomposite $\ln$ $(1-\varepsilon_\kappa)$ in the series:

$$\ln (1-\varepsilon_\kappa) \simeq -\varepsilon_\kappa - \frac{\varepsilon_\kappa^2}{2} - \dots \tag{11}$$

Hence

$$D \ln (1-\varepsilon_\kappa) \simeq D \varepsilon_\kappa + \frac{1}{4} D \varepsilon_\kappa^2 + 2 \rho D \varepsilon_\kappa D \varepsilon_\kappa^2 \tag{12}$$

Becouse $\rho \geq 0$, for $0 \leq \varepsilon_\kappa \leq 1$

$$D \ln (1-\varepsilon_\kappa) \geq D \varepsilon_\kappa \geq D \ln \varepsilon_\kappa \tag{13}$$

That is $D \ln$ $(1-\varepsilon_\kappa) \geq D \ln \varepsilon_\kappa$

Consequently $D \ln \eta_\kappa \leq D \ln \eta_{\kappa+1}$ (14)

Thus the variances of the logorithms of contents, of minerals crystallizating succession from magmatic melt, increase in accordance of the increasing their numbers in the sequence. This perculiarity may be used for the development of statistical criterium for the revealing of mineral crystallization sequence.

Let in researching rock n calculations of m mineral contents were made. They may be the calculations on the integrator in the thin sections or the results of mineral contents definition in the samples of rocks and oth. As the result, the initial data will be represented as the table, containing n rows and m colomns, that is

$$X = \begin{pmatrix} x_{11} & x_{12} & \cdots & x_{1k} & \cdots & x_{1m} \\ x_{21} & x_{22} & \cdots & x_{2k} & \cdots & x_{2m} \\ \cdots & \cdots & \cdots & \cdots & \cdots & \cdots \\ x_{t1} & x_{t2} & \cdots & x_{tk} & \cdots & x_{tm} \\ \cdots & \cdots & \cdots & \cdots & \cdots & \cdots \\ x_{n1} & x_{n2} & \cdots & x_{nk} & \cdots & x_{nm} \end{pmatrix} \qquad (15)$$

Each row of this table is corresponding to concrete sample of rock, and the colomn - to the mineral.

These initial data are transformated into the logarithms (it is possible to use any logarithm base) and the matrix (15) will have next form:

$$\ln X = \begin{pmatrix} \ln x_{11} & \ln x_{12} & \cdots & \ln x_{1k} & \cdots & \ln x_{1m} \\ \ln x_{21} & \ln x_{22} & \cdots & \ln x_{2k} & \cdots & \ln x_{2m} \\ \cdots & \cdots & \cdots & \cdots & \cdots & \cdots \\ \ln x_{t1} & \ln x_{t2} & \cdots & \ln x_{tk} & \cdots & \ln x_{tm} \\ \cdots & \cdots & \cdots & \cdots & \cdots & \cdots \\ \ln x_{n1} & \ln x_{n2} & \cdots & \ln x_{nk} & \cdots & \ln x_{nm} \end{pmatrix} \qquad (16)$$

The estimation S^2 of unknown variance $D \ln \eta_k$ is calculated for each colomn by next formula:

$$S^2 = \frac{1}{n-1} \sum_{t=1}^{n} (\ln x_{tk} - \overline{\ln x}_k)^2, \quad k = 1, 2, \ldots, m \qquad (17)$$

where

$$\overline{\ln x}_k = \frac{1}{n} \sum_{t=1}^{n} \ln x_{tk}, \quad k = 1, 2, \ldots, m \qquad (18)$$

is arithmetic mean of mineral contents logarithms for

the mineral with the number κ . It is necessary to remember, that under conditions of our problem the sequence of crystallization is unknown and we have to estimate it. Because the arrangement of the colomns in the tables (15) and (16) are arbitrary and after the estimations of logarithm variances are calculated we must regulate them. Thus, arranging the estimations S^2_κ as the new sequence in accordance with requirement

$$S^2_{\kappa 1} \leq S^2_{\kappa 2} \leq \ldots \leq S^2_{\kappa j} \leq \ldots \leq S^2_{\kappa m} \qquad (19)$$

we shall obtain the approximate characteristic of true sequence of mineral crystallization from melt for the researching rock.

However, because in the sequence (19) we have only the estimations $S^2_{\kappa_j}$ of unknown variances (let us design them σ^2_1 , σ^2_2 , ..., σ^2_κ , ..., σ^2_m), it is possible these estimations will distinguish unsignificantly. It is impossible to make any inferences about the sequence of mineral crystallization. Besides if even some estimations in the sequence (19) distinguish significantly, it is possible that there are the distinguishing groups of estimations, which contain undistinguish estimations inside each group. Hence, the obtained increasing sequence S^2_1 , S^2_2 ,..., S^2_i ,..., S^2_m requires more careful research, namely, the test of hypothesis about its homogenity. If this hypothesis will be reject we have to look for the boundary between distinguishing statistically homogeneous groups.

Let S^2_1 , S^2_2 ,..., S^2_i ,..., S^2_m is regulated increasing sequence of estimations of variances σ^2_1,..., σ^2_m which were obtained by n_1 , n_2 ,..., n_i ,..., n_m observations. It is required by these data to test hypothesis

$$H_0 : \sigma^2_1 = \sigma^2_2 = \ldots = \sigma^2_m = \sigma^2_0$$

by alternative $H_1 : \sigma_i^2 \neq \sigma_0^2$ although for one $i = 1, 2, \ldots, m$.

It would be possible to use Barthlett criterium for the test of this hypothesis. However this criterium permits only to accept H_0 or to reject it, and in case of rejection this criterium does not permit to find the point where the homogenity is broken. Therefore it is better for testing of H_0 to use as criterium some function F_κ which is defined on the set of dividings of regulate sequence $s_1^2, \ldots, s_m^2$ into two parts of κ and $m-\kappa$ items, $\kappa = 1, 2, \ldots, m-1$. Let us design S_κ^2 generalized estimation of variance for the first κ items of sequence, calculated under condition that $\sigma_1^2 = \sigma_2^2 = \ldots = \sigma_\kappa^2$. Let us design also $S_{m-\kappa}$ similar estimation for other $m-\kappa$ items of sequence.

$$S_\kappa^2 = \frac{1}{\left(\sum_{i=1}^{\kappa} n_i\right) - 1} \sum_{i=1}^{\kappa} (n_i - 1)\, s_\kappa^2 \tag{20}$$

$$S_{m-\kappa}^2 = \frac{1}{\left(\sum_{i=\kappa+1}^{m} n_i\right) - 1} \sum_{i=\kappa+1}^{m} (n_i - 1)\, s_i^2 \tag{21}$$

The ratio

$$V_\kappa = \frac{S_{m-\kappa}^2}{S_\kappa^2} \tag{22}$$

is corresponding to each $\kappa = 1, 2, \ldots, m-1$. If H_0 is true, the ratio will be the value of F-distributed random variate with

$$f_{m-\kappa} = \left(\sum_{i=\kappa+1}^{m} n_i\right) - 1 \quad \text{and} \quad f_\kappa = \left(\sum_{i=1}^{\kappa} n_i\right) - 1 \tag{23}$$

freedom of degrees.

Thus hypothesis H_0 is accepted by the level of significance α if

$$\max_{\kappa} V_{\kappa} \leq F_{\alpha}\binom{f_{m-\kappa}}{f_{\kappa}}, \quad \kappa = 1,2,\ldots,m \qquad (24)$$

where $F_{\alpha}\binom{f_{m-\kappa}}{f_{\kappa}}$ is value F-distribution for level of significance α and $f_{m-\kappa}$ and f_{κ} freedom of degrees.

If $$\max_{\kappa} V_{\kappa} > F_{\alpha}\binom{f_{m-\kappa}}{f_{\kappa}}, \qquad \kappa = 1,2,\ldots,m-1 \qquad (25)$$

H_0 is rejected and the alternative H_1 is accepted. In this case the sequence is divided into two parts in the point corresponding to $\max_{\kappa} V_{\kappa}$. The procedure is repeated for each of these parts. This dihotomical process is continuing until for each of revealed group will be true unique

$$\max_{\kappa} V_{\kappa} \leq F_{\alpha}\binom{f_{m-\kappa}}{f_{\kappa}} \qquad (26)$$

As a result it will be obtained a set of statistically homogeneous groups of variance estimations by logarithms of mineral contens. Naturally, the inference about crystallization sequence inside that homogeneous group will be nongrounded and it will be necessary to consider these minerals as crystallized approximately simultaneously. However the sequence of mineral groups will be grounded and it will correspond to sequence of their formation in time.

Let us consider as the example the results of mineral research of gabbro samples of Polar Ural gabbrotonalite complex by S.F.Sobolev (I). The table 1 contains the variance estimations of logarithms of mineral contents which are regulated as increasing sequence. As a result of data shawn in table I the regulated sequence of mineral was divided into 4 homogeneous pacts.

Table 1

Results of mineral content analysis in gabbro

NN	Mineral	N_i of calculations	Estimation of variance	Estimation of variance in group	$\frac{F_{m-k}}{F_k}$	$F_{0,05}$	$\left(\frac{f_{m-k}}{f_k}\right)$
1	Plagioclase	12	0,025	0,029			
2	Magnetite	6	0,042		3,65	2,19	$\binom{17}{23}$
3	Pyroxene	12	0,102	0,107			
4	Quartz	12	0,123		2,99	2,64	$\binom{5}{23}$
5	Pyrite	6	0,320	0,320	7,17	4,50	$\binom{29}{5}$
6	Sphene	6	0,804				
7	Chalcopyrite	6	2,414				
8	Limonite	6	3,060	2,296			
9	Apatite	6	3,290				
10	Zircon	6	3,750				

The test of hypothesis about equality of generalized variances for adjacent groups had shown this hypothesis is rejected everywhere and it permits to infer about sequence of crystallization of revealed mineral groups.

Reference

1. Sobolev S.F. 1965. Polar Ural gabbro-tonalite complex. "Nauka", Moscow, p.161.

APPLICATIONS OF STOCHASTIC GEOMETRY IN GEOLOGY

Dietrich Stoyan
Sektion Mathematik, Bergakademie Freiberg, Freiberg, 9200, G.D.R.

INTRODUCTION

Stochastic geometry deals with point fields, fibre fields, domain fields, random tessellations, and other geometrical structures, see Stoyan et al.(1986). In geology frequently irregular patterns have to be studied which have a behaviour similar to these structures and which can be described by models of stochastic geometry, see Deffeyes et al.(1982). Nevertheless, until now stochastic geometry has had no intensive application in mathematical geology; here random fields or regionalized variables dominate. A traditional field of application of stochastic geometry is, of course, the stereological analysis of stones and minerals.

In order to be concrete, this paper discusses in detail two particular cases: (1) point patterns with interaction and evolution and (2) cross-correlation analysis of geological line systems.

POINT PATTERNS WITH INTERACTION AND EVOLUTION

There are a lot of geological patterns which can be modelled as planar point fields (or point processes), e.g. systems of small ore deposits, see Agterberg(1976), epicentres of earthquakes, see Vere-Jones and Ozaki(1984), and centres of sinkholes. It is possible that the instants at which the points appeared are known, as in the study of earthquakes. If not, then the description of spatial interaction is the main problem.

Probably, so-called Gibbs or Markov point processes are suitable models for geological point patterns with interaction. Their application is demonstrated here by means of a very small data set, see fig.1. The 31 points are centres of sinkholes in an area of North Harz region (G.D.R.). Geologists say that Karstification processes in the upper Buntsandstein (in depth of ca. 400 m there)

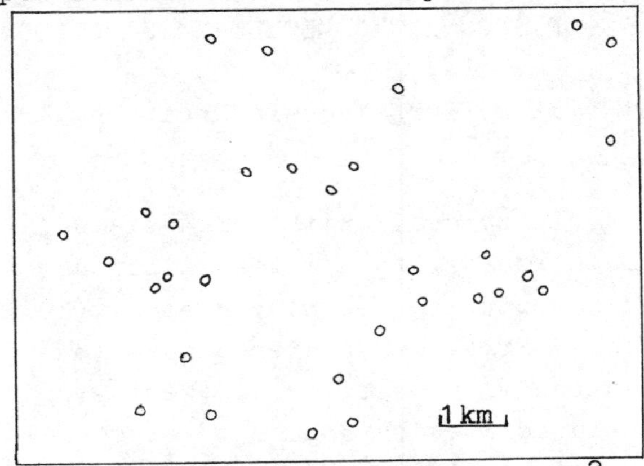

Figure 1. 31 sinkholes in a $6.9 \times 9.2\ km^2$ window.

cause the sinkholes in that area. Nearly 2...4 sinkholes arise in 100 years there; only the data of the last five events are exactly known.

By means of a simulation test it could be shown that the pattern should not be interpreted as a sample of a binomial process (n independent points uniformly distributed in the window). In analogy to Ripley's K-function approach (Ripley, 1977; Stoyan et al., 1986) the function s was used, where

$s(r)$ = number of interpoint distances $< r$ in the pattern.

It is assumed now that the joint probability density of n points in the window W has the form

$$f_n(x_1,\ldots,x_n) = \exp\Big(-\sum_{1\le i<j\le n} \theta(\|x_i - x_j\|)\Big)/Z,$$

where Z is a norming constant.

The corresponding point process is called Gibbs or Markov point process, the function θ is called pair-potential. It describes the interaction of the points.

The empirical form of s and asymptotic properties of

sparse point patterns (Saunders et al., 1982) suggest for θ the form

$$\theta(r) = \begin{cases} \infty, & r \leq r_0 \\ -v, & r_0 < r \leq r_1 \\ 0, & r > r_1. \end{cases}$$

If the point pattern is interpreted as a part of a sample of a stationary Gibbs process with such a pair-potential, then the "cusp method" could be used: In Hanisch and Stoyan(1983) it is shown that the K-function (which is, up to a factor, close to s) vanishes for $r \leq r_0$ and has a cusp at $r = r_1$ with

$$\lim_{r \uparrow r_1} K'(r) / \lim_{r \downarrow r_1} K'(r) = e^v .$$

This leads to the estimates $\hat{r}_0 = 0.25$ = minimal interpoint distance, $\hat{r}_1 = 1.1$ (here the ascent of s(r) begins to be smaller than for smaller r), and $\hat{v} = 0.62$ for the sinkhole pattern.

For more general pair-potentials and samples of stationary Gibbs point processes, the Takacs-Fiksel method should be used, see Stoyan et al.(1986).

Since the sinkhole sample is small, methods adopted to stationary point processes may be inappropriate here. As an alternative approach the maximum likelihood method is known, see Ogata and Tanemura(1981) and Penttinen(1984). For the pair-potential above the log likelihood has the form

$$\ln L(r_0, r_1, v; x_1, \ldots, x_n) = v \cdot s(r_1) - \ln Z.$$

The known statistical methods use approximations of Z, which are of acceptable accuracy for "sparse" patterns without strong clustering, but may be rather poor otherwise, see Gates and Westcott(1986). Two approximations of Z are

$$Z_{OT} = A^n (1 - a/A)^{n(n-1)/2}$$

and

$$Z_P = A^n \exp(\pi n^2 r_1^2 (b - 1)/(2A))$$

with A = area of window W, $a = \pi[r_1^2 - e^v(r_1^2 - r_0^2)]$, and

$$b = e^v(r_1^2 - r_0^2)/r_1^2 .$$

They lead to the estimates $\hat{v} = 0.50$ and $\hat{v} = 0.43$ if $\hat{r}_0$ and $\hat{r}_1$ are taken as above.

Now the point of view is changed: While Gibbs processes are connected with the idea of an equilibrium state, now a dynamic element is included into the considerations, since the original aim of the statistical analysis was a prediction of positions of next sinkholes. For doing this, an assumption on the order of occurrence of sinkholes is necessary. A promising assumption comes from the known close relation of hard-core Gibbs processes to the so-called simple sequential inhibition process, see Ripley(1977). Consider a Markov process $\{X_k\}$ of the following form:

$\{Y_i\}$ is a sequence of random points in W, where Y_1 is uniformly distributed in W. Given the positions $y_1,\ldots,$ y_m of $Y_1,\ldots, Y_m$, the probability density of Y_{m+1} has the form

$$h_{m+1}(y) = \exp\left(-\sum_{i=1}^{m} \Theta(\|y - y_i\|)\right)/z_{m+1}, \quad m = 2,3,\ldots,$$

where Θ is the function above and z_{m+1} is a norming constant (which depends on Θ and $y_1,\ldots,y_m$), $y \in W$. Put

$$X_k = \{Y_1,\ldots,Y_k\} .$$

A prediction of the position of the next or (n+1)th point is possible if one assumes that the evolution of the pattern follows indeed the rules of the Markov process, at least for values near n. Then the probability density of the (n+1)th point is h_{n+1}. It enables to determine areas with high or low probability of occurrence of the next point.

Starting from the estimates of the Gibbs approach, the parameters v, r_0 and r_1 of the Markov process were determined by the Monte Carlo estimation method, were s(r) was used. It produced the values $\hat{v} = 0.4$, $\hat{r}_0 = 0.25$, and $\hat{r}_1 = 1.1$.

For an assessment of goodness-of-fit, fig.2 shows a graphical comparison between the data and 20 simulations of the fitted model. As a summary description of the da-

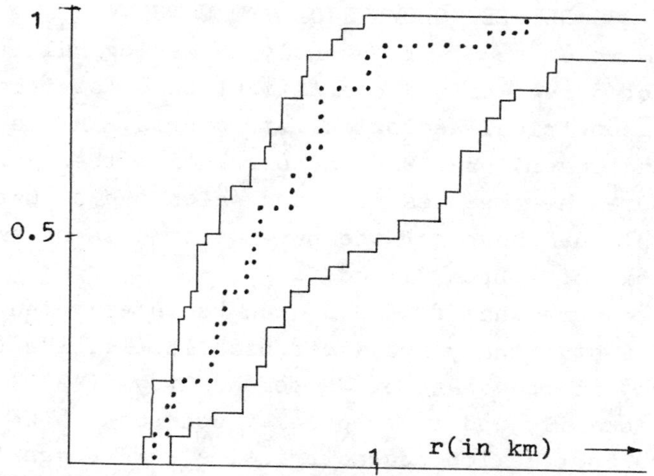

Figure 2. Empirical distribution function of distances from points to nearest neighbour. ... sinkhole data; — upper and lower extremes from 20 simulations.

ta, the empirical distribution function of distances from each of the 31 points to its nearest neighbour is used. Since the empirical curve lies well between the upper and lower extremes and these curves are not used in the parameter estimation, some confidence in the statistical adequacy of the model is justified.

For the example of 31 sinkholes, h_{32} has the simple form

$h_{32}(y) = 0$ if y has a distance less than 0.25 to one of the 31 points of the pattern,

$h_{32}(y) = \exp(v \cdot \text{number of neighbours of } y \text{ in the pattern}) / z_{32}$ if the distance is greater.

(A neighbour is a point with distance between 0.25 and 1.1 km to the reference point.)

In the papers Fiksel(1984) and Fiksel and Stoyan(1983) other Markov process models were discussed and applied to other data. These models are closely related to so-called self-exciting point process models, which were used in the study of earthquake processes, see Ogata and Akaike(1982) and Vere-Jones and Ozaki(1984).

CORRELATIONS BETWEEN GEOMETRICAL STRUCTURES

A problem which occurs frequently in geological studies is that of detection and quantification of correlations between geometrical-geological structures. In the papers Stoyan and Ohser(1982,1985) a correlation theory for random measures has been established which can be used here It is explained here for the particular case of two planar systems of fibres f_1 and f_2.

Let us assume that f_1 and f_2 can be interpreted as samples of stationary and isotropic fibre fields (or processes) in the sense of Mecke and Stoyan(1980) which are stationarily and isotropically connected. (Notice that also correlation characteristics were suggested for the instationary case.)

Each of the fibre processes F_1 and F_2 can be associated with a random measure Φ_i, where, for all Borel A,

$\Phi_i(A)$ = length of all fibre pieces in A.

We have

$$E\,\Phi_i(A) = L_i\,\nu(A),$$

where ν is the Lebesgue measure. L_i is the mean fibre length of F_i per unit area.

The correlation between F_1 and F_2 or Φ_1 or Φ_2 can be characterized by the cross-moment measure μ_{12},

$$\mu_{12}(A_1 \times A_2) = E(\Phi_1(A_1)\cdot\Phi_2(A_2)),\ A_1, A_2 \text{ Borel.}$$

It can be disintegrated and takes then the form

$$\mu_{12}(A_1 \times A_2) = L_1\,L_2 \int_{A_1} \mathcal{K}_{12}(A_2 - x)\,dx.$$

Here $\mathcal{K}_{12}$ is a measure on R^2 called reduced cross-correlation measure. The quantity $L_2\,\mathcal{K}_{12}(A)$ can be interpreted as the mean length of fibre pieces of F_2 in A under the condition that the origin o lies on a fibre of F_1. (A precise definition uses the theory of Palm distributions) In case of isotropy it suffices to consider the particular $A = b(o,r)$ (= disc with radius r centred at o) and to use the function K_{12},

$$K_{12}(r) = \mathcal{K}_{12}(b(o,r)),\ r = 0,$$

or the cross-pair-correlation function g_{12},

$$g_{12}(r) = \frac{d}{dr}K_{12}(r)/(2\pi r).$$

The latter function has the following interpretation: Let C_1 and C_2 be infinitesimal discs of areas dA_1 and dA_2 r length units apart. Then $L_1L_2 \cdot g_{12}(r) \cdot dA_1dA_2$ is the mean of the product of lengths of pieces of fibres of F_1 in C_1 and F_2 in C_2 respectively.

Of course, in the case of independence

$$L_2K_{12}(r) = L_2\pi r^2 \quad \text{and} \quad g_{12}(r) \equiv 1.$$

Values of $g_{12}(r)$ greater than 1 indicate attraction between the members of both fibre systems at distance r, etc..

Let us consider a simple model of two correlated fibre processes. F_1 is a Poisson line process of intensity L_1, while F_2 is a process of lines which are parallel to the lines of F_1: a fraction p of lines of F_2 coincide with their partner in F_1 while the other lines of F_2 have their partner in a random distance. So $L_1 = L_2$. If the distances are distributed uniformly on $[a,b]$ $(a > 0)$ then

$$L_2K_{12}(r) = 2pr + \pi r^2L_2 + (1-p)\cdot\begin{cases} 0 & , r \leqq a \\ 2\int_0^{b_r}\sqrt{r^2 - x^2}\,dx, & r > a \end{cases}$$

with $b_r = \min(b,r)$. Fig. 3 shows g_{12} for $L_1 = 1$, $p = 0.2$, $a = 1$, and $b = 2$.

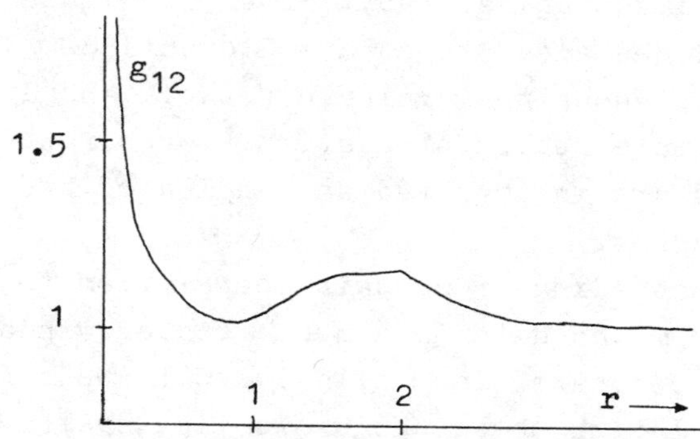

Figure 3. Cross-pair-correlation function for two correlated line processes.

If a sample of F_1 and F_2 is given in a window of observation W then K_{12} can be estimated: An unbiased estimator of $L_2K_{12}(r)$ is $\kappa_{12}(r)$,

$$\kappa_{12}(r) = \int_{W^{(r)}} k(x,r)\, \Phi_2(S(x,r,W^{(r)})\, \Phi_1(dx),$$

where

$W^{(r)} = \{x \in W: \partial b(x,r) \cap W \neq \emptyset\}$, $s(x,r,A)$ = union of sectors of the circle $b(x,r)$ lying completely in A, $k(x,r) = 2\pi/(\alpha_x(r)L_1\nu(W^{(r)}))$, where $\alpha_x(r)$ is the sum of angles corresponding to the sectors of $S(x,r,W^{(r)})$, cf. Stoyan et al.(1986), p.246.

If the samples are in the memory of a computer then the use of $\kappa_{12}(r)$ can be recommended. A similar method is to approximate the fibres by sequences of points and to use methods of statistical analysis of bivariate point patterns.

The papers Stoyan and Ohser(1982,1985) suggest so-called stereological methods which can be used if $K_{12}(r)$ has to be determined manually.

Until now the statistical properties of these estimators are not known; the situation is similar to that in the much simpler case of spatial point processes.

The cross-pair-correlation function g_{12} can be determined by numerical differentiation of K_{12}.

To give an example, the correlation between drainage networks and geological fault lines (or lineaments) is considered, see also Stoyan and Stoyan(1982,1983). Fig.4 on next page shows the drainage network and a system of lineaments in a region of Eastern Erzgebirge (G.D.R.). These lines are interpreted as samples of two correlated fibre processes.

The corresponding cross-pair-correlation function g_{12} obtained from the data in a much greater window is shown in Fig. 5 next page. The solid curve there reflects the close correlation between the two line systems. The minimum at a distance of ca 0.3 km seems to correspond to

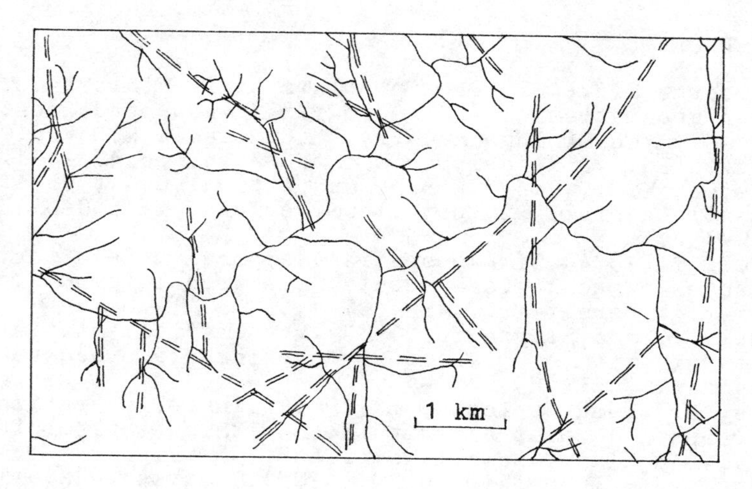

Figure 4. Drainage network and lineaments in a region of Eastern Erzgebirge.

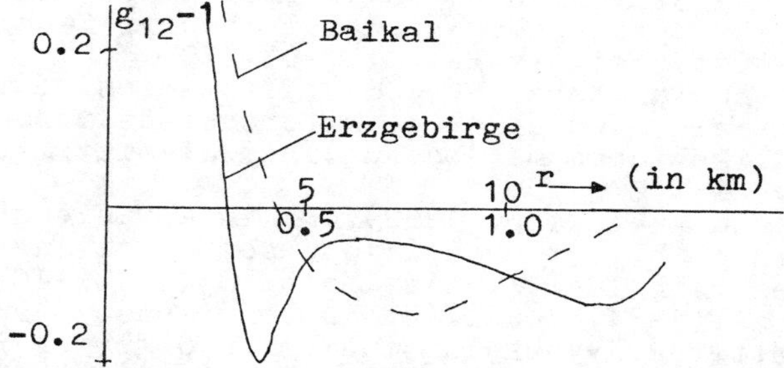

Figure 5. Cross-pair-correlation functions for two drainage networks and systems of lineaments.

the width of valleys.

Fig.5 also shows the cross-pair-correlation function g_{12} for the drainage network and a system of fault lines in a region at the Baikal Lake (U.S.S.R.).

ACKNOWLEDGMENTS

The author thanks Dr.T.Fiksel, Dr.J.Ohser, and Dr.A.Penttinen for their help and stimulating discussions. He is grateful to Prof. H.Bankwitz who kindly offered the data for the Baikal Lake example, which were obtained by remote sensing during the Interkosmos experiment.

REFERENCES

Agterberg, F.P.(1976).New problems at the interface between geostatistics and geology. In: Advanced Geostatistics in the Mining Industry, M.Guarascio,M.David,C.Huijbregts(Ed.).D.Reidel,Dortrecht/Boston, pp.403-421.
Deffeyes, K.S., Ripley, B.D. and Watson, G.S.(1982).Stochastic geometry in petroleum geology. Math.Geol. 14, 419-432.
Fiksel, T.(1984).Simple spatial-temporal models for sequences of geological events. Elektron.Informationsverarb.Kybernet. 20, 480-487.
Fiksel, T. and Stoyan, D.(1983).Mathematical-statistical determination of risk areas for sinkhole processes. Z.angew. Geol. 29, 455-459.
Gates, D.J. and Westcott, M.(1986).Clustering estimates for spatial point distributions with unstable potential. Ann.Inst.Statist.Math. A38, 55-67.
Hanisch, K.-H. and Stoyan, D.(1983).Remarks on statistical inference and prediction for a hard-core clustering model.Math.Operationsf.Statist.,ser.statist.,14,559-567.
Mecke, J. and Stoyan, D.(1980).Formulas for stationary planar fibre processes - general theory. Math.Operationsf.Statist.,ser.statist.,12,267-279.
Ogata, Y. and Tanemura, M.(1981).Estimation of interaction potentials of spatial point processes through the maximum likelihood method.Ann.Inst.Statist.Math. B33, 315-338.
Ogata, Y. and Akaike, H.(1982).On linear intensity models for mixed doubly stochastic and self-exciting point processes. Ann.Inst.Statist.Math. B34, 102-107.
Penttinen, A.(1984).Modelling interactions in spatial point patterns.Jyväskyla Studies in Computer Science, Economics and Statistics, 7.University of Jyväskyla.
Ripley, B.D.(1977).Modelling spatial patterns (with discussion). J.Roy.Statist.Soc. B39, 172-212.
Saunders, R., Kryscion, R.J. and Funk, G.M.(1982).Poisson limits for a hard-core clustering model. Stoch.Proc. Appl. 12, 97-106.
Stoyan, D., Kendall, W.S. and Mecke, J.(1986). Stochastic Geometry and Its Applications. J.Wiley, Chichester.
Stoyan, D. and Ohser, J.(1982).Correlations between planar random structures. Biom.J. 24, 631-647.
Stoyan, D. and Ohser, J.(1985).Cross-correlation measures of weighted random measures. Theor.Probab.Appl. 29, 345-355.
Stoyan, D. and Stoyan, H.(1982).Quantification of correlations between geometrical structures on geological maps. Z. angew.Geol. 28, 238-244.
Stoyan, D. and Stoyan, H.(1983).On a method of quantification of correlations between geological line systems. Z. angew.Geol. 29, 512-517.
Vere-Jones,D. and Ozaki, T.(1984). Some examples of statistical estimation... Ann.Inst.Statist.Math.,B 34,189-207

Bernoulli, Vol. 2, pp. 573-577

CLASSIFICATION AND PARTITIONING OF IGNEOUS ROCKS

E. H. Timothy Whitten

Michigan Technological University, Houghton, Michigan, 49931, U.S.A.

Although rock-naming and classification schemes are integral to virtually all igneous petrology and petrography text books, in most cases the terminology, definitions, and concepts used are imprecise for quantitative evaluations. Many different qualitative and quantitative variables have been used by petrographers to identify and describe igneous-rock types and suites.

Sedimentary and metamorphic rocks can be defined either *descriptively* in terms of known present physical and chemical composition and spatial variability, or *genetically* in terms of conditions during formation and/or of distinctive parental materials. For example, "marine beach" can be characterized either descriptively by features that permit present-day or fossil marine beaches to be identified, or genetically by environmental conditions (waves, currents, sediment transport, etc.) known (or predicted) to result in the formation of beaches. In most traditional and modern classification schemes for plutonic igneous rocks, only descriptive bases have been employed.

From among the plutonic igneous materials, consider granitic rocks. Several hundred variables can be observed for each individual granite sample. Using numerous quantitative (*e.g.*, K_2O%, Sr ppm) and qualitative (*e.g.*, texture) variables for individual samples, petrologists have recognized suites (mega-groups) within major granitoid batholiths (*e.g.*, Lachlan Fold Belt, SE. Australia, Chappell, 1984; Coastal Batholith, Peru, Pitcher, *et al.*, 1985; Okhotsk-Chuckchi belt, Soviet Far East).

Expert field geologists, probably correctly, claim that suites can be recognized with ease in the field (*e.g.*, that mega-groups of granite plutons can be identified within major batholiths). For such purposes, easily observed and subjectively integrated, but rarely quantified or measured, characteristics (variables) are included routinely. There is urgent need for such important characteristics to be identified explicitly and for their mode of assessment (measurement) to be standardized.

For scientific purposes, it is important to establish whether such suites (mega-groups) are real and significant. What wholly objective quantitative tests, for example, prove the reality of granite suites that have been identified previously by petrographers? Attempting to determine what objective criteria would permit unequivocal identification of analogous suites in a new terrane poses fundamental definitional, mathematical, and petrological issues. These same issues are also critical in the classification and partitioning of ALL igneous rocks (Whitten, *et al*., 1986a, 1986b) and indicate that current practices need revision.

From Vandermonde's Theorem (a polynomial of degree *n* can always be found that fits exactly (*n*+1) observed values with distinct abscissae) it follows that, provided enough variables are measured for a sampled population of granitic rocks, some variable (or set of variables) can always be found that effectively discriminates between any truly distinct pair of sets into which rocks are separated (*cf*., Li, *et al*., 1986). In general, one variable (or set of variables) can be expected to discriminate a different pair of groups of samples (descriptive suites or mega-groups) from those discriminated by another variable (or set of variables). Hence, a granite suite (*e.g*., as identified by a cluster-analysis algorithm) has no meaning apart from the specific variable/s by which it is defined. Consistent use of the same variable/s will yield comparable real descriptive suites in different regions (although dissimilar variable weighting will, and variance at the sample level could, yield different suites for interpenetrating sample sets from the same terrane [sampled population]). Hence, different sets of descriptive suites (mega-groups) coexist in the same terrane (*e.g*., a granitoid batholith). A genetic scenario for one such suite may not be useful unless it embraces concomitantly all other possible coexisting interpenetrating descriptive suites. In consequence, such suites remain arbitrary and, in general, of undetermined petrogenetic significance unless the variable/s used were prescribed by a genetic model.

In classification (identification of canonical, natural classes), variables are chosen on the bases of real (or supposed) petrogenetic understanding (*e.g.*, normative Q-Ab-Or-feldspathoid relationships; Vistelius´ [1972] mineral-grain transition probabilities for identifying ´ideal granites´; Rogers, *et al.* [1984] differentiation of volcanic rocks on the bases of detailed chemistry associated with observed tectonics). In partitioning (identification of arbitrary, conventional, symbolic classes), variables are specified by the class definitions and are uninfluenced by *a priori* considerations (*e.g.*, the cluster-analysis ´classification procedures´ of mathematical petrologist LeMaitre [1982, p. 170]); a different set of variables (or variable weightings) will result in different classes. Whitten, *et al.* (1986b) gave a simplified outline of a rigorous foundation for heuristic cluster-analysis methods; procedures and theorems with higher cluster-resolving powers are subjects of ongoing research.

Such classification and partitioning relationships also apply to granitoid suites, thus:

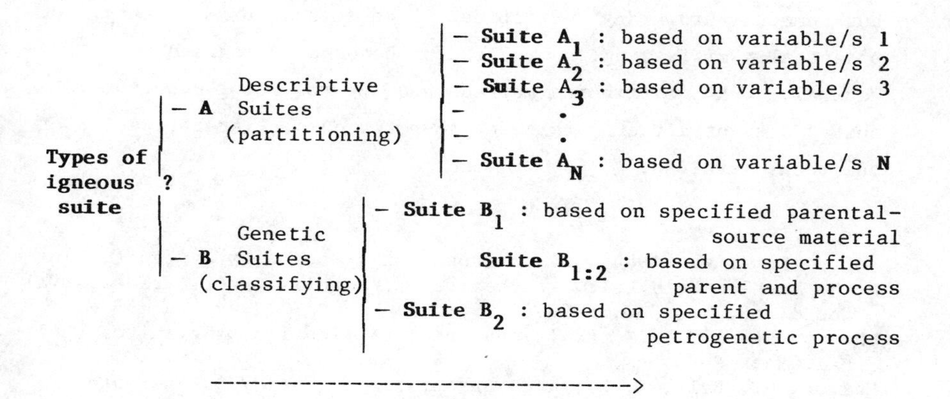

Sedimentary and metamorphic rocks are now commonly classified on the bases of variables that have genetic significance. Traditionally, this has not been the case for igneous rocks (*e.g.*, Shand, 1947; Johannsen, 1931). With few exceptions, partitioning, rather than classification, procedures are still widely used for igneous rocks (but rarely for sedimentary and metamorphic rocks), thereby evolving arbitrary classes and descriptive terminology ("classification") (*e.g.*, Chayes, 1957; Streckeisen, 1976; Cox, *et al.*, 1979, Fig. 2.2).

If the desired results are specified, pattern-recognition techniques permit parameter selection in order to recognize features (suites, natural classes) of interest. Pattern recognition involves filtering (interpretation of) observational data, which is an ill-posed problem because of dimensional changes in the mapping process and of entropy inherent in data-acquisition processes. Petrogenetic events that result in plutonic igneous rocks are commonly not directly observable; hence, interpretation of available observations to develop a chemically and petrographically significant classification is almost always an ill-posed problem.

CONCLUSION

The extensive use of partitioning, rather than classification, for describing igneous rocks (particularly plutonic igneous rocks) has resulted in descriptive classes or suites based on specified, but commonly arbitrary, variables. Such methods permit ready, but potentially misleading, communication between geologists. In general, such partitioning methods and the resulting descriptive suites are not directly relevant to petrogenetic interpretation of the rocks involved.

REFERENCES

Chappell, B. W. (1984). Source rocks of I- and S-type granites in the Lachlan Fold Belt, southeastern Australia. Phil. Trans. Royal Soc. Lond., A. 310, 693-707.

Chayes, F. (1957). A provisional reclassification of granite. Geol. Mag. 94, 58-68.

Cox, K. G., Bell, J. D., and Pankhurst, R. J. (1979). The interpretation of igneous rocks. George Allen & Unwin, London, 450 p.

Johannsen, A. (1931). A descriptive petrography of the igneous rocks. Univ. Chicago Press, Chicago, (Four Vols.)

LeMaitre, R. E. (1982). Numerical petrology: statistical interpretation of geochemical data. Elsevier Scient. Pub. Co., Amsterdam, 281 p.

Li, G., Bornhorst, T. J., Whitten, E. H. T., and Hicks, D. L. (1986). The risk of using discriminant analyses to support a classification hypothesis. (In preparation).

Pitcher, W. S.,, Atherton, M. P., Cobbing, E. J., and Beckinsale, R. D. (Ed.). (1985). Magmatism at a plate edge: The Peruvian Andes. Blackie Halstead Press, Glasgow, 352 p.

Rogers, J. J. W., Suayah, I. B., and Edwards, J. M. (1984). Trace elements in continental-margin magmatism: Part IV. Geochemical criteria for recognition of two volcanic assemblages near Auburn, western Sierra Nevadas, California. Geol. Soc. Amer. Bull. 95, 1437-1445.

Shand, S. J. (1947). Eruptive rocks. Thos. Murby & Co., London, 488p.

Streckeisen, A., (1976). To each plutonic rocks its proper name. Earth Sci. Rev. 12, 1-33.

Vistelius, A. B. (1972). Ideal granite and its properties. Journ. Internat. Assoc. Math. Geol. 4, 80-102.

Whitten, E. H. T., Li, G., Bornhorst, T. J., Christenson, P., and Hicks, D. L. (1986a). Quantitative recognition of granitoid suites within batholiths and other igneous assemblages. In: Use and abuse of statistical methods in the earth sciences, McCammon, R. (Ed.). Oxford Univ. Press, NY, (In press).

Whitten, E. H. T., Bornhorst, T. J., Li, G., Hicks, D. L., and Beckwith, J. P. (1986b). Suites, subdivision of batholiths, and igneous-rock classification: geological and mathematical conceptualization. Amer. Journ. Sci. 286 (In press).

APPLICATION OF FUZZY SETS THEORY TO THE SOLUTION OF PATTERN RECOGNITION PROBLEMS IN OIL AND GAS GEOLOGY

Bagirov B.A.,Djafarov I.S.,Djafarova N.M.
Institute ofSolving Problems of Deep Oil and Gas Deposits of the Azerbaijan SSR Academy of Sciences, Baku,USSR

Information used in complicated problems on making decisions is subjective as a rule,and hence it bears a fuzzy character.The present paper suggests a method for solving pattern recognition problems in oil field geology and geophysics based on the approach developed by S.A.Orlovsky /2/.

When solving recognition problems a target function may be formulated,for example,as follows: it is desirable to allocate the given pattern characterized by a number of features to one of the known classes with the possible highest degree of accuracy or in other words, to set the given pattern the most preferable corresponding alternative from a set of the alternatives assumed.

The analysis of problems with a fuzzy described target function is made by means of fuzzy relations of preference based on the theory of fuzzy sets.

Let us give the necessary determinations.The fuzzy relation on the set X is called a fuzzy subset of the direct product $X \times X$ characterized by the membership function

$$\mu_R : X \times X \longrightarrow [0,1] \qquad (1)$$

The meaning $\mu_R(x,y)$ of this function is understood as a subjective measure or degree of the relation accomplishment xRy.

The fuzzy preference relation R on the set X is described by a membership function of type (1) having a reflexivity property, i.e. $\mu_R(x,x)=1$ at any $x \in X$. If μ_R is a fuzzy preference relation on the set of alternatives X, then for any pair of alternatives $x, y \in X$ the meaning $\mu_R(x,y)$ is understood as a degree of preference accomplishment "x is not worse than y" or $x \succsim y$.

The fuzzy relation of strict preference R^s corresponding to the fuzzy preference relation R has a membership function of the following type :

$$\mu_R^s = \begin{cases} \mu_R(x,y) - \mu_R(y,x) & \text{at } \mu_R(x,y) > \mu_R(y,x) \\ 0 & \text{at } \mu_R(x,y) \leq \mu_R(y,x) \end{cases}$$

$\mu_R^s(x,y)$ is said to be a degree of domination of the alternative x over the alternative y.

The membership function of type

$$\mu_R^{n.d.}(x) = 1 - \sup_{y \in X} \mu_R(y,x) = 1 - \sup[\mu_R(y,x) - \mu_R(x,y)], \quad x \in X \qquad (2)$$

represents a degree of domination of the alternative x over the other alternatives of the set X, or a degree of non-dominance of the alternative x. Non-dominated alternatives are to a certain extent, unimproved within the set (X, R), and their choice in the problem of making decisions shoud be naturally considered rational in the limits of the available information.

Up to now we have considered fuzzy preference relations between the alternatives, i.e. among the elements of the set. This notion may be generalized into relations among fuzzy sets. The function

$$\eta(\nu_1, \nu_2) = \sup_{z, y \in Y} \min\{\nu_1(y), \nu_2(z), \mu_R(y,z)\} \qquad (3)$$

describes fuzzy preference relations among fuzzy subsets $\nu_1, \nu_2 \subset Y$; $\nu_1(y), \nu_2(z)$ are membership functions of fuzzy sets ν_1, ν_2 ; $\mu_R(y,z)$ are fuzzy preference relations among elements Y.

Above mentioned determinations (1)-(3) are used with a number of features, when solving the problem of rational alternatives selection. In this case on the set of alternatives there arise several preference relations (according to the number of features).

Let X be a set of alternatives, and P - a set of features. Each alternative $x \in X$ in this or that degree is characteristic of each feature $p \in P$. Fuzzy preference relations φ on the set of alternatives X are known for any fixed feature $p \in P$, or in other words, we know the membership function $\varphi: X \times X \times P \to [0,1]$, the meaning $\varphi(x_1, x_2, p)$ of which is understood as a degree of preference of the alternative x_1 to the alternative x_2 according to the feature p. Thus φ describes a family of fuzzy preference relations on the set X according to the parameter p.

Generally speaking, elements of the set P differ according to their importance. Let $\mu: P \times P \to [0,1]$ be the given fuzzy relation of preference (importance) of features. The value $\mu(p_1, p_2)$ is understood as a degree with wich the feature p_1 is considered to be not less important than the feature p_2. The rational choice of alternatives has the following stages.

By analogy with (2) we determine $\varphi^{n.d.}(x, p)$, i.e. a fuzzy subset of non-dominated alternatives corresponding to the fuzzy preference relation $\varphi(x_1, x_2, p)$ at fixed $p \in P$:

$$\varphi^{n.d.}(x, p) = 1 - \sup_{y \in X} [\varphi(y, x, p) - \varphi(x, y, p)]$$

If we realized the choice of alternatives taking into account one feature p only, then one should consider rational the choice of alternatives giving the possible greatest meaning of the membership function $\varphi^{n.d.}(x, p)$ (the degree of non-dominance) on the set X. In this case it is necessary to realize the choice considering the whole complex of features differing according to their importance.

At fixed $x^0 \in X$ the function $\varphi^{n.d.}(x^0, p)$ describes a fuzzy subset of features according to which the alternative x^{o} is non-dominated. Two fuzzy sets $\varphi^{n.d.}(x_1, P)$ and $\varphi^{n.d.}(x_2, P)$ correspond to two alternatives x_1 and x_2. Result (3) is used for their comparison. Considering that if the fuzzy set of features $\varphi^{n.d.}(x_1, P)$ is "not less important" than the fuzzy set of features $\varphi^{n.d.}(x_2, P)$, the alternative x_1 should be assumed to be not less preferable than the alternative x_2. Finally we get the following fuzzy preference relation on the set X induced by the function $\varphi^{n.d.}(x, P)$ and fuzzy preference relation μ :

$$\eta(x_1, x_2) = \sup_{P_1, P_2 \in P} \min\{\varphi^{n.d.}(x_1, P_1), \varphi^{n.d.}(x_2, P_2), \mu(P_1, P_2)\}$$

By forming the fuzzy preference relation the initial problem of choice is led to the problem of choice with a single preference relation. To solve it one should determine a fuzzy subset of non-dominated alternatives

$$\tilde{\eta}^{n.d.}(x) = 1 - \sup_{x' \in X} [\eta(x', x) - \eta(x, x')]$$

and choose an alternative giving the maximum $\tilde{\eta}^{n.d.}(x)$.

If not one but several alternatives have the greatest degree of non-dominance, a person who makes a decision can specify the result guided by some geuristics.

A number of problems in oil and gas geology has been solved according to the algorithm suggested /1/.

R e f e r e n c e s

1. Djafarov I.S., Djafarova N.M., Efendiev G.M. (1985) Application of fuzzy sets theory to the solution of pattern recognition problems in oil and gas geology and geophysics. M.: VIEMS, 34 p. (in Russian)

2. Orlovsky S.A. (1981) Problems on making decisions at fuzzy initial information. M.: Nauka, 206 p. (in Russian)

THE APPLICATION OF MULTIDIMENSIONAL RANDOM FUNCTIONS FOR STRUCTURAL MODELLING OF THE PLATFORM COVER

Jan Harff; Günther Schwab

Central Institute for Physics of the Earth, Academy of Sciences of the GDR, Berlin, GDR

INTRODUCTION

A main task of the geological interpretation of indirect geophysical measuring data and drilling data from the platform cover is the identification of homogeneous regions and their buried boundaries. The boundaries represent inhomogeneity zones resulting from vertical crustal movement and they are important for the process of formation of mineral resources deposits.

Numerical methods and computers become increasingly important for the localisation of inhomogeneity zones on the base of digital recorded multidimensional measuring data.

The application of an appropriate model of the investigated area is one of the prerequisites for successful numerical data interpretation.

MODEL

The investigated plane is indicated by $r \in R$, whereby r denotes a two-dimensional vector of plane coordinates.

An instationary random function is applied as a model for an n-dimensional geologic-geophysical feature field.

$$\vec{X}(r) = \vec{m}(r) + \vec{Y}(r), \quad \forall\, r \in R.$$

Plane R is subdivided into homogeneous regions $R_i \subset R$, $i \in \{1,\ldots,H\}$. In these regions moment functions of $\vec{X}(r)$ are constant:

n-dimensional expected value function

$$E[\vec{X}(r_i)] = \vec{m}(r_i) = \vec{m}_i, \quad \forall\, r_i \in R_i, \quad i \in \{1,\dots,H\},$$

$$\mathfrak{M} = \{\vec{m}_1, \vec{m}_2, \dots, \vec{m}_K\}, \quad K \leq H,$$

$$\vec{m}_i = (m_{i1}, m_{i2}, \dots, m_{in}),$$

$$\vec{m}_i \neq \vec{m}_j, \quad i,j \in \{1, \dots, K\},$$

Covariance matrix

$$C(r_i) = (E[(X_k(r_i) - m_{ik})\,(X_l(r_i) - m_{il})]) = \Sigma_i,$$

$$k, l \in \{1, \dots, n\},\; r_i \in R_i,\; i \in \{1, \dots, H\}.$$

A distance matrix D reflects the substantial differences of homogeneous regions

$$D = (d(i,j)),\; i, j \in \{1,\dots,H\},$$

$$d(i,j) = tr(\Delta_{ij}),$$

$$\Delta_{ij} = (\delta_{kl}(i,j)),\; k, l \in \{1,\dots, n\},$$

$$\delta_{kl}(i,j) = ((m_k(r_i) - m_k(r_j))(m_l(r_i) - m_l(r_j)).$$

A generalized variance is also constant in homogeneous regions R_i

$$\varrho(r_i) = \min_j \varrho_j(r_i) = tr(\Sigma_i), \quad \forall r_i \in R_i, \quad j \in \{1,\dots,H\},$$

$$\varrho_j(r_i) = tr\,(C_j(r_i)),$$

$$C_j(r_i) = (E[(X_k(r_i) - m_{jk})\,(X_l(r_i) - m_{jl})]),$$

$$k,l \in \{1,\dots,n\}.$$

The geological boundary between homogeneous regions R_i, R_k, ..., R_l can be expressed in the following form

$$\Lambda_{ik\dots l} = \{ r \in R: \varrho_i(r) = \varrho_k(r) = \dots = \varrho_l(r) \}.$$

METHOD

Primary data for numerical analysis are elements of a finite set of measuring vectors

$$\mathcal{R} = \{\vec{x}(r_i)\},\; i \in \{1,\dots,N\}.$$

These vectors are distributed in the plane of investigation and form the random sample.
The method elaborated here is called REGIONALIZED CLASSIFICATION and is to be associated with the main methodical steps of analysis:

(1) Typification by classification
In the step of classification measuring vectors are agglomerated by the cluster method. The vectors form the classes

$$\mathcal{R}_i \subset \mathcal{R}, \mathcal{R}_i = \{\vec{x}(r_j)\}, \ j \in \{1,\ldots,N_i\}, \ i \in \{1,\ldots,K\}.$$

The conditions for classification are
- $tr(G) \longrightarrow$ MIN (intra group covariance matrix G),
- $tr(H) \longrightarrow$ MAX (between group covariance matrix H),
- Rejection of hypothesis $H_o: \vec{m}_i = \vec{m}_j, \quad i,j \in \{1,\ldots,K\}, \quad i \neq j.$

A strategy by WARD (1963) is used for agglomeration, and a method by AHRENS & LÄUTER (1981) is applied for test statistics.
Distance matrix D^* (expressed by a dedrogramm), the expected value vectors and the covariance matrixes S_i are evaluated by the standard methods.
The result is the geological type model

$$TM = \{\mathfrak{M}^*, D^*, S_i\}, \quad i \in \{1,\ldots,K^*\}.$$

(2) Regionalization by interpolation
The regionalization is carried out by evaluation of the discontinuous n-dimensional expected value function for the whole plane of investigation as the solution of an multivariable interpolation task:

$$\vec{m}^*(r_p) = \{m_i^* \in \mathfrak{M}^*: \ \varrho_i^*(r_p) = \min_j \varrho_j^*(r_p) \quad, r_p \in R.$$

The base is the type model TM. In addition an experimental generalized variance function called Regionalization Function $\varrho^*(r)$ is evaluated with

$$\varrho^*(r) = \min_j \varrho_j^*(r),$$

$$\varrho_j^*(r) = \sum_{l=0}^{L} a_l^* f^l(r), \quad \forall r \in R', R' \subset R.$$

The coefficients a_l^* are determined by the least squares method. Geological boundaries are marked by salient points of function $\varrho^*(r)$. These salient points can be determined as experimental geological boundaries

$$\Lambda_{ik\ldots l}^* = \{r \in R: \varrho_i^*(r) = \varrho_k^*(r) = \ldots = \varrho_l^*(r)\}.$$

The results are summarized by the experimental space model SM including the experimental type model TM

$$SM = \{TM, \varrho^*(r)\}, \quad \forall r \in R.$$

EXAMPLE

For a case study the RECLAS method was applied to the northern part of the GDR. Here, in the Northern German Polish Depression the basement is covered by sedimentary sequences of different thickness.

Faults are the result of ancient vertical crustal movement. These faults are buried by pleistocene sediments and can be identified only indirectly. Measurements of magnetic and gravimetric anomalies and of the recent vertical crustal movement at 57 measuring points distributed in the plane of investigation were used for numerical analysis by the RECLAS method.

Figure 1 shows the results of typification. The dendrogram represents two main classes each of which is subdivided into two subclasses that describe four different types of crustal blocks.

The results of regionalization are shown in Figure 2. The boundaries identified by the RECLAS method coincide with the main tectonical elements of the investigated area. The boundaries determine tectonical gradient zones which are interpreted as buried faults (depth fracture). Therefore the boundaries also deter-

mine the block structure of the investigated part of the Earth's crust.
The identified homogeneous regions are characterized by specific vertical movements, depth of basement and geophysical features of the basement.

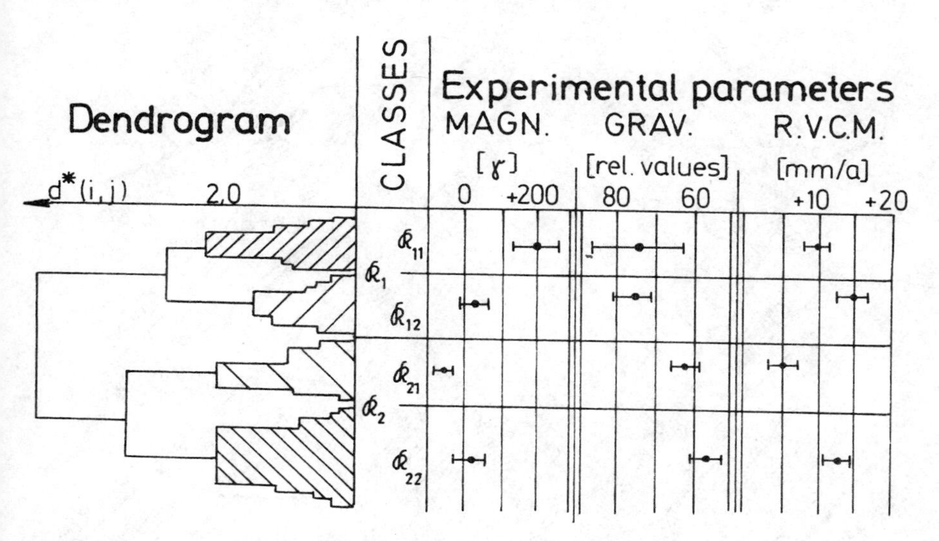

Figure 1 Typification of geophysical measuring vectors

MAGN Magnetic anomalies vertical field intensity
GRAV Gravimetric Bouguer anomalies
RVCM Recent vertical crustal movement related to Level Warnemünde

⊢•⊣ Grafic representation of m^*_{ik} and s_{ik} for Variable $x_k(r_i)$ and class $\mathcal{K}_i$

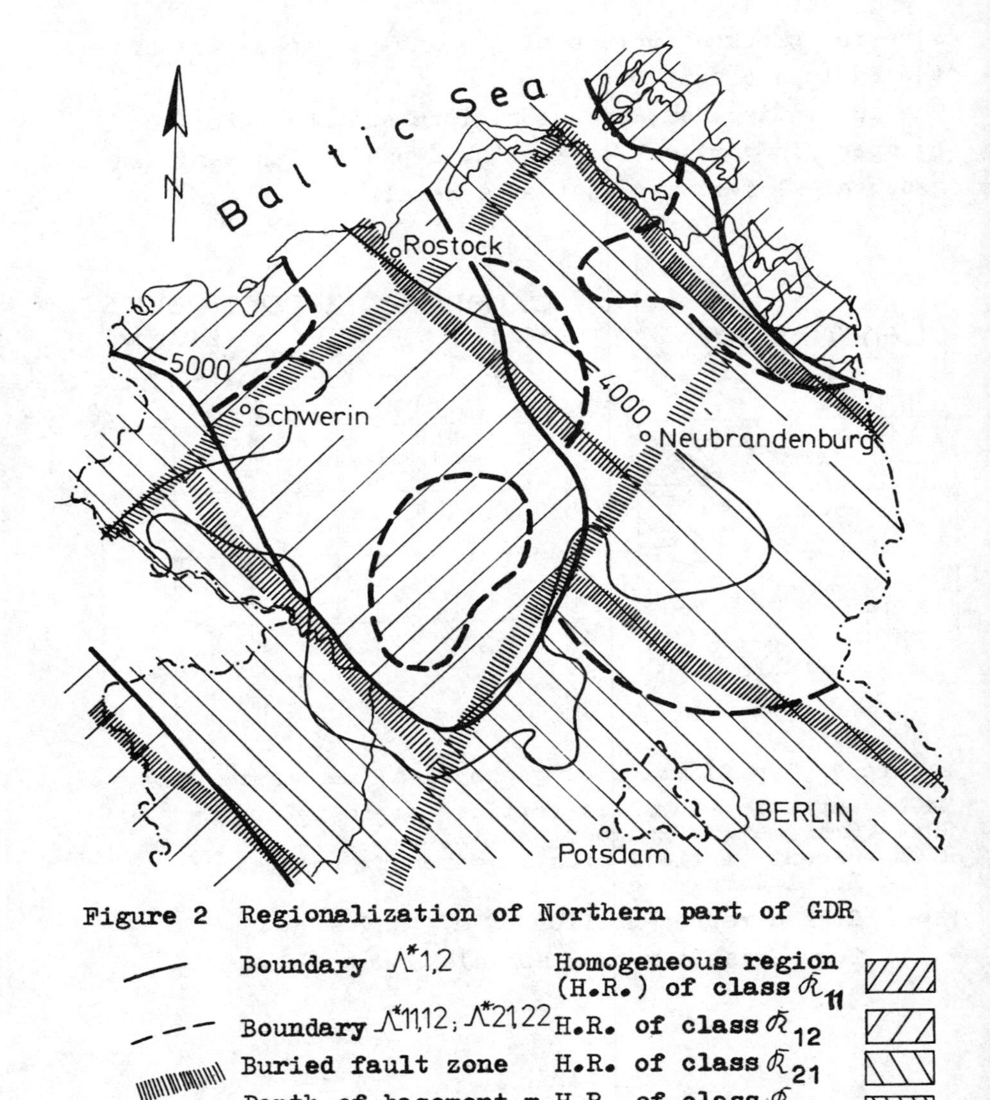

Figure 2 Regionalization of Northern part of GDR

Symbol	Meaning	Class
solid line	Boundary $\Lambda^*1{,}2$	Homogeneous region (H.R.) of class $\mathcal{K}_{11}$
dashed line	Boundary $\Lambda^*11{,}12$; $\Lambda^*21{,}22$	H.R. of class $\mathcal{K}_{12}$
hatched band	Buried fault zone	H.R. of class $\mathcal{K}_{21}$
—4000—	Depth of basement m	H.R. of class $\mathcal{K}_{22}$

REFERENCES

AHRENS, H.; LÄUTER, J.: Mehrdimensionale Varianzanalyse. - Berlin: Akademie Verlag 1981

WARD, J.H.: Hierarchical grouping to optimize an objective function. - J. Amer.Statist.Assoc., 58 (1963), p. 236-244

PREDICTION OF ROCK TYPES IN OIL WELLS FROM LOG DATA.

HOMLEID,M. , BERTEIG,V. , BØLVIKEN,E. , HELGELAND,J. , MOHN,E.
Norwegian Computing Centre, Oslo, Norway.

The determination of lithofacies from wireline log data may be viewed as a pattern recognition problem. Because of the dependencies from one depth to the next in the log curves and in the occurrence of lithofacies, contextual classification methodes are developed. Various models for the relationship between the log variables and lithofacies are considered. A comparison of the corresponding models are performed through a simulation experiment.

1. GEOLOGICAL DESCRIPTION

To estimate hydrocarbon in place and recoverable reserve in new fields, one needs knowledge about the geology of the formation. The geology can be described by factors such as mineral content, texture, fossils etc. It is assumed that these factores at a particular depth can be summarized in one of a number of classes, called lithofacies. One way of getting knowledge of the lithofacies in the formation is to drill bore-holes. Two kinds of data from bore-holes are considered here : core data and log data. When drilling, it is possible to take out a string of the rock, called a core sample, an then at a series of points observe variables as dominating grain size, sorting index, content of carbonate and degree of consolidation. Lithofacies can be determined from core data. However, coring is rather expensive and often no core is taken out. Information is also obtained by lowering a measuring probe in the bore-hole, recording various radioactive, acoustic and electric signals at different depths. Examples of such log data are natural gamma ray and resistivity. They give an indication of the geology in the bore hole surroundings. The problem is to predict lithofacies in bore-holes from log data. This can be considered as a classification problem.

2. CONTEXTUAL CLASSIFICATION IN A SIMPLE MODEL

Discriminant analysis may be used for classification. But there are dependencies from one depth to the next that we would like to bring into the models and methods, both dependencies in the log registrations and in the occurences of lithofacies. The methods to be presented here are based on contextual classification. They require statistical models for the log variables, the lithofacies and their inter-relationship. In choosing the statistical models we must strike a compromise between realism and complexity. Complex models may imply large training data sets and heavy computations. We assume that the general statistical model for the log response can be written: $\underline{X}(t)=\underline{S}(t)+\underline{Y}(t)$, $t=1,..,T$, where

$\underline{X}(t)$ - log registrations in depth t

$\underline{S}(t)$ - signal process, describing underlying geology

$\underline{Y}(t)$ - noise process, caused by e.g. measurement inaccuracies

In the simple model the log registrations are supposed to vary around a level determined by the lithofacies in the actual depth. This means that the signal from depths with the same lithofacies are equal : $\underline{S}(t)=\mu_{C(t)}$ where $\mu_{C(t)}$ is the conditional expectation of the log vector $\underline{X}(t)$ given true lithofacies $C(t)$. $C(t)\in\{1,..,k\}$ The noise $\underline{Y}(t)$ is assumed to follow a 1st order autoregressive process: $\underline{Y}(t)=\Lambda\underline{Y}(t-1)+\underline{\varepsilon}(t)$, where $\underline{\varepsilon}(t)$ is white noise with covariance matrix dependent on lithofacies.

To model the sequence of lithofacies a 1st order Markov chain is used. Before prediction can take place, the system must be "trained". The training set is data from wells with both core data and log data from which all parameters of the models are estimated. The details of the estimation will not be described here.

The posterior probabilities for each lithofacies may be calculated according to : $\pi(k|D)=\text{const.}\pi_k f_k(\underline{x})R_k(D)$, where

$D=\{\underline{x}(t),t=t1,..,t2\}$ - the log reg. in a window surrounding t

π_k - the prior probability for class k ,

$f_k(\underline{x})$ - the conditional density function of the logs

$R_k(D)$ - a certain contextual correction factor

When the loss function equals the number of wrong classifications, the following rule is optimal : for each depth, choose the lithofacies that maximises the above posterior probability.

We have carried out prediction by this contextual method on data from a certain well in the North Sea and compared the results with prediction based on traditional discriminant analysis. The contextual method gives no significant improvement. The main reason is that the model is far to simple to give a realistic description of the data. Two models that take care of geological properties that where ignored in the simple model, are presented in the next sections.

3. FILTER MODEL

One property of the log registrations that is not handled by the simple model is due to the finite resolution of the logs. The signal is also dependent on lithofacies in neighbouring depths. This is modeled by : $\underline{S}(t)=\sum_{u=-m}^{m} H(u)\ \underline{S}(t+u)$, $H(-u) = H(u)$.

The noise follows a simultaneous autoregressive process :

$\sum_{u=-n}^{n}\alpha(u)\ \underline{Y}(t+u) = \underline{\varepsilon}(t)$, where $\underline{\varepsilon}(t)$ is white noise with covariance matrix dependent on lithofacies.

The posterior probabilities given log data from all depths are calculated. The computational burden can be very high and depends critically on m, n and k. One way to simplify could be hierarchic classification. Another simplifications in the algorithm may be obtained by performing local maximisation, e.g. by the type proposed by Besag (1985).

4. LEVEL MODEL

Empirical data studies indicate that the assumption of same mean and variation for log registration from depths with the same lithofacies is not realistic. There are significant differencies from one lithological unit to another within the same lithofacies.

In the level model the signal S(t) is stocastic, but constant for each lithological unit. For each occurence of a new lithological unit the signal is supposed to be drawn from a normal distribution with mean and covariance matrix dependent on the actual lithofacies, and then to be constant within that layer. The noise is assumed to be white. A special case is treated by Yao (1984).

The lithofacies are supposed to follow a semi-Markov model. This assumption is favorable with regard to the computation of posterior probabilities. There are also empirical support for a semi-Markov model for the lithofacies.

5. SIMULATION EXPERIMENT

Training and classification with the non-contextual method and the three contextual ones described in Sec. 2-4 have been performed on simulated and real data. In the table below error rates from two different simulations are compared. Log registrations from 4 logs are simulated in 500 depths according to the filter model; with 7 lithofacies following a 1st order Markov chain. In the case H0 there are no smoothing. In H1 a 1st order filter is applied; H(0) and H(1) are diagonal matrixes with diagonal elements 0.4 and 0.3.

	non-contextual	SIMPLE	FILTER	LEVEL
H0	9.4	5.8	7.0	8.2
H1	27.6	22.2	10.4	27.2

The difference between the methodes in case H0 is partly due to fluctuations. When the signal is smoothed (H1), the filter model shows significant lower error rates than the other ones. Real data are smoothed and the filter model seems to capture these features.

No matter the degree of improvement, all the models are interesting, and we believe that such models may prove useful for other purposes in statistical well log analysis as well.

References

1. Berteig,V., Helgeland,J., Mohn,E., Langeland,T. and van der Wel,D. Lithofacies predictions from well data. SPWLA 26th Logging symposium, Dallas, June 17-20,1985.
2. Yao, Y. : Estimation of a noisy discrete-time step function : Bayes and Empirical-Bayes approaches. Annals of Statistics 12, 1984, pp 1434-1447.
3. Besag,J.: On the statistical analysis of dirty pictures. Workshop on "Statistics and pattern recognition", Edinburgh, July 4 - 12, 1985.

ON TESTS FOR OUTLYING OBSERVATIONS

Pagurova V.I.
Moscow State University, Moscow;
Rodionov K.D.
Institute of Ore Deposits, Petrography, Mineralogy and Geochemistry of the USSR Academy of Sciences, Moscow, USSR.

The method of testing outlying obeservations concerned the mean of normal population proposed by Bol'sev (1974), Bol'sev, Ubaidullaeva (1974), Ubaidullaeva (1974), is extended to more general situation and the asymptotic behaviour of a testing procedure is investigated.

Let $X_1,\dots,X_n$ be independent identically distributed variables with common continuous distribution function $F((x-\theta_1)/\theta_2)$, a location parameter θ_1 and a scale parameter θ_2 are unknown, $\hat{\theta}_1$ and $\hat{\theta}_2$ are consistent estimators based on our sample. Let $T_i=(X_i-\theta_1)/\theta_2$, $Y_i=(X_i-\hat{\theta}_1)/\hat{\theta}_2$, $i=1,\dots,n$, $X^{(1)}\leqslant X^{(2)}\leqslant\dots\leqslant X^{(n)}$ and $Y^{(1)}\leqslant Y^{(2)}\leqslant\dots\leqslant Y^{(n)}$ be corresponding order statistics. For arbitrary fixed $t>0$ we define $x(t)$ as a solution of the equation

$$F(x(t)) = 1-t/n \quad . \tag{1}$$

Let $y(t)$ be such a value that

$$\begin{aligned} &P\{Y_1 \geqslant y(t)\} = t/n + o(n^{-1}) ,\\ &P\{Y_1 \geqslant y(t), Y_2 \geqslant y(t)\} = t^2/n^2 + o(n^{-2}) ,\\ &P\{Y_1 \geqslant y(t), T_1 \geqslant x(t)\} = t/n + o(n^{-1}) ,\\ &P\{Y_1 \geqslant y(t), T_2 \geqslant x(t)\} = t^2/n^2 + o(n^{-2}) \end{aligned} \tag{2}$$

as $n\to\infty$. Now we introduce a counting process $N(t)=\sum_{i=1}^{n} I\{Y_i \geqslant y(t)\}$ and a Poisson process $L(t)$ with unit intensity. The level of significance α and a natural number k are given. Let $t_i=i/(1+c)$, $i=1,\dots,k$, $c>0$,

$Z_\kappa(c)=\sup_{0<t\leq t_\kappa}[L(t)-t]/t$, $Z_{\kappa,n}(c)=\sup_{0<t\leq t_\kappa}[N(t)-t]/t$. Let $c=c(\kappa,\alpha)$ denote the solution of the equation

$$P\{Z_\kappa(c)<c\}=1-\alpha . \tag{3}$$

Theorem 1. Let us assume, that the conditions (1)-(3) hold. Then $P\{Z_{\kappa,n}(c)<c\}=1-\alpha+o(1)$ as $n\to\infty$.

Statistical applications. Let $X_1,\dots,X_n$ be independent observations, X_i has a distribution function $F((x-\theta_{1i})/\theta_2)$, $i=1,\dots,n$. Consider the testing problem $H_0:\theta_{1i}=\theta_1$, $i=1,\dots,n$, against H_1 : there are no more κ observations with $\theta_{1i}>\theta_1$, the remaining observations have a common distribution function $F((x-\theta_1)/\theta_2)$. For this aim we consider the vector$(Y^{(n)},Y^{(n-1)},\dots,Y^{(n-\kappa+1)})$. If the event $\{Y^{(n-i+1)}\geq y(t_i)\}$ has taken place for at least one $i\in\{1,2,\dots,\kappa\}$ then we reject the hypothesis H_0 and accept H_1 and the corresponding value $X^{(n-i+1)}$ is announced to be an outlier. If follows from Theorem 1 that a level of significance of the test is close to α for n large enoug

In the case when $F(x)$ is a standard normal distribution function the test is a generalization of tests proposed by Pearson and Chandra Sekar (1936), Smirnov (1941), Grubbs (1950) when a number of outliers does not exceed a given value κ . It is known, that when $\kappa=1$ and a distribution is normal this test maximizes the probability of making a correct decision among a class of testing procedures invariant under the addition of any constant to all the observations and when all the observations are multiplied by any positive constant (Kudo (1956)).

Theorem 2. Let $F(x)$ be a standard normal distributior function and $\hat\theta_1=\sum_{i=1}^n X_i/n$, $\hat\theta_2^2=\sum_{i=1}^n (X_i-\hat\theta_1)^2/n$ then $P\{Z_{\kappa,n}(c)<c\}=1-\alpha+O(n^{-1}\ln^2 n)$ as $n\to\infty$.

Theorem 3. Let $F(x)=1-e^{-x}$, $x\geq 0$, $\hat\theta_1=X^{(1)}$, $\hat\theta_2=\sum_{i=1}^n X_i/n-\hat\theta_1$, then $P\{Z_{\kappa,n}(c)<c\}=1-\alpha+O(n^{-1}\ln^2 n)$ as $n\to\infty$.

When $F(x)$ is a standard normal distribution functior or $F(x)=1-e^{-x}$, $x\geq c$, the exact distribution of $Z_{\kappa,n}(c)$

is obtained. The critical values of the test are presented in the table for $\alpha = 0{,}1$; $0{,}05$; $0{,}02$; $0{,}01$; $k=1(1)4$; $n=4(1)30(2)50(5)100$ with the error no more them one unit of the third decimal digit. We give some recommendations for computations of critical values lying outside of the table .

We have used this technique to solve a detection outlier problem in general linear regression models and in two-way analysis of variances.

REFERENCES:

I. Большев Л.Н. (1974) Обнаружение грубых ошибок в результатах наблюдений. Междунар. летняя школа по теории вероятн. и матем. статистике. Варна, 8-4I.
2. Большев Л.Н., Убайдуллаева М. (1974) Критерий Шовенэ в классической теории ошибок. Теор. вероятн. и примен. I9, № 4, 7I4-723.
3. Убайдуллаева М. (1974) Об отбраковке резко выделяющихся наблюдений. Теор. вероятн. и примен. I9, № 4, 864-868.
4. Pearson E.S., Chandra Sekar C. (1936) The efficiency of statistical tools and a criterion for the rejection of outlying observations. Biometrika, 28, 308-320.
5. Смирнов Н.В. (194I) Об оценке максимального члена в ряду наблюдений. ДАН СССР (новая серия). 33, 346-349.
6. Grubbs F.E. (1950) Sample criteria for testing outlying observations. Ann. Math. Statist., 21, 27-58.
7. Kudo A. (1956) On the testing of outlying observations. Sankhya, 17, 67-76.
8. Пагурова В.И., Родионов К.Д. (1986) Об асимптотических свойствах критерия обнаружения выбросов. Теор. вероятн. и примен. 3I, № 4, 798-80I.

THE IDEAS OF PERCOLATION THEORY IN GEOPHYSICS AND FAILURE THEORY

V.F.Pisarenko and A.Ya.Reznikova.
Institute of Physics of the Earth, U.S.S.R., Ac.Sci., Moscow, U.S.S.R.

One of the most appealing qualities of the percolation theory (from the solid state physics and geophysics view point) consists in the possibility of abrupt qualitative reorganizations via accumulation of microcracks. An important example is the concentration failure theory of S.N.Zurkov (see Zurkov et al (1977)). This theory is based on percolation ideas and has many experimental corroborations. However, the classical percolation approach in geophysics has many essential difficulties. The most important difficulty is a multiscale property of the geophysical media. The authors of this report have studied a class of the multiscale models of the percolation theory in failure theory. The central part of these considerations is the renormalization group approach. This method has been studied in the work of Allegre C.J et al. (1982).

We now describe the simpest class of the multiscale (or hierarchical) models. As always in the percolation theory we will study the global topological properties of some random set in R^d. We will interpret this set as a set of cracks, microdefects and so on, and we will call it a "defect" set.

Multiscale percolation models are described by three parameters: the number of scales N+1, the scale step (scale unit) and the intensity p of the microdefects of each scale.

The defect set can be constructed by induction. At the initial step (0-th step) the d-dimensional space R^d is divided into unit cubes (scale $\ell_0 = 1$) and a cube is assigned a color (black or white) using Bernuolli trials with the probabilities p or q= 1-p resp. The black color will be interpreted as indicating the defect set. The set of all black cubes of the 0-th rank (scale) we denote $\mathcal{D}_0$, it has the volume density p (in the standard ergodic sense). In the next step each white cube of 0-th rank is divided into ν^d equal cubes of 1-st ($\ell_1 = \frac{1}{\nu}$) scale and for each such cube is assigned a black or a white color with the same probabilities p and q. All cubes of the 1-st rank form the defect set $\mathcal{D}_1$, its density being pq.

By induction we construct the defect sets $\mathcal{D}_2, \ldots, \mathcal{D}_N$ consisiting of black cubes of the scales $\ell_2 = \frac{1}{\nu^2}, \ldots, \ell_N = \frac{1}{\nu^N}$. The volume densities of these sets are $pq^2, \ldots, pq^N$ resp., the concentration of the total defect set $\mathcal{D} = \bigcup_{i=0}^{N} \mathcal{D}_i$ equals

$$\rho = p + pq + pq^2 + \ldots + pq^N = 1 - q^{N+1} \quad (1)$$

The "normal" part of the space has the volume concentration q^{N+1}.

The method of investigation for multiscale models is based on the central idea of coding, see Allegre C.J et al (1982).

We explain the essence of this method. We consider a cube Q_{N-1} of (N-1)-th rank. It consists of cubes of minimal scale (if Q_{N-1} is not black with the probability q). The colors of the elementary cubes of N-th rank define some configuration σ . The number of all possible configurations is 2^{ν^d}. We introduce two classes $\mathcal{K}$ and $\bar{\mathcal{K}}$ of all configurations. We say that $\mathcal{K}$ is a class of failure configurations.

We define the generating function of $\mathcal{K}$:

$$f(p) = P\{\sigma \in \mathcal{K}\} = \sum_{\sigma \in \mathcal{K}} p^{\nu(\sigma)} q^{\nu^d - \nu(\sigma)}, \quad (2)$$

where $\nu(\sigma)$ is the number of black cubes from the configuration σ .

Let us code the cubes of (N-1)-th rank. We declare a cube of (N-1)-th rank to be a defect cube if it was black at (N-1)-th scale or it was white at (N-1)-th scale but contained the failure (in the sense of class $\mathcal{K}$) configuration.

Thus, we eliminate all the cubes of N-th rank and the cubes of (N-1)-th rank are now the minimal cubes in our model.

The probability of a defect of the (N-1)-th rank is

$$p_{N-1} = p + q f(p) = F(p). \tag{3}$$

We repeat the procedure with the cubes of (N-2)-th scale (rank). Now we have the cubes of (N-2)-th rank as minimal with the probability

$$F_2(p) = p_{N-1} + q_{N-1} f(p_{N-1}) = F(F(p)). \tag{4}$$

Repeating the procedure of the coding N times we have the one-scale model of black and white cubes with the probability of the defect cube

$$p_0 = F_N(p) = \underbrace{F(\ldots F}_{N}(p)). \tag{5}$$

We will say that the system has failed if $p_0 - p_{cr}^{(1)}(d)$, where $p_{cr}^{(1)}(d)$ is the percolation threshold in the one-scale model.

We can construct various classes $\mathcal{K}$ with various mechanical or percolation meanings.

1. Allegre C.J. et al (1982) have offered an iterative multiscale failure scheme. The definition of $\mathcal{K}$ in that case is very natural: a configuration $\sigma \in \mathcal{K}$ if two opposite sides of a cube may be connected by the skeleton of white cubes of the next scale. In this work the special case $d = 3$, $\nu = 2$ was considered and the iterations of function F were analysed in the com-

puter. We have generalized these results in that we consider any d, ν, N. The theoretical analysis of $F_N(p)$ has borne out the qualitative results of the abovementioned work: at the failure threshold (in the parameter p) the volume concentration of defect cubes is much greater than that in the one-scale models. The same effect is also known in multiscale percolation models.

2. The sufficient conditions for the percolation.

We will consider a class K^{ℓ}. A configuration $\sigma \in K^{\ell}$ if black cubes Q_N form a connected set connecting any two opposite sides of Q_{N-1} a number of black cubes on each side of Q_{N-1} and is more than half of all the cubes of (N-1)-th rank forming the side. Thus, from the percolation of the cubes of (N-1)-th scale after coding it follows that the percolation takes place in the initial system.

We obtain the condition

$$F_N^{K^{\ell}}(p) > p_{cr}^{(1)}(d) \Leftrightarrow p < F_N^{-1}(K^{\ell})(p_{cr}^{(1)}(d)) \quad (6)$$

($p_{cr}^{(1)}(d)$ is the percolation threshold in d-dimensional one-scale model) is the sufficient condition of percolation for the initial model.

3. The necessary condition of percolation.

The class K_H consists of configurations σ such that there exists percolation (diagonal percolation) at least for one pair of opposite sides of the cubes of largest scale.

It is obvious that for such weak failure criteria the failure follows from the percolation and we have the necessary condition of percolation:

$$p > F_N^{-1(K_H)}(p_{cr}^{(1)}(d)). \quad (7)$$

Above we have supposed that the concentration p is the same for all scales (the uniformity of scales). This condition permits us to use the iterations of a function n of one variable. Now we will analyse a nonuniform model where the probabilities of the defects of 0-th, 1-st

..., N-th ranks are p_o, P_1,...p_N resp.

The corresponding formulas for the coding have the form:

$$\begin{aligned} \tilde{p}_{N-1} &= p_{N-1} + (1 - p_{N-1}) f(p_N), \\ \tilde{p}_{N-2} &= p_{N-2} + (1 - p_{N-2}) f(\tilde{p}_{N-1}), \\ &\cdots\cdots \\ \tilde{p}_0 &= p_0 + (1 - p_0) f(\tilde{p}_1). \end{aligned} \quad (8)$$

References

Allegre C.J., Le Monel J.L. and Provost A.(1982). Scaling rules in rock fracture and possible implications for earthquake prediction. Nature. v.297, 5861, 47-49

Molcanov S.A., Pisarenko VàF. and Reznikova A.Ya. On the percolation approach in the failure theory. In: Matematcheskie metody v seismologii i v geodinamike (Vychislitelnaya seismologia, 19)

Zurkov S.N., Kuksenko V.S., Petrov V.A. et al.(1977). On prediction of rock failure. Izvestiya AN SSSR, Fizika Zemli (Solid Earth), 6, 8-16

HYDROLOGY AND METEOROLOGY
(Session 15)

Chairman: A.H. Murphy

SOME STOCHASTIC MODELS OF RAINFALL WITH PARTICULAR REFERENCE TO HYDROLOGICAL APPLICATIONS

D.R. Cox
Imperial College of Science and Technology, London, England

I. Rodriguez-Iturbe
Universidad Simon Bolivar, Caracas, Venezuela

1. INTRODUCTION

Mathematical models of rainfall range from the complex deterministic systems of dynamic meteorology to very simple empirical models concerning, for example, the marginal distribution of daily rainfall.

In the present paper we review briefly some previous work on stochastic models and then outline recent results of the authors and their colleagues. We focus primarily on models that are reasonably simple but which contain some link with underlying physical processes. The models are intended partly as inputs into hydrological schemes and partly as a framework for condensing and comparing large bodies of empirical data.

2. REQUIREMENTS

Reasonable requirements for the kinds of model we have in mind include the following:

(i) There should be a modest number of adjustable parameters

(ii) The component parameters should have a physical interpretation

(iii) The model should be capable of representing with the same parameter values phenomena on a range of time scales, e.g. hourly, daily, . . . totals

(iv) It should be possible to determine preferably analytically a number of different aspects of the systems

To achieve all or most of these aims fairly drastic simplifying assumptions are unavoidable and these are to some extent dictated by mathematical convenience.

3. OUTLINE OF SOME PREVIOUS WORK

Detailed stochastic models of rainfall seem to date from the paper of Le Cam [1]. He essentially considered point impulses of rain occurring in a clustered point prouss, e.g. of Neyman-Scott type, and with amounts per cluster independently distributed. Waymire and Gupta [8] and Smith and Carr [4,5,6] have developed these ideas, the last three papers considering especially formal statistical estimation and fitting procedures; most practical work in fitting to data has been based on equating first and second moment features.

Stern and Coe [7] concentrated on daily rainfall, representing the pattern of dry and wet days via a Markov chain and describing separately the amounts of rain per wet day. Their paper and its discussion provides also a valuable review of the literature.

4. SPATIAL ASPECTS

A full treatment of the topic requires a spatial-temporal stochastic model and to provide a fully satisfactory treatment of the spatial aspects is perhaps the most challenging open aspect. Two simplified ways of dealing with the spatial aspects are:

(i) To concentrate on a finite set of sites, typically those where there are rain gauges, to label the sites 1,2. . . and to disregard their precise geographical location. This requires the formulation of multivariate generalizations of single-site models and is usually not too difficult

(ii) To eliminate the temporal aspect by integration through time over the passage of a storm system, thus leading to a purely spatial process of total rainfall. From some hydrological viewpoints this may be adequate.

A treatment via route (ii) has been given by Rodriguez-Iturbe, Cox and Eagleson [2]. In their simplest model cell centres form a spatial Poisson process and around each centre is a rain cell of random depth and shape. When various functions are elliptical and Gaussian the whole model is reasonably flexible and tractable mathematically.

5. A TEMPORAL MODEL BASED ON A POISSON PROCESS

We now outline some results for a purely temporal process. A quite widely used model is in its simplest form in continuous time as follows. There is a Poisson process of storm origins. Following each storm origin is a 'rectangular' pulse of rain of constant random depth X (mm per unit time) and random duration L. The total rainfall intensity at time t, Y(t) say, is the sum of such contributions from all storms 'active' at time t. The main properties of {Y (t)} can be determined, in particular detail if the distributions of X and L are exponential. Further the properties of the aggregate process of hourly, daily, . . . rainfall can also be found, e.g. the proportion of periods of length h that are dry. The limitation of this model appears to be that it is unsuccessful in accommodating effects on various time scales.

6. MODEL BASED ON BARTLETT-LEWIS CLUSTER PROCESS

Rodriguez-Iturbe, Cox, Isham and Febres de Power [3] have aimed to overcome the last-mentioned difficulty by replacing the Poisson process by a special form of Bartlett-Lewis cluster process in which the storm origin is followed by cell origins in a finite Poisson process. The main properties of such a process are analytically accessible.

Comparison with a long series of hourly rainfall data from Denver, Colorado shows that this last model, but not that of Section 5, is capable of giving a good fit to hourly, daily and two-daily totals using the same parameter values.

REFERENCES

1. Le Cam, L. (1961). A stochastic description of precipitation. Proc. 4th Berkeley Symp. 3, 165-186.

2. Rodriguez-Iturbe, I., Cox, D.R., Isham, V. and Febres de Power, B. (1986). Some models for rainfall based on stochastic point processes. Proc. Roy. Soc. (London) A, to appear.

3. Rodriguez-Iturbe, I., Cox, D.R. and Eagleson, P.S. (1986). Spatial modelling of total storm rainfall. Proc. Roy. Soc. (London)

A, 403, 27-50.

4. Smith, J.A. and Karr, A. (1983). A point process model of summer season rainfall occurrences. Water Resources Research 19, 95-103.

5. Smith, J.A. and Karr, A. (1985). Statistical inference for point process models of rainfall. Water Resources Research 21, 73-79.

6. Smith, J.A. and Karr, A. (1985). Parameter estimation for a model of space-time rainfall. Water Resources Research 21, 1251-1257.

7. Stern, R.D. and Coe, R. (1984). A model fitting analysis of daily rainfall data (with discussion). J.R. Statist. Soc. A 147, 1-34.

8. Waymire, E. and Gupta, V.K. (1981). the mathematical structure of rainfall representations. Water Resources Research 17, 1261-1272.

CANONICAL CORRELATIONS FOR RANDOM PROCESSES AND THEIR METEOROLOGICAL APPLICATIONS

A.M.Obukhov, M.I.Fortus, and A.M.Yaglom
Institute of Atmospheric Physics, USSR Academy of Sciences, Moscow, USSR

The method of empirical orthogonal functions (EOF's) was first developed and applied to meteorological problems by Lorenz (1956) and Obukhov (1960), and now it is widely used for a simplified description of meteorological fields (see, e.g., Kutzbach, 1967, and Fortus, 1980a). When applied to finite collections of random variables, this method coincides with the well-known principal component analysis introduced by Hotelling in the early 30's and now presented in all the text-books on multivariate statistical analysis (see, e.g., Anderson, 1984, Chap. 11). The EOF's enable one to replace a large collection of irregularly fluctuating variables by a few most varying linear combinations containing practically all information about the statistical variability of the initial data set.

In practice, however, the meteorological data are often not interesting by themselves but only as the input data (predictors) for the approximate determination (prediction) of the values of some other unobservable meteorological parameters (predictands). In such cases it is not the most variable linear combinations of predictors that are of main interest but those linear combinations which are most closely related to the predictands. The appropriate mathematical technique is then the canonical correlation analysis introduced independently by Hotelling (1936) and Obukhov (1938, 1940) (see also Anderson, 1984, Chap. 12). Let $\underset{\sim}{X}=(X_1,X_2,\ldots,X_n)$ and

$\underset{\sim}{Y}=(Y_1,Y_2,\ldots,Y_m)$ be two arbitrary random vectors whose components have zero mean values and finite variances. (The assumption about zero means is not a serious restriction but it only means that we deal with the deviations of the considered parameters from their mean values.) According to the general theory of canonical variables, one can always find a transformation of coordinates in the spaces of these vectors such that all the components of the compound vector

$$(U_1,U_2,\ldots,U_n,V_1,V_2,\ldots,V_n)$$

(where $U_i=\sum_{k=1}^{n}\alpha_{ik}X_k$, $i=1,\ldots,n$, and $V_j=\sum_{l=1}^{m}\beta_{jl}Y_l$, $j=1,\ldots,m$, are the components of $\underset{\sim}{X}$ and $\underset{\sim}{Y}$ in the new coordinate systems) will be pairwise uncorrelated with the exception only of pairs (U_i,V_i), $i=1,\ldots,p$ where $p\leq\min(n,m)$. Without any restriction of generality we can assume that $EU_i^2=EV_j^2=1$ for all i and j, where E is the mean value symbol; then $EU_iU_k=EV_iV_k=EU_iV_j=0$ for $i\neq j$ while $EU_iV_i=\rho_i$, where $\rho_1\geqslant\rho_2\geqslant\ldots\geqslant\rho_p>0$ and $\rho_i=0$ for $i>p$.

Let $X_1,\ldots,X_n$ be the known predictors and $Y_1,\ldots,Y_m$ the unknown predictands. We shall consider only the linear prediction of Y_j's (i.e.prediction by linear combinations of predictors). Then U_1 is clearly the most informative (i.e.most useful for prediction) linear combination of predictors, U_2 is the most informative combination uncorrelated with U_1, U_3 is the most informative combination uncorrelated with both U_1 and U_2, and so on. Similarly, V_1 is the best predictable linear combination of predictands, V_2 is the best predictable linear combination uncorrelated with V_1 and so on. The best linear prediction of V_i is $\hat{V}_i=\rho_iU_i$ and the corresponding mean square error of prediction is

$$\sigma_i^2=E|V_i-\hat{V}_i|^2=1-\rho_i^2.$$

Correlation coefficients $\rho_1,\ldots,\rho_p$ are called the <u>canonical correlations</u> of the pair of vectors $(\underset{\sim}{X},\underset{\sim}{Y})$, and the random variables $U_1,\ldots,U_p,V_1,\ldots,V_p$ are called the canonical variables. (ρ_i is the ith canonical correlation and (U_i,V_i) is the ith pair of canonical variables). The determination of all canonical correlations and variables can be reduced to a simple algebraic eigenvalue problem. Let $B_{xx}=\|EX_iX_k\|$ and $B_{yy}=\|EY_jY_l\|$ be the covariance matrices of vectors $\underset{\sim}{X}$ and $\underset{\sim}{Y}$ while $B_{xy}=\|EX_iY_j\|$ is the cross-covariance matrix. Then it is easy to see that the canonical correlations ρ_i and the vectors $(\alpha_{i1},\ldots,\alpha_{in})=\underset{\sim}{\alpha}_i'$, $(\beta_{i1},\ldots,\beta_{im})=\underset{\sim}{\beta}_i'$ (here and also below prime always symbolizes a transposition) can be found from the equations

$$B_{xy}\underset{\sim}{\beta}-\rho B_{xx}\underset{\sim}{\alpha}=0,\quad B_{xy}\underset{\sim}{\alpha}-\rho B_{yy}\underset{\sim}{\beta}=0, \tag{1}$$

supplemented by normalization conditions $\underset{\sim}{\alpha}'B_{xx}\underset{\sim}{\alpha}=\underset{\sim}{\beta}'B_{yy}\underset{\sim}{\beta}=1$ (see,e.g.,Anderson,1984). If $m\leq n$ and the matrices B_{xx} and B_{yy} are non-singular, then according to (1) $\rho_1,\ldots,\rho_p$ coincide with the non-zero eigenvalues of the $(m\times m)$ matrix $B_{yy}^{-1}B_{xy}'B_{xx}^{-1}B_{xy}$.

In practice, the covariances EX_iX_k, EX_iY_j and EY_iY_l are usually unknown and should be estimated from the observations. The important problem of statistical accuracy and reliability of the resulting estimates of canonical correlations and variables is considered, in particular, by Anderson (1984), but still requires further investigation. Some practical methods for the increase of reliability by some artificial coarsening of the initial data are discussed by Gandin (1967),pp.19-20, but we shall not pursue this topic here.

Seceral attempts to apply the canonical correlation analysis to meteorological problems can be found in the available literature (see,e.g.,Gandin,1967, Glahn,1968, Olevskaya,1968, Brinkmann,1980, Lawson and Cerveny,1985, Tatarskaya and Fortus,1985). The results of the last paper will be discussed later; now we shall only consider quite typical examples from two independent but closely related papers by Glahn and Olevskaya. Here the computations of several first canonical correlations and variables are presented for the case where $\underset{\sim}{Y}=(Y_1,\ldots,Y_m)$ are air pressures (more precisely, the heights of an isobaric surface closely related to the pressure values) observed at a given day at some fixed meteorological stations while $\underset{\sim}{X}=(X_1,\ldots,X_n)$ are the same data for the previous day (or for two previous days). It was found in both works that the data X_i allow to predict some linear combinations of the observations with much greater accuracy than that which can be achieved when the observations Y_j at only one point are predicted. Note also that the results related to the stations in the USA and in the USSR turn out to be qualitatively rather similar.

It is known that the functional time or/and spatial variability of the meteorological parameters is often of most interest for meteorological practice. Therefore the canonical analysis of **infinite** (functional) sets $\underset{\sim}{X}$ and $\underset{\sim}{Y}$ of random variables should be of value in the study of many meteorological situations. The generalization of the canonical correlation analysis to the case where $\underset{\sim}{X}=\{X(t),t\in T\}$ and $\underset{\sim}{Y}=\{Y(s),s\in S\}$ are two random functions (having zero mean values and finite variances) given on arbitrary sets T and S was first studied by Gelfand **and** Yaglom (1957). Consider the Hilbert space H_{xy} of random variables Z (with the inner product $(Z_1,Z_2)=EZ_1\bar{Z}_2$) consisting of all linear combinations of the form $\sum_i \alpha_i X(t_i) + \sum_j \beta_j Y(s_j)$ (where $t_i\in T$, $s_j\in S$, and α_i,β_j are complex numbers) and all mean square limits of sequences of such combinations. Let now H_x and H_y be the linear subspaces of H_{xy} spanned by the vectors $X(t)$, $t\in T$, and $Y(s)$, $s\in S$, respectively, while P_x and P_y be the linear projection operators in H_{xy} which project elements of this space onto H_x and H_y. Then, $A_1=P_xP_y$ and $A_2=P_yP_x$ are linear nonnegative self-ajoint operators in the subspaces H_x and H_y with norm not greater than unity. (Both these operators can be extended to the whole space H_{xy} by setting $A_1=P_xP_yP_x$ and $A_2=P_yP_xP_y$.) It is easy to show that operators A_1 and A_2 will have common spectrum disposed within the closed interval $0 \leq \rho \leq 1$ (see,e.g.,Gelfand and Yaglom,1957; Hannan,1961; Yaglom,1965). The points ρ of this spectrum play now the role of canonical correlations and if the spectrum is purely discrete then the eigenvectors of A_1 and A_2 are the corresponding canonical variables. However, the spectrum $\rho=\rho(k)$ of A_1 and A_2 can be also continuous (see Hannan,1961, and Olevskaya,1974). In this case there exist no eigenvectors U(k) and V(k) which correspond to the point $\rho(k)$ of continuous spectrum and $\rho_1=\sup\rho(k)=\sup\{EUV/(EU^2EV^2)^{1/2}\}$ where the supremum (least upper bound) on the right-hand side is taken over all vectors $U\in H_x$ and $V\in H_y$ but is attained for no pair of such vectors. Hence, in such a case canonical variables must be understood in some generalized sense (see,e.g., an example studied by Olevskaya,1974).

Let us assume that T and S are two intervals of the real axis $-\infty < t < \infty$ so that X(t), $t\in T$ and Y(s), $s\in S$ are two continuous time random processes. If $B_{xx}(t,t)=EX(t)X(t')$, $B_{yy}(s,s')=EY(s)Y(s')$, and $B_{xy}(t,s)=EX(t)Y(s)$, $B_{yx}(s,t)=B_{xy}(t,s)=EY(s)X(t)$ are the covariance and cross-covariance functions of these processes, then the determination of canonical correlations and canonical variables for $\{X(t),t\in T\}$ and $\{Y(s),s\in S\}$ can formally be reduced to the solution of the following eigenvalue problem related to (1)

$$\begin{aligned} -\rho\int_T B_{xx}(t,t')\phi(t')dt'+\int_S B_{xy}(t,s')\psi(s')ds'=0,\ t\in T,\\ \int_T B_{yx}(s,t')\phi(t')dt'-\rho\int_S B_{yy}(s,s')\psi(s')ds'=0,\ s\in S, \end{aligned} \quad (2)$$

where ϕ (t) and ψ (s) are often generalized functions (Yaglom,1966). The eigenvalue problem (2) can be effectively solved only in some exceptional cases. In particular, an explicit solution can be often found if X(t) and Y(s) are stationary and stationary correlated random processes with rational spectral densities $f_{xx}(\lambda)$, $f_{yy}(\lambda)$, and $f_{xy}(\lambda)$. This is due to the fact that the system (2) can be reduced in this case to a system of linear differential equations with constant coefficients and some special boundary conditions at the ends of intervals T and S. See in this respect a quite typical example studied by Gelfand and Yaglom (1957) where it is assumed that T=S is a finite interval, X(t) is a stationary random process with rational spectral density $f(\lambda)$, and $Y(t)=X(t)+N(t)$ where X(t) and N(t) are uncorrelated and N(t) is a "white noise", i.e. a generalized stationary process with a constant spectral density. In this case the canonical correlation spectrum is discrete and all canonical correlations $\rho_1, \rho_2, \rho_3, \ldots$ are the roots of some simple transcendental equation.

If, however, the prediction of the future values of some meteorological processes is of primary interest, then we must consider the case where $\underset{\sim}{X}=\underset{\sim}{X}_0^-=\{X(t), -\infty < t < 0\}$ is the "past" of a given process X(t) and $\underset{\sim}{Y}=\underset{\sim}{X}_\tau^+= X(t)$, $\tau \leqslant t < \infty$ (where τ is a given positive number) is its "future". Under the assumption that X(t) is a stationary random process with rational spectral density $f(\lambda)$, the canonical correlation of past and future of X(t) were found by Yaglom (1965). In this case there exists only a finite number n (where 2n is the degree of the polynomial in the denominator of $f(\lambda)$) of non-zero canonical correlations $\rho_1, \rho_2, \ldots, \rho_n$ and these correlations are the roots of a simple algebraic equation of degree n. When ρ_k is evaluated, the corresponding canonical variables U_k, V_k can be easily found from some system of 2n linear algebraic equations with 2n unknowns. A more general canonical correlation problem for $\underset{\sim}{X}=\underset{\sim}{X}_{0,T_1}^-=\{X(t), -T_1 \leqslant t \leqslant 0\}$ and $\underset{\sim}{Y}=\underset{\sim}{X}_{\tau,T_2}^+=\{X(t), \tau \leqslant t \leqslant \tau+T_2$ where τ ,T_1

and T_2 are three arbitrary positive numbers (and X(t) is the same stationary process as above) is only slightly more complicated. Here again there exist only n-zero canonical correlations which are the roots of some algebraic equation of degree n (see Yaglom,1966).

In what follows we shall restrict our discussion to the case where $T_1=T_2=\infty$ (i.e. $\underset{\sim}{X}=\underset{\sim}{X}_0^-$ and $\underset{\sim}{Y}=\underset{\sim}{Y}_\tau^+$). A particular example, where X(t) is a model of velocity fluctuations of a Brownian particle in a turbulent flow was considered by Glukhovsky (1971). In this case $B(\tau)= EX(t+\tau)X(t)=A^2\exp(-|\tau|/T_L)+B^2\exp(-|\tau|/T_m)$, where T_L is the Lagrangian time scale introduced by Obukhov (1959) and $T_m \ll T_L$ is the molecular collision time scale. There are only two non-zero canonical correlations $\rho_1(\tau)$ and $\rho_2(\tau)$ for $\underset{\sim}{X}_0^-$ and $\underset{\sim}{X}_\tau^+$, and $\rho_1(\tau)\approx\exp(-\tau/T_L)$ while $\rho_2(\tau)$ is much smaller than $\rho_1(\tau)$. The best predictable characteristic of the future behavior of X(t) is very close here to the quantity

$$V_1(\tau)= \int_0^\infty e^{-s/T_K}X(\tau+s)ds$$

where T_K is the so-called Kolmogorov time scale of turbulence which is much smaller than T_L (see,e.g., Monin and Yaglom,1975,p.348).

Later Fortus (1980b) studied a number of more general processes X(t) modeling climatic time series. Note that all the best short-range meteorological forecasts are at present based on numerical integration of dynamical equations; therefore the canonical correlation analysis of the meteorological fields at rather close time moments (of the type presented by Glahn,1968, and Olevskaya,1968) is only theoretically interesting but hardly useful for meteorological practice. However, the situation related to large-scale climatic processes is quite different. There is no reliable dynamic theory of such processes and only the long-range forecasting is of interest here; therefore the statistical forecast of any future feature of the climatic processes, even having only a moderate accuracy, can prove to be of practical importance. Unfortunately, the available data about the correlation functions and spectra of climatic time series and fields are till now very scarce and unreliable and such data are quite necessary for canonical correlation computations. Nevertheless some useful climatic data have begun to appear lately (see,e.g., Mitchell,1976, and Polyak, 1979). The power spectrum of Earth's yearly mean air temperatures calculated by Mitchell (1976) is especially interesting. This spectrum includes a number of sharp peaks (whose statistical significance requires further justification in some cases) and several flat plateaus separated by narrow regions of rapid variations of $f(\lambda)$ (or by narrow peaks). It was shown by Fortus (1980) that spectra of such a type can be often modelled by simple rational functions. Fortus studied also the canonical correlations and variables of past and future for time series with such spectral densities. She showed that all canonical correlations can be rather closely approximated in such cases by the exponential functions $\rho_k(\tau)= \exp(-\theta_k\tau)$, where $\theta_1 < \theta_2 < \ldots < \theta_n$. Moreover, the correlation coefficient between the random variable $X(\tau)$ and its best linear predictor $\hat{X}(\tau)$ obtained by using the entire past $\underset{\sim}{X}_0^-$ decreases here as $\exp(-\theta_n\tau)$, which is much faster than the correlation coefficient $\exp(-\theta_1\tau)$ between the best predictable smoothed variable $V_1(\tau)$ and its best linear predictor $U_1(\tau)$.

It is known that the spectral densities $f(\lambda)$ of many climatic processes $X(t)$ increase with the decrease of frequency λ (such spectra $f(\lambda)$ and the corresponding processes $X(t)$ are often called in the applied literature <u>red noise spectra</u> and <u>red noises</u>). This fact makes clear why the time averaging of the future values of a climatic time series $X(t)$ which supresses poorly predictable high -frequency components of the series produces often a much better predictable characteristic of the future behavior of X(t) than the value of X(t) at a fixed time moment. Note now that the empirical climatic spectral density of Mitchell (1976) provides even some grounds for the assumption that the true spectral density $f(\lambda)$ tends in this case to infinity as $\lambda \to 0$. If this is true, then this fact can drastically change the behavior of the canonical correlations $\rho_k(\tau)$ of past $\underset{\sim}{X}_0^-$ and future $\underset{\sim}{X}_\tau^+$ at large values of τ. In fact, as it was shown by Yaglom (1965), all canonical correlations $\rho_k(\tau)$ decrease exponentially as $\tau \to \infty$ if the spectral density $f(\lambda)$ is rational (and hence f(0) is bounded). However, for stationary processes X(t) such that $f(\lambda) \to \infty$ as $\lambda \to 0$ this latter statement can be wrong. Thus more than 25 years ago Rosenblatt (1961) gave an example of a discrete time stationary process X(t) with a spectral density $f(\lambda)$ which is proportional to $1/\lambda^\gamma$ at small values of λ where $1/2 < \gamma < 1$ and then proved that the first canonical correlation $\rho_1(\tau)$ does not tend to zero as $\tau \to \infty$ in this case.

One can expect that random processes X(t) with spectral densities increasing like $\lambda^{-\gamma}$, where $\gamma > 0$, as $\lambda \to 0$ will prove to be useful as models of some climatic time series. Among such processes "self-similar red noises" with power spectral density of the form $f(\lambda)=C/\lambda^{2\alpha}$, where $1/2 < \alpha < 3/2$, play an important role. Self-similar red noises were studied by many authors (beginning with Kolmogorov, 1940) and they have several interesting geophysical applications. Such a noise X(t) is a non-stationary random process with stationary increments. The self-similarity of the increments implies that the canonical correlation spectrum of past increments $\underset{\sim}{X}_0^-=\{X(0)-X(t),\ t<0\}$ and future increments $\underset{\sim}{X}_\tau^+=\{X(t)-X(0),\ t>\tau\}$, where $\tau > 0$, does not depend on τ in this case. This spectrum was evaluated by Fortus (1983) who showed that it is continuous and coincides with the interval $0 \leq \rho \leq \rho_1$

where $\rho_1=|\sin\pi\alpha|$. The least upper bound $\rho_1=|\sin\pi\alpha|$ of the correlation coefficients between a linear function V of the future values X(t), $t \geqslant \tau$, and a linear function U of past values X(s), $s \leqslant 0$, is, for any α, considerably higher than the value $\rho_0=[1-\Gamma(2-\alpha)/\Gamma(3-2\alpha)\cdot\Gamma(\alpha)]^{1/2}$ (found by Yaglom, 1955) of the correlation coefficient between the increment X(t)-X(0) and its optimal linear prediction. In particular, if $\alpha \rightarrow 1/2$, then $\rho_1 \rightarrow 1$ while $\rho_0 \rightarrow \sqrt{2}/2 \approx 0.7$.

Let us consider in conclusion an example of application of canonical correlation analysis to a genuine climatic time series, namely, the time series X(t) of mean yearly air temperatures at the Earth's surface averaged over the zonal belt between 30 and 80°N. An estimate of the corresponding correlation function from the observations during the years 1892-1976 was evaluated by Tatarskaya and Fortus (1984) by the equation

$$B(\tau)=\frac{1}{N}\sum_{t=1}^{N-\tau}X'(t+\tau)X'(t),\quad X'(t)=X(t)-\frac{1}{N}\sum_{t=1}^{N}X(t),$$

where N=86. Unfortunately the accuracy of the estimate was not sufficient to obtain reliable analytic approximations for the correlation function $B(\tau)$ and its Fourier transform $f(\lambda)$. Therefore the numerical values of $B(\tau)$ were only used to calculate the first two canonical correlations $\rho_1^{(n)}$ and $\rho_2^{(n)}$ and corresponding canonical variables for finite segments of "past" $\{X'(t),X'(t-1),\ldots,X'(t-n+1)\}$ and of "future" $\{X'(t+1),X'(t+2),\ldots,X'(t+n)\}$ where n=1,2,3,4, and 5. It was found by Tatarskaya and Fortus that $\rho_1^{(n)}$ increases monotonically with n from $\rho_1^{(1)}=0.60$ to $\rho_1^{(5)}=0.80$ while the best predictable linear combination of X'(t+1),...,X'(t+5) is quite different from X'(t+1), but rather close to the arithmetic mean of the five successive yearly temperatures. Thus when the time evolution of the yearly mean temperatures of large regions of the Earth is forecasted, then apparently the accuracy of prediction can be often raised substantially if, instead of just the mean temperature for the next year, the averaged temperature for several future years is predicted.

References

Anderson,T.W.(1984). An Introduction to Multivariate Statistical Analysis. 2nd Edition. Wiley, New York.

Brinkmann,W.A.R.(1980). Lake Superior area temperature variations. Ann.Assoc.Amer.Geogr. 70, 17-30.

Fortus,M.I.(1980a) Method of empirical orthogonal functions and its meteorological applications. Meteor. i Gidrol. No.4, 113-119.

Fortus,M.I.(1980b). A stochastic model related to predictability problem for climatic processes. In: Atmospheric Physics and the Problem of Climate, Golitsyn, G.S., and Yaglom,A.M.(Eds.). Nauka, Moscow, pp.139-161.

Fortus,M.I.(1983).Canonical correlations and canonical functionals for a stationary random process with a power spectral density. Dokl.Akad.Nauk SSSR 271, 13 1328. English transl.: Soviet Math.Dokl. 28, 280-284.

Gandin,L.S.(1967). On applications of the method of canonical correlations in meteorology. Trudy Glavn.Meteor. Observ. No.208, 5-22.

Gelfand,I.M., and Yaglom,A.M.(1957). Calculation of the amount of information about a random function contained in another such function. Uspekhi Mat.Nauk 12, No.1, 3-52. English transl.: Amer.Math.Soc.Translations 12, 192-246, 1959.

Glahn,H.R.(1968). Canonical correlation analysis and its relationship to discriminant analysis and multiple repression. J.Atmosph.Sci. 25, 23-31.

Glukhovsky,A.B.(1971). On statistical description of Brownian particle motion in a turbulent flow. Izv. Akad.Nauk SSSR, Ser.Fiz.Atmosf. i Okeana 7, 1039-1044.

Hannan,E.J.(1961). The general theory of canonical correlation and its relation to functional analysis. J.

Austr.Math.Soc. 2, 229-242.
Hotelling,H.(1936). Relation between two sets of variables. Biometrika 28, 321-377.
Kolmogorov,A.N.(1940). Wiener's spiral and some other interesting curves in Hilbert space. Dokl.Akad.Nauk SSSR 26, 115-118.
Kutzbach,J.E.(1967). Empirical eigenvectors of sea-level pressure, surface temperature, and precipitation complexes over North America. J.Appl.Meteor. 6, 791-802.
Lawson, M.P., and Cerveny,R.S.(1985). Seasonal temperature forecasts as products of anticedent linear and spatial temperature arrays. J.Climate Appl.Meteor. 24, 848-859.
Lorenz,E.N.(1956). Empirical orthogonal functions and statistical weather prediction. Sci.Rep.No.1, Statistical Forecasting Project, Mass.Inst.Technology, Cambridge (Mass.), 49pp.
Mitchell,J.M.(1976). An overview of climatic variability and its mechanisms. Quarternary Res. 6, 3-13.
Monin,A.S., and Yaglom,A.M.(1975). Statistical Fluid Mechanics, vol.2. MIT-Press, Cambridge (Mass.).
Obukhov,A.M.(1938). Normal correlation of vectors. Izv. Akad.Nauk SSSR, Ser.Mat. i Estestv.Nauk No.3, 339-370.
Obukhov,A.M.(1940). Theory of correlation for vectors. Uchen.Zap.Mosk.Gos.Univ., Ser.Matem. No.45, 73-92.
Obukhov,A.M.(1959). Description of turbulence in terms of Lagrangian variables. Adv. in Geophys. 6, 113-115.
Obukhov,A.M.(1960). On statistically orthogonal decompositions of empirical functions. Izv.Akad.Nauk SSSR, Ser.Geofiz. No.3, 432-439.
Olevskaya,S.M.(1968). Application of canonical correlation method to the analysis of a geopotential height field. Izv.Akad.Nauk SSSR, Ser.Fiz.Atmosf. i Okeana 4, 1149-1159.
Olevskaya,S.M.(1974). Canonical correlations for random fields. In: Probability Theory and Mathematical Sta-

tistics No.8, Skorokhod,A.V. et al. (Eds.), Izd.Kiev Gos.Univer., Kiev, pp.38-45.

Polyak,I.I.(1979). Methods for analysis of climatic random processes and fields. Gidrometeoizdat, Leningrad.

Rosenblatt,M.(1961). Independence and dependence. In: Proc. 4th Berkeley Symp. on Math.Stat. and Probab., vol.2, Neyman,J.(Ed.), Univ.Calif.Press, Berkeley-Los Angeles, pp.431-443.

Tatarskaya,M.S., and Fortus,M.I.(1984). An application of the canonical correlation method to the analysis of climatic time series. Izv.Akad.Nauk SSSR, Ser.Fiz. Atmosf. i Okeana 20, 1027-1034.

Yaglom,A.M.(1955). Correlation theory of processes with stationary random increments of order n. Mat.Sbornik 37, 141-196. English transl.: Amer.Math.Soc.Translations, Ser.2, 8, 87-141, 1958.

Yaglom, A.M.(1965). Stationary Gaussian processes satisfying the strong mixing condition and best predictable functionals. In: Bernoulli, Bayes, and Laplace Anniversary Volume, LeCam,L.M., and Neyman,J.(Eds.). Springer, Berlin, pp.241-252.

Yaglom,A.M.(1966). Outline of some topics in linear extrapolation of stationary random processes. In: Proc. 5th Berkeley Symp.Math.Stat. and Probab., vol. II, part 1, LeCam,L.M., and Neyman,J.(Eds.), Univ.Calif.Press, Berkeley-Los Angeles, pp.259-278.

STATISTICAL DECISIONS AND PROBLEMS OF THE OPTIMUM USE OF METEOROLOGICAL INFORMATION

Zhukovsky, E.E., Agrophysical Institute, Leningrad, USSR

1. GENERAL

Methods of the probability theory and of mathematical statistics are applied long ago and quite successfully for a description of the atmospheric turbulence, for generalization of results of long-term meteorological observations and for the problems of weather prediction. There is one more field of meteorology based upon probabilistic-statistical methods, i.e. the optimization problems of utilization and estimation of meteorological information (MI) efficiency.

At present these investigations have been formed as an independent field in the applied meteorology based on general approaches developed in statistical decisions and applied for the analysis of "weather - MI - user" systems (Zhukovsky, 1984). A construction of mathematical models of such systems includes:

- description of the set Ω_F of meteorological situations F to the changes of which the MI user is sensitive;
- description of the set Ω_d of economic decisions d of the MI user;
- determination of u(F,d) function showing what benefits or losses occur in case of various combinations of F and d variables (for convenience, the losses are further considered);
- description of the set Ω_s of admissible strategies S, every of which establishes a certain rule for decision making;
- specification of a criterion allowing to compare

various strategies and to select the optimum one.

Within the framework of the Bayesian approach used as the basis for the theory of statistical decisions, the optimum strategy $S_o \in \Omega_s$ is usually determined from the condition of attaining the minimum of average losses. These losses are defined as $U_s = Eu(F,d)$, where E is the averaging with respect to probabilities of the coupled events (F,d) dependent on the selected strategy S.

The potential efficiency of MI is estimated as a difference $\Delta U = U_o^- - U_o^+$ or by means of a dimensionless index $\lambda_o = \Delta U/U_o^-$, where U_o^+ and U_o^- are U_s values corresponding to optimum strategies of decision making when the appropriate MI is available (U_o^+) and when it is missing (U_o^-). Therefore, it means that the problem of the assessment of the MI potential efficiency is closely connected with the search of the optimum strategies of the user in cases of his different knowledge of the meteorological situation.

Some problems on the analysis of meteorological and economic models have been discussed in (Omshansky,1933; Thompson, 1950, 1976; Obukhov, 1955; Monin, 1962; Murphy, 1966, 1977; et al.). These problems are described in detail in (Zhukovsky, 1981) with the attached bibliography, containing publications on this subject by 1980's. The results given below reflect some further development in this field.

2. THE PROBLEM OF DECISION MAKING FOR CLIMATICALLY NON-HOMOGENEOUS AREAS

Most of meteorological factors are highly variable in space; moreover, not only particular values of meteorological parameters usually vary, but long-range characteristics as well which reflect the climate peculiarities in various points of the area under investigation. Proceeding from this it is necessary to discover the following: what effect may be obtained due to a transition from planning of different economic arrangements relative to mean climate conditions to spatial differentiation of decisions; what should be the level of this differentiation; how to utilize best the information on spatial variability of climatic parameters, etc. Some of these problems have been already discussed relative to the case of the account of non-homogeneous area with respect to climatological mean of the affecting meteorological factor (Zhukovsky, 1983). Here we consider one more case which often occurs.

It is assumed below that the weather is characterised by two possible states - F_1 and F_2, and, consequently, two economic decisions d_1 and d_2 control the economic activity of the MI user. The matrix of losses $\| u_{ij} = u(F_i, d_j);\ i,j=1,2 \|$ is assumed to be known.

In the similar alternative situation the selection of the best decision usually follows the rule

$$\left.\begin{array}{l} p_1 \geqslant p^\circ \rightarrow d_1 \\ p_1 < p^\circ \rightarrow d_2 \end{array}\right\} \qquad (1)$$

where p_1 is climatic frequency of meteorological conditions F_1; p° is optimum threshold value of p_1 equal to $\beta/(1+\beta)$; β is a dimensionless economic parameter determined as $\beta = (u_{21}-u_{22})/(u_{12}-u_{11})$.

Suppose n climatically homogeneous zones are selected within the geographical region under consideration, and every of these zones is characterised by its relative area g_l($\sum_{l=1}^{n} g_l = 1$) and by the particular value of

$p_1 = p_{1,l}$. Proceeding from this let us compare two strategies, i.e. non-differential economic planning based on the mean frequency $\bar{p}_1$ equal to

$$\bar{p}_1 = \sum_{l=1}^{n} g_l \, p_{1,l} \tag{2}$$

and differentiated planning with the account of particular peculiarities of the climate over individual zones. It may be shown that the mean gain for unit area due to differentiation of decisions for this alternative model is equal to

$$\Delta\bar{U} = \begin{cases} (r_1 + r_2) \sum\limits_{p_{1,l} \leq p^\circ} (p^\circ - p_{1,l}) g_l & \text{if } \bar{p}_1 \geq p^\circ \\ (r_1 + r_2) \sum\limits_{p_{1,l} > p^\circ} (p_{1,l} - p^\circ) g_l & \text{if } \bar{p}_1 < p^\circ \end{cases} \tag{3}$$

where r_1 and r_2 are "meteorological losses" related to the loss matrix elements $\|u_{ij}\|$ by equations $r_1 = u_{12} - u_{11}$ and $r_2 = u_{21} - u_{22}$. The maximum effect is attained if $\bar{p}_1 = p^\circ$ and is equal to

$$\Delta\bar{U}_{max} = \frac{r_1 + r_2}{2} \sum_{l=1}^{n} |p_{1,l} - \bar{p}_1| \, g_l \tag{4}$$

When the variable p_1 has a wide spectrum of values, it may be interpreted as a continuous random value described by some distribution density $g(p_1)$. Every particular type of this function depends on the climate variability over the non-homogeneous area under consideration. It should be kept in mind, however, that p_1 values are always between the physical limits of 0 and 1. Therefore it may be assumed that beta-distribution may be applied to approximate the function $g(p_1)$:

$$g(p_1) = \frac{1}{B(a,b)} \, p_1^{a-1} (1-p_1)^{b-1} \tag{5}$$

where a and b are positive parameters; $B(a,b) = \int_0^1 p_1^{a-1}(1-p_1)^{b-1}dp_1$ is the complete beta-function (Euler integral of the 1st order). In this case equation (3) may be written in the form

$$\Delta\bar{U} = \begin{cases} (r_1+r_2)\,[p^\circ I_{p^\circ}(a,b)-\bar{p}_1 I_{p^\circ}(a+1,b)] \text{ at } \bar{p}_1 \geqslant p^\circ \\ (r_1+r_2)\,[\bar{p}_1\hat{I}_{p^\circ}(a+1,b)-p^\circ\hat{I}_{p^\circ}(a,b)] \text{ at } \bar{p}_1 < p^\circ \end{cases} \quad (6)$$

where $\bar{p}_1=a/(a+b)$, I_{p° is the incomplete beta-function ratio determined as

$$I_{p^\circ}(v,w)=\frac{1}{B(v,w)}\int_0^{p^\circ} p_1^{v-1}(1-p_1)^{w-1}dp_1 \quad (7)$$

tabulated by K.Pearson, and $\hat{I}_{p^\circ}(v,w)$ is the addition of function $I_{p^\circ}(v,w)$ up to 1.

For the case when $\bar{p}_1= p^\circ$, it follows from (6) that if spatial variations of p_1 are approximated by beta-distribution, tha value of $\Delta\bar{U}_{max}$ is calculated from:

$$\Delta\bar{U}_{max}= (r_1+ r_2)\bar{p}_1\,[\,I_{\bar{p}_1}(a,b)-I_{\bar{p}_1}(a+1,b)] \quad (8)$$

Let us consider the situation when a=b. A family of functions $g(p_1)$ symmetric relative to the point $p_1=\bar{p}_1= =1/2$ corresponds to this situation. In particular, when a=b=1, this is the simplest uniform distribution $g(p_1)= =1$, for which equation (6) gives

$$\Delta\bar{U} = \begin{cases} \frac{r_1+ r_2}{2}(p^\circ)^2 & \text{if } p^\circ \leqslant 1/2 \\ \frac{r_1+ r_2}{2}(1-p^\circ)^2 & \text{if } p^\circ > 1/2 \end{cases} \quad (9)$$

This dependence is shown in Figure 1, where $\Delta\bar{U}_{max}= = 0.125(r_1+r_2)$. The case of differentiation of alternative economic decisions at uniform distribution of

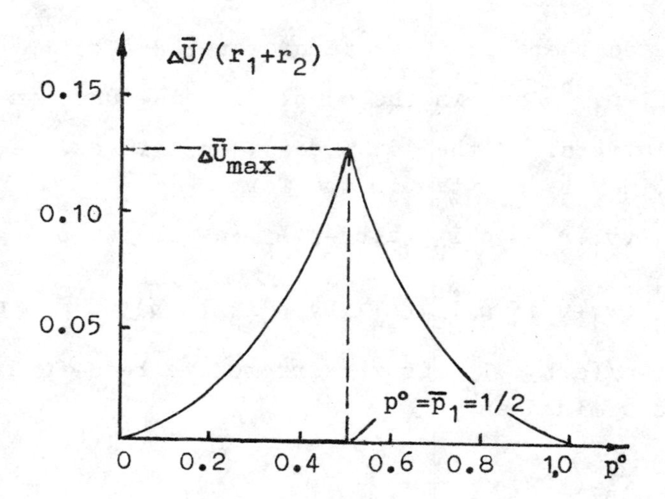

Figure 1. Dependence of $\Delta\bar{U}(p^{\circ})$ in case of uniform p_1 distribution within the range [0,1].

p_1 within an arbitrary range was investigated earlier (Zhukovsky, 1984). It was obtained for that problem, that

$$\Delta\bar{U} = (r_1 + r_2)\,\frac{l_{p_1}}{4}\left[\frac{p^{\circ}-\bar{p}_1}{l_{p_1}} - \mathrm{sign}(p^{\circ}-\bar{p}_1)\right]^2 \qquad (10)$$

where $\mathrm{sign}(p^{\circ}-\bar{p}_1) = \pm 1$ to the sign of difference $(p^{\circ}-\bar{p}_1)$ respectively; l_{p_1} is half-range of p_1 changes.

It is easy to show that at $\bar{p}_1 = 1/2$ and $l_{p_1} = 1/2$ the above equations (9) are obtained from the general equation (10). In (Zhukovsky, 1984) a formula for $\Delta\bar{U}$ has been also obtained for the case when p_1 is a Gaussian random variable. In this case

$$\Delta\bar{U} = (r_1 + r_2)\sigma_{p_1}\left[\, t\Phi(t) + \varphi(t) - |t|/2 \,\right] \qquad (11)$$

where σ_{p_1} is the standard deviation of p_1; t is a dimensionless parameter equal to $(p^{\circ}-\bar{p}_1)/\sigma_{p_1}$; $\varphi(t)$ and $\Phi(t)$ are density of Gaussian distribution and probability integral corresponding to a particular t.

3. UTILIZATION AND POTENTIAL EFFICIENCY OF ALTERNATIVE FORECASTS

As in the above problems, let us assume that weather conditions are characterised by two possible states (F_1, F_2) and, as a result, there are two economic decisions (d_1, d_2), and a matrix of losses $\| u_{i,j}; i,j=1,2 \|$ determines the consequences of these decisions realization in case of a certain weather. Then let us assume that the above decisions may be differentiated according to a categorical alternative forecast, which may be formulated in two mutually exclusive ways, i.e. Π_1 ("Weather F_1 is expected") and Π_2 ("Weather F_2 is expected"). A complete information on the reliability of this forecast is provided by matrix $\| p_{ij}=P(F_i, \Pi_j); \ldots i,j=1,2 \|$, the elements of this matrix are probabilities of various combinations of forecasted and observed meteorological situations.

The described situation has been properly studied (Thompson, 1950; Obukhov, 1955; Monin, 1962; Bagrov, 1966; Zhukovsky, 1979; et al.). In particular, it is shown that it is reasonable to apply the alternative forecast in cases when the two-sided inequality $æ_2 < ß < æ_1$ is valid, where $æ_1 = p_{11}/p_{21}$, $æ_2 = p_{12}/p_{22}$, and, consequently, the forecast should not be used if $ß < æ_2$ or $ß > æ_1$. Irrespective of the predicted weather, the weather conditions of F_1 should be orientated to, in the first case, and those of F_2 - in the second case. In case of a fixed ß the inequality

$$p_{1|1} > ß/(1 + ß) \quad \text{if } ß > p_1/p_2 \qquad (12)$$

or

$$p_{2|2} > 1/(1 + ß) \quad \text{if } ß < p_1/p_2 \qquad (12a)$$

is the necessary and sufficient condition of the economic utility of the alternative forecast (Zhukovsky, 1979,

1981). Here $p_2=1-p_1$ is a climatic probability of conditions F_2; $p_{1|1}=p_{11}/(p_{11}+p_{21})$ and $p_{2|2}=p_{22}/(p_{12}+p_{22})$ are probabilities of a correct prediction of conditions F_1 and F_2, respectively.

Inequalities $\beta > p_1/p_2$ and $\beta < p_1/p_2$ are not compatible. It follows from here that the requirements of any "alternative" user (i.e. having only two decisions) to the categorical forecast are always one-sided. In other words, depending on the ratio of parameters β and $\beta^o = p_1/p_2$ eigher conditions of F_1 or F_2 should be forecasted with a sufficient reliability. The maximum gain will be obtained from the use of the alternative categorical forecast if $\beta=p_1/p_2$. Relative to such users index λ_o of the potential efficiency of the prognostic information may be computed by equation

$$\lambda_o = \frac{(u_{12}-u_{11})(u_{21}-u_{22})}{u_{12}u_{21}-u_{11}u_{22}} Q \qquad (13)$$

where $Q=p_{11}/p_1+p_{22}/p_2-1$ is the alternative-forecast successfullness (according to A.M.Obukhov).

Let us consider a more complicated case, assuming that the forecast is used not by one user but by many users simultaneously, and every user has his own matrix of losses. Let us introduce a term "usefulness on average", admitting that the forecast is valid on average if, assuming that all the potential users completely rely on it, the economic benefit is obtained compared with the results which might be obtained in case of prognostic information. Let us evaluate from this viewpoint the categorical forecasts of dangerous meteorological events. Here we assume that all the users of such a forecast have their loss matrices of the same type

$$\| u_{ij} \| = \begin{Vmatrix} C & L \\ C & 0 \end{Vmatrix} \qquad (14)$$

but different by the values of their elements. Here L is damage resulted from nonpredicted dangerous event; C is the cost of protective measures (Thompson, 1950; Cringorten, 1951; Murphy, 1966; et al.). Moreover, let us assume that L is constant for the considered ensemble of users and C/L ratio is homogeneously distributed within the range of [0, 1]. It may be shown that in this particular case the forecast will satisfy the usefulness on average if point $(p_{1|1}, p_{1|2})$ ($p_{1|2}=1-p_{2|2}$ is the probability of "missing" the occurred dangerous event) in the Cartesian coordinates (x,y) appears in the area between the abscissa axis x, perpendicular x=1 and the curve

$$y = \frac{(2-p_1)x-1}{2x-1-p_1^2} p_1 \qquad (15)$$

The appropriate areas for different p_1 values are shown in Figure 2.

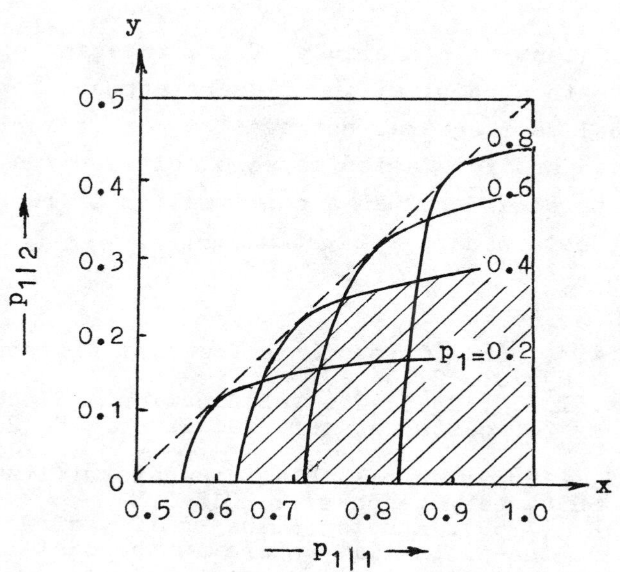

Figure 2. Areas within which the alternative forecast with specified $p_{1|1}$ and $p_{1|2}$ is useful on average. Shading shows appropriate area for $p_1 = 0.4$.

Unlike one-sided requirements to the alternative forecast by every user, the two possible weather situations, i.e. a dangerous event and its absence, should be predicted with a sufficient reliability to provide the conditions of the forecast usefulness on average. In order to estimate the potential efficiency of the prognostic information it is possible to compute $\tilde{\lambda}_o$ by the analogy of λ_o, as $\tilde{\lambda}_o=(MU_o^- - MU_o^+)/MU_o^-$ (M is the operator of the probabilistic averaging of U_o^+ and U_o^- over the ensemble of users). For the alternative forecast of the dangerous event $\tilde{\lambda}_o$ is computed from

$$\tilde{\lambda}_o = \frac{1 - p_1}{2 - p_1} R^2 \qquad (16)$$

where R is the coefficient of qualitative correlation of binary variables F and Π .

Thus, the potential efficiency of the forecast estimated relative to a combination of users appears to be proportional to a squared correlation coefficient of the observed and forecasted meteorological situations, which may be considered as a confirmation of results previously obtained by M.I.Yudin (1977).

REFERENCES

Bagrov, N.A. (1966). On economic forecast usefulness. Meteor. a.Hydrol. 2, 3-12.

Gringorten, I.I. (1951). The verification and scoring of weather forecasts. J.Amer.Stat.Assoc. 48, 255, 279-296.

Monin, A.S. (1962). On reliable forecasts utilization. Izv. AN SSSR, ser. geophys. 2, 218-228.

Murphy, A.H. (1966). A note on the utility of probabilistic and the probability score in the cost-loss ratio decision situation. J. Appl. Met. 5, 4, 534-537.

Murphy, A.H. (1977). The value of climatological, categorical and probabilistic forecasts in the cost-loss ratio situation. Mon. Wea.Rev.105,7,803-816.

Obukhov, A.M. (1955). On the problem of assessment of alternative forecasts successfulness. Izv. AN SSSR, ser. geophys., 4, 72-81.
Omshansky, M.A. (1933). On the account of forecasts accuracy and forecasts utilization. J.Geophys. 3, 4, 489-499.
Thompson, J.C. (1950). A numerical method for forecasting rainfall in the Los Angeles area. Mon. Wea. Rev. 78, 7, 113-124.
Thompson, J.C. (1976). Economic and social impact of weather forecasts. - Weather Forecasting and Weather Forecasts: Models, Systems and Users. Boulder, Colo., NCAR, 2, 525-578.
Yudin, M.I. (1975). Long-range forecast as a means for man's activity control. In: Application of statistical methods in meteorology. Gidrometeoizdat, Leningrad, pp.5-10.
Zhukovsky, E.E. (1979). Alternative weather forecasts: economic efficiency and requirements to forecast successfulness. Scient.-Techn. Bull. on Agronom. Physics, 36, 29-36.
Zhukovsky, E.E. (1981). Meteorological information and economic decisions. Gidrometeoizdat, Leningrad, 304p.
Zhukovsky, E.E. (1983). Statistical models for economically optimum application of climatological data. II Int. Meet. on Statist. Climatology, Sept. 26-30, Lisboa, p.9.2.1-9.2.8.
Zhukovsky, E.E. (1984). Theory and methods of the optimum utility of hydrometeorological information. Autoabstract of theses, Leningrad, 24 p.

APPLICATION OF DATA ANALYSIS METHODS FOR THE EVALUATIONS OF EFFICIENCY OF WEATHER MODIFICATIONS EXPERIMENTS

G. Der Megreditchian

Direction de la Météorologie. EERM/CRMD
PARIS - FRANCE

INTRODUCTION : The combinaison of field and laboratory experiments with theory and numerical modeling has allowed cloud physicists to confirm the usefulness of cloud seeding in certain meteorological conditions. This was the begining of a big wave of cloud seeding experiments to prevent catastrophic hailstorms, disperse the fog on the airports, or increase the precipitations. Unfortunately, in operational conditions postive results, scientifically approved remain rare and often much debated. Nevertheless the member countries of the World Meteorological Organisation have maintened stong interest to the possibilities of successful Weather Modification Experiments (W.M.E.). It is not surprising considering the adverse effects of weather in many parts of the world. Two orientations characterize the main purpose of control of W.M.E.
A) Confirmatory analysis : to verify if a perceptible effect, statistically significant appears for the unique parameter chosen for this confirmatory test.
B) Exploratory analysis : to study which parameters seem to be altered and how much. In terms of Pattern Recognition the question will be : Do the differences appearing on the parameters allow to recognize the type of situation (natural or treated) ? Here we have a delicate statistical problem due to the non repetitivity of experimentation which leads to the necessity of randomization of the seeding procedure for an objective evaluation of efficiency. But first at all in order to concretize our subject let us consider a particular case of W.M.E., the problem of hail suppression or attenuation.
In our study we assimilate each hailstorm $O_j(j=1,L)$ to a concrete meteorological situation characterised by a set of parameters :

$$X_{(j)} = \left\{x_1(j),\ldots,\ x_i(j),\ldots,\ x_n(j)\right\} .$$

They describe as well the general circulation of the atmosphere (local or synoptic variables measured on the ground or in altitude) as the specificity of the hailstorm (radar measurements and various parameters obtained from an hailpads

network). The parameters $x_i(j)$ are associated with a given spatio-temporal area characterising the cell (in our model the index j of the hailstorm).
The realisation of a W.M.E. leads to a set of data defining the data file **X** , having all the available inquires about the N stormy cells analysed :

$\mathbf{X} = [x_i(j)]$, where i=1, n ; j=1, L.

In the frame of the randomized model [8] of the prevention experiment two kind of hailstorms are considered : the "natural" hailstorms $O_N(j)$ and the treated (seeded) one's $O_T(j)$. One of the variable x_i is the indicator variable of the type of hailstorm. Assimilating the data generation to the realization of a random experiment we consider the vector X as fully defined by his multidimensional distribution function. The influence of W.M.E. will express itself by means of a more or less important differences between the distribution function of the "natural" cells $F_N(X)$ and of the "treated" cells $F_T(X)$. Here two possibilities arise : the procedure is without any effect or on the contrary, it has significative influence on some parameters. In this connection two contradictory hypothesis can be formulated :
H_o : $F_N(X) = F_T(X)$, the W.M.E. is not efficient
H_1 : $F_N(X) \neq F_T(X)$, the W.M.E. is efficient.

The evaluation of the efficiency of any prevention method is then equivalent the following statistical problem : to test the H_o hypothesis against the alternative hypothesis H_1.

1/ APPLICATION OF PARAMETRIC AND NO PARAMETRIC TESTS

A) Testing the evolution of mean values

We test in this case the hypothesis H_o of equality of mean values :

H_o : $M_N = M_T$, against the alternative hypothesis.
$H_1 = M_N \neq M_T$.

We distingwish here the unidimensional and multidimensional cases. In the unidimensional case it's possible to apply either the parametric Student test [6], or the non parametric Mann-Witney test [6], and in the multidimensional case the Hotelling test [6].

B) Alteration of the variability

Any modification in the variability of vector X will be expressed by a difference between the covariance matrix $V_{XX}(N)$ and $V_{XX}(T)$, respectively calculated for the "natural" or the "treated" cases. So we have to verify the hypothesis H_o : $V_{XX}(N) = V_{XX}(T)$, against the alternative hypothesis H_1 : $V_{XX}(N) \neq V_{XX}(T)$. In the unidimensional case (the covariance matrix becomes simply the variance) we can apply the classical Fischer's test [10] and in the multidimensional case its generalisation given by the Box test [1].

C) Global modification of the distribution law

It's possible to verify the more general hypothesis $H_0 : F_N(X) = F_T(X)$ of homogeneity of the two populations of "natural" and "treated" cells. Either the observed values $X_N(j)$ and $X_T(j)$ are extracted from the same population if W.M.E. is not efficient or on the contrary from two different populations (if W.M.E. is efficient). Here we are not interested by the concrete form of F(X), but only by the fact $F_N(X) = F_T(X)$ or $F_N(X) \neq F_T(X)$. Various tests can be applied here : the khi square test [3], the informational test [3] the Kolmogorov-Smirnov test [10], the Renyi test [10], etc.

2. STRUCTURAL ANALYSIS OF THE HAILPADS DATA

The hailpads network gives some picture of the structure of hailstorms. Is it possible to detect in this structure some difference for the "natural" and "treated" cases ? The answer is given by a structural analysis mainly based on the study of the corresponding structure functions (variogramms) $\gamma(d)$ and the spatial correlation function r(d) giving some synthetical knowledge about the statistical cohesion of each hailstorm. Then some "distance" is introduced between variograms of various cells :

$$d[\gamma_N(d),\ \gamma_T(d)],\ d[\gamma_N(d),\ \gamma_N(d)]\ ;\ d[\gamma_T(d),\ \gamma_T(d)].$$

and it leads to the problem of the statistical analysis of the matrix D of distances between variograms for the whole set of cells.

3. PROXIMITY ANALYSIS OF THE FILE OF HAILSTORMS DATA

After appropriate quantification of the difference between the "natural" and "treated" hailstorms we get some distances matrix D from which we would like to extract corresponding synthetic information about reciprocal similarity or dissimilarity of concerned objects (variograms or correlation functions). It's not easy to perform that directly, so it's convenient to apply the specific technic of Proximity Analysis [4, 9]. For any distances matrix $D = (d_{ij})$, where $d_{ij} = d[e_i, e_j]$ measure the distance between some objects of arbitrary nature e_i, e_j considered as points in the space R^N, the basic principle consists to realize appropriate projections $\tilde{e}_i$ of the e_i in some subspace $R^n (n < N)$ in order to minimize the deformation of the cluster of real points e_i. Choosing n = 2, thanks to the human property of visual perception, the spatial organisation of the figurations points appears clearly on the factorial plane. The coordinate $Z_s[\tilde{e}_i]$ of the "optimal" figurative points are given by the formula :

$Z_s[e_i] = \sqrt{\lambda_s}\, c_{is}$, where λ_s is the S-eigenvalue and $C_s = \{\ldots, c_{is}, \ldots\}$ the corresponding eigenvector of the matrix $H = -1/2\ \overset{\circ}{D}{}^2$, and $\overset{\circ}{D}{}^2$ is the double centered (either on the rows, as on the columns) matrix D^2.

4. DISCRIMINANT ANALYSIS APPLIED TO THE IDENTIFICATION OF THE TYPE OF HAILSTORMS.

The influence of the W.M.E. on the parameters x_i corresponds to a more or less important modification of their values. Then the more the prevention method is efficient, the more it will be easier to identify the type ("N" or "T") of the hailstorms through the values of x_i. So the efficiency of the W.M.E. can be assimilated to the capacity of discrimination of the type by means of "predictors" x_i. In linear discrimination such index of efficiency of W.M.E. will be the classical index of separability of the two populations the Mahalanobis distance [1, 4] :

$$\Delta^2_N\,[N,T] = M_X'(-)\ V^{-1}{}_{XX} M_X(-).$$

Foreward or Backward Progressive Selection will permit to reveal the relative order of the amount of alteration of the parameters x_i by the prevention method. Appropriate Fisherian tests will allow to estimate the degree of the statistical signification of prevention effect. Then the quality of the identification procedure will be also a synthetic index of the efficiency of the prevention method. For that usually we use the percentage of success :

$$P[S] = 1/\sqrt{2\pi}\int_{-\infty}^{\Delta/2} e^{-z^2/2}\,dz$$

directly linked with the separability criteria Δ^2 by such a simple formula.

5. APPLICATION OF THE CLUSTER ANALYSIS

Cluster Analysis will allow to answer the question : Is the cluster of the points $X_{(j)}$ characterized either by a total absence of structure, or by the existence of some specific configurations ? In particular have we two kinds of cluster : the cluster of the points "N" and the cluster of the points "T". ? we can use both the method of agregation around mobil centers (Etalons), or the hierarchical clustering [2, 4, 7]. After that a frequential analysis of the type of cells ("N" or "T") will answer to the main question of an eventual efficiency of the W.M.E.

6. FACTORIAL ANALYSIS OF CORRESPONDANCES F.A.C.

F.A.C. allows to reveal the statistical interrelations between elements of arbitrary nature characterized by a set of quantitative or qualitative parameters. Here it will be possible to investigate the correspondance between the hailstorms and the qualitative items of their parameters. The introduction of Khi square distance between qualitative items $d_{\chi^2}(A_i, A_j)$ followed by an application of Proximity Analysis of the matrix $D_{\chi^2}[A,A]$ allow detailed synthetical investigation of the efficiency of $\bar{W}$.M.E. [2,7].

7. PREFERENCE ANALYSIS OF THE FILE OF DATA

Preference Analysis deals generally with the problem of preferential ranking of a set of objects O_i realized by a group of experts E_i. Here we have an obvious interpretation for our problem, if each parameter of the hailstorm is assimilated to an "expert" and the hailstorm to the O_j. So we obtain some preferential ranking of the hailstorm that in some order of influence of the prevention method on the whole set of hailstorms and after that we have to see if such ranking is in some concordance with the definition of "natural" or "treated" cases.

8. MODEL OF SIMULATION OF A PREVENTION EXPERIMENT

Conclusions about the efficiency of W.M.E. are often negative because of their enormous cost and the small number of avalaible cases. So the random simulation of a long serie of W.M.E. is of great interest. The model inputs are : the climatological characteristics of hail for a given area, the degree of efficiency of the prevention method, the characteristics of hail damages for various types of culture, the distribution of cultures on a given area, the cost of the prevention operation, the cost associated to various strategies of the user. Then a random simulation of W.M.E. for 10,50 or 100 years will aid to define the "optimal" strategy for a given type of users.

9. OBJECTIVE METHODS FOR THE EVALUATION OF W.M.E. EFFICIENCY (Jacknife, Bootstrap).

Each method for the evaluation of W.M.E. leads to some quality index Q, a random variable for which the W.M.E. gives only one realisation : $\hat{Q} = Q[\mathbf{X}]$. It will be much better if we could have another test sample $\mathbf{X}$(test) for an objective verification $\tilde{Q} = Q[\mathbf{X}(\text{test})]$ of the quality given by the learning sample $\hat{Q} = Q[\mathbf{X}(\text{learn})]$. The Jacknife method allows to perform such objective evaluation on a single learning sample by means of a "slipping" control [4,5]. The best situation will occur in the case (fully utopical) we would have a great number of W.M.E., that is a great number of data samples $\mathbf{X}[k]$. Thus we could compute Q and $\sigma[Q]$, or even $F_Q(Z)$, a much more complete information than a single realisation $\hat{Q}$ of Q. So the Bootstrap method [4] allows to do that through a random simulation of a series of samples $\mathbf{X}[k]$ starting from the initial sample $\mathbf{X}$.

10. CONCLUSIONS

Actual methods of multidimensional data analysis allow a multiform evaluation of the efficiency of W.M.E. Thus it is possible to reveal the preferential or selective character of W.M.E. effects, to quantify their intensity and even to built confidence intervalls for the unknow theoretical value.

REFERENCES

1. ANDERSON : An introduction to multivariate statistical analysis. Wiley. 1970.
2. BENZECRI : L'analyse des données. Tomes I, II. Dunot. 1979.
3. CRAMER : Mathematical methods of statistics. Princeton. 1946.
4. DER MEGREDITCHIAN : Le traitement statistique des données multidimensionnelles. Tomes I, II. Toulouse. 1983.
5. DIACONIS, EFRON : Méthodes de calculs statistiques intensifs sur ordinateurs. Pour la Science. Juillet 1983.
6. KENDALL, STUART : The advanced theory of statistics. Griffin 1966.
7. LEBART, MORINEAU, FENELON : Traitement des données statistiques. Dunod. 1977.
8. MEZEIX, DORAS : La prévention de la grêle. Réalité, échec ou espoir ? GNEFA. 1983.
9. NEYMAN : A statistician's view of weather modification. Proc. Natl. Acad. Sci. USA. 74 (1977).

PREDICTOR-COUNTING CONFIDENCE INTERVALS FOR THE VALUE OF EFFECT IN RANDOMIZED RAINFALL ENHANCEMENT EXPERIMENTS

Kudlaev E.M.
Moscow Lomonosov University, Mosdow, USSR

Suppose that each $\mathcal{O}_k$-th object under seeding is related to a pair of quantities, (ε_k, X_k), $k=1,\dots,n$, where X_k is the precipitation fallen, $\varepsilon_1,\dots,\varepsilon_n$ are random quantities relating to the randomization procedure. Let (ε_k, X_k), $k=1,\dots,n$ be a sequence of independent random pairs taking two values for each k, $(1, b_k)$ and $(0, a_k)$ with probabilities $0<p<1$ and $q=1-p$ respectively, where a_k and b_k $(k=1,\dots,n)$ are constants; if $\varepsilon_k=1$ then an object $\mathcal{O}_k$ is seeded, if $\varepsilon_k=0$ then an object $\mathcal{O}_k$ is not seeded. By the results of observation of all the pairs we estimate the difference

$$C(n) = B(n) - A(n) \quad \text{where } A(n)=\sum_{k=1}^{n} a_k, \quad B(n)=\sum_{k=1}^{n} b_k.$$

The Horvitz-Tompson estimates for $A(n)$ and $B(n)$ and their variances are equal to

$$\hat{A}(n)=q^{-1}\sum_{k=1}^{n} a_k(1-\varepsilon_k), \quad \hat{B}(n)=p^{-1}\sum_{k=1}^{n} b_k\varepsilon_k,$$

$$\mathbb{D}\hat{A}(n)=pq^{-1}\sum_{k=1}^{n} a_k^2, \quad \mathbb{D}\hat{B}(n)=qp^{-1}\sum_{k=1}^{n} b_k^2.$$

respectively. Zhurbenko and Kudlaev (1984) contains the estimate for the lower bound of the probability of $C(n)$ hitting the one-sided confidence interval, that is based on the Höfding inequality. Moreover, for $x>0$

$$\mathbb{P}(\hat{C}(n) - x < C(n)) \geqslant \varphi(x),$$

$$\varphi(x) = 1-\exp\{-2pqx^2[\sqrt{\mathbb{D}\hat{A}(n)}+\sqrt{\mathbb{D}\hat{B}(n)}]^{-2}\} \quad (1)$$

instead of the non-biased variance estimates, $\mathbb{D}\hat{A}(n)$ and $\mathbb{D}\hat{B}(n)$, the corresponding Horvitz-Tompson estimates are taken:

$$\hat{\mathbb{D}}\hat{A}(n)=pq^{-2}\sum_{k=1}^{n} a_k^2(1-\varepsilon_k), \quad \hat{\mathbb{D}}\hat{B}(n)=qp^{-2}\sum_{k=1}^{n} b_k^2\varepsilon_k. \quad (2)$$

Suppose that we have forcasts a_k and b_k for each $k=1,\dots,n$, so that

$$a_k = \bar{a}_k + a'_k \quad , \qquad b_k = \bar{b}_k + b'_k .$$

Denote

$$\hat{\varphi}_\Pi(x) = 1 - exp\left\{-2pqx^2\left[\sqrt{pq^{-2}\sum_{k=1}^{n}(a'_k)^2(1-\varepsilon_k)} + \sqrt{qp^{-2}\sum_{k=1}^{n}(b'_k)^2\varepsilon_k}\right]^{-2}\right\} \quad (3)$$

and suppose that there is a positive constant M such that

$$|a'_k| \leq M, \quad |b'_k| \leq M \quad (k=1,\dots,n). \quad (4)$$

Theorem. If $x > 0$, the condition (4) holds, the expression in square brakets in the right-hand part of (3) is distinct from, then for $n \to \infty$ we have

$$P\left(\hat{C}'(n) + \sum_{k=1}^{n}(\bar{b}_k - \bar{a}_k) - x < C(n)\right) \geqslant \hat{\varphi}_\Pi(x) + o_p(1), \quad (5)$$

where

$$\hat{C}'(n) = p^{-1}\sum_{k=1}^{n} b'_k \varepsilon_k - q^{-1}\sum_{k=1}^{n} a'_k(1-\varepsilon_k).$$

In particular, for

$$\sum_{k=1}^{n} \bar{b}_k = \sum_{k=1}^{n} \bar{a}_k \quad (6)$$

we can rewrite (5)

$$P\left(\hat{C}'(n) - x < C(n)\right) \geqslant \hat{\varphi}_\Pi(x) + o_p(1). \quad (7)$$

The relation (6) holds, if, for example, $\bar{a}_k = \bar{b}_k$ for each value of k ; i.e. forcast is independent of seeding and non-seeding. This supposition testifies to our poor knowledge of the physical basis of the mechanism of seeding clouds. Besides, if it is know that the effect of seeding is insignificant, then it is apparently worth fulfilling the relation (6) obtaining difference (if there is any) through estimation of $C'(n)$.

In the Ukrainian randomized experiment (1973-78) under seeding were one- and multi-cell cumulo-nimbus clouds. On the results of the study of 190 clouds in the experimental meteorological area (1967-76) Kornienko (1982) obtained the regression equation

$$lg\,Q_p = 0{,}08H + 1{,}15\,lg\,S - 0{,}01\,h - 0{,}19, \quad (8)$$

where Q_p is one-cloud precipitation (in thousand tons

$S(km^2)$ is the mean area (in the existence period of a cloud) of the horizontal section of the cloud radioecho at the base, $h\,(km)$ is the height of the lower boundary of the base of the cloud, $H\,(km)$ is the maximum vertical strength of the cloud radioecho in the existence period. We applied formulae (1)-(7) to the data of the Ukrainian experiment and form a table

Table 1

ℓ %	0	10	20	30	40	50	60	67
$\hat{\varphi}_n(x)$	0,977	0,964	0,943	0,914	0,874	0,822	0,755	0,7
$\hat{\varphi}\,(x)$	0,958	0,946	0,93	0,911	0,888	0,86	0,828	-

The values of x were chosen so that they corresponded to increase in precipitation (under seeding) by ℓ % with respect to $\hat{A}(n)$:

$$x = \hat{C}(n) - (\ell/100)\hat{A}(n), \quad x = \hat{C}'(n) - (\ell/100)\hat{A}(n)$$

for the cases with and without predictors respectively. Table 1 presents the calculated results only for one-cell clouds. For multi-cell clouds at $\ell = 0$ the value of $\hat{\varphi}$ defined by (1) through the estimates (2) equals 0,31; the value of $\hat{C}'(n)$ equals -7585,5; hence the value of x at $\ell = 0$ is non-negative and formula (7) is not applicable. Thus, the effect of seeding multi-cell clouds is not clear. To assess it one needs more subtle methods and additional information. For one-cell cloud the effect of seeding is noteworthy.

Suppose that under both seeding and non-seeding random quantities X_k have a gamma-distribution with two parameters, $\Theta_1(j)$ and $\Theta_2(j)$; $j=1$ corresponds to seeding, $j=0$ corresponds to non-seeding. Consider the approaching alternatives:

$$\Theta_1(1) = \Theta_1(0)\exp\{\lambda_1/\sqrt{n}\}, \Theta_2(1) = \Theta_2(0)\exp\{-\lambda_2/\sqrt{n}\}; \lambda_1, \lambda_2 > 0.$$

We can show that the one-sided asymptotic ($n \to \infty$) confidence interval at the level γ for the value of effect

$$d = [\mathbb{E}(X_1 | \varepsilon_1 = 1)] / [\mathbb{E}(X_1 | \varepsilon_1 = 0)] - 1$$

equals $(x_\gamma, +\infty)$. Here

$$x_\gamma = \frac{\Theta_1^*(1)\,\Theta_2^*(0)}{\Theta_2^*(1)\,\Theta_1^*(0)} - 1 - \frac{u_\gamma}{\sqrt{npq\,\Theta_1^*(0)}}, \quad u_\gamma = \Phi^{-1}(\gamma),$$

"asterisk" means a moment estimate calculated from an appropriate sample, Φ is a function of distribution of a standard normal law. The calculated values of x_γ for the Ukrainian experiment are given in the following table

Table 2

γ	0,99	0,975	0,95	0,9	0,85	0,8	0,75
	one-cell clouds						
x_γ	0,835	0,956	1,057	1,174	1,255	1,317	1,372
	multi-cell clouds						
x_γ	-0,236	-0,144	-0,066	0,024	0,087	0,135	0,177

The results of the table do not contradict those of table 1. Besides, within the given parametric model for multi-cell clouds we have succeded in assessing a positive effect with probability close to 0,9 and increase in precipitation (under seeding) by no less than 20% with probability close to 0.7.

REFERENCES

Kornienko, E.E. (1982). The results of the weather modification experiment on cumulo-nimbus clouds to regulate precipitation. In: Trudy Ukrainskogo regionalnogo NII, vyp. 187, 3-25.

Zhurbenko, I.G. and Kudlaev E.M. (1984). On assessment of the effect of seedingin randomised experiments. Uspehi matem. nauk, tom 39, vyp. 2, 3-38.

ON BEHAVIOUR OF SEA SURFACE TEMPERATURE ANOMALIES

Piterbarg L.I., P.P.Shirshov Oceanology Institute,
Sokoloff D.D., Moscow State University, Moscow, U.S.S.R.

Since 1950's interest to large-scale air-sea interactions and sea-surface temperature (SST) anomalies noticably increased because it was found out that an understanding of the coupling between ocean and atmosphere might be a key to long-range weather forecasting and short-time climate predictions. Hasselmann (1976) has proposed a stochastic model of climate variability in which slow changes of climate are explained as the integral response of it to continuous random excitation by shorter time-scale disturbances. Frankignoul and Hasselmann (1977) have applied the model to a simplified atmospheric and oceanic system, the climatic components being represented as SST anomalies which were driven by uncorrelated white-noise atmosperic forcing.

In this paper we consider a more realistic model of the air-ocean interactions. This model includes advection,diffusion and takes into account random fluctuations of the velocity field. Our purpose is to study SST anomalies damping and generation. Note that from theoretical point of view anomalies are high level outliers of a random field.

We start from the heat balance equation

$$\frac{\partial T}{\partial t} + (u \cdot \nabla)T = æ\Delta T + Q - \lambda T,$$

where $T(t,x)$ is the temperature in the point $x \in R^2$, $u(t,x)$, $Q(t,x)$ - the random fields of velocity and distributed sources correspondently, æ - the constant diffusion coefficient, Δ - Laplase operator, λ - the constant feedback factor (Frankignoul,C. and Hasselmann,K. (1977)). Let $T'(t,x) = T(t,x) - \langle T(t.x) \rangle$, where the angle braces indicate ensemble mean, and let $m_n(t,x^{(1)},\dots,x^{(n)}) = \langle T'(t,x^{(1)})..T'(t,x^{(n)}) \rangle$ be the n-points moment of the temperature fluctuations. Suppose that the fields u,Q are stationary and δ-correlated in time. This conjecture bases on the essential difference between atmospheric and oceanic variability time scales (about two orders). From δ-correlation conditions we obtain the following equations for mean temperature and temperature fluctuations moments

$$\frac{\partial \langle T \rangle}{\partial t} = \frac{\partial}{\partial x_i} (æ \delta_{ij} + \frac{\tau_o}{2} B_{ij}(x ,x)) \frac{\partial \langle T \rangle}{\partial x_j} + \langle Q \rangle - \tau_o \langle u' \cdot \nabla Q' \rangle - \lambda \langle T \rangle ,$$

$$\frac{\partial m_n}{\partial t} = \tau_o B_{ij}(x^{(k)}, x^{(1)}) \frac{\partial^2 m_n}{\partial x_i^{(k)} \partial x_j^{(1)}} + \frac{\partial}{\partial x_i^{(k)}} (2 æ \delta_{ij} + B_{ij}(x^{(k)},x^{(k)})) \frac{\partial m_n}{\partial x_j^{(k)}} + \tau_o B_Q(x^{(k)},x^{(1)}) m_{n-2,k,1} - n\lambda \tag{1}$$

where $B_{ij}(x,y) = \langle u_i'(x) \cdot u_j'(y) \rangle$ - the spatial correlators of the velocity field, $B_Q(x,y) = \langle Q_o'(x) \cdot Q_o'(y) \rangle$ - the spatial correlator for the "new" source $Q_o' = Q' - u' \cdot \nabla$ $m_{n-2,k,1}$ - (n-2)-points moment for the collection

$x^{(1)}, \ldots, x^{(k-1)}, x^{(k+1)}, \ldots, x^{(l-1)}, x^{(l+1)}, \ldots, x^{(n)}$ and τ_o is a time-scale. In particular, if the fields u and Q are homogeneous and isotropic in space, div $u = 0$, we obtain the closed equation for temperature spatial correlation function

$$\frac{\partial R}{\partial t} = (2\varkappa + \tau_o F(r))\,\Delta R = \tau_o \frac{\partial F}{\partial r}\frac{\partial R}{\partial r} + \tau_o B_Q(r) - 2\lambda R\,, \quad (2)$$

where $R(t,r) = \langle T'(t,x)\, T'(t,y)\rangle$, $r = |x - y|$, $F(r)$- the velocity longitudinal correlator,(Monin,A. andYaglom,A. (1967)), $\tau_o = l_u / \sigma_u$, σ_u^2 is the velocity fluctuation variance, l_u - Taylor's spatial correlation scale of the velocity field.

It is known (Nosko, V. (1985)) that high level outlier shape of a homogeneous random field is similar to its correlation function shape. Hence, we can study a "typical" anomaly behaviour solving the equation (2) under the initial condition $R|_{t=0} = R_o(r)$. A number of following conclusions can be drawn.

1. On the initial stage the temperature variance increases linearly and the spatial correlation scale, $l_T = \sqrt{-R(t,0)/\Delta R(t,0)}$, decreases exponentially $l^2_T = l^2_u(t/\tau_o)\ (\exp\{t/\tau_o\} - 1)^{-1}$. Fast decreasing of the correlation scale and simultaneous increasing of the variance means growth of a temperature peaks number and their sharpening. Hence, a joint action of a random source and velocity is the cause of the temperature intermittency. If the Peckleut number $Pe = l_u \sigma_u / \varkappa \gg 1$ then the duration of the initial stage is equal to $\tau_o \ln Pe$ (in the order of Pe).

2. Define the lifetime t_N of an axially symmetric anomaly $R_o(r)$, as the N-fold temperature decreasing time in the centre of anomaly. Solving a boundary problem for the operator in the right side of (2) and supposing that $Q = 0$, $\lambda = 0$ we obtain that the main asymptotic member of the lifetime is equal ($Pe \gg 1$)

$$t_N \sim c_N \tau_o \ln(d^2 Pe / l_u^2) \quad (3)$$

where d is the initial spatial scale of the anomaly ($d \lesssim l_u$) and c_N is a constant. Note that (3) gives a more realistic value of anomaly lifetime than the "traditional" estimation $t_N \sim \tau_o d^2/l_u^2$.

3. There is a large number of t_n^* for each n, nonrandom for $Pe \to \infty$, that is sufficient for m_n, the solution of the equation (1) to be near to correspondent to n-point moment of the field $\widetilde{T}$. $\widetilde{T}$ satisfies the following equation

$$\frac{\partial \widetilde{T}}{\partial t} = k \Delta \widetilde{T} + Q - \lambda \widetilde{T}, \qquad (4)$$

where $k \sim l_u \sigma_u$ is the constant "effective" diffusion coefficient. Stationary regime (4) is characterized by a Gaussian multivariative distributions and by the frequency wave spectrum

$$E_T(\omega, p) = E_Q(p)\,(\omega^2 + (\lambda + kp^2)^2)^{-1},$$

where ω -frequency, p -wave number, $E_Q(p)$ - frequency wave spectrum of the source. In stationary regime typical value of anomaly lifetime with the spatial scale p^{-1} is equal to $(\lambda + kp^2)^{-1}$.

4. Let $g' = \nabla T'$ be the temperature fluctuation gradient. If the initial spatial correlation scale of the temperature field satisfies the condition $d > l_u Pe^{1/2}$ then the gradient variance $\sigma_g^2 = \langle |g'|^2 \rangle$ increases exponentially for $t \le t^*$. On this stage the growth rate of m_n is equal $\tau_o^{-1} n(n-1)/2$. It is the evidence of the intense intermittency of a temperature fluctuations gradient. Certainly, σ_g^2 approximates to zero if $t \to \infty$.

ACKNOWLEDGEMENTS

Authors wish to thank S.A.Molchanov, A.A.Ruzmaikin and Ya.B.Zeldowich for helpful discussions.

REFERENCES

Hasselmann,K.(1976).Stochastic climate models: Part1. Theory. Tellus,28, 473-485

Frankignoul,C. and Hasselmann,K. (1977). Stochastic climate models: Part 2. Application to sea-surface temperature anomalies and thermocline variability. Tellus, 29, 289-305.

Monin,A.S. Yaglom,A.M. (1967) Statistical hydrodynamics, Nauka, Moscow

THE SAMPLING VARIABILITY OF THE AUTOREGRESSIVE SPECTRAL ESTIMATES FOR TWO-VARIATE HYDROMETEOROLOGICAL PROCESSES

Privalsky, V.E., Water Problems Institute, Protsenko, I.G., ASUrybproject, Fogel, G.A., Institute of Geography, Moscow, USSR

The probability distribution of autoregressive spectral estimates in the one-variate (1-V) case is known (Koslov and Jones, 1985) but practically no information is available for 2-V time series. In this paper, results of a Monte Carlo study are reported concerning the sampling variability of autoregressive estimates (ARE) of spectral density s(f), coherence C(f), gain H(f), phase P(f), and $F = 0.5 \ln[(1+C)/(1-C)]$ for 2-V AR time series with relatively smooth properties in the frequency domain (Fig. 1). (The dependence on frequency f is dropped in what follows for the sake of brevity.) The aim is to assess quantitatively the variability of ARE and find, if possible, approximate expressions for their confidence bounds. The ARE were computed by using the conditional maximum likelihood approach while the non-parametric estimates were found by using the Blackman-Tukey method with the Parzen spectral window. Ensembles of 5,000 Gaussian time series of length N = 50, 100, or 200 were generated for each AR model and used to compute both ARE and Blackman-Tukey estimates (BTE), the latter possessing 8, 17, or 37 equivalent degrees of freedom respective to N. The variability of BTE was then compared with the approximate theoretical results (Bendat and Piersol, 1986) and with the variability of computed ARE.

It should be noticed that quite a few processes in the atmosphere and hydrosphere possess smooth spectral characteristics while the appropriateness of their AR rather

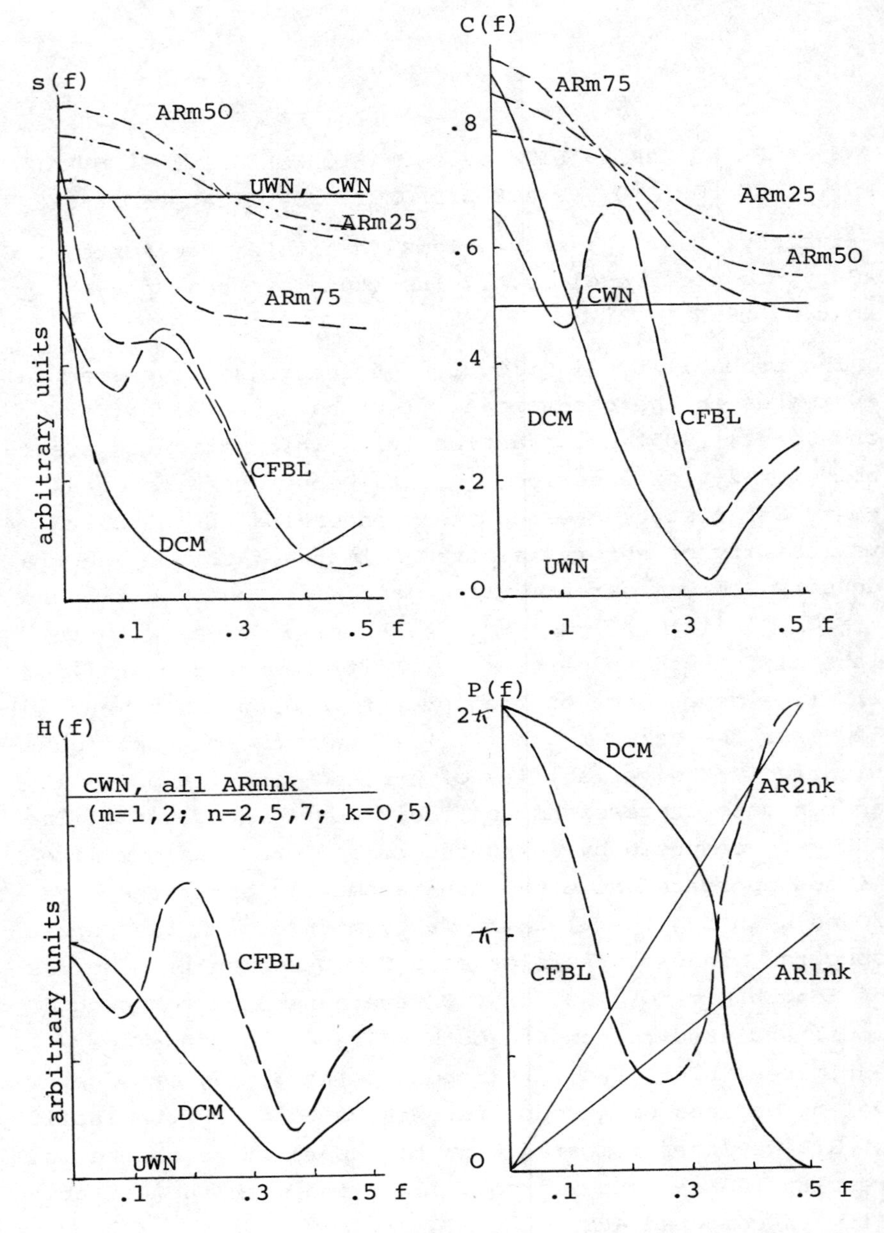

Fig.1. Spectral characteristics of AR processes under the study.

than moving average (MA) representation follows from physical considerations (see, e.g. Privalsky, 1985).

The order p of AR was estimated for each time series according to five criteria: AIC, FPE, BIC, CAT, and HQ but only the true order has been assumed for further computations for all twelve AR processes considered here while estimates p(AIC) have been adopted for MA models.

The following major conclusions can be drawn from our results.

1. AR order criteria. Though BIC and HQ were the best criteria in most cases, AIC and FPE proved to be much better for at least two AR(2) models (DCM and CFBL in Fig.1). Thus, our results agree only partially with those given in Lütkepohl (1985).

2. ARE of s and, especially, C, H, F, and P possess relatively small bias and variance in practically all cases. In the range between $0.2f_N$ and $0.9f_N$ where f_N is the Nyquist frequency, the empirical probability densities (p.d.) of ARE and BTE of C, H, and F do not seem to be markedly dissimilar, the p.d. of BTE and ARE of F and H being rather close to the Gaussian. Thus, it seems reasonable, as a first approximation, to introduce for ARE an equivalent number of degrees of freedom similar to that which is used in non-parametric spectral analysis. Taking into account the results of Kromer (1970), this quantity is defined here as $n = kN/2p$ and k is to be estimated.

3. ARE of s. (a) For models with a constant or monotonously decreasing s, ARE are always better than BTE. The value k = 2 can be taken for AR(0) while for AR(1) and AR(2) k = 1. (b) The approximation k = 1 is rather good for those AR(2) models whose spectra contain smooth peaks though k = 2/3 is safer at the peaks's frequencies.

4. ARE of C and F. (a) The values k = 4 and k = 2 are suggested for AR(0) with C=0 and C≠0 respectively while k=1 for AR(1) models with monotonous C. (b) For AR(2) mo-

dels with smooth but non-monotonous C the approximation k=1 seems to be conservative for all frequencies between $0.1f_N$ and $0.9f_N$ and acceptable at the frequencies where C attains its peak values. The value k=1.2 can be assumed for F.

5. ARE of H. For AR(0) with C≠0 k can be taken equal to 1.8 - 2.0. (b) For AR(1) with monotonous C (AR1nk in Fig.1) k lies between 0.7 and 1.3 and increases as N and C grow. The average value for all frequencies is $\bar{k}$=1.1. (c) For AR(2) with monotonous C (AR2nk in Fig.1) k=1.3, $\bar{k}$=1.5. (d) For AR(2) with smooth but non-monotonous C (CFBL and DCM in Fig.1) $\bar{k}$=1 seems to be conservative.

6. ARE of P. When no significant feedbacks exist between the process"s components, k=1 can be assumed as a conservative approximation while $\bar{k}$ = 1.5.

All in all, when no sharp peaks appear in the spectral characteristics of a 2-V time series, their ARE seem to be distributed more or less similar to respective BTE, possess smaller bias and variance and, as a first and conservative approximation, the equivalent number of degrees of freedom of ARE is n=N/2p; their approximate confidence bounds can be found in the same manner as for non-parametric spectral estimates (Bendat and Piersol, 1986).

Finally, ARE of C, H, F, and P proved to be much better than respective BTE for one MA model of order one and better than BTE for an MA(3) model with non-monotonous s and C

Bendat, J., and Piersol, A. (1986). Random Data. Analysis and Measurement Procedures. John Wiley and Sons, New York, 566 pp.
Koslov, J., and Jones, R. (1986). A unified approach to confidence bounds for the autoregressive spectral estimator. J. Time Series Analysis, 6, 141-161.
Kromer, R. (1970). Asymptotic Properties of the Autoregressive Spectral Estimator. Ph.D.Thesis. Stanford Univ.
Lütkepohl, H. (1985). Comparison of criteria for estimating the order of a vector AR process. J. Time Series Analysis, 6, 35-52.
Privalsky, V. (1985). Climatic Variability. Nauka Publishers, Moscow, 184 pp.

ON THE FORECASTING OF THE FLUCTUATIONS IN LEVELS OF CLOSED LAKES

Zelikin, M.I., Moscow State University, Moscow, USSR;
Zelikina, L.F., Central Institute of the Mathematical Economics, Moscow, USSR; Schulze, J., TH Leuna-Merseburg "Carl Schorlemmer", GDR

A closed lake has, by definition, no outlet. Consequently, the water level of a closed lake is due to random changes caused by various phenomena such as precipitation on and evaporation from the lake surface and the runoff from the sorrounding catchment.

Recently, the stochastic water balance method suggested by Kritskii and Menkel (1946) has been largely used in the long-term water level prediction for closed lakes. This report presents experimental data obtained from testing this method against the statistics of Caspian Sea level behaviour. The experiments show that the reliability of this prediction method is very low. Let's shortly describe the essence of the Kritskii-Menkel method (see Kritskii et al.(1975); Muzylev et al.(1982)). The majority of models used for level prediction involve the assumption that the change in the volume of water in a closed lake is described by the differential equation

$$dW/dt = V(t) + (P(t)-E(t))F(W) - Y(t), \qquad (1)$$

where $V(t)$ is the runoff rate, $P(t)$ is the rate of precipitation (volume per unit time per unit area), $E(t)$ is the evaporation rate (with the dimensions of $P(t)$), and $F(W)$ is the area of the lake, $Y(t)$ is the rate of unrecoverable removal. Also, let $e(t)=E(t)-P(t)$. The method under consideration uses the following discrete approximation of equation (1)

$$W_t = W_{t-1} + V_t - e_t(a(W_t+W_{t-1})/2+b) - Y_t, \qquad (2)$$

where W_t, V_t, e_t, Y_t are the annual average values of the corresponding parameters in the t-th year. It is assumed that $F(W) = aW+b$ and V_t and e_t are stationary Markov Processes, Y_t is a deterministic function. The forecasting consists of estimating the mathematical expectation EW_t of the process W_t and its variance DW_t.

To predict EW_t, the method in question suggests that, rather than equation (2), one should consider equation

$$\hat{W}_t = \hat{W}_{t-1} + \bar{V} - \bar{e}(a(\hat{W}_t+\hat{W}_{t-1})/2+b) - Y_t \qquad (3)$$

derived from (2) by averaging its coefficients as $\bar{V}=EV_t$, $\bar{e}=Ee_t$. Here it is assumed that the solution $\hat{W}$ to equation (3) is an estimate of the mathematical expectation EW_t of the solution to equation (2).

The method also concerns with computing significance levels, but this procedure is omitted here (the significance of a given level is the probability that it is exceeded).

The method in question was programmed in PL/1 on EC-1060 and computation of the Caspian Sea level in 50 years was performed for various initial years.

A hydrological forecasting is assumed to be justified if the deviation of a real process from the estimate EW_t does not exceed 0.67 σ which corresponds to 25-75% of significance levels. Among the 34 50-years predictions of Caspian Sea level with the initial years 1915, 1917, 1919 and so on up to 1981, all predictions fall beyond the interval 25-75% of the significance levels, 23 predictions fall beyond the interval 10-90%, while 11 failed to be even in the interval 1-99%.

In our opinion the major reasons for such poor quality of the forecasting are the following:

1. The assumption that the runoff and evaporation are stationary is not realistic.

2. A linear model of water level behaviour is too rough to adequately describe the real situation of many lakes.

3. The averaging of the coefficients of the stochastic difference equation (2) performed for estimating the mathematical expectation of its solution can yiel a very large error since the corresponding equations are stochastically unstable for an open set of the space of parameters defining the input and the coefficients of the equation.

To construct a more reliable model of the behaviour of a closed lake, one has, first of all, to remove the first two reasons. In what follows the assumption that the runoff and evaporation are stationary is abandoned.

If $F(W) = aW+b$, then equation (1) is transformed as

$$dW/dt = -ae(t)W + V(t) - be(t) - Y(t) \qquad (4)$$

Suppose that the random processes $V(t)$ and $e(t)$ can be represented as $-ae(t) = x(t)+p(t)$, $v(t) = y(t)+q(t)$, where $p(t)$ and $q(t)$ are the Ornstein-Unlenbeck processes; $E(p(t_1)p(t_2)) = \sigma^2\exp(-l|t_1-t_2|)$; the functionis $x(t)$ and $y(t)$ are deterministic components corresponding to the long-term and seasonal variation in the runoff and evaporation processes. By expanding the processes $p(t)$ and $q(t)$ in Wiener measure the mathematical expectation of the solution to equation (4) is shown to be of the form

$$EW(t) = (W_o+b/a)\exp(h^2(t) \quad)\exp(\int_o^t x(s)ds) + \qquad (5)$$

$$+ \int_o^t (y(r)-Y(r)-bx(r))\exp(h^2(t)-h^2(r))\exp(\int_r^t x(s)ds)\, dr,$$

where $h^2(t) = a^{-2}\sigma^2(t/l-l^{-2}+l^{-2}\exp(-lt))/2$. Using (5) it is shown that the averaging of the coefficients of linear differential equation (4) may lead to a large error.

Using the same formula it is also shown that the annual

average values of the runoff and evaporation levelling their seasonal variation lead to systematic errors which grow with the forecasting horizon. A similar phenomenon is observed when averaging is taken over any period.

In conclusion, we would like to say about new phenomena appearing if nonlinearities are taken into account. If we abandon the assumption that F depends on W linearly and take into account the nonlinearities caused by evaporation such as the increase in evaporation over shallow water, then, as can be seen from preliminary investigations, equation (1) has some peculiarities which yield a 'quantizing' effect. In other words, instead of one attraction level there is a set of values $w_1,\ldots,w_k$ such that the limit behaviour of the solution to equation (1) is described by a process of random walk among the corresponding levels.

REFERENCES

Kritskii, S.N., and Menkel', M.F. (1946). Some propositions of the statistical theory of natural reservoir level oscillation and their application to the study of the Caspian Sea level. In: Proceedings of the First Conference on Runoff Control. Academy of Sci. of the USSR, Moscow-Leningrad, 76-97.

Kritskii, S.N., Korenistov, D.V., and Ratkovich, D.Ya. (1975). Caspian Sea level oscillation. Nauka, Moscow, 153 pp.

Muzylev, S.V., Prival'skii, V.E., and Ratkovich, D.Ya. (1982). Stochastic models in engineering hydrology. Nauka, Moscow, 184 pp.

BIOLOGICAL MODELS AND GENETICS
(Session 16)

Chairman: P. Jagers

THE EQUILIBRIUM LAWS AND DYNAMIC PROCESSES IN POPULATION GENETICS

Yu.I. Lyubič
State University, Kharkov, USSR

The evolutionary operator V of panmictic autosomal infinite population without selection has the such form:

$$(Vx)_j \equiv x'_j = \sum_{i,k=1}^{n} p_{ik,j} x_i x_k \quad (1 \leq j \leq n), \quad (1)$$

where $x = (x_1, \ldots, x_n)$ is a state of a generation, and $x' = Vx$ is the state of the next generation. A state is a probability distribution of the types $1, \ldots, n$, into which the population is divided; $p_{ik,j} = p_{ki,j}$ is the probability of j-type offspring of i,k-parents. The space of states is the coordinate simplex $\Delta \subset R^n$. A classical example is the Hardy-Weinberg operator:

$$x'_1 = p^2, \; x'_2 = q^2, \; x'_3 = 2pq,$$

where $p = x_1 + \frac{1}{2}x_3$, $q = x_2 + \frac{1}{2}x_3$. A more general example is given by the multiallele Hardy-Weinberg operator:

$$x'_{ii} = p_i^2, \; x'_{ik} = 2p_i p_k \quad (1 \leq i < k \leq m),$$

where

$$p_i = x_{ii} + \frac{1}{2}\sum_{k \neq i} x_{ik}.$$

Here p_i is the probability of the allele A_i in the genofond; x_{ik} is the probability of the zygote genotype $A_i A_k$. For the genes there exist laws of conservation of probabilities: $p'_i = p_i$, by virtue of which the stationary principle $V^2 = V$ is valid.

In the 20-ties S.N. Bernstein formulated (and solved for $n \leq 3$) the problem of description of all the stationary operators. I think that this problem has a biological meaning only when there exists such a set of invariant linear forms:

$$p_i = \sum_{k=1}^{m} \pi_{ik} x_k \quad (1 \leq i \leq m) \quad (2)$$

that the operator V can be represented in a such form

$$(Vx)_j = \sum_{i,k=1}^{m} c_{ik,j} p_i p_k \quad (1 \leq j \leq n). \quad (3)$$

In that case I say that the population posesses the stationary genic structure (s.g.s.), in this case the condition $V^2 = V$ is valid automatically.

To explain the above-stated we note that the invariant linear forms, that is, the linear forms satisfying the relation $f(Vx) = f(x)$ $(x \in \Delta)$, form the linear space J_V. That space contains the cone C_V of non-negative invariant linear forms. This is a corporal cone with non-empty interior and a finite number of generators $p_1, \ldots, p_m$. It is these generators that must form the set number (2), and having in mind the subsequent probabilistic interpretation we will accept such normalization: $\max_k \pi_{ik} = 1$. The idea is that $p_1, \ldots, p_m$ must be the gene probabilities unchanged from one generation to another. The formulas (2) describe meyosis (that is the formation of genofond) in a state x, and the formulas (3) describe a fertilization by random sample of independent parental genes. Such interpretation of formulas (3) is possible only when $c_{ik,j} \geq 0$. I was lucky to prove the latter inequality.

Finally I obtained the complete description of the stationary genic structure. The visible formulation of that result is connected with the requirement of normality consisting of three conditions:

1) absence of disappearing types that is of such j that $x'_j \equiv 0$; 2) external irreducibility that is absence of such pairs j_1, j_2 that x'_{j_1}, x'_{j_2} are proportional; 3) internal irreducibility that is absence of such α, β that all x'_j depend only on $x_\alpha + x_\beta, x_\gamma$ $(\gamma \neq \alpha, \beta)$. This requirement doesn't restrict generality as it can be provided by the way of a certain procedure of normalization eliminating possible redundancy of population description (for example, formal distinction between the

same genes).

Theorem. If a population possesses the stationary genic structure and is normal then its evolutionary operator belongs to one of the two forms:

I.
$$\begin{cases} x_i' = p_i^2 + 2\sum_{k\neq i} \theta_{ik} p_i p_k \quad (1 \le i \le m), & (4) \\ x_j' = 2\theta_j p_{i_j} p_{k_j}, & (5) \end{cases}$$

where
$$p_i = x_i + \sum_{j=m+1}^{n} \pi_{ij} x_j, \tag{6}$$

(π_{ij}) is a stochastic (by columns) $m\times(n-m)$ matrix, in every column of which exactly two elements are non-zero, namely in the column j: $\pi_{i_j j} \equiv \pi_j > 0, \pi_{k_j j} \equiv \bar{\pi}_j = 1-\pi_j > 0$ $(i_j < k_j)$; and all the pairs (i_j, k_j) $(m+1 \le j \le n)$ are distinct; furthermore, all the $\theta_{ik} \ge 0, \theta_j > 0$; the following relations are valid:

$$\theta_{i_j k_j} + \pi_j \theta_j = \theta_{k_j i_j} + \bar{\pi}_j \theta_j = \frac{1}{2} \tag{7}$$

and
$$\theta_{ik} = \theta_{ki} = \frac{1}{2} \tag{8}$$

for the pairs (i,k) $(i<k)$, distinct from (i_j, k_j) $m+1 \le j \le n$. In this case I say that the population possesses the elementary genic structure (e.g.s.). The identity

$$\sum_{i=1}^{m} p_i = 1 \tag{9}$$

is the criterion of e.g.s. (in the presence of s.g.s.).

II. The coordinates of a state form a rectangular matrix

$$(x_{ik}) \quad (1 \le i \le l, 1 \le k \le r, lr = n, l > 1, r > 1)$$

and
$$x_{ik}' = p_i q_k, \tag{10}$$

where
$$p_i = \sum_{k=1}^{r} x_{ik}, \quad q_k = \sum_{i=1}^{l} x_{ik}. \tag{11}$$

In this case

$$\sum_{i=1}^{l} p_i = 1, \quad \sum_{k=1}^{r} q_k = 1. \tag{12}$$

For $\ell = r = 2$ we obtain the so called Bernstein's quadrille operator.

Let us try to explain in genetic terms the hereditary mechanism in a normal population possessing s.g.s. The following model of normal e.g.s. is proposed. There are m alleles $A_1, \ldots, A_m$ in an autosomal locus which are able to form homozygotas $A_1A_1, \ldots, A_mA_m$. No other genotypes whatsoever exist in the population:

$$n \leq \frac{1}{2} m(m+1).$$

In the process of meyosis the homo- and the heterozygotas disintegrate into constituting them genes, though for the heterozygotas meyotic drive or gamete selection is taking place which results in gene representation in genofond with inequal, generally speaking, probabilities $\pi_j, \bar{\pi}_j$:

$$A_iA_i \to A_i \ , \ A_{i_j}A_{k_j} \to \pi_j A_{i_j} + \bar{\pi}_j A_{k_j} .$$

In the process of fertilization the coupling of genes A_{i_j}, A_{k_j} yields not only heterozygote $A_{i_j}A_{k_j}$ but two homozygotas $A_{i_j}A_{i_j}, A_{k_j}A_{k_j}$ as well. The probability of the heterozygote equals θ_j , the probabilities of the homozygotas equal $\theta_{i_jk_j}, \theta_{k_ji_j}$. Accordingly

$$\theta_j + \theta_{i_jk_j} + \theta_{k_ji_j} = 1 .$$

The above-mentioned non-Mendelian way of homozygotas' origin may be interpreted as the death of one of two coupling genes and the simultaneous duplication of the other.

If genotype A_iA_k $(i<k)$ is forbidden (that is the pair (i,k) is different from every (i_j, k_j)) then in the process of coupling of genes A_i, A_k a heterozygote is not formed at all, and both homozygotes arise with equal probabilities $(\theta_{ik} = \theta_{ki} = \frac{1}{2})$.

The coupling of the same genes A_i, A_i always brings forth a homozygote A_iA_i .

The deviations from the Mendelian model in the processes of meyosis and fertilization are balanced in such a way that gene probabilities in the genofond remain constant: $p_i' = p_i$ $(1 \leq i \leq m)$. This is provided by the relations (7) and (8) which are naturally to be called the equations of gene balance. According to these equations the deficiency of one gene in the genofond which is due to meyotic probability shift is compensated by duplication in fertilization. The combination $\theta_{i_j k_j} + \pi_j \theta_j$ is contribution to the probability A_{i_j} in the next generation due to the individuals resulting from coupling of genes A_{i_j}, A_{k_j}. The analogous meaning has the combination $\theta_{k_j i_j} + \bar{\pi}_j \theta_j$. According to the equations of gene balance these probabilities are equal. Therefore the ratio of probabilities for genes A_{i_j}, A_{k_j} (that is the conditional distribution supported by this pair of genes) doesn't change from generation to generation. The analogous interpretation may be given for the remaining pairs of genes A_i, A_k.

The multiallele Hardy-Weinberg operator is obtained if there are no forbidden genotypes, that is $n = \frac{1}{2}m(m+1)$ (m is number of alleles), alternate genes are equally probable in meyosis, that is $\pi_j = \bar{\pi}_j = \frac{1}{2}$ $(m+1 \leq j \leq n)$; only heterozygotas are formed in coupling of different genes, that is $\theta_j = 1$ $(m+1 \leq j \leq n)$, $\theta_{ik} = \frac{1}{2}$ $(1 \leq i, k \leq m)$.

The model of non-elementary normal s.g.s. is as following:

there are l male genes $A_1, \ldots, A_l$ and r female ones $B_1, \ldots, B_r$.

They give rise to $n = lr$ bisexual (hermaphrodite) zygotas

$$A_i B_k \ (1 \leq i \leq l, \ 1 \leq k \leq r).$$

According to the formulas (10) describing the procedure of random independent sampling of genes A_i, B_k out of male and female sections of the genofond in the process of the coupling of genes belonging to the different sexes A_i and B_k forms the zygota $A_i B_k$. In meyosis the zygota $A_i B_k$ breaks up into genes A_i and B_k entering the male and female sections of genofond respectively. Each gene enters its own section with the probability 1.

The general Bernstein's problem, as I have already told appears to go beyond the framework of biological interpretations though purely mathematically it is interesting enough. There are some results in that direction but the complete solution has not yet been obtained.

From the biological point of view the following hypothesis appears to be important: if a stationary population is ultranormal (that is normal together with all its subpopulations), then it possesses s.g.s. Until recently that is proved just for some particular cases.

Let us note that some special class of non-associative algebras having been called Bernsteinian or stationary algebras is the essential apparatus in studying of the Bernstein's problem. Those are algebras with an identity $(x^2)^2 = \sigma^2(x)x^2$ where σ is the distinguished character.

Every evolutionary operator V is connected with a non-associative non-commutative algebra in the space R^n with the structural constants $p_{ik,j}$, that is with an evolutionary algebra. The stationary principle $V^2 = V$ is the equivalent to the stationarity of the algebra. Another important class of genetic algebras was introduced by Etherington in the 30-ties. These algebras occur naturally in studying of dynamic processes.

Probability properties of structural constants

$$p_{ik,j} \geq 0, \quad \sum_j p_{ik,j} = 1 \tag{13}$$

leaves its mark in the nature of evolutionary algebra. I call an algebra satisfying the condition (13) the stochastic algebra. The finite Markov chains are included in the structure by defining

$$p_{ik,j} = \frac{a_{ji} + a_{jk}}{2}$$

where (a_{ji}) is the matrix of the chain.

Dynamics of population with an evolutionary operator V is described by its iterations (by the consequent squari in the terms of algebra). The equilibrium is achieved in the first generation of offsprings for every stationary population. Among the infinite processes a purely recombination process in a multiallele autosomal population is most thoroughly studied. A Reiersöl algebra corresponds to an evolutionary operator of recombination process on the level of gametas. Every such algebra is connected with a table of genes:

$$(a_{ik})_{1 \le i \le l,\ 1 \le k \le m_i}, \tag{14}$$

whose i -line is formed by the i-locus alleles. A distribution of probabilities $\tau(U|V)$ is given on the set of partitions $U|V$ of the set of loci $L = \{1,\dots,l\}$. Statistically it describes all possible crossingovers and hence is called linkage distribution.

The formula (14) defines a set of gametas

$$g = a_{1k_1} a_{2k_2} \dots a_{lk_l}$$

that is the basis of algebra. Each partition $U|V$ defines subgametas

$$g_U = \prod_{i \in U} a_{ik_i}, \quad g_V = \prod_{i \in V} a_{ik_i}.$$

The mappings $g \to g_U,\ g \to g_V$ are naturally extended to formal linear gamete combinations

$$G = \sum_g \xi(g) g$$

And after this Reiersöl algebra is introduced as:

$$G \times H = \sum_{U|V} \tau(U|V) \cdot \frac{1}{2} (G_U H_V + G_V H_U)$$

(the multiplication in the right side is verbal).

The above-described technique allows to prove convergence and equilibrium pretty easy and allows to cover equilibrium state manifold (Reiersöl, 1962). The equilibrium is characterized by the stochastic independence of loci. The subsequent development of Reiersöl method resulted in exact solution of the non-linear equation describing a recombination process with arbitrary number of loci and arbitrary linkage distribution (Lyubič, 1971). I also managed to obtain the exact lower estimate of the rate of convergence to an equilibrium in exponential scale:

$O(n^{\alpha} \varkappa^{n})$, where n is the number of generation,

α - an integer ≥ 0,

$$\varkappa \geq \left[\frac{l-1}{2}\right] / \left(2\left[\frac{l-1}{2}\right]+1\right). \qquad (15)$$

Here l is the number of loci as before. The right side of estimate (15) doesn't depend on the linkage distribution τ. There exists a formula for a given τ

$$\varkappa = \max_{K \subset L, |K|=2} \tau(K), \qquad (16)$$

where

$$\tau(K) = \sum_{U \supset K} \tau(U|V) \qquad (17)$$

is the probability of subset $K \subset U$ to be not parted during the crossingover. As the right side of the exact estimate (15) is less than $\frac{1}{2}$ so the trivial estimate $\varkappa \geq \tau(L)$ is better than (15), if $\tau(L) \geq \frac{1}{2}$. From the biological point of view the value $\varkappa = \frac{1}{2}$ divides the domains of positive and negative interference of locus pair For two loci i, k under positive interference the "linkage phase" is more probable the "repulsion phase", that is, $\tau(ik) > \tau(i|k)$, which apparently more often are met in real populations. In particular, this is the case if the crossingovers between the neighbouring loci in every chromosome are independent (such a linkage distribution is called a linear one). Coming back to the explicit evolution formula I'd like to add that by its structure the formula resembles the perturbation theory formulae in the theoretic physics, though it is not a series but a finite sum of rather great number of summands. These summands can be graduated by the special parameters, namely non-equilibrium measures of the 1st, 2nd and the following powers characterizing the order of approximity to equilibrium at the beginning of the process. If the non-equilibrium measures of high powers are not taken into account, the formula is considerably simplified.

The involved non-equilibrium measures enter the explicit formula through the linear combinations with the coefficients dependent on the linkage distribution only. The expressions for those coefficients have a complicated combinatorial structure determined by the dichotomic trees.

These methods and results can be extended to quite general genetic situations excluding selection (Kiržner-Lyubič, 1973). However, the general approaches are not known to the systems where recombination probability is dependent not just on the set of partition of loci but on the presence of one or another allele as well. In particular cases yielding to analysis one manages to employ contraction principle.

The evolutionary operator form is considerably influenced when natural selection is taken into account. The standard mathematical model which goes back to Fisher describes the selection in single m-allele locus. This model is a superposition of Hardy-Weinberg operator and Baies formulae. Namely, if p_i is a probability of A_i-allele $(1 \leq i \leq m)$ then in the next generation

$$p_i' = p_i W_i / W \quad (1 \leq i \leq m), \tag{18}$$

where

$$W = \sum_{i,k=1}^{m} \lambda_{ik} p_i p_k , \quad W_i = \sum_{k=1}^{m} \lambda_{ik} p_k , \tag{19}$$

λ_{ik} is a coefficient of genotype $A_i A_k$ fitness (therefore W_i is mean fitness of allele A_i and W is mean fitness of whole population).

The Fundamental Fisher Theorem states that mean fitness increases in the evolutionary process: $W' > W$, if the initial state is non-equilibrium. Modernizing proof of this theorem proposed by Kingman (1961) - Lyubič, Maistrovski and Ol'chovski established the inequality

$$W' \geq W + C\sigma^2, \tag{20}$$

where

$$\sigma^2 = \sum_{i=1}^{m} p_i (W_i - W)^2 \tag{21}$$

is alleles' mean fitness dispersion, $C = \frac{3}{4} (\max_{i,k} \lambda_{ik})^{-1}$.

This and some upper dispersion estimates of non-equilibrium allowed us to prove the general theorem of convergence of every trajectory to its equilibrium state. Let us note that relaxation processes' technique that had been developed earlier for the needs of computation mathematics was used here (1970). In general, the convergence rate is estimated as $O(\frac{1}{\sqrt{t}})$ (t is being time measured by the generations), however, the rate proves to be exponential, if in the limit state all the alleles with mean fitness equal to that of the population have the positive probabilities (we call such equilibrium states the regular ones).

Later Lyubič and Kun (1970) extended the described technique to multiallele autosomal system with additive selection. As it was proved by Evens (1969) the Fundamental Theorem is valid in this case but to prove convergence here we had to consider another increasing function, namely, gametas' probability distribution enthropy on the limit set of trajectories. Let us note that H-theorem already proved by the same authors is valid also in the case of recombination process which leads to the new proof of its convergence (Kun and Lyubič, 1980).

The selection system deviation from additivity may lead to arising of cyclic trajectories. The question of existence of more complicated attractors is still remaining open.

All the reported results are explicitly considered in the monograph by Lyubič [1] . The English translation is being prepared at the Springer publishing house.

REFERENCES

1. Lyubič Yu.I. The Mathematical Structures in Population Genetics. - Kiev, "Naukova Dumka", 1983.

COMMUNITY SIZE AND AGE AT INFECTION: HOW ARE THEY RELATED?

Angela R. McLean
Parasite Epidemiology Research Group
Department of Pure and Applied Biology
Imperial College
Prince Consort Road
London SW7 2BB
England

INTRODUCTION

This paper investigates the impact of changes in community size upon the epidemiology of an infectious disease. Throughout the paper measles is used as an example of a directly transmitted infectious disease. When planning a vaccination programme against measles in a developing country account must be taken of the so called 'window problem'. This is the problem caused by the fact that most newborns cannot be successfully immnised against measles because of the presence of transplacentally derived antibodies. These antibodies persist for 3 - 12 months. In many developing countries the average age at infection is 2 years or even younger, so a large proportion of a cohort will already have experienced measles before all of a cohort have lost their maternal antibody derived protection and can be successfully immunised. It is therefore difficult to find the optimal age for vaccination. The work described here forms part of an effort to build a mathematical model of disease transmission that could be used to assess the impact of different policies of mass vaccination in a developing country.

The type of model presented is a deterministic compartmental model which describes the changes in age prevalence of disease that take place over the course of time. Such models have already been extensively studied, and have been usefully applied to questions about optimal vaccination policy in developed countries (for example Anderson & May 1985 and Schenzle 1985). However, existing models make two assumptions that, although quite reasonable when modelling events in developed countries are harder to defend when trying to mimic events in the developing world. These two assumptions are (1) that there are negligible disease related deaths, and (2) that the population is of fixed size. There is ample evidence that case fatality rates are far from negligible in developing countries. For example Williams (1983) found a case fatality rate of 64% amongst Gambian infants, falling to a rate of 4% amongst 6 - 8 year olds. United Nations data (United Nations, 1983) shows that in Kenya the population is currently growing at an annual rate of 39 per 1000 contrasting sharply with an annual growth rate in the United Kingdom of 1.1 per 1000. Thus whilst it may be reasonable to assume that the population of the U.K. is of fixed size, the same cannot be said of Kenya, Thailand or many other developing countries.

The layout of the paper is along the following lines. Following description of the model attention focuses upon the definition of the force of infection - the element of the model which governs the rate at which new cases are generated. There then follows a section containing data on how age prevalence of measles is observed to change in growing communities. The results of the investigation then follow.

MODEL DESCRIPTION

As the model is compartmental the approach is to split the population under consideration into a number of classes and then describe the rates at which people move from one class to the next. This model splits the living community into the following five groups.

- $M(a,t)$ - Infants protected by antibodies derived across the placenta from their mothers. This protection lasts for 3 - 12 months. It is assumed that individuals in this class can neither be infected nor immunised.
- $X(a,t)$ - Susceptibles; people who are no longer protected by maternal antibodies, but have not yet had measles.
- $H(a,t)$ - Incubators; infected but not yet infectious.
- $Y(a,t)$ - Infectives; capable of passing the disease onto others.

Z(a,t) - Immunes; recovered from the disease and assumed to be immune to reinfection for the rest of their lives.

It has also proved useful to keep track of a subset of those individuals who have died.

E(a,t) - Excess Deaths; individuals who have died as a result of having had measles, who we would not yet expect to have died from some other cause.

The total population, consisting of the sum of the first five classes is denoted by N(a,t).

N(a,t) - Total population.

Bearing in mind the definitions of these 6 compartments, the differential equations describing the flows from one class to another are as follows:

$$\frac{\partial M}{\partial a} + \frac{\partial M}{\partial t} = -(\mu(a) + \delta)\, M(a,t) \qquad \text{Maternal Antibody Protected}$$

$$\frac{\partial X}{\partial a} + \frac{\partial X}{\partial t} = \delta M(a,t) - (\mu(a) + \lambda(a,t))\, X(a,t) \qquad \text{Susceptible}$$

$$\frac{\partial H}{\partial a} + \frac{\partial H}{\partial t} = \lambda(a,t)\, X(a,t) - (\mu(a) + \sigma)\, H(a,t) \qquad \text{Incubating}$$

$$\frac{\partial Y}{\partial a} + \frac{\partial Y}{\partial t} = \sigma H(a,t) - (\mu(a) + \alpha(a) + \gamma)\, Y(a,t) \qquad \text{Infectious}$$

$$\frac{\partial Z}{\partial a} + \frac{\partial Z}{\partial t} = \gamma Y(a,t) - \mu(a)\, Z(a,t) \qquad \text{Immune}$$

$$\frac{\partial E}{\partial a} + \frac{\partial E}{\partial t} = \alpha(a) Y(a,t) - \mu(a)\, E(a,t) \qquad \text{Excess Deaths}$$

$$\frac{\partial N}{\partial a} + \frac{\partial N}{\partial t} = -\mu(a)\, N(a,t) - \alpha(a)\, Y(a,t) \qquad \text{Total Population}$$

Boundary Conditions

$$M(0,t) = \int_0^L m(a')\, N(a',t)\, da'$$

$$X(0,t) = H(0,t) = Y(0,t) = Z(0,t) = E(0,t) = 0$$

The death rate $\mu(a)$ represents deaths from all causes other than measles and applies to all groups. An additional death rate $\alpha(a)$ represents case fatalities and applies to the infectious class alone. The other parameters then determine the speed at which individuals progress from one class to the next. People leave the class protected by maternal antibody to enter the susceptible class at a rate δ, thus the average duration of protection by maternal antibody is approximately $1/\delta$. The most important of these rates is $\lambda(a,t)$, the rate at which susceptibles become infected, which is called the force of infection. Then $1/\sigma$ and $1/\gamma$ are, respectively, the average duration of the incubation period and the average duration of infectiousness. The inclusion of the case fatality rate $\alpha(a)$ also affects the dynamics of the total population $N(a,t)$. Population growth is acknowledged in the boundary condition for the class protected by maternal antibody, and also in the definition of the force of infection.

The force of infection is the driving force of the model and it is in the definition of the relationship between the force of infection, the number of infectious people and the total size of the population that this paper's emphasis lies. Many of the studies which assume that the host population is of fixed size have used the age dependent version of the mass action principle and set lambda at age a and time t to be a weighted sum of all the infectious people.

$$\lambda(a,t) = \int_0^L \beta(a,a')\, Y(a',t)\, da' \tag{1}$$

May and Anderson (1985) have applied this definition to growing populations. Using this definition for a growing population implies the assumption that the force of infection rises linearly with increasing community size. In earlier work (McLean, 1985) the following relationship was used;

$$\lambda(a,t) = \frac{\int_0^L \beta(a,a')\, Y(a',t)\, da'}{\int_0^L N(a',t)\, da'} \tag{2}$$

This assumes that the force of infection is determined by the proportion of the community that are infectious. So the force of infection is independent of community size. It is clear that the true relationship between disease transmission and community size lies somewhere between the two extremes that have already been studied. There are of course a variety of ways to construct a function to act as a continuum between these two definitions, but the one chosen to be studied here is as follows:

$$\lambda(a,t) = \frac{\int_0^L \beta(a,a')\, Y(a',t)\, da'}{\left[\int_0^L N(a',t)\, da'\right]^{\varrho}} \tag{3}$$

where ϱ lies between 0 and 1. Setting $\varrho = 0$ recaptures definition (1), and setting $\varrho = 1$ recaptures definition (2), where the force of infection is independent of community size. So what does the parameter ϱ measure?

Existing studies using these types of models have tended to define quantities (for example infectives, $Y(a,t)$ as densities. As they have dealt with populations that are of fixed size, occasional use of the words 'number' and 'density' as interchangeable has not caused confusion. However when dealing with a growing population it becomes necessary to be quite clear about the distinction between the two. In this paper Y(a,t) is defined to be the number of infectious people of age a at time t. ρ, then, is a composite measure of two quantities; one physical and one sociological . The physical measurement describes the relationship between community size and population density. It is a measure of the way in which the area of a city changes as the number of people increases. The sociological component is more subtle, and is a measure of changes in peoples lifestyles that come about as their village grows into a city.

DATA PRESENTATION

Before going on to show the repertoire of behaviour that can be generated by varying this parameter ρ, data is considered that gives an indication of the range of phenomena observed. Because the parameter is a measure of the way in which age prevalence changes over the long term, it is hard to find data that sheds much light on the issue. However in the years 1883 to 1902 the Town Council of Aberdeen ruled that the reporting of cases of measles should be compulsory. The age stratification of the records kept is fine enough to show that over the course of the twenty years there was no change in the age distribution of cases, even though the population of the city increased by 50% (figure 1). This is evidence for the assigning of the value 1 to the parameter ρ, and having $\lambda(a,t)$ independent of community size N(a,t). However this is incompatible with the wealth of data showing that in urban areas the average age at infection is lower than in rural areas. Figure 2 shows data from two such studies that are from developing countries. They compare Dakar (Senegal) with a fishing village; and the flatter, more densely populated areas of Nepal with the Hills. Both data sets show that denser population leads to greater disease transmission.

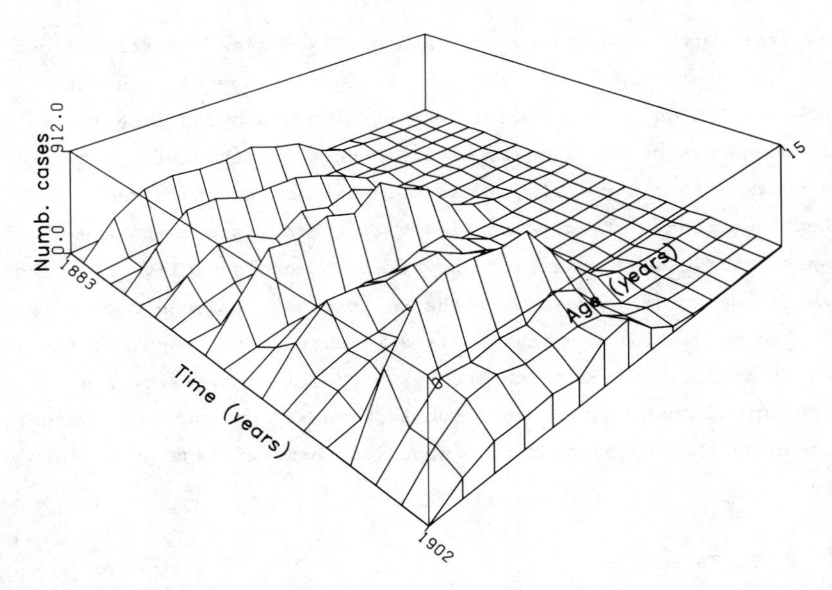

Figure 1. Age incidence of measles in Aberdeen 1883 to 1902. Data are from Wilson (1904)

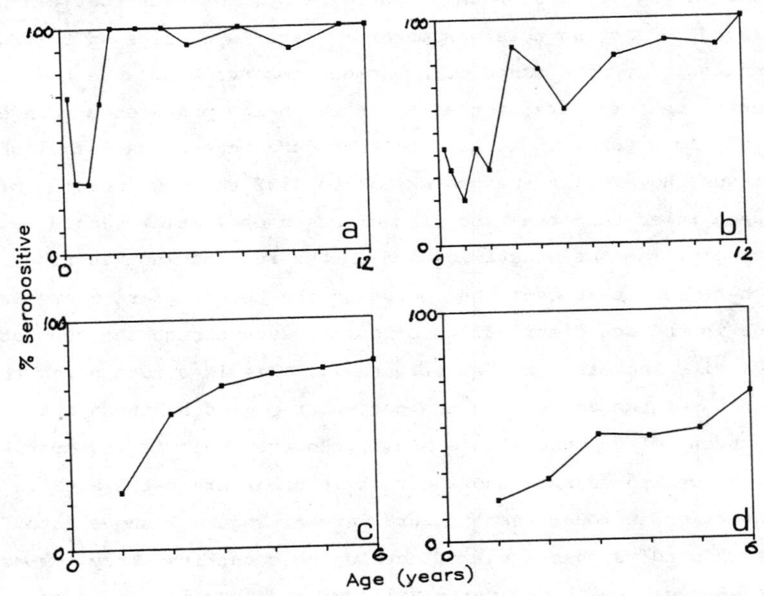

Figure 2. The urban/rural dichotomy in age prevalence of measles. (a) Dakar, Senegal and (b) Popenguine (a small fishing village also in Senegal), data are from Boue (1964). (c) More densely populated Terai areas of Nepal, (d) Hill areas of Nepal, data are from Brink and Nakano (1978)

RESULTS

How, then, can this model accommodate both these phenomena? Figures 3 and 4 show the different consequences of population growth when the parameter ρ is assigned two different values. Figures 3(a) and 3(b) are from simulations where the force of infection was defined with $\rho = 1$. Figure 3(a) shows the total number of cases over twenty years under two different rates of population growth. The population modelled with a lower birth rate grows at a rate of 25 per thousand per year; the faster growing population grows at a rate of 41 per 1000 per year. The only effect of the greater rate is to increase the number of cases. The inter-epidemic period is unaffected, and the serological profile at the end of the twenty years (figure 3(b)) illustrates that the age distribution of cases is also unaffected. If, however, the parameter ρ is given a value less than 1 different results are obtained. In figure 4 the same two sets of demographic data are used to determine the size of the underlying population, but the parameter ρ has been set to a value of 0.5. Here the rate of population growth does affect the inter-epidemic period. Figure 4(a) shows that the faster rate of population growth leads to a shortening of the time between epidemics. Taking slices through the serological profile at the peaks of the final epidemics of the 20 year period, figure 4(b) shows that the greater rate of population growth leads to a lower

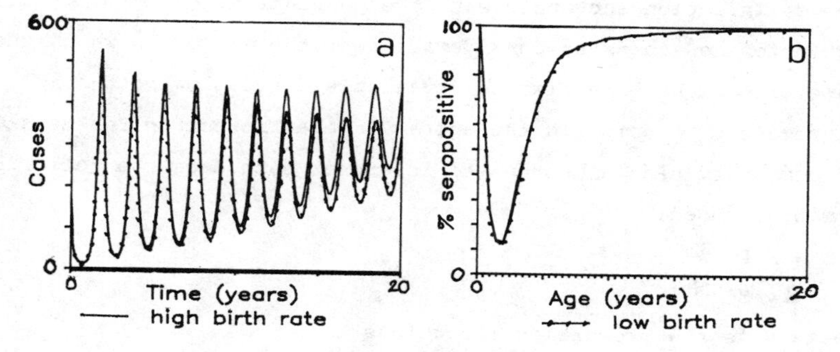

Figure 3. (a)Total cases through time for two different birth rates. The figure shows that when $\rho=1$ changes in birth rate do not change the inter-epidemic period. (b)Serological profiles at the end of 20 years for two different birth rates. The figure illustrates the fact that when $\rho=1$ the rate of population growth does not affect the age distribution of cases.

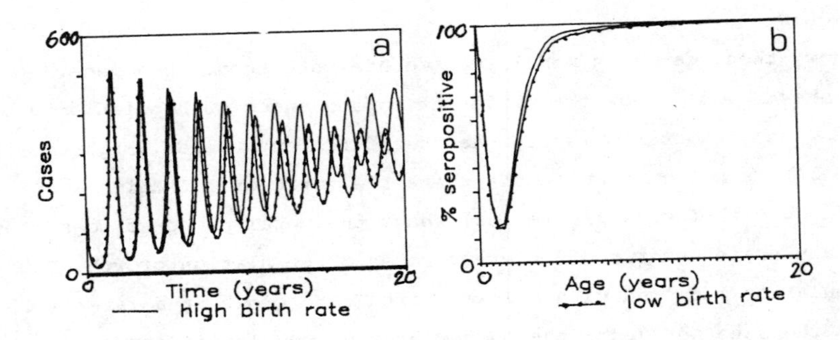

Figure 4. (a) Total cases through time for two different birth rates with $\rho=0.5$. The figure shows that when ρ is less than 1 changes in the birth rate do change the inter-epidemic period. (b)Serological profiles at the end of twenty years for two different birth rates. The figure illustrates the fact that when ρ is less than 1 the rate of population growth does affect the age distribution of cases.

average age at infection: that is, the force of infection has increased. Thus when ρ is set to 1 a situation like that seen in Aberdeen at the end of last century can be generated, where all that happens as a result of growing population is that the number of cases grows, with a fixed age distribution. However when ρ is less than one the model can generate something akin to to the situation where population growth shortens the inter-epidemic period and steepens the serological profile. This is a situation where the force of infection increases with the population.

Attention now turns to consideration of the manner in which the parameter ρ might be estimated. Available data allowing the measurement of ρ compares the average age at infection for measles with community size. In order to interpret such data the following argument is used:

$$\lambda \simeq 1 / A$$

$$\lambda = \beta y N^{(1-\rho)}$$

where y is the proportion infectious

$$\log \simeq \log (\beta y) + (1 - \rho) \log N$$

Figure 5(a) shows some such data collected in New York State at the beginning of this century. Figure 5(b) plots the log of 1/A against the log of community size. The best estimate for the slope of a line

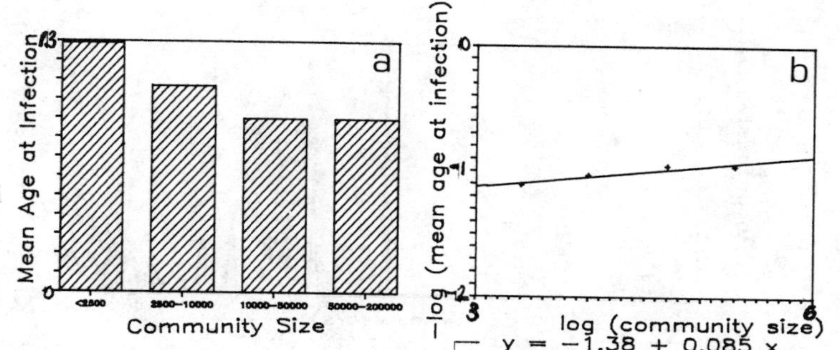

Figure 5. (a) Mean age at infection for different sizes of community in New York State 1918 - 1919. Data are from Fales (1928). (b) The same data under log transformations.

through the points is 0.085, giving an estimated value of ρ of 0.915. This data would imply that changes in community size have only a small effect upon the rate of disease transmission.

The final section of this paper presents some predictions made over 60 years, and considers the impact of vaccination. Figure 6 shows the results of an experiment that considers the consequence of variation in the parameter ρ to long term model predictions. The two

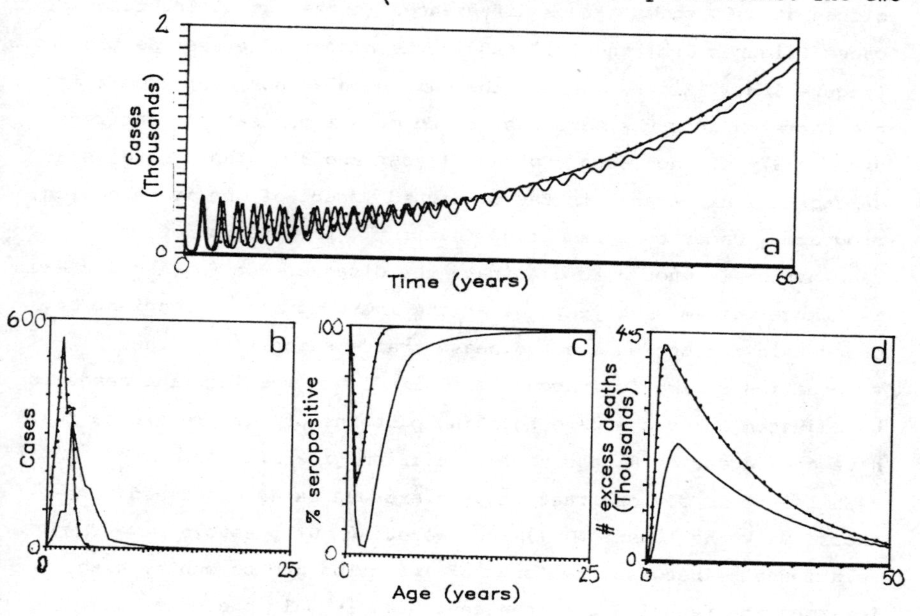

Figure 6. (a) Total cases through time over 60 years for two values of parameter ; — ρ = 1 and —•— ρ = 0. (b), (c) & (d) Age incidence, serology and numbers of excess deaths at time t = 60 years.

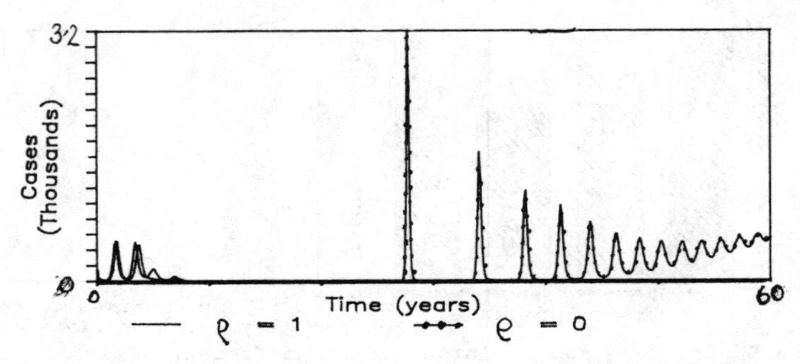

Figure 7. Total cases through time over the course of 60 years in a population subjected to immunisation of 97% of susceptibles at age 1 year 3 months.

extreme cases $\rho = 1$ and $\rho = 0$ are compared over a 60 year period. The number of cases is largely unaffected, but when ρ is zero the inter-epidemic period gets shorter and shorter and the rate of damping of the oscillations is quickened (figure 6(a)). Taking slices at t=60 shows gross differences in the age distribution of cases (figures 6(b) and (c)) and in the number of excess deaths (figure 6(d)). Figure 7 shows the same populations, but from time t=4 years onwards they are subject to mass immunisation. The regime used is 97% of each cohort at age 1 year and 3 months. There is a substantial difference in the predicted impact of the immunisation programme. Under the assumption that $\rho = 1$ this regime of vaccination is enough to eradicate the disease. But for $\rho = 0$ there is a huge epidemic 24 years after the introduction of vaccination. Under this assumption about disease transmission (ρ=0), the vaccination regime is inadequate to eradicate measles, and needs to be adjusted to accommodate shifting patterns of age prevalence that have come about as a result of population growth.

Whilst it is obvious that the two extreme cases discussed here should mark the 'best' (ρ=1) and 'worst' (ρ=0) possible relationships between the force of infection and community size, the intermediate function that has been used is only one of a whole range of possible families of functions that could encompass these two definitions. Dietz (1982) has suggested that a saturation function should be used for the definition of λ. When the population was small such a function would behave like this model does when $\rho=0$ (reflecting the urban/rural dichotomy), and once the population had grown very large it would behave as this model does when ρ is set to 1 (mimicking the behaviour that was recorded in Aberdeen).

It is hard to see a way in which to interpret field data so as to give an indication of which would be the best functional relationship to use. However having chosen a relationship between and N it should be possible to collect data that would allow the estimation of the value of the parameter. Ideally such data would be in the form of a series of serological profiles taken over the course of time in the same place. However this serology would only be meaningful if drawn from an unvaccinated population, and for measles such populations are disappearing fast. However, in the late 1950's and early 1960's quite a lot of serological profiles were collected in developing countries. It would be of great value to return to those communities now and take another serological profile so as to see if age prevalence has changed in the past twenty years. The alternative is to repeat the data collection exercise performed in New York State early this century and collate age prevalence data from a variety of unvaccinated communities of different size. For example this could be done in India where there is as yet no mass vaccination against measles. However such a study would need to be done now, before EPI Geneva persuades the Indian Health Authorities to introduce mass vaccination.

ACKNOWLEDGMENTS

The work presented here was supported in full by the Rockefeller Foundation to whom I am most grateful. I would also like to thank Professor R.M. Anderson and Professor R.M. May for useful discussions concerning this work.

BIBLIOGRAPHY

Anderson, R.M., & May, R.M. (1985) Vaccination and herd immunity to infectious diseases. Nature 318 323 -329

Boue, A. (1964) Contribution a l'etude serologique de l'epidemiologie de la rougeole au Senegal. Bull. Soc. Med. d'Afrique Noire. 9 253 -254

Brink E.W. (1978) Naturally acquired measles immunity in Nepal and Sri Lanka. Trop Geogr. Med. 30 103 - 113

Dietz K. (1982) Overall population patterns in the transmission cycle of infectious disease agents. In Population biology of Infectious diseases (R.M. Anderson & R.M. May) Springer-Verlag.

Fales, W.T. (1928) Am. J. Hyg. 8 759

May, R.M., & Anderson, R.M. (1985) Endemic infections in growing populations. Math. Biosci. 76 1 - 16

McBean, A.M. (1976) Evaluation of a mass measles immunization campaign in Yaounde, Cameroun. Transactions of the Royal Society of Tropical Medicine and Hygeine. 70 206 - 212

McLean, A.R. (1985) Dynamics of childhood infectious diseases in high birthrate countries. Lect. Notes. Biomath. 65 171 - 197

Schenzle, D. (1985) Control of virus transmission in age structured populations. Lect. Notes Biomath 57 171 - 178

United Nations (1983) U.N. Demographic yearbook, 1983.

Williams, P.J. (1983) Status of measles in The Gambia. Rev. Infect. Dis. 5 391 - 394

Wilson, G.N. (1904) Measles: its prevalence and mortality in Aberdeen. Report of the Medical Office of Health Aberdeen 41 - 50

BRANCHING PROCESSES AND NEUTRAL MUTATIONS

NERMAN O., Department of Mathematics, Chalmers University of Technology and University of Göteborg, Göteborg, Sweden

Supported by Swedish Natural Science Research Council

1. The Model

Consider the following scheme of single type asexual population multiplication. Independently of everything else each individual born into the population leads her *life* $\omega \in \Omega$, chosen according to a *life law* P on the *life space* $(\Omega, \mathcal{A})$.

Assume that ω tells us all the *reproduction ages*

$$\tau(1,\omega) \leq \tau(2,\omega) \leq \ldots \leq \infty,$$

and the *life span*

$$\lambda(\omega) \in [0,\infty]$$

of an individual with life ω. Let ξ denote the *point process of reproduction ages* having points at $\tau(1)$, $\tau(2)$,..., and let μ be the *reproduction intensity*

$$\mu = E[\xi].$$

Hitherto we have just described a single type *Crump-Mode-Jagers branching population*, cf. Jagers [-75] and Jagers & Nerman [-84]. The lives ω in this model are flexible entities that can be varied quite freely. If, for example, the individuals are supposed to be biological cells, ω can inform about the way an individual synthesizes its DNA, the duration of its cell-phases etc.

Here we suppose that ω informs about the genetic status of the offspring,

$$\rho(1,\omega), \rho(2,\omega), \ldots \in \{0,1\}.$$

The interpretation intended is that $\rho(i,\omega) = 1$ if the i:th child carries the same allele at one fixed chromosome locus as her mother, while $\rho(i,\omega) = 0$ means that the i:th child is a mutant and carries an entirely new allele, different from all previous alleles in the population.

The aim of this talk is to sketch how general branching process theory can be applied to study the asymptotics of the genetical composition of exponentially growing populations of this kind.

All our results are from a forthcoming Ph.D. thesis by Ziad Taib [to appear]. Some closely related references are Pakes [-84], Sawyer [-77].

For the actual genetical background and some arguments for infinite allele models, see Kimura [-83], see also Kingman [-75] and Tavaré [-84] for more traditional neutral mutation models and for more references.

Certainly the role of this kind of oversimplified population models in genetics should not be overemphasized. Populations do not grow exponentially for ever, all mutations are not neutral, individuals do not lead their lives totally independent of each other a.s.f.. On the other hand more realistic models tend to be very complicated and their mathematics hard.

A nice feature of our model, which is not so common in mathematical genetic theory of similar kind, is that the mutation probability may vary with the age of the mother, as well as with the birth-rank of the individual. Observe also that the occurence of mutations among siblings may be mutually dependent in a completely general fashion.

Let us continue with more preliminaries. The individual reproduction point process ξ can naturally be viewed as a sum

$$\xi = \xi_m + \xi_n$$

of *mutants* ξ_m and *non-mutants* $_n\xi$ Similarly, with notation $\mu_m = E[\xi_m]$ and $\mu_n = E[\xi_n]$,

$$\mu = \mu_m + \mu_n.$$

The main purpose of the following assumptions is to guaarantee an asymptotic exponential population growth and composition stabilization, as time passes.

Basic assumptions

There exists an $\alpha \in (0,\infty)$, such that

$$\int_0^\infty e^{-\alpha u}\mu(du) = 1,$$

and

$$\int_0^\infty ue^{-\alpha u}\mu(du) < \infty.$$

μ is not concentrated on $\{0,h,2h,\ldots\}$ for any $h > 0$.

The individual reproduction has a finite variance

$$E[\xi([0,\infty))^2] < \infty.$$

2. A Measure of Evolutionary Speed.

The notion of a *stable pedigree law* $\tilde{P}$ corresponding to a supercritical branching law P was introduced in Nerman & Jagers [-84] and further discussed in Jagers & Nerman [-84]. The idea is that just like the age compositions are known to stabilize with time, so should the proportions of different family constellations. The limiting deterministic proportions turn out to be conveniently understood as $\tilde{P}$ - probabilities of events in a natural *space of pedigrees*.

Think of a branching population initiated by a finite number of individuals at time $t = 0$, properly constructed on a suitably complex sample space.

For each realisation of such a population, we can consider the proportion of individuals born during time $[0,t]$, which have the following properties

(1) their age at time t is in interval $[0,u]$.

(2) their lives are in some subset $A_0 \in \mathcal{A}$ of the life space.

(3) their birth ranks are j_1, i.e. they are the j_1:th oldest child of their sibships.

(4) their mothers' lives belong to another subset $A_1 \in \mathcal{A}$, their grandmothers' to $A_{-2} \in \mathcal{A}$, and so on up to

their (n-2)th great grandmothers, whose lives are in A_{-n}.

(5) their mothers' birth ranks are $j_2 \in N$, their grandmothers' j_3, and so on, up to the birth rank, j_n of their (n-3)th great grandmothers in their sibships.

(6) All other individuals x stemming from these (n-2)th great grandmothers similarly lead their lives in some $A_x \in \mathcal{A}$.

This proportion is then an almost well-defined measurable function of the realisation of the population, the only problem, which we skip here, is what to do with the asymptotically negligible group of individuals in the n first generations. It can be shown, Nerman & Jagers [-84], that almost surely on the set of non-extinction, the sketched proportion converges, as time t passes, to the constant

$$(1-e^{-\alpha u})P(A_0)E[e^{-\alpha\tau(j_1)};A_{-1}]\ E[e^{-\alpha\tau(j_2)};A_{-2}] \ldots . E[e^{-\alpha\tau(j_n)};A_{-n}] \prod_{\substack{x \neq \{0,-1,\ldots,-n\} \\ x \text{ being individuals} \\ \text{stemming from } -n}} P(A_x)$$

Variation of $u, j_1,\ldots,j_n$ and all A_x, for n fixed, in this expression determines a probability measure on a space which contains vectors with coordinates

a = the age of *ego*, the sampled individual,

ω_0 = the life of ego,

ω_{-1} = the life of ego's mother

⋮

ω_{-n} = the life of ego's n-2:th great grandmother,

i_1 = birth rank of ego,

⋮

i_n = the birth rank of ego's n-3:th great grandmother,

and life coordinates of anyone else stemming from ego's n-2:th great grandmother. It is convenient to think of these measures for different n as projections of one single probability measure $\tilde{P}$ on a space with an infinite line of ancestral lives and lives of any relative stemming from anyone of them.

The product structure of our expression shows that $\tilde{P}$ renders the age a, the extended ancestral lives (ω_{-1},i_1), $(\omega_{-2},i_2),\ldots$, and the lives of ego and her other relatives independent random quantities. Here a is exponentially distributed, all ancestral lives have distribution determined by

$$\tilde{P}(\omega_{-k} \in A_k; i_k = j_k) = E[e^{-\alpha\tau(j_k)}; A_k],$$

while other lives follow the original branching law P. The probability that ego or any of her direct ancestors is a mutant can be calculated from this:

$$\tilde{P}(\rho(i_k,\omega_k) = 0) = \sum_{j=1}^{\infty} \tilde{P}(\rho(i_k,\omega_{-k}) = 0,\ i_k = j) =$$
$$\sum_{j=1}^{\infty} E[e^{-\alpha\tau(j)}; \rho(j) = 0] = \int_0^{\infty} e^{-\alpha u}\mu_m(du)$$

for any k. Similarly the mean of any ancestral generation span is

$$\tilde{E}[\tau(i_k,\omega_{-k})] =$$
$$= \sum_{j=1}^{\infty} E[e^{-\alpha\tau(j)}\tau(j)]$$
$$= \int_0^{\infty} ue^{-\alpha u}\mu(du).$$

This (and the independences) suggest that the ratio

$$\int_0^{\infty} e^{-\alpha u}\mu_m(du)/\int_0^{\infty} ue^{-\alpha u}\mu(du)$$

is a natural measure of the speed of evolution. For details, and strict limit theorems we refer to Taib [to appear].

Let us just make two short remarks.

The sampling procedure is not so essential, if for example sampling is performed among living individuals or newborns, ancestral lives will behave as sketched but of course the asymptotics of the age- and life-distribution of the sampled individual changes.

It is possible to show (without stable pedigree arguments) that if M_t is the number of mutant ancestors of a randomly sampled individual at time t, then, as $t \to \infty$,

$$\frac{M_t}{t} \longrightarrow \int_0^\infty e^{-\alpha u}\mu_m(du) / \int_0^\infty ue^{-\alpha u}\mu(du),$$

as $t \to \infty$, provided the population does not die out.

3. A Measure of Relatedness

Think of an old exponentially growing stably composed branching population of the kind outlined. Suppose that we sample two individuals at random from the population born up to the sampling time, t say. Let φ_t be the conditional probability that they carry the same allele, given the evolution of the branching process. I shall try to sketch, how, given some extra conditions, φ_t can be approximated by a natural constant divided by the total population size, y_t, at time t.

Suppose that the sampling is performed in two steps. Then, in the second step we choose between $y_t - 1 \approx y_t$ individuals. Define φ to be the number of individuals carrying the same allele as the ego individual (not including himself) born before the sampling time. Thus φ is a function of pedigrees. In the average, averaging over the first sampling step, approximately $\tilde{E}[\varphi]$ of those sampled among second will be genetically identical to the first sampled, if sampling time is large, and if $\tilde{E}[\varphi] < \infty$. After some, not so trivial calculations it turns out that

$$\tilde{E}[\varphi] = (\int_0^\infty e^{-\alpha u}\mu_m(du))(\alpha\int_0^\infty e^{-\alpha t}E[y_{n,t}^2]dt) - 1,$$

where $E[y_{n,t}^2]$ is the mean square of the total population born up to time t in a branching process initiated by a single newborn ancestor at time $t = 0$, but with individual reproduction ξ_n instead of ξ.

This approximation can be rigorously performed with the

help of branching processes counted by general characteristics, a tool that is actually used in the very derivation of the stable pedigree law $\tilde{P}$, Nerman & Jagers, [-84]. A necessary and sufficient extra condition for the finiteness of $\tilde{E}[\varphi]$ is that

$$\int_0^\infty e^{-\frac{\alpha}{2}u} \mu_n(du) < 1.$$

Thus mutations must not be to rare. Again we refer to Taib [to appear] for details and just make two remarks.

It is possible to use urn-ball type arguments for heuristic derivation of the approximation above.

If sampling is performed among living individuals and z_t and $z_{n,t}$ denote the number of living individuals in the original ξ-generated and the new ξ_n-generated branching population respectively, then the probability of sampling the same allele approximately equal

$$\frac{(\int_0^\infty e^{-\alpha u}\mu_m(du))(\int_0^\infty e^{-\alpha u}E[z_{n,t}^2]dt)}{z_t\int_0^\infty e^{-\alpha u}P(\lambda>u)du},$$

as can be rigorously proved by the same technique (and best understood by urn-ball arguments?).

4. The Old Alleles

For some purposes, for example for the study of when the first k-mutant appeared, a mutant with exactly k-1 mutant ancestors, or for the study of the very oldest alleles present in the population, it is convenient to think of the population in terms of a sort of macro individuals. By a *macro individual* we mean the entire set of individuals that carry one and the same allele, and we think of the life of a macro individual as the set of their lives. Certainly it is natural to let the creation time of an allele be the birthtime of the corresponding macro individual and we can naturally think of a macro life span - Λ say, as the time elapsed between the creation of an allele and its disappeareance from the population.

The subject now considered is the asymptotics of the times when the oldest alleles in the population first appeared.

Here the tail of the distribution of Λ certainly plays a key role. Since Λ can be identified as the extiction time of an ordinary branching process with ξ_n-reproduction and λ-life spans, we can lean on well-known results (cf. Asmussen & Hering [-83])

$$P(\Lambda > u) \sim c_1/u, \text{ as } u \to \infty,$$

under

<u>Condition A</u>.

$$\mu_n(\infty) = E[\xi_n([0,\infty))] = 1,$$

$$c_1 = 2(\int_0^\infty u\, \mu_n(du))/\mathrm{Var}[\xi_n[0,\infty)] \in (0,\infty),$$

and

$$u^2 P(\lambda>u) \longrightarrow 0, \text{ and } u^2\mu_n[u,\infty) \to 0, \text{ as } u \to \infty.$$

Moreover, in the case $\mu_n(\infty) < 1$,

<u>Condition B</u>

There is a $\gamma < 0$; $\int_0^\infty e^{-\gamma u}\mu_n(du) = 1$, $\int_0^\infty ue^{-\gamma u}\mu_n(du) < \infty$,

$\int_0^\infty e^{-\gamma u}P(\lambda>u)du < \infty$ and $E[\int_0^\infty e^{-\gamma u}\xi_n(du) \log^+(\int_0^\infty e^{-\gamma u}\xi_n(du))] < \infty$,

another nice tail behaviour of Λ holds true:

There is then a $c_2 > 0$ such that

$$P(\Lambda > u) \sim c_2 e^{\gamma u}, \text{ as } u \to \infty.$$

The exponential increase in the inflow rate of new alleles and the sketched tails of their survival times now cooperate so that under Condition A the point process of all 'ages' of the present alleles, translated by the population size determined quantity

$$\frac{\log z_t - \log\log z_t}{\alpha} + \frac{\log((\int_0^\infty e^{-\alpha u}\mu_m(du))c_1\alpha \;/\int_0^\infty e^{-\alpha u}P(\lambda>u)du)}{\alpha}$$

can be approximated by a Poissonprocess with intensity density function $u \to e^{-\alpha u}$. While under Condition B the appropriate random translation,

$$u = \frac{\log z_t}{\alpha-\gamma} + \frac{\log(c_2 \int_0^\infty e^{-\alpha u}\mu_m(du)/\int_0^\infty e^{-\alpha u}P(\lambda>u)du)}{\alpha-\gamma}$$

renders the ages of the oldest alleles an approximate Poisson process with intensity density function $u \to e^{-(\alpha-\gamma)u}$.

These results are more or less direct consequences of an extension of example c in Jagers & Nerman [-84B] applied to the sketched macro process. Again Taib [to appear] will give the arguments needed.

References

Asmussen, S., Hering, H. Branching Processes. - Birkhäuser, Boston, 1983.

Jagers, P., Nerman, O. The growth and composition of branching populations. Adv. Appl. Prob. 1984, p. 221-259.

Jagers, P., Nerman, O. Limit theorems for sums determined by branching processes and other exponentially growing processes. Stoch. Proc. Appl. 17. 1984B p. 47-71.

Kimura, M. The Neutral Theory of Molecular Evolution. - Camebridge University Press. - Cambridge, 1983.

Nerman, O., Jagers, P. The stable doubly infinite pedigree process of supercritical branching populations. Zeitschrift fur Wahrscheinlichkeitsteorie 65. 1984. p. 445-460.

Pakes, T. Coloured branching processes. (Unpublished). 1984.

Sawyer, S. Asymptotic properties of the equilibrium probability of identity in a geographical structured population. - Adv. Appl. Prob. 9. 1977. p. 268-282.

Taib, Z. Forthcoming Ph.D. Thesis. Dept. of Mathematics, University of Göteborg. - Göteborg.

Tavaré, S. Lines of descent and genealogical processes and their applications in population genetics. Theor. Pop. Biol. 26. 1984. p. 119-164.

Kingman, J.F.C. The Mathematics of Genetic diversity. SIAM - Washington. 1980.

Jagers, P. Branching Processes with Biological Applications. Wiley, Chichester. 1975.

LIMIT THEOREM FOR SOME STATISTICS OF MULTITYPE GALTON-WATSON PROCESS.

Badalbaev I.S., Mukhitdinov A.
Institute of Mathematics AS UzSSR, Tashkent, USSR.

Let $\{Z^i(t)=(Z^i_1(t),\dots,Z^i_r(t)) / Z^i(0)=e_i, t\geq 0\}$, $r\geq 2$, be a r-type indecomposable, non-periodic Galton-Watson (G-W) process with the matrix of means $A=\|EZ^i_j(1)\|$, and let m be the Frobenius-Perron root of A with associated left and right eigenvectors v, u. It is well known, that under the condition of finiteness of second moments if $m>1$ then $Z^i_j(t)m^{-t}\rightarrow w^i v_j$, $j=1\div r$, with probability 1 (wp1) and the distribution of w^i has a jump of magnitude $q^i<1$ at the origin and a continuous density function on the set of positive real numbers. This fact indicates that asimptotical behavior of $Z^i(t)$ process trajectory is mainly regulated by the m and corresponding eigenvectors. Therefore, the problem of statistical estimation of this parameters is very important. Estimation of Perron root for twotype processes was first considered by Badalbaev (1976). At the present time, there are many works concerned with the estimation of parameters in multitype processes.

Introduce statistics:

$$\Delta Z^i(t)=(\Delta Z^i_2(t),\dots,\Delta Z^i_R(t)),\quad \Delta Z^i_j(t)=v_j^{-1}Z^i_j(t)-v_1^{-1}Z^i_1(t) \qquad (1)$$

$$\Delta_t Z^i=(\Delta_t Z^i_1,\dots,\Delta_t Z^i_r) \qquad \Delta_t Z^i_j=Z^i_j(t+1)-mZ^i_j(t) \qquad (2)$$

$$\hat{v}^i(t)=(\hat{v}^i_2(t),\dots,\hat{v}^i_r(t)),\quad \hat{v}^i_j(t)=Z^i_j(t)/(Z^i_1(t)+1) \qquad (3)$$

$$\hat{m}^i(t)=(\hat{m}^i_1(t),\dots,\hat{m}^i_r(t)),\quad \hat{m}^i_j(t)=Z^i_j(t+1)/(Z^i_j(t)+1) \qquad (4)$$

This statistics of $\hat{v}^i(t)$ is the estimation of v vectors direction i.e. the vector $(1,v_2/v_1,\dots,v_r/v_1)$, and $\hat{m}^i_j(t)$

of m. The necessity of observation of this functional (1) outcomes from the study of $\hat{v}^i(t)$, but it has also an independent interest: for supercritical Galton-Watson processes limit distribution for functional $(\bar{a}, Z^i(t))$, where vector $\bar{a}$ satisfies the condition $(\bar{a}, v)=0$, $\bar{a}\neq 0$, were obtained by Kesten H. and Stigum B.P. (1966). But here the j-th component of the vector $Z^i(t)$ is scalar product $(\bar{a}^j, Z^i(t))$, where $\bar{a}^j=(-1/v_1, 0.., 1/v_j, 0..)$, $(\bar{a}^j, v)=0$, $j=2\div r$ and set $\{\bar{a}^j\}_{j=1}^{r}$ is the base of space orthogonal to the direction of v, i.e. the joint distribution of $(\bar{a}^j, Z^i(t))$, $j=2\div r$, is studied therefore the result of Kesten H. and Stigum B.P. (1966) folloves from our results.

For the functional (1) and (2) asimptotical behaviors of the first and second moments are established (we are not giving them its becouse they are very cumbersome) which depend on all eigenvalues of A and on their eigenvectors and affiliated vectors (see (5)).

Let $m=m_1>|m_2|\geqslant|m_3|\geqslant\ldots\geqslant|m_n|$ be eigenvalues of matrix A. We will also denote by $N(k)=(N_1,\ldots,N_k)$ the k-dimensional random vector with normal distribution and means equel to zero.

Theorem 1. Let $m>1$, $E|Z^i(1)|^2<\infty$, $Z^i(t)\neq 0$, $i=1\div r$ and $t\to\infty$.

1°. If $m\geqslant|m_2|^2$ then

$$(\Delta Z^i_j(t) m^{-t/2} f_j^{-1}(t),\ j=2\div r) \xrightarrow{d} N(r-1)\sqrt{w^i}$$

$$((\hat{v}^i_j(t)-v_j/v_1) m^{t/2} f_j^{-1}(t), j=2\div r) \xrightarrow{d} (N_1 v_2, .., N_{r-1} v_r)/v_1\sqrt{w^i}$$

where

$$f_i(t)=\begin{cases}1 & \text{if} \quad |m_2|^2<m \\ t^{k-1/2} \quad 0\leqslant k\leqslant r-1 & \text{if} \quad |m_2|^2=m\end{cases}$$

2°. If $m<|m_n|^2$ then

$$(\Delta Z^i_j(t)|m_{l(j)}|^{-t} t^{-k-1},\ j=2\div r) \longrightarrow (\Delta Z^i_j,\ j=2\div r) \quad \text{wp1}$$

where limit random vector $\Delta Z^i=(\Delta Z^i_j,\ j=2\div r)$ has finite second moments and

$$m^t((\hat{v}^i_j(t)-v_j/v_1)|m_{l(j)}|^{-t}t^{k-1},j=2\div r)\longrightarrow(\Delta v^i_j,j=2\div r) \quad \text{wp1}$$

besides $\Delta v^i_j=v_j\Delta Z^i_j/(v_1w^i)$. Here $m_{l(j)}$- the eigenvalue of A defining the asimptotic behavior $E(\Delta Z^i_j(t))^2$.

Let us divide the set of types into two subsets like that:

$$F=\left\{j:\ E(\Delta_t Z^i_j)^2=O(|m_{l(j)}|^{2t}t^{2k-2})\right\}$$
$$M=\left\{j:\ E(\Delta_t Z^i_j)\ ^2=O(m^t t^{2k-1})\right\} \tag{5}$$

Let denote the number of elements in F and in M by the same letters.

Theorem 2. Let conditions of Th.1 be hold. Then

$$(|m_{l(j)}|^{-t}t^{-k-1}\Delta_t Z^i_j,\ j\in F)\longrightarrow(\Delta \dot{Z}^i_j,\ j\in F) \quad \text{wp1}$$

where the limit random veztor $\Delta \dot{Z}^i=(\Delta \dot{Z}^i_j,\ j\in F)$ has finite sekond moments, and

$$m^t((\hat{m}^i_j(t)-m)|m_{l(j)}|^{-t}t^{-k-1},\ j\in F)\longrightarrow(\Delta\hat{m}^i_j,\ j\in F) \quad \text{wp1}$$

here $\Delta\hat{m}^i_j=\Delta\dot{Z}^i_j/w^i$, $j\in F$ and

$$(\Delta_t Z^i_j m^{-t/2}g^{-1}_j(t),\ j\in M)\xrightarrow{d} N(M)\sqrt{w^i}$$

$$((\hat{m}^i_j(t)\ -m)m^{t/2}g^{-1}_j(t),\ j\in M)\xrightarrow{d} N(M)/\sqrt{w^i}$$

where $g_j(t)$- some polynomial of order less than r. For any $j\in F$ and $s\in M$ under $t\to\infty$

$$E((\hat{m}^i_s(t)-m)m^{t/2}g^{-1}_s(t)(\hat{m}^i_j(t)-m)m^t|m_{l(j)}|^{-t}t^{-k-1})\longrightarrow 0$$

$$E(\Delta_t Z^i_s m^{-t/2}g^{-1}_s(t)\Delta_t Z^i_j|m_{l(j)}|^{-t}t^{-k-1})\longrightarrow 0$$

Let spectrum eigenvalue of A contain m_l: $|m_l|^2>m$ and m_s: $|m_s|^2\leqslant m$. Then, as it follows from Th.2, the set of types is divided into two subsets, such that limit distribution of estimations which are construeted by meanings of processes using types belonging to one of these classes, are essentially differented from the similar estimates, constracted by using number of particals, type of which belongs to the other class; covariance between these statistics converges to zero. It is proved that if $|m_l|^2 > m$ for some l, $2 \leq l \leq n$, then there exist at least one type such, that $E(\Delta Z^i_j(t))^2 = O(|m_2|^{2t} t^{2k-2})$ (similar is true also for $\Delta_t Z^i_j$). This shows that in the asimptotical behavior of the trajektory of the process a big role is played also by eigenvalue m_l: $|m_l|^2 > m$ and it would be very usefull to estimate these values using the meanings of the process $Z^i(t)$. Since the general proposition is too big we shall give the result for twotype processes.

Theorem 3. Let $Z^i(t)$ be twotype, supercritical Galton-Watson process satisfing the conditions of Th.1, and $m_2^2 > m$. Then $\Delta Z^i_2(t+1)/\Delta Z^i_2(t)$ is consistent estimate of m_2.

REFERENCES:

1. Badalbaev I.S.(1976). On one estimate of the eigenvalue of means matrix of twotype branching process. Izvestiya AN UzSSR, 8, №5, 8-13.
2. Kesten H., Stigum B.P.(1966). Additional limit theorem for indecomposable multidimensional Galton-Watson process. Ann. Math. Stat., 37, №6, 1463-1481.

THE REGULARITY OF METAPHASE CHROMOSOMES ORGANIZATION IN CEREALS

N.L.Bolsheva, N.S.Badaev, E.D.Badaeva, O.V.Muravenko
Institute of Molecular Biology, USSR Acad. of Sci., Moscow, USSR
Yu.N.Turin
Moscow State University, Faculty of Mathematics and Mechanics, Moscow, USSR

A metaphase chromosome is an oblong body, divided into two arms by centromeric constriction. The specific heterochromatic segments more intensively stained are arranged along chromosomes. Generally heterochromatic segments consist of highly repeated DNA sequences. The functions of heterochromatin is not yet known. It assumed that heterochromatin may influence upon regulation of genome function.

Each individual chromosome may be described from the morphological point of view by its sizes, the arms length ratio and the arrangement of the heterochromatic segments along the chromosome.

As it has been shown by Lima-de-Faria (1980) a lot of specific regions have a definite location with respect to the ends of an arm. We noticed that the arrangement to heterochromatic segments in some chromosome arms in cereals is highly similar. This observation and evidences of Lima-de-Faria suggest that the distribution of the heterochromatin along chromosome arms is non-random. To clarify this problem we investigated the distribution of the heterochromatin along chromosomes in cereals.

The chromosomes of durum and aestivum wheats, rye, barley, hexaploid and octoploid triticales were studied. 20-60 chromosomes of each morphological type where taken from the cells of different plant varieties and measured. In order to determine the localization of particular heterochromatic segment with respect to the ends of arm we calculated the mean ratio between a distance from the centromeric constriction to the heterochromatic segment and the length of an arm. Thus, the localization of heterochromatic segments in chromosome arms was expressed in numbers from 0 to 1, which made it possible to compare the distribution of heterochromatic segments in chromosome arms of a different length.

Figure 1 presents the positions of 126 heterochromatic segments in chromosome arms in cereals.

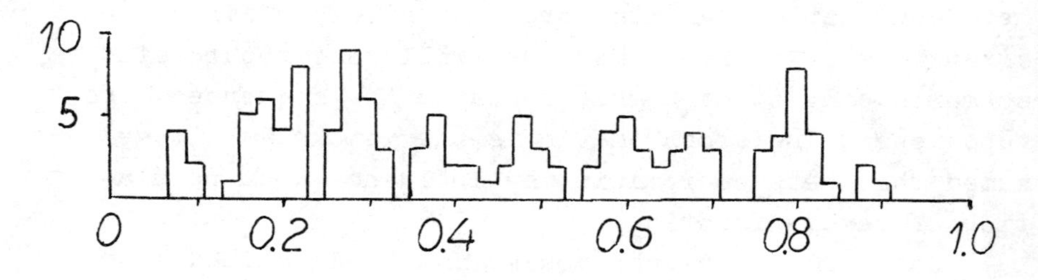

Denying the existence of regularity in the detected positions of heterochromatic segments means to advance the statistical hypothesis H_o that these values are random and distributed uniformly with the entire range from 0 to 1. We checked H_o using the David's (1950) criterion of "empty boxes". The region of observation was subdivided into "m" equal intervals. We used m = 50. Let us calculate a random number of those intervals which contain no observations. In the case of H_o the distribution of a random number of those intervals which contain no observations almost obeys the normal law with the mean "A", where "A" is a following expression: $A = me^{-n/m}$ and the dispersion "D" is a following expression: $D = me^{-n/m} - e^{-2n/m}$. In these expressions "n" is a total number of heterochromatic segments and "m" is a number of intervals. The normalized random value was 6.46 in our case, which rejected H_o with a high significance. Therefore, the distribution of heterochromatic segments along chromosome arms is not random.

A visual evaluation of the histogram presented on second slide suggests a periodicity in the distribution of heterochromatic segments along chromosome arm. The following approach to check this hypothesis was used. Let us represent the position "x" of each heterochromatic segment in the chromosome arm as: $X_i = K_i \Delta + \mathcal{E}_i$ where "k" is a total number of periods Δ and $\mathcal{E}$ is a residue. If there is a periodicity in the distribution under study and the length of a period is taken correctly, then residues "$\mathcal{E}_i$" will not be distributed uniformly along Δ, but will be accumulated at the edges of this segment. We tested various Δ with a small step in calculations with computer. A distinct difference from the uniformity was found at $\Delta = 0.1$, ($\chi^2_{10} = 39.5$), which rejects the hypothesis about the uniform distribution of residues with high significance. Figure 2 presents the distribution of residues "$\mathcal{E}_i$" for this case.

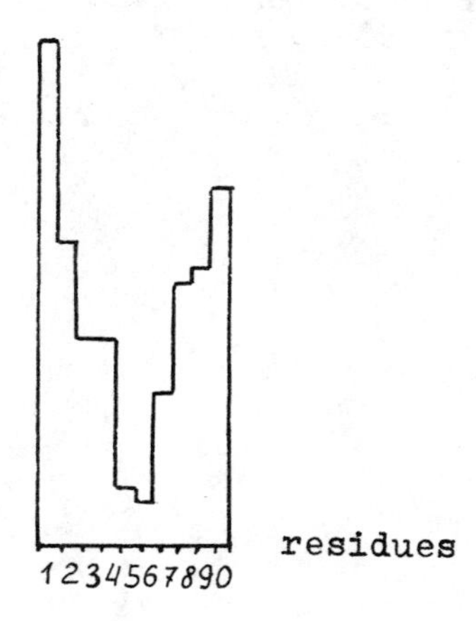

It may be concluded therefore that in the studied cereals, intersticial heterochromatin is most often located in 9 positions in chromosome arms. These positions reside at an identical distance from one another. At the same time, in an individual arm, heterochromatin may be revealed not in all of the positions, but only in some of them. It has been thus shown, that there exists a common principle of metaphase chromosome organization in studied cereals.

It should be noticed, that in the course of evolution the chromosome size and morphology were changed. Thus the chromosomes of A and D wheat genomes are much smaller than chromosomes of rye, barley and chromosomes of wheat B-genome. It's well known, that in the course of rye evolution there was structural rearrangement between 4th, 6th and 7th chromosomes. In spite of these changes a unique principle of the heterochromatin distribution along chromosome arms was preserved in the course of evolution. This fact indicates the functional significance of such chromosome organization.

David F.N. Two combinatorial tests of whether the sample has come from a given population. Biometrika, 1950, v. 37, p. 97-110.

Lima-de-Faria A. Classification of genes rearrangements and chromosomes according to the chromosome field. Hereditas, 1980, v. 93, p. 1-46.

THE GENEALOGY OF THE INFINITE ALLELES MODEL

Peter Donnelly
Department of Statistical Science
University College
Gower Street
London WC1E 6BT, U.K.

Simon Tavaré
Department of Mathematics
University of Utah
Salt Lake City
Utah 84112, U.S.A.

INTRODUCTION

Our purpose here is to describe a particular stochastic process associated with the genealogy of a (hypothetically infinite) population evolving according to the infinite alleles model of population genetics (on the so called diffusion time scale). The process represents a generalization of both Kingman's (1982b) coalescent, in accounting for mutation, and the n-coalescent with ages of Donnelly and Tavaré (1986) (see also Kingman 1982a and Watterson 1984) to the infinite case. We content ourselves with an outline of the dynamics and the distribution of the process. For a more detailed exposition and several applications see Donnelly and Tavaré (1987). For more of the genetic background and in particular a description of the infinite alleles model see for example Ewens (1979). Our process is associated with any of the models within the domain of attraction of the usual (Wright Fisher) diffusion approximation (see for example Griffiths 1980, to which our work is closely allied). Beyond this we make no specific assumptions about the nature of the reproductive mechanism.

THE PROCESS

Fix $t > 0$ and consider the genealogy of the population at the present time (time 0) with respect to the ancestral population at time $-t$. A certain proportion of the population, x_1 say, may share a common ancestor, without intervening mutation, at time $-t$. A further proportion x_2 say, may share a different common ancestor, again without intervening mutation between time 0 and time $-t$, and so on. Suppose there are k such groups (for $t > 0$, k will be a.s. finite) with frequencies $x_1,\ldots,x_k$. Each such group will be of the same genetic type as the corresponding ancestor. There will also be a number of types in the current population which do not appear in the ancestral population. Each new type will correspond to individuals who share a common ancestor, without intervening mutation, at some time $-r$, $0 < r < t$, that ancestor being a mutant. By assumption each new type will arise at a different time, so order (and label) the new types by age (i.e. time since first appearance), and denote the frequencies of the oldest, second oldest, ..., new type by $x_{k+1},x_{k+2},\ldots$. (For $t > 0$ and non zero mutation rate, there will be an infinite number of new types). For this value of t, describe the above genealogy by the point $(k;x_1,x_2,\ldots,x_k,x_{k+1},x_{k+2},\ldots)$ of $N_\infty \times \Delta$ (where $N_\infty = N\cup\{0,\infty\}$ and $\Delta = \{(x_1,x_2,\ldots) : 0 \leqslant x_i \leqslant 1,\ \sum_{i=1}^{\infty} x_i \leqslant 1\})$. As t varies this gives rise to a stochastic process $\{M_t;t \geqslant 0\}$, and we put $M_o = (\infty,0,0,\ldots)$.

The process $\{M_t\}$ has a special structure and particularly simple dynamics. It inherits these from the finite n-coalescents with ages of Donnelly and Tavaré (1986) which (as in Kingman 1982b) may be embedded (in two sorts of ways) in $\{M_t\}$. Denote by D_t the number of alleles in common between the population at time 0 and the ancestral population at time $-t$ (so that D_t is the value of k above). Then the process $\{D_t;t \geqslant 0\}$ is a death process with entrance boundary at ∞ and death rates $k(k + \theta - 1)/2$ from state k. (As usual θ denotes the scaled mutation rate.) Furthermore we may construct $\{M_t\}$ by putting $M_o = (\infty;0,0,\ldots)$ and $M_t = \mathcal{M}_{D_t}$ for

$t > 0$, where $\{\mathcal{M}_k, k = 0,1,\ldots\}$ is a discrete time Markov chain on $N_\infty \times \Delta$, independent of $\{D_t; t \geq 0\}$ with

$P(\mathcal{M}_{k-1} = (k - 1; x_1,\ldots,x_{i-1},x_i + x_j,\ldots,x_{j-1},x_{j+1},\ldots,x_k,x_{k+1},\ldots)|$
$\mathcal{M}_k = (k; x_1,\ldots,x_k,x_{k+1},\ldots)) = 2/k(k + \theta - 1)$, $1 \leq i < j \leq k$,

$P(\mathcal{M}_{k-1} = (k - 1; x_1,\ldots,x_{i-1},x_{i+1},\ldots,x_k,x_i,x_{k+1},x_{k+2},\ldots)|$
$\mathcal{M}_k = (k; x_1,\ldots,x_k,x_{k+1},\ldots)) = \theta/k(k + \theta - 1)$, $1 \leq i \leq k$.

Thus the times between changes of $\{M_t; t \geq 0\}$ depend only on the number of "old" types, and the changes take one of two simple forms: a "coalescence" of two old types or one of the old types becoming "new". The existence of the Markov chain $\{\mathcal{M}_k, k = 0,1,\ldots\}$ is established in Donnelly and Tavaré (1987).

The distribution of $\{D_t; t \geq 0\}$ is well known. See for example Tavaré (1984, equation (5.5)). The distribution, ν_k say, of $\mathcal{M}_k$ is concentrated on $\{k\} \times \Delta^1$ (where $\Delta^1 = \{(x_1,x_2,\ldots) : 0 \leq x_i \leq 1 , \sum_{i=1}^{\infty} x_i = 1\} \subseteq \Delta$) and has the following representation. Let V_k be a random variable having the beta density f_k given by

$$f_k(x) = \frac{\Gamma(k + \Theta)x^{k-1}(1 - x)^{\Theta-1}}{\Gamma(\Theta)(k - 1)!} , \quad 0 \leq x \leq 1 .$$

Let $(U_1,U_2,\ldots,U_k)$ be a random k-vector having a uniform distribution on the simplex $\{(u_1,u_2,\ldots,u_k); u_i \geq 0, u_1 + u_2 + \ldots + u_k = 1\}$ and let $Z_1,Z_2,\ldots$ be independent and identically distributed random variables with density f given by

$$f(x) = \Theta(1 - x)^{\Theta-1}, \quad 0 \leq x \leq 1 ,$$

and choose $(U_1,U_2,\ldots,U_k)$, V_k , and $\{Z_j\}$ to be mutually independent. Finally define the random element $X^{(k)}$ of Δ^1 by

$$X^{(k)} = (V_kU_1,V_kU_2,\ldots,V_kU_k,(1-V_k)Z_1,(1-V_k)(1-Z_1)Z_2,$$
$$(1-V_k)(1-Z_1)(1-Z_2)Z_3,\ldots) .$$

When $k = 0$, put

$$X^{(o)} = (Z_1,(1-Z_1)Z_2,(1-Z_1)(1-Z_2)Z_3,\ldots) .$$

Now for any Borel subset A of Δ , define ν_k by

$$\nu_k(\{k\} \times A) = P(X^{(k)} \in A) , \quad k = 0,1,2,\ldots .$$

This form lends itself readily to calculation and is exploited in Donnelly and Tavaré (1987), to recover the transition density of the usual K-allele diffusion models.

ACKNOWLEDGEMENTS

The authors were supported in part by NSF grants DMS 85-01763 and DMS 86-08857.

REFERENCES

Donnelly, P., Tavaré, S. (1986), The ages of alleles and a coalescent, *Adv. Appl. Prob.* **18**, 1-19.

Donnelly, P., Tavaré, S. (1987), The population genealogy of the infinitely many neutral alleles model. To appear.

Ewens, W.J. (1979), *Mathematical Population Genetics*, Springer, Berlin.

Griffiths, R.C. (1980), Lines of descent in the diffusion approximation of neutral Wright-Fisher models. *Theor. Popn. Biol.* **17**, 37-50.

Kingman, J.F.C. (1982a), On the genealogy of large populations, *J. Appl. Probl* **18**, 27-43.

Kingman, J.F.C. (1982b), The Coalescent. *Stoch. Proc. Appln.* **13**, 235-248.

Tavaré, S. (1984), Line of descent and genealogical processes and their application in population genetic models. *Theor. Popn. Biol.* **26**, 119-164.

Watterson, G.A. (1984), Lines of descent and the coalescent, *Theor. Popn. Biol.* **26**, 77-92.

THE RELATIONSHIP BETWEEN THE STOCHASTIC AND DETERMINISTIC VERSION OF A MODEL FOR THE GROWTH OF A PLANT CELL POPULATION

De Gunst, Mathisca C.M.
Department of Mathematics
University of Leiden
P.O. Box 9512
2300 RA LEIDEN
The Netherlands

INTRODUCTION.

When a population of plant cells in batch suspension culture is followed in time, it turns out that its growth depends on the substrate and hormone concentration in the medium. Moreover, the individual cell cycles appear to be of variable duration. A stochastic model for the growth of a cell population with the observed properties is presented.

THE MODEL.

In batch suspension cultures usually two types of cells are found: cells which are participating in the cell division process, denoted by "type-A cells", and cells which are differentiating and therefore will not divide, the "type-B cells". For $n = 1,2,\ldots$ we consider a process starting at $t = 0$ with n cells, which are all of type-A. Let $N(t)$ be the random number of cells at time t, $N(0) = n$, and let $N_A(t)$ be the numer of type-A cells at time t, $N_A(0) = n$. The question is: can these stochastic processes be approximated to first order by their deterministic counter parts? To answer this question we investigate the behaviour of
$X_n(t) = (N(t)-n)/n$ as $n \to \infty$.

Let $\tau_i = \inf\{t:N(t) = n+i\}$

denote the time of the i-th division, $i = 1,2,\ldots$. The idea is that a cell dividing at τ_i has started its mitotic cycle at time $\tau_i - c$, because it has received a stimulus at this time. Here c is a constant, independent of n. A type-A cell receives this stimulus at the time of the first event in a counting process starting at the birth of the cell and having a variable, random rate $Q(t)$. For different cells the processes are only coupled through this rate, but otherwise independent. The two cells originating at time τ_i independently become type-A cells with probability P_i .

A fixed quantity $1/y_s$ of substrate is used up in providing a stimulus. When the available amount of substrate at $t = 0$ is

$$S(0) = n\, b_s\, y_s^{-1} ,$$

where b_s is fixed, the amount of substrate at time t is given by

$$S(t) = (S(0) - i\, y_s^{-1}) \vee 0 \quad \text{for} \quad \tau_i - c \le t < \tau_{i+1} - c,\ i = 1,2,\ldots$$

It is assumed that the rate $Q(t)$ depends on S in the following way.

$$Q(t) = \begin{cases} \dfrac{S(t)}{d(S(t)+k_s)} = \dfrac{nb_s - i}{d(nb_s + na_s b_s - i)} = Q_i & \text{for } \tau_i - c \le t < \tau_{i+1} - c , \\ & i = 1,2,\ldots,[nb_s]-1 = n_s - 1, \\ 0 & \text{otherwise,} \end{cases}$$

where d and $a_s = k_s/S(0)$ are fixed. This means that the total number of stimuli never exceeds n_s . Furthermore the amount of hormone is given by

$$H(c) = nb_H\, y_H^{-1} ,$$

$$H(\tau_i) = (H(c) - (i-1)y_H^{-1}) \vee 0 ,\ i = 1,2,\ldots ,$$

where b_H and y_H are constants. The two cells originating at time τ_i independently become type-A cells with probability P_i depending on the amount of hormone present:

$$P_i = \frac{H(\tau_i)}{H(\tau_i)+k_H} ,$$

where $k_H/H(c) = a_H$ is fixed. Note that the processes may continue after the hormone has run out. Then only type-B cells are produced.

THE ASYMPTOTIC BEHAVIOUR OF X_n .

Given for each division the number of newly born A-cells, i.e. given the sequence $Z = (Z_1, Z_2, \ldots)$, where $Z_1, Z_2, \ldots$ are independent and $Z_j \sim \text{Bin}(2, P_j)$, $j = 1,2,\ldots$, $N(t) - n$ can be considered as a stopped counting process with intensity process $\Lambda_z(t)$ given by

$$\Lambda_z(t) = (2n-N(t) + \sum_{j=1}^{N(t-c)-n} Z_j) Q_{N(t)-n} \cdot 1_{[c,\infty)}(t) .$$

This expression depends on $N(t)$ and $N(t-c)$. When $\sum_{j=1}^{N(t-c)-n} Z_j$ is replaced by a smoothed version of its expectation, we get a non-random expression, which we denote by $n\,\psi(t,X_n)$. Here

$$X_n(t) = (N(t)-n)/n ,$$

$$\psi(t,x) = \{(1-x(t)+2(x(t-c)\wedge b_H)-2a_H b_H \log\left(\frac{b_H+a_H b_H}{b_H+a_H b_H-(x(t-c)\wedge b_H)}\right)) \vee 0\} \cdot$$

$$\cdot \left(\frac{b_s-x(t)}{d(b_s+a_s b_s-x(t))} \vee 0\right) 1_{[c,\infty)}(t) .$$

Now, let X be the unique solution of

$$\begin{cases} x(t) = 0 & t \leq c \\ x(t) = \int_0^t \psi(s,x)ds & t \geq c . \end{cases}$$

Then X is a deterministic, differentiable function of t and under certain conditions X_n satisfies

$$\limsup_{n\to\infty} \sup_{t\geq 0} |X_n(t) - X(t)| = 0 \quad \text{a.s.}$$

In fact for every constant $M > c$ we have the following rate of convergence.

$$\sup_{0\leq t\leq M} |X_n(t) - X(t)| = O\left(\left(\frac{\log n}{n}\right)^{\frac{1}{2}}\right) \quad \text{a.s.}$$

GENE ACTION FOR AGRONOMIC CHARACTERS IN WINTER WHEAT

W. Lonc, Institute of Plant Breeding and Seed Production
Agricultural University of Wrocław, Poland

INTRODUCTION

The purpose of the studies was to determine gene action for quantitatively inherited characters in winter wheat. In quantitative characters, only type of gene action is generally studied without estimating the number of genes because the effect of a separate gene is small and influenced by the environment. With additive gene action the phenotype is a good indicator of the genotype and selection can be practiced in early hybrid generations. When large dominance and epistatic effects are present in the inheritance of a character, selection should be delayed.

MATERIALS AND METHODS

Seven lines derived from the following winter wheat cultivars, differing morphologically, were crossed in a half-diallel fashion: Oregon 394 /USA/ - O, selection 642 Carsten 102 x Mex. x Mex. /France/ - C, Atlas 66 /USA/ - A, Marksman /Great Britain/ - M, selection Kr 118/78 /Poland/ - K, Arminda /The Netherlands/ - Ar, Liwilla /Poland/ - L. F_2 hybrids and parental lines were space planted in a randomised complete block design and harvested in 1984. Graphical analysis of Hayman (1954) and Jinks (1954) was applied. The regression of Wr /covariance/ on Vr /variance/ was used to test the adequacy of a simple additive-dominance model. Whenever the simple model was inadequate epistatic lines were removed from the analysis until a subgroup showing no interaction was obtained. Correlations between parental means /y_r/ and Wr + Vr were also estimated.

RESULTS

Regression of Wr on Vr for plant height gave a coefficient which differed from zero, but not from unity, indicating the adequacy of a simple additive-dominance model /fig.1/. The regression line cut Wr axis above the origin, indicating partial dominance of genes. Parental lines Liwilla and Atlas possessed a slight excess of dominant alleles whereas Arminda and Marksman had more recessive alleles. The other lines were not genetically differentiated and had aproximately equal numbers of recessive and dominant alleles. Correlation between y_r and Wr + Vr was -0,72, very close to the theoretical value /$r_{0,05}$ = 0,75/, demonstrating that dominance was acting in the direction of high expression of the character.

Productive tillering was also governed by partial dominance /fig.2/. Lines Carsten and Arminda had an excess dominant alleles while Marksman carried most recessive alleles. No significant correlation was found between y_r and Wr + Vr /r = 0,12/.

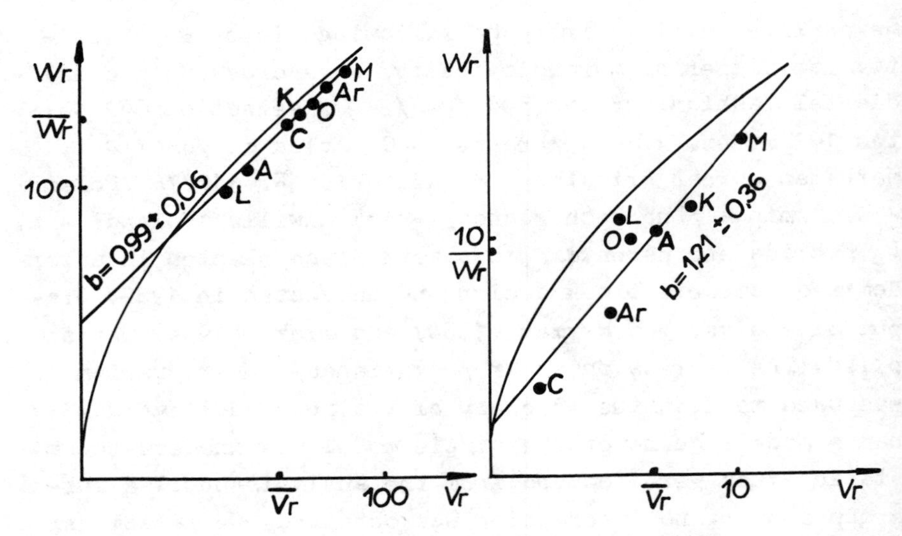

Fig.1. Plant height

Fig.2. Productive tillering

Non allelic interaction was observed in the expression of no. of grains/ear. When the F_2 data, after omitting Atlas and Kr 118/78, were reanalyaed, a slope not differing from unity was obtained and partial dominance was indicated in the inheritance of the character /fig.3/. Marksman carried a considerable excess of recessive alleles while Liwilla, Arminda and Oregon had a slight excess of dominant alleles. A high correlation was found for y_r and Wr + Vr /r = 0,88/ indicating a relationship between recessive alleles and high no. of grains/ear.

Grain yield/ear was determined by partial dominance. Kr 118/78 and Arminda contained a slight excess of dominant alleles while Carsten, Marksman and Liwilla carried more recessive alleles. Atlas and Oregon had aproximately equal numbers of recessive and dominant alleles /fig.4/. Correlation between y_r and Wr + Vr was r = 0,63, and although not significant, showed a tendency for recessive alleles to be associated with high grain yield/ear.

Similar results were discussed in a previous paper Lonc (1985).

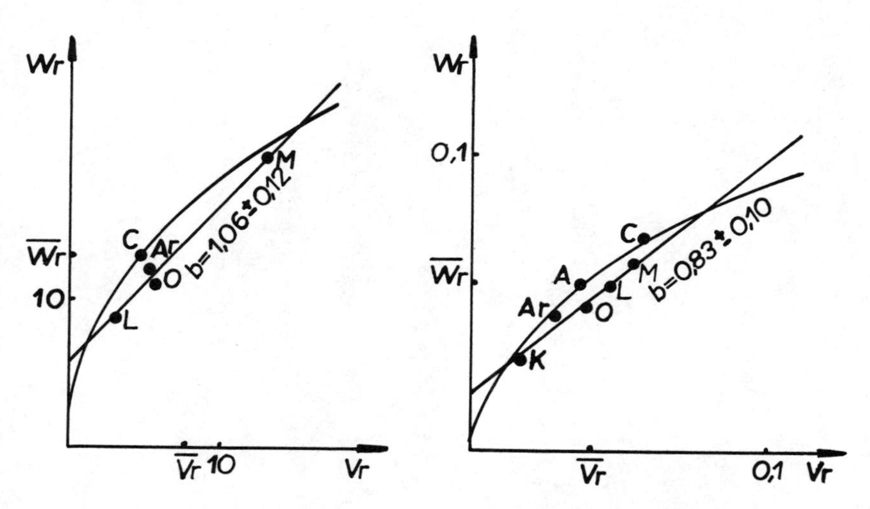

Fig.3. No. of grains/ear

Fig.4. Grain yield/ear

CONCLUSIONS

Line Kr 118/78 and Atlas 66 caused epistasis in the inheritance of no. of grains/ear. Varying degrees of partial dominance were operating in the inheritance of the studied character. Recessive genes increased no. of grain/ear

REFERENCES

Hayman, B.I. (1954). The theory and analysis of diallel crosses. Genetics 39, 789-809.

Jinks, J.W. (1954). The analysis of continuous variation in a diallel cross of Nicotiana rustica varieties. Genetics 39, 767-788.

Lonc, W. (1985). Sposoby działania genów warunkujących cechy ilościowe pszenicy ozimej. Hod. Rośl. Aklim. 29, 3-4.

TOTAL PROGENY OF A CRITICAL BRANCHING PROCESS

Sagitov S.M.

Institute of Mathematics and Mechanics, Academy of Sciences of the Kazakh SSR, Alma-Ata, USSR

1.Main result. We consider a general critical branching process see, e.g., Jagers (1969). Let l be the life-span and ν be the number of offspring of the initial individual. We will assume

(1) $Es^{\nu} = s + (1-s)^{1+\alpha} L(1-s)$, $\alpha \in (0,1]$,

(2) $P\{l > t\} = o(t^{-1-1/\alpha} \chi(t))$, $t \to \infty$,

where $L(x), \chi(1/x)$ are slowly varying as $x \to 0$ and

$$\chi^{\alpha}(t) L(t^{-1/\alpha} \chi(t)) \to 1, \ t \to \infty.$$

Let $Z(t)$ be the number of individuals alive at time t, and $Y(t)$ be the total number of individuals, who was born by time t. Denote

$$Q(t) = P\{Z(t) > 0\}, \ a = E(\tau_1 + \ldots + \tau_{\nu}),$$

where τ_j is the age of original parent at the birth of the j-th offspring. Next theorem generalizes the analogous results obtained for Galton-Watson and Bellman-Harris branching processes by Sagitov, Vatutin (1986) and Vatutin (1986) respectively.

Theorem 1. If (1), (2) hold and $0 < a < \infty$, then

$$E(\exp\{-Q(t)[\lambda_1 Y(t)/EY(t) + \lambda_2 Z(t)/EZ(t)]\} \mid Z(t) > 0) \to$$

$$\to h(\lambda_1, \infty) - h(\lambda_1, \lambda_2), \ t \to \infty,$$

where $h(\lambda_1, \lambda_2)$ is the unique solution of the equation

$$(1+\alpha)\lambda_1 \frac{\partial h}{\partial \lambda_1} + \lambda_2 \frac{\partial h}{\partial \lambda_2} = h - h^{1+\alpha} + \alpha \lambda_1,$$

$$h(0,0)=0, \quad \frac{\partial h}{\partial \lambda_i}(0,0)=1, \quad i=1,2.$$

2. Multidimentional generalization. Assume that each individual I is associated with a nonnegative multidimentional stochastic process χ^I. Suppose $\chi^I=(\chi_1^I,\dots,\chi_n^I)$ to be independent copies of $\chi=(\chi_1,\dots,\chi_n)$. Let us consider the process

$$X=\sum_{I:\,\sigma_I\le t}\chi^I(t-\sigma_I), \quad X=(X_1,\dots,X_n),$$

where σ_I is the time of birth of I. This is a natural extention of the onedimentional process, that was introduced by Jagers (1974).

Theorem 2. Let (1), (2) hold and $0<\alpha<\infty$. Suppose that

(i) $EX_i(t)$ is regularly varying as $t\to\infty$ with parameter $\beta_i>0$, $i<n$,

(ii) $E\chi_i^{1+\varepsilon}(t)=O((E\chi_i(t))^{1+\varepsilon})$, $t\to\infty$, $\varepsilon>0$, $i<n$,

(iii) $X_n\equiv Z$, $\beta_n=0$.

Then

$$E(\exp\{-Q(t)\sum_{i=1}^{n}\lambda_i X_i(t)/EX_i(t)\}\mid Z(t)>0)\to$$

$$\to h(\lambda_1,\dots,\lambda_{n-1},\infty)-h(\lambda_1,\dots,\lambda_n), \quad t\to\infty,$$

where $h(\lambda_1,\dots,\lambda_n)$ is the unique solution of the equation

$$\sum_{i=1}^{n}(1+\alpha\beta_i)\lambda_i\frac{\partial h}{\partial\lambda_i}=h-h^{1+\alpha}+\alpha\sum_{i=1}^{n}\beta_i\lambda_i$$

$$h(0,\dots,0)=0, \quad \frac{\partial h}{\partial\lambda_i}(0,\dots,0)=1, \quad i=1,\dots,n.$$

Remark 1. It is useful to compare this equation with (2.9) from Foster, Ney (1978). The statement of theorem 2 may be extended to $\beta_i\geqslant 0$.

<u>Corollary</u>. Let (1),(2) hold and $0 < a < \infty$. Suppose that

(i) $n = 1$,

(ii) $E\chi(\cdot)$ is directly Riemann integrable,

(iii) $E\chi^{1+\alpha}(t)L(\chi(t)) \to 0, \quad t \to \infty.$

Then

$$E(\exp\{-\lambda X(t)t^{-1/\alpha}\chi(t)\} \mid Z(t) > 0) \to$$

$$\to 1 - [1 + a\alpha^{1+\alpha} - 1(\lambda \int_0^\infty E\chi(t)dt)^{-\alpha}]^{-1/\alpha}, \quad t \to \infty.$$

<u>Remark 2</u>. Here we don't suppose that

$$P\{\chi(t) = 0 \mid l < t\} = 1, \quad t \geqslant 0$$

as distinct from Sagitov (1983, 1986).

REFERENCES

1 Foster,J., Ney,P. (1978). Limit laws for decomposable critical branching process. Z.Wahrscheinlichkeitstheor. und verw. Geb. 46, 13-43.

2 Jagers, P. (1969) A general stochstic model for population development. Scand. Aktuarietidskr. 52, 84-103.

3 Jagers, P. (1974). Convergence of general branching processes and functionals thereof. J.Appl. Pribability, 11, 471-478.

4 Sagitov, S.M. (1983). Limit theorem for general critical branching process. Math. Notes 34.

5 Sagitov, S.M. (1986) Limit behaviour of the general branching processes. Math. Notes 39.

6 Sagitov, S.M., Vatutin,V.A. (1986). Decomposable critical branching process with two types of particles. Proc. Steklov Inst. Math. 177.

7 Vatutin, V.A. (1986). Bellman-Harris critical branching process with final type. Theor. Probability Appl. 31.

LIMIT THEOREMS FOR A CRITICAL BRANCHING CRUMP-MODE-JAGERS PROCESSES

Topchii V.A., Computer Centre of the Siberian Branch of the USSR Academy of Sciences, Omsk, USSR

Consider a population with one ancestor, which develops in time t. The evolution of each individual is determined by the two-dimensional stochastic process $(\eta, N(t); t \in [0,\infty))$, where η - is interpreted as a life-length, and point process N(t) - is a general number of offsprings generated by a particle in time t from its birth. The evolution processes for all particles are independent and identically distributive. The number of particles alive in time t is denoted by $\xi(t)$, $t \geq 0$ and these processes are called the general branching or Crump-Mode-Jagers. For details of definition of $\xi(t)$ see Jagers (1975).

Define: R(t) - the total number of individuals that have been born up to time t; $Q(t) = P(\xi(t) > 0)$; $m_n(t) = M\{R^n(t); \xi(t) = 0\}$; $q(t) = P(\eta > t)$; $N = \lim_{t\to\infty} N(t)$; $A = MN$; $B = M(N-1)N$; $h(z) = Mz^N$; $a(t) = MN(t)$; $a = \int_0^\infty t\,da(t)$; $\overline{Q}(t) = aB^{-1}t^{-1} + \sqrt{a^2B^{-2}t^{-2} + 2B^{-1}q(t)}$; $G(y) = h(1-y) - 1 + y$; $F(x) = G^{-1}(x)$, i.e. $G(F(x)) = x$.

The condition $q(t) \in \mathcal{O}\!\mathcal{l}$ denotes that $\lim_{t\to\infty, d\to 1} (q(t) + t^{-2})(q(td) + t^{-2}d^{-2})^{-1} = 1$ and $(q(t)t^2 + 1)^{-1} + 1$ - is a slowly varying function.

<u>Theorem 1</u>. Let $A = 1$, $0 < B < \infty$, $0 < a < \infty$, $q(0) = 1$, $q(t) \in \mathcal{O}\!\mathcal{l}$ then, as $t \to \infty$

$$Q(t) \sim \overline{Q}(t).$$

Theorem 2. Under the conditions of theorem 1

$$m_n(t) \sim v_n(t)Q^{-2n+1}(t),$$

where $v_1(t) = Q(t)(BQ(t) + at^{-1})^{-1}$, and for $n > 1$ $v_n(t) = BQ(t)\sum_{i=1}^{n-1} C_{n-1}^i v_i(t)v_{n-i}(t)(BQ(t) + (2n - 1)at^{-1})^{-1}$.

In the case where the function $q(t)t^2$ has finite or infinite limit $\lim_{t\to\infty} q(t)t^2 = a^2B^{-1}c/2$ we have $v_n(t) \to v_n$ as $t\to\infty$, where for $c = \infty$ $v_n = \sum_{i=1}^{n-1} C_{n-1}^i v_i v_{n-i}$ as $n > 1$, $v_1 = B^{-1}$, and for $c \in [0,\infty)$ $v_1 = B^{-1}(1 - (\sqrt{1 + c} + 2)^{-1})$, $v_n = (1 + \sqrt{1 + c})\sum_{i=1}^{n-1} C_{n-1}^i v_i v_{n-i}(\sqrt{1 + c} + 2n)^{-1}$ as $n > 1$.

From theorem 2, Taylor's formula and continuity theorem for Laplace transforms applying to $W_t(u) = \int_0^\infty e^{-ux} \cdot w_t(x)dx$, where $w_t(x) = Q^{-1}(t)P(R(t)Q^2(t) > x;\ \xi(t)=0)$, we can prove the results

Theorem 3. Under the conditions of theorem 1 in the case $\lim_{t\to\infty} t^2q(t) = \infty$ there exists $w(x) = \lim_{t\to\infty} w_t(x)$ and $W(u) = \lim_{t\to\infty} W_t(u) = \int_0^\infty e^{-ux}w(x)dx$. For $\Phi(x) = 2\pi^{-1/2} \cdot \int_0^x \exp\{-t^2\}dt$ the functions have the form

$$w(x) = \sqrt{2\pi^{-1}x^{-1}B^{-1}}\exp\{-Bx/2\} + \Phi(\sqrt{Bx/2}) - 1 \text{ and}$$

$$W(u) = u^{-1}(\sqrt{1 + 2B^{-1}u} - 1).$$

Theorem 4. Let $A = 1$, $0 < B < \infty$, $0 < a < \infty$, $q(0) = 1$ and $\lim_{t\to\infty} t^2q(t) = a^2B^{-1}c/2 \in [0,\infty)$, then

$$W(u) = \lim_{t\to\infty} W_t(u) = \sqrt{2B^{-1}u^{-1}}\, I_p(b\sqrt{B^{-1}u/2})I_{p-1}^{-1}(b\sqrt{B^{-1}u/2}),$$

where $b = 1 + \sqrt{1 + c}$, $p = (1 + b)/2$, and $I_p(z) = (z/2)^p\sum_{k=0}^{\infty}(z/2)^k\Gamma^{-1}(p + k + 1)/k!$ is the modified Bessel function.

Corollary 1. Let $A = 1$, $0 < B < \infty$, $0 < a < \infty$, $q(0) = 1$ and $q(t) = o(t^{-2})$, as $t \to \infty$, then

$w(2xB^{-1}) = (2\sum_{n=1}^{\infty} \exp\{-n^2/x\} + 1)\sqrt{\pi^{-1}x^{-1}} - 1$ and

$W(u) = \sqrt{2B^{-1}u^{-1}} \operatorname{cth} \sqrt{2uB^{-1}} - u^{-1}$.

The condition $B < \infty$ is necessary for asymptotic investigation of $m_n(t)$. The results of theorems 3 and 4 may be generalized for $G(y) = y^{\alpha+1}l(y)$, where $0 < \alpha \leq 1$ and $l(y)$ - is a slowly varying function. Such functions $G(y)$ are called $(\alpha + 1)$ - regular. In theorem 4 (case $\alpha = 1$) $W(u)$ is a solution of a some differential equation. In the case $\alpha < 1$ we may write the differential equation for $W(u)$, but we do not know its solution.

Suppose that $G(y)$ and $Q(t)$ have no restriction on behaviour except for

$$\lim_{t \to \infty, d \to 1} Q(t)Q^{-1}(td) = 1 \qquad \text{(i)}$$

For sufficient conditions for (i) see Topchii (1986).

The condition $f(t) \in \mathcal{L}$ denotes that for some $d > 0$ $\lim_{t \to \infty} f(t + d)f^{-1}(t) = 1$, $f(t)$ is monotonous and continuous, $f(t) \to \infty$ as $t \to \infty$ and $G(Q(t))f(t) \to 0$ as $t \to \infty$.

Let $u(t,x) = f_1(t)P(\, R(t)f(t)G(Q(t)) > x;\ \xi(t) = 0\,)$.

Theorem 5. If $A = 1$, $P(\,\eta = 0\,) = 0$, (i) holds and $f(t) \in \mathcal{L}$ is fixed, the limit $\lim_{t \to \infty} u(t,x) \neq \text{const}$ exists if and only if there exists

$$\lim_{t \to \infty} f_1(t)F(f(t)G(Q(t))) = c > 0$$

and $G(y)$ - is $(\alpha + 1)$ - regular for some $\alpha \in (0,1]$.

If $c = 1$ and $G(y)$ is $(\alpha + 1)$ - regular

$$\lim_{t \to \infty} u(t,x) = \Gamma^{-1}(\alpha/(\alpha + 1))x^{-1/(\alpha + 1)}.$$

The results of theorems 3 and 4 and corollary 1 may be written as the asymptotic behaviour of large deviation probability. Under corresponding assumptions and notations is true

$$P(R(t) > x(t); \xi(t) = 0) \sim u(x(t)Q^{-2}(t))Q(t)$$

for x(t) such that $\limsup_{t\to\infty} x(t)Q^{-2}(t)<\infty$, $\liminf_{t\to\infty} x(t)\cdot Q^{-2}(t) > 0$.

In the last interpretation theorem 5 has the form

Corollary 2. Let A = 1, (i) holds, G(y) is (α + 1) - regular for $\alpha\in(0,1]$, $x(t)\to\infty$ and $x(t) = o(1/G(Q(t)))$, then

$$P(R(t) > x(t); \xi(t) = 0) \sim \Gamma^{-1}(\alpha/(\alpha+ 1))F(x(t)).$$

The previous results have also anothe natural interpretation as behaviour of total number of particles that have been born up to time t on the set $\{\xi(t) = 0\}$ for branching process with $1/Q(t)$ or $f_1(t)$ ancestors (case scheme of series).

References

Jagers P. (1975). Branching processes with biological applications. Wiley, London, 268p.

Topchii V.A. (1986). The extinction probability of critical branching processes. Teorija Verojatnostej i jejo Primenenija. V.31, №1, pp.175-176. (Russian).

THE BEHAVIOUR OF THE PRAY-EDITOR SYSTEM IN THE NEIGHBOURHOOD OF STATISTIC EQUILIBRIUM

Yevgenij F.Tsarkov

Special Mathematics Department, Riga Polytechnical Institute,
Faculty of Automation and Computing Techniques, Riga,
Latvian SSR, USSR

Walterr's classical model of n ecological associations is represented in the form:

$$\frac{dx_i}{dt} = x_i(b_i + \sum_{j=1}^{n} a_{ij}x_j), \quad i = 1,2,\dots,n, \tag{1}$$

where b_i is the rate of the natural increase or mortality of i-spaces in the absence of all the rest species, a_{ij} is the indicator of the insidespecies interaction of i-species and coefficients a_{ij} $(i \neq j)$ show the character and intensivity of j-spacies influence upon i-species. However, as it was mentioned by many authors, we ought to considert the influence of random factors of various kinds and the fact that the intensivity of the propagation and death is different in various age groups to make more complete description in mathematical models of the laws of the ecological systems functioning. Therefore, we must add to the system of equations (1) random perturbations and the summands, containing the meaninh of solution in the preceding moments of time. Apparently, the system of functional-differential equations

$$\frac{dx_i}{dt} = x_i(t)(b_i + \sum_{j=1}^{n} a_{ij} \int_0^h x_j(t-\tau)dF_{ij}(\tau) + \varepsilon f_i(t,x_t,\omega,\varepsilon), \tag{2}$$

where ε is a small parameter, $F_{ij}(\tau)$ are monotonous functions, fixed by a condition $F_{ij}(h)-F(0)=1$ and the summands $f_i(t,x_t,\omega,\varepsilon)$ reflect the random factors, depending on time t and the part of vector-function solution, determined by the correlation

$x_t \overset{\Delta}{=} \{(x_1(t+\theta), x_2(t+\theta), \dots, x_n(t+\theta))^T, \theta \varepsilon [-h,0]\}$,

is rather good mathematical model as it allows to take into account all above-mentioned factors. Already in the analysis of an isolated population the consideration of the age structure even in the determined variant allows to discover a new and interesting effect: there's no stable equilibrium in the population with the determined correlation among the parameters and it's size may change irregularly, remaining near some constant. It is rather easy to receive this result using the method of bifurcation theory [1], expanding the solution (2) in a series on the fractional degrees of a small parameter ε in the neighbaurhood of the equilibrium. Unfortunatelly the necessity of taking into consideration the random perturbations in the equations of the type (2) makes the process of analysis more complicated. The method described below is proposed by the author [6]. He uses the averaging principle of R.Chasminsky [3] and some results received by A.Ventzeletal [4]. This method allows to simplity the analysis of the pray-editor system fluctations, considering the age structure in the neighbourhood of the statistic equilibrium.

Suppose that the equation (2) may be rewritten in the neighbourhood of the studied equilibrium in a quasilinear form:

$$\frac{dy_i(t)}{dt} = \sum_{j=1}^{n} r_{ij} \int_0^h y_j(t-\tau)\,dF_{ij}(\tau) + \varepsilon f_i(t, y_t, \omega, \varepsilon) \tag{3}$$

so, that the linear equation

$$\frac{dZ_i(t)}{dt} = \sum_{j=1}^{n} r_{ij} \int_0^h Z_j(t-\tau)\,dF_{ij}(\tau) \tag{4}$$

would be in the indifferent equilibrium. The natural phase space for (3) and (4) will be the space $C \overset{\Delta}{=} C([-h,0] \to R^n)$. The equation (4) generates in C the semigroup of the displacement operators $\{T(t), t \geq 0\}$ [5] where due to the made assumptions the infinitesimal operator has some simple eigenvalues $\lambda_1, \lambda_2, \dots, \lambda_m$ on the imaginary axis and the rest part of spectrum is situated in the halfplane $\{\lambda \varepsilon C : R\,\lambda < -\rho < 0\}$. No we may decompose C in the direct sum of subspaces

$C = G \oplus Q$, where G is an invariant root subspace of the operator A, corresponding to the eigenvalues $\lambda_1, \lambda_2, \ldots, \lambda_m$. If we denote by E the matrix of basis in and determine the matrix B with the equality $T(t)E = E\exp\{Bt\}$ we may show from some additional suppositions that the solutions of (3) is well approximated by the vector-function $E\exp\{Bt\}u(t)$, where u(t) is the solution of the equation

$$\frac{du(t)}{dt} = \varepsilon e^{-Bt} Sf(t, Ee^{Bt}u(t), \omega, \varepsilon), \tag{5}$$

the matrix S has m lines and n colomns and is rather easy determined from the system (4). The methods [3] and [4] are ready for analysis (5).

The methods described above allow to study the pray-editor system stability of equilibrium in the accidental environment. It's possible to use a quasi-potential of A. Ventcel [3] build for the system of the type (5) and the Lyapunoff-Krasovsky quadratic functionals [6] found with the help of the solution of the analoque of the Lyapunoff's equation in $C^* \otimes C^*$ for the equation (4). In the last case we ought to give up the consideration of the age structure in the general form and to suppose, that this factor's consideration is the introduction to the mathematical model of a constant time delay in the argument of the solution to be found.

Let's describe the results of one of the most simple variants of the pray-editor system mathematical model taking into account the age structure;

$$\frac{dx_1(t)}{dt} = (a+\varepsilon\xi(t))x_1(t) - b_1x_1(t)x_2(t) - cx_1(t)x_1(t-h),$$

$$\frac{dx_2(t)}{dt} = -gx_2(t) + b_2x_1(t)x_2(t) + H \tag{6}$$

This model differs from the simplest Walterr's model by a stochastic perturbation $\varepsilon\xi(t)$ of the natural birth rate α of the pray and time delay in the argument. Moreover, a constant summand H is put into the second equation indicating a constant increase of the editors from autside. For example, the association of that kind takes place in the case of using biological ways of fighting with the pests in agriculture. If $b_1 = 0, \varepsilon = 0$, than it follows from [7] results that the prays' number makes periodical oscillations, and their amplitude depending on different correlations among parameters either decreases or converges to some stable state. In the first case a regime appears in the system close to either periodical or stacionary one. In the second case ascillations occur and small perturbations of the parameter a ($\varepsilon \neq 0$) lead to small fluctuations of these oscillations' amplitude. As it is clear from Walterr's model (1) this regime appears to be rough. In the system (6) the regimes may appear close to periodical with a frequency different from pray's ascillations' frequency.

REFERENCES

1. Hassard, A., Kazarinoff, N., and Wan, Y. (1981). Theory and Applications of Hopf Bifurcation, Cambridge University Press, Cambridge.
2. Tsarkov, Ye. (1984). The Random Perturbations of the Systems with Delay. IX YCNO. Abstracts of Communications. Vol.1. Naukova Dumka. Kiev. pp.401-403.
3. Ventcel, A., and Friedlin, M. (1979). The Fluctuations in the Dynamical Systems under the Influence of the small Random Perturbations. Nauka. Moscow.
4. Hasminsky, R. (1956). On the Random Processes determined by the Differential Equations with the small Parameter. Prob. Theor. and Appl. 11 : 2. pp. 240-259.
5. Hale, J. (1977). Theory of Functional Differential Equations. Springer-Verlag. New York.
6. Tsarkov, Ye., and Engelson, L. (1981). On the statistic Solutions of the Linear Systems of the Functionsl Differential Equations. The Topological Space and its reflections. Latv. State University. pp. 142-151.
7. Kolesov, Yu., and Shvitra, D. (1979). The Autooscillations in the Systems with Delay. Mokslas. Vilnius.

BELLMAN-HARRIS BRANCHING PROCESSES AND DISTRIBUTIONS OF MARKS IN PROLIFERATING CELL POPULATIONS.

A.Yu.Yakovlev[1], M.S.Tanushev[2], N.M.Yanev[2].

[1]Central Research Institute of Roentgenology and Radiology of the Ministry of Health of the USSR, Leningrad, USSR,

[2]Mathematical Institute of the Bulgarian Academy of Sciences, Sofia, Bulgaria.

INTRODUCTION

The microscopic study of a radioautograph allows one to count not only the fractions of cells in periods S and M of the mitotic cycle at the given instant but the number of grains (marks) in each labeled single cell as well. The time-varying distribution of the number of marks per cell contains useful information for the analysis of cell proliferation kinetics. This distribution for the cell population obeying the Bellman-Harris branching process has been derived in this work.

BASIC ASSUMPTIONS AND INTRODUCTORY FORMALISM

We need the following assumptions: (a) the cell population dynamics is considered within the model of Bellman-Harris age-dependent branching process; (b) the pulse labeling of cells is interpreted as the attachment to them of a random number of discrete marks; (c) the initial distribution of marks is Poisson with parameter θ ; (d) when a cell dies, its marks disappear; when a cell splits, each of the marks attached to the mother cell transfers independently and with probability 1/2 to one of daughter cells.

If $Z_j(t)$ is the number of cells bearing j marks at the moment $t \geqslant 0$, then $\mathbf{Z}(t) = (Z_0(t), Z_1(t),\ldots, Z_j(t),\ldots)$ formes an infinite-type Bellman-Harris branching process with a distribution function G(t) for the mitotic cycle duration and an offspring p.g.f. for type k = 0,1,2,...

(1) $h_n(S) = 1-p_n + p_n \sum_{i=0}^{n} \binom{n}{i} 2^{-n} S_i S_{n-i}$, $\mathbf{S}=(S_o, S_1, \ldots, S_k, \ldots)$.

We assume that G(0) = 0, and

(2) $p_n = pa^n$, $0 < a \leqslant 1$, $0 < p \leqslant 1$, n=0,1,2,... ,

where p is a probability of the successful binary splitting and 1-p is a probability of the reproductive death of a cell. The dependence (2) is introduced because of some different biological applications. Namely, the cell damages induced by cytotoxic agents may be interpreted as discrete marks and in this case it is natural to assume that the value of p depends on the number of such marks. The distribution of marks is defined as the ratio of expectations of the processes $Z_j(t)$ and $Z(t) = \sum_{j=0}^{\infty} Z_j(t)$, i.e. $\Pi_j(t) = N_j(t)/N(t)$, where $N_j(t) = E\{Z_j(t)\}$, $N(t) = E\{Z(t)\}$.

THE DISTRIBUTION $\Pi_j(t)$ WITHIN BRANCHING PROCESS FRAMEWORK

Introduce p.g.f. $F(t; \mathbf{S})$ of the process $\mathbf{Z}(t)$ with components

$F_k(t; \mathbf{S}) = E\left\{ \mathbf{S}^{\mathbf{Z}(t)} | \mathbf{Z}(0) = \mathbf{e}_k \right.$, $|\mathbf{S}| \leqslant \mathbf{1}$, k=0,1,2...,

where $\mathbf{S}^{\mathbf{Z}(t)} = \prod_{K=0}^{\infty} S_k^{Z_k(t)}$, and $\mathbf{e}_k$ is infinite vector with the k th component equal to 1 and all other components equal to 0. The generating functions $F_k(t, S)$ satisfy the following system of integral equations (Athreya and Ney, 1972):

$$(3)\ F_k(t; \mathbf{S}) = S_k \bar{G}(t) + \int_0^t h_k[\mathbf{F}(t-u; \mathbf{S})]\, dG(u), \quad |\mathbf{S}| \leqslant \mathbf{1},$$

where $\bar{G} = 1-G$.

Denote the expectation of the number of particles of the j th type originating from one particle of the k th type by

$$A_{kj} = E\{ Z_j(t) | Z_k(0) = 1\}.$$

From (1) and (3) we obtain the following system of equations

$$(4)\quad \begin{aligned} A_{kj}(t) &= \delta_{kj}\bar{G}(t) + p_k 2^{1-k}\sum_{i=0}^{k}\binom{k}{i}\int_0^t A_{ij}(t-u)dG(u), k \geqslant j, \\ A_{kj}(t) &= 0, \quad k < j, \end{aligned}$$

where $\delta_{kj}=0$ when $j \neq k$, and $\delta_{kj} = 1$ when $j = k$.

Consider the situation when at the moment $t = 0$ there exists a random number of particles with mathematical expectation equal to M. Then on the strength of assumption (c)

$$(5)\quad N_j(t,\theta) = E\{Z_j(t)\} = M\, e^{-\theta}\sum_{k=j}^{\infty}\frac{\theta^k}{k!}A_{kj}(t).$$

By substituting of (5) into the system (4) and using (2) we obtain

$$(6)\quad N_j(t,\theta) = M\, e^{-\theta}\frac{\theta^j}{j\,!}\,\bar{G}(t) + 2pe^{-\theta(1-a)}\int_0^t N_j(t-u;\frac{a\theta}{2})\, dG(u).$$

The relationship (6) may be treated as a recurrent one in regard to the parameter θ. Hence iterating (6) n times, we obtain

$$(7)\quad N_j(t,\theta) = M\frac{e^{-\theta}\theta^j}{j\,!}\sum_{k=0}^{n}(2p(\tfrac{a}{2})^j)^k[\bar{G}*G_k](t)\exp\left\{\frac{\theta a(1-a^k/2^k)}{2(1-a/2)}\right\} + I(n),$$

where

$$I(n) = (2p)^{n+1}\exp\left\{-\theta(1-a)\frac{1-(a/2)^{n+1}}{1-a/2}\right\}\int_0^t N_j(t-u;\,\theta(\tfrac{a}{2})^{n+1})\,dG_{n+1}(u)$$

symbol $*$ designates onefold convolution, and $G_{k+1}(t) = G_k * G(t)$ for $t \geqslant 0$.

In consequence of the regularity of the process $Z(t)$ it is not difficult to show that $I(n) \to 0$ as $n \to \infty$ for any fixed value of $t \geqslant 0$. Therefore, letting n go to infinity, from (7) we find

$$(8)\quad N_j(t,\theta)= \frac{M\, e^{-\theta}\theta^j}{j\,!}\sum_{k=0}^{\infty}(2p(\tfrac{a}{2})^j)^k[\bar{G}*G_k](t)\exp\left\{\frac{\theta a}{2}\cdot\frac{1-(\frac{a}{2})^k}{1-a/2}\right\}$$

$j = 0,1,2,\ldots$.

The expected total number of cells, $N(t) = \sum_{j=0}^{\infty} N_j(t,\theta)$, is

$$(9)\quad N(t)=M\exp\left\{-\theta\frac{1-a}{1-a/2}\right\}\sum_{k=0}^{\infty}(2p)^k[\bar{G}*G_k]\,(t)\; e^{\frac{\theta(1-a)}{1-a/2}(\frac{a}{2})^k}$$

From (8) and (9) we derive finally the expression

$$(10)\quad \Pi_j(t)=\frac{\theta^j}{j\,!}\;\frac{\sum_{k=0}^{\infty}[2p(\frac{a}{2})^j]^k\exp\left\{-\frac{\theta}{1-a/2}(a/2)^{k+1}\right\}\bar{G}*G_k(t)}{\sum_{k=0}^{\infty}(2p)^k\exp\left\{\frac{\theta(1-a)}{1-a/2}(\frac{a}{2})^k\right\}\;\bar{G}*G_k\;(t)}$$

In the special case a=1 the expression (10) has the form

$$(11)\quad \Pi_j(t)=\frac{e^{-\theta}\theta^j}{j!}\;\frac{1+\sum_{k=1}^{\infty}(2^{1-j}p)^k[1-2^{j-1}p^{-1}\exp(-\theta/2^k)]\exp(-\theta/2^k)\;G_k(t)}{1+(1-1/2p)\sum_{k=1}(2p)^k\;G_k(t)}\;.$$

If assumption (a) is replaced by the postulates of a linear homogeneous birth-death process (Markov case), the distribution (11) is reduced to the one Williams has discovered under the same conditions imposed on the evolution marks in a cell population. Some modified versions of the distribution $\Pi_j(t)$ corresponding to special states of cell proliferation kinetics have been considered by Yanev and Yakovlev (1985).

REFERENCES

Athreya, K., and Ney, P. (1972). Branching Processes, Elsevier, New York.

Williams, T. (1969). The distribution of inanimate marks over a nonhomogeneous birth-death process. Biometrica 56, 225-227.

Yanev, N.M., and Yakovlev, A.Ya. (1985). On the distribution of marks over a proliferating cell population obeying the Bellman-Harris branching process. Math. Biosci. 75, 159-173.

STOCHASTIC SIMULATION
(Session 18)

Chairman: G.A. Mikhailov

Methods in Quantum Monte Carlo

M.H. Kalos
Courant Institute of Mathematical Sciences, New York University, New York, USA

1. Introduction

Monte Carlo methods are widely used in the computer study of systems of condensed matter. The study of "classical systems", that is those governed by Boltzmann statistics is now very widespread [1,2]. The basic approach is that Monte Carlo quadrature is used to evaluate expectations with respect to the Boltzmann distribution exp $[-\beta H]$, with $\beta = 1/(K_B T)$, T the absolute temperature.

The algorithm of Metropolis et al. [3] has generally been used to sample the (otherwise extremely difficult) probability density function.

Partly because the computations are more time consuming, partly because the understanding of the necessary algorithms was not so straightforward, the corresponding study of quantum systems has not been so widespread. Nevertheless, in recent years the interest and applications have been growing rapidly. For a recent survey, see reference [4].

If one expresses the results as averages with respect to Brownian walks and discretizes the walk in time, then methods similar to classical statistical mechanics can again be used with the additional complication that interacting replicas of the system for each "time slice" must be considered. An alternative is "Green function Monte Carlo". This is potentially free of truncation errors and admits of a sophisticated transformation for variance reduction, a form of "importance sampling". These may be of some general interest.

2. Solution to the Schrodinger Equation

We consider a system of N particles interacting by a potential energy function of the coordinates of all particles, V. Let R denote a point in $\mathbb{R}^{3n}$, the configuration space. Then in dimensionless units, the Hamiltonian is

$$H = -\Delta + V(R) \tag{1}$$

and Schrodinger's equation is

$$H\psi(R) = E\psi(R) \tag{2}$$

Let us shift the energy scale by Vo to obtain

$$(H+V_0)\psi(R) = (E+V_0)\psi(R) \tag{3}$$

and define the Green function for H+Vo:

$$(H+V_0)G(R,R') = \delta(R-R') \tag{4}$$

It is not difficult to show that if Eo is the lowest eigenvalue and

$$(E_0+V_0) > 0, \tag{5}$$

then

$$G(R,R') \geq 0 \tag{6}$$

We may now integrate equation (3) formally to obtain an equivalent integral equation:

$$\psi(R) = (E+V_0)\int G(R,R')\psi(R')\,dR' \tag{7}$$

We may consider the solution by iteration: let $\varphi^{\circ}(R)$ be some approximation to $\psi(R)$ and define

$$\varphi^{n+1}(R) = (E+V_0)\int G(R,R')\varphi^{n}(R')\,dR' \qquad (8)$$

It is easy to show that this converges to the lowest eigenfunction of (1) not orthogonal to φ°.

Monte Carlo methods can be used because G(R,R') is positive and integrable.

Thus, if a population of points $R_1, R_2, R_3, \ldots, R_m$ is sampled from $\varphi^{\circ}(R)$ and for each of these new points are sampled from $(E+V_0)G(R,R_k)$ then the new population has density $\varphi^{1}(R)$. Thus a random walk generates successive densities φ^{n} which converge to ψ. In general we will seek the lowest eigenfunction ψ_0 and lowest eigenvalue Eo.

However, $(E+V_o)\int G(R',R)\,dR' \neq 1$ for every R, so that the random walk is in fact a birth-death process. This presents only minor technical difficulty; it implies that a population of replicas of the system having fluctuating size must be considered.

It is assumed that G(R,R') can be sampled for arbitrary R,R' and potential function V. A crucial technical step is how that may be done. We use interated first passage probabilities.

Let Ω be a subdomain of $\mathbb{R}^{3N}$, and

$U \geq V(R)$ for every R in Ω ; U constant.

Let G_U be the solution of

$$(-\Delta + U)\,G_U(R,R') = \delta(R-R') \qquad (9)$$

on Ω with $G_U = 0$ on and outside $\partial\Omega$. R' is contained in Ω . Then multiply eq. (4) by G_U, equation (9) by G, subtract and integrate over Ω. There results the integral equation

$$G(R,R') = G_U(R,R') + \int_{\partial\Omega} G(R,R'')\left[-\nabla_n G_U(R'',R')\right] dS'$$

$$+ \int_{\Omega} G(R,R'')\left[1 - V(R'')/U(R'')\right] U(R'') G_U(R'',R')\, dR'' \qquad (10)$$

which describes a random walk starting at R' that can move to R'' in $\partial\Omega$ with p.d.f. $-\nabla_n G_U$(R'', R') or to R'' in Ω with p.d.f. G_U(R'', R') followed by a renewal with pr [1 - V/U]. If one can find and sample G_U for any class of finite domains Ω and any U, then a practical algorithm follows. We have used both hyperspheres in $\mathbb{R}^{3n}$ and (now more usually) cartesian products of 3-spheres with different radii.

3. Importance Sampling for the Schrodinger Equation

Algorithms that derive directly from the methods discussed above work very well for few-body problems if high accuracy is not required. For many-body problems such as fluids and crystals of the light elements, especially helium, and for the great accurary required for quantum chemistry, one must investigate appropriate variance reduction methods.

It is a general principle of Monte Carlo that an "importance function" can sometimes be defined which may play a role in variance reduction. The importance function has the significance of the contribution of a random walker at a particular position, in this problem, at some Ro, to the quantity of primary interest. We are concerned here with the asymptotic population φ^n for large n. What is the relative contributions of a walker at Ro? To answer the question let

$$\varphi^0(R) = \delta(R - R_0) = \sum_k \psi_k(R)\psi_k(R_0) \qquad (11)$$

where we expand in the normalized eigenfunctions of H. Then

$$Q^n(R) = \sum_k \left[\frac{E+V_0}{E_K+V_0}\right]^n \psi_k(R)\psi_k(R_0) \underset{n\to\infty}{\longrightarrow} \left[\frac{E+V_0}{E_0+V_0}\right]^n \psi_0(R)\psi_0(R_0) \quad (12)$$

thus the importance function is ψ_0 (Ro) - not surprising result given that H is self-adjoint.

That suggests that a population with density $\psi_0(R)\psi(R)$ rather than $\psi(R)$ should be generated. Assume for the moment that ψ_0 is known. Then equation (7) may be multipled through by $\psi_0(R)$ to get

$$\psi_0(R)\psi(R) = (E+V_0)\int \left[\psi_0(R)G(R,R')/\psi_0(R')\right]\psi_0(R')\psi(R')dR' \quad (13)$$

To generate the new transition matrix, equation (10) may be replaced by

$$\psi_0(R)G(R,R')/\psi_0(R') = \psi_0(R)G_U(R,R')/\psi_0(R')$$

$$+\int_{\partial\Omega} \left[\psi_0(R)G(R,R'')/\psi_0(R'')\right]\left[-\psi_0(R'')\nabla_n G_U(R'',R')/\psi_0(R')\right]dR' \quad (14)$$

$$+\int_{\Omega} \left[\psi_0(R)G(R,R'')/\psi_0(R'')\right]\left[1-V(R'')/U(R'')\right]\left[\psi_0(R'')U(R'')G_U(R'',R')/\psi(R')\right]dR'$$

To see what effect this has on the random walk that generates G from G, and thence on the sampling of the successive populations drawn from we multiply equation (3) by G_U, equation (10) by ψ, subtract and integrate over Ω. The result is

$$1 = (E_0+V_0)\int_{\Omega}\left[\psi_0(R)G_U(R,R_0)/\psi_0(R_0)\right]dR$$

$$+\int_{\partial\Omega}\left[-\psi_0(R)\nabla_n G_U(R,R_0)/\psi_0(R_0)\right]dS \quad (15)$$

$$+\int_{\Omega}\left[\psi_0(R)U(R)G_U(R,R_0)/\psi_0(R_0)\right]\left[1-V(R)/U(R)\right]dR$$

This produces a conservation of the number of walkers when the p.d.e.'s are modified by the ratio ψ_0 (R)/ ψ_0 (Ro). That is, the number of walkers in population $n+1$ which results from a walker in population n is exactly one. What was a birth-death process becomes a random walk by one walker at a time. In this ideal circumstance, the energy of the physical system can be estimated from the random walk with zero variance, and other properties with very low variance.

Of course, the exact eigenstate ψ_0 (R) is not known - or else we would not have been doing this calculation! What we have done in practise is to replace ψ_0 (R) by ψ_T (R) an approximate trial function known in advance.

Such functions are the input to the more usual approximate numerical and analytical theories of physical systems. For example a widely used and reasonable trial function for a homogeneous boson fluid in its ground state is a product of "pair functions"

$$\psi_T(R) = \psi_T(r_1, r_2 \cdots r_n) = \prod_{i<j} f(|r_i - r_j|) \tag{16}$$

It is satisfactory to choose a physically reasonable form of f(r) (satisfying the differential equation in the limit of small r and approaching one for large r) and adjust parameters to minimize the Rayleigh quotient for H. In quantum chemistry the structure of is generally well, if approximately, known.

With good approximate ψ_T instead of ψ_0 the walkers again undergo birth-death processes, but the populations are very nearly conserved. The variance of the Monte Carlo estimates of Eo are reduced by many orders of magnitude and the accurate quantitative study of large quantum systems becomes possible.

References

1. Monte Carlo methods in "Statistical Physics", K. Binder, Editor, Second Edition. Springer-Verlag, Berlin, 1986.

2. Applications of the Monte Carlo Method in "Statistical Physics",

K. Binder, Editor, Springer-Verlag, Berlin, 1984.

3. N. Metropolis, A.W. Rosenbluth, M.N. Rosenbluth, A.M. Teller, E. Teller: J. Chem. Phys. 21, 1087 (1953)

4. Monte Carlo Methods in Quantum Problems, M.H. Kalos, Editor, Reidel, (1984)

THE MONTE-CARLO METHOD AND ASYNCHRONIC CALCULATIONS

Ermakov S.M.

The Monte-Carlo method is a very important means which enables to utilize theoretic and probabilistic methods in many various branches of knowledge. Below the prospects of development of this method for the solution of equation (1) are briefly discussed. Let

$$\varphi = \mathcal{K}\varphi + f, \qquad (1)$$

where φ is an unknown function, $\mathcal{K}$ is in general case a nonlinear operator and its connection with asynchronic calculations. Let first $\mathcal{K}$ be linear $\mathcal{K}\varphi[x] = \int k(x,y)\varphi(y)\mu(dy)$ where μ is σ-finite measure on $\mathcal{X}$, $x \in \mathcal{X}$, k is a given kernel on $\mathcal{X} \times \mathcal{X}$ and $|\mathcal{K}|\varphi[x] = \int |k(x,y)|\varphi(y)\mu(dy)$. It is known the classical Neumann-Ulam scheme comparing with equation (1) to any markovian process is used for its solution only at that case when "majorant" equation

$$\bar{\varphi} = |\mathcal{K}|\bar{\varphi} + |f| \qquad (2)$$

has an iteration solution, Ermakov et al.(1984). The case when (1) has iteration solution but (2) has not is of a great interest for the investigation. In Wagner,(1983) a question about representation in this case corresponding functional from solution (1) as a main value of generalized integral on trajectories is discussed. In Ermakov (1983) vector algorithms and their generalizations are applied to evaluate the main values of integrals on trajectories. Majorant equations connecting with vector algorithms have an iteration solution under more weaker restrictions.

As it was mentioned in Ermakov (1983) the Neumann-Ulam scheme is exclusively convenient for applying on contemporary multiprocessor systems and its vector generalizations are so convenient if multiprocessor computers are supplied with a matrix processor.

The connection between branching random processes and nonlinear equations enabled to construct an analog of the Neumann-Ulam scheme for the solution of equations with polynomial nonlinearity, Ermakov (1979). The use of linearization methods treated to Karlemann made us to realize duality between sampling algorithms of branching processors and markovian processors with interaction, Nekrutkin (1979). Due to this fact algorithms of solution nonlinear Boltzmann equations have been carried out. These algorithms possess mentioned advantadges of the Neumann-Ulam scheme. The restrictions connecting with a majorant equation generate restrictions on the length of time interval where the solution may be obtained without remembering at an intermediate time interval. Studying of these restrictions similar to conditions of steability difference scheme makes possible designing of more economical in the sense of volume memory multiprocessor computer algorithms. The question of representation functional from the solution of nonlinear equation as the main value integral on trajectories has been studied in Wagner and Ermakov (1982). As in linear case to evaluate such integrals one may use vector algorithms and projection estimators. Studying these problems has begun comparatively not so late, Ermakov et al.(1984).

Thus, the problem of constructing Monte-Carlo for the solution of equation type (1) is connected first of all with considering of corresponding majorant equation (majorant iteration process).

It is natural to expect that analogue problems must

be arisen in another branches connected with calculations. It is really to turn out that the majorant equatiins are connected with significant branch of computional mathematics, especially asynchronic calculations.

It is known that for the realization of advantadges of modern multiprocessor systems in many cases the elaboration of special numerical methods are demanded. In particular when solving equations by the iteration method after discretization a part of processors may calculate values of unknown function before anothers. The methods which suppose sufficient on n iteration of all values of the function for receiving $n+1$ iteration (synchronic) cause nonproductive loss of time on expectation. Expenditures of such kind may be very large, Marchuk and Nesterenko (1983). If $n+1$ iteration will be calculated by free processrs at once (that is it will be used some values obtained in the result n, $n-1$ etc. iteration) loss on expectations won't be. It may be some cases when the corresponding values are calculated by groups (bloc iterations). The connection between asynchronic calculations and the Monte-Carlo method, in particular, is formally represented by the next result. Let $\mathcal{X}$ be $\mathcal{X} = \mathcal{Y} \times \mathcal{Z}$. Suppose a weak convergence of a process $\varphi_n = \mathcal{K}\varphi_{n-1} + f$. If $\{J_j\}_{j=0}^{\infty}$ is a chaotic sequence of sets integers from Marchuk and Nesterenko (1983) with maximal sediment and $\{1:N\}$, $Y = \bigcup_{t=1}^{N} Y_t$ is a subdivision on the intersecting subsets then iterations

$$\varphi_n(x) = \sum_{t \in J_n} \{\mathcal{K}\varphi_{n-1}[x]\}_{Y_t} + \sum_{t \in J_n} \{\varphi_{n-1}(x)\}_{Y_t} + f(x) \quad (3)$$

are called $\mathcal{Z}$ -bloc asynchronic. Here subdivision may depend from iteration number n, and $\{\psi(x)\}_{Y_t}$ for any function ψ on $\mathcal{X}$ is $\psi(x)$, if $y \in Y_t$ and zero otherwise.

If the Neumann series for some linear functional

from $\varphi(\mathcal{X})$ is calculated by the Monte-Carlo method with the help of simplest $\mathcal{Z}$-vector estimator (according to exact integration of the Neumann series on all variables corresponding $\mathcal{Z}$ in representation $\mathcal{X} = \mathcal{Y} \times \mathcal{Z}$) then the next theorem is valid.

THEOREM For the convergence of $\mathcal{Z}$-bloc iteration process it is sufficient and nessesary that simplest $\mathcal{Z}$-vector estimator is integrable on trajectories of sampling Markov chain.

The main idea of proving is in comparison of corresponding majorant processes for convergence (3) and integrability. Analogue results may be obtained in more complicated cases.

REFERENCES

I.Вагнер В. (I983) О несмещенности некоторых оценок метода Монте-Карло в знакопеременном случае. ЖВМ и МФ, 32, №3, I25-I38.

2.Вагнер В. и Ермаков С.М. (I982) О представлении решений нелинейных уравнений континуальными интегралами. ДАН СССР, 267, №6, I346-I350.

3.Ермаков С.М. (I979). Метод Монте-Карло для итерации нелинейных операторов. ДАН СССР, 204, №2, 27I-274.

4.Ермаков С.М. (I983). О суммировании рядов, связанных с интегральными уравнениями. Вестник ЛГУ, матем.,мех., астрон. I, 48-53.

5.Ермаков С.М., Некруткин В.В., Сипин А.С. (I984). Случайные процессы для решения классических уравнений математической физики. М., Наука.

6.Марчук В.А. и Нестеренко Б.Б. Введение в асинхронные методы решения краевых задач. Препринт 83.I7. Институт математики АН УССР, Киев.

7.Некруткин В.В. (I979). Несмещенные оценки решений нелинейных интегральных уравнений. В кн.:Исследование операций и статистическое моделирование. Ленинград, вып. 5, I2I-I59.

INCREASING THE EFFICIENCY OF STATISTICAL SAMPLING WITH THE AID OF INFINITE-DIMENSIONAL UNIFORMLY DISTRIBUTED SEQUENCES

Sobol' I.M.
Keldysh Institute of Applied Mathematics, Moscow, USSR

1. Introduction.

Let $X=(x_1,x_2,\ldots)$ be an infinite-dimensional point in the unit cube I^∞ and $P=(x_1,\ldots,x_n)$ - its projection in the n-dimensional unit cube I^n.

Definition. The sequence of points $X_1,X_2,\ldots$ is called an infinite-dimensional uniformly distributed sequence (i.u.d.s.) if the sequences $P_1,P_2,\ldots$ are uniformly distributed in I^n (in the sense of H. Weyl) at all $n\geq 1$.

It can be easily verified that generalized LP_τ-sequences and generalized Halton sequences investigated in my book (Sobol', 1969) are i.u.d.s.

Coordinates of $X_1,X_2,\ldots$ may be used as substitutes for random numbers $\gamma_1, \gamma_2, \ldots$ in various computations. Limit theorems that are true for $\gamma_1, \gamma_2, \ldots$ can be proved for these coordinates too. The following two theorems are analogues of the central limit theorem and of Kolmogorov's limit theorem. We assume that $X_1,X_2,\ldots$ is a i.u.d.s. and $X_k=(x_1^k,x_2^k,\ldots)$.

Theorem 1. If the normalized sum of the coordinates of X_k is

$$z_n^k=\sqrt{\frac{12}{n}}\sum_{i=1}^{n}\left(x_i^k-\frac{1}{2}\right)$$

and $S_n^N(u,v)$ is the number of z_n^k with $1\leq k\leq N$ satisfying $u\leq z_n^k<v$, then

$$\lim_{n\to\infty}\lim_{N\to\infty}\frac{S_n^N(u,v)}{N}=\frac{1}{\sqrt{2\pi}}\int_u^v \exp\left(-\frac{t^2}{2}\right)dt.$$

Theorem 2. Let $F_n^k(x)$ be the empirical distribution function of the values $x_1^k,\dots,x_2^k$ and $K_n^k=\sqrt{n}\sup_{0<x<1}|F_n^k(x)-x|$ – the Kolmogorov statistic. Let $S_n^N(z)$ be the number of K_n^k satisfying $1\leqslant k\leqslant N$ and $K_n^k<z$. Then

$$\lim_{n\to\infty}\lim_{N\to\infty}(S_n^N(z)/N)=\mathcal{K}(z),$$

where $\mathcal{K}(z)$ is the well-known Kolmogorov function.

2. Calculation of infinite-dimensional integrals.

We consider in I^∞ the product measure (dX) induced by Lebesgue measures on all coordinate axes and assume that f(X) is integrable in I^∞ with respect to dX. Quadrature formulas for computation of such integrals have been constructed (Sobol', 1969, 1974) – a problem suggested by N.N. Čencov. The convergence of these formulas depends on the distribution of the sequence of nods.

Definition. A sequence of functions $f_n(x_1,\dots,x_n)$ define and Riemann-integrable in I^n at $n\geqslant n_1$ is called a finite-dimensional approximation to f(X) if

$$\lim_{n\to\infty}\int_0^1\cdots\int_0^1 f_n(P)\,dx_1\cdots dx_n=\int_{I^\infty}f(X)\,dX.$$

Without loss of generality the definition of f_n can be extended to I^∞ so that $f_n(X)=f_n(x_1,\dots,x_n)$.

Theorem 3. The relation

$$\lim_{n\to\infty}\lim_{N\to\infty}\frac{1}{N}\sum_{k=1}^{N}f_n(X_k)=\int_{I^\infty}f(X)\,dX$$

holds for all integrable f(X) and arbitrary finite-dimensional approximations $f_n(X)$ if and only if the sequence of nods $X_1,X_2,\dots$ is a i.u.d.s.

3. Implementation of Monte Carlo algorithms.

Let J be a quantity that has to be evaluated and η is a random variable satisfying $M\eta=J$. A Monte Carlo algorithm for estimating J is specified by two relations (Sobol', 1973):

$$J \approx \frac{1}{N}\sum_{k=1}^{N}\eta_k, \qquad \eta = g(\gamma_1, \gamma_2, \dots).$$

We may introduce a random point $\Gamma = (\gamma_1, \gamma_2, \dots)$ in I^∞ with probability density p(X)=1. Then $J = Mg(\Gamma) = \int_{I^\infty} g(X)\,dX$.

It follows from theorem 3 that (theoretically) one may choose an arbitrary finite-dimensional approximation $g_n(X)$ to g(X) and points $X_1, X_2, \dots$ of a i.u.d.s. as substitutes for random points $\Gamma_1, \Gamma_2, \dots$: if N and n are sufficiently large then

$$J \approx \frac{1}{N}\sum_{k=1}^{N} g_n(X_k).$$

In various problems a remarkable increase in the rate of convergence has been attained with the aid of LP_τ-sequences.

Example. The mathematical expectation $M\eta = 1$ of the random variable

$$\eta = (2\gamma_1+1)(2\gamma_2+2)\cdots(2\gamma_9+9)/10!$$

has been estimated. The probable error of the Monte Carlo approximation is $0.30\ N^{-1/2}$; using a 9-dimensional LP_τ-sequence errors $2.0\ N^{-1}$ were obtained (at $N=2^5, 2^6, \dots, 2^{20}$).

4. Simulation of Markov chains.

Consider a homogeneous Markov chain $\xi_1 \to \xi_2 \to \dots \to \xi_s \to \dots$ with a finite or infinite state space $\{\omega^j\}$ and assume that 1° the initial state ω^{i_0} is recurrent nonnull; 2° a stationary (final) probability distribution p_j^* exists; 3° a function $f(\xi_s)$ is given and its final expectation $M^*f = \sum_j f(\omega^j)p_j^*$ is finite. Due to the law of large numbers as $T \to \infty$

$$\frac{1}{T}\sum_{s=1}^{T} f(\xi_s) \xrightarrow{P} M^*f, \qquad (1)$$

so M^*f can be estimated simulating one chain.

Let us define τ as the length of a cycle between two repetitions of the state ω^{i_0} and Q - as the sum of $f(\xi_s)$ over a cycle. Then $M^*f = MQ/M\tau$.

Denote by $\tilde{\xi}_1 \to \tilde{\xi}_2 \to \cdots \to \tilde{\xi}_s \to \cdots$ an auxiliary chain that obeys the same rules as the original chain with one extra rule: if n successive states are different from ω^{i_0} then (by definition) the next state is ω^{i_0}.

We now apply the standard Monte Carlo procedure to simulate the auxiliary chain but while constructing the k-th cycle we use for random numbers coordinates of X_k (Sobol', 1974). The lengths of the cycles of the auxiliary chain and the corresponding sums of $f(\tilde{\xi}_s)$ are in fact finite-dimensional approximations to τ and Q.

<u>Theorem 4</u>. Assume that 1°-3° are fulfilled and a segment of the auxiliary chain consisting of N cycles has been constructed with the aid of points of a i.u.d.s. Let $T_{N,n}$ be the length of the segment. Then

$$\lim_{n\to\infty} \lim_{N\to\infty} \frac{1}{T_{N,n}} \sum_{s=1}^{T_{N,n}} f(\tilde{\xi}_s) = M^* f . \qquad (2)$$

Consider (1) with $T = \tau_1 + \cdots + \tau_N$ - a sum of N cycles. As $N \to \infty$ the probable error of (1) is always $O(N^{-1/2})$. However the rate of convergence on N in (2) may be $O(N^{-1})$

List of references.

Sobol' I.M. (1969). Multidimensional quadrature formulas and Haar functions, Nauka, Moscow, 1 - 288.

Sobol' I.M. (1973). Numerical Monte Carlo methods, Nauka, Moscow, 1 - 312.

Sobol' I.M. (1974). On pseudo-random numbers for constructing discrete Markov chains by the Monte Carlo method. Ž. Vyčisl. matem. i mat. fiz., Vol. 14, N 1, 36 - 44.

Sobol' I.M. (1974). Infinite-dimensional uniformly distributed sequences in numerical mathematics. Preprint, Inst. Appl. Mathem., USSR Acad. Sciences, Moscow, N 22, 1 - 42.

CONTROLLED UNBIASED ESTIMATORS FOR CERTAIN FUNCTIONAL INTEGRALS

WAGNER, W., Akademie der Wissenschaften der DDR, Karl-Weierstrass - Institut für Mathematik, DDR-1086 Berlin, Mohrenstrasse 39, PF 1304

1. Introduction

We consider functional integrals of the form

$$I_c^{(t,x)}(t_o,x_o) := E \exp\left(\int_{t_o}^{t} c(s,w(s))\,ds\right) ,$$

where w is the Brownian bridge from x_o at the time t_o into x at the time t. The symbol E denotes the mathematical expectation. Let C be the space of continuous functions $v:[t_o,t]\rightarrow R^d$ such that $v(t_o)=x_o$ and $v(t)=x$. Let m_o denote the conditional Wiener measure on C. The function $c:[t_o,t]\times R^d\rightarrow R$ is supposed to be such that $\int_{t_o}^{t} c(s,v(s))ds$ is finite for almost all $v\in C$ with respect to m_o and that I_c is finite.

Functional integrals of this kind are important in mathematics and physics because of their connection with Green's function for certain partial differential equations (cf. the Feynman-Kac formula). Various parameters of quantum-mechanical systems (like the lowest energy level) can be evaluated numerically with the help of the functional integral representation of Green's function. The first papers in this direction are that by Donsker/Kac [3] and by Gel'fand/Chentsov [6], where the functional integral is evaluated by means of averages over independent samples of approximate Wiener trajectories.

Dealing with stochastic numerical algorithms for functional integrals two main problems have to be considered;

the approximation problem: to replace the functional of a continuous path by a functional of a discrete trajectory and to estimate the corresponding systematic error of the algorithm;

the variance reduction problem: to choose the parameters of the algorithm in order to reduce the statistical error depending on the variance of the estimator.

We present a numerical algorithm that unifies and generalizes several known algorithms and that solves the approximation problem, since no systematic error is involved. First we outline the algorithm with a simple choice of parameters in order to explain the main idea. Later we show, how to introduce more general parameters in order to deal with the variance reduction problem.

2. A solution of the approximation problem

We consider a sequence of fixed moments (t_i), $i=1,\dots,n$, $t_o<t_1<\dots<t_n<t=:t_{n+1}$. For any $i=0,1,\dots,n$, let $(t_{i,j})$, $j=0,1,\dots,k_i$, be a Markov chain of random moments, where $t_{i,o}:=t_i$ and $t_{i,j+1}$ is generated on $[t_{i,j},t_{i+1}]$ with the density proportional to $\exp(-g_i t_{i,j+1})$, $j=0,1,\dots,k_i-1$. The length k_i is random with Poisson distribution with the parameter $g_i(t_{i+1}-t_i)$. The (g_i) are positive constants. Denote $x_i:=w(t_i)$ and $x_{i,j}:=w(t_{i,j})$, $i=0,1,\dots,n$, $j=0,1,\dots,k_i$. The Wiener trajectory and the random sequences of moments are to be generated independently. Let $(d_i:=d_i(t_i,x_i,t_{i+1},x_{i+1}))$, $i=0,1,\dots,n$, be a set of measurable functions to be chosen later.

Theorem 1. Consider the estimator

$$\xi := \exp\Big(\sum_{i=o}^{n} d_i(t_{i+1}-t_i)\Big)\prod_{i=o}^{n}\frac{\exp(g_i(t_{i+1}-t_i))}{g_i^{k_i}}\prod_{j=1}^{k_i}(c(t_{i,j},x_{i,j})-d_i) .$$

If $E|\xi|<\infty$, then $E\xi = I_c^{(t,x)}(t_o,x_o)$.

Let us consider some special cases.

In the case $n=0$, $d_o=0$ we obtain the estimator

$$\xi = \frac{\exp(g_o(t-t_o))}{g_o^{k_o}}\prod_{j=1}^{k_o} c(t_{o,j},x_{o,j}) .$$

This is the "estimator by absorption" from the von Neumann-Ulam scheme for solving linear integral equations. It has been used in [4]. More general estimators of this kind were developed in [9]. These estimators use the Wiener trajectory exclusively at random moments $(t_o<t_{o,1}<\dots<t_{o,k_o}<t)$ and will be refered to as uncontrolled unbiased estimators.

In the case $d_i=(c(t_i,x_i)+c(t_{i+1},x_{i+1}))/2$, $g_i=0$, $i=0,1,\dots,n$, we obtain the estimator

$$\xi = \exp\Big(\sum_{i=o}^{n}(c(t_i,x_i)+c(t_{i+1},x_{i+1}))(t_{i+1}-t_i)/2\Big) .$$

This is the direct realization of Chorin's formula [2], which has been investigated and developed in [8], [1], [7]. This formula uses the Wiener trajectory only at fixed moments $(t_o<t_1<\dots<t_n<t)$. Its systematic error is n^{-2}.

Our estimators from Theorem 1 use the Wiener trajectory at some fixed moments (t_i) and at some random moments $(t_{i,j})$. They will be called controlled unbiased estimators and can be interpreted (for the corresponding (d_i)) as Chorin's formulas with correction terms, which make the approximation formulas unbiased.

3. On the variance reduction problem

The estimators from Theorem 1 are derived using Fosdick's mixed integration theorem [5]:

$$I_c^{(t,x)}(t_o,x_o) = \int_{(R^d)^n} p_{tr}^{(t,x)}(t_o,x_o,t_1,x_1) \dots p_{tr}^{(t,x)}(t_{n-1},x_{n-1},t_n,x_n) \, I_c^{(t_1,x_1)}(t_o,x_o) \dots I_c^{(t,x)}(t_n,x_n)\, dx_n \dots dx_1 ,$$

where $p_{tr}^{(t,x)}$ is the transition density of the Brownian bridge w. The usual way to apply mixed integration theorems to the construction of numerical algorithms is to approximate the inner functional integrals (e.g. , in Chorin's formula, the integrals $I_c^{(t_{i+1},x_{i+1})}(t_i,x_i)$ are approximated by the terms $\exp((c(t_i,x_i)+c(t_{i+1},x_{i+1}))(t_{i+1}-t_i)/2)$). Instead of this we use unbiased estimators for the inner functional integrals and combine them with an estimator for the finite-dimensional integral over $(R^d)^n$.

Thus, general parameters to be chosen in order to reduce the variance of the estimator are:

(a) the distribution of $(x_1,\dots,x_n)$,

(b) the distribution of $(t_{i,j},x_{i,j})$, $i=0,1,\dots,n$, $j=0,1,\dots$.

The basic principle of the importance-sampling technique is not to generate the Wiener trajectory itself, but to use the function c or some information about I in the choice of these distributions (cf. [9]). Several results on optimal parameters were obtained.

We mention only some results concerning the basic process itself. Consider a function c_o such that $I_{c_o}^{(t,x)}(t_o,x_o) < \infty$, and define a measure m_{c_o} on C:

$$dm_{c_o}[v] := \exp\Big(\int_{t_o}^{t} c_o(s,v(s))ds\Big)\, dm_o[v]\; I_{c_o}^{(t,x)}(t_o,x_o)^{-1}.$$

<u>Theorem</u> 2. The measure m_{c_o} corresponds to a Markov process with the transition density

$$p_{tr,c_o}^{(t,x)}(s_1,y_1,s_2,y_2) = p_{tr}^{(t,x)}(s_1,y_1,s_2,y_2)\frac{I_{c_o}^{(t,x)}(s_2,y_2)\,I_{c_o}^{(s_2,y_2)}(s_1,y_1)}{I_{c_o}^{(t,x)}(s_1,y_1)},$$

$t_o \le s_1 < s_2 < t$, $y_1,y_2 \in R^d$.

Fosdick's mixed integration theorem can be generalized to these Markov processes. We can use them for importance sampling in the infinite-dimensional space C, if we are able to generate the distributions given by their transition density p_{tr,c_o} .

Example. Consider $c_o(s,y)=-a^2y^2/2$, $y\in R$, $a>0$. The transition density $p^{(t,x)}_{tr,c_o}(s_1,y_1,s_2,y_2)$ is Gaussian with

mean $y_1\sinh(a(t-s_2))/\sinh(a(t-s_1)) + x\sinh(a(s_2-s_1))/\sinh(a(t-s_1))$ and variance $\sinh(a(t-s_2))\sinh(a(s_2-s_1))/(a\sinh(a(t-s_1)))$. The process corresponding to m_{c_o} should be called the harmonic oscillator process because of its relation to the quantum-mechanical harmonic oscillator having the potential c_o.

References

1. Blankenship G.L., Baras J.S. Accurate evaluation of stochastic Wiener integrals with applications to scattering in random media and to nonlinear filtering.-SIAM J.Appl.Math., 1981, 41, 3, 518-55
2. Chorin A.J. Accurate evaluation of Wiener integrals.-Math.Comput., 1973, 27, 121, 1-15.
3. Donsker M.D., Kac M. A sampling method for determining the lowest eigenvalue and the principal eigenfunction of Schrödinger's equation.-J.Res.Nat.Bur.Standards, 1950, 44, 5, 551-557.
4. Dyatkin I.G., Zhukova S.A. On the problem and the algorithm of solving Schrödinger's equation by the Monte Carlo method.(in Russian)-Zh.Vychisl.Mat.iMat.Fiz., 1968, 8, 1, 222-229.
5. Fosdick L.D. Approximation of a class of Wiener integrals.-Math. Comput., 1965, 19, 90, 225-233.
6. Gel'fand I.M., Chentsov N.N. On the numerical evaluation of functional integrals.(in Russian)-Zh.Ehksper.Teor.Fiz., 1956, 31, 6, 1106-1107.
7. Hald O.H. Approximation of Wiener integrals.- University of California-Preprint, 1985, PAM-303, 10p.
8. Maltz F.H., Hitzl D.L. Variance reduction in Monte Carlo computations using multi-dimensional Hermite polynomials.-J.Comput.Phys., 1979, 32, 3, 345-376.
9. Wagner W. Unbiased Monte Carlo evaluation of certain functional integrals.-AdWderDDR, Institut für Mathematik-Preprint, 1985, P-MATH-11/85, 21p.

KAC'S MODEL FOR A GAS OF N PARTICLES AND MONTE-CARLO SIMULATION IN RAREFIED GAS DYNAMICS

Yanitskii V.E.

Computing Centre. Academy of Sciences. USSR, Moscow

1. The approach involved is simulation of the Boltzmann gas by a system of a finite number of N particles ($N \sim 10^3 \div 10^4$). The laws of elastic collisions of molecules with indices i and j are defined by differential $d\sigma_{ij}$ and full σ_{ij} cross-sections, $\sigma = \int d\sigma$. Consider one of the gas models containting N particles. Its Monte-Carlo simulation allows to calculate the kinetics of space-nonhomogeneous gas flows. In physical space Ω we introduce a sufficiantly fine mesh that devides Ω into cells of volume V. Continuous time t is substituted by discrete $t_\alpha = \alpha \cdot \Delta t$. At the initial instant space Ω is filled with particles. Their coordinates x and velocities C are chosen by Monte-Carlo methods in accordance with densities $n_0(x) = \int f_0(x,c)dc$ and $\varphi_0(C/x) = f_0(x,c)/n_0(x)$ of initial distributions x and C . Calculation of evolution of this particle system over small time interval Δt is divided into the following two stages: first, pair particle collisions are simulated in cells, only velocities changing here. Second, a free flight of particles from cell to cell and their collisions with boundary are simulated. This approach was first realized by Bird (1976). The main peculiarities of the model considered are in the scheme of collision simulation and in the approach to designing these schemes, Janitskii (1975), Belotserkovskii and Janitskii (1976).

We assume that at time instant t_α ($\alpha = 0,1,\dots$) in a cell with centre $x_j (j = 1, 2,\dots,J)$ there are $N(\alpha,j)$ particles with velocities $c_1,\dots,c_{N(\alpha,j)}$. State $\{N(\alpha), C(\alpha)\}$ is defined by the succession out of J points, with appearance $\{N(\alpha,j);\ c_1,\dots,c_{N(\alpha,j)}\}$. Alternation of the collision stages mentioned above and particle shifts gives trajectory $\{N(\alpha)C(\alpha)\};\ \alpha = 0,1\dots$. Having a set of realized trajectories we can calculate any gas macroparametre ψ at instant t_α at point x_j using estimates of the corresponding integrals.

2. To construct simulation algorithms of N particle collisions in a cell at time interval Δt we use $3N$ dimensional Kac's (1967) model approximating the Boltzmann equation; vector $\mathbf{C} = \{c_1,\dots,c_N\}$ being its state. Let $\mu_t(\mathbf{C})$ be the density distribution of $\mathbf{C}(t)$ over the hypersphere of constant energy E and impulse P of system $\{c_1,\dots,c_N\}$. The basic equation of this model $\partial\mu_t/\partial t = A\mu_t$ is linear

$$A = \sum_{1\le i<j\le N}' \omega_{ij}(T_{ij} - I)\ ,\ T_{ij}\varphi = \int \varphi(\mathbf{C}_{ij})\, d\sigma_{ij}/\sigma_{ij}\ ,$$

$$\omega_{ij} = |c_i - c_j|\sigma_{ij}/V,\ \mathbf{C}_{ij} = \{c_1,\dots,c_i',c_{i+1},\dots,c_j',c_{j+1},\dots,c_N\}.$$

The connection of Kac's model with the Boltzmann equation is known. A chain of connected equations, with

$$\partial f^{(1)}(t,c_1)/\partial t = \int [f^{(2)}(t,c_1',c_2') - f^{(2)}(t,c_1,c_2)]\, g_{12}\, d\sigma_{12}\, dc_2$$

is valid for s-particle "kinetic distribution functions"

$$f^{(s)}(t,c_1,\dots,c_s) = \frac{N!}{(N-s)!\,V^s} \int \mu_t(\mathbf{C}) \prod_{i=s+1}^{N} dc_i$$

The latter coincides with the Boltzmann equation with accurasy up to the value of statistical particle dependence. In flow simulating this error is negligible, if average concentration $\bar{N}/V$ of model particles is great enough.

Conversion of densities $\mu_{t+\Delta t} = G(\Delta t)\mu_t$ corresponds to accurate realization of Kac's model, and

$$G(\Delta t) = EXP[\Delta t \cdot \sum_{1 \le i < j \le N} \omega_{ij}(T_{ij} - I)]$$

The corresponding algorithm is quite elaborate, if a special case of "maxvellian molecules", for which ω_{ij} = =const, is not considered. In a general case of frequencies ω_{ij} depending on $|c_i - c_j|$ more effective is an approximating scheme analogous to the succession of Bernulli tests, with different probabilities of success. An approximating operator corresponds to this scheme:

$$G(\Delta t) = \prod_{1 \le i < j \le N} [(1 - \Delta t \cdot \omega_{ij})I + \Delta t \cdot \omega_{ij} T_{ij}]$$

THE SCHEME OF COLLISION SIMULATION BY SUCCESSION OF BERNULLI TESTS

For every velocity pair (c_i, c_j) of N particles in a cell the following procedure is carried out.

Step 1. Collision of pair (c_i, c_j) is simulated with probability equal to

$$p_{ij} = \Delta t \cdot g_{ij} \sigma_{ij} / V, \quad \sigma_{ij} = \sigma(g_{ij}), \quad g_{ij} = |c_i - c_j|.$$

If the end of this test is "success", the following step is performed.

Step 2. The values of velocities c_i and c_j are changed into c_i' and c_j', respectively.

$$c_i' = \frac{1}{2}[(c_i + c_j) + g_{ij} e], \quad c_j' = \frac{1}{2}[(c_i + c_j) - g_{ij} e]$$

here e - random single vector, distributed over the sphere with density

If there is "failure" at step 1 values c_i and c_j are not changed.

The second stage of model evolution at step t which is a collisionless particle transfer can be realized by the scheme of Bernulli tests, Janitskii (1974). For the sake of simplicity consider one--dimensional flow in coordinate space. In this case calculation of the system state at instant $t + \Delta t$ with state $\{\{N(j); c_1, \ldots, c_{N(j)}\}$; $j = 1, \ldots, J\}$ at instant t is reduced to simulating tests of all particles in each cell with index j for an output with probabilities $|c_{x_i} \Delta t / \Delta x|$ where i=1,...

..., N(j). If there is "success" vector c_i of the tested particle is transferred into the set of the neighbouring cell minding the sign c_{xi}.

3. The described model was applied to calculate the shock wave structure and heat transfer in rare gas between two plates. Comparison was made and good agreement with the result of direct numerical intergration of the Boltzmann equation was achieved, (Janitskii(1975),Belotserkovskii and Janitskii(1976). Application of weight multiples to the scheme of collision simulation permitted to enlarge the method to kinetics problems of neutral and chemically reacting mixtures, Radev et al.(1984), Korolev and Janitskii (1985).

REFERENCES

Bird, G. (1976). Molecular Gas Dynamics, Clarendon Press, Oxford.

Belotserkovskii, O.M., Janitskii, V.E. (1976). Numerical investigation of rarefied gas flows by a statistical particle-in-cell method. Lect. Not. Phys. 59,105-113.

Kac, M. (1967). Probability and related topics in physical sciences. Interscience Publishers Ltd. London - New York.

Korolev, A.E., Yanitskii, V.E.(1985). Development of particle Monte-Carlo method for physice-chemical kinetics problems. Zh.vychisl.Mat.mat.Fiz.,(in Russian), 25, 431-441.

Radev,S.P.,Stephanov,S.K.,Janitskii,V.E.(1984),Particle-in-cell statistical method and application to problems of gas kinetics. Proc.14th Int.Symp.on Rare.Gas Dynam.,Univ.Tokyo Press. pp. 207-214.

Janitskii, V.E.(1974) Application of random walk processes for Monte-Carlo simulation free molecular gas flows. Zh.vych.Mat.mat.Fiz., (in Russian) 14,259-262.

Janitskii, V.E.(1975). Statistical model of an ideal gas flow and its peculiarities. In: Numerical Methods in mechanics of a continuous medium. Novosibirsk (in Russian) 4, 139-150.

STATISTICAL COMPUTING
(Session 21)

Chairman: S. Mustonen

MATHEMATICAL PROGRAMMING IN STATISTICS : an overview

Yadolah Dodge
Université de Neuchâtel
Groupe d'Informatique et de Statistique
Pierre-à-Mazel 7
2000 Neuchâtel
Switzerland

ABSTRACT: This is a review of mathematical programming applications in some areas of statistics, concentrating basically on the L_1-regression estimators. It begins with a brief exposition of the classical and the mathematical optimization methods. The linear programming formulation of L_1 regression along with its properties is reviewed. An integer programming formulation of clustering is given. The last application concerns optimally choosing a design of experiment.
KEYWORDS: Mathematical programming, optimization, L_1-norm, design of experiment, regression analysis.

1. INTRODUCTION

In the development of theories underlying statistical methods, one is often faced with an optimization problem. Whether it is designing a scientific experiment, planning a survey for collection of data, or drawing inference from the available data, one has to choose an objective function and maximize or minimize it subject to given constraints such as costs or unknown parameters.

For example, to determine an estimator we need a set of criteria by which its performance can be judged. Intuitively, by an estimator of parameter θ we mean a function T of the observations $(x_1,\ldots,x_n)$ which is closest to the true value in some sense. In laying down criteria of estimation one attempts to provide a measure of closeness of an estimator to the true value of a parameter and impose suitable restriction on the class of estimators. An optimum estimator in the restricted class is determined by minimizing the measure of closeness. Some restrictions on the class of estimators seem to be necessary to avoid trivial estimators.

Regression analysis is concerned with the problem of predicting a variable, called "dependent variable", on the basis of information provided by certain other variables, called "independent variables". A function $f(x_1,...,x_k)$ of the independent variables $X=(x_1,...,x_k)$ is called a predictor of a dependent variable Y that is considered. Here again, different criteria are considered for optimization. For instance, the classical approach is to minimize the mean square error.

Examples of such can be found in almost every branch of statistics. The techniques for solving such problems can be classified as **classical, numerical, variational methods**, and **mathematical programming**.

In many situations, the classical optimization methods based on differential calculus are too restrictive and are either inapplicable or difficult to apply. Moreover, the lack of suitable numerical algorithms for solving such problems has placed severe limitations on the choise of objective functions and constraints and in most cases leads to development and use of some inefficient statistical procedures.
In what follows, we try to give some examples to clarify why it is necessary to introduce the mathematical programming techniques in solving statistical problems.

2. SOME TYPICAL EXAMPLES

2.1 Regression analysis

One of the most extensively and exhaustively discussed

methods among the statistical tools available for analysis of data is regression.

Suppose we assume that the relationship is linear between the two variables. That is, the functional relationship of **Y** and **X** is of the following form

$$\mathbf{Y} = \mathbf{X}\boldsymbol{\beta} + \boldsymbol{\varepsilon} \qquad (3.1)$$

where Y is an nx1 vector of response variables, **X** an nxk matrix of predictor variables, $\boldsymbol{\beta}$ a kx1 vector of unknown parameters, and $\boldsymbol{\varepsilon}$ an nx1 vector of random errors.

We would like the predicted value of **Y** to be close to the observed value of **Y**. This closeness can be stated, in general, in terms of a L_p-norm. Here the L_1-norm problem can be stated as : find a vector that minimizes

$$\Sigma \; |Y_i - \mathbf{x}_i'\boldsymbol{\beta}|^p \; ,$$

for $p \geq 1$, where Y_i is the ith element of the vector Y and $\mathbf{x}_i$ is the ith row of the matrix **X**. The values of p most widely studied in the literature are p=1,2, and ∞.

For p=2, the criterion is the familiar least squares, and computational techniques for obtaining the estimators can be found in almost every existing text book in statistics (**everyone born should know the least squares method**). It is used extensively since it was discovered because of its analytical tractability and its highly developed theory and widespread literature. The other values of p that are commonly used are 1 and ∞. For p=1, the criterion is least absolute values ; and for $p=\infty$, it is the minimax criterion.

2.1.1 L_1 Estimation

The L_1 criterion of estimation has been refered to in the literature by a variety of names : minimum or least sums of absolute errors ; minimum or least absolute deviations, or errors, or residuals ; L_1; and so on.

The criterion of minimizing the L_1-norm is preferable to that of least squares in the presence of large disturbances (outliers) or when the classical assumption of a normal distribution or error is violated due to "contamination" or "heavy tails". Furthermore, if the errors are independently and identically distributed as the double exponential distribution, then it is known that the application of the maximum likelihood method for estimating β would imply minimization of sum of absolute deviations.

Nevertheless, the use of the least absolute criterion has been restricted because of (i) the lack, until recently, of good statistical inference procedures, and (ii) the need for specialized computer algorithms, to which statisticians and many practitioners often do not have access.

The L_1-norm estimator was studied as early as 1757 by Boscovich (see Eisenhart 1961, 1962). Laplace (1793) also suggested and studied this estimator (see Stigler 1973). Edgeworth (1887) presented a method for the simple regression, with L_1-norm estimator. However, Turner (1887) questioned Edgeworth'claim of his method's computational superiority over the least-squares method, and also pointed out the nonuniqueness of the L_1-norm estimator.

2.1.2 Linear programming formulation of L_1 regression

One can distinguish two general types of algorithms for computing the L_1 estimate. One type makes explicit use of the fact that the minimum occurs at one of the simplex points. These algorithms employ linear programming or projection methods to search among the simplex vertices. The other type of algorithm uses iterative descent methods. The algorithm which work only with vertices are generally more efficient and more accurate.

Consider the problem of minimizing $\Sigma|d_i|$ with respect to $\boldsymbol{\beta}$ where d_i is the deviation from the observed, and predicted values of Y_i the ith observation.

The problem can be stated as follows :

$$\text{Minimize } \Sigma|d_i|$$
$$\text{subject to } \mathbf{X\beta+d=Y}$$
$$\mathbf{d,\beta} \text{ unrestricted in sign}$$

Noting the fact that $d_i=d_{1i}+d_{2i}$ where d_{1i} and d_{2i} are nonnegative, and $d_i=d_{1i}-d_{2i}$, we can reformulate the problem as :

$$\text{Minimize } \Sigma d_{1i}+\Sigma d_{2i}$$
$$\text{subject to } \mathbf{X\beta+d_1-d_2=Y} \qquad (3.2)$$
$$\boldsymbol{\beta} \text{ unrestricted in sign}$$
$$\mathbf{d_1,d_2 \geq 0}$$

While it is possible to apply the standard simplex algorithm to (3.2) or to its dual, this formulation generally requires an unacceptable amount of computation time and storage. Barrodale and Roberts (1972, 1973) present a modification of the linear programming simplex method that saves storage and gains efficiency in computation by skipping over simplex vertices. The problem can be stated as

$$\text{Minimize } \Sigma d_{1i}+\Sigma d_{2i}$$
$$\text{subject to } \mathbf{X(\beta_1-\beta_2)+d_1-d_2=Y}$$
$$\mathbf{\beta_1,\beta_2,d_1,d_2 \geq 0}$$

For details on computation aspect of the L_1 see Kennedy and Gentle (1980) pages 513-525, and Arthanari and Dodge (1981) pages 32-88.

2.2 Cluster Analysis

One of the problems in multivariate data analysis is the problem of cluster analysis, problems which involve grouping a certain number of entities into a certain number of groups, in some sense **optimally**, arise in various fields of scientific inquiry.

2.2.1 Clustering methods

Methods available for clustering can be divided into two categories, those that use probability distribution assumptions on the characterization vectors and those that do not. The most common methods of the second category are the hierarchical ones.

In hierarchical, one may find among other methods nearest neighbor, forthest neighbor, centroid, and Ward's methods.

The primary charm of these methods is their computational ease. However, these methods suffer from a basic deficiency : they generally search for locally optimal clustering ; once a cluster is formed we do not remove any element or elements to see whether that can produce different clusters at different threshold levels.

These advantage and disadvantage are shared by other procedures available for solving clustering problems. Thus we should consider methods that can produce optimal solutions to clustering problems.

In regard to these methods, Everitt (1974) states :

"The number of clustering techniques available is large, as is the number of problems in applying them. The greater part of the mushrooming literature on classification is concerned with new techniques which in general suffer from many of the problems of those methods already in use. Perhaps the problem is that mentioned by Johnson (1968), namely that anyone who is prepared to learn quite a deal of matrix algebra, some classical mathematical statistics, some advanced geometry, a little set theory, perhaps a little information theory and graph theory and some computer techniques and who has access to a good computer and enjoy mathematics (as he must if he gets this far) will probably find the development of new taximetric methods much more rewarding, more up-to-date, more "general", and hence more presti-

gious than merely classifying plants or animals."
The need for algorithms that can find the global optimal cluster has been filled to some extent by the applications of integer and dynamic programming methods and partial-enumeration techniques, known as branch-and-bound methods.

2.2.2 Integer programming and cluster median problem

Let N={1,2,...,n} be the set of objects or elements that are
to be clustered into m clusters. We are given for each $i \varepsilon N$, a vector $X_i = (x_{1i},...,x_{zi}) \varepsilon R^z$.

Let n_k denote the number of elements in the kth cluster, k=1,2,...,m. So

$$\sum_{k=1}^{m} n_k = n.$$

Let m_0 denote the given upper bound on n_k. If there is no such restriction, m_0=n. Even though we wish to have only m clusters we define n clusters fictiously, n-m of which will have no elements at all. Further, for each nonempty cluster we have a median. Thus there should be m medians in all. We call the cluster for which element j is the median, j-cluster. Let

$$x_{ij} = \begin{cases} 1 & \text{if the ith element belongs to the j-cluster} \\ 0 & \text{otherwise,} \end{cases}$$

for i,j=1,...,n.

Let $((d_{ij}))$ be any distance matrix. An integer programming formulation can be given as :

Minimize $\quad \Sigma \Sigma x_{ij} d_{ij}$

subject to $\quad \Sigma x_{ij}=1 \;, \; i=1,...,n$

$\Sigma x_{jj}=m$

$x_{jj} \geqslant x_{ij} \;, \; i=1,...,n, \; j=1,...,n$

each x_{ij} is either 0 or 1.

The cluster-median problem can thus be solved as a 0-1 integer programming problem.

2.3. Optimal designs

Different optimality criteria are considered for finding optimal designs. For an extensive bibliography on the theory of optimal designs see Ash and Hedayat (1978). Gribik and Kortanek (1977) identify the mathematical-programming problems arising in such situation and provide a cutting-plane algorithm to solve the problems.

Consider the model,

$$y|x = \sum_{r=1}^{p} \theta_r f_r(x)+\varepsilon(x)$$

where x belongs to the factor space, $E(\varepsilon(x))=0$, and Var $(\varepsilon(x))=\sigma_x{}^2$.

Let the n_i independent observations of y be made at x_i, $i=1,...,m$. The observations are made independently at x_i and x_j, $i=j$. Let y_{ij} denotes the value of the jth observations made at point x_i. Let

$$n = \sum_{i=1}^{m} n_i \quad \text{and} \quad M = \sum_{i=1}^{m} (n_i/n)f(x_i)f(x_i)'$$

where $f(x_i) = (f_1(x_i),...,f_p(x_i))$. M corresponds to the information matrix (X'X) in the linear model.

Assume that M is nonsingular. Then, because $1/nM^{-1}$ is the covariance matrix of the least squares estimator θ of θ, it would be desirable to choose M for a given n, n_i, x_i, so that the covariance matrix of θ is "best" in terms of some criterion.

Suppose the criterion function is $C(M^{-1})$. Then the problem can be formulated as

Minimize $\quad C(M^{-1})$

subject to $\quad M = \Sigma\, p_i f(x_i)f(x_i)'$, which is nonsingular,

$\Sigma\, p_i=1$

$p_i \geqslant 0$, $i=1,...,m$

$p_i n$ is an integer , $i=1,\dots,m$

m is an arbitrary positive integer.

This problem corresponds to finding an exact design. An approximate design, $\xi(.)$, is a probability measure over the factor space. A similar problem for approximate designs can be stated as :

Minimize $C(M^{-1})$

subject to $M \epsilon \Omega = \{M(\xi) \epsilon R^{pxp} |$

$M(\xi) = \int f(x) f'(x) \xi(dx)$

for $\xi(.)$, a probability measure on the factor space$\}$

and M nonsingular.

4. Concluding remarks

Mathematical programming problems have received the attention of researchers in mathematics, economics, and operations research for over three decades. Since the development of the simplex method for efficiently solving the linear programming problem, both the theory and the methods of mathematical programming have seen unprecedented growth. Also, the emphasis has turned for solving certain problems, toward finding efficient methods suitable for computers.

Thus it is encouraging to attempt to use the available literature on mathematical programming to formulate and solve several of the statistical problems that might not be solved efficiently with classical methods alone.

With the recently developed asymptotic normal theory for L_1 estimators in the linear regression model and the availability of a number of computationally efficient algorithms for calculating L_1 estimates, it is reasonable to expect that L_1 estimation procedures will be more commonly applied to data sets than before in regression experiments. For the error distributions such as Cauchy and Laplace for which the median is superior to the mean as an estimator of location, L_1 estimation is certainly preferred to least squares and should be used in these cases.

Inference techniques are now developed for L_1 estimation in the regression model based on the asymptotic normal theory. These techniques are relatively easy to use.

Even now that the strength and utility of the L_1-norm has been established by many authors during the past to decades, it is unfortunate to say that no traces are to be found in classical textbooks.

Texts on regression dating of 1985 suprisingly contain no information of this norm, and I feel that the time is long due for the establishment of the L_1-norm, accessible in all textbooks in statistics at the undergraduate level.

Generalized versions of the Neyman-Pearson Lemma are obtained through duality results in mathematical programming. Mathematical programming can be applied to ensure nonnegative estimates of Rao's MINQUE approach. These are a few of the applications of mathematical programming in statistics.

All the literature cited in this paper can be found in the following references.

References

1. Arthanari T.S. and Dodge Y. (1981). Mathematical Programming in Statistics. John Wiley and Sons, New York.
2. Gentle J.E. and Kennedy W.J. (1980). Statistical computing. Marcel Dekker, New York.
3. Koenker R. and Bassett G. (1982). Tests of linear hypotheses and L_1 estimation. Econometrica, 50, 1577-1583.
4. Zanakis S.H. and Rustagi J.S. (Editors) (1982). Optimization in Statistics. North Holland, Amsterdam.

MODEL SEARCH IN LARGE MODEL FAMILIES

Tomáš Havránek
Institute of Computer Science, Czechoslovak Academy of Sciences,
Pod vodárenskou věží 2, 182 07 Prague, Czechoslovakia

INTRODUCTION

In the present paper we wish to summarize a rather general and abstract framework for formulation and solving a data analytic task both from computational and statistical point of view. Consider the following situation: let RQ be a family of models (hypotheses, relevant questions), let $\underline{M}$ be a corresponding sample space, and let d be a decision rule, giving for each $M \in \underline{M}$ value 'accept' (d(m,M)=a) or 'reject' (d(m,M)=r). The task is now, for a given data set M to find the set A = {m ∈ RQ; d(m,M) = a} or equivalently R = {m ∈ RQ; d(m,M)=r} and, moreover, to find a comprehensive representation of these sets. Note, that d is not necessarily a statistical test. If d corresponds to a test, say a goodness-of-fit test on a given significance level α, then clearly the nature of the decision 'accept' should be taken into account. Moreover, even if d is a test, then sometimes the set of interest is A, sometimes R or our interest is focused to both of these sets. C.f. Havránek (1984), Hájek (1984) and Edwards and Havránek (1985) respectively.

If we speak about large families of models, we suppose RQ to have from hundreds to millions of members (say 128 to 4.2×10^6 as in Table 1 of Havránek (1981a) or even 2.2×10^{12} members). The last number appears if one wish to consider all 2x2 tables derived by collapsing rows and columns from a 25x18 contingency table). Clearly, in such a situation it is not possible to inspect all members of RQ to obtain A. It is necessary to consider techniques inspecting only some small proportion of RQ and enabling to decide about the whole RQ (i.e. to find A and R). Typically, such a technique will use, for example, 17 models from 128 (speed-up factor 7.53) or 9853 models from 4.2×10^6

(speed-up factor 425.69) or 7672979 models from 2.2×10^{12} (speed-up factor 2.86×10^{5}); the efficiency depends on data.

The most straightforward way is to use some partial ordering on RQ defined by d, say $m_1 \leqq_0 m_2$ if for each $M \in \underline{M}$, $d(m_1,M)=a$ implies $d(m_2,M)=a$ (we are using subscript 0 for observational - data level ordering). This way was consequently choosen in GUHA methods developed gradually from a very special case of simple dichotomous data to a general theory in Hájek et al. (1966), Hájek and Havránek (1977, 1978). This approach was oriented mainly towards multiway contingency tables. The principle is that if we find an m_1 such that $d(m_1,M)=a$, we shall without evaluating to put into A each m_2, $m_2 \geqq {}_0 m_1$. A good representation of A is then the set of its minimal elements with respect to the ordering $\leqq_0$. In practice the situation is much more complicated; for example, the question to establish for each pair m_1, m_2 whether $m_1 \leqq_0 m_2$ could be computationally expensive, hence only a quickly recognisable part of the ordering is used. Closely related to this approach is the approach used in Pokorný and Havránek (1978) and Pokorný (1982) for the search of sources of dependence in two-way contingency tables; the numerical example above refers to this case. Here the set RQ is adapted during a run of a recursive hierarchical procedure. Orderings of this kind are used in variable (regressor) selection procedures, like Furnival and Wilson (1974), but usually not for finding a whole set A.

Another way is to use some "simplicity" partial ordering between models, say $m_1 \leqq_T M_2$ if the parametric space of m_1 is embedded in that of m_2 (subscript T refers here to theoretic, data independent ordering). Here, if we find for some m_1 that $d(m_1,M)=a$, we put into A without evaluation each $m_2, m_2 \geqq_T m_1$ (and if we find $d(m_1,M)=r$, we put into R each $m_2, m_2 \leqq_T m_1$). A general algorithm of this kind using the lattice structure generated by $\leqq_T$ was developed in Edwards and Havránek (1987).

We shall now introduce three particular instances of partial ordered sets (posets) of models:

(i) Family of hierarchical log-linear models, RQH, for multidimensional contingency tables. Let the dimension of the table, say p, be given. Variables are then denoted by 1,2,...,p and we shall suppose that in the log-linear representation all main effects are present. Then the models in question can be described by its generating expressions including superscripts of maximal log-linear parameters present in the model. For p=5 we have, for example, ({1,2,3,4},{1,2,3,5}) expressing the conditional independence of variables 4 and 5. Usually a shortened notation is used, namely (1234,1235). Clearly (1234,145, 345) < (1234,1345), (1234,135,345) < (1245,2345,1234) and (1234,1345) and (1234,1245,2345) are incomparable. We write, as usual < for ≦ and ≠. The world clearly above refers to the fact that this ordering is just the natural simplicity ordering corresponding to the log-linear parameters involved (nesting of models). From the computational point of view it is important that this ordering can be recognised directly from generating expressions: generally $m_1=(A_1,\ldots,A_k)\leq m_2=(B_1,\ldots,B_l)$ if each $A_i \subseteq B_j$ for some j=1,...,l.

(ii) Family of graphical models RQG for multidiomenional contingency tables. This family is a proper subset of the previous one: in the present context see Edwards and Havránek (1985). There is one-to-one correspondence between members of this family and graphs. For p=3 e.g. 1-3-2 corresponds to (13,23), $\widehat{1\text{-}2\text{-}3}$ corresponds to (123). The model (12,13,23) is not graphical.

(iii) The third family RQP is inspired by one-way ANOVA, but it can be defined more generally (Edwards and Havránek, 1987) consider a set of parameters $\{\mu_1,\ldots,\mu_p\}$ and partitions of the set of indices {1,..., p}. E.g. for p=5, (13,24,5) codes such a partition. If now $m=(A_1,\ldots, A_k)$ is defined as $\mu_i=\mu_j$ if $i,j\in A_r$ for some r=1,...,k. The "natural" simplicity ordering can be here defined as follows: $m_1=(A_1,\ldots,A_k)\leq m_2(B_1,\ldots B_l)$ if each $A_i \supseteq B_j$ for some j=1,...,l; i.e. if m_2 is a refinement of the partition m_1.

In the data analytic task discussed above, we can consider two kinds of partial orderings. Let us suppose a family RQ be given and that there is a one-to-one correspondence between elements of RQ and decision functions (m<->d(m,.)) and hence we restrict ourselves to orderings on RQ only. Let now an ordering ≦ on RQ be given; we have then a poset (RQ,≦). Such an ordering can be quite arbitrary and we can discuss some properties of appropriate orderings. The ordering ≦ is 0--sound if it is embedded into the ordering $\leq_0$, i.e. if $m_1\leq m_2$ then $m_1 \leq {}_0 m_2$. The notion of soundness is a logical notion joining a deduction rule (as a formal syntactic relation) with underlying semantics.

One semantics here is just the semantics linking models with data via the decision function d. Another semantics on the theoretical level links models as formal expressions with probability distributions. We can suppose that each model $m \in RQ$ defines a subset P_m of probability distributions and we can say that a probability distribution p satisfies a model m if $p \in P_m$. In such a case, ordering $\leq_T$ is defined by $m_1 \leq_T m_2$ if $P_{m_1} \subseteq P_{m_2}$. Now we can say that the ordering $\leq$ from the beginning of this paragraph is T-sound if it is embedded into $\leq_T$.

Now we can define in the present framework the fundamental notion of coherence: A partial ordering $\leq$ on RQ is coherent if it is both O-sound and T-sound. A coherent ordering is completely coherent if each coherent ordering is embedded into it. There are two principles or rules for a treatment the data analytic task in question (Edwards and Havránek, 1987):

The *upward rule:* if for a given data M and a model m,d(m,M)=A then consider each $m_1, m \leq m_1$ as accepted (for M). We can call such a model weakly accepted.

The *downward rule:* If for a given data M and a model m,d(m,M)=r, then take each $m_2 \leq m$ as rejected. We can call such a model weakly rejected.

Using these rules one can evaluate not the whole set RQ but in particular instances only its small proportion. These rules are perfectly rational if (i) the ordering $\leq$ is (completely) coherent and (ii) the decision whether $m \leq m'$ is for each $m,m' \in RQ$ computationally inexpensive. A comment to (ii): Computationally inexpensive means in comparison with evaluating of d(m,M). In fact usually $\leq$ is so defined to be evaluated immediately. Just for this reason we distinguish between $\leq$ and $\leq_T$. Some comments to (i): Coherence guarantees that the result of applying a procedure using both rules is in principle the same as the result of evaluating all models in RQ. Completeness guarantees highest effectivity of applying the rules in terms of reduction of the number of evaluated d(m,M).

The above rules are used even in the case of non-coherent ordering $\leq$; then they give only some "approximative" results c.f. Edwards and Havránek (1985). The quality of such results depends on the degree of non-coherence. Usually only non-coherence due to a violation of

0-soundness is considered, but non-coherence due to a violation of T--soundness is worth of an attention as well. Particular case is if $\leqq_0$ or $\leqq_T$ is embedded into $\leqq$, for example $\leqq_0 \subsetneqq \leqq \subseteqq \leqq_T$.

LATTICES

To make the above rules really effective needs to consider more deeply the structure of the ordering on RQ. Remember now that RQ is a finite set. For each two elements m_1, m_2 of RQ we can define their meet $m_1 \wedge m_2$ as the greatest lower bound (i.e. such $m \in RQ$ that $m \leq m_1$, $m \leqq m_2$ and there is no $m' < m$ with this property). Clearly in a general (RQ,$\leq$) the existence of the meet for arbitrary two elements is not guaranteed. If for each $m_1, m_2 \in RQ$, $m_1 \wedge m_2 \in RQ$ exists, we speak about $\wedge$-semilattice (lower semilattice). This corresponds to the notion of closed set of models, c.f. Hommel (1986), Marcus et al. (1976) when $\leqq \subseteqq \leqq_T$. There are reasonable classes of models that do not form a semilattice - such a class is the class of decomposable models as a subclass of the class of graphical models for contingency tables. The meet can be viewed as an idempotent, commutative and associative operation between elements of RQ. An element $m \in RQ$ is called $\wedge$-irreducible if $m=m_1 \wedge m_2$ implies that $m_1=m$ or $m_2=m$. An element m is expressible as an irredundant meet of $\wedge$-irreducible elements $e_1,\ldots,e_k$ if $m=e_1\wedge \ldots \wedge e_k$ and $m \neq e_{i_1} \wedge \ldots \wedge e_{i_1}$ for each $\{i_1,\ldots,i_1\} \neq \{1,\ldots,k\}$.

In connection with closed sets of hypotheses it is worth to mention two facts: (a) In each finite $\wedge$-semilattice RQ there is the least element $\underline{0}$ (i.e. for each $m \in RQ$, $\underline{0} \leqq m$). The connection with the "overall" hypothesis will be discussed in Sect. 5. (b) If a finite $\wedge$-semilattice RQ contains a greatest element $\underline{I}$, then RQ is a lattice. It means that the closedness property has further important consequences.

Now we have to define a lattice. A poset (RQ,$\leq$) forms a lattice if it is a $\wedge$-semilattice and a $\vee$-semilattice, i.e. for each two elements $m_1, m_2 \in RQ$ the set RQ contains not only meet $m_1 \wedge m_2$ but moreover their join $m_1 \vee m_2$ defined as the least upper bound. Both operations are linked by the absorption: $m_1 \vee (m_1 \wedge m_2) = m_1$ and $m_1 \wedge (m_1 \vee m_2) = m_1$. Important is that each element of a finite lattice RQ can be expressed as an irredundant meet of $\wedge$-irreducible elements and as an irredundant join $\vee$-irreducible elements defined dually. Both of this representatations are not necessarily unique.

A lattice RQ is distributive lattice if for each $m_1.m_2,m_3 \in RQ$, $m_1 \wedge (m_2 \vee m_3) = (m_1 \wedge m_2) \vee (m_1 \wedge m_3)$. In a finite distributive lattice the above meet and join representation of elements is unique and moreover there is one-to-one correspondence between $\wedge$ and $\vee$ irreducible elements enabling immediately transform the meet representation to the join representation and vice versa. If we exclude $\underline{0}$ and $\underline{I}$ elements then $\wedge$-irreducible elements can be considered as "elementary" hypotheses and $\vee$-irreducible elements as "atomic" hypotheses in the terminology discussed in Hommel (1986).

From the computational point of view it is important that operations $\wedge$ and $\vee$ can be defined in many cases as formal (syntactic) operations over some strings of symbols (expressions):

(i) Hierarchical log-linear models RQH: the meet operation can be expressed as follows: $(A_1,\ldots,A_k) \wedge (B_1,\ldots,B_l) = (A_1 \cap B_1, A_1 \cap B_2, \ldots, A_k \cap B_l)$ after removing redundant intersections (a set is redundant if it is a subset of another set in this expressions; from two equal set clearly only one is redundant). For example (p=5): (1234, 1345) $\wedge$ (1234,1245,2345) = (1234,145,345). The join operation is even simpler: $(A_1,\ldots,A_k) \vee (B_1,\ldots,B_l) = (A_1,\ldots,A_k, B_1,\ldots,B_l)$ after removing redundant sets. The $\wedge$-irreducible elements are just $\underline{I}$=(1...p) and $(A_1,\ldots,A_k)$ where each A_i is of cardinality p-1 (one variable missing from each A_i). Then the above example is just the meet representation for (1234,145,345). This lattice is distributive.

(ii) If we consider a subclass of RQH containing $\underline{I}$ and all models represented as meet of the above $\wedge$-irreducible elements containing just two subsets (i.e. (A_1,A_2)), we obtain the lattice of graphical models RQG with the same ordering $\leq$. The meet operation is the same as above, the $\wedge$-irreducible elements are (1...p) and (A^i,A^j) for i=j, where $A^i = \{1,\ldots,p\} - \{i\}$, similarly A^j. This characterisation was considered in Havránek (1982) where a complexity hierarchy within RQH was discussed based on various complexity of meet representations. In RQG the corresponding $\vee$-irreducible elements are $\underline{0}$=(1,...,p) and m^{ij}= $=(B_1,\ldots,B_{p-1})$ where one B_r is (ij) and other are one element sets. This lattice is the least lattice with respect to the considered ordering containing the class of decomposable models. It is not only distributive lattice but a boolean lattice. The join is e.g. (12.3) $\vee_g$ (13,23) = (123). This join operation can be defined in the language of graphs; applying the join operation for RQH to $m_1, m_2 \in$ RQG we obtain $m_1 \vee m_2$ (possibly $\notin$ RQG) described by a generating expression, say $m=(C_1,\ldots,C_k)$. Such expression defines a graph with vertices $1,\ldots,p$ and edges $\{(i,j);\ (ij) \subseteq C_l$ for some $l=1,\ldots,k\}$. The least $m' \in$ RQG such that $m \leq m'$ is given by cliques of this graph. Clearly if $m \in$ RQG, then $m = m'$. The join $\vee_g$ is defined as $m_1 \vee_g m_2 = (m_1 \vee m_2)'$.

(iii) Similarly as in the previous cases (RQP,$\leq$) forms a lattice. The join of $m_1 = (A_1,\ldots,A_k)$ and $B_1,\ldots,B_l)$ is defined as $m_1 \vee m_2 = (A_1 \cap B_1, A_1 \cap B_2,\ldots,A_k \cap B_l)$ after removing empty intersections. The meet $m_1 \wedge m_2$ can be obtained in two steps. First, $(A_1,\ldots,A_k, B_1,\ldots, B_l)$ is formed and, second, non disjoint sets in it concatenated. The $\vee$-irreducible models are the models (A_1,A_2), meet irreducible models are $(A_1,\ldots,A_{p-1})$ (if we do not consider $\underline{0}=(1\ldots p)$ and $\underline{I}=(1,\ldots,p)$). As an example, see $(123,4,5)=(12,3,4,5) \wedge (1,23,4,5)=(12,3,4,5) \wedge (13,2,4,5)$. Hence this lattice is not distributive.

LATTICES IN EFFECTIVE MODEL SEARCH METHODS

For a poset (RQ,$\leq$) and a decision rule d we can describe the following algorithm for performing the considered data analytic task (Edwards and Havránek, 1985, 1987); first we define the notion of duals for an incomparable set S = RQ (for no models $m_1,m_2 \in S$ $m_1 \leq m_2$ or $m_2 \leq m_1$). The a-dual of S is the set of simplest models in RQ that are not contained in any model in S, formally: $D_a(S)=\min\{a\colon a \nleq s$ for each $s \in S\}$. Similarly r-dual of S is defined as the set of maximal models in RQ that does not contain any models in S, i.e. $D_r(S) = \max\{r\colon s \nleq r$ for each $s \in S\}$. Duals can be interpreted as follows: if models in S are rejected, then, using coherence, we can say that $d_a(S)$ consists of the simplest models in RQ that could concievably be accepted. Similarly if models in S are accepted, we can say, using coherence, that $D_r(S)$ consists of the most complex models in RG that could concievably be rejected.

Suppose now that we have in a stage of evaluating models a (incomparable) set A = RQ of accepted models and a (incomparable) set R = RQ of rejected models. Consider the set T of models whose status as accepted or rejected has not yet been determined, i.e. $T = \{m \in RQ: a \nleq m$ for all RQ $a \in A$ and $m \nleq r$ for all $r \in R\}$. Then $\max(T)=D_r(A)-R$ and $\min(T)=D_a(R)-A$. This is the base of the proposed procedure, since clearly evaluating either max(T) or min(T) can give an effective gain to the status of not yet evaluated models.

Suppose that $(RQ,\leq)$ and d are given. For a set of models S denote $S^r = \{m:m \in S,d(m,M)=r\}$, and similarly for S^a. Now we can describe the general algorithm for model search:

Input: initial incomparable set of models S_o, data $M,A:=R:=\emptyset$.

(1) $A:S^a$, $R:=S^r$,

(2) If $A=\emptyset$ go to (3), if $r=\emptyset$ go to (4), else choose between (3) and (4),

(3) evaluate $S:=(D_r(A)-R)$. If $S^r=S$ then stop, else go to (5),

(4) evaluate $S:=(D_a(R)-A)$. If $S^a=S$ then stop, else go to (5),

(5) $A:=\min\ (A \cup S^a)$, $R:=\max\ (R \cup S^r)$, go to (2).

The initial set S_o can be choosen freely. For the choice in (2), we suggest to compute number models in $D_r(A)-R$ and $D_a(R)-A$ and to evaluate the smalest set.

If $\leq$ is 0-sound then the algorithm gives a unique solution independent off the starting set S and choices in the step (2). The A is the set of minimal accepted models and R is the set of maximal rejected models; weak acceptance and rejection equals to acceptance and rejection.

The algorithm can be applied but with a cautin in situations in that $\leq$ is T-sound but not 0-sound. On the coherence or non-coherence (or a degree of non-coherence) depends the statistical properties of the algorithm.

The critical point from the computational point of view is there the search for duals. This search can be made effectively if $(RQ,\leq)$ forms a lattice and it is particularly effective if this lattice is distributive where the one-to-one correspondence between $\wedge$-irreducible and $\vee$-irreducible elements is used. Less effective is the search in partition lattices. This task and its solution for a general finite lattices as well as for both particular cases is described in Edwards and Havránek (1987); in Edwards and Havránek (1985) it is applied to contingency tables. Further applications can be found in Antoch (1986).

LATTICES IN SIMULTANEOUS INFERENCE

In the previous section we stressed the lattice properties in connection with computational aspects, now we can formulate in the lattice language some statistical properties in conncection to Hommel (1986), Marcus et al. (1976). Consider first finite $\wedge$-semilattices. As we know this semilattice contains $\underline{0}$. If $P_{\underline{0}} \neq \emptyset$ we can say that this model is satisfiable and we can speak about the overall hypothesis. Recall that if RQ contains $\underline{I}$ (the saturated model) then RQ forms a lattice. Note that it does not mean automatically that $\underline{I}$ is taken into account for an evaluating process. Meet irreducible elements distinct from $\underline{I}$ can be considered as elementary hypotheses. There if $m_1 \leq m_2$ then m_2 is called to be an implication of m_1; elementary hypotheses in the present context are not generally identical with maximal hypotheses of Hommel (1976). As we know for a distributive lattice, the representation of hypotheses by elementary hypotheses is unique - this is a condition for one to one relation between the elements of $\underline{\tilde{P}}$ and $\underline{\tilde{H}}$ of Hommel (1976).

In the context of simultaneous inference methods it is usual to assume that $\leq$ is T-sound. Let (RQ,$\leq$) be given. Suppose now that each d is a goodness-of-fit test on a (local) significance level α. Consider the algorithm deciding in any case in (2) in favour of (4) and using as the starting set the minimal satisfiable elements of RQ. Such a procedure can be called a purely upward procedure as using consequently the upward rule of Sect. 2. Now an important fact mentioned in Marcus et al. (1976) can be rather generally formulated:

If (RQ,$\leq$) is a $\wedge$-semilattice and $\leq$ is T-sound, then the purely upward procedure keeps the multiple α-level.

For multiple α-level see Hommel (1976); it is necessary to generalize its definition: if $\underline{0}$ is satisfiable, the multiple α-level refers to probability of at least one rejection under $\underline{0}$, if not then it refers to this probability under each minimal satisfiable $m \in$ RQ (the case of weakly closed class of hypotheses is covered; it is corresponds to the case where $P_{\underline{0}}=\emptyset$). Clearly the fact remains true if all local levels are less equal to α. Noticeable is the fact that we do not need here coherence. See the example from Marcus et al. (1976), pp. 658-659. On the other hand, if we consider a procedure using both rules, coherence is in general necessary to keep the multiple α-levels.

If we take a slightly different point of view, then the above fact is based on a construction of a new decision rule: $d'(m,.)=r$ if $d(m_1,.) = r$ for each $m_1 \leq m$ (in the lattice terminology it means that each m is substituted by the principal ideal defined by it: $m^* = \{m_1;\ m_1 \leq m\}$). This leads to the idea that if we have a general poset $(RQ,\leq)$ to consider as new "models" its subsets closed from the bottom (a set $T = RQ$ is closed from the bottom if the following holds: if $m \in T$ and $m_1 \leq m$ then $m_1 \in T$). Since RQ is finite the class of all such subsets of RQ, say RQ' forms a distributive lattice (with operations $\cap$ and $\cup$). A decision rule for RQ' can be defined as $d'(T,.)=r$ if $d(m,.)=r$ for each $m \in T$. A distributive lattice is an ideal field for an application of the algorithm. Moreover, if is T-sound then the ordering of the new lattice is coherent. Each $T \in RQ'$ is characterized by its maximal elements, say $\max(T) = \{m_1, \ldots, m_k\}$; sets with only one maximal element are $\cup$-irreducible. Hence considering elements of RQ' corresponds to considering incomparable sets of hypotheses in RQ. Note that even if RQ is a lattice, then T with $\max(T) = \{m_1, m_2\}$ is in general distinct from $(m_1 \vee m_2)^*$. Since due to the definition of d' the downward rule is ineffective, this construction leads again to purely upward procedure.

But if we define in some way a new coherent or non-coherent decision rule using only maximal elements of sets from RQ', we can obtain a really new but strange procedure, statistical and interpretional merits of which are to be discussed elsewhere.

There are in general some open questions linked with model search procedures: we can say (even in the non-coherent case) that the upward rule is related to keeping the overal significance level. In which way is the downward rule related to power? With which mode of measuring the quality of statistical procedures are related procedures using both downward and upward rule substantially?

In Aickin (1983) a poset $(RQ,\leq)$ of models is used. The procedure suggested there is a stepwise procedure working downward along chains in RQ (and using in a way the upward rule). Again the procedure can be discussed in the lattice framework, e.g. the notion of protection is linked with meet of models and its izomorphism with set intersection in the space of distributions.

LATTICES FOR DECOMPOSITIONS

In Whittaker (1984a,b) lattices of models are considered, but not for constructing, say automatic, model search procedures as above. The aim is there to use lattice properties, namely the Möbius inversion, to decompose easily computable information statistics into additive elements related to particular models enabling their direct evaluation and interpretation and the construction of appropriate models supported by data.

Partition lattices and Möbius inversion were recently and fruitfully used in the theory of cummulants, k-statistics and their generalizations (see e.g. Speed 1986 for a series of papers) leading in consequence to computing formulae for variances and covariances for mean squares in ANOVA under broadest possible assumptions (Speed and Silcock, 1986).

REFERENCES

Aickin, M. (1983). Serial tests of multiple hypotheses. Commun. Statist. Theor. Meth. 12, 1535-1551.

Antoch, J. (1986). Algoritmic development in variable selection procedures. In: COMPSTAT 86. Physica-Verlag, Wien, 83-90.

Edwards, D. and Havránek, T. (1985). A fast procedure for model search in multidimensional contingency tables. Biometrika 72, 339-351.

Edwards, D. and Havránek, T. (1987). A fast model selection procedure for large families of models. J. Amer. Statist. Assoc. 82 (in print).

Furnival, G.M. and Wilson, R.W.J. (1974). Regression by leaps and bounds. Technometrics 16, 499-511.

Hájek, P. (1984). The new version of the GUHA procedure ASSOC (generating hypotheses on associations). In: COMPSTAT'84. Physica-Verlag, Wien, 360-365.

Hájek, P., Havel, I. and Chytil, M. (1966). The GUHA method of automatic hypotheses determination. Computing 1, 293-303.

Hájek, P. and Havránek, T. (1977). On generation of inductive hypotheses. Int.J.Man-Machine Studies 9, 415-438.

Hájek, P. and Havránek, T. (1978). Mechanising Hypothesis Formation-Mathematical Foundations for a General Theory. Springer-Verlag, Heidelberg.

Havránek, T. (1981). The GUHA method in the context of data analysis. Int.J.Man-Machine Studies 15, 265-285.

Havránek, T. (1982). Some complexity considerations concerning hypotheses in multidimensional contingency tables. In: Trans. 9th Prague Conf. on Inf. Theory, Decision Functions and Random Processes. Reidel and Academia, Dordrecht and Prague, 281-286.

Havránek, T. (1984). A procedure for model search in multidimensional contintency tables. Biometrics 40, 95-100.
Hommel, G. (1986). Multiple test procedures for arbitrary dependence structures. Metrika (in print).
Marcus, R., Peritz, E. and Gabriel, K.R. (1976). On closed testing procedures with special reference to ordered analysis of variance- Biometrika 63, 655-660.
Pokorný, D. (1982). Procedures for optimal collapsing in two-way contingency tables. In: COMPSTAT'82. Physica Verlag, Wien, 96-102.
Pokorný, D. and Havránek, T. (1978). On some procedures for identifying sources of dependence in contingency tables. In: COMPSTAT'78. Physica Verlag, Wien, 221-227.
Speed, T.P. (1986). Cummulants and partition lattices III-multiply--indexed arrays. J. Austral.Math.Soc. (Series A) 40, 161-182.
Speed, T.P. and Silcock, H.L. (1986). Cummulants and partition lattices VI-variances and covariances of mean squares (submitted).
Whittaker, J. (1984a).Fitting all possible decomposable models to multiway contingency tables. In: COMPSTAT'84. Physica Verlag, Wien, 401-406.
Whittaker, J. (1984b). Model interpretation from additive elements of the likelihood function. Appl.Statist. 33, 52-64.

PROGRAMMING LANGUAGES AND OPPORTUNITIS THEY OFFER TO THE STATISTICAL COMMUNITY

Naeve, Peter
University of Bielefeld

At first I would like to modify the title of my paper to "programing languages and opportunities they should offer to the statistical community". By adding this one word "should" I want to express two things:

- Firstly a somewhat critical review of the field of accessible programming languages (in contrast to the unnumbered body of proposals for programming languages burried in the literature).
- Secondly a plea for a shift in attitude of the so called user from a supply oriented to a demand oriented point of view.

The term programming languages stands for what computer scientists call high level languages or general purpose languages (e.g. FORTRAN, PASCAL), but I will let it include for the moment dedicated problem oriented languages (like GLIM).

It is claimed that high level languages free the user from the burden of machine dependencies, bit and byte manipulations etc. and allow him to concentrate on the formulation of the solution to his problem. Valuable tools are provided to ease the users task. High level programming languages usually incorporate facilities to support

- structured programming
- user supplied data types (and structures)
- recursion.

But unfortunately it turns out that the effort to avoid machine orientation resulted in a kind of splendid isolation for the high level languages. They usually do not provide reasonable interfaces to

- other high level languages
- new hardware features such as graphics devices and array processors
- the operating system.

This is an especially sad finding with respect to graphics. Either you leave it or dig deep into escape sequences and this kind of stuff. Some may consider this a little bit unjustified but I think for most users it comes close to the true description of their situation.

To proceed with our theme let us investigate who makes up the statistical community. In alphabetic order there is the consultant, researcher and the teacher. The computer influences the work of all of them.

When looked at in detail we may conclude that although with different weights statisticians use mathematics, logic and deal with data which usually have their unique kind of structure. The language of statisticians resembles these different roots. Turning to a computer the statistician has to use a programming language to convey his ideas and thoughts. What are the requirements a programming language should fulfil to suit the statistician's needs?

1. It should allow him both:
   - not to bother about the computer (i.g. he does not want to be a computer specialist)
   - to bring as much special knowledge about computers and computer programming he has into action (i.e. with respect to the computer the language should be open).
2. It must be as close to mathematics as possible - at least to the 'useful and essential' part of mathematics. Consider the well known formula from the field of linear models

$$SS = \hat{B}'C(C'GC)^{-1}\ C'\hat{B}\ .$$

   Would it not be nice to have a programming language which allows to evaluate this formula with the statement

   SS ← *CB* +.× (⌹ (⍉*C*) +.× *G* +.× *C*) +.× *CB* ← (⍉*C*) +.× *B* .
3. It should allow to deal with logic in a condensed form. For instance: to speak about the non-negative values in a data vector X should be as easy as

   (X ≥ 0) / X   or   X[X ≥ 0] .
4. It should contain statistical concepts, i.e. mean, variance etc., for instance in the form mean X, var X .
5. It should allow to deal with data objects in an easy way of name, attribute(s), value(s). To give an example

```
res  ← reg(y,X)        result of regression of y on X
plot res$resid         plot of the residuals
regsum res             summary of regression statistics
plot y, res$resid      plot of residuals versus y
regprt res             print report of regression
```

where the result res of the regression of y on X is a structure with components such as residuals, which are accessed in an obvious way. The structure may be processed in various ways (i.e. regprt, regsum).

6. It should not only allow to speak about analysis of data but also about the form one wants the result to be presented.
7. It should support the interactive style of work a statistician is used to. This calls for a kind of protocol feature.
8. It should fit to the user's way of thinking and working and not vice versa.
9. It should support the statistician in using as much of the hardware and other software available at his site.

Think for a moment how your favourite programming language fulfils these requirements.

Here are some statements to back your findings:

"We learned several things from this experiment, one of the earliest in the Computational Probability Project. The programming for computing the functions mentioned was easy, but still took a good deal of time to do, debugging and running a program in batch mode is not very efficient in term of the user's time. It was more time consuming to program the CALCOMP and, more important, it was psychologically unattractive to run the plotter off-line." Grenander (5, p. 14).

"It is quite common that when writing a research report containing numerical tables the output from the computer cannot be used as such, but the results have to be retyped manually. This may happen even if the computer output is well designed, since the needs of the user may change during the reporting phase." Mustonen (9, p. 337).

Perhaps you will sympathize with the following two quotes from a panel discussion on "programming languages issues for the 1980's"(10)

"On the whole programming languages are overrated. ... What's really important in the programming game is not the language but the tools you have to work with." "... there's an intermediated sort of thing which profession oriented language is the buzzword for ..."

In 1984 a software catalog (13) listed 252 statistical software packages. Many of these packages claim to offer a kind of statistical language for ease of use too. Do these packages fulfil our scheme of requirements? My answer is no. They suffer from the same kind of isolation as do programming languages. They are rather fixed in scope too.

There is still another reason to be a bit reluctant with respect to statistical packages. There are some doubts regarding their standards.

The paper by J.L. Longley: An appreisal of least square programs for the user (8) dates 19 years back but its message is not yet obsolate. The same is true with respect to the subject dealt with in papers by Ling (7) or Wilkinson and Dallal (14) on sample moments calculation to mention just two. To give some evidence to this claim I will present some citations form "Statistical Computing Software Reviews" found in the last two volumes of The American Statistician.

1. "Results of the analyses cannot be stored for analysis. For example one cannot store means and standard deviations to plot in checking for possible variance-stabilizing transformations. Similary, residuals from regressions can be printed but not stored to plot, for example against predicted values." (12, p. 165)
2. "When the median of an odd sample size was programmed incorrectly, one wonders about computations for more complicated procedures." (12, p. 166)
3. "The variance function in the command FUN, however, failed the "Anscombe test" dismally; Anscombe (1967, p. 3) discussed what could go wrong in the computation of the three readings 9,000, 9,001, and 9,003. The function VAR in FUN produced 1.33 for the variance of these values (should be 2.33) and a negative variance for 9,997, 9,998, and 9,999, suggesting the inappropriate use of the "desk calculator" formula." (11, p. 218f.)

I intendently did not mention the name of the packages reviewed. I do not want to turn those packages down, for I strongly believe that the packages not yet reviewed are not free of bugs too.

What should be learned from these citations is that packages may suffer from everything starting with numerical inaccuracy and going to inflexible and inconsistent management of statistical analysis.

To make explizit that we are looking for more let us coin a new name for it. I think statistical environment would be the right name. By this I mean a computing environment made up through an integrated collection of tools for exploiting as much of the hardware and software at hand (not calling for special devices) in an easy and natural way. This should be embedded in an concise consistent concept of a language well suited to the needs of statisticians. The language should have a syntax strongly influenced by mathematics and logic, with basic statistical terms such as mean, median etc. but also - and this is essential - it must contain tools to enlarge the language for personal or common usage.

At the moment I just see two such systems which are more than experimental prototypes. I speak of the S-system and APL.

Let them introduce themselves.

"S is an interactive environment for data analysis and graphics with two components: a language and a support system. The S language is a very high-level language for specifying computations. S is also a system in that it provides a total environment for the user, including data management, documentation, and graphics.

The primary goal of the S environment is GOOD DATA ANALYSIS. The facilities in S are directed toward this goal. S encourages the iterative, interactive style of data analysis which leads to unterstanding. In this way, S is quite unlike most statistical 'packages'.

S provides the user with interactive computation, both simple and complex, graphical displays on a wide vairety of graphics devices, data management and structuring." (2, p. i)

The S-system was developed at the Bell Telephone Laboratories by Becker and others. It runs on UNIX machines (for instance I made my experiences with S on a HP 9000). The book by Becker and Chambers: S An interactive environment for data analysis and graphics describes the system - it is the reference manual. The quote given was from this book.

Turning to APL let me quote Dr. Iverson (the father of APL) "Nearly all programming languages are rooted in mathematical notation, employing such fundamental notions as functions, variables, and the decimal (or other radix) representation of numbers ... APL has, in its development, remained much closer to mathematical notation, ..." (4, p. 39)

"The primitive objects of the language are arrays. ... The syntax is simple: there are only three statement types (name assignment, branch, or neither), there is no function precedence hierarchy, functions and defined functions ... are treated alike.

The semantic rules are few: the definition of primitive functions are independent of the representations of data to which they apply, all scalar functions are extended to other arrays in the same way (that is item - by - item) ...

The utility of the primitive functions is vastly enhanced by operators which modify their behavior in a systematic manner. ...

External communication is established by means of variables which are shared between APL and other systems. These shared variables are treated both syntactically and semantically like other variables. A subclass of shared variables, called system variables, provides convenient communication between APL programs and their environment." (5, p. 40f)

APL was first developed by Iverson, Falkoff and others within IBM. But it began to spread out into other companies and universities. As a result we have different implementations of APL grouped around a core of the language which is standardized.

It is impossible to give a complete review of both systems within this short paper. Nor do I intended to rank one against the other.

As I consider them both to be advanced prototypes of a statistical environment we should look for I will present certain of their features in a number of small examples to exemplify some essential points

Just two further remarks before I start:

i) all "lines of code" presented so far were written down either in S or APL.

ii) Although being unique S and APL have some things in common. Both are interpretative languages - so interactive work is eased. Both have a concept of workspace, allowing the user to tailor his own special environment without loosing contact to the overall system or other users if he wishes to cooperate.

Example 1: Closeness to mathematical notation

APL offers matrices as primitive objects and a set of primitive functions and operators to deal with them. Talking about requirements we presented the formula

$$SS = \hat{B}'C(C'GC)^{-1}\ C'\hat{B}\ .$$

and its transformation into an APL expression

```
SS ← CB +.× (⌹ (⍉C) +.× G +.× C) +.× CB ← (⍉C) +.× B .
```

This is almost a one-to-one mapping of our well acquainted mathematical notation. For instance "⍉" mirrors matrix transpose, "+.×" tells exactly what a scalar product of vectors is. If you learn that "←" stands for assignment and become acquainted with the (at first somewhat peculiar) right to left anti-hierachic style of formula evaluation unique in APL you will read and unterstand such expressions as easy as you do with normal mathematical notation.

Probably someone knows Conway's game of life. Let a rectangular grid divide the plane into cells. A cell can be dead or alive. A living cell will survive, if its neighbourhood is neither overcroweded (i.e. 4 or more living cells) nor it is isolated (i.e. 0 or 1 living cell) otherwise it dies. A cell will be born if exactly 3 cells in its neigbourhood are alive.

A simple APL solution - exploiting direct function definition is shown below. The plane is mapped on a bit-matrix, where 1 stands for a living cell.

```
GENER : HH ∨ H RULE2 HH ← (H ← NB ω) RULE1 ω
RULE1 : ω ∧ α ∊ 2 3
RULE2 : (~ω) ∧ 3 = α
NB    : (HOR H) + (VERT H) + DIAG H ← BORD ω
BORD  : 0 ,[1] (0 , ω , 0) ,[1] 0
DIAG  : (¯2 2 ↓ ω)+(2 ¯2 ↓ ω)+(¯2 ¯2 ↓ ω)+2 2 ↓ ω
VERT  : (¯2 1 ↓ H) + 2 1 ↓ H ← 0 ¯1 ↓ ω
HOR   : (1 ¯2 ↓ H) + 1 2 ↓ H ← ¯1 0 ↓ ω
```

Besides easily identified logical notation the main features of APL applied in this solution are the "take" and "drop" function. L↑R ≙ pull L items from one end or corner of R, L↓R ≙ wipe L items from one end or corner of R. Note that you may specify which dimensions of your objects are envolved such as in L,[I]R ≙ join values of L and R together into a larger array by catenating along L's or R's I-th coordinate.

Example 2: Extensibility of the language

Look at those two stem and leaf display done with the S function stem.

```
> stem  income, depth=T
N = 50  Median = 695.665
Quartiles = 287.77, 1813.93
Decimal points is 3 places to the right of the colon
   19   19   0 : 1112222222333333444
        11   0 : 56667778889
   20    3   1 : 134
   17    6   1 : 557789
   11    5   2 : 12234
    6    3   2 : 556
    3    2   3 : 03
    1    0   3 :
    1    1   4 : 0
> stem  income, depth=T, twodig=T
N = 50  Median = 695.665
Quartiles = 287.77, 1813.93
Decimal points is 3 places to the right of the colon
   20   20   0 : 09,12,14,15,19,21,22,23,24,25,25,28,29, ...
        10   0 : 57,58,60,65,66,73,77,77,81,87
   20    4   1 : 14,26,39,49
   16    5   1 : 51,68,74,81,90
   11    6   2 : 11,21,23,33,45,46
    5    3   2 : 50,63,98
    2    1   3 : 30
    1    0   3 :
    1    1   4 : 00
```

The only difference should be the difference in leaf size, i.e. 1 or digits. But more has happened. Some items changed their stem. The reason for this is, that the stem function rounds the data. There was no way to avoid it. So I decided to extend the abilities of the S-function. The next lines show what has to be done. The subroutine stemw as delivered by Bell reads as follows.

```
ROUTINE(stemw,produce stem and leaf portion for stem)
subroutine stemw(a,n,fc,kkl,kku,ksl,nl,wd,twodig,depth)
real a(n)
integer n,fc,kkl,kku,ksl,nl,width,wd
.
.
.
   if (twodig) {#... ditto above, but for 2 digit leaves.
      t = s*a(k)+0.0499
      if (a(k) < 0.0) t = t-10.1
   }
   else {#...adjusts round-off
      t = s*a(k)+0.499
      if (a(k) < 0.0) t = t-11.0
   }
.
.
.
```

This has to be changed to

```
ROUTINE(stemw,produce stem and leaf portion for stem)
subroutine stemw(a,n,fc,kkl,kku,ksl,nl,wd,twodig,depth,ritup)
real a(n)
integer n,fc,kkl,kku,ksl,nl,widt,wd,ritup
.
.
.
   if (twodig) {# ...ditto above, but for 2 digit leaves.
      t = s*a(k)+(0.0499*ritup)
      if (a(k)< 0.0) t = t-10.0-(0.1*ritup)
   }
   else {# ...adjusts round-off
      t = s*a(k)+(0.499*ritup)
      if (a(k)< 0.0) t = t-10.0-(1.0*ritup)
  }
.
.
.
```

The essential calculations are done by function stem.

The old version

```
FUNCTION stem(
              x/REAL,NAOK/
             nl/INT,1,0/
          scale/INT,1,1000/
        twodig/LGL,1,FALSE/
         fence/REAL,1,2./
          head/LGL,1,TRUE/
         depth/LGLG,1,FALSE/
)
INCLUDE(option,io)
if(nl!0&nl!=2&nl!=5&nl!=10)
  FATAL(number of leaves per stem is not 0 2 5  or  10)
n=LENGTH(x)
NAOUT(`x') # remove NAs
if(n!=LENGTH(x) && head)
FPRINT(OUTFC,"Contained",I(n-LENGTH(x)),"NAs")
call stems(x,LENGTH(x),OUTFC,0,lwidth,nl,2.,FALSE,
          twodig,!head,scale,fence,FALSE,depth)
END
```

has to be changed to

```
FUNCTION STEM(
              x/REAL,NAOK/
             nl/INT,1,0/
          scale/INT,1,1000/
        twodig/LGL,1,FALSE/
         fence/REAL,1,2./
          head/LGL,1,TRUE/
         depth/LGL,1,FALSE/
         round/LGL,1,FALSE/
)
INCLUDE(option,io)
```

```
if(nl!=0&nl!=2&nl!=5&nl!=10)
  FATAL(number of leaves per stem is not 0 2 5  or  10)
n=LENGTH(x)
NAOUT(`x') # remove NAs
if(n!=LENGTH(x) && head)
FPRINT(OUTFC,"Contained",I(n-LENGTH(x)),"NAs")
if(round)
   iritup=1
else
   iritup=0
call stems(x,LENGTH(x),OUTFC,0,lwidth,nl,2.,FALSE,
          twodig,!head,scale,fence,FALSE,depth,iritup)
END
```

I do not present the compilation - and binding steps necessary to include the new version in S. Your are well supported by the S-system (and the operating system UNIX without which S would not be so powerful) in going through this procedures. ●

Example 3: It is easy to make a plot

Imagine you want to build a plot function for histogramms. Here are some ideas how the result should look like.

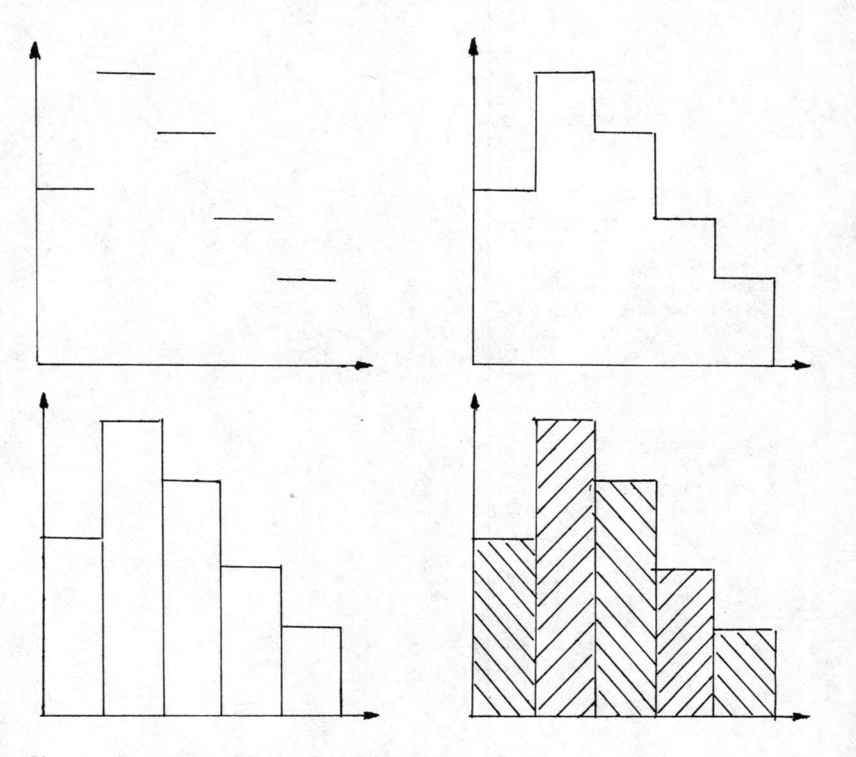

figure 1: Some ideas for histogram plots

In the APL implementation APL*PLUS you are provided with a set of system functions which give you control over your plotting device.

Some system functions for graphics in APL*PLUS:

⎕GINIT	initializes the graphics display for the particular adapter in use
⎕GVIEW	selects the rectangular area on the screen onto which graphics displays are projected
⎕GLINE	plots individual points, draws straight line segments, and fills rectangular bars
⎕GCIRCLE	creates displays of entire circles or ellipses, or arcs from them, or pie-shaped wedges formed by such arcs and two radii
⎕WRITE	writes text on the graphics screen

One thing more you have to learn is that different people (or graphics devices) have different coordinate systems to structure a picture.

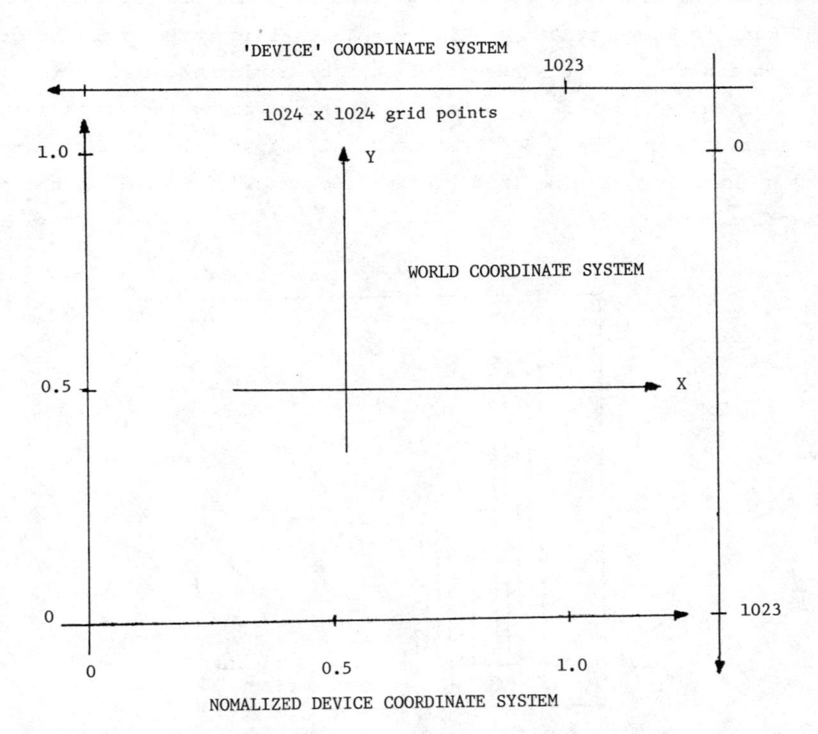

figure 2: Different coordinate systems

Taking this into account the simple plot function is easyly coded.

```
∇ HIST XY
   ⎕ELX←⎕ALX←' → END ◊ ⎕DM
   0 0 ρ0 ⎕GINIT 'IBMCOLOR
   XY←NORM XY
  ⍝ NEXT LINE FOR SOLID BARS AND BOXES ONLY
  ⍝ XY←(¯1 0 ↓XY),(1↓XY[;1]),[1.5]⊖ρ¯1↑XY[;2]
  ⍝ SOLID BARS
  ⍝ 1 ⎕GLINE XY
  ⍝ BOXES
  ⍝ 1 ⎕GLINE((1↑ρXY), 5 2)ρXY[; 1 2 1 4 3 4 3 2 1 2]
  ⍝ HORIZONTAL LINES
  ⍝ 1 ⎕GLINE((¯1+1↑ρXY),2 2)ρ(¯1 0↓XY),(1↓XY[;1]),[1.5]¯1↓XY[;2]
  ⍝ SKY LINE
  ⍝ 1 ⎕GLINE((¯1+1↑ρXY),3 2)ρ(¯1 0↓XY),(1↓XY[;1]),(¯1↓XY[;2]),1 0↓XY
  ⍝ ⎕GLINE 1 2 2 ρXY[1;],XY[1;1],¯1↑XY[;2]
   0 0 ρ⎕INKEY
  END: 0 0 ρ3 ⎕INT 16
 ∇
```

If you are willing to put somewhat more effort into the plotting business you may produce a result like that shown in the following figure 3

I want to demonstrate by this example that is rather easy to create a plot function for the essential part of a picture. But it is always this essential part that matters if you try to use the graphical mode to express your ideas. So if a graphical output is just for your sake why spent a lot of time in making it look nice or in making the plot function foolproof? ●

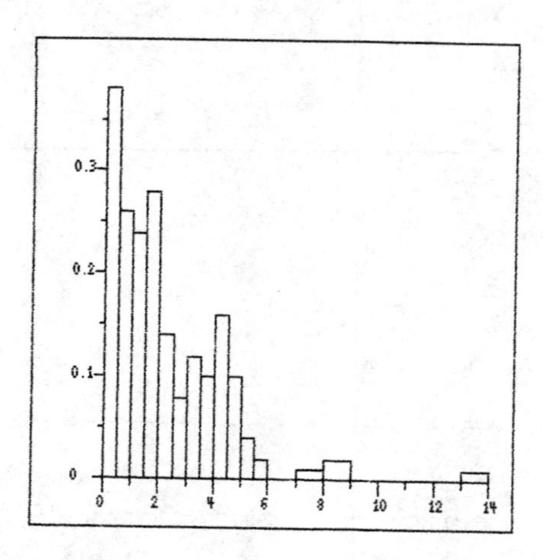

figure 3: One nice histogram plot

I hope this set of examples has made the vision of a computing environment for statistician a little bit more concrete. As I said I do not want to rank those two candidates. At the moment you may praise yourself if you can lay your hand on either of them. But I think it is fair to confess that I myself still are biased towards APL.

And it is also fair to report that both systems are not free of bugs. The three APL systems are all implemented on the same computer (an IBM compatible PC with a 8088 and 8087 processor).

'Inconsistencies' in the implementation of the Gamma function and the Binomial coefficient in APL on an IBM PC

LIMITS for X					
! X	SHARP	: ≤ 56.5452	1 ! X	SHARP	: ≤ 3.602 E 16
	APL*PLUS	: ≤ 170.6243		APL*PLUS	: ≤ 255
	IBM 1.0	: ≤ 170.6243		IBM 1.0	: ≤ 1.796 E 308
0 ! X	SHARP	: ≤ 3.602 E 16	X ! X	SHARP	: ≤ 3.602 E 16
	APL*PLUS	: ≤ 255		APL*PLUS	: ≤ 1.796 E 308
	IBM 1.0	: ≤ 1.796 E 308		IBM 1.0	: ≤ 1.796 E 308

Ironically enough when we had implemented S on our HP 9000 we got a run time error when calling the gamma function with argument 6, i.e. the result should be 120. The reason was an inconsistent handling of the machine dependent number ranges.

To bring this paper to an end let me summarize. Not what a computer language offers but what a computer language should offer should be the point of concern. One might put it in one sentence. The computer language must fit to the statistician's way of thinking and working. In some detail this plea was worked out in requirements 1 - 9 and examples 1 - 3. The history of APL reveals that user's demands can be fruitfully incorporated into the development of a language. So do not be a devote beggar grasping what will be given to you by the computer scientists. It is you who has the real problems to solve not they. And S and APL show that it can be done.

R e f e r e n c e s

1. Becker, R.A. Chambers, J.M. (1984). Design of the S System for data analysis. Communications of the ACM, vol 27.
2. Becker, R.A., Chambers, J.M. (1984). S an interactive environment for data analysis and graphics. Wadsworth Inc., Belmont.
3. Becker, R.A., Chambers, J.M. (1985). Extending the S System. Wadsworth Inc., Belmont.
4. Falkoff, A.D., Iverson, K.E. (1978). The evolution of APL. APL Quote Quad, vol 9, no 1.
5. Falkoff, A.D., Iverson, K.E. (1978). APL language. APL Quote Quad, vol 9, no 1.
6. Grenander, U. (1982). Mathematical experiments on the computer. Academic Press, New York.
7. Ling, R.F. (1974). Comparison of several algorithmus for computing sample means and variances. Journal of the American Statistical Association, vol 69.
8. Longley, J.W. (1967). An appraisal of least squares programs for the electronic computer from the point of view of the user. Journal of the American Statistical Association, vol 62.
9. Mustonen, S.(1981). Statistical computing with text editor. In: Computational Statistics, Naeve, P., Büning, H. (Ed.). de Gruyter, Berlin.
10. Programming languages issuess for the 1980's (1984). SIGPLAN-notices, vol 19, no 8.
11. The American Statistician (1985), vol 39, no 3.
12. The American Statistician (1986), vol 40, no 2.
13. The Software Catalog (1984). Science and Engineering. Elsevier, New York.
14. Wilkinson, L., Dallal, G.E. (1977). Accuracy of sample moment calculations among widely used statistical programs. The American Statistician, vol 31.

STATISTICAL SOFTWARE FOR MICRO-COMPUTERS

R. GILCHRIST
Polytechnic of North London, London N7 8DB, UK.

ABSTRACT

Recent developments in computer hardware means that the established British and US 'main-frame' statistical packages can now be run on desk-top machines. At the same time, a new breed of statistical software is appearing, written specifically for micro-computers. This paper reviews the field in the light of the author's evaluations of some two dozen packages.

INTRODUCTION

The most popular main-frame statistical packages (BMDP, GENSTAT GLIM, MINITAB, P-STAT, SAS, SPSS) have now been down-loaded to run on the IBM-PC series of machines. (This paper concentrates on developments in statistical software in the non-Communist countries. The author apologises to colleagues from COMICON countries for lack of reference to software developed by them). In fact, over the last 5 or 6 years, about 100 micro-based packages have appeared, about half in IBM-PC format (Gilchrist, 1985). The micro-user is confronted by a wide range of choices; many packages appear to offer the facilities required by the user. How should s/he choose?

User Choice

The main-frame user has a limited range of established software; should the micro-user purchase the down-loaded main-frame package? This has several attractions; firstly, user-familiarity is very important; moreover, the main-frame based packages are well-supported, with at least adequate and sometimes good documentation. They are (generally) accurate and bug-free. On the other hand they tend to be quite expensive ($250 to $1000), without substantial discounts for multiple copies (although site licences are now appearing at a more realistic price). This clearly leads the user to consideration of micro-based software.

However, the packages specifically designed for micros tend to be more restricted than the established main-frame-based package, some offering little more than descriptive statistics, regression and simple anova. Most are quite accurate, some notably more so than those developed on main-frames. Most do, moreover, make some limited use of micro technology, for example being menu-driven or giving access to user directories. This contrasts with the majority of the 'ex-main-frame' packages. In fact, with the notable exception of SPSS and SAS, the main-frame packages have been converted to run on micros without attempting to make use of micro technology. (An extreme example of this is BMDP, which still produces 132 character-wide output in the micro version, with attendant non-readable word-wrap).
Some micro packages are noticeably quicker than others; both programming skill and host language seem important here. Interpretive BASIC and Pascal appear slow! And speed is not necessarily lost in the quest for accuracy. (As we shall remark later, MICROSTAT is both highly accurate yet it is not slow). Menu-based systems should be carefully (better) designed; they can be very tedious for the expert user. Some packages are rather unfriendly, either landing the user deeper and deeper into a hierarchy of menus and/or abruptly dumping the user back to the operating system. Documentation varies greatly; most is adequate to get started; some is good. But very few packages provide any algorithmic information. It would be nice to know how the regression equations are solved (Givens, Gram-Schmidt, etc might increase user confidence in results); the precision of storage and calculations should surely be stated. Users should be able to insist on the updating algorithm for means and SSP matrices. At the very least, it seems odd that no (?) packages provide information on simple matters such as whether divisor n or n-1 is used.

Accuracy

Standard benchmark tests of accuracy can be carried out easily. The most familiar data set is that of Longley (1967); this consists of highly correlated econometric data and has been used to test regression. More illuminating tests can be carried out using the ideas of Greenfield and Siday (1980). Data with differing stiffness (number of common digits) and of differing collinearity can quickly show a package's limitations. For example, what appears to be single precision storage and calculation means that some packages cannot even calculate simple statistics for data such as 7.001. With values such as 7.000001, precision difficulties mean that most micro packages cannot calculate even simple statistics.

The so-called Wampler data provides another way of testing regression (see Lessage & Simon, 1984), by introducing an increasingly ill-conditioned X'X matrix. The number of digits accuracy of parameter estimates in regression is calculated as the stability measure gets increasingly smaller. MICROSTAT again does remarkably well, giving results 2 decimal places better than any other package tested and greatly more accurate than most others. (4 or more decimal places more accurate in estimating what are true values of 1.0). Clearly, a well-designed micro-package can be very accurate; a badly designed one can fail with bizarre data.

General Facilities

A useful checklist of facilities of packages is provided by Carpenter et al (1984). Students working with the author have used these in evaluation of about 20 packages. It is useful to supplement this list with considerations of the user interface. For example, the number of floppy disks required is an important consideration for the user with a non-hard disc machine. Items such as how often the system dumped the user; error handling, in general, and speed should be noted.

A notable omission is the facility to write routines in the host language which can then be linked in with the package. This has been a feature in BMDP for some years; it is now available in GLIM3.77. Packages are unlikely to provide all the facilities the user may potentially require, so extensibility is very desirable. Unfortunately, most micro software is provided in compiled form, so extensibility may be a problem (even with BMDP and GLIM).

CONCLUSION

Many stats packages developed for micros are quite good at doing statistical calculations. Some, like MICROSTAT are remarkably accurate and easy to use. (The author has heard good reports of SYSTAT, too). They may outperform the main-frame based packages in many ways. Some are reasonable at dealing with difficult data, including missing values, but are less user friendly. (INSTAT seems worthy of mention; as a newcomer, it has potential).

Having noted that some packages are quite effective, it is unfortunate that most are very boring. SAS and SPSS-PC are totally rewritten versions and really look like micro-packages (at the expense of compatibility with the 'old' main-frame version). But no package (except SAS and to some extent SPSS) yet uses the latest ideas in micro software; e.g. icons, pull down menus, windowing or touch-screens. Few even have good graphical output. And even less have 2.5-D, with 3-D unheard of. Yet data-base management software, spreadsheets, etc are moving rapidly into the provision of statistical facilities. They have full-screen editing, good file management, interfaces to other software, colour graphics, etc. Their popularity could lead to the demise of (generally reliable) statistical software produced by statisticians. Thus, because of the limitations of statistical packages, there is the possibility that users will prefer to use exciting, novel software which may have unstable or incorrect statistical procedures.
Is there a danger of moving back to the accuracy type of problems which occurred 20 years ago with some popular 'main-frame' packages?

References

Carpenter, J, Deloria, D and Morgenstein, D (1984). Statistical analysis for microcomputers. BYTE, April, 234-264.

Greenfield, T and Siday, S (1980). Statistical Computing for Business and Industry. The Statistician, 29, 1, 33-55.

Gilchrist, R (1985). Statistical packages for the IBM PC. In the Research and Academic Users Guide to the IBM PC. Chichester: IBM UK Ltd, 111-129.

Lessage, J P and Simon, S D (1985). Numerical accuracy of statistical algorithms for microcomputers. Computational Statistics and Data Analysis, 3, 47-57.

Longley, J W (1967). Appraisal of least squares programs from the point of view of the user. J. Amer. Stat. Ass., 62, 819-841.

DATA CLASSIFICATION BY COMPARING OF COMPUTER IMPLEMENTATIONS OF STATISTICAL ALGORITHMS

N.N.Lyashenko
Lrningrad Institute of informatic and automatisation
Fourteenth line 39, Leningrad, 199178, USSR

M.S.Nikulin
Leningrad Branch of Steklov Mathematical Institute,
Fontanka 27, Leningrad, 191011, USSR

Using any statistical algorithm we deal with it's constructive implementation rather than it's ideal form. In particular, the discreteness of computing devices and physical constraints force us to restrict ourselves to finite sets of number values. If a theoretical model (statistical procedure) uses sets of higher power (for example, infinite sets) then any implementation is necrssary connected with approximations and therefore with computational errors.

Let $\mathbb{X}_n = (X_1, \ldots, X_n)$ be a sample, and $T_n = T_n(\mathbb{X}_n)$- statistics under investigation, $G = \{g_1, \ldots, g_s\}$ a set of basic vector-functions corresponding to a given computer device (G may consist of arithmetical functions some standard functions like sqrt, sin etc.).

Consider a representation of T_n in terms of the basic functions

$$T_n(\mathbb{X}_n) = (g_{i_1} \circ \ldots \circ g_{i_m})(\mathbb{X}_n), \qquad (1)$$

where $g_{i_k} \in G$. If $\mathcal{R}$ is an operator transforming into it's computer implementation (very often $\mathcal{R}g_i = g_i$), then the implementation of T_n may be written in the form

$$\mathcal{R}T_n^* = \mathcal{R}g_{i_1} \circ \mathcal{R}g_{i_2} \circ \ldots \circ \mathcal{R}g_{i_m},$$

where T_n^* is a particular representation of T_n in

terms of g_i ($\mathbf{T}_n^*$ takes into account an algorithm for calculating T_n). It is important to remark that in general

$$\mathcal{R}\mathbf{T}_n^* \neq \mathcal{R}\mathbf{T}_n^{**}$$

for two representations T_n^*, T_n^{**}

$$\mathbf{T}_n = g_{i_1} \circ \dots \circ g_{i_m} = g_{j_1} \circ \dots \circ g_{j_r} ,$$

where $g_{i_\ell}, g_{j_k} \in G$.

An operator $\mathcal{R}$ may have different interpretations. For example, $\mathcal{R} = fl_d$ (rounding floating point operation with d binary digits) or $\mathcal{R} = Pr_d$ (a projection on a linear subspace generated by functions $1, x, \dots, x^{d-1}$ in systems of analytical calculations) and so on.

Further we consider the case $\mathcal{R} = fl_d$ which is very important in Statistical calculations.

If $\mathbf{T}_n(\mathbb{X}_n)$ is a statistic then $(fl_d \mathbf{T}_n^*)(\mathbb{X}_n)$ is it's computer realization in d -digit floating point arithmetic. From the definition of fl_d (see Wilkinson (1963), Lyashenko, Nikulin (1986)) for any n we have

$$\lim_{d \to +\infty} fl_d \, \mathbf{T}_n^*(\mathbb{X}_n) = \mathbf{T}_n(\mathbb{X}_n) .$$

For a large n we can use properties of $\mathbf{T}_n$ described by asymptotic theory. On the other hand, for any real computer d is necessary bounded and asymptotical properties of $fl_d \mathbf{T}_n$ may be essentially different from those proved in asymptotic theory. Further more, in general

$$\lim_{n \to \infty} \lim_{d \to \infty} (fl_d \mathbf{T}_n^*)(\mathbb{X}_n) \neq \lim_{d \to \infty} \lim_{n \to \infty} (fl_d \mathbf{T}_n^*)(\mathbb{X}_n).$$

For example, if $X_1, \dots, X_n$ are i.i.d.r.v., $\mathbf{E}|X_1| < \infty$,

$$\mathbf{T}_n(\mathbb{X}_n) = \frac{1}{n}(X_1 + \dots + X_n)$$

and, in particular,

$$\mathbf{T}_n^*(\mathbb{X}_n) = (\cdots(X_1 + X_2) + X_3) + \cdots + X_n)/n,$$

then by virtue of Khinchin's theorem

$$\lim_{n\to\infty} \lim_{d\to\infty} (fl_d \mathbf{T}_n^*)(\mathbb{X}_n) = \mathbf{E}X_1,$$

but

$$\lim_{d\to\infty} \lim_{n\to\infty} (fl\, \mathbf{T}_n^*)(\mathbb{X}_n) = 0,$$

see Lyashenko, Nikulin (1984, 1985, 1986).

Such phenomena lead us to investigate the stability of statistical algorithms $\mathbf{T}_n^*$ with respect to the operator fl_d. The problem of an optimal representation of type (1) for a statistics $\mathbf{T}_n$ arises.

It is convenient to have a program of automatic classification of data $(\mathbb{X}_n \in R^n)$ with the help of "preference regions":

$$M(\mathbf{T}_n^*) = \{\mathbb{X}_n : |fl_d \mathbf{T}_n^*(\mathbb{X}_n) - \mathbf{T}_n(\mathbb{X}_n)| \le |fl_d \mathbf{T}_n^{**}(\mathbb{X}_n) - \mathbf{T}_n(\mathbb{X}_n)|\},$$

$$M(\mathbf{T}_n^{**}) = R^n \setminus M(\mathbf{T}_n^*).$$

We propose to use the special program system generating program texts of $fl_d \mathbf{T}_n^*$, $fl_d \mathbf{T}_n^{**}$ etc. under the given d, calculating floating point errors and constructing preference regions. To obtain the (approximate) boundary surfaces of preference regions one can use methods of discriminant analysis.

The report contains as an example a comparison of two different algorithms for calculation of $\Phi(x)$ (the standard normal distribution function): Hastings formula and the Chebyshev approximation.

From general point of view our problem may be considered as a problem of choice of a robust algorithm T_n^* for T_n. The operator fl_d is an interesting example of nontrivial contamination of both the sample X_n and the statistics T_n, physically realized by a computer. An investigation of properties of the operator fl_d in this sense is equivalent to the investigation of the influence of computer rounding on a statistical inference process (see Lyashenko, Nikulin (1986)).

REFERENCES

Wilkinson, J.H. (1963). Rounding errors in algebraic processes. London, 160 p.

Lyashenko, N.N. and Nikulin, M.S. (1984). On computer version of statistical procedures. Zapiski Nauchn. Semin. LOMI. 136, pp.142-152.

Lyashenko, N.N. and Nikulin, M.S. (1984), On constructive versions of decomposable statistics. In: Abstracts of the Colloquium on Goodness-of-Fit, Debrecen: Janos Bolyai Mathematical Society, p.21.

Lyashenko, N.N. and Nikulin, M.S. (1985). On constructive versions of statistical procedures. Theor. Veroyatn. Primen. 30, pp.215-216.

Lyshenko, N.N. and Nikulin, M.S. (1986). Computer and some statistical inference. Tech. report N 32 of the University of British Columbia (Canada), 67 p.

TIME DISCRETE APPROXIMATION OF ITO PROCESSES

PLATEN, E., Karl-Weierstraß-Institut für Mathematik,
AdW der DDR, DDR-1086 Berlin, Mohrenstraße 39, GDR

Let be given an Itô process which is defined by the equation:

$$X_t = X + \int_0^t a(X_s)ds + \int_0^t b(X_s)dW_s,$$

$t \in [0,T]$, where x denotes the initial value at time 0 and $W = \{W_t\}_{t \in [0,T]}$ the standard Wiener process.

Simulation studies and theoretical investigations show that one can not simply apply well-known "deterministic" numerical methods to simulate approximate trajectories or functionals of diffusion processes. The paper considers time discrete approximations of Itô processes which can be used for the construction of recursive Monte-Carlo simulation algorithms for Itô processes. Several authors, for instance Milstein [1], Rao Borwankar, Ramakrishna [2], Wagner, Platen [3], Clark, Cameron [4], Rümelin [5], Talay [6] and Platen [7] developed and investigated time discrete approximations.

If one needs pathwise approximate trajectories, then one can apply the mean square criterion to estimate the order of convergence with respect to the step size of the time discretization. By the use of the stochastic Taylor formula derived in [3], [8] and [9] it is shown in [3] and [10] that for any desired order of mean square convergence there exists a corresponding so called mean square Taylor approximation. In [7] conditions are described for derivative free approximations to converge with a given mean square order. From the approach in [10] it becomes clear that one needs for higher algorithms more and more mixed multiple stochastic integrals which are considered in[11] . The time discrete mean square approximation of the Itô process with jump component is treated in [12] .

In many applications one wishes to approximate functionals of the diffusion. In this case one can use the so called mean criterion to obtain some order of convergence.

This type of convergence was studied by Milstein [13] Talay [6] and also in [7] . In [14] the order of mean convergence of the Euler approximation under Hölder conditions is studied. Talay proposed in [6] a class of second order mean approximations. For any order of mean convergence corresponding mean Taylor approximations are introduced in [7] by the use of the stochastic Taylor formula see [8] . Corresponding results for the Itô process with jump component are proved in [15] . A convenient derivative free second order mean approximation is proposed in [7] , where one can find also conditions for quite general higher order mean approximations.

Simulation studies in [16] show that derivative free approximations are convenient and useful.

Open problems are concerning the step size control and numerical stability of simulation algorithms.

References

[1] Milstein, G.N.: Approximate integration of stochastic differential equations. Theor. Prob. Appl., XIX, 3, (1974), 583-588.

[2] Rao, N.J.; Borwankar, J.D.; Ramakrishna, D.: Numerical solution or Itô integral equations. SIAM J. Control, 12, 1, (1974), 124-139.

[3] Wagner, W.; Platen, E.: Approximation of Itô integral equations. Preprint ZIMM, AdW der DDR, Berlin (1978).

[4] Clark, J.M.C.; Cameron, R.J.: The maximum of convergence of discrete approximations for stochastic differential equations. Lect. Notes in Control and Inform. Sc. 25, Springer Verlag (1980), 162-171.

[5] Rümelin, W.: Numerical treatment of stochastic differential equations. Report Nr. 12, Univ. Bremen (1980).

[6] Talay, D.: Efficient numerical schemes for the approximation of expectations of functionals of the solution of a S.D.E. and applications. Lect. Notes in Control and Inf. Sc., 61, Springer Verlag (1984).

[7] Platen, E.: Zur zeitdiskreten Approximation von Itôprozessen, Diss. B, AdW der DDR, IMath, (1984).

[8] Platen, E.; Wagner, W.: On a Taylor formula for a class of Itô processes. Probab. and Math. statistics, Vol. 3, Fasc. 1, (1982), 37-51.

[9] Platen, E.: A generalized Taylor formula for solutions of stochastic equations. Sankhya, A, 44, 2, (1982), (163-172).

[10] Platen, E.: An approximation method for a class of Itô processes. Lietuvos Matem. Rink, XXI, 1, (1981), 121-133.

[11] Liske, H.; Platen, E.; Wagner, W.: About mixed multiple Wiener integrals. Preprint, P-Math-23/82, IMath, AdW der DDR, Berlin (1982).

[12] Platen, E.: An approximation method for a class of Itô processes with jump component. Lietuvos Matem. Rink., XXII, 2, (1982), 124-136.

[13] Milstein, G.N.: A second order method for integration of stochastic differential equations. Theor. Probab. Appl. XXIII, 2, (1976), 414-419.

[14] Mikulevicius, R.; Platen, E.: Rate of convergence of the Euler approximation for diffusion processes. Preprint, P-Math/86, IMath, AdW der DDR, Berlin (1986).

[15] Mikulevicius, R.; Platen, E.: Time discrete Taylor approximations for Itô processes with jump component. Preprint, P-Math/86, IMath, AdW der DDR, Berlin (1986).

[16] Liske, H.; Platen, E.: Simulation studies on time discrete diffusion approximations. To appear.

FAST ALGORITHM OF PEAK LOCATION IN SPECTRUM

Surina I.I.
I.V.Kurchatov Institute of Atomic Energy, Moscow, USSR

Among the problems of spectroscopic data processing there exists a well-known problem of peak location in a spectrum. A great number of computer codes are available to solve this problem. Some codes are based on the Fourier method and the rest, on polynomial methods (see Briggs and Seah, Eds. (1983) and references therein). The existing codes are efficient, but their foundation in theoretical statistics is not sufficient enough.

We suggest a new approach to its solution. Employing the observed values $\{z_i\}$, $i=1,\ldots,N$ constituting the spectrum (figure 1), the least square method (LSM) line segments are constructed under nonstandard conditions when mean values of observables do not lie on the straight line. Then a statistical analysis of the line slopes is made. On its basis conclusions are drawn on the number and location of peaks.

Let us consider the spectrum $\{z_i\}$. The observed z_i values are normally independently distributed $N(\lambda_i, \sigma_i^2)$ with unknown mean $\lambda_i = \lambda(x_i)$ values and known variances σ_i^2. The function $\lambda(x)$ is usually known to belong to a wide parametric family. The x_i values are held fixed.

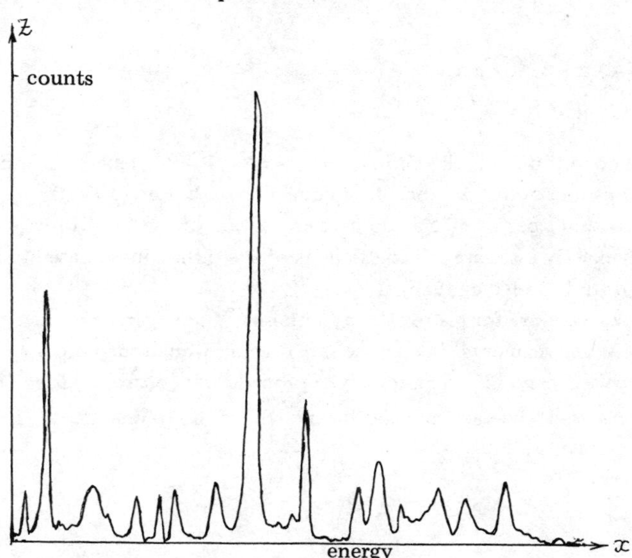

Figure 1. The observed (neutron) spectrum.

A procedure to test a hyothesis of an isolated peak presence in some region is constructed. First of all, one simple <u>statement</u> is formulated. Let $\{\lambda_j\}$ be an arbitrary function, $j=1,...,l$ (figure 2). A LSM line $\tilde{y}=\tilde{A}+\tilde{B}x$ is drawn where $\tilde{A}=\sum a_j\lambda_j$, $\tilde{B}=\sum b_j\lambda_j$. Then $A=\sum a_j z_j$ and $B=\sum b_j z_j$ are unbiased estimates with a minimum variance for $\tilde{A}$ and $\tilde{B}$.

<u>To prove</u> the statement, note, firstly, that LSM coefficients satisfy the ratios

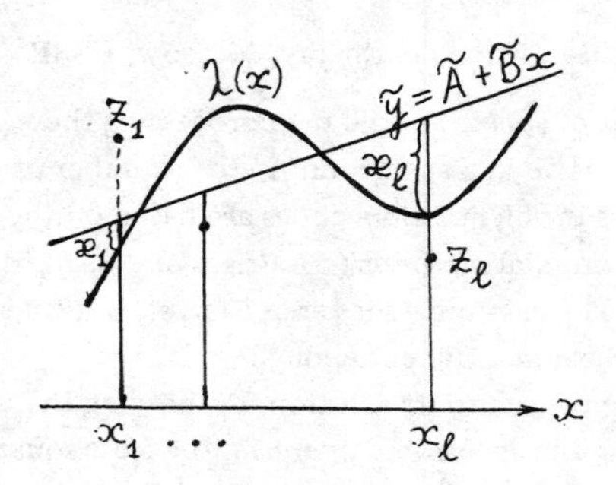

Figure 2

$$\sum a_j=1, \quad \sum a_j x_j=0, \quad \sum b_j=0, \quad \sum b_j x_j=1. \qquad (1)$$

Secondly, instead of the initial spectrum $\{z_j\}$ with the mean values lying on the curve $\{\lambda_j\}$, consider the spectrum $\{z_j+\varkappa_j\}$ the mean values of which lie on $\tilde{y}=\tilde{A}+\tilde{B}x$. Here $\varkappa_j=\tilde{y}_j-\lambda_j$. Using this spectrum, the LSM line $\tilde{\tilde{y}}=\tilde{\tilde{A}}+\tilde{\tilde{B}}x$ is constructed. From

$$\tilde{\tilde{A}}=A+\tilde{A}\sum a_j-\tilde{A}+\tilde{B}\sum a_j x_j ,$$

$$\tilde{\tilde{B}}= B+\tilde{A}\sum b_j+\tilde{B}\sum b_j x_j-B$$

and the conditiion (1) it follows that $\tilde{\tilde{A}}=A$, $\tilde{\tilde{B}}=B$. Hence, A and B are the best estimates of $\tilde{A}$ and $\tilde{B}$ (compare whith Rao (1973)).

Thus, without performing a compete estimation of the $\{\lambda_i\}$ function parameters, one can exactly estimate the coefficients of the LSM lines related to the function (if l is high enough).

Now a procedure for testing the hypothesis: $\tilde{B}>0$ can be constructed. The LSM line segments $\tilde{y}=\tilde{A}+\tilde{B}x$ are drawn through the points $\lambda_i,..., \lambda_{i+l-1}$ for variouus i and l . Accordingly, a procedure is constructed for $\tilde{B}<0$.
$B_l(i)$ is used to denote the coefficient B of LSM line $y=A+Bx$ plotted by the points $z_i,...,z_{i+l-1}$:

$$B_l(i)=\sum_{j=i}^{i+l-1} b_j z_j .$$

Hereinafter, we assume $x_i = i$, $\sigma_i = 1$. Then $b_j = 6(2j - l - 1)/(l(l^2 - 1))$.
The $B_l(i)$ values are normally distributed. Let $\beta_l(i)$ denote a mean of $B_l(i)$. The higher is l, the lower is the variance of $B_l(i)$. Figure 3 and the Table present critical values for $B_l(i)$ in testing the hypothesis: $\beta_l(i) > 0$. The significance level is $\alpha = .01$.

Table

Critical region for $B_l(i)$

l	2	3	4	5	6
$\sigma_B^2 = \frac{12}{l(l^2-1)}$	2.	.5	.2	.1	.057
$B_{cr.} = 2{,}33\sigma_B$	3.29	1.64	1.04	.74	.56

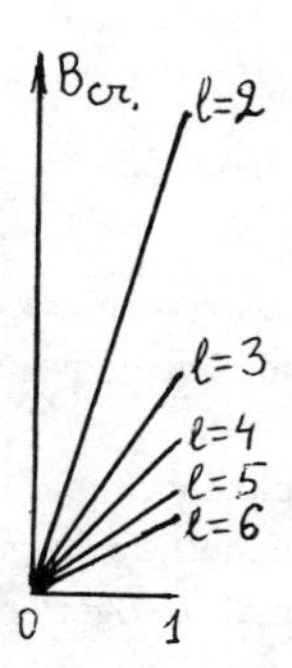

Figure 3.

A testing procedure for the hypothesis of the peak presence in the spectrum is constructed. Note that the values $B_l(i)$, $B_l(i+k)$ are correlated at $0<k<l$. Their correlation coefficient has the form:

$$\rho(B_l(i), B_l(i+k)) = (l - k)(l^2 - 2k^2 - 2lk - 1)/(l(l^2 - 1)).$$

We assume, by definition, that there is at least one peak between the points x_{i_1} and x_{i_2+l-1} in the spectrum $\{z_i\}$ if $\beta_l(i_1)$ and $\beta_l(i_2)$ with a preset significance level are in the fourth quadrant: $\beta_l(i_1) > 0$, $\beta_l(i_2) < 0$. Figure 4 presents the critical region for testing this hypothesis. Here u is determined from the condition:

$$\int_{S_u} \varphi(B_1, B_2)\,ds = 1 - \alpha,$$

where φ is the probability density function for a joint distribution of $B_1 = B_l(i_1)$ and $B_2 = B_l(i_2)$. The level lines of φ are shown in figure 5. The region of S_u is shaded in figure 5.

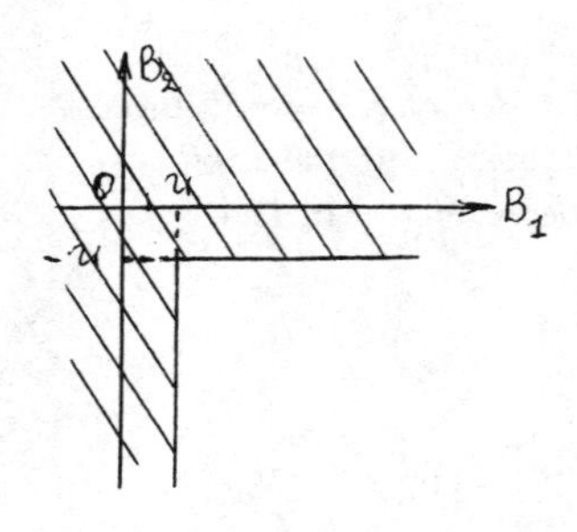

Figure 4.

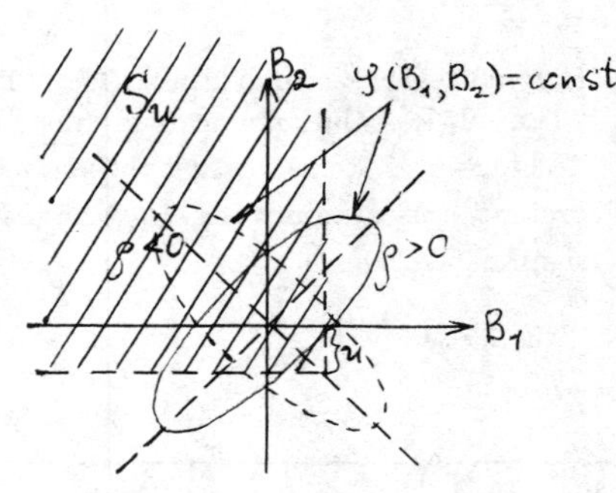

Figure 5.

We have developed a computer code VIS realizing partially the above-mentioned algorithm. The code VIS is successfully applied to analyze neutron and electron spectra.

In conclusion, there are some remarks.

(1) The method suggested is applicable not only to detect isolated peaks but also to analyze overlapping peaks and various features of the spectrum: slopes, bumps, etc. The method permits narrow peaks to be detected in a "true" spectrum $\{\lambda_i\}$.

(2) Using the LSM line segments, one can considerably speed up the procedures of the peak location since the $\{b_j\}$ coefficients calculated in advance can be stored externally and only linear expressions are calculated in the direct processing of the spectra.

(3) Here we have suggested the procedure for testing the hypothesis of isolated peak presence in the spectrum. It would be interesting to find $B_l(i)$-dependent statistics for a simultaneous estimation of the number and location of all the peaks in the spectrum.

It would be desirable to get rid of the assumption of the $\{z_i\}$ distribution normality.

References

Practical surface analysis by Auger and X-ray photoelectron spectroscopy.(1983). Briggs, D., and Seah, M.P. (Eds). New York: Wiley, 528p.

Rao, C.R. (1973). Linear statistical inference and its applications. New York: Wiley,625p.

EMPIRICAL PROCESSES
(Session 23)

Chairman: E. Gine

APPROXIMATIONS OF WEIGHTED EMPIRICAL PROCESSES WITH APPLICATIONS TO EXTREME, TRIMMED AND SELF-NORMALIZED SUMS

Sándor Csörgő, Szeged University, Hungary;
David M. Mason, University of Munich, W. Germany

1. *Introduction.* In [3], a probability space is constructed carrying a sequence $U_1, U_2, \ldots$ of independent rv's uniformly distributed on $(0,1)$ and a sequence $B_1, B_2, \ldots$ of Brownian bridges such that for all $0 \leq \nu_1 < 1/2$,

$$(1) \qquad \sup_{1/(n+1) \leq s \leq n/(n+1)} |\sqrt{n}(s-U_n(s))-B_n(s)|/(s(1-s))^{1/2-\nu_1} = O_P(n^{-\nu_1})$$

and for all $0 \leq \nu_2 < 1/4$,

$$(2) \qquad \sup_{0 \leq s \leq 1} |\sqrt{n}(G_n(s)-s)-\bar{B}_n(s)|/(s(1-s))^{1/2-\nu_2} = O_P(n^{-\nu_2}),$$

where, with $U_{1,n} \leq \ldots \leq U_{n,n}$ denoting the order statistics of $U_1, \ldots, U_n$, $U_n(s)=U_{k,n}$ for $(k-1)/n < s \leq k/n$, $k=1,\ldots,n$, $G_n(s)= =n^{-1}\#\{1 \leq k \leq n : U_k \leq s\}$, and $\bar{B}_n(s)=B_n(s)$ for $1/n \leq s \leq 1-1/n$ and 0 elsewhere. The relation in (1) is in fact a consequence of an improvement of the Komlós-Major-Tusnády type exponential inequality of Csörgő and Révész from 1978 for $\sqrt{n}(s-U_n(s))$, where the improvement is near 0 and 1, while (2) follows from (1). The dual approximation, in which the original Komlós-Major-Tusnády inequality for $\sqrt{n}(G_n(s)-s)$ is improved in the same manner, with (2) following for any $0 \leq \nu_2 < 1/2$ and (1) following from (2) for any $0 \leq \nu_1 < 1/4$, is done in [21].

The approximations (1) and (2) have proved to be very powerful tools in handling many problems in empirical-process theory itself and in statistics in general (cf. [3, 4, 5, 20, 6, 2] and the preprints of M. Csörgő and Horváth in No. 42, 55, 60, 61 and 71 of the Techn. Rep. Ser. Lab. Res. Statist. Probab., Carleton Univ., Ottawa, Canada). Here we shall concentrate only on probabilistic applications concerning the asymptotic distribution of sums of independent and identically distributed rv's and the associated trimmed and extreme sums. Section 2 containes a condensed review of results of this type. As an illustration of the technique, we prove in Section 3 half of a conjecture of Logan, Mallows, Rice and Shepp [19] on the asymptotic normality of Student's t statistic.

2. *A condensed survey of some of the probabilistic applications.* Let $X_{1,n} \leq \ldots \leq X_{n,n}$ be the order statistics of n independent rv's with the common (right-continuous) distribution function F and corresponding quantile function $Q(s) = \inf\{x : F(x) \geqq s\}$, $0 < s \leqq 1$, $Q(0) = Q(0+)$. Essential to our approach is the link between sums and the empirical process given by the formula

$$S_n(m,k) - \mu_n(m,k) = \sum_{i=m+1}^{n-k} X_{i,n} - n\int_{m/n}^{1-k/n} Q(u)\,du \overset{D}{=} \sum_{i=m+1}^{n-k} Q(U_{i,n}) - \mu_n(m,k)$$

(3)

$$= n\Big\{\int_{m/n}^{U_{m,n}} \Big(G_n(s) - \frac{m}{n}\Big)dQ(s) + \int_{m/n}^{1-k/n} (s - G_n(s))\,dQ(s) + \int_{U_{n-k,n}}^{1-k/n} \Big(G_n(s) - \frac{n-k}{n}\Big)dQ(s)$$

for any integers m and k, $0 \leqq m < n-k \leqq n$, $U_{0,n} = 0$, where $\overset{D}{=}$ denote distributional equality.

Using (3), it is shown in [3,4] that the sufficiency of the classical criterion for F to belong to the domain of attraction of the normal law, written $F \in D(2)$, i.e. for the asymptotic normality of (the properly centered and normalized) $S_n(0,0)$ follows from (2). The method used in [4] yielded the following analogue of the classic Hinchin-Feller-Lévy necessary and sufficient condition expressed through the quantile function: Whenever F is non-degenerate $F \in D(2)$ if and only if

(4) $$\lim_{s \downarrow 0} \{s^{1/2}|Q(s)| + s^{1/2}|Q(1-s)|\}/\sigma(s) = 0$$

and

(5) $\sigma^2(s) = \int_s^{1-s}\int_s^{1-s} (u \wedge v - uv)\,dQ(u)\,dQ(v)$ is slowly varying at 0

Using the asymptotic Poisson behaviour of $nG_n(s/n)$ and of $n(1 - G_n(1-s/n))$, $s \geqq 0$, and their asymptotic independence, a purely probabilistic proof is given in [4] for the sufficiency of the other classical criterion of Doeblin and Gnedenko for F to belong to the domain of attraction of a stable law with exponent $0<\alpha<2$, written $F\in D(\alpha)$. Of course the quantile version of this condition is also needed in the proof, and this is formulated in [4] as well. When $F \in D(\alpha)$, the asymptotic distribution of $S_n(m,k)$ with fixed $m\geq 0$ and $k\geq 0$ is also derived in [4] for the case $\alpha<2$ and in [12] for $\alpha=2$. Also, in the case $\alpha<2$, related further results for the extreme sums $S_n(0,n-m)$ and $S_n(n-k,0)$ and for the self-normalized sums $S_n(0,0)/(\sum_{i=1}^{n}|X_i|^p)^{1/p}$ of [19], where $p>\alpha$, are derived by the same approach in [7].

Let now $m=m_n\to\infty$ and $k=k_n\to\infty$ such that $m_n/n\to 0$ and $k_n/n\to 0$ as $n\to\infty$. Assuming $F\in D(\alpha)$, $\alpha\leq 2$, it is shown in [12] that (2) implies the asymptotic normality of $S_n(k_n,k_n)$ via (3), and that, when $\alpha<2$, then the extreme sums $S_n(0,n-k_n)+S_n(n-k_n,0)$ have the same stable limiting distribution as $S_n(0,0)$. Note that $m_n\equiv k_n$ here, and see what follows below.

For any fixed finite constant c consider the extreme-value distribution functions

$$\Lambda_c(x)=\begin{cases}\exp(-x^{-1/c}), & x>0 \quad (c>0)\\ \exp(-\exp(-x)), & -\infty<x<\infty \quad (c=0)\\ \exp(-(-x)^{-1/c}), & x<0 \quad (c<0).\end{cases}$$

We shall say that F is in the domain of attraction of Λ_c, written $F\in\Lambda_c$, if there exist sequences of constants a_n and b_n such that $a_n^{-1}(X_{n,n}-b_n)\xrightarrow{\mathcal{D}}Y_c$ as $n\to\infty$, where $\xrightarrow{\mathcal{D}}$ denotes convergence in distribution and Y_c is a rv with distribution function Λ_c. Using results from [16], it is straightforward to establish that the classical criterion for $F\in\Lambda_c$ has the following quantile version: $F\in\Lambda_c$ for some c if and only if for the same c

$$\lim_{s\downarrow 0}\frac{Q(1-sx)-Q(1-sy)}{Q(1-sv)-Q(1-sw)}=\frac{x^{-c}-y^{-c}}{v^{-c}-w^{-c}}$$

for all distinct positive numbers x,y,v,w, where $(x^0-y^0)/(v^0-w^0)=(\log x-\log y)/(\log v-\log w)$. In a slightly different form this was also given in [22].

The above mentioned result for extreme sums in [12] and the results in [15] and [18] (the latter being a generalization of a result in [14]) together say that whenever $F\in\Lambda_c$ then there exist constants $A_n>0$ and B_n such that

$$A_n^{-1}\{\sum_{i=1}^{k_n} X_{n+1-i,n}-B_n\}\xrightarrow{D}Y,$$

where Y is a standard normal rv N(0,1) if $c\leq 1/2$ and is a stable rv with exponent 1/c if c>1/2. (The cases $c\neq 0$ are in [12] and [15], while the case c=0 is in [18].) For $c\leq 1/2$ these results are proved by using (2) and (3).

Using (1)-(3) and greatly extending the first result in [12] mentioned above, a necessary and sufficient condition for the asymptotic normality of $S_n(m_n,k_n)$ is derived in [9]. Setting

$$\Psi_{1,n}(x)=\begin{cases}\sqrt{\frac{m_n}{n}}\{Q(\frac{m_n}{n}+x\frac{\sqrt{m_n}}{n})-Q(\frac{m_n}{n})\}/b_n, & -\sqrt{m_n}\leq 2x\leq\sqrt{m_n},\\ \Psi_{1,n}(-\sqrt{m_n}/2), & -\infty<2x<\sqrt{m_n},\\ \Psi_{1,n}(\sqrt{m_n}/2), & \sqrt{m_n}<2x<\infty,\end{cases}$$

and

$$\Psi_{2,n}(x)=\begin{cases}\sqrt{\frac{k_n}{n}}\{Q(1-\frac{k_n}{n}+x\frac{\sqrt{k_n}}{n})-Q(1-\frac{k_n}{n})\}/b_n, & -\sqrt{k_n}\leq 2x\leq\sqrt{k_n},\\ \Psi_{2,n}(-\sqrt{k_n}/2), & -\infty<2x<-\sqrt{k_n},\\ \Psi_{2,n}(\sqrt{k_n}/2), & \sqrt{k_n}<2x<\infty,\end{cases}$$

where

$$b_n^2=\int_{m_n/n}^{1-k_n/n}\int_{m_n/n}^{1-k_n/n}(u\wedge v-uv)\,dQ(u)\,dQ(v),$$

this result is the following: There exist sequences of normalizing and centering constants $A_n>0$ and B_n such that $T_n^*=A_n^{-1}\{S_n(m_n,k_n)-B_n\}\xrightarrow{D}N(0,1)$ if and only if $\Psi_{i,n}(x)\to 0$ for all x, i=1,2 (in which case A_n can be chosen to be $a_n=\sqrt{n}\,b_n$ and B_n to be $\mu_n=\mu_n(m_n,k_n)$). Moreover, it is shown in [9] that $a_n^{-1}\{S_n(m_n,k_n)-\mu_n\}$ is stochastically compact if and only if $\limsup_{n\to\infty}|\Psi_{i,n}(x)|<\infty$ for all x, i=1,2. In fact, the necessary and sufficient condition for the existence of $A_n>0$ and B_n such that T_n^* be stochastically compact is also given in [9] in terms of the behaviour of the two Ψ_n functions, i.e.

the behaviour of the underlying quantile function Q, but this condition is a little bit more complicated. Also, the direct probabilistic proofs based on (1)-(3) at once give the complete description of the possible subsequential asymptotic distributions of T_n^*.

In the case of heavy trimming when $m_n/n \to \alpha$ and $k_n/n \to 1-\beta$, $0<\alpha<\beta<1$, a variant of Stigler's theorem for $S_n(m_n,k_n)$ is also derived in [9] as a by-product by the same method.

In some statistical applications sums of certain weighted k_n extremes play a role, where, as above, $k_n \to \infty$ and $k_n/n \to 0$. Such sums are treated by the above methodology in [8] and [13].

When $m \geq 0$ and $k \geq 0$ are any fixed integers, our empirical process-quantile function probabilistic approach is used in [10] to give necessary and sufficient conditions concerning domains of attraction and partial attraction and stochastic compactness for the lightly trimmed sums $S_n(m,k)$, containing the classical theory for $S_n(0,0)$, with the description of all possible subsequential limiting distributions. Results on the same level of generality are given in [11] for the extreme sums $S_n(n-k_n,0)$ when $k_n \to \infty$. Thus [9,10,11] together provide a complete asymptotic distribution theory of sums with respect to the natural ordering of the summands.

3. *Asymptotic normality of Student's t.* Let $X, X_1, X_2, \ldots$ be independent with common distribution and quantile functions F and Q as above and assume EX=0 whenever $E|X|<\infty$. We consider the self-normalized sums

$$R_n = \sum_{i=1}^{n} X_i / \left(\sum_{i=1}^{n} |X_i|^2 \right)^{1/2}, \quad n=1,2,\ldots .$$

As remarked in [17] and [19], with $\bar{X}_n = n^{-1} \sum_{i=1}^{n} X_i$ standing for the sample mean, for Student's statistic T_n we have

$$T_n = \sum_{i=1}^{n} X_i / \left(\frac{n}{n-1} \sum_{i=1}^{n} (X_i - \bar{X}_n)^2 \right)^{1/2} = R_n \left((n-1)/(n-R_n^2) \right)^{1/2}.$$

So whenever R_n has a limiting distribution, T_n has a limiting distribution, and the two coincide.

The asymptotic distribution of R_n is investigated in [19 when $F\in D(\alpha)$, $\alpha<2$ (cf. also [7] mentioned above).

If $EX^2<\infty$, then, obviously, by the central limit theorem and the law of large numbers, $R_n \xrightarrow{D} N(0,1)$. Citing from [19], "One is tempted to conjecture that R_n is asymptotically normal if [and perhaps only if] X_i are in the domain of attraction of the normal law." As an illustration of our techniques, we prove here the 'if' part of this conjecture

Assume $F\in D(2)$ and that $EX=0$. We can also assume that $EX^2=\infty$, otherwise there is nothing to prove. Since by Corollary 2.2 in [4]

$$\sum_{i=1}^{n} X_i/(\sqrt{n}\sigma(1/n)) \xrightarrow{D} N(0,1),$$

and by item (A.26) in [4]

$$(6)\quad \sigma^2(1/n)/s^2(1/n)=\sigma^2(1/n)/\int_{1/n}^{1-1/n} Q^2(u)du\to 1$$

as $n\to\infty$, to show that $R_n \xrightarrow{D} N(0,1)$ it is sufficient to prove that

$$(7)\quad \sum_{i=1}^{n} X_i^2/s^2(1/n) \stackrel{D}{=} \sum_{i=1}^{n} Q^2(U_i)/s^2(1/n) \xrightarrow{P} 1,$$

where $\xrightarrow{P}$ denotes convergence in probability as $n\to\infty$. By (6) this will follow if we show that

$$(8)\quad \Delta_n=\frac{1}{n\sigma^2(1/n)}\{\sum_{i=1}^{n} Q^2(U_i)-ns^2(1/n)\} \xrightarrow{P} 0.$$

Using (3), with Q replaced by Q^2, we have

$$\Delta_n=\frac{2}{\sigma^2(1/n)}\int_{1/n}^{U_{1,n}} (G_n(s)-\tfrac{1}{n})Q(s)dQ(s)$$

$$+\frac{2}{\sigma^2(1/n)}\int_{1/n}^{1-1/n} (s-G_n(s))Q(s)dQ(s)$$

$$+\frac{2}{\sigma^2(1/n)}\int_{U_{n-1,n}}^{1-1/n} (G_n(s)-\tfrac{n-1}{n})Q(s)dQ(s)$$

$$+\frac{1}{n\sigma^2(1/n)}(Q^2(U_{1,n})+Q^2(U_{n,n}))$$

$$=D_{1,n}+M_n+D_{2,n}+C_n.$$

Since both $nU_{1,n}$ and $n(1-U_{n,n})$ converge in distribution to an exponential random variable with mean 1, it is easy to verify using (4) and (5) that $C_n \xrightarrow{P} 0$.

Next, writing $Q=Q^+-Q^-$ and noticing that $2QdQ=2(Q^+dQ^+ - Q^-dQ^-)$, we obtain

$$D_{1,n} \leq [n\{G_n(\frac{1}{n})+\frac{1}{n}\}][\frac{1}{n\sigma^2(1/n)}\{Q^2(U_{1,n})+Q^2(\frac{1}{n})\}].$$

Since the first factor is bounded in probability, $D_{1,n} \xrightarrow{P} 0$ by the above argument for C_n. The proof that $D_{2,n} \xrightarrow{P} 0$ is completely analogous.

Finally, defining M_n^+ and M_n^- in the obvious way through $QdQ=Q^+dQ-Q^-dQ$, we have $M_n=M_n^+-M_n^-$. Now for any $0<\nu<1/2$,

$$|M_n^+| \leq \int_0^{1-1/n} n^{-\nu}(1-s)^{1-\nu}Q^+(s)dQ(s) \sup_{0<s<1} n^{\nu}\frac{|s-G_n(s)|}{(1-s)^{1-\nu}}$$

and by item (3) in [20] the sup factor here is bounded in probability. Integrating by parts and using (4) and (5) and an argument as on pages 9 and 10 of [12], we see that the integral factor converges to zero. Therefore $M_n^+ \xrightarrow{P} 0$. Since the proof that $M_n^- \xrightarrow{P} 0$ is similar, we have established (8) and hence the direct half of the Logan-Mallows-Rice-Shepp conjecture.

The more cautiously stated 'only if' part of the conjecture appears to be difficult.

We point out that a version of (7) has been proved earlier by Athreya [1]. He must have used different methods and also asked if (7) holds almost surely. The answer to this question is negative, since it is easily checked that for the choice $Q(1-s)=(s\log(1/s))^{-1/2}$, $0<s<1$, (4) and (5) hold true while, as $n\to\infty$, the limsup of (7) equals to infinity almost surely.

References

1. Athreya, K.B. (1985). Personal communication to S. Csörgő at the 191st Eastern Regional IMS Meeting, June 12-14, Stony Brook, N.Y.
2. Burke, M.D. and Horváth, L. (1966). Estimation of influence functions. *Statist. Probab. Letters* 4, 81-85.

3. Csörgő, M., Csörgő, S., Horváth, L., and Mason, D.M. (1986a). Weighted empirical and quantile processes. *Ann. Probab.* 14, 31-85.
4. Csörgő, M., Csörgő, S., Horváth, L., and Mason, D.M. (1986b). Normal and stable convergence of integral functions of the empirical distribution function. *Ann. Probab.* 14, 86-118.
5. Csörgő, M., Csörgő, S., Horváth, L., and Mason, D.M. (1986c). Sup-norm convergence of the empirical process indexed by functions and applications. *Probab. Math. Statist.* 7, No. 1.
6. Csörgő, M., and Mason, D.M. (1985). On the asymptotic distribution of weighted uniform empirical and quantile processes in the middle and on the tails. *Stochastic Process. Appl.* 21, 119-132.
7. Csörgő, S. (1986). Notes on extreme and self-normalised sums from the domain of attraction of a stable law. Preprint.
8. Csörgő, S., Deheuvels, P., and Mason, D.M. (1985). Kernel estimates of the tail index of a distribution. *Ann. Statist.* 13, 1050-1077.
9. Csörgő, S., Haeusler, E., and Mason, D.M. (1986). The asymptotic distribution of trimmed sums. Submitted.
10. Csörgő, S., Haeusler, E., and Mason, D.M. (1987a). A probabilistic approach to the asymptotic distribution of sums of independent, identically distributed random variables. Submitted.
11. Csörgő, S., Haeusler, E., and Mason, D.M. (1987b). The asymptotic distribution of extreme sums. In preparation.
12. Csörgő, S., Horváth, L., and Mason, D.M. (1986). What portion of the sample makes a partial sum asymptotically stable or normal? *Probab. Theory Rel. Fields* 72, 1-16.
13. Csörgő, S., and Mason, D.M. (1984). Simple estimators of the endpoint of a distribution. Submitted.

14. Csörgő, S., and Mason, D.M. (1985). Central limit theorems for sums of extreme values. *Math. Proc. Cambridge Philos. Soc.* 98, 547-558.

15. Csörgő, S., and Mason, D.M. (1986). The asymptotic distribution of sums of extreme values from a regularly varying distribution. *Ann. Probab.* 14, 974-983.

16. De Haan, L. (1975). *On Regular Variation and its Application to the Weak Convergence of Sample Extremes.* Amsterdam; Mathematical Centre.

17. Efron, B. (1969). Student's t-test under symmetry conditions. *J. Amer. Statist. Assoc.* 64, 1278-1302.

18. Lo, G.S. (1985). A note on the asymptotic normality of sums of extreme values. Submitted.

19. Logan, B.F., Mallows, C.L., Rice, S.O., and Shepp L.A. (1973). Limit distributions of self-normalized sums. *Ann. Probab.* 1, 788-809.

20. Mason, D.M. (1985). The asymptotic distribution of generalized Rényi statistics. *Acta Sci. Math.* (*Szeged*). 48, 315-323.

21. Mason, D.M., and van Zwet, W.R. (1987). A refinement of the KMT inequality for the uniform empirical process. *Ann. Probab.* 15, to appear.

22. Meizler, D.G. (1949). On a theorem of B.V. Gnedenko. *Sb. Trudov Inst. Mat. Akad. Nauk Ukrain. S.S.R.* 12, 31-35. (in Russian)

MINIMIZATION OF EXPECTED RISK BASED ON EMPIRICAL DATA

Vapnik V.N., Chervonenkis A.Ja. Institute of Control Sciences, Moscow, USSR

The report deals with the problem of expected risk minimization on empirical data, the study of two principles of expected risk minimization: the principle of empirical risk minimization and the principle of structural risk minimization - and the application of these principles to the problem of the estimation of probability distribution density.

I°. <u>Minimizing the expected risk on empirical data.</u> Let $(X,\mathcal{F},P)$ be the probability space, $Q(x,\alpha), \alpha\in\Lambda$ the set of measurable functions. It is required to minimize the expected risk functional

$$\Phi(\alpha)=\int Q(x,\alpha)\,d P(x) \tag{1}$$

if the probability measure $P(x)$ is unknown but a random independent sample

$$x_1,\ldots,x_\ell\ldots \tag{2}$$

obtained according to $P(x)$ is given.

This statement embraces a wide range of statistical problems and particularly the problem of estimation of probability distribution density.

Let the required density $p(x,\alpha_0)$ belong to the set of densities $p(x,\alpha), \alpha\in\Lambda$ (Radon-Nikodim derivatives with respect to σ -finite measure ν : $p(x,\alpha)=\frac{d P_\alpha(x)}{d\nu}$) and let $p(x,\alpha_0)$ have a finite Shannon entropy $H_{\alpha_0}=-\int \ln p(x,\alpha_0)dP_{\alpha_0}(x)$.

Then the problem of density estimation is equivalent to minimization of the functional

$$H_{\alpha_0}(\alpha)=-\int \ln p(x,\alpha)\,d P_{\alpha_0}(x) \tag{3}$$

on empirical data (2).

2°. <u>The principle of minimizing the empirical risk.</u> We shall minimize the empirical functional

$$\Phi_\ell(\alpha)=\frac{1}{\ell}\sum_{i=1}^{\ell} Q(x_i,\alpha) \tag{4}$$

constructed by means of the sample (2). The function $Q(x,\alpha_\ell)$ yielding the minimum to the functional (4) on $Q(x,\alpha), \alpha\in\Lambda$ is chosen as the required function $Q(x,\alpha_0)$ minimizing (I).

The method of minimizing (I) by means of minimizing the empirical functional (4) is called the method of empirical risk minimization. For the problem of estimating the density the principle of minimizing the empirical risk makes the maximum likelihood method: minimization of the functional (3) is done by means of minimizing the empirical functional

$$H_\ell(\alpha) = -\frac{1}{\ell}\sum_{i=1}^{\ell} \ln p(x_i,\alpha).$$

3°. The conditions of consistency of the method of empirical risk minimization. Introduce some definitions.

Definition. The method of minimizing the empirical risk on the set $Q(x,\alpha), \alpha\in\Lambda$ is strictly consistent if for any C on the set of functions $\Lambda(C)=\{Q(x,\alpha): \alpha\in\Lambda$ and $\int Q(x,\alpha)dP(x) > C\}$ the convergence

$$\inf_{Q(x,\alpha)\in\Lambda(C)} \frac{1}{\ell}\sum_{i=1}^{\ell} Q(x_i,\alpha) \xrightarrow[\ell\to\infty]{P} \inf_{Q(x,\alpha)\in\Lambda(C)} \int Q(x,\alpha)dP(x) \quad (5)$$

takes place.

Remark. The convergence (5) implies the convergence

$$\int Q(x,\alpha_\ell)dP(x) \xrightarrow[\ell\to\infty]{P} \inf_{\alpha\in\Lambda} \int Q(x,$$

Definition. We shall say that on the set $Q(x,\alpha), \alpha\in\Lambda$ the modulo uniform convergence of means to their mathematical expectations occurs if

$$\sup_{\alpha\in\Lambda} \left| EQ(x,\alpha) - \frac{1}{\ell}\sum_{i=1}^{\ell} Q(x_i,\alpha)\right| \xrightarrow[\ell\to\infty]{P} 0.$$

is valid.

We shall say that on the set $Q(x,\alpha), \alpha\in\Lambda$ one-sided uniform convergence occurs if

$$\sup_{\alpha\in\Lambda} \left(EQ(x,\alpha) - \frac{1}{\ell}\sum_{i=1}^{\ell} Q(x_i,\alpha)\right)_+ \xrightarrow[\ell\to\infty]{P} 0, \quad (z) = \begin{cases} z, & z>0, \\ 0, & z\le 0. \end{cases}$$

Theorem I. Let $-\infty < a \le EQ(x,\alpha) \le A < \infty$. Then the following two assertions are equivalent:

I) the method of empirical risk minimization is strictly consistent on $Q(x,\alpha), \alpha \in \Lambda$,

2) on the set $Q(x,\alpha), \alpha\in\Lambda$ one-sided uniform convergence exists.

4°. The uniform convergence on the set of uniformly-bounded functions. Consider the set of uniformly-bounded functions $|Q(x,\alpha)| \le B, \alpha\in\Lambda$. Introduce the set of ℓ -dimensional vectors

$$q(\alpha) = (Q(x_1,\alpha), \dots, Q(x_\ell,\alpha)), \alpha\in\Lambda.$$

using (2). Let $N^\Lambda(\varepsilon, x_1,\dots,x_\ell)$ be a number of elements in the minimal ε -net of the set $q(\alpha), \alpha\in\Lambda$ in the metric C : $\rho(q_1,q_2) = \sup_{1\le i\le \ell} |Q(x_i,\alpha_1) - Q(x_i,\alpha_\ell)|$.

Let the functions $Q(x,\alpha), \alpha\in\Lambda$ satisfy the appropriate conditions of measurability.

Denote $H^\Lambda(\varepsilon,\ell) = E \ln N^\Lambda(\varepsilon; x_1,\dots,x_\ell)$.

Theorem 2 [I] For a modulo uniform convergence on the set $Q(x,\alpha), \alpha\in\Lambda$ it is necessary and sufficient that for any $\varepsilon>0$ the equality

$$\lim_{\ell\to\infty} \frac{H^\Lambda(\varepsilon,\ell)}{\ell} = 0$$

hold.

Theorem 3. For one-sided uniform convergence on the set $Q(x,\alpha), \alpha\in\Lambda$ it is necessary and sufficient that for any positive ε, δ and η there can be found the set of bounded functions $|R(x,\beta)| \le B, \beta\in\mathcal{B}$ satisfying the conditions of measurability such that for any function $Q(x,\alpha^*)$ $\alpha^*\in\Lambda$ the function $R(x,\beta^*), \beta^*\in\mathcal{B}$ exists satisfying the conditions

$$Q(x,\alpha^*) \ge R(x,\beta^*),\ EQ(x,\alpha^*) - ER(x,\beta^*) \le \delta \quad (6)$$

and for the set $R(x,\beta), \beta\in\mathcal{B}$ the inequality

$$\lim_{\ell\to\infty} \frac{H^{\mathcal{B}}(\varepsilon,\ell)}{\ell} < \eta \quad (7)$$

is fulfilled.

Corollary. [2] If uniform convergence exists it exists almost sure.

The proof of theorem 3 is based on the following two theorems.

Theorem 4. For one-sided uniform convergence on the set

$Q(x,\alpha), \alpha\in\Lambda$ it is necessary that for any $\varepsilon>0$ on $Q(x,\alpha)$ $\alpha\in\Lambda$ there exist a finite ε-net in the metric $L_1(P)$: $\rho_{L_1}(Q_1,Q_2)=\int|Q(x,\alpha_1)-Q(x,\alpha_2)|dP(x)$.

Theorem 5. Let for the set of functions $Q(x,\alpha), \alpha\in\Lambda$ there exist such $\varepsilon_0>0$ that

$$\lim_{\ell\to\infty}\frac{H^\Lambda(\varepsilon_0,\ell)}{\ell}=C\neq 0.$$

Then such functions $\Psi_1(x)\geq\Psi_0(x)$ and the number $\tau(\varepsilon_0)$ satisfying the inequality

$$\int(\Psi_1(x)-\Psi_0(x))dP(x)\geq\tau(\varepsilon_0)$$

can be found that for any $\delta>0$, any binary sequence $\omega_1,\ldots,\omega_\ell, (\omega_i=0,1)$ and for almost any sample $x_1,\ldots,x_\ell$ the function $Q(x,\alpha^*)$ can be found for which the inequalities

$$|Q(x_i,\alpha^*)-\Psi_{\omega_i}(x_i)|<\delta, \quad i=1,\ldots,\ell$$

hold.

5°. Conditions of consistency of maximum likelihood method for the set of uniformly bounded and uniformly separated from zero densities.

Definition. The maximum likelihood method is consistent on the set of densities $p(x,\alpha),\alpha\in\Lambda$ if for any density $p(x,\alpha_0), \alpha_0\in\Lambda$ the convergence in probability

$$\inf_{\alpha\in\Lambda}\frac{1}{\ell}\sum_{i=1}^{\ell}-\ln p(x_i,\alpha)\xrightarrow[\ell\to\infty]{P}-\int\ln p(x,\alpha_0)dP_{\alpha_0}(x)$$

takes place.

In this section we shall study densities satisfying the condition

$$0<a\leq p(x,\alpha)\leq A<\infty, \ \alpha\in\Lambda, \ x\in(-\infty,\infty). \tag{8}$$

For such densities the theorem holds:

Theorem 6. In order that the maximum likelihood method will be consistent it is necessary and sufficient that on the set of (uniformly bounded) functions $-\ln p(x,\alpha),\alpha\in\Lambda$ one-sided uniform convergence with respect to any measure $P_{\alpha_0}(x), \alpha_0\in\Lambda$ take place.

6°. Generalization on the set of unbounded functions.

Now let $Q(x,\alpha),\alpha\in\Lambda$ be the set of unbounded functions

Define the functions

$$[Q(x,\alpha)]_c = \begin{cases} Q(x,\alpha), & \text{if } |Q(x,\alpha)| \le c, \\ c \,\mathrm{Sgn}\, Q(x,\alpha), & \text{if } |Q(x,\alpha)| > c. \end{cases}$$

<u>Theorem 7.</u> [5] In order that on the set of unbounded functions $Q(x,\alpha)$, $\alpha\in\Lambda$ one-sided uniform convergence take place it is sufficient that for any positive ε,δ,η there can be found the set of functions $R(x,\beta)$, $\beta\in\mathcal{B}$ satisfying the conditions of measurability and the following three conditions:

I) $|R(x,\alpha)| \le K(x)$, $\int K(x)\,dP(x) < \infty$;

2) for any function $Q(x,\alpha^*)$ the function $R(x,\beta^*)$ satisfying the inequalities (6) exists,

3) for any C for the set $[R(x,\beta)]_c$, $\beta\in\mathcal{B}$ the inequality (7) is true.

To find the conditions of the consistency of the maximum likelihood method for the set of unbounded and unseparable from zero densities introduce the following definitions.

<u>Definition.</u> We shall say that for the set $p(x,\alpha)$, $\alpha\in\Lambda$ the estimator of the maximum likelihood exists if for any $P_{\alpha_0}(x)$, $\alpha_0\in\Lambda$ such finite $K(\alpha_0)$ may be found that

$$\inf_{\alpha\in\Lambda}\left(-\sum_{i=1}^{K(\alpha_0)} \ln p(x_i,\alpha)\right) \geqslant L(x_1,\dots,x_{K(\alpha_0)}),\quad EL(x_1,\dots,x_{K(\alpha_0)}) > -\infty,$$

where mathematical expectation of the function of $K(\alpha_0)$ variables $\bar{x}=(x_1,\dots,x_{K(\alpha_0)})$ is taken according to measure

$$P_{\alpha_0}(x_1,\dots,x_{K(\alpha_0)}) = \prod_{i=1}^{K(\alpha_0)} P_{\alpha_0}(x_i).$$

<u>Theorem 8.</u> [5] In order that the maximum likelihood method will be consistent in L_1 it is sufficient that on $p(x,\alpha)$ $\alpha\in\Lambda$ there exist anestimator of maximum likelihood and for any $P_{\alpha_0}(x)$, $\alpha_0\in\Lambda$ and arbitrary $c>0$ on the set of functions $[-\sum_{i=1}^{K(\alpha_0)} \ln p(x_i,\alpha)]_c$ ($K(\alpha_0)$ variables $\bar{x}=(x_1,\dots,x_{K(\alpha_0)})$) one-sided uniform convergence with respect to measure $P_{\alpha_0}(\bar{x}) = \prod_{i=1}^{K(\alpha_0)} P_{\alpha_0}(x_i)$ take place.

7°. <u>The principle of structural risk minimization.</u> Now consider the new method of minimizing the expected risk

(I) on empirical data (2) that is the generalization of the method of minimizing the empirical risk. Let on the set S of functions $Q(x,\alpha)$, $\alpha \in \Lambda$ the structure

$$S_1 \subset S_2 \subset \ldots \subset S_n \subset \ldots \qquad (9)$$

be given where $S_n = \{Q(x,\alpha) : \alpha \in \Lambda_n\}$ and the law $n = n(x_1,\ldots,x_\ell)$ be given defining the element S_n of the structure (9) depending on the sample $x_1,\ldots, x_\ell$ We shall choose as the approximation $Q(x,\alpha_\ell)$ to the required function $Q(x,\alpha_0)$ such a function that yields the conditional minimum to the empirical risk functional (4) on the set S_n.

We shall call such method of minimizing the expected risk (I) on empirical data (2) the method of structural risk minimization.

We shall define the conditions that the structure (9) and the law $n = n(\ell)$ should satisfy in order that the method of structural risk minimization provide the convergence of the estimator to the required density. To formulate these conditions introduce the following concepts,

Let on X the system of sets $\Omega_\alpha(X)$, $\alpha \in \Lambda$ be defined. Call the index of the system $\Omega_\alpha(X)$, $\alpha \in \Lambda$ with respect to the collection $x_1,\ldots,x_\ell$ the number of different sets $(x_{i_1},\ldots,x_{i_k}) = (x_1,\ldots,x_\ell) \cap \Omega_\alpha(X)$ when $\alpha \in \Lambda$. Denote the index $\Delta^\Lambda(x_1,\ldots,x_\ell)$. Call the function

$$m^\Lambda(\ell) = \max_{x_1,\ldots,x_\ell} \Delta^\Lambda(x_1,\ldots,x_\ell)$$

the function of the growth of the system $\Omega_\alpha(X)$, $\alpha \in \Lambda$.

<u>Theorem 9.</u>[3] The estimate is valid:

$$m^\Lambda(\ell) \begin{cases} \text{either} \equiv 2^\ell; \\ \text{or} < 1.5\,\ell^h, \quad h = \max(i : m^\Lambda(i) = 2^i) \end{cases}.$$

<u>Definition.</u> We shall say that the set of functions $Q(x,\alpha)$, $\alpha \in \Lambda$ has a finite capacity h if the growth functions $m_c^\Lambda(\ell)$ of the systems of the sets

$$\Omega_\alpha^c(X) = \{x : Q(x,\alpha) \geq c, \ \}, \ \alpha \in \Lambda$$

are such that

$$h = \sup_{-\infty \leq c \leq \infty} \lim_{\ell\to\infty} \left(\frac{\lg m_c^\Lambda(\ell)}{\ell} \right) < \infty.$$

8°. <u>Theorems of the convergence of the method of struc-</u>

tural risk minimization. Consider the structure $\mathcal{A}$. Let the elements S_n of the structure (9) satisfy the following three conditions:

I) $\tau_n = \inf_{\alpha \in \Lambda_n} EQ(x,\alpha) - \inf_{\alpha \in \Lambda} EQ(x,\alpha) \xrightarrow[n\to\infty]{} 0$;

2) the set of functions $Q(x,\alpha), \alpha \in \Lambda_n$ has capacity h_n ;

3) S_n contains such functions that $|Q(x,\alpha)| \leq B_n, \alpha \in \Lambda_n$.

For the structure $\mathcal{A}$ the following theorem is valid.

Theorem IO. [5].The method of structural risk minimization provides the convergence

$$\rho(\alpha_\ell, \alpha_0) = EQ(x,\alpha_\ell) - EQ(x,\alpha_0) \xrightarrow[\ell\to\infty]{} 0$$

almost sure with asymptotic rate

$$V(\ell) = \tau_{n(\ell)} + \sqrt{\frac{h_{n(\ell)} B^2_{n(\ell)} \ln \ell}{\ell}}$$

(i.e. $P\{\overline{\lim_{\ell\to\infty}} V^{-1}(\ell)\rho(\alpha_\ell,\alpha_0) = Const\}$),
if the law $n = n(\ell)$ is such that

$$\frac{h_{n(\ell)} B^2_{n(\ell)} \ln \ell}{\ell} \xrightarrow[\ell\to\infty]{} 0$$

holds.

Now let $Q(x,\alpha) \geq 0, \alpha \in \Lambda$ Consider the structure $\mathcal{B}$ the element of which S_n satisfies the following three conditions:

I) $\tau_n = \inf_{\alpha \in \Lambda_n} EQ(x,\alpha) - \inf_{\alpha \in \Lambda} EQ(x,\alpha) \xrightarrow[n\to\infty]{} 0$;

2) the set of functions $Q(x,\alpha), \alpha \in \Lambda_n$ has capacity h_n;

3) for the functions S_n the pair $\nu_n > 1$ and $0 < C_n < \infty$ is defined for which the inequality

$$\sup_{\alpha \in \Lambda_n} \frac{(EQ^{\nu_n}(x,\alpha))^{1/\nu_n}}{EQ(x,\alpha)} \leq C_n$$

holds.

For the structure $\mathcal{B}$ the theorem is valid.

Theorem II. [5] Let $\forall n: \nu_n \geq a > 2$ Then the method of structural risk minimization provides the convergence

$$\rho(\alpha_\ell, \alpha_0) = EQ(x,\alpha_\ell) - EQ(x,\alpha_0) \xrightarrow[\ell\to\infty]{} 0 \qquad (10)$$

almost sure with asymptotic rate

$$V(\ell)=\tau_{n(\ell)}+\sqrt{\frac{C^2_{n(\ell)}\, h_{n(\ell)} \ln \ell}{\ell}}$$

if the function $n=n(\ell)$ is such that

$$\frac{C^2_{n(\ell)}\, h_{n(\ell)} \ln \ell}{\ell} \xrightarrow[\ell\to\infty]{} 0$$

But if $\exists\, n : \nu_n \le 2$ the method of structural risk minimization provides the convergence almost sure with asymptotic rate

$$V(\ell)=\tau_{n(\ell)}+\ell^{-1+1/\nu_{n(\ell)}}\, C_{n(\ell)} h^{1/2}_{n(\ell)} (\ln \ell)^{3/2-1/\nu_{n(\ell)}},$$

if the function $n=n(\ell)$ is such that

$$h_{n(\ell)}\,\ell^{-2+2/\nu_{n(\ell)}}\, C^2_{n(\ell)} (\ln \ell)^{3-2/\nu_{n(\ell)}} \xrightarrow[\ell\to\infty]{} 0$$

holds.

9°. The estimation of densities by the method of maximum conditional (structural) likelihood. Let us apply the method of structural risk minimization for the estimation of density in the class of densities with finite Shannon entropy. For this define on the set of functions $Q(x,\alpha)=-\ln p(x,\alpha), \alpha\in\Lambda$ the structure of the type $\mathcal{A}$ (it is possible) to which we shall apply the method of structural risk minimization. With the help of theorem II in this case the fact of the convergence of the estimate to the required density in Kulbac metric is proved and the rate of this convergence is estimated.

Consider now the general case: the densities $p(x,\alpha), \alpha\in\Lambda$ are arbitrary (may have no Shannon entropy). Construct the structure $\mathcal{A}_c$ the elements of which $S_n=\{p(x,\alpha): \alpha\in\Lambda_{n(\ell)}\}$ satisfy the following four conditions:

I) $\tau_n=\inf\limits_{\alpha\in\Lambda_n} \int |p(x,\alpha)-p(x,\alpha_0)|\, d\nu \xrightarrow[n\to\infty]{} 0$;

2) the set of functions $-\ln p(x,\alpha), \alpha\in\Lambda_n$ has capacity h_n

3) S_n contains such functions that $|\ln p(x,\alpha)|\le B_n, \alpha\in\Lambda_n$

4) the set of functions $S_n=\{p(x,\alpha):\ \alpha\in\Lambda_n\}$ is convex.
For the structure $\mathcal{A}_e$ the theorem is valid:

Theorem I2. [5] The method of maximum conditional (structural) likelihood provides the convergence in the metric $\rho(\alpha_\ell,\alpha_0)=\int|p(x,\alpha_\ell)-p(x,\alpha_0)|dx$ almost sure with asymptotic rate

$$V(\ell)=\tau_{n(\ell)}+e^{-B_{n(\ell)}}+\sqrt[4]{\frac{h_{n(\ell)}B^2_{n(\ell)}\exp\{8B_{n(\ell)}\}\ln\ell}{\ell}},$$

if the function $n=n(\ell)$ is such that

$$\frac{h_{n(\ell)}B^2_{n(\ell)}\ln\ell\exp\{8B_{n(\ell)}\}}{\ell}\xrightarrow[\ell\to\infty]{}0$$

holds.

IO°. Direct methods of estimating the density. According to the definition the density of the probability distribution is the solution of the integral equation

$$\int_{-\infty}^{x}p(t,\alpha)dt=P_\alpha(x) \qquad (11)$$

The estimation of the density on empirical data $x_1,\dots,x_\ell$ is the solution of the equation (II) in conditions when instead of the cumulative function $P_\alpha(x)$ its approximation is given, e.g.

$$\widehat{P}_\alpha(x)=\frac{1}{\ell}\sum_{i=1}^{\ell}\theta(x-x_i),\quad \theta(z)=\begin{cases}1, & z>0;\\ 0, & z\le 0.\end{cases}$$

Consider the equation

$$Af(t)=F(x)$$

defined by the linear non-degenerated operator A mapping the elements of Gilbert space Γ into elements of metric space M_* (the solution of the equation (I2) makes an ill-posed problem).

Let $F_\ell(x)$ be a sequence of random functions and $f_\ell(t)$ be a sequence of functions minimizing the functional

$$\Phi(f)=\rho^2_{M_*}(Af,F_\ell)+\gamma\|f\|^2,$$

where $\gamma_\ell\xrightarrow[\ell\to\infty]{}0$ is a decreasing positive sequence.

Theorem I3. 6 For any $\varepsilon>0$ there exists a positive number $n(\varepsilon)$ such that for every $\ell>n(\varepsilon)$ the inequality

$$P\{\|f_\ell(t)-f(t)\|^2>\varepsilon\}<2P\{\rho^2_{M_*}(F,F_\ell)>\frac{\varepsilon}{2}\gamma_\ell\}$$

holds.

Corollary. Let $p(t,\alpha_\ell)$ minimize the functional

$$\Phi(\alpha)=\rho_0(F,F_\ell)+\gamma_\ell\|p(t,\alpha)\|^2 \qquad (14)$$

where $F(z)=BP_\alpha(x)$, $F_\ell(z)=B\hat{P}_\alpha(x)$, $B: M_*\to M$ is a limited non-degenerated linear operator, let

$$\rho_0(F,F_\ell)\le R(F)\sup_z|F(z)-F_\ell(z)|. \qquad (15)$$

Then using Kolmogorov-Smirnov inequality we obtain

$$P\{\|p(t,\alpha_\ell)-p(t,\alpha_0)\|^2>\varepsilon\}<2\exp\left\{-\frac{\varepsilon\ell\gamma_\ell}{R^2(F)\|B\|^2}\right\}. \qquad (16)$$

It follows from (I6) that if

$$\gamma_\ell\xrightarrow[\ell\to\infty]{}0,\quad \ell\gamma_\ell\xrightarrow[\ell\to\infty]{}\infty$$

the convergence $\|p(t,\alpha_\ell)-p(t,\alpha_0)\|\xrightarrow[\ell\to\infty]{}0$ in probability takes place; but if for any $\mu>0$ the inequality

$$\sum_{\ell=1}^{\infty}\exp\{-\mu\ell\gamma_\ell\}<\infty$$

fulfills the convergence takes place almost sure.

Obtain the estimators of densities for the case when the functions $F(z)=BP_\alpha(x)$ belong to the space with the metric

$$\rho(F,F_\ell)=\sqrt{\int(F_\ell(z)-F(z))^2dz}$$

satisfying (I5).

I. Let $p(t,\alpha)\in L_2(-\pi,\pi)$ and let B be the identical operator. We shall seek the approximation $p(t,\alpha_\ell)$ as the expansion

$$p(t,\alpha_\ell)=\frac{1}{2\pi}+\sum_{k=1}^{\infty}(a_k\sin kt+b_k\cos kt).$$

The minimum of the functional (I4) is attained in case when

$$b_k=\frac{\frac{1}{\ell}\sum_{i=1}^{\ell}\cos kx_i}{\pi(1+\gamma_\ell k^2)},\quad a_k=\frac{1}{\pi(1+\gamma_\ell k^2)}\left(\frac{1}{\ell}\sum_{i=1}^{\ell}\sin kx_i+\frac{(-1)^{k+1}k}{1+2\Gamma_\gamma}(\bar{x}+2R_\gamma\right),$$

where $\bar{x}=\frac{1}{\ell}\sum_{i=1}^{\ell}x_i$,

$$\Gamma_\gamma=\sum_{r=1}^{\ell}\frac{1}{1+\gamma_\ell r^2}\approx\frac{\pi}{\sqrt{\gamma_\ell}}\ ;\quad R_\gamma=\sum_{r=1}^{\infty}\frac{(-1)^r\frac{1}{\ell}\sum_{i=1}^{\ell}\sin r x_i}{(1+\gamma_\ell r^2)r}.$$

2. Let $p(t,\alpha)\in L_2(-\infty,\infty)$ and $p(t,\alpha)$ differs from zero on finite support. Let operator B be defined in the form

$$BP_\alpha(x)=\int K(z-x)P_\alpha(x)dx$$

In this case the minimum of the functional (I4) (obtained by variation of (I4) and the afterwords Fourier transform of the obtained equation) makes Parzen-Rosenblatt estimator [4]

$$p(t,\alpha_\ell)=\frac{1}{\ell\sqrt{\gamma_\ell}}\sum_{i=1}^{\ell}g\left(\frac{t_i-x_i}{\sqrt{\gamma_\ell}}\right),$$

where the kernel function $g(z)$ is defined by the kernel $K(z)$ of the operator B :

$$\hat{K}(\omega)=\int_{-\infty}^{\infty}K(z)e^{-i\omega z}dz,\quad \tilde{g}(\omega)=\frac{\tilde{K}(\omega)\hat{K}(-\omega)}{\gamma_\ell\omega^r+\hat{K}(\omega)\hat{K}(-\omega)},$$

$$g(z)=\frac{1}{2\pi}\int_{-\infty}^{\infty}\tilde{g}(\omega)e^{i\omega z}d\omega.$$

Observe that estimating the equation solution (II) on empirical data $x_1,\dots,x_\ell$ by method of minimization (I4) is equivalent to the application of the method of structural risk minimization. Here the empirical functional $\rho^2(BP_\alpha(x),B\hat{P}_\alpha(x))$ is minimized on the element of the structure

$$S_n=\{p(t,\alpha):\|p(t,\alpha)\|^2\le C_n\}.$$

Using different metrics $\rho_M(F,F_\ell)$ in (I4) is a source of inventing various algorithms of density estimation.

References

I. Вапник В.Н., Червоненкис А.Я. (I98I) Необходимые и достаточные условия равномерной сходимости средних к их математическим ожиданиям. Теория вероятн. и ее примен. XXII, 3, 543-563.

2. Вапник В.Н., Червоненкис (I974) Теория распознавания образов. М., Наука, 4I5с. (Wapnik W.N., Tscherwonenkis A.Ja. Theorie der Zeichenerkennung. Akademie-Verlag, Berlin,

1979).
3. Вапник В.Н., Червоненкис А.Я. (1971) О равномерной сходимости частот появления событий к их вероятностям. Теория вероятн. и ее примен. XУI, 2, 264-280.
4. Вапник В.Н. Восстановление зависимостей по эмпирическим данным. М., Наука. 1979, 448с. (Vapnik V. Estimation of Dependences Based on Empirical Data. Springer-Verlag, N.Y.,1982).
5. Vapnik V.N. Maximum Likelihood Principle and its Generalization.The 5-th International Summer School on Probability Theory and Mathematical Statistics. Varna, 1985.
6. Вапник В.Н., Стефанюк А.Р. (1978) Непараметрические методы восстановления плотности вероятностей. Автоматика и телемеханика, 8.

RATES OF CONVERGENCE IN THE INVARIANCE PRINCIPLE FOR EMPIRICAL MEASURES

I.S. Borisov
Institute of Mathematics, Novosibirsk, USSR, 630090

Let $X_1, X_2, \ldots$ be a sequence of independent identically distributed random variables taking values in an arbitrary measurable space $(Y, \mathcal{A})$ and with the distribution $P(\cdot)$. Let $P_m(\cdot)$ be the empirical distribution based on the sample $X_1, \ldots, X_m$. Denote

$$S_n(t,A) = n^{-1/2}[nt](P_{[nt]}(A) - P(A)), \quad 0 \leq t \leq 1.$$

Let $W_n(t,A)$ be a centered Gaussian random field with the same covariance as $S_n(t,A)$. Consider the following distance between $S_n(t,A)$ and $W_n(t,A)$:

$$d_n(\mathcal{A}_o) = \inf_{S_n, W_n} \inf(x: Pr^*(\sup_{0 \leq t \leq 1} \sup_{A \in \mathcal{A}_o} |S_n(t,A) - W_n(t,A)| \geq x) \leq x) \tag{1}$$

where the first infimum is taken on all pairs of $S_n(\cdot)$ and $W_n(\cdot)$ defined on a common probability space, Pr^* is the outer probability, $\mathcal{A}_o \subseteq \mathcal{A}$.

Let $\mathcal{A}_o$ be an arbitrary subclass of $\mathcal{A}$. Suppose that for every $x > 0$ there are the sets $A_i \in \mathcal{A}$, $i=1,2,\ldots,N(x)$, with the following properties: for any $B \in \mathcal{A}_o$ there are A_i, A_j, $i,j \leq N$, such that

$$A_i \subseteq B \subseteq A_j, \qquad P(A_j - A_i) \leq x. \tag{2}$$

The logarithm of the minimal number of (A_i) satisfying (2) is called the metric entropy with inclusion (see

Dudley, 1978) and is denoted by $H_1(x,@_o,P)$.

Let the family of sets $@(N)=(A_i \in @;\ i=1,\dots,N)$ satisfy (2). Denote by $@_g(N)$ some finite class (B_j) generating the family $@(N)$, that is

$$A_i = \bigcup_{k=1}^{M_i} \bigcup_{j=l(i,k)}^{l(i,k-1)} B_j, \quad i=1,\dots,N, \tag{3}$$

where $B_i \cap B_j = \emptyset$, $i \neq j$; $(l(i,k))$ is the increasing on k subsequence of natural numbers.

Let $M(@(N),@_g(N))=\max_{i\le N} M_i$, where M_i are defined in (3).

It is obvious that $M(\cdot)$ depends on $@(N)$, $@_g(N)$ and on the enumeration of elements from $@_g(N)$. Denote

$$r(x,@_o,P) = \inf M(@(N),@_g(N)) \tag{4}$$

where the infimum in (4) is taken on all families $@(N)$ satisfying (2) for $N \ge \exp(H_1(x,@_o,P))$, on all classes $@_g(N)$ and all enumerations of subsets $B_j \in @_g(N)$.

Further the symbols C, C_k denote some absolute positive constants. The dependence on some parameters is fixed by the corresponding arguments from $C(\cdot)$.

The below statements may be considered as some corollaries from the author's results(1985a,1986,1983,1985b).

Theorem 1. Let for some $@_o$ and P the R.Dudley's condition be fulfilled:

$$\int_0^1 (H_1(x^2,@_o,P))^{1/2}dx < \infty. \tag{5}$$

Then

$$d_n(@_o) \le C \inf_{m:H_1(2^{-m},@_o,P)\le n2^{-m}} R_n(m), \tag{6}$$

where

$$R_n(m) = n^{-1/2}(\log n)^2 r(2^{-m},@_o,P) +$$

$$+ \left(1 + \frac{\log n}{H_1(2^{-m}, @_o, P)+1}\right) \sum_{k>m} (H_1(2^{-k}, @_o, P)2^{-k})^{1/2}. \quad (7)$$

Theorem 2. Let $A_1, A_2, \ldots$ be a sequence of pairwise disjoint non-empty measurable subsets. Suppose that $\bigcup_i A_{k_i} \in @_o$ for any subsequence (k_i) of natural numbers. Then

$$d_n(@_o) \geq$$

$$\sup_{m\geq 0} \min((2\pi)^{-1/2}\sum_{k>m} p_k^{1/2} - 2n^{1/2}\sum_{k>m} p_k,\ 1-3\exp(-n\sum_{k>m} p_k/8))$$

where $p_k = P(A_k)$.

As an example, let us consider the case when $Y = [0,1]^2$, $@$ is the class of all Borel subsets of Y, P is an absolute continuous distribution in Y with bounded density, $@_1$ is the class of all closed convex subsets of Y,

$$@_2 = @_2(a,L) = ((z,y) \in Y: y \leq f(z);\ f \in \mathrm{Var}(a,L)), \quad 0 < a \leq 1,$$

where $\mathrm{Var}(a,L)$ is the family of all absolute continuous on $[0,1]$ functions for which $\sup|df/dz| + \int_0^1 |d(df/dz)|^{1/a} \leq L$.

Theorem 3. For $Y, @_1$ and P above

$$H_1(x, @_1, P) \leq C(P)x^{-1/2}, \quad r(x, @_1, P) \leq C_1(P)x^{-2},$$

$$d_n(@_1) \leq C_2(P)(\log n)^2 n^{-1/18};$$

$$H_1(x, @_2, P) \leq C(L,P)x^{-1/(1+a)},\ r(x, @_2, P) \leq C_1(L,P)x^{-3/(1+a)},$$

$$d_n(@_2) \leq C_2(L,P)(\log n)^2 n^{-a/(12+2a)}.$$

Theorem 4. Let the assumptions of Theorem 3 be fulfilled and moreover, the density of P is separated from zero on a rectangle $[z_1, z_2]\times[y_1, y_2] \subseteq Y$. Then

$$d_n(@_1) \geq C(P)n^{-1/6}, \quad d_n(@_2) \geq C(L,P)n^{-a/(4+2a)}.$$

Denote $L_1(x) = \log\max(x,2)$, $L_k(x) = L_1(L_{k-1}(x))$, $k \geq 2$.

Theorem 5. For every natural k there exist $(Y,@), P = P^{(k)}$ and $@_o \subseteq @$ such that

$$C_1(k)L_k^{-1}(n) \leq d_n(@_o) \leq C_2(k)L_k^{-1}(n).$$

Remark 1. The statement of Theorem 5 will hold if we consider, instead of (1), the distance

$$\overset{\circ}{d}_n(@_o) = \sup_x |Pr(\sup_{A\in @_o} |S_n(1,A)| < x) - Pr(\sup_{A\in @_o} |W_n(1,A)| < x)|$$

where it is supposed that the corresponding measurability conditions are fulfilled (see Borisov (1985b), (1985c)).

Remark 2. By the results of Borisov, 1985a, 1986, 1983 it is easy to show that the estimates for $d_n(\cdot)$ in Theorems 3 and 4 will be true if $P = vP^{(1)} - (1-v)P^{(2)}$, $v \in [0, 1]$, where $P^{(1)}$ satisfies the conditions of Theorem 3 or 4 and $P^{(2)} = (p_k)$ is an arbitrary discrete distribution in Y for which $p_k = O(k^{-2(s+1)})$ as $k \longrightarrow \infty$ and $s > \max(a/6, 1/8)$. Moreover, the constants $C(\cdot)$ do not depend on v.

REFERENCES

Borisov, I.S. (1985a). Rate of convergence in central limit theorem for empirical measures. In: Limit Theorems for Sums of Random Variables, Optimization Software Inc., New York, 186-213 (English translation).

Borisov, I.S. (1986). A new approach to the approximation problem of sum distributions of independent random variables in linear spaces. In: Limit Theorems of Probability Theory, Optimization Software Inc., New York (English translation, to appear).

Borisov, I.S. (1983). Problem of accuracy of approximation in the central limit theorem for empirical measures. Siberian Math.J., v.24, No 6, 833-843.(English transl.)

Borisov, I.S. (1985b). A remark on the speed of convergence in the central limit theorem in Banach spaces. Siberian Math.J., v.26, No 2, 180-185.(English transl.)

Borisov, I.S. (1985c). Upper and lower estimates of convergence rate in the invariance principle for empirical measures. Proceedings of the 4-th Vilnius Conference on Probab.Theory and Math.Statist. (to appear).

Dudley, R.M. (1978). Central limit theorem for empirical measures. Ann.Probab., v.6, No 6, 899-929.

ALMOST SURE BEHAVIOUR OF WEIGHTED EMPIRICAL PROCESSES IN THE TAILS

John H.J. EINMAHL[1] and David M. MASON[2]
[1] Dept. of Med. Informatics and Statistics, University of Limburg, P.O. Box 616, 6200 MD Maastricht, The Netherlands.
[2] Dept. of Mathematical Sciences, 501 Ewing Hall, University of Delaware, Newark, DE 19716, USA.

1. RESULTS

Let $X_1, X_2, \ldots$ be a sequence of independent random vectors, each uniformly distributed over $[0,1]^d$, $d \in \mathbb{N}$. The first n random vectors determine the empirical df F_n in the usual way:

$$F_n(t) = n^{-1}\#\{1 \le i \le n : X_i \le t\}, \qquad t \in [0,1]^d.$$

Writing $t = \langle t_1,\ldots,t_d\rangle$ and $|t| = \Pi_{j=1}^d t_j$ we define for $\alpha \in [0,1]$

$$\Delta_{n,\alpha} = \sup_{t\in[0,1]^d} \frac{|F_n(t)-|t||}{(|t|(1-|t|))^\alpha} \tag{1}$$

and for d=1 and fixed $k \in \mathbb{N}$

$$\Delta_{n,\alpha}^{(k)} = \sup_{X_{k:n}\le t\le X_{n-k+1:n}} \frac{|F_n(t)-t|}{(t(1-t))^\alpha}, \tag{1'}$$

where $X_{i:n}$ is the i-th order statistic of $X_1,\ldots,X_n$. For a non-decreasing sequence of numbers $k_1,k_2,\ldots$ with $0 < k_n \le n$ define also

$$D_{n,\alpha}(k_n) = \sup_{0<|t|\le\frac{k_n}{n}} \frac{|F_n(t)-|t||}{|t|^\alpha} \tag{2}$$

and for d=1 and fixed $k \in \mathbb{N}$

$$D_{n,\alpha}^{(k)}(k_n) = \sup_{X_{k:n}\le t\le \frac{k_n}{n}} \frac{|F_n(t)-t|}{t^\alpha}. \tag{2'}$$

In this paper the a.s. behaviour of the random variables in (1)-(2') will be determined for all $\alpha \in [0,1]$, i.e. for a rather general class of weight functions the a.s. behaviour of weighted empirical processes in the tail(s) will be obtained.

We first consider the rv's in (1) and (1'). In James (1975, d=1) and Alexander (1982, $d \in \mathbb{N}$) it is shown that a law of the iterated logarithm holds for $\alpha \in [0,\frac{1}{2})$. Therefore we may restrict ourselves to the case $\alpha \in [\frac{1}{2},1]$. Let $a_1, a_2, \ldots$ be a sequence of positive numbers.

<u>Theorem</u> 1. Let $d \in \mathbb{N}$ and $\alpha \in [\frac{1}{2},1]$. Then we have

$$\text{(A)} \quad \Sigma a_n (\log(1/a_n))^{d-1} = \infty \Rightarrow \limsup_{n\to\infty} n a_n^{\alpha} \Delta_{n,\alpha} = \infty \quad \text{a.s.}$$

$$\text{(B)} \quad \Sigma a_n (\log(1/a_n))^{d-1} < \infty,\ n a_n \downarrow \Rightarrow \lim_{n\to\infty} n a_n^{\alpha} \Delta_{n,\alpha} = 0 \quad \text{a.s.}$$

<u>Corollary 1</u>.

$$\limsup_{n\to\infty} \frac{\log n^{1-\alpha} \Delta_{n,\alpha}}{\log\log n} = \alpha d \quad \text{a.s.}$$

<u>Theorem 1'</u>. Let d=1, $\alpha \in [\frac{1}{2},1]$ and $k \in \mathbb{N}$ fixed. Then we have

$$\text{(A)} \quad \Sigma n^{k-1} a_n^k = \infty,\ a_n \downarrow \Rightarrow \limsup_{n\to\infty} n a_n^{\alpha} \Delta_{n,\alpha}^{(k)} = \infty \quad \text{a.s.}$$

$$\text{(B)} \quad \Sigma n^{k-1} a_n^k < \infty,\ n a_n \downarrow \Rightarrow \lim_{n\to\infty} n a_n^{\alpha} \Delta_{n,\alpha}^{(k)} = 0 \quad \text{a.s.}$$

<u>Corollary 1'</u>.

$$\limsup_{n\to\infty} \frac{\log n^{1-\alpha} \Delta_{n,\alpha}^{(k)}}{\log\log n} = \alpha/k \quad \text{a.s.}$$

Next we consider the rv's in (2) and (2'). From Theorems 1 and 1' and their proofs it readily follows that $D_{n,\alpha}(k_n)$ and $D_{n,\alpha}^{(k)}(k_n)$ have the same a.s. behaviour as $\Delta_{n,\alpha}$ and $\Delta_{n,\alpha}^{(k)}$. Therefore we need only consider $\alpha \in [0,\frac{1}{2})$. For the sake of brevity we omit the unweighted case $\alpha=0$, which shows a behaviour which is slightly different from the other values of α in $[0,\frac{1}{2})$. For any $\alpha \in (0,\frac{1}{2})$ write

$$b_n = \left(n\, k_n^{(1-2\alpha)/(2\alpha)} (\log\log n)^{1/(2\alpha)}\right)^{-1}.$$

<u>Theorem 2</u>. Let $d \in \mathbb{N}$ and $\alpha \in (0,\frac{1}{2})$. Then we have

(A) $\quad \Sigma b_n(\log(1/b_n))^{d-1} = \infty \Rightarrow \limsup_{n\to\infty} \dfrac{n^{1-\alpha}D_{n,\alpha}(k_n)}{k_n^{\frac{1}{2}-\alpha}(\log\log n)^{\frac{1}{2}}} = \infty \quad$ a.s.

(B) $\quad \Sigma b_n(\log(1/b_n))^{d-1} < \infty\ k_n/n\downarrow 0 \Rightarrow \limsup_{n\to\infty} \dfrac{n^{1-\alpha}D_{n,\alpha}(k_n)}{k_n^{\frac{1}{2}-\alpha}(\log\log n)^{\frac{1}{2}}} \leq (2d)^{\frac{1}{2}} \quad$ a.s.,

with equality almost surely for the case d=1.

(C) If in addition to the conditions in (B) we have $\log\log(n/k_n)/\log\log n \to \gamma$, then

$$\limsup_{n\to\infty} \frac{n^{1-\alpha}D_{n,\alpha}(k_n)}{k_n^{\frac{1}{2}-\alpha}(\log\log n)^{\frac{1}{2}}} = (2(1+\gamma(d-1)))^{\frac{1}{2}} \quad \text{a.s.}$$

<u>Theorem 2'</u>. Let d=1, $\alpha \in (0,\frac{1}{2})$ and $k \in \mathbb{N}$ fixed. Then we have

(A) $\quad \Sigma n^{k-1}b_n^k = \infty \Rightarrow \limsup_{n\to\infty} \dfrac{n^{1-\alpha}D^{(k)}_{n,\alpha}(k_n)}{k_n^{\frac{1}{2}-\alpha}(\log\log n)^{\frac{1}{2}}} = \infty \quad$ a.s.

(B) $\quad \Sigma n^{k-1}b_n^k < \infty,\ k_n/n\downarrow 0 \Rightarrow \limsup_{n\to\infty} \dfrac{n^{1-\alpha}D^{(k)}_{n,\alpha}(k_n)}{k_n^{\frac{1}{2}-\alpha}(\log\log n)^{\frac{1}{2}}} = 2^{\frac{1}{2}} \quad$ a.s.

2. DISCUSSION OF THE RESULTS

Theorem 1, contained in Einmahl & Mason (1985a), generalizes various results in the literature. In the one dimensional case this theorem was proved for $\alpha=\frac{1}{2}$ in Csáki (1975) and for $\alpha=1$ in Shorack & Wellner (1978). Mason (1981) connected these two results and proved the theorem for every $\alpha \in [\frac{1}{2},1]$; he (1982) also considered $\alpha=1$ for arbitrary $d \in \mathbb{N}$. Theorem 1 shows that for the given weight functions no law of the iterated logarithm type result holds, i.e. there is no standardization which will yield a finite positive value for the limsup almost surely. However, by Corollary 1 we see that for $\log n^{1-\alpha}\Delta_{n,\alpha}$ a law of the iterated logarithm does hold. Theorem 1' is contained in Einmahl et al. (1985). It shows another way of generalizing Theorem 1 with d=1.

Theorems 2 and 2' are established in Einmahl & Mason (1985b). As far as we know, this is the first time that results of this type have been considered. These theorems show that for $\alpha \in (0,\frac{1}{2})$ a law of the iterated logarithm in the tail holds, provided the sequence $\{k_n\}_{n=1}^{\infty}$

increases sufficiently rapidly.

Theorems 1' and 2' are likely to have a wide variety of applications in probability theory, in particular in the theory of extreme values. Theorem 1' is already applied in the study of the almost sure stability of sums of extreme values in Einmahl et al. (1985). Theorem 2', in turn, has already proved to be a valuable tool in establishing laws of the iterated logarithm for sums of extreme values, see Haeusler & Mason (1985) and Deheuvels et.al. (1986).

For more details and other results concerning the (a.s.) behaviour of weighted empirical processes the reader is referred to Shorack & Wellner (1986 , d=1) and Einmahl (1986, $d \in \mathbb{N}$).

REFERENCES

Alexander, K.S. (1982). Some limit theorems for weighted and non-identically distributed empirical processes. Ph.D. dissertation, M.I.T., Cambridge.

Csáki, E. (1975). Some notes on the law of the iterated logarithm for empirical distribution function. In: Colloq. Math. Soc. János Bolyai 11, Révész, P. (Ed.). North-Holland, Amsterdam, pp. 47-58.

Deheuvels, P., Haeusler, E., and Mason, D.M. (1986). On the almost sure behaviour of sums of extreme values from a distribution in the domain of attraction of a Gumbel law. Preprint.

Einmahl, J.H.J. (1986). Multivariate empirical processes. CWI Tract, Amsterdam, to appear.

Einmahl, J.H.J., Haeusler, E., and Mason, D.M. (1985). An extension of a theorem of Csáki with application to the study of the almost sure stability of sums of extreme values. Preprint.

Einmahl, J.H.J., and Mason, D.M. (1985a). Bounds for weighted multivariate empirical distribution functions. Z. Warsch. Verw. Gebiete 70, 563-571.

Einmahl, J.H.J., and Mason, D.M. (1985b). Laws of the iterated logarithm in the tails for weighted uniform empirical processes. Ann. Probab., to appear.

Haeusler, E., and Mason, D.M. (1985). A law of the iterated logarithm for sums of extreme values from a distribution with a regulary varying upper tail. Ann. Probab., to appear.

James, B.R. (1975). A functional law of the iterated logarithm for weighted empirical distributions. Ann. Probab. 3, 762-772.

Mason, D.M. (1981). Bounds for weighted empirical distribution functions. Ann. Probab. 9, 881-884.

Mason, D.M. (1982). Some characterizations of almost sure bounds for weighted multidimensional empirical distributions and a Glivenko-Cantelli theorem for sample quantiles. Z. Warsch. Verw. Gebiete 59, 505-513.

Shorack, G.R., and Wellner, J.A. (1978). Linear bounds on the empirical distribution function. Ann. Probab. 6, 349-353.

Shorack, G.R., and Wellner, J.A. (1986). Empirical processes with applications to statistics. Wiley, New York.

CONVERGENCE OF THE EMPIRICAL CHARACTERISTIC FUNCTIONALS

Kolčinskii V.I.
Kiev University, Kiev, U.S.S.R.

The asymptotic properties of the empirical characteristic functionals (e.c.f.) in Banach spaces are considered in the paper. The similar results in finite dimensional case are well known (see Marcus (1981)).

Let F be a separable Banach space, ξ - a random element, taking values in F with distribution μ, $\{\xi_n\}_{n\geq 1}$ - the sequence of independent copies of ξ.

Let F^* denote the dual space of F, U_{F^*} - the unit ball of F^*.

Let us define the characteristic functional of measure μ

$$\varphi_\mu(f) = \int_F \exp\{if(x)\}\,\mu(dx)$$

and its estimate - e.c.f.

$$\varphi_{\mu_n^*}(f) = n^{-1}\sum_1^n \exp\{if(\xi_k)\},\ f\in U_{F^*}.$$

The accuracy of the estimation will be determined by the value

$$\Delta_n = \sup\{|\varphi_{\mu_n^*}(f) - \varphi_\mu(f)| : f\in U_{F^*}\}.$$

Given F separable Banach space, the $*$ - weak topology in U_{F^*} is metrizable. Let W denote the corresponding metric. Then (U_{F^*}, W) is compact metric space. Let us denote $(C_W(U_{F^*})$ the space of continuous complex valued functions. Obviously, $\varphi_\mu\in C_W(U_{F^*})$ and $\varphi_{\mu_n^*}\in C_W(U_{F^*})$, $n\geq 1$. Therefore, Δ_n is random variable for each $n\geq 1$.

It follows from the law of large numbers in separable Banach space $C_W(U_{F^*})$ that $\Delta_n \to 0$, $n \to \infty$ almost surely (a.s.) and in the mean.

Let

$$Z_n(f) = n^{1/2}[\varphi_{\mu_n^*}(f) - \varphi_\mu(f)],\ f \in U_{F^*}.$$

By the multidimensional central limit theorem the finite dimensional distributions of random functions Z_n converge weakly to the corresponding distributions of Gaussian random function $G_\mu(f), f \in U_{F^*}$ with mean zero and convariance.

$$M G_\mu(f)\overline{G_\mu(g)} = \varphi_\mu(f-g) - \varphi_\mu(f)\overline{\varphi_\mu(g)},\ f, g \in U_{F^*}.$$

Let E be a separable Banach space, $T: E \to F$ - a bounded linear operator, $\xi = T\eta$, where η is random element, taking values in E, $1 < p \leq 2$.

<u>Theorem I.</u> If T is operator of type p, $M\|\eta\|^p < +\infty$, then

(I) for $p < 2$ $\Delta_n = O(n^{1/p-1})$, $n \to \infty$ a.s. and in the mean;

(2) for $p = 2$ the laws of random elements Z_n in space $C_W(U_{F^*})$ converge weakly to the law of Gaussian random element G_μ.

There exist a random element ξ in space ℓ_p such that for all $r < p$ $M\|\xi\|^r < +\infty$ and with some constant $C > 0$

$$M\Delta_n > Cn^{1/p-1} \quad \text{for} \quad p \in (1,2);$$
$$M\Delta_n > Cn^{-1/2}\ln n \quad \text{for} \quad p = 2.$$

Let now consider the characterization of the operators of type p in terms of metric entropy, see Kolčinskii (1986). For probability measure ν on E and $p \in (1,2]$ we define

$$d_{\nu,p}(f,g) = \left(\int_E |f(x) - g(x)|^p \nu(dx)\right)^{1/p},\ f, g \in E^*.$$

Let $\mathcal{M}_p$ denote the set of probability measures on E, satisfying the condition

$$\int_E \|x\|^p \nu(dx) \leq 1.$$

Let $q = p(p-1)^{-1}$.
Given (S, ρ) is totally bounded metric space, $H_\rho(S,\varepsilon)$ denote ε - entropy of S with respect to metric ρ, $\varepsilon > 0$. T^* denote the dual operator to T.

Theorem 2. (1) If

$$\sup_{\nu \in \mathcal{M}_p} \int_0^\infty H^{1/q}_{d_{\nu,p}} (T^* U_{F^*}; \varepsilon)\, d\varepsilon < +\infty,$$

then T is operator of type p. (2) If T is operator of type p, then

$$\sup_{\nu \in \mathcal{M}_p} \sup_{\varepsilon > 0} \varepsilon H^{1/q}_{d_{\nu,p}} (T^* U_{F^*}; \varepsilon) < +\infty .$$

The next statement refines the result of theorem I. Let $W_2(E)$ denote the space of measures ν on E, having finite total variation and satisfying the condition

$$\|\nu\| = \int_E \max(1, \|x\|^2)\, |\nu|(dx) < +\infty .$$

Obviously, $(W_2(E), \|\cdot\|)$ is linear normed space. Let us define the bounded linear operator $\hat{T} : W_2(E) \to C_W(U_{F^*})$

$$(\hat{T}\nu)(f) = \int_E \exp\{if(Tx)\}\, \nu(dx), \; f \in U_{F^*} .$$

Theorem 3. If $T: E \to F$ is operator of type 2, then $\hat{T} : (W_2(E), \|\cdot\|) \to (C_W(U_{F^*}), \|\cdot\|_\infty)$ is operator of type 2.

The proofs of the results are based on the use of general limit theorems for empirical processes obtained last years in the works of R.Dudley, D.Pollard, E.Gine, J.Zinn, M.Talagrand and the author (see Gine and Zinn (1986)). In particular, the comparison theorem of Talagrand is essentially used. The proof of theorem 2 uses the results of Marcus and Pisier (1983).

REFERENCES

Gine,E. and Zinn,J. (1986). Lectures on the central limit theorem for empirical processes.Preprint.

Kolčinskii, V.I. (1986). Operators of type p and metric entropy. Teor.Verojatnost.Mat.Statist (Kiev) to appear.

Marcus,M.B. (1981). Weak convergence of the empirical characteristic function. Ann.Probab.9, 194-201.

Marcus,M.B. and Pisier,G. (1983). Characterizations of almost surely continuous p - stable random Fourier series and strongly stationary processes. Preprint.

SAMPLE APPROXIMATION OF THE DISTRIBUTION BY MEANS OF k POINTS: A CONSISTENCY RESULT FOR SEPARABLE METRIC SPACES

Pärna K.
Faculty of Mathematics, Tartu State University, Estonia, USSR

ABSTRACTS

Empirical measure P_n, corresponding to the theoretical measure P on separable metric space, is approximated by k points that minimize the given approximation criterion. Conditions are found that ensure the almost sure convergence of the empirical minimum of the criterion to the theoretical minimum. Analogous result for Euclidean spaces has been proved in Pollard (1981).

1. PRELIMINARIES AND NOTATIONS

Let $\{x_1, x_2, \ldots, x_n\}$ be a random sample from the unknown probability distribution P on a separable metric space T with metrics d. Our aim is to approximate the distribution P by some discrete distribution, concentrated at k points of T (k is a given number), with only the sample $\{x_1, x_2, \ldots, x_n\}$ known. Such an approximation problem may be interpreted as a problem of optimal coding, optimal location of resources, k-means clustering and so on.

Introduce the measure of goodness of approximation of the distribution P by a finite set A, $A \subset T$, as follows:

$$W(A,P) = \int_T \min_{a \in A} \phi(d(x,a))P(dx). \quad (1)$$

For our purposes, the function ϕ must satisfy some regularity conditions. We shall need ϕ defined on the interval $[0, \infty)$, being continuous and increasing, with $\phi(0) = 0$ and $\phi(r) \to \infty$ as $r \to \infty$. In order to control the growth of ϕ in the tails, assume that there exists a constant λ such that $\phi(2r) \leq \lambda \cdot \phi(r)$ for every $r > 0$. For example, any function of the form $f(r) = r^s$ with positive s will go.

Further, let for j = 1, 2, ...

$$W_j(P) = \inf \{W(A,P): A \text{ contains } j \text{ points}\}, \quad (2)$$

from which the inequalities

$$W_1(P) \geq W_2(P) \geq \ldots \geq W_k(P) \geq \ldots$$

follow (addition of a point to the given set A never increases the approximation criterion).

It must be said that in metric spaces these infimum values are possibly not attainable on any set A, $A \subset T$. Thereby, we avoid the term "optimal set" in this paper.

Let P_n be the empirical measure obtained from the sample by placing mass 1/n at each of $x_1, x_2, \ldots, x_n$. Define the sample analogue for the criterion (1) by

$$W(A,P_n) = 1/n \sum_{i \leq n} \min_{a \in A} \phi(d(x_i,a)) \tag{3}$$

and the same for the infimum value (2) by

$$W_j(P_n) = \inf \{W(A,P_n): A \text{ contains j points}\}.$$

This is the criterion (3) and not the criterion (1), which can be minimized actually, when only the sample $\{x_1, x_2, \ldots, x_n\}$ from the distribution P is known. We do not discuss the practical problems of minimizing of the criterion (3) here.

For each fixed A, the strong law of large numbers argument shows that $W(A,P_n) \to W(A,P)$ almost surely, as n increases. It may be expected, therefore, that under some conditions the convergence of $W_k(P_n)$ (i.e. sample infimum) to $W_k(P)$ (i.e. population infimum) holds, too. This is the problem studied in this paper. In a more popular form, the convergence $W_k(P_n) \to W_k(P)$ ensures that a set A, which is a "good" set of approximating points for a large sample, is nearly as good for the population. Still, it doesn't ensure the convergence (in some suitable sense) of the optimal approximating set for the sample to that set for the population, provided their existence.

2. RESULTS

The main result of this work is contained in the following theorem.

<u>Theorem</u>. Suppose that $\int_T \phi(d(x,z))P(dx) < \infty$ for some $z \in T$ and strict inequalities

$$W_1(P) > W_2(P) > \ldots > W_k(P) \tag{4}$$

hold. Then

$$W_k(P_n) \to W_k(P) \tag{5}$$

almost surely.

This theorem generalizes one of the assertions from Pollard (1981), where the case of finite-dimensional Euclidean space is considered. In that paper the convergence of the optimal sample sets A_n^* to the optimal population set A^* has been proven, too (provided that A^* is unique). Unfortunately, the compactness argument, used in the paper mentioned, does not work in general metric spaces. In these spaces the question of the convergence $A_n^* \to A^*$ remains to be opened. Sverdrup-Thygeson (1981) considers the case of $k = 1$ for compact metric spaces. Some pioneer works in this area are MacQueen (1967) and Hartigan (1978).

We briefly sketch our method of proof. Say a set A_ε, containing k points, is ε-optimal with respect to the measure P, if $W(A_\varepsilon,P) < W_k(P) + \varepsilon$, $\varepsilon > 0$. Let A_{ε_n} be an ε_n-optimal set with respect to P_n. First it is shown, that there exists a sequence $\varepsilon_n \to 0$ such that A_{ε_n} will be contained in some closed ball $B \subset T$, eventually. The second step of the proof involves showing that, almost surely, $W(A,P_n) - W(A,P)$ converges to zero uniformly over those subsets of B containing k points.

More precisely, let

$$E(B) = \{A : A \subset B, \text{ A contains k points}\}.$$

Then the following lemma holds.

Lemma 1. For arbitrary closed ball B from the separable metric space T the uniform convergence

$$\lim_n \sup_{A \in E(B)} |W(A,P_n) - W(A,P)| = 0$$

holds almost surely.

The proof of this uniform strong law of large numbers is based on a theorem given in Ranga Rao (1962).

As the 3-rd step, the asymptotic equivalence of minimizing $W(\cdot,P_n)$ and $W(\cdot,P)$ - the desired result - is shown.

The detailed proof of this theorem one can find in Pärna (1986).

In addition, we give a sufficient condition to ensure the inequalities (4) be satisfied.

Lemma 2. Let P be not concentrated at any k points of T and let ϕ be strictly increasing (besides other restrictions given above). Suppose that for each j = 1, 2, ..., k-1 there exists an P-optimal set A(j), containing exactly j points, i.e. $W(A(j),P) = W_j(P)$. Then inequalities (4) hold.

REFERENCES

Hartigan, J.A. (1978). Asymptotic distributions for clustering criteria. Ann. Statist. 6, 117-131.

MacQueen, J.B. (1967). Some methods for classification and analysis of multivariate observations. Proc. Fifth Berkeley Symp. Math. Statist. Prob. 1, 281-297.

Pollard, D. (1981). Strong consistency of k-means clustering. Ann. Statist. 9, 135-140.

Pärna, K. (1986). Strong consistency of k-means clustering criterion in separable metric spaces. Acta et Commentations Universitatis Tartuensis 733, 86-96.

Ranga Rao, R. (1962). Relations between weak and uniform convergence with applications. Ann. Math. Statist. 33, 659-680.

Sverdrup-Thygeson, H. (1981). Strong law of large numbers for measures of central tendency and dispersion of random variables in compact metric spaces. Ann. Statist. 9, 141-145.

AUTHOR INDEX*

* Volume:Session:Page number